Yakov Berkovich and Zvonimir Janko

**Groups of Prime Power Order**

De Gruyter Expositions in Mathematics

Volume 61

Yakov Berkovich and Zvonimir Janko

# Groups of Prime Power Order

Volume 4



DE GRUYTER

**Mathematics Subject Classification 2010**
20-02, 20D15, 20E07

**Authors**
Prof. Yakov G. Berkovich
18251 Afula
Israel
berkov@math.haifa.ac.il

Prof. Dr. Zvonimir Janko
Ruprecht-Karls-Universität Heidelberg
Mathematisches Institut
Im Neuenheimer Feld 288
69120 Heidelberg
Germany
janko@mathi.uni-heidelberg.de

ISBN 978-3-11-028145-3
e-ISBN (PDF) 978-3-11-028147-7
e-ISBN (EPUB) 978-3-11-038155-9
Set-ISBN 978-3-11-028148-4
ISSN 0938-6572

**Library of Congress Cataloging-in-Publication Data**
A CIP catalog record for this book has been applied for at the Library of Congress.

**Bibliographic information published by the Deutsche Nationalbibliothek**
The Deutsche Nationalbibliothek lists this publication in the Deutsche Nationalbibliografie; detailed bibliographic data are available on the Internet at http://dnb.dnb.de.


Typesetting: le-tex publishing services GmbH, Leipzig
Printing and binding: CPI Books GmbH, Leck
♾ Printed on acid-free paper
Printed in Germany

www.degruyter.com

# Contents

# List of definitions and notations

## Set theory

- $|M|$ is the cardinality of a set $M$ (if $G$ is a finite group, then $|G|$ is called its order).
- $x \in M$ ($x \notin M$) means that $x$ is (is not) an element of a set $M$. $N \subseteq M$ ($N \nsubseteq M$) means that $N$ is (is not) a subset of the set $M$; moreover, if $M \neq N \subseteq M$ we write $N \subset M$.
- $\emptyset$ is the empty set.
- $N$ is called a nontrivial subset of $M$, if $N \neq \emptyset$ and $N \subset M$. If $N \subset M$, we say that $N$ is a proper subset of $M$.
- $M \cap N$ is the intersection and $M \cup N$ is the union of sets $M$ and $N$. If $M, N$ are sets, then $N - M = \{x \in N \mid x \notin M\}$ is the difference of $N$ and $M$.

## Number theory and general algebra

- $p$ is always a prime number.
- $\pi$ is a set of primes; $\pi'$ is the set of all primes not contained in $\pi$.
- $m, n, k, r, s$ are, as a rule, natural numbers.
- $\pi(m)$ is the set of prime divisors of $m$; then $m$ is a $\pi$-number if $\pi(m) \subseteq \pi$.
- $n_p$ is the $p$-part of $n$, $n_\pi$ is the $\pi$-part of $n$.
- $\mathrm{GCD}(m, n)$ is the greatest common divisor of $m$ and $n$.
- $\mathrm{LCM}(m, n)$ is the least common multiple of $m$ and $n$.
- $m \mid n$ should be read as: $m$ divides $n$.
- $\mathrm{GF}(p^m)$ is the finite field containing $p^m$ elements.
- $F^*$ is the multiplicative group of a field $F$.
- $\mathcal{L}(G)$ is the lattice of subgroups of a group $G$.
- $\mathcal{L}_N(G)$ is the lattice of normal subgroups of a group $G$.
- If $n = p_1^{\alpha_1} \dots p_k^{\alpha_k}$ is the standard prime decomposition of $n$, then $\lambda(n) = \sum_{i=1}^k \alpha_i$.

## Groups

We consider only finite groups which are denoted, with a pair of exceptions, by upper case Latin letters.

- If $G$ is a group, then $\pi(G) = \pi(|G|)$.
- $G$ is a $p$-group if $|G|$ is a power of $p$; $G$ is a $\pi$-group if $\pi(G) \subseteq \pi$.
- $G$ is, as a rule, a finite $p$-group.
- $H \leq G$ means that $H$ is a subgroup of $G$.

- $H < G$ means that $H \leq G$ and $H \neq G$ (in that case $H$ is called a *proper* subgroup of $G$). $\{1\}$ denotes the group containing only one element.
- $H$ is a nontrivial subgroup of $G$ if $\{1\} < H < G$.
- $H$ is a maximal subgroup of $G$ if $G > \{1\}$, $H < G$ and it follows from $H \leq M < G$ that $H = M$.
- If $H$ is a proper normal subgroup of $G$, then we write $H \triangleleft G$. Expressions 'normal subgroup of $G$' and '$G$-invariant subgroup' are synonyms.
- A normal subgroup $H$ of $G$ is nontrivial provided $G > H > \{1\}$.
- $H$ is a minimal normal subgroup of $G$ if (a) $H$ normal in $G$; (b) $H > \{1\}$; (c) $N \triangleleft G$ and $N < H$ implies $N = \{1\}$. Thus, the group $\{1\}$ has no minimal normal subgroup.
- A group $G$ is *metabelian* if $G/A$ is abelian for some abelian $A \triangleleft G$. Groups of class 2 are metabelian but the converse is not true.
- A $p$-group $G$ is said to be *Dedekindian* if all its subgroups are normal.
- A group $G$ is said to be *minimal nonabelian* if it is nonabelian but all its proper subgroups are abelian.
- $\mathcal{A}_1(G)$ is the set of all minimal nonabelian subgroups of a $p$-group $G$.
- A group $H$ is said to be *capable* if there exists a group $G$ such that $G/Z(G) \cong H$.
- A $p$-group $G$ is said to be *meta-Hamiltonian* if all its minimal nonabelian (so all nonabelian) subgroups are normal.
- $H \leq G$ is quasinormal if it is permutable with all subgroups of $G$.
- A $p$-group is said to be *modular* if all its subgroups are quasinormal.
- $G$ is simple if it is a minimal normal subgroup of $G$ (so $|G| > 1$).
- $H$ is a maximal normal subgroup of $G$ if $H < G$ and $G/H$ is simple.
- The subgroup generated by all minimal normal subgroups of $G$ is called the *socle* of $G$ and denoted by $\mathrm{Sc}(G)$. We put, by definition, $\mathrm{Sc}(\{1\}) = \{1\}$.
- $\mathrm{N}_G(M) = \{x \in G \mid x^{-1}Mx = M\}$ is the normalizer of a subset $M$ in $G$.
- $\mathrm{C}_G(x)$ is the centralizer of an element $x$ in $G$ : $\mathrm{C}_G(x) = \{z \in G \mid zx = xz\}$.
- $\mathrm{C}_G(M) = \bigcap_{x \in M} \mathrm{C}_G(x)$ is the centralizer of a subset $M$ in $G$.
- If $A \leq B$ and $A, B$ are normal in $G$, then $\mathrm{C}_G(B/A) = H$, where $H/A = \mathrm{C}_{G/A}(B/A)$.
- $H < G$ is a TI-subgroup if $T \cap T^x = \{1\}$ for all $x \in G - \mathrm{N}_G(H)$.
- $A \operatorname{wr} B$ is the wreath product of the 'passive' group $A$ and the transitive permutation group $B$ (in what follows we assume that $B$ is a regular permutation group, as a rule, a $p$-group); $B$ is called the active factor of the wreath product). Then the order of that group is $|A|^{|B|} \cdot |B|$.
- $\mathrm{Aut}(G)$ is the group of automorphisms of $G$ (the automorphism group of $G$).
- $\mathrm{Inn}(G)$ is the group of all inner automorphisms of $G$.
- $\mathrm{Out}(G) = \mathrm{Aut}(G)/\mathrm{Inn}(G)$ is the outer automorphism group of $G$.
- $\mathcal{N}(G)$ is the norm of $G$, the intersection of normalizers of all subgroups of $G$.
- If $a, b \in G$, then $a^b = b^{-1}ab$ is a conjugate with $a$ in $G$.
- $a \in G$ is real if it is conjugate with $a^{-1}$.
- An element $x \in G$ inverts a subgroup $H \leq G$ if $h^x = h^{-1}$ for all $h \in H$.
- If $M \subseteq G$, then $\langle M \rangle = \langle x \mid x \in M \rangle$ is the subgroup of $G$ generated by $M$.

- $M^x = x^{-1}Mx = \{y^x \mid y \in M\}$ for $x \in G$ and $M \subseteq G$.
- $[x, y] = x^{-1}y^{-1}xy = x^{-1}x^y$ is the *commutator* of elements $x, y$ of $G$. If $M, N \subseteq G$ then $[M, N] = \langle [x, y] \mid x \in M, y \in N\rangle$ is a subgroup of $G$.
- $o(x)$ is the order of an element $x$ of $G$.
- An element $x \in G$ is a $\pi$-element if $\pi(o(x)) \subseteq \pi$.
- $G$ is a $\pi$-group, if $\pi(G) \subseteq \pi$. Obviously, $G$ is a $\pi$-group if and only if all of its elements are $\pi$-elements.
- $G'$ is the subgroup generated by all commutators $[x, y]$, $x, y \in G$ (i.e., $G' = [G, G]$), $G^{(2)} = [G', G'] = G'' = (G')'$, $G^{(3)} = [G'', G''] = (G'')'$ and so on. $G'$ is called the *commutator* (or *derived*) subgroup of $G$.
- $Z(G) = \bigcap_{x \in G} C_G(x)$ is the center of $G$.
- $Z_i(G)$ is the $i$-th member of the upper central series of $G$; in particular, $Z_0(G) = \{1\}$, $Z_1(G) = Z(G)$.
- $K_i(G)$ is the $i$-th member of the lower central series of $G$; in particular, $K_2(G) = G'$. We have $K_i(G) = [G, \dots, G]$ ($i \geq 1$ times). We set $K_1(G) = G$.
- If $G$ is nonabelian, then $\eta(G)/K_3(G) = Z(G/K_3(G))$.
- $\mathcal{M}(G) = \langle x \in G \mid C_G(x) = C_G(x^p)\rangle$ is the Mann subgroup of a $p$-group $G$.
- $\mathrm{Syl}_p(G)$ is the set of $p$-Sylow subgroups of an arbitrary finite group $G$.
- $S_n$ is the *symmetric* group of degree $n$.
- $A_n$ is the *alternating* group of degree $n$.
- $\Sigma_{p^n}$ is a Sylow $p$-subgroup of $S_{p^n}$.
- $\mathcal{H}_{2,p}$ is a nonabelian metacyclic $p$-group of order $p^4$ and exponent $p^2$.
- $\mathrm{GL}(n, F)$ is the set of all nonsingular $n \times n$ matrices with entries in a field $F$, the $n$-dimensional *general linear* group over $F$, $\mathrm{SL}(n, F) = \{A \in \mathrm{GL}(n, F) \mid \det(A) = 1 \in F\}$, the $n$-dimensional *special linear* group over $F$.
- If $H \leq G$, then $H_G = \bigcap_{x \in G} x^{-1}Hx$ is the *core* of the subgroup $H$ in $G$ and $H^G$, the intersection of all normal subgroups of $G$ containing $H$, is the *normal closure* or *normal hull* of $H$ in $G$. Obviously, $H_G$ is normal in $G$.
- If $G$ is a $p$-group, then $p^{b(x)} = |G : C_G(x)|$; $b(x)$ is said to be the *breadth* of $x \in G$, where $G$ is a $p$-group; $b(G) = \max\{b(x) \mid x \in G\}$ is the *breadth* of $G$.
- If $H \leq G$ and $|G : N_G(H)| = p^{\mathrm{sb}(H)}$, when $\mathrm{sb}(H)$ is said to be the *subgroup breadth* of $H$. Next, $\mathrm{sb}(G) = \max\{\mathrm{sb}(H) \mid H \leq G\}$.
- $\Phi(G)$ is the Frattini subgroup of $G$ (= the intersection of all maximal subgroups of $G$), $\Phi(\{1\}) = \{1\}$, $p^{\mathrm{d}(G)} = |G : \Phi(G)|$.
- $\Gamma_i = \{H < G \mid \Phi(G) \leq H, |G : H| = p^i\}$, $i = 1, \dots, \mathrm{d}(G)$, where $G > \{1\}$.
- If $H < G$, then $\Gamma_1(H)$ is the set of all maximal subgroups of $H$.
- $\exp(G)$ is the exponent of $G$ (the least common multiple of the orders of elements of $G$). If $G$ is a $p$-group, then $\exp(G) = \max\{o(x) \mid x \in G\}$.
- $k(G)$ is the number of conjugacy classes of $G$ (= $G$-classes), the class number of $G$.
- $K_x$ is the $G$-class containing an element $x$ (sometimes we also write $ccl_G(x)$).
- $C_m$ is the cyclic group of order $m$.
- $A \times B$ is the direct product of groups $A$ and $B$.

- $G^m = G \times \dots G$ ($m$ times) is the direct product of $m$ copies of a group $G$.
- $A * B$ is a central product of groups $A$ and $B$, i.e., $A * B = AB$ with $[A, B] = \{1\}$. In particular, the direct product $A \times B$ is a central product of groups $A$ and $B$.
- $\mathrm{E}_{p^m} = \mathrm{C}_p^m$ is the elementary abelian group of order $p^m$. $G$ is an elementary abelian $p$-group if and only if it is a $p$-group $> \{1\}$ and $G$ coincides with its socle. Next, $\{1\}$ is elementary abelian for each prime $p$, by definition.
- A group $G$ is said to be *homocyclic* if it is a direct product of isomorphic cyclic subgroups (obviously, elementary abelian $p$-groups are homocyclic).
- $\mathrm{ES}(m, p)$ is an *extraspecial* group of order $p^{1+2m}$ (a $p$-group $G$ is said to be extraspecial if $G' = \Phi(G) = \mathrm{Z}(G)$ is of order $p$). Note that for each positive integer $m$, there are exactly two nonisomorphic extraspecial groups of order $p^{2m+1}$.
- $S(p^3)$ is a nonabelian group of order $p^3$ and exponent $p > 2$.
- A *special* $p$-group is a nonabelian $p$-group $G$ such that $G' = \Phi(G) = \mathrm{Z}(G)$ is elementary abelian. Direct products of extraspecial $p$-groups are special.
- $\mathrm{D}_{2m}$ is the *dihedral* group of order $2m$, $m > 2$. Some authors consider $\mathrm{E}_{2^2}$ as the dihedral group $\mathrm{D}_4$.
- $\mathrm{Q}_{2^m}$ is the *generalized quaternion* group of order $2^m \geq 2^3$.
- $\mathrm{SD}_{2^m}$ is the *semidihedral group* of order $2^m \geq 2^4$.
- $\mathrm{M}_{p^m}$ is a nonabelian $p$-group containing exactly $p$ cyclic subgroups of index $p$ (see Theorem 1.2).
- $\mathrm{cl}(G)$ is the *nilpotence class* of a $p$-group $G$.
- $\mathrm{dl}(G)$ is the *derived length* of a $p$-group $G$.
- $\mathrm{CL}(G)$ is the set of all $G$-classes.
- A $p$-group of *maximal class* is a nonabelian group $G$ of order $p^m$ with $\mathrm{cl}(G) = m-1$.
- A $p$-group is s-self dual if every one of its subgroups is isomorphic to a quotient group.
- A $p$-group is q-self dual if every one of its quotient groups is isomorphic to a subgroup.
- $\mathrm{GL}(n, F)$ ($\mathrm{SL}(n, p)$) the general (special) linear group of degree $n$ over the field $F$.
- $\Omega_m(G) = \langle x \in G \mid o(x) \leq p^m \rangle$, $\Omega_m^*(G) = \langle x \in G \mid o(x) = p^m \rangle$ and $\mho_m(G) = \langle x^{p^m} \mid x \in G \rangle$.
- A $p$-group $G$ is said to be *regular*, if for any $x, y \in G$ there exists $z \in \langle x, y \rangle'$ such that $(xy)^p = x^p y^p z^p$.
- A $p$-group is *absolutely regular* if $|G/\mho_1(G)| < p^p$. Absolutely regular $p$-groups are regular.
- A $p$-group is *thin* if it is either absolutely regular or of maximal class.
- $G = A \cdot B$ is a *semidirect product* with *kernel* $B$ and *complement* $A$.
- A group $G$ is an extension of a normal subgroup $N$ by a group $H$ if $G/N \cong H$. A group $G$ splits over $N$ if $G = H \cdot N$ with $H \leq G$ and $H \cap N = \{1\}$ (in that case, $G$ is a semidirect product of $H$ and $N$ with kernel $N$).
- $H^\# = H - \{e_H\}$, where $e_H$ is the identity element of the group $H$. If $M \subseteq G$, then $M^\# = M - \{e_G\}$.

- An automorphism $\alpha$ of $G$ is *regular* (= *fixed-point-free*) if it induces a regular permutation on $G^{\#}$ (a permutation is said to be *regular* if it has no fixed points).
- An *involution* is an element of order 2 in a group.
- A group $G$ is said to *metacyclic* if it contains a normal cyclic subgroup $C$ such that $G/C$ is cyclic.
- A group $G$ is said to be *minimal nonmetacyclic* if it is nonmetacyclic but all its proper subgroups are metacyclic.
- A subgroup $A$ of a group $G$ is said to be *soft*, if $\mathrm{C}_G(A) = A$ and $|\mathrm{N}_G(A) : A| = p$.
- A *section* of a group $G$ is an epimorphic image of some subgroup of $G$.
- If $F = \mathrm{GF}(p^n)$, then we usually write $\mathrm{GL}(m, p^n)$, $\mathrm{SL}(m, p^n), \ldots$ instead of $\mathrm{GL}(m, F)$, $\mathrm{SL}(m, F), \ldots$.
- $\mathrm{c}_n(G)$ is the number of cyclic subgroups of order $p^n$ in a $p$-group $G$.
- $\mathrm{c}(G)$ is the number of nonidentity cyclic subgroups of $G$.
- $\mathrm{s}_n(G)$ is the number of subgroups of order $p^n$ in a $p$-group $G$.
- $v_n(G)$ is the number of normal subgroups of order $p^n$ in a $p$-group $G$.
- $e_n(G)$ is the number of subgroups of order $p^n$ and exponent $p$ in $G$.
- A group $G$ is said to be s-self dual (q-self dual) if any its subgroups (quotient group) are isomorphic to some quotient group (subgroup).
- $M_p(m, n) = \langle a, b \mid o(a) = p^m,\ o(b) = p^n,\ a^b = a^{1+p^{m-1}} \rangle$ is the metacyclic minimal nonabelian group of order $p^{m+n}$.
- $M_p(m, n.1) = \langle a, b \mid o(a) = p^m,\ o(b) = p^n,\ [a, b] = c,\ [a, c] = [b, c] = c^p = 1 \rangle$ is the nonmetacyclic minimal nonabelian group of order $p^{m+n+1}$.
- An $\mathcal{A}_n$-group is a $p$-group $G$ all of whose subgroups of index $p^n$ are abelian but $G$ contains a nonabelian subgroup of index $p^{n-1}$. In particular, $\mathcal{A}_1$-group is a minimal nonabelian $p$-group for some $p$.
- $\alpha_n(G)$ is the number of $\mathcal{A}_n$-subgroups in a $p$-group $G$.
- $\mathcal{D}_n$-group is a 2-group all of whose subgroups of index $p^n$ are Dedekindian, containing a non-Dedekindian subgroup of index $p^{n-1}$ and that is not an $\mathcal{A}_n$-group.
- $\mathcal{MA}(G)$ is the set of minimal nonabelian subgroups of a $p$-group $G$.
- $\mathcal{MA}_k(G) = \{H \in \mathcal{MA}(G) \mid \Omega_k(H) = H\}$.
- $\mathrm{D}_k(G) = \langle \mathcal{MA}_k(G) \rangle = \langle H \mid H \in \mathcal{MA}_k(G) \rangle$.
- $L_n = |\{x \in G \mid x^n = 1\}|$.
- If $G$ is a metacyclic $p$-group and $w(G) = \max\{i \mid |\Omega_i(G)| = p^{2i}\}$, then $R(G) = \Omega_{w(G)}(G)$. In that case, $G/R(G)$ is either cyclic or a 2-group of maximal class.

## Characters and representations

- $\mathrm{Irr}(G)$ is the set of all *irreducible* characters of $G$ over complex numbers.
- A character of degree 1 is said to be *linear*.
- $\mathrm{Lin}(G)$ is the set of all *linear* characters of $G$ (obviously, $\mathrm{Lin}(G) \subseteq \mathrm{Irr}(G)$).

- $\mathrm{Irr}_1(G) = \mathrm{Irr}(G) - \mathrm{Lin}(G)$ is the set of all *nonlinear* irreducible characters of $G$; $\mathrm{n}(G) = |\mathrm{Irr}_1(G)|$.
- $\chi(1)$ is the *degree* of a character $\chi$ of $G$.
- $\chi_H$ is the *restriction* of a character $\chi$ of $G$ to $H \le G$.
- $\chi^G$ is the character of $G$ induced from the character $\chi$ of some subgroup of $G$.
- $\bar{\chi}$ is a character of $G$ defined as follows: $\bar{\chi}(x) = \overline{\chi(x)}$ (here $\bar{w}$ is the complex conjugate of a complex number $w$).
- $\mathrm{X}(G)$ is a character table of $G$.
- $\mathrm{Irr}(\chi)$ is the set of irreducible constituents of a character $\chi$ of $G$.
- $1_G$ is the principal character of $G$.
- $\mathrm{Irr}^{\#}(G) = \mathrm{Irr}(G) - \{1_G\}$.
- If $\chi$ is a character of $G$, then $\ker(\chi) = \{x \in G \mid \chi(x) = \chi(1)\}$ is the *kernel* of a character $\chi$.
- $\mathrm{Z}(\chi) = \{x \in G \mid |\chi(x)| = \chi(1)\}$ is the *quasikernel* of $\chi$.
- If $N$ is normal in $G$, then $\mathrm{Irr}(G \mid N) = \{\chi \in \mathrm{Irr}(G) \mid N \not\le \ker(\chi)\}$.
- $\langle \chi, \tau \rangle = |G|^{-1} \sum_{x \in G} \chi(x)\tau(x^{-1})$ is the *inner product* of characters $\chi$ and $\tau$ of $G$.
- $\mathrm{I}_G(\phi) = \langle x \in G \mid \phi^x = \phi \rangle$ is the *inertia subgroup* of $\phi \in \mathrm{Irr}(H)$ in $G$, where $H \triangleleft G$.
- $1_G$ is the *principal character* of $G$ ($1_G(x) = 1$ for all $x \in G$).
- $\mathrm{M}(G)$ is the *Schur multiplier* of $G$.
- $\mathrm{cd}(G) = \{\chi(1) \mid \chi \in \mathrm{Irr}(G)\}$.
- $\mathrm{mc}(G) = \mathrm{k}(G)/|G|$ is the *measure of commutativity* of $G$.
- $\mathrm{T}(G) = \sum_{\chi \in \mathrm{Irr}(G)} \chi(1)$.
- $\mathrm{f}(G) = \mathrm{T}(G)/|G|$.

# Preface

This is the fourth volume of the series devoted to elementary parts of finite $p$-group theory. The material presented here is appearing in book form for the first time.

Below we list some new characterizations of certain classes of $p$-groups and results presented in this volume:

(1) groups all of whose maximal subgroups, except one, have cyclic derived subgroups,
(2) subgroup structure of metacyclic groups,
(3) groups with many minimal nonabelian subgroups,
(4) characterizations of Dedekindian $p$-groups, minimal nonabelian, metacyclic and $p$-groups of maximal class,
(5) classification of $p$-groups $G$ all of whose subgroups containing $\Phi(G)$ as a subgroup of index $p$, are minimal nonabelian,
(6) classification of nonabelian $p$-groups with exactly one noncyclic maximal abelian subgroup,
(7) $p$-groups with cyclic intersections of any two distinct conjugate subgroups,
(8) $p$-groups all of whose noncyclic abelian subgroups are normal,
(9) $p$-groups in which the intersection of any two distinct conjugate subgroups is (a) either cyclic or generalized quaternion, (b) either cyclic or of maximal class,
(10) $p$-groups in which the intersection of any two distinct conjugate cyclic subgroups is trivial,
(11) $p$-groups with small subgroups generated by two conjugate elements,
(12) 2-groups in which every cyclic subgroup has index $\leq 4$ in its normal closure,
(13) $p$-central $p$-groups,
(14) some generalizations of 2-central 2-groups,
(18) another proof of Baer's theorem about 2-groups with nonabelian norm,
(19) $p$-groups all of whose nonnormal abelian subgroups lie in centers of their normalizers
(20) the structure of $G/Z(G)$, where the two-generator $p$-group $G$ possesses an abelian maximal subgroup,
(21) $p$-groups all of whose subgroups not contained in Frattini subgroup are quasinormal,
(22) the centralizer equality subgroup in a $p$-group,
(23) extraspecial normal subgroups of $p$-groups
(24) $p$-groups with given intersections of certain subgroups.

For further information, see Contents.

The problems from the section 'Research problems and themes', Part IV, are posed, if it is not stated otherwise, by the first author. That section contains more than 500 items, some of which are solved by the second author. Many problems from the lists in the previous three volumes were solved mainly by the second author and mathematicians from Shanxi Normal University.

There are in the text many hundreds of exercises, most of which are solved (these exercises, if it is not stated otherwise, are written by the first author).

Sections and appendices written by the authors are listed in the Author index. Avinoam Mann wrote § 150. § 163 is combined from two independent texts, presented by Mann and the second author. In Appendix 53 the letters of Qinhai Zhang were used. Appendix 46 is written by Moshe Roitman. § 178 and Appendix 56 were inspired by the letters of Martin Isaacs. Some sections were discussed with Isaacs and Mann. The problems were discussed with Mann and, of course, with the second author. We are indebted to all the named mathematicians.

We added in the Bibliography of Volume III those items that are cited in this volume and omitted those items that are not cited here.

# § 145 $p$-groups all of whose maximal subgroups, except one, have derived subgroup of order $\le p$

The class of all $p$-groups $G$ with the property that the derived subgroup of every maximal subgroup of $G$ is of order at most $p$ was characterized in § 137. But there is no way to completely determine the structure of such $p$-groups.

It is quite surprising that we can completely determine (in terms of generators and relations) the title groups, where exactly one maximal subgroup has the commutator subgroup of order $> p$. We shall prove our main result (Theorem 145.8) starting with some partial results about the title groups. However, Propositions 145.4 and 145.5 are also of independent interest.

**Proposition 145.1.** *Let $G$ be a title $p$-group. Then we have $\mathrm{d}(G) \le 3$, $\mathrm{cl}(G) \le 3$, $p^2 \le |G'| \le p^3$ and $G'$ is abelian of exponent $\le p^2$. Also, $G$ has at most one abelian maximal subgroup.*

*Proof.* Let $H$ be the unique maximal subgroup of $G$ with $|H'| > p$. This gives $|G'| \ge p^2$. Let $K \ne L$ be maximal subgroups of $G$ that are both distinct from $H$. We have $|K'| \le p$, $|L'| \le p$ and so $K'L' \le Z(G)$ and $|K'L'| \le p^2$. By a result of A. Mann (Exercise 1.69), we get $|G' : (K'L')| \le p$. This implies that $|G'| \le p^3$, $G'$ is abelian and $G$ is of class $\le 3$. Since $K'L'$ is elementary abelian, we also get $\exp(G') \le p^2$. If $G$ would have more than one abelian maximal subgroup, then (by the above argument) $|G'| \le p$, a contradiction. Hence $G$ has at most one abelian maximal subgroup.

Note that each nonabelian $p$-group $X$ has exactly 0, 1 or $p + 1$ abelian maximal subgroups and in the last case $|X'| = p$ (Exercise 1.6 (a)). Suppose that $\mathrm{d}(G) \ge 4$. Then $G$ has at least $1 + p + p^2 + p^3$ distinct maximal subgroups and so the set $\mathcal{S}$ of maximal subgroups of $G$ with the commutator group of order $p$ has at least $p + p^2 + p^3 - 1$ elements. Since $G'$ has at most $p^2 + p + 1$ pairwise distinct subgroups of order $p$ (and the maximum is achieved if $G' \cong \mathrm{E}_{p^3}$), it follows that there are $K \ne L \in \mathcal{S}$ such that $K' = L'$. By the above argument (using a result of A. Mann), we get $|G'| = p^2$ and so $G'$ has at most $p + 1$ pairwise distinct subgroups of order $p$ (where the maximum is achieved if $G' \cong \mathrm{E}_{p^2}$). If $M \in \mathcal{S}$, then considering $G/M'$, we see that there are at most $p + 1$ elements $N \in \mathcal{S}$ such that $N' = M'$. This gives

$$p + p^2 + p^3 - 1 \le (p + 1)^2, \quad \text{and so} \quad p^3 - p \le 2 \quad \text{or} \quad p(p^2 - 1) \le 2\,,$$

a contradiction. Our proposition is proved. □

**Proposition 145.2.** *Let $G$ be a title $p$-group. Then the subgroup*

$$H_0 = \langle M' \mid M \text{ is any maximal subgroup of } G \text{ with } |M'| \le p\rangle$$

*is noncyclic and so $H_0$ is elementary abelian of order $p^2$ or $p^3$ and $H_0 \le Z(G)$.*

*Proof.* Suppose that $H_0$ is cyclic. Then we have $|H_0| = p$ and so $|G'| = p^2$ because (by Exercise 1.69) $|G' : H_0| \le p$ and Proposition 145.1 implies that $|G'| \ge p^2$. This gives $H' = G'$, where $H$ is the unique maximal subgroup of $G$ with $|H'| > p$. Consider the nonabelian factor group $G/H_0$. In this case $G/H_0$ has exactly one nonabelian maximal subgroup $H/H_0$. Since $\mathrm{d}(G/H_0) = 2$ or 3, the last statement would imply that the nonabelian $p$-group $G/H_0$ would have exactly $p$ or $p+p^2$ abelian maximal subgroups, a contradiction (by Exercise 1.6 (a)). □

**Proposition 145.3.** *Let $G$ be a title $p$-group. Then we have* $\mathrm{d}(G) = 2$.

*Proof.* Assume that $\mathrm{d}(G) = 3$ and we use the notation from Proposition 145.2.

First suppose that $H_0 = G'$ so that $G$ is of class 2 with an elementary abelian commutator subgroup. For any $x, y \in G$, we get $[x^p, y] = [x, y]^p = 1$ and this implies that $\mho_1(G) \le \mathrm{Z}(G)$. It follows $\Phi(G) = \mho_1(G)G' \le \mathrm{Z}(G)$ and $G/\Phi(G) \cong \mathrm{E}_{p^3}$. Let $X$ be any maximal subgroup of $G$ so that $X/\Phi(G) \cong \mathrm{E}_{p^2}$ and all $p+1$ maximal subgroups of $X$ which contain $\Phi(G)$ are abelian. This implies $|X'| \le p$. But then each maximal subgroup of $G$ has its derived subgroup of order $\le p$, contrary to our assumption.

Now assume $H_0 \ne G'$. In this case $H_0 \cong \mathrm{E}_{p^2}$, $H_0 \le \mathrm{Z}(G)$ and $|G'| = p^3$. There are exactly $p + p^2$ maximal subgroups $M_i$ of $G$ such that $|M_i'| \le p$, $i = 1, 2, \dots, p+p^2$. Since $H_0$ has exactly $p+1$ subgroups of order $p$, it follows that there exist the indices $i \ne j \in \{1, 2, \dots, p+p^2\}$ such that $M_i' = M_j'$ is of order $p$. Again by Exercise 1.69, we have $|G' : (M_i'M_j')| \le p$ and this gives $|G'| \le p^2$, a contradiction. Our proposition is proved. □

**Proposition 145.4.** *Let $G$ be a two-generator $p$-group, $p > 2$, with $G' \cong \mathrm{C}_{p^2}$. Then each maximal subgroup of $G$ is nonabelian.*

*Proof.* Assume that $G$ has an abelian maximal subgroup $M$ so that $|M/\Phi(G)| = p$. Take an element $a \in M \setminus \Phi(G)$ and an element $b \in G \setminus M$ so that we have $G = \langle a, b\rangle$ and $G' = \langle [a, b]\rangle$. Since $G'$ is cyclic, Theorem 7.1 (c) implies that $G$ is regular. We have $b^p \in \Phi(G) < M$ and so $[a, b^p] = 1$. Hence

$$(a^{-1}b^{-p}a)b^p = ((b^{-1})^a)^p b^p = 1 \quad \text{and so} \quad (b^a)^p = b^p\,.$$

By Theorem 7.2 (a) (about regular $p$-groups), the last relation gives $((b^{-1})^a b)^p = 1$ or equivalently $[a, b]^p = 1$, a contradiction. □

**Remark 1.** The assumption $p > 2$ in Proposition 145.4 is essential. This shows a 2-group of maximal class and order 16.

**Proposition 145.5.** *Let $G$ be a two-generator $p$-group, $p > 2$, with $G' \cong \mathrm{E}_{p^2}$. Then $G$ has an abelian maximal subgroup.*

*Proof.* By Proposition 137.4, each proper subgroup of $G$ has the derived subgroup of order at most $p$. Then we may apply Proposition 137.5 and so for each $x, y \in G$, we get $[x^p, y] = [x, y]^p = 1$. This gives $\mho_1(G) \le \mathrm{Z}(G)$ and therefore $\Phi(G) = \mho_1(G)G'$ is abelian.

Let $M \in \Gamma_1$ centralize $G'$. We have $|M : \Phi(G)| = p$ and $M$ centralizes $\mho_1(G)$ and $G'$ so that $\Phi(G) \leq Z(M)$, and so that $M$ is abelian. □

**Remark 2.** The assumption $p > 2$ in Proposition 145.5 is essential. Let $G$ be a faithful and splitting extension of an elementary abelian group of order 8 by a cyclic group of order 4. Then we have $d(G) = 2$ and $G' \cong E_4$ but $G$ has no abelian maximal subgroup.

**Proposition 145.6.** *Let $G$ be a title $p$-group and $\Gamma_1 = \{H_1, H_2, \dots, H_p, H\}$ be the set of all maximal subgroups of $G$, where $|H'| > p$. Then $G'$ is abelian of order $p^3$, $H' \cong E_{p^2}$, $H' \leq Z(G)$ and $H_1', H_2', \dots, H_p'$ are pairwise distinct subgroups of order $p$ contained in $H'$. If $G = \langle x, y\rangle$ for some $x, y \in G$, then $[x, y] \in G' \setminus H'$ and $[x, y] \notin Z(G)$ so that $G$ is of class 3. Finally, $G/H'$ is nonmetacyclic minimal nonabelian and so if $a \in G \setminus G'$ is such that $a^p \in G'$, then $a^p \in H'$.*

*Proof.* Let $H_0$ be the subgroup of $G'$ as defined in Proposition 145.2. Then $H_0 \leq Z(G)$ and $H_0$ is elementary abelian of order $p^2$ or $p^3$. Suppose for a moment that $H_0 = G'$. We have $G = \langle x, y\rangle$ for some $x, y \in G$ and $[x, y] \in H_0$ so that $G/\langle [x, y]\rangle$ is abelian and $G' = \langle [x, y]\rangle$ is of order $p$, a contradiction. It follows that $H_0 \neq G'$ which gives that $H_0 \cong E_{p^2}$, $|G' : H_0| = p$ and $G'$ is abelian of order $p^3$. Since $d(G/H_0) = 2$ and $|G'/H_0| = p$, it follows that $G/H_0$ is minimal nonabelian (see Lemma 65.2 (a)). In particular, we have $H' \leq H_0$, which together with $|H'| > p$ implies $H' = H_0 \cong E_{p^2}$. If $G/H'$ is metacyclic, then a result of N. Blackburn (see Lemma 44.1 and Corollary 44.6) gives that $G$ is also metacyclic. This is a contradiction because $G'$ is noncyclic. Hence $G/H'$ is nonmetacyclic minimal nonabelian so that Lemma 65.1 gives that $G'/H'$ is a maximal cyclic subgroup of $G/H'$. Thus for each element $a \in G \setminus G'$ such that $a^p \in G'$, we get $a^p \in H'$.

We have $G = \langle x, y\rangle$ for some $x, y \in G$. It is clear that $\langle [x, y]\rangle$ is not normal in $G$. Indeed, if $\langle [x, y]\rangle \trianglelefteq G$, then $G/\langle [x, y]\rangle$ is abelian and so $\langle [x, y]\rangle = G'$ is of order $\leq p^2$ (noting that $\exp(G') \leq p^2$), a contradiction. We have proved that $\langle [x, y]\rangle$ is not normal in $G$. In particular, $[x, y] \notin Z(G)$ and so $[x, y] \in G' - H'$ and $G$ is of class 3.

If $\Gamma_1 = \{H_1, H_2, \dots, H_p, H\}$ is the set of all maximal subgroups of $G$, then we have $H_i' \leq H_0 = H'$ for all $i = 1, 2, \dots, p$. We claim that $H_1', H_2', \dots, H_p'$ are pairwise distinct subgroups of order $p$. Indeed, if $|H_i'H_j'| \leq p$ for some $i \neq j$, $i, j \in \{1, 2, \dots, p\}$, then a result of A. Mann (see Exercise 1.69) implies $|G' : (H_i'H_j')| \leq p$ and so $|G'| \leq p^2$, a contradiction. Our proposition is proved. □

**Remark 3.** If $X$ is a two-generator $p$-group of class 2, then it is well known that $X'$ is cyclic. Hence if $G$ is any two-generator $p$-group, then $G'/K_3(G)$ is cyclic, where $K_3(G) = [G', G]$.

**Proposition 145.7.** *If $G$ is a title $p$-group, then $p = 2$.*

*Proof.* Assume that $p > 2$ and we use Proposition 145.6 together with the notation introduced there.

First suppose that $G'$ is not elementary abelian. Then we have $o([x, y]) = p^2$ and $\langle [x, y]^p \rangle$ is a subgroup of order $p$ contained in $H'$. Let $H_i'$, $i \in \{1, 2, \dots, p\}$, be such that $H_i' \neq \langle [x, y]^p \rangle$ which gives $G' = H_i' \times \langle [x, y] \rangle$. We consider the factor group $\bar{G} = G/H_i'$. Since $d(\bar{G}) = 2$, $p > 2$, and $\bar{G}' \cong C_{p^2}$, we may use Proposition 145.4 saying that each maximal subgroup of $\bar{G}$ is nonabelian. But $\bar{H}_i = H_i/H_i'$ is an abelian maximal subgroup of $\bar{G}$, a contradiction.

We have proved that $G'$ is elementary abelian of order $p^3$. Let $\{H_1', H_2', \dots, H_p', K\}$ be the set of all $p + 1$ subgroups of order $p$ in $H'$ and consider the factor group $G/K$. All $p + 1$ maximal subgroups of $G/K$ are nonabelian, $d(G/K) = 2$, $p > 2$, and $(G/K)' = G'/K \cong E_{p^2}$. By Proposition 145.5, $G/K$ possesses an abelian maximal subgroup, a contradiction. We have proved that we must have $p = 2$. □

**Theorem 145.8.** *Let $G$ be a $p$-group with exactly one maximal subgroup $H$ such that $|H'| > p$. Then we have $d(G) = 2$, $p = 2$ and $G'$ is abelian of order 8 and type $(4, 2)$. Also, $[G', G] = \Omega_1(G') \le Z(G)$, $\Phi(G) = C_G(G')$ is abelian and $\mho_2(G) \le Z(G)$. Let $\{H_1, H_2, H\}$ be the set of maximal subgroups of $G$. Then $H_1' = \langle z_1 \rangle$ and $H_2' = \langle z_2 \rangle$ are both of order 2, $\langle z_1, z_2 \rangle = \Omega_1(G') = H' \cong E_4$, $d(H) = 3$ and $\mho_1(G') = \langle z_1 z_2 \rangle$. Finally, $H$ is the unique maximal subgroup of $G$ that contains an element inverting $G'$. We have the following two possibilities:*

(i) $d(H_1) = d(H_2) = 2$ *in which case $H_1$ and $H_2$ are minimal nonabelian. In this case either $H_1$ and $H_2$ are both metacyclic and $G$ is isomorphic to one of the groups of Theorem 100.3 (a) and (b) or $H_1$ and $H_2$ are both nonmetacyclic and $G$ is isomorphic to one of the groups of Theorem 100.3 (c).*

(ii) $d(H_1) = d(H_2) = 3$ *and the group $G$ is given with:*

$$G = \langle a, b \mid [a, b] = v,\ v^4 = 1,\ [v, a] = z_1,\ [v, b] = z_1^{\epsilon} z_2,\ z_1^2 = z_2^2 = 1,\ v^2 = z_1 z_2,$$
$$[z_1, a] = [z_1, b] = [z_2, a] = [z_2, b] = 1,\ a^{2^m} = z_1^{\alpha} z_2^{\beta},\ b^{2^n} = z_1^{\gamma} z_2^{\delta} \rangle,$$

*where $m \ge 2$, $n \ge 2$ and $\alpha, \beta, \gamma, \delta, \epsilon \in \{0, 1\}$. We have here $|G| = 2^{m+n+3} \ge 2^7$, $G' = \langle v, z_1 \rangle \cong C_4 \times C_2$, $[G', G] = \langle z_1, z_2 \rangle = \Omega_1(G') \le Z(G)$ and the Frattini subgroup $\Phi(G) = \langle G', a^2, b^2 \rangle$ is abelian. Finally, if $\epsilon = 0$, then $H = \Phi(G)\langle ab \rangle$ and if $\epsilon = 1$, we have $H = \Phi(G)\langle b \rangle$.*

*Conversely, all groups stated in parts (i) and (ii) of this theorem are $p$-groups all of whose maximal subgroups, except one, have their derived subgroups of order $\le p$.*

*Proof.* We use Proposition 145.6 together with the notation introduced there. By Proposition 145.7, we have in addition $p = 2$.

Let $X$ be a maximal subgroup of $G$. By Schreier's inequality (Theorem A.25.1), we have

$$d(X) \le 1 + |G : X|(d(G) - 1),$$

and so $d(X) \le 3$. Since $H' \cong E_4$ and $H' \le Z(H)$, the maximal subgroup $H$ cannot be two-generator (see Remark 3). It follows that we have $d(H) = 3$. Since $G$ is a nonmetacyclic two-generator 2-group, we may use Theorem 107.1 saying that such a group has an even number of two-generator maximal subgroups. It follows that we have either $d(H_1) = d(H_2) = 2$ or $d(H_1) = d(H_2) = 3$.

Set $H_1' = \langle z_1 \rangle, H_2' = \langle z_2 \rangle$ so that we have $H' = \langle z_1 \rangle \times \langle z_2 \rangle \cong E_4$ and $\Phi(G) = H_1 \cap H_2$. Since $(\Phi(G))' \leq \langle z_1 \rangle \cap \langle z_2 \rangle = \{1\}$, it follows that $\Phi(G)$ is abelian and so $\Phi(G)$ is a maximal normal abelian subgroup of $G$ (containing $G'$). Take elements $h_1 \in H_1 \setminus \Phi(G)$ and $h_2 \in H_2 \setminus \Phi(G)$ so that we have $G = \langle h_1, h_2 \rangle$, $[h_1, h_2] = v \in G' \setminus H'$ and $o(v) \leq 4$. If $v$ commutes with both $h_1$ and $h_2$, then we get $v \in Z(G)$, a contradiction. Without loss of generality we may assume that $[v, h_1] \neq 1$ and so we get $[v, h_1] = z_1$.

Assume for a moment that $G' \cong E_8$ so that $v$ is an involution. We compute

$$[h_1^2, h_2] = [h_1, h_2]^{h_1}[h_1, h_2] = v^{h_1}v = (vz_1)v = v^2 z_1 = z_1 .$$

This is a contradiction since $h_1^2 \in \Phi(G)$ and $\langle h_1^2, h_2 \rangle \leq H_2$, where $H_2' = \langle z_2 \rangle$. We have proved that $G'$ is abelian of type $(4, 2)$ and so $o(v) = 4$ and $1 \neq v^2 \in H'$.

We have $K_3(G) = [G', G] \geq \langle z_1 \rangle$. Since $d(G) = 2$, it follows by Remark 3 that $G'/K_3(G)$ is cyclic. Suppose that $[v, h_2] = 1$ so that in this case we have $K_3(G) = \langle z_1 \rangle$. We compute

$$[h_1, h_2^2] = [h_1, h_2][h_1, h_2]^{h_2} = vv^{h_2} = v^2 \neq 1 .$$

We have $\langle h_1, h_2^2 \rangle \leq H_1$ and so $v^2 = z_1$. But then we have $G'/K_3(G) = G'/\langle z_1 \rangle \cong E_4$, a contradiction. We have proved that $[v, h_2] \neq 1$ and so $[v, h_2] = z_2$. This gives

$$K_3(G) = \langle z_1, z_2 \rangle = H' \leq Z(G)$$

and $G$ is of class 3.

We get

$$[h_1^2, h_2] = [h_1, h_2]^{h_1}[h_1, h_2] = v^{h_1}v = (vz_1)v = v^2 z_1 ,$$

and since $\langle h_1^2, h_2 \rangle \leq H_2$, it follows that $v^2 z_1 \in \langle z_2 \rangle$ and so $v^2 \in \{z_1, z_1 z_2\}$. Similarly, we get

$$[h_1, h_2^2] = [h_1, h_2][h_1, h_2]^{h_2} = vv^{h_2} = v(vz_2) = v^2 z_2 ,$$

and since $\langle h_1, h_2^2 \rangle \leq H_1$, it follows that $v^2 z_2 \in \langle z_1 \rangle$ and so $v^2 \in \{z_2, z_1 z_2\}$. As a result, we get $v^2 = z_1 z_2$ and so $\mho_1(G') = \langle z_1 z_2 \rangle$. Note that $H = \Phi(G)\langle h_1 h_2 \rangle$ and

$$v^{h_1 h_2} = (vz_1)^{h_2} = v(z_1 z_2) = vv^2 = v^3 = v^{-1}$$

and so $h_1 h_2$ inverts $G'$. It follows that $\Phi(G) = C_G(G')$ and $H$ is the unique maximal subgroup of $G$ that contains an element inverting $G'$.

Let $x, y \in G$. Then $\langle x^2, y \rangle$ is contained in one of the maximal subgroups $X_i$ of $G$, where $X_i'$ is elementary abelian of order $\leq 4$ and $\mathrm{cl}(X_i) = 2$ $(i = 1, 2, 3)$. It follows

$$[x^4, y] = [(x^2)^2, y] = [x^2, y]^2 = 1 ,$$

and so we get $\mho_2(G) \leq Z(G)$.

Now suppose that $d(H_1) = d(H_2) = 2$. In this case both $H_1$ and $H_2$ are minimal nonabelian (see Lemma 65.2 (a)) and $H$ is neither abelian nor minimal nonabelian.

Since $\mathrm{d}(G) = 2$ and $H_1' \neq H_2'$ such 2-groups are completely determined in Theorem 100.3, which gives the groups quoted in part (i) of our theorem.

It remains to consider the case $\mathrm{d}(H_1) = \mathrm{d}(H_2) = 3$. By Theorem 107.2 (a), a nonmetacyclic two-generator 2-group $G$ has the property that every maximal subgroup of $G$ is not generated by two elements if and only if $G/G'$ has no cyclic subgroup of index 2. Thus $G/G'$ is abelian of type $(2^m, 2^n)$, where $m \geq 2$, $n \geq 2$ and so $|G| = |G'|2^{m+n} = 2^{m+n+3} \geq 2^7$. There are normal subgroups $A$ and $B$ of $G$ such that $G = AB$, $A \cap B = G'$, $A/G' \cong \mathrm{C}_{2^m}$, $B/G' \cong \mathrm{C}_{2^n}$, $m \geq 2$, $n \geq 2$. Let $a \in A' - G'$, $b \in B - G'$ be such that $\langle a \rangle$ covers $A/G'$ and $\langle b \rangle$ covers $B/G'$. Since $G/H'$ is nonmetacyclic minimal nonabelian, we know that (see Lemma 65.1) $G'/H'$ is a maximal cyclic subgroup of $G/H'$ and so we have $a^{2^m} \in H'$ and $b^{2^n} \in H'$. We have $G = \langle a, b \rangle$ and so $[a, b] = v$ is an element of order 4 contained in $G' \setminus H'$.

Maximal subgroups of $G$ are $M_1 = A\langle b^2 \rangle$, $M_2 = B\langle a^2 \rangle$ and $M_3 = \Phi(G)\langle ab \rangle$, where $\Phi(G) = G'\langle a^2 \rangle\langle b^2 \rangle$ is abelian. Since $\Phi(G) = \mathrm{C}_G(G')$ and $\Omega_1(G') = H' \leq \mathrm{Z}(G)$, we see that $G/\Phi(G) \cong \mathrm{E}_4$ acts faithfully on $G'$ stabilizing the chain $G' > H' > \{1\}$. Interchanging $A$ and $B$ (if necessary), we may assume that $|M_1'| = 2$ and so we may set $v^a = vz_1$ which gives that $[v, a] = z_1$ and $M_1' = \langle z_1 \rangle$, where $z_1 \in H' - \langle v^2 \rangle$. Set $z_2 = z_1 v^2$ so that we have $v^2 = z_1 z_2$. Then we have two possibilities.

(1) We assume $v^b = vz_1z_2 = v^{-1}$ or equivalently $[v, b] = z_1z_2$ so that the element $b$ inverts each element in $G'$. Since the maximal subgroup $H$ is the unique maximal subgroup of $G$ which contains an element inverting $G'$, we have in this case $M_2 = B\langle a^2 \rangle = H$, where we should have $H' = \langle z_1, z_2 \rangle$. Indeed, we have

$$[a^2, b] = [a, b]^a[a, b] = v^a v = (vz_1)v = v^2 z_1 = (z_1z_2)z_1 = z_2\,,$$

and so we get $H' = \langle z_1, z_2 \rangle$. In this case $M_3 = \Phi(G)\langle ab \rangle$ has the property $M_3' = \langle z_2 \rangle$. Indeed, here we have

$$[a^2, ab] = [a^2, b] = z_2\,,$$
$$[ab, b^2] = [a, b^2]^b = ([a, b][a, b]^b)^b = (vv^b)^b = (vv^{-1})^b = 1\,,$$

and

$$v^{ab} = (vz_1)^b = (vz_1z_2)z_1 = vz_2 \quad \text{and so } [v, ab] = z_2\,.$$

(2) Now we suppose $v^b = vz_2$ or equivalently $[v, b] = z_2$. In this case we get $M_2' = \langle z_2 \rangle$ since

$$[a^2, b] = [a, b]^a[a, b] = v^a v = (vz_1)v = v^2 z_1 = (z_1z_2)z_1 = z_2\,.$$

Also, we have here $M_3 = H$ because

$$v^{ab} = (vz_1)^b = (vz_2)z_1 = v(z_1z_2) = vv^2 = v^3 = v^{-1}$$

and so $ab$ inverts $G'$. We have $[v, ab] = z_1z_2$ and

$$[a^2, ab] = [a^2, b] = [a, b]^a[a, b] = v^a v = (vz_1)v = v^2 z_1 = (z_1z_2)z_1 = z_2\,,$$

and so we have here $M_3' = \langle z_1, z_2 \rangle$.

In both cases (1) and (2), we may set $v^b = z_1^{\epsilon} z_2$, where in case (1) we have $\epsilon = 1$ and in case (2) we have $\epsilon = 0$. Thus, if $\epsilon = 0$, then $H = \Phi(G)\langle ab \rangle$ and if $\epsilon = 1$, we have $H = \Phi(G)\langle b \rangle$.

Also, we may set

$$a^{2^m} = z_1^{\alpha} z_2^{\beta}, \quad b^{2^n} = z_1^{\gamma} z_2^{\delta},$$

where $\alpha, \beta, \gamma, \delta \in \{0, 1\}$ since we know that $a^{2^m}, b^{2^n} \in H' = \langle z_1, z_2 \rangle$.

Conversely, by inspection of groups given in parts (i) and (ii) of our theorem, we see that all these groups have the title property. Our theorem is proved. □

# § 146 $p$-groups all of whose maximal subgroups, except one, have cyclic derived subgroups

In this section written by the second author, the solution of Problem 2248 is presented.

In § 145 we completely determined $p$-groups $G$ that possess exactly one maximal subgroup $H$ such that $|H'| > p$. It turns out that in this case we must have $p = 2$, $d(G) = 2$, $G'$ is abelian of type $(4, 2)$, $H'$ is a four-group and $G$ is of class 3.

Here we generalize this result by assuming only that a $p$-group $G$ has exactly one maximal subgroup $H$ such that $H'$ is noncyclic (and so the derived subgroups of all other maximal subgroups of $G$ are cyclic). It is surprising that here we must also have $p = 2$, $G'$ is abelian of rank 2, $|G' : H'| = 2$ and $d(G) \leq 3$. Some more information about such 2-groups $G$ are given in Theorem 146.6 for the case $d(G) = 2$ and in Theorem 146.8 for the case $d(G) = 3$ (where in this case the nilpotence class of $G$ is not bounded). It follows from the hypothesis that the title group $G$ is nonmetacyclic (indeed, its derived subgroup is noncyclic).

We shall prove our results starting with a series of propositions about the title groups. The following result (see Exercise 1.69) is used very often in this section: If $M_1$ and $M_2$ are two distinct maximal subgroups of a $p$-group $G$, then $|G' : (M_1'M_2')| \leq p$.

**Proposition 146.1.** *Let $G$ be a title $p$-group. Then $G'$ is noncyclic abelian of rank $\leq 3$.*

*Proof.* Let $H$ be the unique maximal subgroup of $G$ such that $H'$ is noncyclic. Let $K \neq L$ be two maximal subgroups of $G$ which are both distinct from $H$. Then $K'$ and $L'$ are both cyclic normal subgroups of $G$. By Exercise 1.69, $|G' : (K'L')| \leq p$ and so $d(G') \leq 3$. On the other hand, the cyclic subgroups

$$K', L' \lhd G, \mathrm{Aut}(K'), \mathrm{Aut}(L')$$

are abelian which infers that $G'$ centralizes both $K'$ and $L'$ and so $K'L' \leq Z(G')$. This implies that $G'$ is abelian. □

**Proposition 146.2.** *Let $G$ be a title $p$-group. Then* $d(G) \leq 3$.

*Proof.* Since $G'$ is noncyclic, there exists a $G$-invariant subgroup $G_0$ contained in $G'$ such that $G'/G_0 \cong E_{p^2}$. We want to show that $d(G/G_0) \leq 3$ (in which case we would also have $d(G) \leq 3$ since $G_0 \leq \Phi(G)$), and so we may assume that $G_0 = \{1\}$. Thus, we may suppose that $G$ is a $p$-group with $G' \cong E_{p^2}$ and each maximal subgroup of $G$ with at most one exception $H$ has its derived subgroup of order $\leq p$. If $|H'| \leq p$, then, by Theorem 139.A, $d(G) \leq 3$. If $H' \cong E_{p^2}$, then by Theorem 145.8, we get $d(G) = 2$ and we are done. □

**Proposition 146.3.** *Let $G$ be a title $p$-group. Then the characteristic subgroup:*

$$H_0 = \langle M' \mid M \text{ is any maximal subgroup of } G \quad \text{with } M' \text{ cyclic}\rangle$$

*is noncyclic and has index $p$ in $G'$.*

*Proof.* Suppose that $H_0$ is cyclic; then $H_0 < G'$ so that, by Exercise 1.69, $|G' : H_0| = p$. Let $H < G$ be maximal with noncyclic $H'$; then $G' = H'H_0$. It follows that the group $G/H_0$ is nonabelian with exactly one nonabelian maximal subgroup $H/H_0$. In this case the number of abelian maximal subgroups in $G/H_0$ is $> 0$ and divisible by $p$, contrary to Exercise 1.6 (a). □

**Proposition 146.4.** *Let $G$ be a $p$-group with exactly one maximal subgroup $H$ with a noncyclic derived subgroup $H'$. Assume in addition that* $\mathrm{d}(G) = 2$. *Then the (noncyclic) characteristic subgroup $H_0$ of $G$ defined in Proposition 146.3 is of index $p$ in $G'$ and $G/H_0$ is nonmetacyclic and minimal nonabelian. For any two distinct maximal subgroups $K, L$ of $G$ one has $K'L' = H_0$. In particular,* $\mathrm{d}(H_0) = 2$ *and $H' \leq H_0$ with $H' \geq \Omega_1(H_0) \cong \mathrm{E}_{p^2}$.*

*Proof.* By Proposition 146.3, there are maximal subgroups $M \neq N$ such that $M'$ and $N'$ are cyclic and $K_0 = M'N'$ is noncyclic. By Exercise 1.69, $|G' : K_0| \leq p$.

Suppose for a moment that $G' = K_0$. Let $X$ be a $G$-invariant subgroup contained in $G' = K_0$ such that $M' \leq X < K_0$ and $|K_0 : X| = p$; then $N' \nleq X$. Since $\mathrm{d}(G/X) = 2$ and $(G/X)' = K_0/X$ is of order $p$, it follows that $G/X$ is minimal nonabelian (Lemma 65.2 (a)). But then $N/X$ is abelian and so $N' \leq X$, a contradiction. Thus, $|G' : K_0| = p$. Therefore, by Lemma 65.2 (a) again, $G/K_0$ is minimal nonabelian. Hence for each maximal subgroup $L$ of $G$, $L/K_0$ is abelian and so $L' \leq K_0$. In particular, $K_0 = H_0$ (where $H_0$ is defined in Proposition 146.3) and $H' \leq H_0$ with $H' \geq \Omega_1(H_0) \cong \mathrm{E}_{p^2}$. For any two distinct maximal subgroups $K, L$ of $G$, we get, by Exercise 1.69, $K'L' = H_0$. Also, since $G$ is nonmetacyclic, it follows from Theorem 36.1 that $G/H_0$ is nonmetacyclic. □

Thus, the subgroup $H_0$ of Proposition 146.3 contains derived subgroups of all maximal subgroups of $G$.

**Proposition 146.5.** *Let $G$ be a $p$-group with exactly one maximal subgroup $H$ with noncyclic derived subgroup $H'$. Assume, in addition, that* $\mathrm{d}(G) = 2$. *Then the characteristic subgroup $H_0$ of $G$, defined in Proposition 146.3, is equal to $H'$.*

*Proof.* Assume, by way of contradiction, that $H' < H_0$. In this case, each maximal subgroup of $G/\Phi(H_0)$ has a derived group of order $\leq p$ and $(G/\Phi(H_0))'$ is noncyclic of order $p^3$. By Theorem 137.7,

$$p > 2, \quad \mathrm{cl}(G/\Phi(H_0)) = 3, \quad G'/\Phi(H_0) \cong \mathrm{E}_{p^3}.$$

Hence $\mathrm{d}(G') = 3$ and $\Phi(H_0) = \Phi(G')$. If $\Phi(H_0) = \Phi(G') = \{1\}$, then $H_0 \cong \mathrm{E}_{p^2}$ and, by assumption, $H' < H_0$ is of order $p$, a contradiction. Thus, $\Phi(H_0) = \Phi(G') > \{1\}$.

Let $L$ be a fixed maximal subgroup of $G$ such that $L'$ is cyclic. Then there is another maximal subgroup $K \neq L$ of $G$ such that $K'$ is cyclic and $L'K' = H_0$ (Proposition 146.4). In particular, $H_0/L'$ is cyclic. On the other hand, $G'/L'$ is noncyclic since $\mathrm{d}(G') = 3$ and so we may apply Proposition 139.4 on the factor group $G/L'$. It follows that $G'/L' \cong \mathrm{E}_{p^2}$. This gives $|H_0 : L'| = p$, where $|L'| = p^m$, $m \geq 2$, $H_0$ is of type $(p^m, p)$ and $\Phi(H_0) =$

$\Phi(G')$ is cyclic of order $p^{m-1}$. Also, for each maximal subgroup $M$ of $G$ which is distinct from $H$, $M' \cong C_{p^m}$ is a cyclic maximal subgroup of $H_0$ (here we use Exercise 1.69). Hence $H_0$ has exactly $p$ cyclic maximal subgroups that are all derived subgroups of $p$ pairwise distinct maximal subgroups of $G$, which are all distinct from the maximal subgroup $H$.

Note that there are elements of order $p$ in $G' - H_0$ and so $G' \cong C_{p^m} \times E_{p^2}$. Also, $H_0/\Omega_1(H_0)$ is cyclic of order $\geq p$ and $G'/\Omega_1(H_0)$ is noncyclic (since there are elements of order $p$ in $G'-H_0$) and so by Proposition 139.4 applied on the factor group $G/\Omega_1(H_0)$, we obtain $G'/\Omega_1(H_0) \cong E_{p^2}$. This implies $m = 2$ and so $H' = \Omega_1(H_0) \cong E_{p^2}$.

Consider $G/\Phi(H_0)$, where $\Phi(H_0) = \Phi(G') \leq Z(G)$ is of order $p$, $G'/\Phi(H_0) \cong E_{p^3}$, $p > 2$, $d(G) = 2$, and each maximal subgroup of $G/\Phi(H_0)$ has a derived subgroup of order $p$. Since $K_3(G) = [G, G'] \leq H_0$ and $G'/K_3(G)$ is cyclic, it follows that $K_3(G)$ covers $H_0/\Phi(H_0)$ and so $K_3(G) = H_0$. We can find elements $a \in G$ and $c \in G'$ such that $k = [a, c] \in K_3(G) - H'$, where $H' = \Omega_1(H_0)$. Hence $o(k) = p^2$ and $k^p \in \Phi(H_0) = \Phi(G') \leq Z(G)$. Let $L$ be a maximal subgroup of $G$ which contains $a$ and $c$. Then $L' = \langle k \rangle$, which implies that $\langle k \rangle$ is normal in $G$. It follows that $L_0 = \langle a, c \rangle$ is a nonabelian two-generator $p$-group, $p > 2$, such that

$$LO_0' = \langle [a, c] \rangle = \langle k \rangle \cong C_{p^2} .$$

By Lemma 139.1, $\mho_1(L_0') = \langle [a, c]^p \rangle = \langle [a, c^p] \rangle$. But $c^p \in \Phi(G') \leq Z(G)$, and so $[a, c^p] = 1$, a contradiction. The proof is complete. □

Let $G$ be a title $p$-group. By Proposition 146.2, $d(G) \leq 3$. Here we suppose $d(G) = 2$ and then we use Proposition 146.4 together with the notation introduced there. By Proposition 146.5, $H' = H_0$. Considering the factor group $G/\Phi(H_0)$ we see that each maximal subgroup of $G/\Phi(H_0)$ except $H/\Phi(H_0)$ has its derived subgroup of order $\leq p$. By Theorem 145.8, we get $p = 2$ and $G'/\Phi(H_0)$ is of type (4, 2). Hence $d(G') = 2$ and so $G'$ is metacyclic. By Proposition 146.4, $|G : H'| = 2$ and, if $\{H_1, H_2, H\}$ is the set of maximal subgroups of $G$, then $H_1'H_2' = H'$. Thus, the following result is proved:

**Theorem 146.6.** *Let $G$ be a $p$-group with exactly one maximal subgroup $H$ such that $H'$ is noncyclic. Then $d(G) \leq 3$. Assume here $d(G) = 2$. In this case we have $p = 2$, $G'$ is abelian of rank 2 and $|G' : H'| = 2$. If $\{H_1, H_2, H\}$ is the set of maximal subgroups of $G$, then $H_1'H_2' = H'$.*

It remains to investigate the title groups $G$ with $d(G) = 3$. For this purpose we need the following auxiliary result:

**Lemma 146.7.** *Let $G = \langle a, b, c \rangle$ be a $p$-group with $d(G) = 3$, $cl(G) = 2$, $G' \cong E_{p^2}$, and $\Phi(G) = Z(G)$. Then $G$ has exactly one abelian maximal subgroup $A$ and for each subgroup $X_i$ of order $p$ in $G'$ ($i = 1, 2, \ldots, p + 1$) there are exactly $p$ pairwise distinct maximal subgroups $L_{ij}$ ($j = 1, 2, \ldots, p$) of $G$ such that $L_{ij}' = X_i$.*

*Proof.* We may choose the generators $a, b, c$ of $G$ so that setting $[a, b] = k$ and $[b, c] = l$, one obtains $G' = \langle k, l\rangle$. For each $i = 0, 1, \dots, p-1$,

$$[ac^{-i}, b] = [a, b][c, b]^{-i} = kl^i$$

and so each subgroup of order $p$ in $G'$ is a commutator subgroup of a maximal subgroup $\Phi(G)\langle b, c\rangle$ or $\Phi(G)\langle ac^{-i}, b\rangle$ $(i = 0.1, \dots, p-1)$ of $G$.

We find an abelian maximal subgroup $A$ of $G$. If $[a, c] = 1$, then we may set $A = \Phi(G)\langle a, c\rangle$. If $[a, c] = l^j$ with $j \not\equiv 0 \pmod p$, then

$$[ab^{-j}, c] = [a, c][b, c]^{-j} = l^j l^{-j} = 1$$

and so we may set $A = \Phi(G)\langle ab^{-j}, c\rangle$. If $[a, c] = (kl^i)^j$ for some $i \in \{0, 1, \dots, p-1\}$, where $j \not\equiv 0 \pmod p$, then we compute:

$$[ac^{-i}, b^{-j}c] = [ac^{-i}, b]^{-j}[a, c] = (kl^i)^{-j}(kl^i)^j = 1\,,$$

and so we may set $A = \Phi(G)\langle ac^{-i}, b^{-j}c\rangle$. Indeed, $\langle a, b^{-j}\rangle$ (with $j \not\equiv 0 \pmod p$) covers $G/(\Phi(G)\langle c\rangle)$ and so $\Phi(G)\langle ac^{-i}, b^{-j}c\rangle$ is a maximal subgroup of $G$. By Exercise 1.69, $A$ is a unique maximal abelian subgroup of $G$.

Let $X$ be any given subgroup of order $p$ in $G'$. By the above, there is at least one maximal subgroup $L \neq A$ of $G$ such that $L' = X$. By considering the factor group $G/X$, we see that $A/X$ and $L/X$ are two distinct abelian maximal subgroups of $G/X$ and so $G/X$ has exactly $p+1$ abelian maximal subgroups. Note that a nonabelian $p$-group has exactly 0, 1, or $p+1$ abelian maximal subgroups (see Exercise 1.6 (a)). Hence there are exactly $p$ maximal subgroups $L_i$ of $G$ such that $L_i' = X$. Our lemma is proved. □

**Theorem 146.8.** *Let $G$ be a $p$-group with exactly one maximal subgroup $H$ such that $H'$ is noncyclic. Assume here $\mathrm{d}(G) = 3$. In that case, $p = 2$, $G'$ is abelian of type $(2^n, 2)$, $n \geq 2$, and $G$ is of class $n+1$. Let*

$$\Gamma_1 = \{N_1, N_2, N_3, N_4, M, A, H\}\,.$$

*Then*

$$N_1' = N_2' = \langle y\rangle \cong C_{2^n}\,, \quad M' = \langle x\rangle\,,$$

*where $x \in G' - \langle y\rangle$ with*

$$x^2 \in \langle y^4\rangle\,, \quad A' = \langle y^2\rangle = \Phi(G')\,, \quad N_3' = N_4' = \langle yx\rangle \cong C_{2^n}\,, \quad H' = \langle y^2, x\rangle \cong C_{2^{n-1}} \times C_2\,.$$

*Finally,*

$$K_3(G) = [G, G'] = \langle y^2\rangle = \Phi(G') = [G, \Phi(G)]$$

*and $K_i(G) = \langle y^{2^{i-2}}\rangle$ for $3 \leq i \leq n+2$ and so $G$ is of nilpotence class $n+1$ $(n \geq 2)$.*

*Proof.* Let $H_0$ be the characteristic subgroup of $G$ which is contained in $G'$ as defined in Proposition 146.3. Assume at the moment that $G' \neq H_0$. Let $M \neq N$ be any to maximal subgroups of $G$ such that $M'$ and $N'$ are cyclic. Then $M'N' \leq H_0$ and, by Exercise 1.69, we get $|G' : (M'N')| \leq p$. It follows that $M'N' = H_0$ and $|G' : H_0| = p$. There are elements $x, y \in G$ such that $[x, y] \in G' - H_0$. Since $\langle x, y\rangle < G$, we must have $\langle x, y\rangle \leq H$ and so $H'$ covers $G'/H_0$. This implies that $G/H_0$ is a nonabelian $p$-group (since $|G'/H_0| = p$) all of whose maximal subgroups, except $H/H_0$, are abelian and so $G/H_0$ has exactly $p^2 + p$ abelian maximal subgroups, a contradiction (see Exercise 1.6 (a)).

We have proved that we must have $H_0 = G'$, where $H_0$ is defined in Proposition 146.3. By Proposition 146.1, $G'$ is abelian with $G'/\Phi(G') \cong E_{p^2}$ or $E_{p^3}$. If $|H'/(H' \cap \Phi(G'))| > p$, then $G/\Phi(G')$ is a $p$-group all of whose maximal subgroups, except one, have their derived subgroups of order $\leq p$. Then, by Theorem 145.8, $d(G/\Phi(G')) = d(G) = 2$, a contradiction. Hence $|H'/(H' \cap \Phi(G'))| \leq p$ and so $\Phi(G') \neq \{1\}$ (since $H'$ was assumed to be noncyclic). All maximal subgroups of $G/\Phi(G')$ have their derived subgroups of order $\leq p$, $G'/\Phi(G')$ is noncyclic, and $d(G/\Phi(G')) = 3$. Thus $G/\Phi(G')$ is a group appearing in part (c) of Theorem 139.A. In particular, $\Phi(G)/\Phi(G') = Z(G/\Phi(G'))$.

Let $S$ be a $G$-invariant subgroup contained in $H' \cap \Phi(G')$ such that $|H' : S| = p$. Indeed, if $H' \nleq \Phi(G')$, then we may set $S = H' \cap \Phi(G')$, and if $H' \leq \Phi(G')$, then $S$ could be any $G$-invariant subgroup of index $p$ in $H'$. By considering the factor group $G/S$, we see that $G/S$ is a $p$-group satisfying the assumption of Theorem 139.A. It follows that each maximal subgroup of $G/S$ has its derived subgroup of order $\leq p$. Hence for each maximal subgroup $M_i$ of $G$ such that $M_i' = \langle x_i\rangle$ is cyclic, $i = 1, 2, \ldots, p^2 + p$, one has $x_i^p \in S$. On the other hand,

$$G' = H_0 = \langle x_i \mid i = 1, 2, \ldots, p^2 + p\rangle$$

and so

$$\Phi(G') = \langle x_i^p \mid i = 1, 2, \ldots, p^2 + p\rangle \leq S\,.$$

We have proved that $H' > \Phi(G')$ and $|H' : \Phi(G')| = p$.

Assume that $G'/\Phi(G')$ is elementary abelian of order $p^3$, i.e., $d(G') = 3$. By Exercise 1.69, we see that all $1 + p + p^2$ maximal subgroups of $G/\Phi(G')$ have $1 + p + p^2$ pairwise distinct derived subgroups of order $p$ contained in $G'/\Phi(G')$. Let $X$ be any maximal subgroup of $G'$ so that $X/\Phi(G') \cong E_{p^2}$. There are exactly $p + 1$ subgroups $X_i/\Phi(G')$ of order $p$ in $X/\Phi(G')$ and so there are at least $p \geq 2$ of them that are distinct from $H'/\Phi(G')$. Hence there exist two distinct maximal subgroups $M_1$ and $M_2$ of $G$ so that $M_1' = \langle x_1\rangle$, $M_2' = \langle x_2\rangle$ are cyclic, $x_1, x_2 \in X - \Phi(G')$, and $\langle x_1, x_2\rangle$ covers $X/\Phi(G')$. By Exercise 1.69, $|G' : (\langle x_1\rangle\langle x_2\rangle)| \leq p$ and so we must have $\langle x_1\rangle\langle x_2\rangle = X$. We have proved that each maximal subgroup $X$ of $G'$ is two-generator. On the other hand, $d(G') = 3$ and $\Phi(G') \neq \{1\}$ and so $G'$ is not elementary abelian. We may set

$$G' = \langle x\rangle \times \langle y\rangle \times \langle v\rangle$$

for some $x, y, v \in G'$, where we may assume that $o(x) \geq p^2$. But then

$$Y = \langle x^p\rangle \times \langle y\rangle \times \langle v\rangle$$

is a maximal subgroup of $G'$ which is not two-generator, a contradiction. We have proved that $G'/\Phi(G') \cong E_{p^2}$, i.e., $d(G') = 2$.

We have actually proved that $G/\Phi(G')$ satisfies all assumptions of Lemma 146.7. Using this lemma, we see that there are maximal subgroups $M$ and $N$ of $G$ such that

$$M' = \langle x \rangle, \quad N' = \langle y \rangle, \quad \text{where } x \in H' - \Phi(G'),\ y \in G' - H'.$$

It follows from $\langle x, y\rangle\Phi(G') = G'$ that $\langle x, y\rangle = G'$ and so $\langle y\rangle\langle x\rangle = G'$ and $G'/\langle y\rangle$ is cyclic. Since $y^p \in \Phi(G') \le H'$ and $x \in H'$, we get $H' = \langle y^p\rangle\langle x\rangle$, which is a maximal subgroup of $G'$, and $H'/\langle y^p\rangle$ is cyclic. Note that $G'/\langle y^p\rangle$ is noncyclic because $y^p \in \Phi(G')$ and $d(G') = 2$. Each maximal subgroup of $G/\langle y^p\rangle$ has a cyclic derived subgroup but $(G/\langle y^p\rangle)'$ is noncyclic. By Theorem 139.A, we get $x^p \in \langle y^p\rangle$ and so $\langle y^p\rangle = \Phi(G') \ne \{1\}$ is cyclic and $o(y) = p^n$, $n \ge 2$. Hence $\langle y\rangle$ is a cyclic subgroup of index $p$ in $G'$ so that $G'$ is of type $(p^n, p)$. Since $H'$ is noncyclic, we must have $x^p \in \langle y^{p^2}\rangle$. Again by Lemma 146.7, there is a unique maximal subgroup $A$ of $G$ such that $A' \le \Phi(G') = \langle y^p\rangle$. If $A' \le \langle y^{p^2}\rangle$, then $A'\langle x\rangle < H'$ and so $|G' : (A'M')| > p$, contrary to Exercise 1.69. Hence we must have $\{1\} \ne A' = \langle y^p\rangle = \Phi(G')$.

Now we consider the structure of $G/\Phi(\Phi(G')) = G/\langle y^{p^2}\rangle$ so that we may assume for a moment $\langle y^{p^2}\rangle = \{1\}$. In this case,

$$o(y) = p^2, \quad |\Phi(G')| = p, \quad G' = \langle y\rangle \times \langle x\rangle \cong C_{p^2} \times C_p.$$

Here

$$M' = \langle x\rangle \cong C_p, \quad N' = \langle y\rangle \cong C_{p^2}, \quad A' = \langle y^p\rangle \cong C_p,$$
$$\Phi(G)/\langle y^p\rangle = Z(G/\langle y^p\rangle), \quad H' = \langle y^p\rangle \times \langle x\rangle \cong E_{p^2}.$$

We have $(M \cap N)' \le M' \cap N' = \{1\}$ and so $M \cap N$ is abelian. But $M \cap N > \Phi(G)$ with $|M \cap N : \Phi(G)| = p$ and so $\Phi(G)$ is abelian. Assume that $\Phi(G) \le Z(G)$. Then all $p + 1$ maximal subgroups of $N$ that contain $\Phi(G)$ are abelian, which implies $|N'| \le p$, a contradiction. Hence $\Phi(G) \nleq Z(G)$ and so we get $[G, \Phi(G)] = \langle y^p\rangle$. By Lemma 146.7 and a result of A. Mann, there are exactly $p - 1$ maximal subgroups $M_i$ of $G$ such that $M_i' = \langle x_i\rangle$ are pairwise distinct subgroups of order $p$ contained in $H'$ and $\langle x_i\rangle \ne \langle y^p\rangle$, $i = 1, \dots, p - 1$. In that case,

$$[M_i, \Phi(G)] \le \langle y^p\rangle \cap \langle x_i\rangle = \{1\} \Rightarrow \Phi(G) \le Z(M_i).$$

If $p \ge 3$, then $\langle M_1, M_2\rangle = G$ and so $\Phi(G) \le Z(G)$, a contradiction. We have proved that we must have $p = 2$. In this case, $M = M_1$ with $M' = \langle x\rangle$, where $x = x_1$ is an involution in $H' - \langle y^2\rangle$. By the above, $\Phi(G) \le Z(M)$. Since $\langle y^2\rangle$ and $\langle x\rangle$ are normal in $G$, it follows that

$$H' = \langle y^2\rangle \times \langle x\rangle \le Z(G).$$

There are exactly two maximal subgroups $N \ne N_1$ such that $N' = N_1' = \langle y\rangle \cong C_4$ and there are exactly two other maximal subgroups $L \ne L_1$ such that $L' = L_1' = \langle yx\rangle \cong C_4$.

Also, $C_G(G') \geq M$ since $\Phi(G) \leq Z(M)$. Assume that $C_G(G') = G$, i.e., $G$ is of class 2. Since $\Phi(G)/\langle y^2\rangle \leq Z(N/\langle y^2\rangle)$, it follows that for any $a, b \in N - \Phi(G)$ such that $\langle a, b\rangle$ covers $N/\Phi(G)$, one has $\langle [a, b]\rangle = \langle y\rangle$. Take $a \in (M \cap N) - \Phi(G)$ and $b \in N - M$ so that $[a, b] = y^{\epsilon}$ with $\epsilon = \pm 1$. We compute (noting that $G$ is of class 2):

$$[a, b^2] = [a, b]^2 = y^2 .$$

On the other hand, $b^2 \in \Phi(G)$, and so $\langle a, b^2\rangle \leq M$. This is a contradiction since $M' = \langle x\rangle \neq \langle y^2\rangle$. We have proved that $C_G(G') = M$ and so $[G, G'] = \langle y^2\rangle$ which infers that $G$ is of class 3. Note that $H' \leq Z(G)$ and $\langle y\rangle$ is normal in $G$ and so each element in $G - M$ inverts $G'$.

We return to the general case, where $\langle y^4\rangle \geq \{1\}$ and we know already that we must have $p = 2$. Here we shall use the results from the previous paragraph for the factor group $G/\langle y^4\rangle$. Our title group $G$ is a 2-group with $d(G) = 3$, $G'$ is abelian of type $(2^n, 2)$, $n \geq 2$, $G$ has exactly two maximal subgroups $N_1 \neq N_2$ such that $N_1' = N_2' = \langle y\rangle \cong C_{2^n}$, and $G$ has exactly one maximal subgroup $M$ such that $M' = \langle x\rangle$, where $x \in G' - \langle y\rangle$ and $x^2 \in \langle y^4\rangle$. Then $\langle yx\rangle$ is another cyclic maximal subgroup of $G'$ (of order $2^n$) distinct from $\langle y\rangle$ and there are exactly two maximal subgroups $N_3 \neq N_4$ such that $N_3' = N_4' = \langle yx\rangle \cong C_{2^n}$. The unique maximal subgroup $H$ of $G$ such that $H'$ is noncyclic has its derived group $H' = \langle y^2, x\rangle$ (which is of type $(2^{n-1}, 2)$). Finally, $G$ has a unique maximal subgroup $A$ such that $A' = \langle y^2\rangle = \Phi(G')$. We have

$$Z(G/\langle y^2\rangle) = \Phi(G)/\langle y^2\rangle\,, \quad [G, \Phi(G)] = [G, G'] = \langle y^2\rangle .$$

Each element $g \in G - M$ inverts on $G'/\langle y^4\rangle$. Note that $\langle y\rangle$ is normal in $G$ and so we have $(y^{-1})^g = ys$ with $s \in \langle y^4\rangle$. This gives $(y^{-2^i})^g = y^{2^i}s^{2^i}$ and therefore we get

$$[g, y^{2^i}] = y^{2^{i+1}}s^{2^i}, \quad \text{where } s^{2^i} \in \langle y^{2^{i+2}}\rangle .$$

Since $K_3(G) = [G, G'] = \langle y^2\rangle$, we get $K_4(G) = [G, K_3(G)] = \langle y^{2^2}\rangle$ and so in general for $3 \leq i \leq n+2$, we have $K_i(G) = \langle y^{2^{i-2}}\rangle$ and so

$$K_{n+1}(G) = \langle y^{2^{n-1}}\rangle \neq \{1\} \text{ but } K_{n+2}(G) = \langle y^{2^n}\rangle = \{1\} .$$

This shows that $G$ is of class $n+1$ and our theorem is proved. □

# § 147 *p*-groups with exactly two sizes of conjugate classes

Our main result (Theorem 147.2) was actually proved by K. Ishikawa and L. Verardi (see [Ish]), but our proof is conceptual and different.

Let $G$ be a $p$-group with exactly two sizes of conjugate classes. In particular, $G$ is nonabelian. Let $A$ be a maximal normal abelian subgroup of $G$ so that

$$C_G(A) = A\,, \quad A < G\,, \quad Z(G) < A\,.$$

Let $x \in G - A$ (so that $x$ is noncentral in $G$) and $a \in A$. Then, by Lemma 123.1,

$$[x, a] \in Z(G) \Rightarrow [G, A] \le Z(G)\,.$$

Now

$$[A, G] \le Z(G) \Rightarrow [A, G, G] = [G, A, G] = \{1\}\,.$$

Therefore, by the Three Subgroups lemma, we get $[G, G, A] = [G', A] = \{1\}$. Hence $G' \le A$ and so $G'$ is abelian, $G/A$ is abelian and so $G$ is metabelian. We also get $[G, G'] \le Z(G)$ and so $G$ is of class at most 3. We have proved:

**Proposition 147.1.** *Let $G$ be a title $p$-group and $A$ be a maximal normal abelian subgroup of $G$. Then $[G, A] \le Z(G)$ and $G' \le A$, so $G$ is of class at most 3.*

Let $G$ be a title $p$-group and $A$ be a maximal normal abelian subgroup of $G$. Suppose that $A/Z(G)$ is not elementary abelian. Then there is $a \in A$ such that $a^p \notin Z(G)$ but $a^{p^2} \in Z(G)$. Since $a \notin Z(G)$, one obtains $A \le C_G(a) < G$, where $C_G(a) \trianglelefteq G$. Let $b \in G - C_G(a)$ such that $b^p \in C_G(a)$. By Lemma 123.1, $1 \ne [a, b] \in Z(G)$, where $\langle [a, b] \rangle \trianglelefteq G$ so that $\langle a, b\rangle/\langle [a, b]\rangle$ is abelian and therefore $\langle a, b\rangle' = \langle [a, b]\rangle$ and $\langle a, b\rangle$ is of class 2. We get

$$[a, b]^p = [a, b^p] = 1$$

and so $\langle a, b\rangle$ is minimal nonabelian (see Lemma 65.2 (a)). It follows that

$$a^p \in \Phi(\langle a, b\rangle) = Z(\langle a, b\rangle)$$

and therefore $a^p$ commutes with $b$. But $a^p \notin Z(G)$ and so

$$1 \ne |G : C_G(a^p)| < |G : C_G(a)|\,,$$

contrary to our assumption that $G$ has exactly two class sizes. Thus, $A/Z(G)$ is elementary abelian.

Assume that $G/A$ is not elementary abelian. Then there is an element $g \in G - A$ such that $g^p \notin A$ but $g^{p^2} \in A$. There is an element $a \in A - Z(G)$ such that $[a, g^p] \ne 1$. Then $[a, g] \ne 1$ and, by Lemma 123.1, $[a, g] \in Z(G)$ so that $\langle a, g\rangle' = \langle [a, g]\rangle$ and $\langle a, g\rangle$ is of class 2. We get

$$[a, g]^p = [a^p, g] = 1$$

since $A/Z(G)$ is elementary abelian and so $a^p \in Z(G)$. Hence $\langle a, g\rangle$ is minimal non-abelian (see Lemma 65.2 (a)) which implies $g^p \in Z(\langle a, g\rangle)$ and so $[a, g^p] = 1$, a contradiction. We have proved that $G/A$ is also elementary abelian.

Let $g \in G - A$ be such that $g^p \in A - Z(G)$. Set

$$C = C_G(g) \quad \text{so that } C \geq \langle g, g^p, Z(G)\rangle\,, \; A > (A \cap C)\,.$$

But $C_G(g^p) \geq \langle A, C\rangle < G$ (since $g^p \notin Z(G)$) and $\langle A, C\rangle > C$, contrary to our assumption that $G$ has only two class sizes. We have proved that $\mho_1(G) \leq Z(G)$. This implies at once that $[G, A]$ is elementary abelian. Indeed, let $g \in G$ and $a \in A$. Then $[g, a] \in Z(G)$ (Proposition 147.1) and $\langle g, a\rangle/\langle [g, a]\rangle$ is abelian and so $\langle g, a\rangle' = \langle [g, a]\rangle$ and therefore $\langle g, a\rangle$ is of class $\leq 2$. This implies $1 = [g^p, a] = [g, a]^p$ and our claim is proved.

Assume $p > 2$. Then we claim that for any $a, b \in G$, one has $[a^p, b] = [a, b]^p$. Indeed, we set $g = [a, b] \in A$. If $g$ commutes with $a$, then for each $n \geq 1$ we prove by induction that $[a^n, b] = [a, b]^n$. Indeed, for $n > 1$,

$$\begin{aligned}[a^n, b] &= [aa^{n-1}, b] = [a, b]^{a^{n-1}}[a^{n-1}, b]\\ &= [a, b][a^{n-1}, b] = [a, b][a, b]^{n-1} = [a, b]^n\,.\end{aligned}$$

In particular, we have $[a^p, b] = [a, b]^p$.

We assume now that $[g, a] = z \neq 1$. Since $[A, G] \leq Z(G)$ and $[A, G]$ is elementary abelian, we have $o(z) = p$ and $z \in Z(G)$. We note that

$$g^a = a^{-1}ga = g(g^{-1}a^{-1}ga) = g[g, a] = gz \quad \text{and so } g^{a^i} = gz^i$$

for all $i \geq 1$. We have

$$\begin{aligned}[a^p, b] &= [a \cdot a^{p-1}, b] = [a, b]^{a^{p-1}}[a^{p-1}, b]\\ &= [a, b]^{a^{p-1}}[a \cdot a^{p-2}, b] = [a, b]^{a^{p-1}}[a, b]^{a^{p-2}}[a^{p-2}, b]\,,\end{aligned}$$

and so continuing we get finally:

$$\begin{aligned}[a^p, b] &= [a, b]^{a^{p-1}}[a, b]^{a^{p-2}} \dots [a, b]^a[a, b]\\ &= g^{a^{p-1}}g^{a^{p-2}} \dots g^a g = (gz^{p-1})(gz^{p-2}) \dots (gz)g\\ &= g^p z^{(p-1)+(p-2)+\dots+1} = g^p z^{(p-1)\frac{p}{2}} = g^p = [a, b]^p\,,\end{aligned}$$

where we have used the fact that $p > 2$. We have proved that in any case we get $[a^p, b] = [a, b]^p$. But $\mho_1(G) \leq Z(G)$ and so $[a^p, b] = [a, b]^p = 1$ and therefore we have proved that in the case $p > 2$ the subgroup $G'$ is elementary abelian.

If $p = 2$, then the above results show that $G/Z(G)$ is elementary abelian and so $G$ is of class 2. In this case for any $x, y \in G$ we get $[x, y]^2 = [x^2, y] = 1$ and so $G'$ is also elementary abelian. We have proved the following result:

**Theorem 147.2.** *Let $G$ be a $p$-group with exactly two sizes of conjugate classes. Then $G$ is of class at most 3, $\mho_1(G) \leq Z(G)$, $G'$ is elementary abelian, and for any maximal normal abelian subgroup $A$ of $G$ we have that $A/Z(G)$ and $G/A$ are elementary abelian. If $p = 2$, then $G$ is of class 2.*

# § 148 Maximal abelian and minimal nonabelian subgroups of some finite two-generator $p$-groups especially metacyclic

*Contents.*



*$1^o$ Introduction.* A group $G$ is said to be *metacyclic* if it contains a normal cyclic subgroup $C$ with cyclic quotient group $G/C$. It is easy to show that, provided $G$ is a noncyclic $p$-group, then this $C$ is a maximal cyclic subgroup of $G$ (otherwise, $C \leq \Phi(G)$ so $G/C$ is noncyclic). All sections of metacyclic groups are metacyclic and so two-generator. There are a great number of papers devoted to this important class of groups. For example, Isaacs and the first author have proved, independently, that the representation group of a nonabelian metacyclic $p$-group is metacyclic (see Theorem 47.2). Also, automorphism groups of such groups are described (see [Kur] and [Mal6]). However, there is surprisingly little known about the subgroup structure of metacyclic $p$-groups in spite of the fact that such groups are classified (i.e., some 1–1-correspondence is established between such groups and certain sets of parameters; see, for example, [Kin3], [XZ1]); see the list of problems).

A number of interesting properties of metacyclic $p$-groups are proved in [HK]. Below, some results obtained in that paper for a nonabelian metacyclic $p$-group $G$ with $|G'| = p^n$, $p > 2$, are presented. (a) One has $|G| = |Z(G)|\,|G'|^2$ (see also the text preceding Theorem 148.20). (b) There are exactly $(p-1)|G : G'|/p^{k+1}$ irreducible characters of degree $p^k$ for $k = 1, \dots, \log_p(|G'|)$ (another proof of the existence of such characters is also contained in the paragraph preceding Theorem 148.20).

There are a number of challenging problems concerning the subgroup structure of metacyclic $p$-groups, especially in the case $p = 2$ (see, for example, $10^o$).

In Subsections $3^o$–$9^o$ a number of new results concerning the subgroup structure of finite metacyclic $p$-groups is proved. More full results are obtained for $p > 2$ (this is due to the fact that, for $p > 2$, metacyclic $p$-groups are regular and there is only one nonabelian $p$-group with cyclic subgroup of index $p$).

Some results of the section were inspired by the following observation of the second author: if the derived subgroup of a two-generator $p$-group $G$, $p > 2$, is cyclic of order $p^2$, then all maximal subgroups of $G$ are nonabelian (see Theorem A.40.30 and Remark 148.9, below); this easily proved result is fairly unexpected. It follows from this and Lemma 148.3 (a), below, that if $p > 2$ and a nonabelian metacyclic $p$-group $G$ possesses an abelian maximal subgroup, then $G$ is minimal nonabelian (Corollary 148.2 (b)). The last assertion is not true for $p = 2$ (for example: $G = D_{2^n}$, $n > 3$).

Let us consider the following three sets of nonabelian (not necessary metacyclic) $p$-groups $G$:

- $\mathcal{G}_* = \{G \mid d(G) = 2\}$,
- $\mathcal{G}_{**} = \{G \mid |G'| = p^2\}$,
- $\mathcal{G}_{***} = \{G \mid G$ has an abelian subgroup of index $p\}$.

Clearly, it is impossible to classify $\mathcal{G}_*$-groups. The $p$-groups with derived subgroup of order $p$ are classified in [Blac2] (see also Lemma 4.2). It is doubtful that a classification of $\mathcal{G}_{**}$-groups may be obtained with the help of elementary methods. A $\mathcal{G}_{**}$-group has at most one abelian subgroup of index $p$.

If $G \in \mathcal{G}_* \cap \mathcal{G}_{**}$ and $L \leq G' \cap Z(G)$ is of order $p$, then $G/L$ is minimal nonabelian (Lemma 148.3 (a)) so any maximal subgroup of our group $G$ has the derived subgroup of order $\leq p$; such groups are classified in § 137. Note that if groups of some family are classified in a usual sense, this does not mean that their subgroup structure is known.

Defining relations of $\mathcal{G}_{***}$-groups are presented in [NRSB]. However, there are many unsolved problems for such groups, for example, description of their subgroups. It is doubtful that the use of the mentioned result from [NRSB] allows us to shorten the proof of Theorem 148.1.

In Theorem 148.1 the groups $G \in \mathcal{G}_* \cap \mathcal{G}_{**} \cap \mathcal{G}_{***}$ are considered. Using that theorem, some new results on the subgroup structure of metacyclic $p$-groups, $p > 2$, and some metacyclic 2-groups are obtained (the last results can be proved directly without appealing to Theorem 148.1, however, that theorem presents an independent interest). For example, if $G$ is such a (metacyclic) $p$-group, $p > 2$, then all its (i) minimal nonabelian subgroups have equal order, say $p^n$, and so all subgroups of order $p^{n-1}$ are abelian, and (ii) all its maximal abelian subgroups have order $p^{n-1}$. The number of minimal nonabelian subgroups in such groups $G$ is computed and it is shown that this number depends on $|G'|$ only.

In what follows $G$ denotes a finite $p$-group. By $\alpha_n(G)$ the number of $\mathcal{A}_n$-subgroups in $G$ is denoted. It is easy to prove, using Lemma 148.3 (f), that a minimal nonabelian $p$-group $G$ of order $> p^3$ has a cyclic center $\iff G \cong M_{p^n}$.

Let $G$ be a metacyclic $p$-group and

$$w(G) = w = \max\{i \geq 0 \mid |\Omega_i(G)| = p^{2i}\}.$$

If $|G| > p^3$, then our $G$ has a normal abelian subgroup of type $(p, p) \iff \Omega_1(G) \cong E_{p^2}$. Write $R(G) = \Omega_w(G)$ (see § 124, where that subgroup is important). By Lemma 148.3 (i), $G/R(G)$ is either cyclic or a 2-group of maximal class. A metacyclic $p$-group $G$ of exponent $p^e$ equals $R(G) \iff |G| = p^{2e}$.

Given a nonabelian $p$-group $G$, there is a positive integer $k$ such that $G$ is an $\mathcal{A}_k$-group. A $p$-group $G$ has an $\mathcal{A}_n$-subgroup $\iff$ $G$ is an $\mathcal{A}_k$-group for some $k \geq n$.

In Subsection 2° the following result is proved:

**Theorem 148.1.** *Suppose that a $p$-group $G \in \mathcal{G}_* \cap \mathcal{G}_{**} \cap \mathcal{G}_{***}$. Then $G/Z(G) \in \{S(p^3), D_8\}$. If $L \leq G' \cap Z(G)$ is of order $p$, then the quotient group $G/L$ is minimal nonabelian. There is $R < Z(G)$ such that $Z(G/R)$ is cyclic and $G'R/R \cong G'$. Write $|G/R| = p^{n+3}$, $n > 1$; then $Z(G/R) \cong C_{p^n}$.*

*Next we assume that $R = \{1\}$; then $Z(G)$ is cyclic of order $p^n$, $|G| = p^{n+3}$, $n > 1$.*

(a) *If $p > 2$, then the following holds:*
  (a1) *$G' \cong E_{p^2}$ (in particular, $G$ is nonmetacyclic),*
  (a2) *One has $G = C\Omega_1(G)$, where $\Omega_1(G)$ is of order $p^3$ and exponent $p$ and $C$ is nonnormal cyclic of order $p^{n+1}$, $Z(G) = \mho_1(G) = \mho_1(C)$,*
  (a3) *The set $\Gamma_1 = \{M_1, \dots, M_p, A = Z(G)\Omega_1(G)\}$, where $M_i$ has a cyclic subgroup of index $p$, all those $M_i$, which are nonabelian, are $\cong M_{p^{n+2}}$.*

(b) *If $p = 2$ and $G$ is nonmetacyclic*[1]*, then the following holds:*
  (b1) *$G' \cong C_4$ is a maximal cyclic subgroup of $G$. Next, $\Gamma_1 = \{A, B, C\}$, where $A/Z(G) \cong E_4 \cong B/Z(G)$, $C/Z(G) \cong C_4$ and $C$ is abelian of type $(2^{n+1}, 2)$,*
  (b2) *$A \cong M_{2^{n+2}}$, $B = DZ(G)$, where $D \cong D_8$ is $G$-invariant, $\Omega_1(G) = \Omega_2(G) = D\Omega_2(Z(G)) = \Omega_2(B)$,*
  (b3) *$G/D$ is cyclic of order $2^n$, $\Phi(G) = G'Z(G)$ is abelian of type $(2^n, 2)$,*
  (b4) *$G = DZ$, where $Z$ is a non-$G$-invariant cyclic subgroup of order $2^{n+1}$, $Z(G) < Z$.*

The above observation of the second author is an immediate consequence of Theorem 148.1 (a). Indeed, suppose that a $p$-group $G$, $p > 2$, satisfies $d(G) = 2$ and $|G'| = p^2$. Assume that $A \in \Gamma_1$ is abelian. Then $G \in \mathcal{G}_* \cap \mathcal{G}_{**} \cap \mathcal{G}_{***}$. In that case, by Theorem 148.1 (a), $G' \cong E_{p^2}$.

The second author has also proved (see Proposition 145.5) the following nice result which, in some sense, is converse to Theorem 148.1 (a): A two-generator $p$-group $G$, $p > 2$, with $G' \cong E_{p^2}$ possesses an abelian maximal subgroup, say $A$. The proof of this result is short but nontrivial (for this $G$ it follows from Lemma 148.3 (e), below, that $G/Z(G)$ is nonabelian of order $p^3$ and exponent $p$). By Lemma 148.2 (c), if the derived subgroup of a two-generator 2-group $G$ is noncyclic, then all members of the set $\Gamma_1$ are nonabelian.

Suppose that a $p$-group $G \in \mathcal{G}_* \cap \mathcal{G}_{**} \cap \mathcal{G}_{***}$. Then all members of the set $\Gamma_1$ have derived subgroups of order $\leq p$ (see the third sentence of the statement of The-

---

1 Note that $D_{16} \in \mathcal{G}_* \cap \mathcal{G}_{**} \cap \mathcal{G}_{***}$.

orem 148.1). The groups with the last property are classified in § 137; for more general result, see § 139. All maximal subgroups of our group $G$, except one, are metacyclic; the $p$-groups with that property are classified in § 87.

**Corollary 148.2.** *Suppose that a nonabelian p-group $G \in \mathcal{G}_*$, i.e., $\mathrm{d}(G) = 2$.*
(a) *If $p > 2$ and there is in $G'$ a $G$-invariant subgroup $L$ such that $G'/L \cong \mathrm{C}_{p^2}$, then all members of the set $\Gamma_1$ are nonabelian.*
(b) *Suppose that $p > 2$ and there is an abelian $A \in \Gamma_1$ (in that case, $G \in \mathcal{G}_* \cap \mathcal{G}_{***}$).*
   (b1) *If $|G'| > p$, then $G'$ is noncyclic.*
   (b2) *If, in addition, $G'$ is cyclic, then $G$ is minimal nonabelian (in that case, $|G'| = p$).*
(c) *If $p = 2$ and $G'$ is noncyclic, then all members of the set $\Gamma_1$ are nonabelian.*

*Proof.* We use the notation introduced in the statement.

(a) Without loss of generality, one may assume that $L = \{1\}$; then $G' \cong \mathrm{C}_{p^2}$. Assume that $H \in \Gamma_1$ is abelian. Then $H/\mathrm{Z}(G) \cong G' \cong \mathrm{C}_{p^2}$ (Lemma 148.3 (e)) so that $G/\mathrm{Z}(G)$ has two distinct cyclic subgroups, say $Z_1/\mathrm{Z}(G)$ and $Z_2/\mathrm{Z}(G)$, of index $p$. Then $Z_1, Z_2$ are abelian so that $Z_1 \cap Z_2 = \mathrm{Z}(G)$ is of index $p^2$. In That case, By Lemma 148.3 (e), $|G'| = p$, a contradiction. Thus, the set $\Gamma_1$ has no abelian member.

(b) Part (b1) follows immediately from (a). Now suppose that $\mathrm{d}(G) = 2$, $G'$ is cyclic and $A \in \Gamma_1$ is abelian. Then $A/\mathrm{Z}(G) \cong G'$ is cyclic of index $p$ in $G/\mathrm{Z}(G)$ (Lemma 148.3 (e)). The group $G/\mathrm{Z}(G)$ has two distinct cyclic subgroups $A/\mathrm{Z}(G)$ and $B/\mathrm{Z}(G)$ of index $p$ (Lemma 148.3 (l), below). Then $A, B$ are abelian of index $p$ in $G$ so $A \cap B = \mathrm{Z}(G)$ is of index $p^2 = |G : \Phi(G)|$ in $G$; hence $\mathrm{Z}(G) = \Phi(G)$ so that $G$ is minimal nonabelian, proving part (b2).

(c) Now let $p = 2$ and let $G'$ be noncyclic. There is a $G$-invariant $L < G'$ such that $G'/L \cong \mathrm{E}_4$. Passing to $G/L$, one may assume that $G' \cong \mathrm{E}_4$. Assume that $A \in \Gamma_1$ is abelian. Then $A/\mathrm{Z}(G) \cong G' \cong \mathrm{E}_4$ (Lemma 148.3 (e)) so that $G/\mathrm{Z}(G) \cong \mathrm{D}_8$.[2] If $B/\mathrm{Z}(G) < G/\mathrm{Z}(G)$ is cyclic of index 2, then $B \neq A$ is abelian, and so $A \cap B = \mathrm{Z}(G)$ has index 4 in $G$, a contradiction. Thus, $A$ does not exist so that all members of the set $\Gamma_1$ are nonabelian. □

It follows from part (b2) of Corollary 148.2 that, if $G$ is a nonabelian metacyclic $p$-group, $p > 2$, and $A \in \Gamma_1$ is abelian, then $G$ is minimal nonabelian.

2 Indeed, if $G/\mathrm{Z}(G)$ is abelian, it is of type $(4, 2)$ so $G/\mathrm{Z}(G)$ contains two distinct cyclic subgroups $F/\mathrm{Z}(G), H/\mathrm{Z}(G)$ of order 4; then, as above, $|G'| = 2$, a contradiction.

Most known results used in the sequel are gathered in the following:

**Lemma 148.3.** *Suppose that $G$ is a nonabelian $p$-group.*

(a) *Lemma 65.2 (a). If $\mathrm{d}(G) = 2$ and $G' \le \Omega_1(Z(G))$, then $G$ is minimal nonabelian.*

(b) *Theorem 36.1. The group $G$ is metacyclic $\iff$ $G/R$ is metacyclic for any $G$-invariant subgroup $R$ of index $p$ in $G'$.*

(c) *Exercise 1.69. If $A, B \in \Gamma_1$ are distinct, then $|G' : A'B'| \le p$.*

(d) *Theorem 7.2 (b,c,d). If $G$ is regular and $p^n \ge \exp(G)$, then*

$$\exp(\Omega_n(G)) = p^n , \quad \mho_1(G) = \{x^p \mid x \in G\} , \quad |\Omega_n(G)| = |G/\mho_n(G)| .$$

(e) *Lemma 1.1. If $A < G$ is abelian of index $p$, then $|G| = p|G'|\,|Z(G)|$ and $A/Z(G) \cong G'$.*

(f) *Lemma 65.1. Suppose that $G$ is minimal nonabelian. Then $|G'| = p$ and $|\Omega_1(G)| \le p^3$. If, in addition, $|\Omega_1(G)| < p^3$, then $G$ is metacyclic. If a nonmetacyclic $G$ of order $> p^3$ has a metacyclic maximal subgroup, it contains exactly $p$ metacyclic maximal subgroups.*

(g) *Theorem 69.3 (see also Theorem 13.7). If a $p$-group $G$, $p > 2$, has no normal subgroup of order $p^3$ and exponent $p$, then it is either metacyclic or a 3-group of maximal class.*

(h) *Exercise 10.10. If $H < G$ and every subgroup of $G$ of order $p|H|$, containing $H$, is of maximal class, then $G$ is also of maximal class.*

(i) *Lemma 1.4. If a normal subgroup $N$ of $G$ has no $G$-invariant abelian subgroup of type $(p, p)$, then $N$ is either cyclic or a 2-group of maximal class. In particular, if $N \le \Phi(G)$ and $Z(N)$ is cyclic, then $N$ is also cyclic.*

(j) *Proposition 72.1. If $n > 1$, $G$ is an $\mathcal{A}_n$-group and $G'$ is cyclic of order $\ge p^n$, then $G$ is metacyclic and $|G'| = p^n$. A metacyclic $p$-group $G$ is an $\mathcal{A}_n$-group $\iff$ $|G'| = p^n$.*

(k) *Proposition 10.28. The group $G$ is generated by minimal nonabelian subgroups.*

(l) *Theorem 1.2. If $G$ possesses a cyclic subgroup of index $p$, then $G$ is one of the following groups: $\mathrm{M}_{p^n}$, $\mathrm{D}_{2^n}$, $\mathrm{Q}_{2^n}$, $\mathrm{SD}_{2^n}$. The last three groups exhaust the 2-groups of maximal class and order $2^n$. A 2-group of maximal class $G$ has only one cyclic subgroup of index 2, unless $G \cong \mathrm{Q}_8$.*

(m) *Proposition 10.19. If a metacyclic $p$-group $G$ contains a nonabelian subgroup of order $p^3$, then either $|G| = p^3$ or $G$ is a 2-group of maximal class.*

(n) *Proposition 1.8. If $G$ contains a self-centralizing subgroup of order $p^2$, it is of maximal class.*

(o) *Theorem 7.1 (b,c). If $p > 2$ and the subgroup $\mathrm{K}_{p-1}(G)$, the $(p-1)$-th member of the lower central series of $G$, is cyclic, then $G$ is regular.*

**Remark 148.4.** Let $G$ be a metacyclic $p$-group with $|G'| > p$. (i) If $A < G$ is a minimal nonabelian subgroup, then $A/A' < G/G'$ is maximal abelian. Indeed, $A' = \Omega_1(G') \lhd G$ since $G'$ is cyclic. Assume that $A/A' < B/A' \le G/A'$, where $B/A'$ is abelian. Then, by Lemma 148.3 (*a*), $B$ is minimal nonabelian since $B' = A' \le \Omega_1(Z(B))$ and $\mathrm{d}(B) = 2$, a contradiction since $A < B$ is nonabelian. (ii) Conversely, if $p > 2$ and $A/\Omega_1(G') < G/\Omega_1(G')$ is maximal abelian, then $A$ is minimal nonabelian. Assume, however, that

$A$ is abelian. Let $A < B \le G$, where $|B : A| = p$. Then $B$, by part (ii) of Corollary 148.2 (b), is minimal nonabelian; hence $B/B' = B/\Omega_1(G')$ is abelian, a contradiction since $A/\Omega_1(G')(< B/\Omega_1(G'))$ is a maximal abelian subgroup of $G/\Omega_1(G')$.

Suppose that $G$ is a metacyclic $p$-group with $|G'| > p$. We claim that if all maximal abelian subgroups of $G/\Omega_1(G')$ have equal order, then all minimal nonabelian subgroups of $G$ have equal order as well. Indeed, suppose that $A, B < G$ are minimal nonabelian. Then $A' = \Omega_1(G') = B'$ is of order $p$. By Remark 148.4, $A/\Omega_1(G')$ and $B/\Omega_1(G')$ are maximal abelian subgroups of $G/\Omega_1(G')$; hence they have equal order, by assumption; then $|A| = |B|$.

**Remark 148.5.** Suppose that $G$ is a metacyclic $p$-group such that, whenever $H \le G$, then $\mho_1(H) = \{x^p \mid x \in H\}$ (this is a case provided $p > 2$, by Lemma 148.3 (*d*), since $G$ is regular, as follows from Lemma 148.3 (e)). We claim that, if $G$ is noncyclic and $A < G$ is cyclic of order $p^e = \exp(G)$, then $G = AB$, where $B$ is cyclic with $A \cap B = \{1\}$, unless $G$ is a generalized quaternion group. This will be proved by induction on $|G|$. One may assume, in view of Lemma 148.3 (*l*), that $|G : A| > p$; in this case the subgroup $\Phi = \Phi(G)(= \mho_1(G))$ is noncyclic of index $p^2$ in $G$ and, by Lemma 148.3 (*i*), it is not generalized quaternion. Then $\Phi = UV$, where $U = A \cap \Phi$ is cyclic of order $p^{e-1} = \exp(\Phi)$ (see Lemma 148.40, below), $V = \langle v\rangle$ is cyclic and $U \cap V = \{1\}$, by induction. By assumption, $v = b^p$ for some $b \in G$ since $\Phi = \mho_1(G)$. Then, by the product formula, $G = AB$, where $B = \langle b\rangle$ and $A \cap B \le A \cap V = \{1\}$.

**Remark 148.6.** A nonabelian $p$-group $G = C\Omega_1(G)$, where $\Omega_1(G) \cong S(p^3)$ (so that $p > 2$) and $C$ is cyclic, has an abelian subgroup of index $p$. Indeed, if $R < \Omega_1(G)$ is $G$-invariant of order $p^2$, then $C_G(R)$ is abelian of index $p$ in $G$ (note that $\Omega_1(C_G(R)) = R$ and $C_G(R)/R$ is cyclic). If $p = 2$ and $G = C\Omega_1(G)$, where $\Omega_1(G) \cong E_8$, then $G$ need not necessarily contains an abelian subgroup of index 2.

**Remark 148.7.** If all minimal nonabelian subgroups of a $p$-group $G$ have equal order, then all its $\mathcal{A}_n$-subgroups also have equal order for any $n \le k$, where $G$ is an $\mathcal{A}_k$-group.

**Remark 148.8.** Suppose that $H \in \Gamma_1$, where $G$ is nonabelian two-generator $p$-group. Then $H' < G'$. Indeed, let $L < G'$ be $G$-invariant of index $p$, then $G/L$ is minimal nonabelian (Lemma 148.3 (*a*)). In this case, one has $L < G' \le \Phi(G) < H$ so that $H/L$ is abelian, and hence $H' \le L < G'$.

**Remark 148.9.** Suppose that $G$ is a $p$-group such that $|G'| > p$ and $G/Z(G)$ has a cyclic subgroup, say $A/Z(G)$, of index $p$. Then $p = 2$, $G/Z(G) \cong D_{2^n}$ and $G'$ is cyclic. Indeed, assume that $A/Z(G)$ and $B/Z(G)$ are distinct cyclic subgroups of index $p$ in $G/Z(G)$. Then $A, B \in \Gamma_1$ are distinct abelian subgroups; hence $A \cap B = Z(G)$ has index $p^2$ in $G$; in that case, $|G'| = p$, by Lemma 148.3 (*e*), a contradiction. Thus, $G/Z(G)$ has exactly one cyclic subgroup, namely $A/Z(G)$, of index $p$ so, by Lemma 148.3 (*l*), $p = 2$ and $G/Z(G)$ is of maximal class. Then $A \in \Gamma_1$ is an abelian subgroup. Assume that $B/Z(G) < G/Z(G)$ is cyclic of order 4 such that $B \not\le A$; then $B$ is abelian. In this case, $A \cap B \le Z(G)$, a contradiction since $|A \cap B| = 2|Z(G)|$. Thus, all elements of the set $(G/Z(G)) - (A/Z(G))$ are involutions, and so the quotient group $G/Z(G)$ is dihedral (Lemma 148.3 (l)). By Lemma 148.3 (e), $G' \cong A/Z(G)$ is cyclic of order $\frac{1}{2}|G/Z(G)| = 2^{n-1}$.

**Remark 148.10.** Suppose that $M$ is a minimal nonabelian subgroup of a metacyclic $p$-group $G$. (i) Write $H = MZ(G)$. Assume that $M < H$. Then $|H'| = p$ and $H$ is two-generator as a subgroup of $G$. Therefore, $H$ is minimal nonabelian (Lemma 148.3 (a)), a contradiction. Thus, $Z(G) < M$. In particular, if $G$ contains a subgroup $\cong M_{p^n}$, then $Z(G)$ is cyclic. (ii) Moreover, considering $MC_G(M)$, one obtains $C_G(M) < M$, i.e., the centralizer of every minimal nonabelian subgroup $M$ in a metacyclic $p$-group $G$ coincides with $Z(M)$.

The results of Remark 148.10 are consequences of the following general fact: Suppose that $M$ is a nonabelian two-generator subgroup of a $p$-group $G$ and all containing $M$ subgroups of $G$ of order $p|M|$ are also two-generator. Then $C_G(M) = Z(M)$.

**Remark 148.11.** Suppose that $H \leq G$ is nonabelian of order $p^3$ and $G$ a metacyclic $p$-group. By Remark 148.4, $H/H' \leq G/H'$ is maximal abelian. Then $G/H'$ is of maximal class since $|H/H'| = p^2$ (Lemma 148.3 ($n$)) and, by Remark 148.10, $H' = Z(G)$. Then $G$ is of maximal class since $|Z(G)| = |H'| = p$. Thus, $|G| = p^3$ if $p > 2$. This is a new proof of Lemma 148.3 (m).

**Remark 148.12.** Suppose that $G$ is a nonabelian $p$-group, $p > 2$, with cyclic $G'$. If $G$ has an abelian subgroup $A$ of index $p$, then $|G : Z(G)| = p^2$. Indeed, by Lemma 148.3 ($e$), $A/Z(G) \cong G'$ is cyclic. In that case, $G/Z(G)$ has two distinct cyclic maximal subgroups, say $A/Z(G)$ and $B/Z(G)$ (Lemma 148.3 ($l$)). Then $A, B$ are distinct abelian maximal subgroups of $G$ hence $A \cap B = Z(G)$ has index $p^2$ in $G$ so that $|G'| = p$. If, in addition, $d(G) = 2$, then $G$ is minimal nonabelian (see part (ii) of Corollary 148.2 (b)).

*2° Proof of Theorem 148.1.* If $G$ is a group of order $p^4$, $p > 2$, and $|G'| = p^2$, then $cl(G) = 3$ and $G' \cong E_{p^2}$. Indeed, by virtue of $G' = \Phi(G)$, one has $d(G) = 2$ and so $|Z(G)| = p$ since $G$ is not minimal nonabelian, in view of $|G'| = p^2 > p$ (Lemma 148.3 (f)). The quotient group $G/Z(G)$ is nonabelian so $cl(G) = 3$; then $G' \cong E_{p^2}$ (indeed, a $p$-group of maximal class and order $> p^3$, $p > 2$, has no normal cyclic subgroup of order $> p$, by Exercise 9.1 (c) and Lemma 148.3 (i); in particular, $G$ is nonmetacyclic). In view of this, one may assume in Theorem 148.1 (a), that $|G| > p^4$.

Suppose that $G$ is a $p$-group with $|G'| > p^2 = |G : G'|$. Take a $G$-invariant subgroup $L$ of index $p^2$ in $G'$. Then $G/L$ is of maximal class and order $p^4$, by the previous paragraph.

If $G$ is a nonabelian two-generator $p$-group, then $Z(G) \leq \Phi(G)$. Indeed, $d(G/Z(G)) \geq 2 = d(G)$ so that $d(G/Z(G)) = 2$, and we conclude that $Z(G) \leq \Phi(G)$.

*Proof of Theorem 148.1.* The notation introduced in the statement of the theorem is retained.

As $|G'| = p^2$, the set $\Gamma_1$ has only one abelian member. By Lemma 148.3 (a), $|G : Z(G)| = p|G'| = p \cdot p^2 = p^3$. It follows from $d(G) = 2$ that $Z(G) < \Phi(G)$. As the two-generator group $G/Z(G)$ has at most one cyclic subgroup of index $p$ (see the first sentence of this paragraph), it follows that $G/Z(G) \cong D_8$ if $p = 2$ and $G/Z(G) \cong S(p^3)$ if $p > 2$; in particular, $cl(G) = 3$, $G/Z(G)$ is nonabelian so that $G' \nleq Z(G)$ hence $L = G' \cap Z(G)$ has order $p$. If $p = 2$ and $A/Z(G)$ is cyclic of index 2 in $G/Z(G)$ then $A$ is abelian, and therefore $G' \cong A/Z(G) \cong C_4$ (Lemma 148.3 (e)). If $p > 2$ and $A \in \Gamma_1$ is abelian, one has $G' \cong A/Z(G) \cong E_{p^2}$ as a maximal subgroup in $S(p^3)$. Note that $G/L$ is minimal nonabelian (Lemma 148.3 (a)).

Assume that $Z(G)$ is noncyclic. If $R < Z(G)$ is as large as possible subjecting $R \cap L = \{1\}$ (here $L = G' \cap Z(G)$), then $R > \{1\}$ since $|L| = p < |\Omega_1(Z(G))|$; then $G' \cap R = \{1\}$. Write $\bar{G} = G/R$; then $\bar{G}' \cong G'$ has order $p^2$. If $A < G$ is abelian of index $p$, then $R < Z(G) < A$ whence $\bar{A} < \bar{G}$ is abelian of index $p$. One has $\overline{Z(G)} \leq Z(\bar{G})$ and, as $|\bar{G}'| = p^2$, one obtains $|\bar{G} : Z(\bar{G})| = p^3$ (Lemma 148.3 (e)). Thus, $\overline{Z(G)} = Z(\bar{G})$ (compare indices!). Assume that $Z(\bar{G})$ is noncyclic. Then, arguing as above, one finds in $Z(\bar{G})$ a subgroup $\bar{R}_1 > \{\bar{1}\}$ such that $\bar{R}_1 \cap \bar{G}' = \{\bar{1}\}$. As above, $R_1 < Z(G)$. However, $R < R_1$ and $R_1 \cap G' = \{1\}$, contrary to the maximal choice of $R$. Thus, the quotient group $G/R$ has the cyclic center and satisfies the hypothesis. Next assume that $R = \{1\}$; then $Z(G) \cong C_{p^n}$ is cyclic, $n > 1$, so that $|G| = p^{n+3}$. It remains to prove assertions (a) and (b).

(a) Here $p > 2$, $G' \cong E_{p^2}$. As $\exp(G/Z(G)) = p$, one has $\mho_1(G) \leq Z(G)$. Assume that $\mho_1(G) < Z(G)$. In that case, $\mho_1(G) = \mho_1(Z(G)) > \{1\}$ since $Z(G)$ of order $p^n > p$ is cyclic; then $|G/\mho_1(G)| = p^4$. Write $F = \mho_1(G)$. Since $G' \cap F = G' \cap Z(G) = L$ is of order $p$, one has

$$|(G/F)'| = |G'F/F| = |G'/(G' \cap F)| = |G'/L| = p$$
$$\Rightarrow (G/F)/(G/F)' \cong E_{p^3}, \quad d(G/F) = 3 > 2 = d(G),$$

a contradiction. Thus, $\mho_1(G) = Z(G)$ is cyclic hence $\mho_1(G) < T < G$, where $T$ is cyclic and $|G : T| = p^2$. Then $G$ has no subgroup of order $p^4$ and exponent $p$ (consider its intersection with $T$). As $G$ is neither metacyclic nor of maximal class (indeed, $|G/\mho_1(G)| = p^3$ and $|Z(G)| = p^n > p$), one obtains $G = TE$, where $E \triangleleft G$ is of order $p^3$ and exponent $p$ and $|T \cap E| = p$ (Lemma 148.3 (g)). Assume that $E < \Omega_1(G)$. As $G/E$ is cyclic, it follows that $|\Omega_1(G)| = p^4$. Since $G$ has no subgroup of order $p^4$ and exponent $p$, $\exp(\Omega_1(G)) > p$ which implies that the subgroup $\Omega_1(G)$ is irregular (Lemma 148.3 (d)) so that $p = 3$ and $\Omega_1(G)$ is of maximal class (Theorem 9.5). By Lemma 148.3 (h), $G$ is of maximal class, a contradiction since $G/E$ is cyclic of order $> 3$. Thus, $E = \Omega_1(G)$. Since $G/E$ is cyclic, one has $G' < E$, and so $G' \cong E_{p^2}$ again.

Now all $|\Gamma_1| = p + 1$ maximal subgroups of $G$ will be described. Since $p^2$ does not divide $\exp(\mathrm{Aut}(E))$ (see § 34) and $G/E$ is cyclic, one obtains $A = EC_G(E) = EZ(G) \in \Gamma_1$.

Let $\{M_1, \dots, M_p\} = \Gamma_1 - \{A\}$. One has

$$\Omega_1(M_i) = M_i \cap \Omega_1(G) = M_i \cap E \cong \mathrm{E}_{p^2} \ (i = 1, \dots, p\,.$$

As $M_i/\Omega_1(M_i) \cong G/\Omega_1(G) = G/E$ is cyclic, $M_i$ has a cyclic subgroup of index $p$, $i \le p$.

Suppose that $E = \Omega_1(G)$ is abelian; then $A = EZ(G)$ is (the unique) abelian maximal subgroup of $G$. In that case, $M_i \cong \mathrm{M}_{p^{n+2}}$ for all $i$ (here $p^n = |Z(G)|$).

Now let $E = \Omega_1(G)$ be nonabelian. Then, by hypothesis, one of subgroups $M_1, \dots, M_p$, say $M_1$, is abelian so $M_i \cong \mathrm{M}_{p^{n+2}}$ for $i = 2, \dots, p$.

(b) Here $p = 2$, $G/Z(G) \cong \mathrm{D}_8$ so that $\mathrm{cl}(G) = 3$, $\Gamma_1 = \{A, B, C\}$, where $C/Z(G)$ is cyclic of order 4 so $C$ is metacyclic abelian since $Z(G)$ is cyclic. By Lemma 148.3 (e), $G' \cong C/Z(G) \cong \mathrm{C}_4$. By Lemma 148.3 (b,a), $\bar{G} = G/L$ is nonmetacyclic and minimal nonabelian and $L = \mho_1(G')$. Since $C/L$ is metacyclic and $|G/L| > 2^3$, at least one of subgroups $A/L$, $B/L$, say $A/L$ is also metacyclic (Lemma 148.3 (f)). Now, $Z(G) < \Phi(G)$ since $\mathrm{d}(G) = 2$ (see the paragraph preceding the proof).

As $A/Z(G) \cong \mathrm{E}_4$ and $\mathrm{d}(A) = 2$ (indeed, $L = \Phi(G') \le \Phi(A)$ since $G' < A$), it follows that the cyclic $Z(G) = \Phi(A) = \mho_1(A)$, and so $A$ has a cyclic subgroup of index 2; then $A \cong \mathrm{M}_{2^{n+2}}$ since $A \ne C$ is nonabelian and $n > 1$ (Lemma 148.3 (l)). By Lemma 148.3 (f), $\mathrm{d}(B/L) = 3$. In that case, $G' < \Phi(G) < B$ and $L = \Phi(G') < \Phi(B)$ imply $\mathrm{d}(B) = \mathrm{d}(B/L) = 3$ so it follows from $B/Z(G) \cong \mathrm{E}_4$ and cyclicity of $Z(G)$ that $\Phi(B) = \mho_1(Z(G))$.

Let $U$, $V$, $W$ be all maximal subgroups of $B$ containing $Z(G)$ (as a subgroup of index 2). All these (noncyclic) subgroups are abelian of type $(2^n, 2)$ since $Z(G) \cong \mathrm{C}_{2^n}$. Let $x \in U - Z(G)$ and $y \in V - Z(G)$ be involutions. Then $y \notin U$ since $U \cap V = Z(G)$, and so $\langle y, x, Z(G)\rangle = \langle y, U\rangle = B$; hence $B = DZ(G)$, where $D = \langle x, y\rangle$. As $B$ is nonabelian, one has $xy \ne yx$; and hence $D$ is dihedral. As $\mathrm{cl}(B) = 2$, one has $D \cong \mathrm{D}_8$. Next, the subgroup $\Omega_2(B) = D\Omega_2(Z(G))$ of order 16, being characteristic in $B = DZ(G) \lhd G$, is normal in $G$. By Appendix 16, $\Omega_2(B)$ contains exactly three subgroups isomorphic to $\mathrm{D}_8$ so one of them which will be denoted by $D$ again, is normal in $G$. By Lemma 148.3 (i), $D \nleq \Phi(G)$ so that $\mathrm{d}(G/D) < \mathrm{d}(G) = 2$; and hence $G/D$ is cyclic. Let $Z < A$ be cyclic of index 2; then $|Z| = 2^{n+1}$ and $Z(G) = Z(A) = \Phi(A) < Z$ since $|A : Z| = 2$. One has $G = DZ$ and $Z \cap D = Z(D)$. As $G/D$ is cyclic, we get $\Omega_2(G) = \Omega_2(B)$. As an abelian subgroup $\Omega_2(C) \le \Omega_2(G)$ and $\Omega_2(G)$ is nonabelian, it follows that $|\Omega_2(C)| < |\Omega_2(G)| = 16$, and so $|\Omega_2(C)| = 8$. Then $C/\Omega_1(C)$ is cyclic, and so $C$ is abelian of type $(2^{n+1}, 2)$. □

**Supplement to Theorem 148.1.** *Let $G$ be a (nonmetacyclic) $p$-group of Theorem 148.1 with cyclic $Z(G)$ of order $p^n > p$.*

(a) *If $p > 2$, then a minimal nonabelian subgroup of $G$ is either $\Omega_1(G) \cong S(p^3)$ (provided $\Omega_1(G)$ is nonabelian) or has a cyclic subgroup of index $p$.*
(b) *If $p = 2$, then a minimal nonabelian subgroup of $G$ has a cyclic subgroup of index 2.*

*Proof.* Let $H < G$ be minimal nonabelian of order $> p^3$. The notation of Theorem 148.1 is retained.

Let $p > 2$ and $H \notin \{M_1, \dots, M_p\}$. Then $H < A = \Omega_1(G)Z(G)$ (Theorem 148.1 (a)) and so $A$ is nonabelian and hence $\Omega_1(G) \cong S(p^3)$. Since $|H \cap \Omega_1(G)| = p^2$ and $H/(H \cap$

$\Omega_1(G)) \cong H\Omega_1(G)/\Omega_1(G)$ is cyclic as a subgroup of $G/\Omega_1(G)$, one has $H \cong \mathrm{M}_{p^k}$, $k \le n+1$.

Now let $p = 2$; then $G = DZ$ is as in Theorem 148.1 (b) (here $Z < G$ is cyclic of index 4 and $D \cong \mathrm{D}_8$). Assume that $H$ has no cyclic subgroup of index 2; then $|H| > 8$. In that case, $H < B = DZ(G)$ (Theorem 148.1 (b)) and $|\Omega_1(H)| = 4$ (since $B$ so $H$ has no subgroup $\cong \mathrm{E}_8$); then $H$ is metacyclic (Lemma 148.3 (f)). In this case, $H$ has exactly six cyclic subgroups of order 4, a contradiction since $\Omega_2(B)$ has exactly four such subgroups. Thus, $H$ has a cyclic subgroup of index 2. □

*3^o Metacyclic p-groups, p > 2.* In this subsection the number $\alpha_n(G)$ of $\mathcal{A}_n$-subgroups in a metacyclic $p$-group $G$, $p > 2$, will be computed. Some structure results will be also proved. For example, it will be proved that, in any such group all maximal abelian subgroups have equal order.

**Lemma 148.13.** *If $G$ is a two-generator nonabelian $p$-group, $p > 2$, with cyclic $G'$, and $H \in \Gamma_1$ is nonabelian, then $H' = \mho_1(G')$, i.e., $|G' : H'| = |G : H|$.*

*Proof.* In view of Lemma 148.3 (a), one may assume that $|G'| > p$ (otherwise, $G$ is minimal nonabelian, contrary to the hypothesis). By Remark 148.8, $H' < G'$. Now let $H$ be such that $|H'|$ is as small as possible. By Corollary 148.2 (b), the nonabelian group $G/H'$ is minimal nonabelian since it contains an abelian subgroup $H/H'$ of index $p$. Since $|G'/H'| = |(G/H')'| = p$ (Lemma 148.3 (f)), one obtains $H' = \mho_1(G')$. □

**Lemma 148.14.** *Let $G$ be a two-generator $p$-group, $p > 2$, Let $G'$ be cyclic and let $H = H_0 < H_1 < \cdots < H_n = G$ be a maximal chain with two-generator nonabelian subgroups $H_1, \ldots, H_n = G$ connecting $H = H_0$ and $H_n = G$.*[3] *Then $|G' : H'| = |G : H| = p^n$.*

*Proof.* If $H = H_0$ is abelian, then $H_1$ is minimal nonabelian (Corollary 148.2 (b)) so that $|H_1'| = p = |H_1 : H|$ (Lemma 148.3 (a)). If $H_0$ is nonabelian, one obtains $|H_1' : H_0'| = |H_1 : H_0| = p$ (Lemma 148.13). Given $i \in \{1, \ldots, n\}$, one has $|H_i' : H_{i-1}'| = |H_i : H_{i-1}| = p$ (Lemma 148.13). Multiplication of these equalities over all $i$ completes the proof. □

Let $G$ be a metacyclic $p$-group with $|G'| = p^k$, $p > 2$. If $k = 1$, then $\alpha_1(G) = 1$ (Lemma 148.3 (a)). If $k = 2$, then $\alpha_1(G) = p + 1$ (Lemmas 148.14, 148.3 (a) and Corollary 148.2 (b)). Our aim is to show that $\alpha_n(G)$, $n \le k$, depends only on $|G'|$ and $n$. By what has just been said, this is true for $k < 3$. By Hall's enumeration principle (Theorem 5.2),

$$\alpha_n(G) = \sum_{H \in \Gamma_1} \alpha_n(H) - p \cdot \alpha_n(\Phi(G)). \tag{1}$$

This identity allows us to use induction in order to find $\alpha_n(G)$.

3 Such a chain exists if $G$ is metacyclic.

**Proposition 148.15.** *Suppose that $G$ is a nonabelian metacyclic $p$-group, $p > 2$, with $|G'| = p^k$, $k > 1$.*

(a) *If $F, H < G$ are nonabelian then $|F'| = |H'| \iff |F| = |H|$. If $n < k$, then all $\mathcal{A}_n$-subgroups of $G$ have index $p^{k-n}$ in $G$.*

(b) *If $G_0$ is a metacyclic $p$-group with $|G_0'| = |G'|$, then $\alpha_1(G) = \alpha_1(G_0)$. Thus, $\alpha_1(G)$ depends only on $|G'|$.*[4]

(c) *If $A, B \le G$, where $A$ is abelian and $B$ is nonabelian, then $|A| < |B|$.*

(d) *If $A < G$ is maximal abelian and $A < B \le G$, where $|B : A| = p$, then $B$ is minimal nonabelian. All maximal abelian subgroups of $G$ have index $p^k$ in $G$ and, conversely, all subgroups of index $p^k$ in $G$ are abelian. If $U < V \le G$, where $V$ is minimal nonabelian and $|V : U| = p$, then $U$ is maximal abelian in $G$. Therefore, all minimal nonabelian subgroups have index $p^{k-1}$ in $G$.*

*Proof.* By Lemma 148.3 (j), $G$ is an $\mathcal{A}_k$-group.

(a) Let $F = F_0 < F_1 < \cdots < F_k = G$ and $H = H_0 < H_1 < \cdots < H_k = G$ be maximal chains. Then, by Lemma 148.14, $|G' : F'| = |G : F|$ so that $|F| = \frac{|G|}{|G':F'|}$ and similarly $|H| = \frac{|G|}{|G':H'|}$. Hence, $|F| = |H| \iff |F'| = |H'|$. The second assertion now follows from Lemma 148.3 (j).

(b) By the paragraph preceding the proposition, the result holds for $k \le 2$. Assume that $k > 2$. We proceed by induction on $|G'|$. By induction, if $H < G$, then $\alpha_1(H)$ depends only on $|H'|$. If $F, H \in \Gamma_1$, then $|F'| = |H'| = p^{k-1}$, by Lemma 148.13, so that, by induction, $\alpha_1(F) = \alpha_1(H)$. Since $k > 2$, then $|\Phi(G)'| = p^{k-2}$, by (a). Application of (1) to $G$ and $G_1$ completes the proof.

(c,d) Let $A, A_1 < G$ be maximal abelian and $A < B \le G$, $A_1 < B_1 \le G$, where $|B : A| = p = |B_1 : A_1|$. Then $B, B_1$ are minimal nonabelian (Corollary 148.2 (b)) so that $|B| = |B_1|$, by (a). It follows that $|A| = \frac{1}{p}|B| = \frac{1}{p}|B_1| = |A_1|$. By (a), $|G : A| = p|G : B| = p^k$. If $H < G$ is of order $|A|$, then $H$ has no minimal nonabelian subgroup so it is abelian. Now let $U < V \le G$, where $V$ is minimal nonabelian, $|V : U| = p$ and $U \le U_1$, where $U_1 < G$ is maximal abelian and $U_1 < V_1 \le G$ with $|V_1 : U_1| = p$. Then $V_1$ is minimal nonabelian so $|V_1| = |V| \Rightarrow U_1 = U \Rightarrow U$ is maximal abelian in $G$. □

By Proposition 148.15 (d), if $A$ is an abelian subgroup of a nonabelian metacyclic $p$-group $G$, $p > 2$, then there is abelian $A_1 \lhd G$ of order $|A|$. Indeed, there is $A_1 \lhd G$ of order $|A|$. This $A_1$ is abelian, by the previous paragraph. This is also a consequence of the following Gillam's theorem (see Theorem 39.7): If $A < G$ is an abelian subgroup of a metabelian $p$-group $G$, then there is an abelian $A_1 \lhd G$ with $|A_1| = |A|$. However, and our partial case of Gillam's result is not obvious.

If $G$ is a nonabelian metacyclic $p$-group, $p > 2$, then $G$ is covered by minimal nonabelian subgroups (Proposition 148.15 (d)); this is not true for $p = 2$ (indeed, the group

4 It is not true that orders of minimal nonabelian subgroups of $G$ and $G_0$ are equal: they depend also on $|G|, |G_0|$; see (a).

$D_{16}$ of exponent 8 has no minimal nonabelian subgroup of exponent 8). The second author has classified those metacyclic 2-groups that are not covered by minimal nonabelian subgroups (see Proposition 154.1; see also § 149). It follows from that proposition that so-called ordinary metacyclic 2-groups (see § 26, Definition 7) are covered by minimal nonabelian subgroups. Moreover, the fairly deep Proposition 149.3 describes the nonabelian 2-groups $G$ such that any maximal abelian subgroup of $G$ is contained in a minimal nonabelian subgroup.

Let $p > 2$, let $C = \langle c \rangle$ be cyclic of order $p^3$ and suppose that a cyclic $B = \langle b \rangle$ of order $p^2$ acts faithfully on $C$. Let $G = B \cdot C$ be the natural semidirect product. Then the nonabelian subgroups $F = \Omega_1(B) \cdot C \cong \mathrm{M}_{p^4}$ and $H = \Omega_2(G)$ are non-isomorphic minimal nonabelian subgroups and both have order $p^4$.

By Lemma 148.3 (j), a metacyclic $p$-group $G$ is an $\mathcal{A}_n$-group $\iff |G'| = p^n$.

Given a $p$-group $G$ with $|G'| = p^k$ as in Proposition 148.15, one writes $\alpha_n(p^k)$ instead of $\alpha_n(G)$. In view of Proposition 148.15 (b), this notation is legitimate. If $k = 2$, then all members of the set $\Gamma_1$ are minimal nonabelian (Corollary 148.2 (b)) so that $\alpha_1(G) = |\Gamma_1| = p + 1$, as it was noted above. If $k > 2$, then $|H'| = p^{k-1} > p$ for all $H \in \Gamma_1$ (Proposition 148.15 (a)) so that $|\Phi(G)'| = p^{k-2}$, by the same proposition. One may rewrite (1) for $1 \leq n < k$ as follows:

$$\alpha_n(p^k) = (p + 1)\alpha_n(p^{k-1}) - p\alpha_n(p^{k-2}). \tag{2}$$

Using (2), one obtains:

$$\alpha_1(p^3) = (p + 1)\alpha_1(p^2) - p\alpha_1(p) = (p + 1)^2 - p = p^2 + p + 1 = \frac{p^3 - 1}{p - 1}.$$

Similarly, one obtains $\alpha_1(p^4) = \frac{p^4-1}{p-1}$.

**Lemma 148.16.** *If $G$ is a metacyclic $p$-group, $p > 2$, $|G'| = p^k \geq p$, then $\alpha_1(p^k) = \frac{p^k-1}{p-1}$.*

*Proof.* As we have noted, this holds for $k \leq 4$. Therefore, in what follows, we assume that $k > 4$. Then $\Phi(G) < H \in \Gamma_1$, where $|H : \Phi(G)| = p$ and $|H'| = p^{k-1} > p$; hence $|\Phi(G)'| = p^{k-2} > p^2$ (Lemma 148.13). Therefore, by induction,

$$\alpha_1(p^{k-1}) = \frac{p^{k-1} - 1}{p - 1}, \quad \alpha_1(\Phi(G)) = \alpha_1(p^{k-2}) = \frac{p^{k-2} - 1}{p - 1}.$$

Now, by (2) with $n = 1$, one obtains:

$$\alpha_1(G) = \alpha_1(p^k) = (p + 1)\frac{p^{k-1} - 1}{p - 1} - p\frac{p^{k-2} - 1}{p - 1} = \frac{p^k - 1}{p - 1},$$

as required. □

**Theorem 148.17.** *Let $k \geq n$ be positive integers. If $G$ is a metacyclic $p$-group, $p > 2$, with $|G'| = p^k$, then $\alpha_n(G) = \alpha_n(p^k) = \frac{p^{k-n+1}-1}{p-1}$.*

*Proof.* In view of Lemma 148.16, one may assume that $n > 1$. We proceed by induction on $n+k$. Since $\alpha_k(p^k) = 1 = \frac{p^{k-k+1}-1}{p-1}$, one may assume that $n < k$; then $k \geq 3$. By Hall's enumeration principle, one has

$$\alpha_n(G) = \alpha_n(p^k) = \sum_{H \in \Gamma_1} \alpha_n(p^{k-1}) - p \cdot \alpha_n(p^{k-2}). \tag{3}$$

Below, Lemma 148.3 (j) and Proposition 148.15 (a) will be used freely.

If $H \in \Gamma_1$, then $|H'| = p^{k-1} < p^k$ so, by induction,

$$\alpha_n(p^{k-1}) = \frac{p^{k-1-(n-1)+1} - 1}{p-1} = \frac{p^{k-n+1} - 1}{p-1}. \tag{4}$$

If $n > k-2$, then $n = k-1$ since $n < k$, and

$$\alpha_n(p^{k-2}) = \alpha_{k-1}(p^{k-2}) = 0 = \frac{p^{(k-2)-(k-1)+1} - 1}{p-1}.$$

Since $\alpha_{k-1}(p^{k-1}) = 1$, one obtains:

$$\alpha_{k-1}(p^k) = (p+1)1 - p0 = p+1 = \frac{p^{k-(k-1)+1} - 1}{p-1} = \frac{p^2-1}{p-1},$$

and this case is complete.

If $n = k-2$, then

$$\begin{aligned}\alpha_{k-2}(G) = \alpha_{k-2}(p^k) &= (p+1)\alpha_{k-2}(p^{k-1}) - p\alpha_{k-2}(p^{k-2}) \\ &= (p+1)^2 - p = p^2 + p + 1 = \frac{p^3-1}{p-1} = \frac{p^{k-(k-2)+1} - 1}{p-1},\end{aligned}$$

completing this case.

If $n < k-2$, then $|\Phi(G)'| = p^{k-2}$ so, by induction,

$$\alpha_n(\Phi(G)) = \alpha_n(p^{k-2}) = \frac{p^{(k-2)-n+1} - 1}{p-1} = \frac{p^{k-1-n} - 1}{p-1}.$$

Since for all $H \in \Gamma_1$, by induction,

$$\alpha_n(H) = \alpha_n(p^{k-1}) = \frac{p^{k-1-n+1} - 1}{p-1} = \frac{p^{k-n} - 1}{p-1}$$

(see Lemma 148.15), one obtains, in view of (3) and (4):

$$\alpha_n(G) = \alpha_n(p^k) = (p+1)\frac{p^{k-n} - 1}{p-1} - p\frac{p^{k-1-n} - 1}{p-1} = \frac{p^{k-n+1} - 1}{p-1},$$

completing the proof. □

Mann [Man33] has computed the number of subgroups of given order in a metacyclic $p$-group, $p > 2$ (see also § 124, where that number is also computed for $p = 2$). Next, the number $c_n(G)$ is computed in a metacyclic 2-group; see Appendix 98.

If $G, k, n$ are as in Theorem 148.17 and $\phi$ is a $p$-automorphism of $G$, then $H^\phi = H$ for some $\mathcal{A}_n$-subgroup $H \leq G$ since $\alpha_n(p^k) \equiv 1 \pmod{p}$.

**Exercise 1.** Let $G$ be a metacyclic $p$-group containing a normal cyclic subgroup $C$ of order $p^e$ such that $G/C$ is cyclic of order $p^f$ (such $G$ is said to be of type $(e, f)$), and let $p > 2$. Mann (see Theorem 124.1) computed the number of subgroups of order $p^m$ in $G$. Using this, Proposition 148.15 and Theorem 148.17, find:
(a) the total number of abelian subgroups in $G$, and
(b) the total number of subgroups in $G$.

**Proposition 148.18.** *If $G$ is a nonabelian metacyclic $p$-group, $p > 2$, then $p + 1$ divides the number of maximal abelian subgroups $A$ of $G$ such that $\mathrm{N}_G(A)/A$ is cyclic.*

*Proof.* This is obvious provided $G$ is minimal nonabelian since then $|\Gamma_1| = p + 1$. Therefore, in what follows, one may assume that $G$ is not minimal nonabelian. Let $\mathcal{A} = \{A_1, \dots, A_r\}$ be the set of maximal abelian subgroups of $G$ and $\mathcal{M} = \{M_1, \dots, M_s\}$ the set of minimal nonabelian subgroups of $G$. Below, Proposition 148.15 (d) is used freely. Let $a_i$ be the number of members of the set $\mathcal{M}$ containing $A_i$; then $a_i \in \{1, p+1\}$, $i \le r$, since all subgroups of $G$ are two-generator. Any subgroup $M_j$ contains exactly $p + 1$ members of the set $\mathcal{A}$. By double counting of the number of pairs $A_i < M_j$, $(i \le r,\ j \le s)$, one has

$$a_1 + \cdots + a_r = (p + 1)s, \tag{5}$$

where $s = \alpha_1(G)$. Set $r_1 = |\{i \mid a_i = 1\}|$. If $a_i = 1$, then $\mathrm{N}_G(A_i)/A_i$ is cyclic (recall that $p > 2$). With this notation, one can rewrite (5) as follows:

$$r_1 + (r - r_1)(p + 1) = (p + 1)s. \tag{6}$$

It follows from (6) that $p + 1 \mid r_1$. □

**Theorem 148.19.** *Let $G$ be a regular $p$-group with $\exp(G) = p^e > p$. Then all minimal nonabelian subgroups of $G$ have equal exponent $\iff$ $\Omega_{e-1}(G)$ is abelian.*

*Proof.* By Theorem 7.2 (b), $\Omega_{e-1}(G) < G$. There is in $G$ a minimal nonabelian subgroup $M \nleq \Omega_{e-1}(G)$ (Lemma 148.3 (k)) so $\exp(M) = p^e$. Thus, all minimal nonabelian subgroups of $G$ have exponent $p^e$ so $\Omega_{e-1}(G)$ is abelian since it has no minimal nonabelian subgroups. The converse assertion is obvious. As metacyclic $p$-group, $p > 2$, is regular, the above result is applicable to such groups. □

The proofs of the following two theorems were produced after acquaintance with Section 1 of Hethelyi-Külshammer's paper [HetK].

Let $G$ be a nonabelian primary $\mathcal{A}_n$-group, $n > 1$, with $|G : \Phi(G)| = p^2$. By Lemma 148.3 (k), there is in $G$ a minimal nonabelian subgroup $F$ that is not contained in $\Phi(G)$; then $F\Phi(G) \in \Gamma_1$. By Lemma 148.3 (k) again, there is in $G$ a minimal nonabelian subgroup $H$ that is not contained in $F\Phi(G)$. Then $G = H(F\Phi(G)) = \langle F, H, \Phi(G)\rangle = \langle F, H\rangle$. In particular, a nonabelian metacyclic $p$-group ($p$ is arbitrary) is generated by two minimal nonabelian subgroups.

Let $G$ be a metacyclic $p$-group and let $|G'| = p^k > p > 2$. By the previous paragraph, $G = \langle F, H\rangle$, where $F, H$ are minimal nonabelian. Then $Z(F) = \Phi(F) < \Phi(G)$ and $Z(H) < \Phi(G)$ hence $Z(F)Z(H) \subseteq \Phi(G)$. Next,

$$|G : Z(H)| = |G : Z(F)| = |G : F|\,|F : Z(F)| = p^2|G : F| = p^2 p^{k-1} = p^{k+1}$$

so that

$$(7) \qquad |\Phi(G) : Z(F)| = \frac{1}{p^2}|G : Z(F)| = \frac{1}{p^2}p^{k+1} = p^{k-1} = |\Phi(G) : Z(H)|.$$

By Remark 148.10, $Z(G) \le Z = Z(F) \cap Z(H)$. As $C_G(Z) \ge \langle F, H\rangle = G$, one has $Z \le Z(G)$, and so $Z(G) = Z$. By (7) and the product formula,

$$(8) \quad |\Phi(G)| \ge |Z(F)Z(H)| = \frac{|Z(F)|\,|Z(H)|}{|Z(G)|} = \frac{|\Phi(G)|^2}{p^{2k-2}|Z(G)|} = p^{-2k}|G : Z(G)|\,|\Phi(G)|\,;$$

hence

$$(9) \qquad |G : Z(G)| \le p^{2k} = |G'|^2.$$

If $\chi \in \mathrm{Irr}(G)$, then $\chi(1)^2 \le |G : Z(G)|$ [Isa1, Corollary 2.30] so, by (9), one has

$$(10) \qquad \chi(1) \le p^k \quad \text{for every } \chi \in \mathrm{Irr}(G).$$

By [BZ, Theorem 4.35], $\bigcap_{\chi \in \mathrm{Irr}_1(G)} \ker(\chi) = \{1\}$. Therefore, there is $\chi \in \mathrm{Irr}_1(G)$ such that $\ker(\chi) \cap G' = \{1\}$ since $G'$ is cyclic. As $G$ is an M-group, it follows that $\chi = \mu^G$ for some $\mu \in \mathrm{Lin}(T)$, where $T < G$ is of index $\chi(1)$. Then

$$\{1\} = G' \cap \ker(\mu^G) = G' \cap \ker(\mu)_G \ge (T')_G = T'$$

(here $H_G = \bigcap_{x \in G} H^x$ for $H \le G$) since $T' \lhd G$, and hence $T' = \{1\}$, $T$ is abelian.[5] In that case, by Proposition 148.15 (c), $p^k = |G : T| = \chi(1) \le p^k$ so that $\chi(1) = p^k$. Next, by (9), $p^{2k} = \chi(1)^2 \le |G : Z(G)| \le p^{2k}$ so that $|G : Z(G)| = p^{2k} = |G'|^2$.

**Theorem 148.20** (Hethelyi-Külshammer [HetK]). *If $G$ is a metacyclic $p$-group, $p > 2$, $|G'| = p^k \ge p$, then:*
(a) $\mathrm{cd}(G) = \{\chi(1) \mid \chi \in \mathrm{Irr}(G)\} = \{1, p, \dots, p^k\}$,
(b) $|G : Z(G)| = |G'|^2$.

*Proof.* (b) is proved. By induction, applied to $G/\Omega_1(G')$, one obtains (a). □

If $G$ is as in Theorem 148.20 and $\chi \in \mathrm{Irr}(G)$, then $\chi(1) = p^k \iff \ker(\chi) \cap G' = \{1\}$. This follows from the proof and the fact that, provided $\ker(\chi) \cap G' > \{1\}$, then $|(G/\ker(\chi))'| < p^k$, and hence all irreducible characters of the group $G/\ker(\chi)$ have degrees $< p^k$, by the above, and this is a contradiction.

5 Inequality (10) follows from Proposition 148.15 (a) and Ito's theorem on degrees [BZ, Theorem 7.7] whose proof depends on Clifford's theory.

**Theorem 148.21.** *Let $G$, $F$, $H$ be as in the paragraph preceding Theorem 148.20. Then:*
(a) $Z(F) \cap Z(H) = Z(G)$,
(b) $Z(F)Z(H) = \Phi(G)$,
(c) *if $U < G$ is maximal abelian, then* $|U| = |Z(G)|\,|G'| > |Z(G)G'|$.

*Proof.* Part (a) is proved in the paragraph preceding Theorem 148.20.
(b) Since, by Theorem 148.20 (b), $|G : Z(G)| = |G'|^2 = p^{2k}$, it follows from (7) and (8) that $|Z(F)Z(H)| = |\Phi(G)|$ hence $Z(F)Z(H) = \Phi(G)$ since $Z(F)Z(H) \subseteq \Phi(G)$.
(c) By Proposition 148.15 (d), $|G : Z(G)| = |G'|^2 \Rightarrow |Z(G)|\,|G'| = |G : G'| = |U|$. Since $Z(G) \cap G' > \{1\}$, it follows that $|U| > |Z(G)G'|$. □

If a $p$-group $G$ has a maximal normal abelian subgroup $A$ with cyclic quotient group $G/A$ of order $p^k$, then (see [Man33, Proposition 11]) $\mathrm{cd}(G) = \{\chi(1) \mid \chi \in \mathrm{Irr}(G)\} = \{1, p, \dots, p^k\}$. Proposition 148.22 and Corollary 148.23 are inspired by this result.

Suppose that $H < G$, where $G$ is an arbitrary finite group. Write $j(H) = \min\{|G : F| \mid F \le G,\ F \cap H = \{1\}\}$. In particular, $j(\{1\}) = 1$, $j(G) = |G|$. If $\{1\} < H < G = Q_{2^n}$, then $j(H) = |G|$. If $\{1\} < H < G \in \{D_{2^n}, SD_{2^n}\}$, then $j(H) = \frac{1}{2}|G|$ if $Z(G) \le H$ and $j(H) = 2$ if $Z(G) \nleq |H|$.

**Proposition 148.22.** *Suppose that $G$ is a nonabelian group and $A \triangleleft G$ is self-centralizing abelian. Write $\bar{G} = G/A$. Let $\bar{H} \le \bar{G}$ be of prime order $p$. Then there is $\chi \in \mathrm{Irr}(G)$ such that $A \nleq \ker(\chi)$ and $\chi(1) \ge j(\bar{H})$.*

*Proof.* By [BZ, Theorem 7.7], $\mathrm{cd}(H) = \{1, p\}$ since $H$ is nonabelian. Let $\psi \in \mathrm{Irr}(H)$ be of degree $p$; then $\psi_A = \phi_1 + \dots + \phi_p$ is the Clifford decomposition, $\phi_1 = \phi$ (see [BZ, Exercise 7.3.7, Theorem 7.3.16]). Let $T = I_G(\phi)$ denote the inertia subgroup of $\phi$ in $G$. Then $H \nleq T$ and $A \le T$ so that $H \cap T = A$, and hence $\bar{H} \cap \bar{T} = \{\bar{1}\}$. If $\theta \in \mathrm{Irr}(T)$ is such that $\phi$ is an irreducible constituent of $\theta_A$, then $\chi = \theta^G \in \mathrm{Irr}(G)$ [Isa1, Theorem 6.11 (a)] so that $\chi(1) \ge |G : T|$. Since $A \nleq \ker(\chi)$ and $|G : T| \ge j(\bar{H})$, in view of $\bar{H} \cap \bar{T} = \{\bar{1}\}$, the proof is complete. □

**Corollary 148.23.** *Suppose that a nonabelian group $G$ has a maximal normal abelian subgroup $A$ such that the quotient group $G/A$ is either a cyclic $p$-group or a generalized quaternion group. Then there is $\phi \in \mathrm{Lin}(A) - \{1_A\}$ with $\phi^G \in \mathrm{Irr}(G)$.*

*Proof.* If $H$ and $\phi \in \mathrm{Lin}(A)$ are such as in Proposition 148.22, then $I_G(\phi) = A$ since $H/A$ is the unique subgroup in $G/A$ of prime order. Therefore, $\chi = \phi^G \in \mathrm{Irr}(G)$. □

If, in Corollary 148.23, $|G/A| = p^k > p$ and $G$ is not a $p$-group, then it is possible that $|\mathrm{cd}(G)| < k + 1$ (for example: $G$ is a Frobenius group with kernel $A$ and cyclic $G/A$; then $|\mathrm{cd}(G)| = 2$).

Let $G$ be a nonabelian metacyclic 2-group with a normal cyclic subgroup $L$ such that $G/L$ is cyclic. Then $C_G(L)$ is a maximal abelian normal subgroup of $G$. Setting $|G : C_G(L)| = 2^k$, it follows from [Man33, Proposition 11] that $\mathrm{cd}(G) = \{1, 2, \dots, 2^k\}$.

$4^o$ *Metacyclic 2-groups without nonabelian sections of order 8.* Most results of Subsection $3^o$ are not true for metacyclic 2-groups. In this section some analogs of theorems from § 3 for the metacyclic 2-groups without nonabelian sections of order 8 are proved or stated.

It is assumed throughout this section that the group $G$ satisfies, if it is not stated otherwise, the following:

**Hypothesis.** $G$ is a metacyclic 2-group of order > 8 which has no nonabelian section of order 8, $G' \cong C_{2^k}$, $k \geq 1$.

If a group $G$ of exponent $2^e$ satisfies the hypothesis, then, using Lemma 148.3 (i), it is easy to prove by induction on $i$ that $\exp(\Omega_i(G)) = 2^i$ for $i \leq e$ and $G/R(G)$ is cyclic (indeed, by Lemma 148.3 (i), $\Omega_1(G) \cong E_4$ and $G/\Omega_1(G)$ satisfies the hypothesis). A metacyclic 2-group $G = R(G)$ can possess a nonabelian section of order 8 (for example: if $G = \mathcal{H}_{2,2} = \langle a, b \mid a^4 = b^4 = 1,\ a^b = a^3\rangle$, then $G/\langle b^2\rangle \cong D_8$ and $G/\langle a^2b^2\rangle \cong Q_8$).

**Lemma 148.24.** *(Compare with Corollary 148.2(b).) If $G$ possesses an abelian subgroup of index 2, then $G$ is minimal nonabelian.*

*Proof.* In view of Lemma 148.3 (a), one may assume that $|G'| = 2^k > 2$. Assume that $A \in \Gamma_1$ is abelian. By Lemma 148.3 (e), $A/Z(G) \cong G'$ so that $G/Z(G)$ has a cyclic subgroup of index 2. Then, by Remark 148.9, $G/Z(G) \cong D_{2^{k+1}}$ hence $G$ has an epimorphic image $\cong D_8$, contrary to the hypothesis. Thus, $k = 1$, so $G$ is minimal nonabelian. □

**Lemma 148.25.** *If $|G'| = 2^k > 2$, then $|G' : H'| = 2$ for all $H \in \Gamma_1$.*

*Proof.* Take $H \in \Gamma_1$ such that $|H'|$ is as small as possible. By Remark 148.8, $H' < G'$. Write $\bar{G} = G/H'$. Then $\bar{G}$ is minimal nonabelian (Lemma 148.24) so that $|G' : H'| = |\bar{G}'| = 2$ (Lemma 148.3 (f)), and the result follows, by the choice of $H$. □

**Theorem 148.26.** *If $2^n < 2^k = |G'|$, then all $\mathcal{A}_n$-subgroups of $G$ have equal order.*

*Proof.* We use induction on $|G| + k + n$. Note that $G$ is an $\mathcal{A}_k$-group and $H < G$ is an $\mathcal{A}_n$-group $\iff$ $|H'| = 2^n$ (Lemma 148.3 (j)).

(i) Assume that $|\Phi(G)'| \geq 2^n$; then there is an $\mathcal{A}_n$-subgroup $A$ in $\Phi(G)$ (Lemmas 148.25 and 148.3 (j)). Let $B < G$ be an $\mathcal{A}_n$-subgroup. If $B \leq \Phi(G)$, then $|B| = |A|$, by induction, applied to $\Phi(G)$. Let $B \nleq \Phi(G)$. Then $B\Phi(G) = H \in \Gamma_1$ so, by induction, applied to $H$, $|B| = |A|$.

(ii) Now assume that $|\Phi(G)'| < 2^n$ and let $F < G$ be an $\mathcal{A}_n$-subgroup; then $F \nleq \Phi(G)$ since $|F'| > |\Phi(G)'|$. Write $H = F\Phi(G)$; then $H \in \Gamma_1$ since $d(G) = 2$, and so $|H : \Phi(G)| = 2$. By Lemmas 148.25 and 148.3 (j),

$$|\Phi(G)'| < 2^n = |F'| \leq |H'| = 2|\Phi(G)'| \leq 2^n = |F'|\,,$$

and so $|H'| = 2^n = |F'|$. Since $F \le H$, one has $F = H$ (Lemma 148.25), and so $\Phi(G) < F$. Then $F \in \Gamma_1$ so that $|F| = \frac{1}{2}|G|$. Since $F$ is an arbitrary $\mathcal{A}_n$-subgroup of $G$, the proof is complete (the above argument also works for $p > 2$). □

If $H < G$ is nonabelian ($G$ is a title 2-group) and $A < H$ is maximal abelian, then $A$ is maximal abelian in $G$. Indeed, assume that $A < A_1 < G$, where $A_1$ is maximal abelian in $G$. Let $A < B \le H$, where $|B : A| = 2$, and $A_1 < B_1 \le G$, where $|B_1 : A_1| = 2$. By Lemma 148.24, $B$ and $B_1$ of distinct orders are minimal nonabelian, contrary to Theorem 148.26.

**Theorem 148.27.** *Suppose that $|G'| = 2^k$, $k > 1$ ($G$ is a title 2-group). Then:*
(a) *If $F, H < G$ are nonabelian, then $|F'| = |H'| \iff |F| = |H|$ and $|G : F| = |G' : F'|$. If $n \le k$, then all $\mathcal{A}_n$-subgroups have index $2^{k-n}$ in $G$.*
(b) *If $A, B \le G$, where $A$ is abelian and $B$ is nonabelian, then $|A| < |B|$ so $|G : A| \ge 2^k$.*
(c) *All maximal abelian subgroups have index $2^k$ in $G$.*
(d) *If $k \ge n$ are natural numbers, then $\alpha_n(2^k) = 2^{k-n+1} - 1$ (compare with Theorem 148.17).*
(e) *Let $\exp(G) = 2^e$. Then all minimal nonabelian subgroups of $G$ have the same exponent $\iff \Omega_{e-1}(G)$ is abelian.*

Theorem 148.27 (a) follows immediately from Theorem 148.26. Let us prove (b). Let $A < K \le G$, where $|K : A| = 2$, and let $B_1 \le B$ be minimal nonabelian. Since $K$ is minimal nonabelian (Lemma 148.24), one has $|K| = |B_1|$ hence $|A| < |B|$, proving (b). Let us prove (e). As it was noted, $\Omega_{e-1}(G) < G$ so there is a minimal nonabelian $A < G$ such that $A \nleq \Omega_{e-1}(G)$ (Lemma 148.3 (k)). It follows that $\exp(A) = 2^e$ so, by hypothesis, all minimal nonabelian subgroups of $G$ have exponent $2^e$. Now the proof is finished as the proof of Theorem 148.19. The proofs of the remaining parts of the theorem are omitted since they repeat, almost word for word, the proofs of corresponding assertions for $p > 2$ (see Subsection 3°).

Let $n > 3$, $\langle a \rangle = A \cong C_{2^n}$ and let $B = \langle b \rangle$ of order $2^{n-2}$ act faithfully on $A$; then one can put $a^b = a^5$. Let $G = B \cdot A$ be the natural semidirect product with kernel $A$. Such $G$ has no nonabelian epimorphic images of order 8. The subgroup $F = \Omega_1(B) \cdot A \cong M_{2^{n+1}}$ is minimal nonabelian. If $H < G$ is minimal nonabelian, then $|H| = |F| = 2^{n+1}$, by Theorem 148.26.

**Theorem 148.28.** *Let $G = \langle A, B \rangle$, where a 2-group $G$ satisfies the hypothesis, $|G'| = 2^k$ and $A, B < G$ are distinct minimal nonabelian. Then:*
(a) $\mathrm{cd}(G) = \{1, 2, \dots, 2^k\}$,
(b) $|G : Z(G)| = |G'|^2$,
(c) $Z(A) \cap Z(B) = Z(G)$,
(d) $Z(A)Z(B) = \Phi(G)$.
(e) *If $U < G$ is maximal abelian, then $|U| = |Z(G)|\,|G'|$.*

The proof is the same as the proofs of Theorems 148.20 and 148.21 (Theorem 148.28 (a) follows, by induction, from Proposition 148.22).

*$5^o$ Metacyclic 2-groups containing an abelian subgroup of index 2.* In this subsection $G$ is a nonabelian metacyclic 2-group of order $> 2^3$ containing an abelian subgroup of index 2.

Let $G$ be a nonabelian $p$-group (not necessarily metacyclic) and let $A \in \Gamma_1$ be abelian. If $B < G$ is maximal abelian and $B \neq A$, then $AB = G$ and $A \cap B = Z(G)$ so that $|B| = p|Z(G)|$, by the product formula. If, in addition, $B < D \leq G$, $|D : B| = p$ and $d(D) = 2$, then $D/Z(G) \cong E_{p^2}$ is noncyclic since $D$ is nonabelian. It follows that $\Phi(D) = Z(G) = Z(D)$ so that $D$ is minimal nonabelian of order $p^2|Z(G)| = p|B|$. Let $M < G$ be minimal nonabelian of order $p^2|Z(G)|$. Then all maximal subgroups, say $A_1$, of $M$ that are $\neq M \cap A$, are maximal abelian in $G$ since all maximal abelian subgroups of $G$ that distinct of $A$, have order $|A_1|$, by the above. Obviously, $Z(G) < M$ (see part (i) of Remark 148.10). It follows that $Z(G) = \Phi(M) = Z(M)$. Let $T < M$ be maximal and $T \neq M \cap A$; then $T$ is maximal abelian in $G$. It follows that if $G$ is a nonabelian metacyclic 2-group with abelian $A \in \Gamma_1$ (for $p > 2$ all that follows in this paragraph, is known; see Subsection $3^o$), then:

(i) all maximal abelian subgroups of $G$ that are $\neq A$, have order $2|Z(G)|$,
(ii) all nonabelian subgroups of $G$ of order $4|Z(G)|$ are minimal nonabelian,
(iii) if $M < G$ is minimal nonabelian and $B < M$ is maximal such that $B \neq M \cap A$, then $B$ is maximal abelian in $G$.

Let us prove (ii). Let $U \leq G$ be nonabelian of order $4|Z(G)|$. By Remark 148.10 and Lemma 148.3 (k), $Z(G) < U$ so $U/Z(G) \cong E_4$, and so $U$ is minimal nonabelian since $d(U) = 2$.

**Lemma 148.29.** *Let $G$ be a metacyclic 2-group, $|G'| = 2^k > 2$ and suppose that $G$ has an abelian maximal subgroup. Then $G/Z(G) \cong D_{2^{k+1}}$. If a nonabelian $H \in \Gamma_1$, then $H' = \mho_1(G')$.*

*Proof.* It follows from Remark 148.9 and Lemma 148.3 (e) that $G/Z(G) \cong D_{2^{k+1}}$. Let $\Gamma_1 = \{A, B, C\}$, where $C/Z(G)$ is cyclic; then $C$ is abelian. Since $k > 1$, $C$ is the unique abelian member of the set $\Gamma_1$. Being noncyclic, maximal subgroups $A/Z(G)$, $B/Z(G)$ of $G/Z(G) \cong D_{2^{k+1}}$ are isomorphic to either $D_{2^k}$ or $E_4$ (provided $k = 2$), so have derived subgroups of order $2^{k-2}$; and hence $|A'|, |B'| \geq 2^{k-1}$ since $A' \cap Z(G) \neq \{1\} \neq B' \cap Z(G)$. Since $|A'|, |B'| < |G'|$ and $G'$ is cyclic, one obtains $|A'| = |B'| = 2^{k-1}$ hence $A' = B' = \mho_1(G')$. □

**Theorem 148.30.** *Suppose that $G$ and $G_0$ are nonabelian metacyclic 2-groups of equal order with $|G'| = |G_0'|$ containing an abelian subgroup of index $p$. If $H \leq G$ and $H_0 \leq G_0$ are minimal nonabelian, then $|H| = |H_0|$.*

*Proof.* Set $|G'| = 2^k$. If $k = 1$, then $G$ and $G_0$ are minimal nonabelian (Lemma 148.3 (a)) so the assertion is true since $|G| = |G_0|$. Now let $k > 1$. We proceed by induction on $k$. Let $F, H < G$ be minimal nonabelian and let $F \le A \in \Gamma_1$, $H \le B \in \Gamma_1$. If $C \in \Gamma_1$ is abelian, then $A \cap C$ and $B \cap C$ are abelian subgroups of index 2 in nonabelian groups $A, B$, respectively. Then, by Lemma 148.29, $|A'| = |B'|$. Therefore, by induction, $|F| = |H|$. The same argument shows that all minimal nonabelian subgroups of $G_0$ have order $|F|$. □

Considering $G/\Omega_1(G)$, $G_0/\Omega_1(G_0)$ and taking into account that $F/\Omega_1(G)$, $H/\Omega_1(G)$, being maximal abelian in $G/\Omega_1(G)$, have the same order, we obtain an alternate proof of Theorem 148.30.

Let us show that $H/Z(G) \cong E_4$, where $H$ is from Theorem 148.30 (it follows from this that $Z(H) = Z(G)$). Let $G' = 2^k > 2$. By Remark 148.9, $G/Z(G) \cong D_{2^{k+1}}$. By Remark 148.10, $Z(G) \le Z(H)$ so that $H/Z(G) < G/Z(G)$. Since $H' \le Z(G)$ and $H/Z(G)$ is abelian and noncyclic, it follows from the structure of $G/Z(G)$ that $H/Z(G) \cong E_4$. This agrees with the result of (ii) in the paragraph preceding Lemma 148.29.

**Proposition 148.31.** *Let $G$ be a metacyclic 2-group containing an abelian subgroup of index 2. If $|G'| = 2^k$ and $n \le k$, then $\alpha_n(G) = 2^{k-n}$.*

*Proof.* Since $\alpha_k(G) = 1 = 2^{k-k}$, one may assume that $n < k$; then $k > 1$ so that $\alpha_1(G) > 1$ (see § 76). Next, induction on $k$ is used. Let $A \in \Gamma_1$ be abelian and let $H \in \Gamma_1$ be nonabelian; then $A$ is a unique abelian member in $\Gamma_1$ so that $F \in \Gamma_1 - \{A, H\}$ is nonabelian. In this case, $H \cap A$ is abelian of index 2 in $H$ and $|H'| = 2^{k-1}$ (Lemma 148.29). Therefore, by induction, $\alpha_n(H) = 2^{k-1-n}$ hence, by Hall's enumeration principle,

$$\alpha_n(G) = \alpha_n(F) + \alpha_n(H) = 2 \cdot 2^{k-1-n} = 2^{k-n}$$

since $\Phi(G)$ is abelian. □

It follows from Proposition 148.31 that if $G$ is a metacyclic 2-group with abelian Frattini subgroup and $|G'| = 2^k > 2$, then $\alpha_n(G) \in \{2^{k-n}, 3 \cdot 2^{k-n-1}\}$ for $k \ge n - 1$.

**Proposition 148.32.** *Let $G$ be a nonabelian metacyclic 2-group with abelian $A \in \Gamma_1$. If $B < G$ is maximal abelian $\ne A$ and $B < D \le G$, where $|D : B| = 2$, then $D$ is minimal nonabelian. All maximal abelian subgroups of $G$ that $\ne A$ have order $|B| = 2|Z(G)|$ and all minimal nonabelian subgroups of $G$ have order $|D| = 4|Z(G)|$.*

*Proof.* See the paragraph preceding Lemma 148.29. □

*6° Metacyclic p-groups with a minimal nonabelian subgroup of order $p^4$.* By Remark 148.8 (see also Lemma 148.3 (m)), if a metacyclic $p$-group $G$ has a minimal nonabelian subgroup of order $p^3$, it is of maximal class. In this subsection, as an application of the results obtained above, the metacyclic $p$-groups possessing a minimal nonabelian subgroup of order $p^4$, are described. There are exactly two nonisomorphic metacyclic minimal nonabelian groups of order $p^4$: $M_{p^4}$ and $\mathcal{H}_{2,p} = \langle a, b \mid a^{p^2} = b^{p^2} = 1,\ a^b = a^{1+p}\rangle$.

Suppose that a metacyclic $p$-group $G$ has a proper minimal nonabelian subgroup of order $p^4$. Assume that $|G| = p^5$. As $G$ is not a 2-group of maximal class, all of its subgroups of order $p^3$ are abelian (Remark 148.8), and hence $G$ is an $\mathcal{A}_2$-group; then $|G'| = p^2$ (Corollary 65.3).

**Remark 148.33.** Let $G$ be a nonabelian metacyclic $p$-group, $p > 2$. Then $\Omega_1(G) \cong E_{p^2}$. Write $\bar{G} = G/\Omega_1(G)$. Let $\bar{A} < \bar{G}$ be maximal abelian. Then $A' < \Omega \cong E_{p^2}$ so that $A' = \Omega_1(G')$ if $A$ is nonabelian, hence $A' \leq \Omega_1(Z(G))$. Assume that $A$ is abelian. Let $A < B \leq G$, where $|B : A| = p$. Then $B$ is minimal nonabelian (Corollary 148.2 (b)); hence $B/\Omega_1(G)$ is abelian, a contradiction since $\bar{A} < \bar{B}$ is maximal abelian in $\bar{G}$.

**Exercise 2.** Let $G$ be a metacyclic $p$-group with $|G'| = p^k$ and $n < k$. If $H < G$ is an $\mathcal{A}_n$-subgroup, then $H/H'$ is a maximal abelian subgroup of $G/H'$.

**Theorem 148.34.** *If a metacyclic p-group G has a proper minimal nonabelian subgroup B of order $p^4$, then one of the following holds:*

(a) *If $|G : A| = p$, then G is an $\mathcal{A}_2$-group.*
*Now let $|G| > p^5$. Then $p = 2$ and $G/\Omega_1(G)$ is a 2-group of maximal class since $A/\Omega_1(G)$ of order $p^2$ is maximal abelian in $G/A$ (Lemma 148.3 (n)). Write $\bar{G} = G/\Omega_1(G)$, $|\bar{G}| = 2^n$ and $\Gamma_1 = \{U, V, T\}$, where $T/\Omega_1(G)$ is cyclic.*

(b) *If T is abelian, then all minimal nonabelian subgroups of G have order $2^4$. If T is nonabelian, then $T \cong M_{2^{n+1}}$ is minimal nonabelian, all other minimal nonabelian subgroups of G have order $2^4$.*

(c) *If $G/\Omega_1(G) \cong D_{2^n}$, then $\Omega_1(G) = Z(G)$ and all minimal nonabelian subgroups of G are isomorphic to $\mathcal{H}_{2,2}$.*

*Proof.* We retain the notation introduced in the statement.

Since $G$ is not minimal nonabelian, one has $|G'| > p$. Since all minimal nonabelian subgroups of a 2-group of maximal class have order 8, $G$ is not a 2-group of maximal class. If $|G| = p^5$, then $G$ is an $\mathcal{A}_2$-group (see the paragraph preceding Remark 148.33), proving (a).

As $|G'| > p$, the quotient group $G/\Omega_1(G)$ is nonabelian so that $d(G/\Omega_1(G)) = 2$, and therefore $\Omega_1(G) < \Phi(G)$.

Now let $|G| > p^5$. We have $\Omega_1(G) = \Omega_1(B) \cong E_{p^2}$ ($B$ is as in the statement). Write $C = C_G(\Omega_1(G))$; then $|G : C| \leq p$. By Remark 148.33, $B/\Omega_1(G)$ (of order $p^2$) is a maximal abelian subgroup of $G/\Omega_1(G)$ so that $G/\Omega_1(G)$ is of maximal class (Lemma 148.3 (n)). As $|G/\Omega_1(G)| > p^3$ and $G/\Omega_1(G)$ is metacyclic, one obtains $p = 2$. Set $|G/\Omega_1(G)| = 2^n$ ($n > 3$). Let $\Gamma_1 = \{U, V, T\}$, where $T/\Omega_1(G)$ is cyclic and $U/\Omega_1(G)$, $V/\Omega_1(G)$ are of maximal class. Then $T$ is either abelian of type $(2^n, 2)$ or $\cong M_{2^{n+1}}$ (Lemma 148.3 (l)) so that all proper subgroups of $T$ are abelian. It follows from $|B| < |T|$ that $B \nleq T$. Let $M < G$ be minimal nonabelian; then $M' < \Omega_1(G) < M$ so that $M/\Omega_1(G)$ is abelian.

Assume that $|M| > 2^4$; then $M/\Omega_1(G)$ is cyclic so that $M = T$ (recall that $G/\Omega_1(G)$ is a 2-group of maximal class).

As $U(\in \Gamma_1)$ (see the statement) is nonabelian, it contains an $\mathcal{A}_1$-subgroup, say $N$. Since $U$ is not of maximal class, $|N| > 8$ (Lemma 148.3 (m)). Therefore, by the previous paragraph, $|N| = 2^4$. It follows that all $\mathcal{A}_1$-subgroups of $G$, except possibly one, have order $2^4$.

If $G/\Omega_1(G) \cong D_{2^n}$ and $M/\Omega_1(G) < G/\Omega_1(G)$ is a four-subgroup, then $\Omega_1(G) \le Z(M)$ since $M$ is metacyclic of exponent 4. Since $G/\Omega_1(G)$ is generated by subgroups $\cong E_4$, it follows that $\Omega_1(G) \le Z(G)$, and we conclude that all $\mathcal{A}_1$-subgroups of $G$ are isomorphic to $\mathcal{H}_{2,2}$. Therefore, $\Omega_1(G) = Z(G)$ (otherwise, $G$ has no subgroup isomorphic $\mathcal{H}_{2,2}$). □

**Corollary 148.35.** *Suppose that a metacyclic $p$-group $G$ of order $> p^4$ possesses a self-centralizing subgroup $A$ of order $p^3$. Let $A < B < G$, where $|B : A| = p$. Then $B$ is minimal nonabelian and so $G$ is one of the groups of Theorem* 148.34.

*Proof.* The subgroup $B$ of order $p^4$, being nonabelian, is either minimal nonabelian or a 2-group of maximal class (Lemma 148.3 (m)). If $B$ is a 2-group of maximal class, then $B$ contains a nonabelian subgroup of order 8, so $G$ is of maximal class (Lemma 148.3 (m)), a contradiction since a 2-group of maximal class and order $> 2^4$ has no self-centralizing abelian subgroup of order $2^3$. Thus, $B$ is minimal nonabelian, so $G$ satisfies the hypothesis of Theorem 148.34. □

If a metacyclic $p$-group $G$ has a proper minimal nonabelian subgroup $A$ of order $p^5$, then $\Omega_1(G) = \Omega_1(A)$ and $A/\Omega_1(G) < G/\Omega_1(G)$ is maximal abelian of order $p^3$ (Remark 148.33), and so $G/\Omega$ is a group from Corollary 148.35.

*7⁰ A property of metacyclic p-groups.* Recall that a group is said to be homocyclic if it is a direct product of cyclic subgroups of equal order.

**Theorem 148.36.** *A metacyclic $p$-group $G$ possesses a composition series all of whose members are characteristic, unless $G$ is homocyclic or* $\cong Q_8$.

*Proof.* Let $G$ be a counterexample of minimal order. Set $|G| = p^m$, $\exp(G) = p^e$. One may assume that $G$ is noncyclic. Then $G$ is a 2-group of maximal class $\iff G \not\cong Q_8$. In the sequel $G$ is not of maximal class; then $m > 3$.

If $G$ has no characteristic subgroup of order $p$, it is abelian (otherwise, $\Omega_1(G')$ is characteristic of order $p$), and so homocyclic (otherwise, $\mho_{e-1}(G)$ of order $p$ is characteristic; an abelian $p$-group $X$ of exponent $p^e$ is homocyclic $\iff d(\mho_{e-1}(X)) = d(X)$). Thus, $G$ has a characteristic subgroup, say $L$, of order $p$. Then $G/L$ has no composition series with all characteristic members so, by induction, it is either homocyclic or $\cong Q_8$.

(i) Assume that $G/L$ is homocyclic; then $L = \mho_{e-1}(G)$ and the subgroup $M = \Omega_1(G) \cong E_{p^2}$ is characteristic. As $G/M$ is nonhomocyclic abelian of rank $\le 2$, it has a composition series all of whose members are characteristic, by induction, completing this case.

(ii) Now let $G/L \cong Q_8$. Then $G$ has no cyclic subgroup of index 2 (Lemma 148.3 (l)). In that case, $G \cong \mathcal{H}_{2,2}$. Then $G/G'$ is abelian of type $(4, 2)$ so nonhomocyclic and it has a composition series all of whose members are characteristic. The proof is complete since $G'$ is characteristic. □

The following corollary is fairly unexpected:

**Corollary 148.37.** *Let $G$ be a nonabelian metacyclic $p$-group, $p > 2$.*
(a) *$G$ has a characteristic minimal nonabelian subgroup.*
(b) *$G$ has a characteristic maximal abelian subgroup.*

*Proof.* Let $p^\nu$ be the order of a minimal nonabelian subgroup $G$ (see Proposition 148.15 (d)). By Theorem 148.36, since $G$ is nonhomocyclic, it has characteristic subgroups, say $U$, $V$, of orders $p^\nu$, $p^{\nu-1}$, respectively. Then $U$ is minimal nonabelian and $V$ is maximal abelian in $G$ (Proposition 148.15). □

Corollary 148.37 is not true for $p = 2$ (for example, $D_{16}$ has no characteristic minimal nonabelian subgroup since it is a subgroup of $H = D_{32}$ and $H$ has no normal nonabelian subgroup of order 8), however that corollary holds for particular types of metacyclic 2-groups considered in Subsections $4^o$, $5^o$, and $8^o$. If $G$ from Corollary 148.37 has the derived subgroup of order $p^k > p$, $p > 2$, then it possesses a characteristic $\mathcal{A}_n$-subgroup for any $n \le k$ (to prove this, use Proposition 148.15).

**Remark 148.38.** In the same way one can prove that if a metacyclic group $G$, which is not necessarily nilpotent, has no chief series all of whose members are characteristic, then $G = Q \times H$, where either: (i) $Q \cong Q_8$ and $H$ is noncyclic homocyclic of odd order, or (ii) $Q = \{1\}$ and $H(= G)$ is homocyclic of odd order.

**Remark 148.39.** Suppose that $G$ is a metacyclic $p$-group. If a prime $q \mid |\mathrm{Aut}(G)|$, then $q \le p$, unless $G$ is either a homocyclic 2-group of rank 2 or $\cong Q_8$. Assume that this is false. Then there is a characteristic $H \in \Gamma_1$ (Theorem 148.36). Let $\phi \in \mathrm{Aut}(G)$ be of prime order $q > p$. Then $\phi$ fixes a subgroup $\bar{H} = H/\Phi(G)$ of the group $\bar{G} = G/\Phi(G)$ of order $p$. By Maschke's theorem, $\bar{G} = \bar{H} \times \bar{F}$, where the subgroup $\bar{F}$ of order $p$ is $\phi$-invariant. Since, by assumption, $q > p$, $\phi$ induces on $G/\Phi(G)$ the identity automorphism. According to Hall's result (Theorem 1.15), in that case, $o(\phi)$ is a power of $p$, a contradiction. If $G \cong Q_8$, then $q = 3 > 2$ since $\mathrm{Aut}(Q_8) \cong S_4$. If $p > 2$ and $G$ is abelian of rank 2, then $q \mid (p-1)p(p+1)$ so $q \le p$.

### $8^o$ *On a class of metacyclic 2-groups.* Let us begin with the following:

**Definition 1.** A group $G$ is said to be a $\Delta_p$-group if it satisfies the following conditions:
($1\Delta_p$) $G$ is a nonabelian metacyclic $p$-group,
($2\Delta_p$) whenever $A < B \le G$, $B' > \{1\}$ and $|B : A| = p$, then $|B' : A'| = p$.

If $p > 2$, then any nonabelian metacyclic $p$-group is a $\Delta_p$-group (Proposition 148.15). All metacyclic 2-groups considered in Subsection $4^0$ are $\Delta_2$-groups. The 2-groups $G$ of maximal class and order $> 8$ are not $\Delta_2$-groups (let $B = G$ and let $A < G$ be cyclic of index 2).

Throughout this section $G$ is a metacyclic $\Delta_2$-group. Below a number of properties of $G$ are proved.

($3\Delta_2$) If $A < B \leq G$, where $A$ is either maximal abelian or nonabelian, then $|B' : A'| = |B : A|$. This follows immediately from the definition.

($4\Delta_2$) If $A < B \leq G$, where $A$ is maximal abelian and $|B : A| = 2$, then $B$ is minimal nonabelian. Indeed, by ($2\Delta_2$), one has $|B'| = 2$ so, by Lemma 148.3 (a), $B$ is minimal nonabelian since $\mathrm{d}(B) = 2$.

($5\Delta_2$) If $U, V < G$ are minimal nonabelian, then $|U| = |V|$. We prove this by induction on $|G'|$. Suppose that $\Phi(G)$ is nonabelian. Take $V \leq \Phi(G)$. Then $U \leq U\Phi(G) < G$ and, by induction, applied to $U\phi(G)$, one has $|U| = |V|$. Now let $\Phi(G)$ be abelian. Then $U\Phi(G), V\Phi(G) \in \Gamma_1$ so these subgroups are minimal nonabelian, by ($4\Delta_2$). In that case, $|U| = |U\Phi(G)| = \frac{1}{2}G = |V\Phi(G)| = |V|$.

($6\Delta_2$) If $B \leq G$ is minimal nonabelian, then all maximal subgroups of $B$ are maximal abelian in $G$. Indeed, let $A < B$ be maximal and $A \leq A_1 < G$, where $A_1$ is maximal abelian. Let $A_1 < B_1 \leq G$, where $|B_1 : A_1| = 2$. By ($4\Delta_2$), $B_1$ is minimal nonabelian so $|B_1| = |B|$, by ($5\Delta_2$). Then $|A| = \frac{1}{p}|B| = \frac{1}{p}|B_1| = |A_1|$ hence $A = A_1$ is a maximal abelian subgroup of $G$.

($7\Delta_2$) If $Z_1$ and $Z_2$ are self centralizing cyclic subgroups of $G$, then $|Z_1| = |Z_2|$. Indeed, if $Z_i < B_i$ with $|B_i : Z_i| = 2$, then $B_i$ is minimal nonabelian, $i = 1, 2$, by ($4\Delta_2$). Now $|B_1| = |B_2|$ by ($5\Delta_2$), and so $|Z_1| = \frac{1}{2}|B_1| = \frac{1}{2}|B_2| = |Z_2|$, completing the proof.

($8\Delta_2$) If $|G'| = 2^k$ and $n \leq k$, then $\alpha_n(G) = 2^{k-n+1} - 1$. The proof is the same as the proof of Lemma 148.16 and Theorem 148.17.

**Exercise 3.** Prove properties ($8\Delta_2$) and ($9\Delta_2$).

**Lemma 148.40.** *If $G$ is a metacyclic $p$-group, then* $\exp(\Phi(G)) < \exp(G)$.

*Proof.* Note that for metacyclic $G$, one has $\Phi(G) = \mho_1(G)$ (indeed, $G/\Phi(G) \cong \mathrm{E}_{p^2} \cong G/\mho_1(G)$ and $\Phi(G)$, $\mho_1(G)$ are incident). Set $\exp(G) = p^e$, $\exp(R(G)) = p^w$, $\exp(G/R(G)) = p^t$, $w, t \geq 0$; then $e = w + t$. If $G$ is regular, then $\mho_1(G) = \{x^p \mid x \in G\}$ so that $\exp(\Phi(G)) = \exp(\mho_1(G)) < \exp(G)$. If $p > 2$, then $G$ is regular, and so the lemma holds in this case. The lemma also holds if $G$ is a 2-group of maximal class (Lemma 148.3 (c,l)). Now let $p = 2$ and $G$ be not a 2-group of maximal class; then $w > 0$.

(i) Let $G = R(G)$; then $\Omega_{e-1}(G) = \Phi(G)$. As $\exp(\Omega_{e-1}(G)) = 2^{e-1} < \exp(G)$, this case is complete.

(ii) Assume that $G/R(G) > \{1\}$ is cyclic. Then $H = \Omega_{e-1}(G) \in \Gamma_1$ so $\Phi(G) < H$ and $\exp(\Phi(G)) \le \exp(H) = 2^{e-1} < \exp(G)$.

(iii) Now let $G/R(G)$ be of maximal class and order $2^{t+1}$. Let $F/R(G) = \Phi(G/R(G))$. Then $\exp(F) = 2^{w+t-1} = 2^{e-1}$, by (ii). Since $F = \Phi(G)$, the proof is complete. □

It is easy to prove, using Lemma 148.3 (i), that if $G$ is a metacyclic $p$-group, then $\exp(\Omega_i(\Phi(G)) = p^i$ for $p^{i+1} \le \exp(G)$.

**Theorem 148.41** (for $p > 2$, see Theorem 148.19). *Let $G$ be a $\Delta_2$-group of order $> 2^3$ and exponent $2^e$. Suppose that $G$ is not minimal nonabelian and all minimal nonabelian subgroups of $G$ have equal exponent. Then $G$ is an $\mathcal{A}_2$-group, all of whose maximal subgroups are minimal nonabelian of exponent $2^e$. Any such $G$ satisfies the hypothesis.*

*Proof.* As we have noted, $G$ is not a 2-group of maximal class. Let $Z < G$ be cyclic of order $2^e$ and $Z \le A < B \le G$, where $A$ is maximal abelian and $|B : A| = 2$. Then $B$ is minimal nonabelian, by ($4\Delta_2$), hence all minimal nonabelian subgroups of $G$ have exponent $2^e$ and so all subgroups of $G$ of exponent $< 2^e$ are abelian. Therefore, by Lemma 148.40, $\Phi(G)$ is abelian of exponent $2^{e-1}$. As $G$ is not minimal nonabelian, all members of the set $\Gamma_1$ are minimal nonabelian, by ($4\Delta_2$), so that $G$ is an $\mathcal{A}_2$-subgroup; then $|G'| = 4$ (Corollary 65.3). As is easily seen, any such $G$ satisfies the hypothesis. □

*$9^o$ Metacyclic p-groups with a self-centralizing cyclic subgroup. Concluding remarks.* Let $A$ be a maximal abelian subgroup of a metacyclic $p$-group $G$. If $G$ is a 2-group of maximal class, then $|A| \in \{4, \frac{1}{2}|G|\}$. If $p > 2$, then, by Proposition 148.15 (d), $|G : A| = |G'|$, i.e., $|A| = |G : G'|$ is determined uniquely.

Let $\mathrm{sc}(G)$ be the number of orders of self-centralizing cyclic subgroups of a nonabelian metacyclic $p$-group $G$. By the previous paragraph, $\mathrm{sc}(G) = 1$ if $p > 2$. We will show, that $\mathrm{sc}(G) \le 2$ if $p = 2$. Note that if $G$ is a 2-group of maximal class, then $\mathrm{sc}(G) < 2 \iff G \cong \mathrm{D}_8$. Of course, the equality $\mathrm{sc}(G) = 0$ is possible (for example: $G \cong \mathcal{H}_{2,2}$).

It is known that $G = R(G)$ has the noncyclic center so it has no self-centralizing cyclic subgroups. Let $p^c$ be the order of a self-centralizing cyclic subgroup of $G/R(G)$ and $|R(G)| = p^{2w}$. Then $p^c = |G/R(G)|$ if $G/R(G)$ is cyclic, and if $G/R(G)$ is a 2-group of maximal class, then $2^c \in \{2^2, \frac{1}{2}|G/R(G)|\}$ (in the last case, $2^c = |G/R(G)|$). This notation is retained in what follows.

**Lemma 148.42.** *If $G = R(G)$ is a metacyclic p-group, then $\Omega_1(G) \le Z(G)$.*

*Proof.* One may assume that $G$ is nonabelian. We proceed by induction on $|G|$. If $G$ is minimal nonabelian, then the result holds since $\Omega_1(G) \le \Phi(G) = Z(G)$ (see Lemma 148.3 (f)). Now let $G$ be not minimal nonabelian; then $w > 2$. Let $H < G$ be minimal nonabelian. Then $\Omega_1(H) = \Omega_1(G)$ and $H/\Omega_1(G) < G/\Omega_1(G)$ is maximal abelian (Remark 148.33) so, by induction, $H/\Omega_1(G)$ is noncyclic. In that case,

$\Omega_1(G) = \Omega_1(H) \le Z(H)$ (Lemma 148.3 (f) again). Then $C_G(\Omega_1(G))$ contains all minimal nonabelian subgroups of $G$ so coincides with $G$ (Lemma 148.3 (k)). □

*Second proof of Lemma 148.42.* There is a cyclic $Z \lhd G$ such that $G/Z$ is cyclic. Since $|Z|, |G/Z| \le \exp(G)$, one has $|Z| = \exp(G) = |G/Z| = p^w$. If $B < G$ is cyclic such that $G = BZ$, then $|B| = p^w$ so $Z \cap B = \{1\}$, and hence $G = B \cdot Z$ is a semidirect product with kernel $Z$. The subgroup $C_G(Z) > Z$ is abelian and normal in $G$. Then $C_G(Z) \cap B \le B_G$ so that $\Omega_1(Z) \times \Omega_1(B_G) \cong E_{p^2}$ is noncyclic and central. □

**Theorem 148.43.** *If $Z$ is a self-centralizing cyclic subgroup of a nonabelian metacyclic $p$-group $G$, $|G| > p^3$, then $Z\Omega_i(G)/\Omega_i(G)$ is self centralizing in $G/\Omega_i(G)$ for any $i \le w$. If $G = R(G)$, then $G$ has no self-centralizing cyclic subgroup. Let $p^c$ be the order of a cyclic self-centralizing subgroup of $G/R(G)$.*

(a) *If $G/R(G)$ is cyclic, then $|Z| = p^{w+c}(= \exp(G))$.*
(b) *If $G/R(G)$ is a 2-group of maximal class, then $|Z| \in \{2^{w+2}, 2^{w+c}(= \exp(G))\}$ ($Z = 2^{w+2}$ can be if $G/R(G)$ is not dihedral).*

*Proof.* By the paragraph preceding the theorem, one may assume that $w > 0$. Put $|Z| = p^v$. If $G$ is minimal nonabelian (in that case, $w = 1$ and $G \cong M_{c+2}$), then $|Z| = p^{1+c} = p^{w+c}$. Let $G$ be not minimal nonabelian; then $|G'| > p$ (Lemma 148.3 (f)). We use induction on $|G|$. Set $\Omega = \Omega_1(G)$, $\bar{G} = G/\Omega$. If $w = 1$ and $\bar{G}$ is cyclic, then $G \cong M_{2^{2+c}}$. One has $\Omega = E_{p^2}$. By Theorem 148.43, below, $Z \not\le R(G)$ since $Z(R(G))$ is noncyclic, so that $G > R(G)$. The subgroup $H = Z\Omega \cong M_{p^{v+1}}$ so the cyclic $\bar{H}$ is maximal abelian in $\bar{G}$ (Remark 148.33). It is easy to prove by induction that $Z\Omega_i(G)/\Omega_i(G)$ is self centralizing for $i \le w$. This is true for $i = 1$: $\bar{Z}$ is self centralizing in $\bar{G}$, by the above. Let $w > 1$ and $|R(\bar{G})| = p^{2\bar{w}}$; then $\bar{w} = w - 1$. In this case, $\bar{Z}\Omega_i(\bar{G})/\Omega_i(\bar{G})$ is self centralizing in $\bar{G}/\Omega_i(\bar{G})$ for $i \le \bar{w}$, by induction, and our claim follows.

As $G$ is not of maximal class, we get $p^v = |Z| > p^2$ (Lemma 148.3 (n)).

By the above, $B = Z\Omega \cong M_{p^{v+1}}$ is minimal nonabelian (Lemma 148.3 (l)). By Remark 148.33, $\bar{B} = B/\Omega$ is maximal abelian in $\bar{G} = G/\Omega$. Note that $\bar{B} \cong Z/(Z \cap \Omega)$ is cyclic of order $\frac{1}{p}|Z| = p^{v-1}$. If $w = 1$, the proof is complete. Indeed, then $\bar{G}$ is a 2-group of maximal class since $G$ is not minimal nonabelian; in that case,

$$\frac{1}{2}|Z| = |\bar{Z}| \in \{4, \frac{1}{2}|\bar{G}| = \frac{1}{8}|G|\} \quad \text{hence } |Z| \in \{2^3, \frac{1}{4}|G|\} = \{2^{w+2}, 2^{w+c}\}.$$

Now let $w > 1$; then $\bar{G}$ is nonabelian in view of $|G'| > p$. Let $\bar{w} = w(G/\Omega)$; then $\bar{w} = w - 1$. If $G/R(G)$ is cyclic, then $|\bar{Z}| = p^{\bar{w}+c}$ so that $|Z| = p \cdot p^{\bar{w}+c} = p^{w+c}$. If $G/R(G)$ is s 2-group of maximal class, then, by induction,

$$\frac{1}{2}|Z| = |\bar{Z}| \in \{2^{\bar{w}+2}, 2^{\bar{w}+c}\} \to |Z| = 2|\bar{Z}| \in \{2^{w+2}, 2^{w+c}\},$$

completing the proof. □

**Corollary 148.44.** *Let $G = R(G)$ be a nonabelian metacyclic $p$-group.*

(a) *G has no minimal nonabelian subgroup* $\cong M_{p^n}$.
(b) *Let* $A < G$ *be maximal abelian. If* $A$ *has a cyclic subgroup of index* $p$, *then* $A$ *is abelian of type* $(p^w, p)$ *(here* $p^w = \exp(G)$*).*

*Proof.* (a) If $B \leq G$ is minimal nonabelian, then $\Omega_1(G) = \Omega_1(B) \cong \mathrm{E}_{p^2}$, therefore, (a) follows from Lemma 148.42.

(b) By hypothesis and Lemma 148.2, $A$ is abelian of type $(p^a, p)$. We have to prove that $a = w$. We use induction on $|G|$. If $w = 2$, the result is obvious since $G$ is minimal nonabelian. Write $\bar{G} = G/\Omega_1(G)$. One has $\Omega_1(G) < A$ Lemma 148.42) so that $\bar{A} < \bar{G} = R(\bar{G})$ is cyclic of order $p^{a-1}$. Let $\bar{A} < \bar{B} < \bar{G}$, where $|B : A| = p$. Then $\bar{B}$ is abelian, by (a), and hence the (nonabelian) subgroup $B$ is minimal nonabelian since $|B'| = p$ (Lemma 148.3 (a)) and $\bar{B} < \bar{G} = R(\bar{G})$ is maximal abelian (Remark 148.33). It follows that $\bar{B}$ is noncyclic (Lemma 148.42). Then, by induction, $\bar{B}$ is abelian of type $(p^{\bar{w}}, p)$, where $p^{\bar{w}} = \exp(\bar{G})$. Then $\bar{w} = w - 1$ and $\exp(B) = p^{1+\bar{w}} = p^w$. The minimal nonabelian subgroup $B$ has no cyclic subgroup of index $p$ so all its maximal subgroups have exponent $p^w$; hence $a = w$. Thus, $A$ is abelian of type $(p^w, p)$. □

**Remark 148.45.** Let $G$ be a nonabelian metacyclic $p$-group and let $Z < G$ be self-centralizing cyclic. Let $Z < B \leq G$, where $|B : Z| = p$. Then either $B \cong \mathrm{M}_{p|Z|}$ is minimal nonabelian or a 2-group of maximal class (Theorem 1.2). In the second case, $B$ has a nonabelian subgroup $R$ of order 8, and then $G$ is of maximal class (Lemma 148.3(m)).

**Remark 148.46.** Suppose that a metacyclic $p$-group $G = R(G)$ has a proper minimal nonabelian subgroup $A$ of order $p^5$. Let us prove that $G$ is an $\mathcal{A}_2$-group of order $p^6$. Set $\bar{G} = G/\Omega_1(G)$. By Lemma 148.3 (*f*), $\Omega_1(A) = \Omega_1(G) \leq \mathrm{Z}(G)$, and $\bar{A}$ is maximal abelian (of type $(p^2, p)$) in $\bar{G} = R(\bar{G})$ (Remark 148.33 and Lemma 148.42). By Corollary 148.44, $\exp(\bar{G}) = \exp(\bar{A}) = p^2$; and hence $|\bar{G}| = (\exp(\bar{G}))^2 = p^4$ so $|G| = p^6$. In that case, $|G'| = p^2$ (by hypothesis, $|G'| > p$, and $|G'| < p^3$ since $G/G'$ is noncyclic and $G$ has no cyclic subgroup of index $p^2$). By Lemma 148.3(j), $G$ is an $\mathcal{A}_2$-group.

**Exercise 4.** Suppose that a metacyclic $p$-group $G$ is not a 2-group of maximal class and $G/\Omega_1(G)$ is an $\mathcal{A}_{k-1}$-group, $k > 1$. Then $G$ is an $\mathcal{A}_k$-group.

**Exercise 5.** Let $A$ be a noncyclic maximal abelian subgroup of a nonabelian metacyclic $p$-group $G$ and write $\bar{G} = G/\Omega_1(G)$. Let $\bar{A} \leq \bar{B} \leq \bar{G}$, where $\bar{B}$ is abelian. Show that $|B : A| \leq p$ with equality $\iff$ $B$ is minimal nonabelian.

**Solution.** Let $A < B$. As the cyclic $B' < \Omega_1(G) \cong \mathrm{E}_{p^2}$, it follows that $|B'| = p$, and Lemma 148.3 (a) implies that $B$ is minimal nonabelian. Clearly, $|B : A| = p$.

**Remark 148.47.** Let a nonabelian metacyclic $p$-group $G = R(G)$ contain a proper maximal abelian subgroup $A$ of order $p^4$. By Lemma 148.42, $\Omega_1(G) < \mathrm{Z}(G) < A$. Let $A < B < G$. where $|B : A| = p$. If $B/\Omega_1(G)$ is abelian, then, by Lemma 148.3(a), $B$ is minimal nonabelian of order $p^5$ since then $|B'| = p$, and $G$ is as in Remark 148.46. Now assume that $B/\Omega_1(G)$ is nonabelian. If $p > 2$, then $|G/\Omega_1(G)| = p^3$ (Lemma 148.3 (m)) so that $|G| = p^5$, a contradiction since $|G| = p^{2w}$. Let $p = 2$. Then $G/\Omega_1(G)$ is of maximal class, a contradiction as $R(G/\Omega_1(G)) \neq G/\Omega_1(G)$. Thus, $B/\Omega_1(G)$ is abelian and $B$ is minimal nonabelian.

**Remark 148.48.** Let $G$ be a metacyclic minimal nonabelian $p$-group of exponent $p^e$ and $U < G$ a nonnormal cyclic subgroup of order $p^e$. Let us prove that there exists a cyclic $V \triangleleft G$ such that $U \cap V = \{1\}$ and $G = UV$. Indeed, there is a cyclic $V \triangleleft G$ such that $G/V$ is cyclic. Since $G' < V$ is of order $p$ and $G' \nleq U$, one has $U \cap V = \{1\}$ since $V$ is cyclic. In particular, $G = UV$, by the product formula since $|G/V| \leq p^e$.

**Remark 148.49.** Let a $p$-group $G = AB$, where $A \leq G$ is abelian and $B \leq G$ is cyclic. Then $C = \mathrm{C}_G(A) = A(C \cap B)$ is abelian since $C \cap B$ is cyclic. Let, in addition, $G$ be metacyclic. Then one can take $A$ and $B$ to be cyclic. In that case, $\mathrm{C}_G(A)$ and $\mathrm{C}_G(B)$ are maximal abelian subgroups of $G$. It follows from $G = \mathrm{C}_G(A)\mathrm{C}_G(B)$ that $\mathrm{C}_G(A) \cap \mathrm{C}_G(B) = \mathrm{Z}(G)$. Thus, the center of a metacyclic $p$-group is the intersection of centralizers of two of its cyclic subgroups.

**Remark 148.50.** Let us prove that if $G$ is a metacyclic group of exponent $p^e$, then the subgroup $\Omega_e^*(G) = \langle x \in G \mid o(x) = p^e \rangle$ has index $\leq p$ in $G$. If $\Omega_{e-1}(G) < G$, then $\Omega_e^*(G) \geq \langle G - \Omega_{e-1}(G) \rangle = G$. If $G = R(G)$, then $\exp(\Omega_{e-1}(G)) < \exp(G)$, and our claim follows. The last inequality holds if $G/R(G) > \{1\}$ is cyclic. Now let $G/R(G)$ be a 2-group of maximal class. If $G/R(G) \cong Q_8$, then $\Omega_e^*(G) = G$ since $\Omega_{e-1}(G) < G$. Otherwise, take $T/R(G) < G/R(G)$, a cyclic subgroup of index 2 in $G/R(G)$. By the above, $\Omega_{e-1}(T)$ has index 2 in $T$. It follows that $\Omega_e^*(T) = \langle T - \Omega_{e-1}(T) \rangle = T$ so that $2 = |G : \Omega_e^*(T)| \geq |G : \Omega_e^*(G)|$. In that case, $T = \Omega_e^*(G)$.

**Definition 2.** A $p$-group $G$ is said to be a *$\Theta_p$-group* if $G'$ is cyclic and all nonabelian subgroups of $G$ are two-generator. The abelian $p$-groups are considered as $\Theta_p$-groups.

The property $\Theta_p$ is inherited by sections. Metacyclic $p$-groups are $\Theta_p$-groups. By Theorem 106.3, nonabelian $\Theta_2$-groups are either metacyclic or minimal nonabelian, so it remains to investigate $\Theta_p$-groups for odd $p$. Many results of Subsections $3^o$ and $5^o$ also hold for $\Theta_p$-groups, $p > 2$.

**Proposition 148.51.** *Suppose that $G$ is a nonabelian $\Theta_p$-group, $p > 2$. Then $G$ is either minimal nonabelian or metacyclic.*

*Proof.* Suppose that $G$ has a nonabelian subgroup $B$ of order $p^3$. Assume that $\mathrm{C}_G(B) \nleq B$. If $K$ is a least subgroup of $\mathrm{C}_G(B)$ not contained in $B$, then $BK$ is nonabelian and $\mathrm{d}(BK) > 2$, contrary to the hypothesis. Thus, $\mathrm{C}_G(B) < B$; and hence $G$ is of maximal class (Proposition 10.17). Then $|G| = p^3$ since $G'$ is cyclic, and so $G$ is minimal nonabelian (here Lemma 148.3 (i) and Exercise 9.1 (b) were used). Now let $G$ have no nonabelian subgroup of order $p^3$.

Let $G$ be a counterexample of minimal order. Then all nonabelian members of the set $\Gamma_1$ are either minimal nonabelian or metacyclic, by induction.

(i) Suppose that a metacyclic $M \in \Gamma_1$ is nonabelian. Since $M$ is absolutely regular and $G$ is neither metacyclic nor of maximal class, there is $E \triangleleft G$ of order $p^3$ and exponent $p$ (Theorem 69.3). By the first paragraph of the proof, $E \cong \mathrm{E}_{p^3}$. Then $E \not\leq M \Rightarrow G = ME$. In that case, $G/E \cong M/(M \cap E)$ is cyclic in view of $G/(M \cap E) \cong (M/(M \cap E)) \times (E/(M \cap E))$ and $\mathrm{d}(G) = 2$. Since $\Omega_1(M) = M \cap E \cong \mathrm{E}_{p^2}$, one has $M \cong \mathrm{M}_{p^n}$ for some $n \geq 3$ (Lemma 148.3 (l)). In that case, there is a cyclic $C < M$ of index $p$; then $G = CE$ with $|C \cap E| = p$, by the product formula. It follows from $\mathrm{d}(G) = 2$ that $C$ is nonnormal in $G$. Therefore, $|G'| > p$. Since $G' < E$, it follows that $G'$ is noncyclic, a contradiction.

(ii) Thus, all nonabelian maximal subgroups of $G$ are minimal nonabelian and so $G$ is a two-generator $\mathcal{A}_2$-group with cyclic $G'$. By Proposition 71.1, and Theorems 71.3 and 71.4, all members of the set $\Gamma_1$ are nonabelian. In that case, by Proposition 71.5, $G' \cong \mathrm{E}_{p^3}$ is noncyclic, a final contradiction. □

Of course, our easy Proposition 148.51 immediately follows from the fairly difficult classification result of [XZA], where the $p$-groups all of whose proper nonabelian subgroups are two-generator, are described.

**Remark 148.52.** Let a $p$-group $G$ be such that, whenever $A, B < G$ are nonabelian and $|A'| = |B'|$, then $|A| = |B|$. It follows that all minimal nonabelian subgroups of $G$ have the same order. Suppose that $A, B < G$ are nonabelian of the same order and let $A$ be minimal nonabelian. Assume that $B$ is not minimal nonabelian. Then there is a minimal nonabelian $B_1 < B$. Since $|A'| = |B_1'|$ and $|B_1| < |B| = |A|$, this is a contradiction. Thus, all nonabelian subgroups of $G$ of order $|A|$ are minimal nonabelian. It follows that all subgroups of $G$ of order $< |A|$ are abelian.

The order of the Schur multiplier of an arbitrary metacyclic group was computed with help of cohomology theory (see [Kar2, pp. 288–292, Section 10.1.C, Theorem 1.25, and Corollary 1.26]). Let us show that for $G = \mathcal{H}_{2,2}$ this is possible to do in elementary way.

**Remark 148.53.** If the group

$$G = \mathcal{H}_{2,2} = \langle a, b \mid a^4 = b^4 = 1,\ b^a = b^3 \rangle$$

and $\mathrm{M}(G)$ is its Schur multiplier, then $|\mathrm{M}(G)| = 2$. Indeed, let $\Gamma$ be a representation group of $G$. Then there is in $\Gamma' \cap \mathrm{Z}(\Gamma)$ a subgroup $M \cong \mathrm{M}(G)$ such that $\Gamma/M \cong G$. By Theorem 47.2, $\Gamma$ is metacyclic. All maximal subgroups of the group $\Gamma/M$ are abelian of type $(4, 2)$. Let $H/M < \Gamma/M$ be maximal. Then $H/M$ has two distinct cyclic subgroups $A/M$ and $B/M$ of index 2. The subgroups $A$ and $B$ are abelian of index 2 in $H$ so $A \cap B \geq \mathrm{Z}(H)$. But $\mathrm{d}(H) = 2$ since $\Gamma$ is metacyclic so that $A \cap B = \Phi(H)$. It follows that all maximal subgroups of $H$ are abelian, and hence $H$ is either abelian or minimal nonabelian. Therefore, $\Gamma$ is either an $\mathcal{A}_1$- or $\mathcal{A}_2$-group (note that $M < \Gamma' < \Phi(\Gamma)$). It follows that $|\Gamma'| \leq 4$ (Lemma 148.3 (j)). Assume that $|\Gamma'| = 2$. Then $\mathrm{M}(G) = \{1\}$ so that $\Gamma = G$. It follows that $\Gamma$ is the unique representation group of $G$. But it will be shown that

$$X = \langle x, y \mid x^8 = y^8 = 1,\ y^x = y^{-1},\ x^4 = y^4 \rangle$$

is a representation group of $G$, which is a contradiction. In that case, $X' = \langle y^2 \rangle \cong C_4$. Let $M = \langle y^4 \rangle$. Then $X/M$ is nonabelian of exponent 4 so that $X/L \cong \mathcal{H}_{2,2}$. Since $M \leq Z(X) \cap X'$. it follows that $M$ is isomorphic to a subgroup of $M(G)$. Thus, $\Gamma$ is an $\mathcal{A}_2$-group of order $2^5$ so that $|M(G)| = 2$. Note that

$$\Gamma^1 = \langle x^4 = y^8 = 1,\ y^x = y^{-1} \rangle$$

is another representation group of $G$. Obviously, $\Gamma \not\cong \Gamma^1$.[6]

**Exercise 6.** All minimal nonabelian subgroups in a nonabelian metacyclic 2-group $G = R(G)$ have the same exponent $\iff$ $\Phi(G)$ is abelian.

**Exercise 7.** Let $G = AB$ be a metacyclic $p$-group and $\{1\} < A < G$, $\{1\} < B < G$ and $A \cap B = \{1\}$. Show that $A$ and $B$ are cyclic.

**Solution.** One may assume that $G$ has no cyclic subgroup of index $p$. Then $\Omega_1(G) \cong E_{p^2}$ and therefore the intersections $A \cap \Omega_1(G) = \Omega_1(A)$ and $B \cap \Omega_1(G) = \Omega_1(B)$ are of order $p$. Since $G$ has no subgroup $\cong Q_8$, $A$ and $B$ are cyclic.

**Exercise 8.** If $G$ is a metacyclic $p$-group and $L < G'$, then there is $\chi \in \operatorname{Irr}(G)$ such that $\ker(\chi) \cap G' = L$.

**Solution.** Let us prove that there is $\chi \in \operatorname{Irr}_1(G)$ such that $\ker(\chi) \cap G' = \{1\}$. This is a case provided $Z(G)$ is cyclic. Otherwise, let $K \lhd G$ be maximal such that $K \cap G' = \{1\}$. Then $Z(G/K)$ is cyclic; therefore, there is $\chi \in \operatorname{Irr}(G/K)$ such that $\ker(\chi) \cap G'K/K = K/K$. Considered as a character of $G$, one obtains $\ker(\chi) \cap G' = \{1\}$ so the result holds if $L = \{1\}$. Now let $L > \{1\}$ and write $\bar{G} = G/L$. Then, by the above, there is $\chi \in \operatorname{Irr}(\bar{G}) = \{\bar{1}\}$, or, considered as a member if $\operatorname{Irr}(G)$, $\chi$ satisfies $\ker(\chi) \cap G' = L$.

**Exercise 9.** Let $G$ be a metacyclic $p$-group, $p > 2$, $|G'| = p^k > p$. Then $G/\Omega_{k-1}(G)$ is a minimal nonabelian group. (**Hint.** Use Theorem 7.2 (b) and Lemma 65.2 (a).)

**Exercise 10.** Is the following assertion true?: If all $\mathcal{A}_k$-subgroups of a metacyclic $p$-group $G$, $|G'| > p^k$, $k > 2$, have a cyclic subgroup of index $p^k$, then $G$ has an abelian subgroup of index $p^k$.

**Exercise 11.** A metacyclic $p$-group $G = R(G)$ has no $\mathcal{A}_2$-subgroup of order $p^5$.

**Solution.** If $H$ is such subgroup, then $H' \cong C_{p^2}$ (Corollary 65.3) so $H/\Omega_1(G)$ is nonabelian of order $p^3$; then either $G = H$ or $G/\Omega_1(G)$ is a 2-group of maximal class, In both cases, $R(G/\Omega_1(G)) \neq G/\Omega_1(G))$ (Lemma 148.3(m)), a contradiction.

6 A similar argument is also applicable to $G = \mathcal{H}_{2,p}$ for $p > 2$. The Schur multiplier of this group has order $p$ since any representation group of $G$ is a metacyclic $\mathcal{A}_2$-group of order $p^5$ and exponent $p^3$.

**Exercise 12.** If a metacyclic $p$-group $G = R(G)$ contains an $\mathcal{A}_2$-subgroup $H$ of order $p^6$, then $|G| = p^6$.

**Solution.** As $H/\Omega_2(G)$ is maximal abelian subgroup of order $p^2$ of $G/\Omega_1(G) = R(G/\Omega_1(G))$, that quotient group is either of order $p^3$ or 2-group of maximal class (Propositions 1.8 and 10.19), which is a contradiction.

**Exercise 13.** Suppose that $G$ is a nonabelian metacyclic $p$-group, $|R(G)| = p^{2w}$. If $G$ has no minimal nonabelian subgroups of exponent $> p^w$, then $G = R(G)$. (**Hint.** Use Lemma 148.3 (k).)

**Exercise 14.** Let $G$ be an arbitrary (not necessarily nilpotent) metacyclic group of order $n^2$ and exponent $n$. Then $G = AB$, where $A, B < G$ are cyclic of order $n$.

**Solution.** There is a cyclic $A \triangleleft G$ such that $G/A$ is cyclic. Clearly, $|A|$ and $|G/A|$ divide $n$ and $|A|\,|G/A| \leq |G| = n^2$. It follows that $|A| = n = |G/A|$. Let $B < G$ be cyclic such that $AB = G$. Then $|B| = n$ since $|B| \mid n$ and $n \mid |B|$. In that case, $|A| = n$ and $A \cap B = \{1\}$, by the product formula.

**Exercise 15.** Let $G$ be a metacyclic $p$-group. Then:
(i) if $G/R(G)$ is cyclic, then $|R(G)| \geq |G'|^2$, and
(ii) if $G/R(G)$ is a 2-group of maximal class, then $\exp(G') = \frac{1}{2}\exp(G)$.

**Hint.** If $G/R(G)$ is cyclic, then $G' \leq R(G)$, and so $|R(G)| \geq |G'|^2$ since $G'$, being a cyclic subgroup of $R(G)$, has order $\leq \exp(R(G)) = p^w$. If $G/R(G)$ is a 2-group of maximal class, then $|G' \cap R(G)| = \exp(R(G))$ and so

$G'R(G)/R(G)$ is cyclic of index 4 in $G/R(G)$, implying the result in this case.

**Exercise 16.** Let $G$ be a metacyclic $p$-group $G$ with $|G'| = p^k > p$. Then for any $k_1 < k$ there is $H < G$ with $|H'| = p^{k_1}$. (**Hint.** Let $H < G$ be an $\mathcal{A}_{k_1}$-subgroup. Use Theorem 72.1)

**Exercise 17.** Classify the metacyclic $p$-groups $G$ with $|G'| > p > 2$ all of whose minimal nonabelian subgroups contain a cyclic subgroup of index $p$.

**Hint.** The centralizer $C_G(\Omega_1(G))$ is abelian (since it has no minimal nonabelian subgroup) of index $p$ in $G$. Conversely, each such $G$ satisfies the condition.

**Exercise 18.** Classify the metacyclic $p$-groups of orders $p^5$ and $p^6$.

**Exercise 19.** Classify the metacyclic $p$-groups $G$ satisfying $|G : Z(G)| = p^3$. (**Hint.** In that case, $G/Z(G) \cong D_8$.)

**Exercise 20.** Classify the metacyclic $p$-groups $G$ such that $G/Z(G)$ is of order $p^4$ and exponent $p^2$.

**Hint.** If $H \in \Gamma_1$, then $|H/Z(H)| \leq p^2$ so that $H$ is either abelian or minimal nonabelian. It follows that $G$ is an $\mathcal{A}_2$-group.

**Exercise 21.** If all $\mathcal{A}_2$-subgroups of a metacyclic $p$-group $G$, $|G'| > p^2$, have a cyclic subgroup of index $p^2$, then $G$ has an abelian subgroup of index $p^2$.

**Solution.** One may assume that $G$ is not a 2-group of maximal class. Let $H/\Omega_1(G) \leq G/\Omega_1(G)$ be minimal nonabelian. Since $H'$ is cyclic, it follows that $H' \cong \mathrm{C}_{p^2}$ so that $H$ is an $\mathcal{A}_2$-subgroup (Lemma 148.3 (j)). Let $Z < H$ be cyclic of index $p^2$. Then $Z\Omega_1(G)/\Omega_1(G)$ is cyclic of index $p$ in $H/\Omega_1(G)$. Thus, each minimal nonabelian subgroup of $G/\Omega_1(G)$ has a cyclic subgroup of index $p$. By Exercise 17, $G/\Omega_1(G)$ has an abelian subgroup $A/\Omega_1(G)$ of index $p$. Since $|A'| \leq p$, $A$ is either abelian or minimal nonabelian (Lemma 1.3 (a)). Any maximal subgroup of $A$ is abelian and has index $p^2$ in $G$.

**Exercise 22.** Is it true that the result of Exercise 8 holds for a $p$-group with cyclic derived subgroup?

**Exercise 23.** Let $A < G$ be a maximal abelian subgroup of index $p^n$ in a $p$-group $G$. Prove that $\alpha_1(G) \geq n$.

**Solution.** We proceed by induction on $|n$. This is true for $n = 1$. Now let $n > 1$. Let $A < M \in \Gamma_1$; then $|M : A| = p^{n-1}$. By induction, $\alpha_1(M) \geq n - 1$. By § 76, $\alpha_1(G) \geq \alpha_1(M) + (p - 1) \geq (n - 1) + (p - 1) \geq n$. Slightly changing the above argument, we get $\alpha_1(G) \geq 1 + (n - 1)(p - 1)$, which is better than the above estimate provided $p > 2$. It is interesting to classify the $p$-groups $G$ satisfying $\alpha_1(G) = 1 + (n - 1)(p - 1)$. For another approach to this theme, see § 250.

**Exercise 24.** Let $M < G$ be a nonabelian subgroup of a metacyclic $p$-group $G$. Study the subgroup $\mathrm{C}_G(\mathrm{Z}(M))$ (see Remark 148.10).

*10º Problems.* Below are stated a number of related problems.

**Problem 1.** Study the $p$-groups $G$ such that, whenever $A, B < G$ are nonabelian and $|A'| = |B'|$, then $|A| = |B|$ (see Remark 148.52).

**Problem 2.** Classify the metacyclic 2-groups $G$ which cannot be presented in the form $G = AB$, where $A, B < G$ with $A \cap B = \{1\}$.

**Problem 3.** Classify the $p$-groups all of whose maximal subgroups, except one, are two-generator.

**Problem 4.** Study the $p$-groups $G$ with cyclic $G'$ and satisfying the condition $(2\Delta_p)$ from $8^o$.

**Problem 5.** Classify the metacyclic groups of exponent $2^e$ that have only one maximal abelian subgroup of exponent $2^e$.

**Problem 6.** Study the two-generator $p$-groups $G$ with cyclic $G'$ such that $|G : H^G| \leq p$ for all minimal nonabelian $H < G$.

**Problem 7.** Classify the metacyclic $p$-groups all of whose maximal abelian subgroups have the same exponent.

**Problem 8.** Given $n$, does there exist a metacyclic 2-group containing exactly $n$ maximal abelian subgroups of pairwise distinct orders? Is it true that $n < 3$?

**Problem 9.** Classify the metacyclic $p$-groups with $\leq p+1$ conjugate classes of minimal nonabelian subgroups.

**Problem 10.** Classify the $p$-groups all of whose maximal metacyclic subgroups are either abelian or minimal nonabelian.

**Problem 11.** Classify the metacyclic $p$-groups all of whose subgroups of order $p$ (maximal subgroups) are characteristic (two problems).

**Problem 12.** Classify the 2-groups $G$ with cyclic $G'$ without a normal minimal nonabelian subgroup.

**Problem 13.** Classify the metacyclic $p$-groups $G$ satisfying $|H^G : H| \leq p$ for all minimal nonabelian $H < G$.

**Problem 14.** Study the $p$-groups $G$ such that, whenever $A < B < G$, where $A$ is maximal abelian in a nonabelian (minimal nonabelian) subgroup $B$, then $A$ is maximal abelian in $G$.

# § 149 $p$-groups with many minimal nonabelian subgroups

The purpose of this section is to demonstrate how the appearance of many minimal nonabelian subgroups in $p$-groups can influence the structure of such groups. We have noted (Proposition 149.1; see also Proposition 148.15 (d)) that metacyclic $p$-groups $G$ with $p > 2$ have the following property:

($*$) Whenever a nonabelian subgroup $H$ of $G$ has an abelian subgroup of index $p$, then $H$ is minimal nonabelian.

It follows from ($*$) that if $A < G$ is maximal abelian, then all containing $A$ subgroups of $G$ of order $p|A|$ are minimal nonabelian.

Therefore it is of interest to classify all $p$-groups $G$ which have the above property ($*$). This will be done in Theorem 149.3. It turns out that not all metacyclic 2-groups have the property ($*$), but only so called "ordinary metacyclic 2-groups" (see § 26, Definition 7) . There are also some nonmetacyclic $p$-groups having the property ($*$).

We shall also consider nonabelian $p$-groups $G$ that are not minimal nonabelian and which possess the following property:

($**$) Whenever $X$ and $Y$ are distinct minimal nonabelian subgroups of $G$ and $x \in X - Y$, $y \in Y - X$, then $\langle x, y\rangle$ is also minimal nonabelian.

It turns out (Theorem 149.4) that $p$-groups $G$ having the property ($**$) also have the property ($*$), and we must have $p = 2$ and $G$ is a special group of order $2^6$, which is isomorphic to a Sylow 2-subgroup of the Suzuki simple group Sz(8). It is surprising that in this case our group $G$ is unique!

In any case, the property ($*$) is much more general than the property ($**$) and actually both properties ($*$) and ($**$) state that in some sense our $p$-groups have 'many' minimal nonabelian subgroups.

One of the most difficult open problems in $p$-group theory is to classify $p$-groups that are covered by its minimal nonabelian subgroups. We see that $p$-groups with the property ($*$), which are studied in this section, are in fact covered by its minimal nonabelian subgroups. This also indicates that the above open problem is really difficult.

The following result was also proved in Proposition 148.15 (d). The presented proof is based on other ideas.

**Proposition 149.1.** *Let $G$ be a nonabelian metacyclic $p$-group with $p > 2$. If $G$ possesses an abelian maximal subgroup $A$, then $G$ is minimal nonabelian.*

*Proof.* We may choose generators $x, y$ of $G$ such that $x \in A$ and $y \in G - A$. Then the cyclic derived subgroup $G' = \langle [x, y] \rangle \neq \{1\}$. By Lemma 139.1, we get

$$\mho_1(G') = \langle [x, y]^p \rangle = \langle [x, y^p] \rangle \Rightarrow [x, y^p] = 1 \quad \text{since } x, y^p \in A .$$

Thus, $\mho_1(G') = \{1\}$. Hence $d(G) = 2$ and $|G'| = p$ so that Lemma 65.2 (a) implies that $G$ is minimal nonabelian. □

We recall that a metacyclic 2-group $G$ is called 'ordinary metacyclic' (with respect to a subgroup $H$) if $G$ has a cyclic normal subgroup $H$ such that $G/H$ is cyclic and $G$ centralizes $H/\mho_2(H)$ (i.e., $[G, H] \leq \mho_2(H)$) (see § 26, Definition 7). For example, $G = M_{2^n}$ is ordinary metacyclic with respect to any of its cyclic subgroups of index 2. However, if $L < G$ is cyclic of order $2^2$ not contained in $\Phi(G)$, then $L \triangleleft G$, $G/L$ is cyclic and $G$ is not ordinary metacyclic with respect to $L$.

**Proposition 149.2.** *Let $G$ be a nonabelian ordinary metacyclic 2-group with respect to a subgroup $H$. Then $[G', G] \leq \mho_2(G')$ and each subgroup $U$ of $G$ is also ordinary metacyclic with respect to $H \cap U$. If $G$ has an abelian maximal subgroup, then $G$ is minimal nonabelian.*

*Proof.* Set $H = \langle h \rangle$ and let $g \in G - H$ be such that $\langle g \rangle$ covers $G/H$ (i.e., $H\langle g \rangle = G$) and so $G = \langle g, h \rangle$. Next,

$$[G, H] \leq \mho_2(H) = \langle h^4 \rangle \Rightarrow h^g = hv \quad \text{with } v \in \langle h^4 \rangle ,$$

and so

$$[h, g] = v , \quad G' = \langle v \rangle , \quad v = h^i \quad \text{with } i \equiv 0 \pmod 4 .$$

We compute:

$$v^g = (h^i)^g = (h^g)^i = (hv)^i = h^i v^i = vv^i ,$$

and so $[v, g] = v^i \in \mho_2(G')$ since $i \equiv 0 \pmod 4$. Thus, $[G', G] \leq \mho_2(G')$.

Let $U$ be any subgroup of $G$ and set $V = U \cap H$ so that $V$ is a cyclic normal subgroup of $G$ and $U/V \cong UH/H$ is cyclic as a subgroup of $G/H$. We have $V = \langle h^j \rangle$ for some integer $j$. We compute:

$$(h^j)^g = (h^g)^j = (hv)^j = h^j v^j = h^j (h^i)^j = h^j (h^j)^i ,$$

and so

$$[h^j, g] = (h^j)^i \in \mho_2(V) \quad \text{where } i \equiv 0 \pmod 4 .$$

It follows that the element $g$ centralizes $V/\mho_2(V)$, and so $U$ centralizes $V/\mho_2(V)$, and we conclude that $U$ is ordinary metacyclic.

Now suppose that $G$ has an abelian maximal subgroup $A$. We may choose generators $x, y$ of $G$ such that $x \in A$ and $y \in G - A$. Then the subgroup $G' = \langle [x, y] \rangle \neq \{1\}$ is cyclic. As $[G', G] \leq \mho_2(G')$, we may use Lemma 139.1 and we get

$$\mho_1(G') = \langle [x, y]^2 \rangle = \langle [x, y^2] \rangle = \{1\} ,$$

since $x, y^2 \in A$, and so $[x, y^2] = 1$. Hence $d(G) = 2$ and $|G'| = 2$, and it follows from Lemma 65.2 (a) that $G$ is minimal nonabelian. □

**Theorem 149.3** (Janko). *Let $G$ be a nonabelian $p$-group that is not minimal nonabelian. Suppose that $G$ has the following property:*

($*$) *Whenever a nonabelian subgroup $H$ of $G$ has an abelian maximal subgroup, then $H$ is minimal nonabelian.*

*Then one of the following possibilities hold:*
(a) *$G$ is any of the nonmetacyclic $p$-groups defined in Proposition 71.5 (a, b),*
(b) *$G$ is a metacyclic $p$-group with $p > 2$ and $|G'| > p$,*
(c) *$G$ is an ordinary metacyclic 2-group with $|G'| > 2$.*
*Conversely, each of the groups given in (a), (b) and (c) satisfies the assumptions of our theorem.*

*Proof.* Let $G$ be a nonabelian $p$-group which is not minimal nonabelian and assume that $G$ has property ($*$). This implies that $G$ has no abelian maximal subgroup.

(i) First assume that $G$ possesses a normal elementary abelian subgroup $E$ of order $p^3$.

Let $A$ be a maximal normal abelian subgroup of $G$ which contains $E$ so that $A$ is of rank $\geq 3$. Since $G$ has no abelian maximal subgroup, $|G/A| > p$. We consider a subgroup $H/A$ of order $p^2$ in $G/A$. Let $H_1/A$ be a subgroup of order $p$ in $H/A$ so that, ($*$), $H_1$ is minimal nonabelian. Then $\Phi(H_1) < A$ and, in view of $\mathrm{d}(H_1) = 2$, one obtains $|A : \Phi(H_1)| = p$, which together with $\Phi(H_1) \leq \Phi(H)$ yields $\mathrm{d}(H) \leq 3$. If $\mathrm{d}(H) = 3$, then $\Phi(H_1) = \Phi(H)$, which implies $H/A \cong \mathrm{E}_{p^2}$. Now suppose that $\mathrm{d}(H) = 2$. In that case, $\Phi(H) < H_1$ so that $\Phi(H)$ is abelian (recall that $H_1$ is minimal nonabelian) and $H/\Phi(H) \cong \mathrm{E}_{p^2}$. As a minimal nonabelian $H_1 < H$, it follows that $H$ is neither abelian nor minimal nonabelian. By ($*$), each maximal subgroup of $H$ is minimal nonabelian. We have $E \leq A$ and so $|H| > p^4$ and $H$ is nonmetacyclic so that one may use Proposition 71.5 (b). It follows that

$$E = \Omega_1(H) = \Phi(H)\,, \quad H/E \cong \mathrm{E}_{p^2}\,, \quad A = E\,.$$

In particular, we get again $H/A \cong \mathrm{E}_{p^2}$.

We have proved that for any subgroup $H/A$ of order $p^2$ in $G/A$, one has $H/A \cong \mathrm{E}_{p^2}$. It follows that $\exp(G/A) = p$. We have either $\Phi(H) = A$ or $\Phi(H) = \Phi(H_i)$, where $H_i/A$ is any subgroup of order $p$ in $H/A$ and, by ($*$), $H_i$ is minimal nonabelian, $i = 1, 2, \dots, p+1$.

Suppose that there is at least one subgroup $H/A$ of order $p^2$ in $G/A$ such that $\Phi(H) = A$ and so $\mathrm{d}(H) = 2$. Since each maximal subgroup $H_i$ of $H$ is nonabelian and $|H_i : A| = p$, it follows by the property ($*$) that $H_i$ is minimal nonabelian for each $i = 1, 2, \dots, p+1$. As $A$ is of rank $\geq 3$, it follows that $|H| > p^4$ and $H$ is nonmetacyclic. By Proposition 71.5 (b), we get

$$p > 2\,, \quad |H| = p^5\,, \quad A = H' = \Omega_1(H) = \Phi(H) \cong \mathrm{E}_{p^3}$$

and $\mathrm{Z}(H) = \mho_1(H) = \mathrm{K}_3(H)$ is a subgroup of index $p$ in $A$.

(i1) Assume that there is at least one subgroup $H/A$ of order $p^2$ in $G/A$ such that $\mathrm{d}(H) = 3$.

We know that in that case $\Phi(H) = \Phi(H_i)$, where $H_i/A$ is any subgroup of order $p$ in $H/A$ and (by the property $(*)$) $H_i$ is minimal nonabelian, $i = 1, 2, \ldots, p+1$. Also $\Phi(H_i) = \mathrm{Z}(H_i)$ for all $i = 1, 2, \ldots, p+1$ (Lemma 65.1) which implies $\Phi(H) \leq \mathrm{Z}(H)$. On the other hand, $H$ has no abelian maximal subgroup (otherwise, the property $(*)$ would infer that $H$ is minimal nonabelian) and so we get $\Phi(H) = \mathrm{Z}(H)$. Let $M$ be any maximal subgroup of $H$ which does not contain $A$. Then $M \cap A = \Phi(H)$ and so $M$ covers $H/A$. Let $X$ be a maximal subgroup of $M$ that contains $\Phi(H)$. Since $\Phi(H) = \mathrm{Z}(H)$ and $|X : \Phi(H)| = p$, we see that $X$ is abelian. But $M$ is nonabelian and so, by the property $(*)$, $M$ is minimal nonabelian. Thus $H$ is a nonmetacyclic $p$-group of order $> p^4$ with $\mathrm{d}(H) = 3$ all of whose maximal subgroups are minimal nonabelian. By Proposition 71.5 (a), we get $p = 2$ and $H$ is a special group of order $2^6$, which is isomorphic to an $\mathrm{S}_2$-subgroup of the simple group $\mathrm{Sz}(8)$. This gives

$$\Phi(H) = \mathrm{Z}(H) = H' = \Omega_1(H) = E \cong \mathrm{E}_8 ,$$

$A$ is abelian of type $(4, 2, 2)$ and each maximal subgroup of $H$ is isomorphic to $\mathcal{H}_{32}$ defined in Theorem 57.5. Since $p = 2$, it follows by the preceding paragraph that for each subgroup $K/A$ of order 4 in $G/A$ we have $\mathrm{d}(K) = 3$ and so, by the above, $K$ is isomorphic to an $\mathrm{S}_2$-subgroup of the simple group $\mathrm{Sz}(8)$. Also, $\exp(G/A) = 2$ gives that $G/A$ is elementary abelian.

Let $X$ be any minimal nonabelian subgroup in $G$. Since $\mathrm{d}(X) = 2$ (see Lemma 65.1), $|X : (X \cap A)| \leq 4$ and so $X$ is contained in a subgroup $H > A$ such that $H/A \cong \mathrm{E}_4$. By the above, $H$ is isomorphic to an $\mathrm{S}_2$-subgroup of the simple group $\mathrm{Sz}(8)$ and so $X \cong \mathcal{H}_{32}$. By Proposition 57.5, $\Omega_1(G) \leq \mathrm{Z}(G)$ and $G$ is of exponent 4 and class 2. It follows from $\mathrm{C}_G(A) = A$ that

$$\Omega_1(G) = \Omega_1(A) = E = \mathrm{Z}(G) \cong \mathrm{E}_8 .$$

Thus $G' \leq E$ and since $H' = E$, we get $G' = E$. In addition (since $\exp(G) = 4$), for each $g \in G - E$, we have $g^2 \in E\{1\}$ and so $\Phi(G) = E$ and therefore $G$ is a special 2-group.

Let $a \in A - E$ so that $a^2 = z$ is an involution in $E$ and $\mho_1(A) = \langle z \rangle$. There are exactly eight elements of order 4 in $A - E$ and $\mathrm{C}_G(A) = A$ implies that $\mathrm{C}_G(a) = A$. This gives that $|G/A| \leq 8$. Suppose that $|G/A| = 8$. In that case all eight elements of order 4 in $A - E$ form a single conjugate class in $G$. In particular, there is an element $g \in G - A$ such that $g^2 \in E$ and $a^g = az = a^{-1}$. By the property $(*)$, $B = A\langle g \rangle$ is minimal nonabelian and so, by the above, $B \cong \mathcal{H}_{32}$ with $B' = \langle z \rangle$. On the other hand, since $\mathcal{H}_{32}$ is nonmetacyclic, it follows (Lemma 65.1) that $B' = \langle z \rangle$ must be a maximal cyclic subgroup in $B$. This is a contradiction since $z = a^2$. We have proved that $|G/A| = 4$ and so $G = H$ is a special group of order $2^6$ which is isomorphic to an $\mathrm{S}_2$-subgroup of the simple group $\mathrm{Sz}(8)$.

(i2) Now we assume that for all subgroups $H/A$ of order $p^2$ in $G/A$, one has $d(H) = 2$.

Then we already know that we must have $p > 2$ and $|H| = p^5$, where $H$ is isomorphic to a Blackburn group from Proposition 71.5 (b). In particular,

$$A = E \cong E_{p^3}\,, \quad \Omega_1(H) = A = H'\,,$$

and so $C_G(A) = A$ implies that either $G = H$ (and $G$ is isomorphic to any group given in Proposition 71.5 (b)) or $|G : H| = p$ and $G/A$ is isomorphic to the nonabelian group $S(p^3)$ of order $p^3$ and exponent $p$. Suppose that $G/A \cong S(p^3)$. Set $K = G'$ so that $A < K < H$ and $K/A = (G/A)'$, which yields $|K : A|$=p. It follows that $K$ is the nonmetacyclic minimal nonabelian group of order $p^4$ given with:

$$K = \langle k, l \,|\, k^{p^2} = l^p = 1\,, \quad [k, l] = m\,, \quad m^p = [m, k] = [m, l] = 1\rangle\,,$$

where $K' = \langle m\rangle \le Z(G)$ and $\mho_1(K) = \langle k^p\rangle \le Z(G)$ since $p > 2$. It follows that:

$$K' \times \mho_1(K) = Z(K) = Z(G) < A \quad \text{with } |A : Z(G)| = p\,.$$

Let $a \in A - Z(G)$ so that $C_G(A) = A$ yields $C_G(a) = A$, and therefore the conjugate class of $a$ in $G$ has the size $p^3$. This implies that

$$|A - Z(G)| = p^3 - p^2 \ge p^3\,,$$

which is a contradiction. Hence we must have $G = H$.

We have proved that if a $p$-group $G$ satisfies the assumptions of Theorem 149.3 and if $G$ possesses a normal elementary abelian subgroup of order $p^3$, then $G$ is isomorphic to the groups defined in Proposition 71.5 (a, b). Conversely, it is clear that each group from Proposition 71.5 satisfies the assumptions of our theorem.

(ii) Now assume that $G$ has no normal elementary abelian subgroup of order $p^3$.

(ii1) First suppose in addition that $G$ is nonmetacyclic.

Let $M$ be a minimal nonmetacyclic subgroup in $G$. Then Theorems 66.1 and 69.1 give us the structure of $M$. If $|M| > p^3$, then $M$ is nonabelian and possesses an abelian maximal subgroup but $M$ is not minimal nonabelian, contrary to the property $(*)$. It follows that $M$ is of order $p^3$ and exponent $p$. Suppose that $M$ is nonabelian. Then $p > 2$ and $M \cong S(p^3)$ (the nonabelian group of order $p^3$ and exponent $p$) that is minimal nonabelian. In that case, $M < G$ and let $V$ be a subgroup of order $p^4$ that contains $M$. Then $V$ has an abelian maximal subgroup but $V$ is not minimal nonabelian, contrary to the property $(*)$. It follows that $M \cong E_{p^3}$. If $p > 2$, then, by Corollary 10.6, the group $G$ has also a normal elementary abelian subgroup of order $p^3$, contrary to the assumption.

We have proved that we must have $p = 2$ and $G$ possesses a subgroup $M \cong E_8$. Let $A$ be a maximal abelian subgroup of $G$ which contains $M$. Let $B > A$ be a subgroup of $G$ such that $|B : A| = 2$. By property $(*)$, $B$ is nonmetacyclic minimal nonabelian and so (Lemma 65.1) $M = \Omega_1(B)$. By the assumption, $B < G$. Set $C = N_G(M)$ so that $B < C < G$. By part (i) of the proof, $C$ is a special group of order $2^6$ that is isomorphic to an $S_2$-subgroup of the simple group Sz(8). But then $M = \Omega_1(C)$ and so $N_G(M) > C$, a contradiction. We have proved that in case (ii) of the proof our group $G$ cannot be nonmetacyclic.

(ii2) The group $G$ is metacyclic.

First we examine the case $p > 2$. Since $G$ is neither abelian nor minimal nonabelian, $|G'| > p$ (see Lemma 65.2 (a)) and these are the groups of part (b) of our theorem. Conversely, let $H$ be a nonabelian subgroup in a metacyclic $p$-group $G$ (with $p > 2$) that possesses an abelian maximal subgroup. Then by Proposition 149.1, $H$ is minimal nonabelian. Hence our groups in part (b) of our theorem have property $(*)$.

We turn now to the case $p = 2$. Let $G$ be a metacyclic 2-group that is neither abelian nor minimal nonabelian, and which has the property $(*)$. Then, by Lemma 65.2 (a), $|G'| > 2$. Let $\langle a \rangle$ be a cyclic normal subgroup of $G$ such that $G/\langle a \rangle$ is cyclic. Let $b \in G$ be such that that $G = \langle a, b \rangle$. We set $o(a) = n2^n$, where $n \geq 4$ since $|G : C_G(a)| > 2$. Indeed, if $|G : C_G(a)| \leq 2$, then $G$ is either abelian or minimal nonabelian (by $(*)$), contrary to the assumption.

Assume that $a^b = a^{-1}v$, where $v \in \langle a^4 \rangle$. Let us consider the subgroup $S = \langle a^{2^{n-3}}, b \rangle$, where $o(a^{2^{n-3}}) = 8$. We have

$$\left(a^{2^{n-3}}\right)^b = \left(a^b\right)^{2^{n-3}} = \left(a^{-1}v\right)^{2^{n-3}} = a^{-2^{n-3}} v^{2^{n-3}},$$

where $o(v^{2^{n-3}}) \leq 2$. Hence $b$ induces on $\langle a^{2^{n-3}} \rangle \cong C_8$ an involutory automorphism such that $S' = \langle a^{2^{n-2}} \rangle \cong C_4$ and $b^2$ centralizes $\langle a^{2^{n-3}} \rangle$. It follows that $S$ is nonabelian but $S$ is not minimal nonabelian (see Lemma 65.1). On the other hand, $S$ has the abelian maximal subgroup $\langle a^{2^{n-3}}, b^2 \rangle$, a contradiction.

Thus, we must have $a^b = av$, where $v \in \langle a^4 \rangle$. Hence $G$ is ordinary metacyclic with respect to the subgroup $\langle a \rangle$, and so $G$ is a group of part (b) of the theorem.

Conversely, let $G$ be an ordinary metacyclic 2-group with $|G'| > 2$. Let $U$ be a nonabelian subgroup in $G$ which has an abelian maximal subgroup. By Proposition 149.2, $U$ is also ordinary metacyclic and so by the same proposition, $U$ is minimal nonabelian. We have proved that the groups in part (c) of our theorem have the property $(*)$. Our theorem is proved. □

**Theorem 149.4.** *Let $G$ be a nonabelian $p$-group that is not minimal nonabelian. We assume that $G$ has the following property:*

$(**)$ *Whenever $X$ and $Y$ are distinct minimal nonabelian subgroups of $G$ and $x \in X - Y$, $y \in Y - X$, then $\langle x, y \rangle$ is also minimal nonabelian.*

*Then $G$ also has the property $(*)$ (see Theorem 149.3) and we must have $p = 2$ and $G$ is a special group of order $2^6$, which is isomorphic to a Sylow 2-subgroup of the Suzuki simple group* Sz(8).

*Proof.* Let $G$ be a $p$-group satisfying the assumptions of our theorem.

(i) First we prove that whenever $B \le G$ is nonabelian and $A$ is an abelian maximal subgroup of $B$, then $B$ is minimal nonabelian (this is the property $(*)$ from Theorem 149.3). Indeed, take an element $x \in B - A$ and set $A_0 = C_A(x)$ so that $x^p \in A_0$, $A_0 = Z(B)$ and $B_0 = A_0\langle x\rangle$ is a maximal abelian subgroup of $B$. Note that $B_0 < B$ and let $B_1 \le B$ be such that $B_1 > B_0$ and $|B_1 : B_0| = p$. Set $A_1 = B_1 \cap A$ so that $|A_1 : A_0| = p$ and $|B_1 : A_1| = p$. It follows that $A_1$ and $B_0$ are two distinct abelian maximal subgroups of the nonabelian group $B_1$. This implies $B_1' = \langle z\rangle$ is of order $p$ and $z \in A_0 = Z(B)$. For each $y \in A_1 - A_0$, one has $\langle [x, y]\rangle = \langle z\rangle$ and so $M = \langle x, y\rangle$ is minimal nonabelian (see Lemma 65.2 (a)). Let $y_1$ be any other element in $A_1 - A_0$ so that $N = \langle x, y_1\rangle$ is minimal nonabelian and assume $M \ne N$. But then $y \in M - N$ and $y_1 \in N - M$ and so, by our property $(**)$, $\langle y, y_1\rangle$ must be minimal nonabelian, contrary to $y, y_1 \in A_1 \le A$. Hence we must have $M = N$ and so $y_1 \in M$. It follows that $A_1 - A_0 \le M$ and so $\langle A_1 - A_0\rangle = A_1 \le M$, which gives that $B_1 = M$ is minimal nonabelian. Since $x$ was an arbitrary element in $B - A$, it follows that each minimal nonabelian subgroup $M_0$ of $B$ contains

$$A_0 = Z(B) = Z(M_0) = \Phi(M_0) \Rightarrow |M_0| = |M| = p^2|A_0|\,.$$

We claim that for each $x \in B - A$, there is a unique minimal nonabelian subgroup $B_1 = M$ of $B$ containing $x$. Indeed, assume that $B_2 \ne B_1$ is another minimal nonabelian subgroup of $B$ containing $x$. Now

$$B_2 \ge A_0\langle x\rangle = B_0 \Rightarrow B_1 \cap B_2 = B_0\,.$$

We have $B_2 \cap A > A_0$ and let $v \in (B_2 \cap A) - A_0$ so that $v \notin A_1$. On the other hand, $B_1 = \langle x, y\rangle$ with $y \in A_1 - A_0$. Then $v \in B_2 - B_1$ and $y \in B_1 - B_2$ and so, by the property $(**)$, $\langle v, y\rangle$ must be minimal nonabelian, contrary to $\langle v, y\rangle \le A$.

Suppose that

$$B \ne B_1, \quad \text{where } B_1 = M = \langle x, y\rangle \quad \text{with } x \in B - A\,, \quad y \in A_1 - A_0\,.$$

Take an element $x_1 \in B - (A \cup B_1)$. Let $M_1$ be a unique minimal nonabelian subgroup of $B$ containing $x_1$ so that $M \ne M_1$. In that case $x \in M - M_1$, $x_1 \in M_1 - M$ and therefore (by the property $(**)$) $\langle x, x_1\rangle$ is minimal nonabelian. By the uniqueness of minimal nonabelian subgroups of $B$ that contain a prescribed element in $B - A$, we get $\langle x, x_1\rangle = M = M_1$, a contradiction. Hence statement (i) is proved.

(ii) If $A$ is a maximal normal abelian subgroup of $G$, then $\exp(G/A) = p$.

Indeed, let $A$ be a maximal normal abelian subgroup of $G$. Suppose that there is a subgroup $H/A$ of $G/A$ such that $H/A \cong C_{p^2}$. Let $h \in H - A$ be such that $\langle h\rangle$ covers $H/A$ and set $k = h^p$. Then, by (i), $B = A\langle k\rangle$ is minimal nonabelian. Let $A^* = C_H(h) = \langle h\rangle C_A(h)$ so that $A^*$ is a maximal abelian subgroup of $H$. Let $B^*$ be a subgroup of

$H$ such that $B^* > A^*$ and $|B^* : A^*| = p$. By (i), $B^*$ is minimal nonabelian. Hence $k = h^p \in \Phi(B^*) = Z(B^*)$ and so $k$ centralizes $B^* \cap A$. This implies that $B^* < H$ and $B^* \cap A < A$. On the other hand, $C_G(A) = A$ and so $k$ does not centralize $A$. Let $a \in A - C_A(k)$ so that $a \in B - B^*$. Also, $h \in B^* - B$ and so (by the property $(**)$) $\langle a, h\rangle$ is minimal nonabelian. In this case, $h^p = k \in Z(\langle a, h\rangle)$ and so $k$ centralizes $a$, which is a contradiction. Our statement (ii) is proved.

(iii) We have $p = 2$ and whenever $A$ is a maximal normal abelian subgroup of $G$, then $A$ is of type $(4, 2, 2)$ and so $G$ is nonmetacyclic.

Let $A$ be any maximal normal abelian subgroup of $G$. By (i) and our assumption, $|G/A| > p$. Let $H/A$ be any subgroup of order $p^2$ in $G/A$. By (ii), $H/A \cong E_{p^2}$.

By (i), for each of the $p + 1$ subgroups $H_i/A$ of order $p$ in $H/A$, $H_i$ is minimal nonabelian, $i = 1, 2, \dots, p + 1$. Let $x, y$ be any elements in $H = A$ such that $\langle x, y\rangle$ covers $H/A$. Set $H_1 = A\langle x\rangle$, $H_2 = A\langle y\rangle$ so that $H_1 \cap H_2 = A$. Since $x \in H_1 - H_2$ and $y \in H_2 - H_1$, it follows (by the property $(**)$) that $\langle x, y\rangle$ is minimal nonabelian. If $\Phi(H) = A$, then $\langle x, y\rangle = H$ is minimal nonabelian, contrary to the fact that $H_1$ is minimal nonabelian and $H_1 < H$. Thus $\Phi(H) < A$ which gives $d(H) \geq 3$ and so, in particular, $H$ is not metacyclic.

Now

$$|A : \Phi(H_1)| = p\,, \quad |A : \Phi(H_2)| = p\,, \quad \Phi(H_1) \leq \Phi(H)\,, \quad \Phi(H_2) \leq \Phi(H)\,.$$

This yields

$$\Phi(H_1) = \Phi(H_2) = \Phi(H)\,, \quad |A : \Phi(H)| = p \Rightarrow d(H) = 3\,.$$

Moreover,

$$\Phi(H_1) = Z(H_1)\,, \quad \Phi(H_2) = Z(H_2) \Rightarrow \Phi(H) \leq Z(H)\,.$$

By (i), $H$ does not possess any abelian maximal subgroup and so we have $\Phi(H) = Z(H)$.

Note that $\langle x, y\rangle$ and $H_1$ are two distinct minimal nonabelian subgroups of $H$. Suppose that $\langle x, y\rangle$ does not contain $\Phi(H) = \Phi(H_1)$ and let $f \in \Phi(H_1) - \langle x, y\rangle$. Then

$$y \in \langle x, y\rangle - H_1\,, \quad f \in H_1 - \langle x, y\rangle$$

so that (by the property $(**)$) $\langle f, y\rangle$ must be minimal nonabelian. This is a contradiction since $f \in Z(H)$. Hence $\langle x, y\rangle > \Phi(H)$ and so $\langle x, y\rangle$ is a maximal subgroup of $H$.

We have proved that each maximal subgroup of $H$ is minimal nonabelian and $d(H) = 3$ so that $H$ is nonmetacyclic. Also we have $|H| > p^4$. Indeed, if $|H| = p^4$, then $H$ would have an abelian maximal subgroup, a contradiction. By Proposition 71.5 (a), $H$ is a special group of order $2^6$ which is isomorphic to a Sylow 2-subgroup of the Suzuki simple group Sz(8). In particular, $A$ is of type $(4, 2, 2)$ and so $G$ is nonmetacyclic. Statement (iii) is proved.

By (i), our group $G$ satisfies the assumptions of Theorem 149.3. In addition, by (iii), $p = 2$ and $G$ is nonmetacyclic. Hence Theorem 149.3 implies that $G$ is isomorphic to a Sylow 2-subgroup of the Suzuki simple group Sz(8). The proof is complete. □

**Problem.** Classify the nonabelian $p$-groups $G$ such that, whenever $A, B < G$ are distinct maximal abelian subgroups and $a \in A - B$ and $b \in B - A$, then the subgroup $\langle a, b \rangle$ is minimal nonabelian (of maximal class, metacyclic).

# § 150 The exponents of finite $p$-groups and their automorphism groups

This short section was written by Mann in a letter dated 21/07/11. We reproduce that letter without changes.

Y. Berkovich asked the following:

**Problem 2336.** Suppose that $G$ is a $p$-group such that $\mathrm{Aut}(G)$ is elementary abelian (of exponent $p$). Is it possible to estimate $\exp(G)$?

The following result answers that question, and a natural generalization of it, affirmatively:

**Theorem 150.1.** *Let $G$ be a finite $p$-group, let $S$ be a Sylow $p$-subgroup of* $\mathrm{Aut}(G)$, *and let* $\exp(S) = q$. *Then* $\exp(G)$ *is bounded by some function of $q$.*

For the proof we need the following result:

**Theorem 150.A.** *Let $G$ be a group for which $G/Z(G)$ is locally finite and of exponent $n$. Then $G'$ is locally finite and of finite exponent, which is bounded by a function of $n$.*

*Proof of Theorem 150.1.* Since $S$ contains a copy of $\mathrm{Inn}(G) \cong G/Z(G)$, we have $\exp(G/Z(G)) \leq q$, and Theorem A shows that $\exp(G')$ is bounded by some function $f(q)$ of $q$. Suppose that $\exp(Z(G)) > q$ and $\exp(G/G') \geq p^2q^2$. We can write $G/G' = C \times D$, where $C$ is a cyclic group, generated by $yG'$, say, of order $\exp(G/G')$. Let $z \in Z(G)$ have order $pq$. Define a homomorphism $\phi : G/G' \to Z(G)$, by:

$$\phi(yG') = z, \ \ \phi(D) = 1 ,$$

regard it as an endomorphism of $G$, and define another endomorphism by:

$$\sigma(x) = x\phi(x) .$$

If $a \in G$ has order dividing $pq$, then $\phi(a) = 1$, and in particular $\phi(z) = 1$. It follows that $\sigma^i(x) = x\phi(x)^i$, for each $i$. If $\sigma(x) = 1$, then $x = \phi(x)^{-1}$ has order dividing $pq$, thus $\phi(x) = 1$ and $x = 1$. Therefore $\sigma$ is an automorphism, and the equality $\sigma(y) = yz^i$ shows that $\sigma$ has order $pq$, a contradiction. Thus either

$$\exp(Z(G)) \leq q , \quad \exp(G) \leq q^2 ,$$

or

$$\exp(G/G') \leq pq^2 , \quad \exp(G) \leq prq^2 . \qquad \square$$

A similar argument proves the following:

**Theorem 150.2.** *Let $G$ be a finite group, and let* $\exp(\mathrm{Aut}(G)) = n$. *Then* $\exp(G)$ *is bounded by a function of $n$.*

*Proof.* Let $k = \exp(G/Z(G))$. By Theorem A, $\exp(G')$ is bounded by a function of $k$. If a prime $p$ does not divide $k$, then a Sylow $p$-subgroup $S$ of $G$ is contained in $Z(G)$; hence $G = S \times R$, for some group $R$. Repeating this argument shows that $G = A \times H$, where $A$ is abelian, $|H|$ is divisible only by primes dividing $k$, and $|A|$ and $|H|$ are relatively prime. Then $\mathrm{Aut}(G) = \mathrm{Aut}(A) \times \mathrm{Aut}(H)$, and for $A$ we have $\exp(A) \leq 2\exp(\mathrm{Aut}(A))^2$. It remains to bound $\exp(\mathrm{Aut}(H))$. Let $p \mid k$, and let a Sylow $p$-subgroup of $\mathrm{Aut}(H)$ have exponent $q$. As in the previous proof, we see that either $\exp(Z(H))$ is not divisible by $pq$, or $\exp(H/H')$ is not divisible by $(pq)^2$. In either case the $p$-contribution to $\exp(H)$ is at most $prq^2$, for $r$ as above, and this bounds $\exp(G)$. □

Since the bound on $r$ in Theorem 150.A depends on the bounds in E. I. Zelmanov's proof of the restricted Burnside problem, the bound on $\exp(G)$ in Theorems 150.1 and 150.2 is very big, but can be reduced under more special assumptions.

# § 151 $p$-groups all of whose nonabelian maximal subgroups have the largest possible center

Minimal nonabelian $p$-groups $G$ are closest to abelian $p$-groups and they have been determined by L. Rédei (see Lemma 65.1). They have the property $|G : Z(G)| = p^2$. But the class of $p$-groups $G$ with $|G : Z(G)| = p^2$ is larger and in that case $G = MZ(G)$, where $M$ is minimal nonabelian. Note that if in a $p$-group $G$ one has $|G : Z(G)| < p^2$, then $G$ is abelian. Therefore it is of interest to classify $p$-groups $G$ such that each nonabelian maximal subgroup $H$ of $G$ has the largest possible center $Z(H)$, i.e., $|H : Z(H)| = p^2$. In that case, $|X : Z(X)| = p^2$ for each nonabelian $X < G$. In fact we shall prove Theorem 151.1, which characterizes the title $p$-groups, and this solves Problem 1404(iii).

**Theorem 151.1** (Janko). *Let $G$ be a $p$-group that is neither abelian nor minimal nonabelian. Then $G$ has the property that each nonabelian maximal subgroup $H$ of $G$ has the largest possible center, i.e., $|H : Z(H)| = p^2$, if and only if one of the following holds:*

(i) $G = M_1Z(G)$, *where $M_1$ is minimal nonabelian and $M_1 < G$,*

(ii) $G = M_1 * M_2$, *where $M_1$ and $M_2$ are minimal nonabelian with $M_1' = M_2'$,*

(iii) $d(G) = 2$ *and* $|G'| = p^2$,

(iv) $p > 2$, $d(G) = 2$, $cl(G) = 3$, $G' \cong E_{p^3}$ *and* $\mho_1(G) \leq Z(G)$,

(v) $d(G) = 3$, $cl(G) = 2$, $G' \cong E_{p^3}$ *or* $E_{p^2}$ *and* $\Phi(G) = Z(G)$.

*Proof.* Let $G$ be a $p$-group that is neither abelian nor minimal nonabelian. Assume that $G$ has the following property:

$(*)$ Whenever $H$ is a nonabelian maximal subgroup of $G$, then $|H : Z(H)| = p^2$.

In that case $H$ has an abelian maximal subgroup and so we may use Lemma 1.1, which implies that $|H'| = p$, and so our group $G$ satisfies the assumptions of Theorem 137.7. Hence our group $G$ must be one of the groups stated in that theorem.

(1) Assume that $G$ is a group in Theorem 137.7 (a). Then we have $|G'| = p$. Let $M$ be a nonabelian maximal subgroup of $G$ and let $M_1$ be a minimal nonabelian subgroup of $M$ so that we have (by the property $(*)$)

$$M = M_1Z(M) \quad \text{with } M_1 \cap Z(M) = Z(M_1) \text{ and } M/Z(M) \cong M_1/Z(M_1) \cong E_{p^2}\,.$$

Since $M_1' = G'$, we get

$$G = M_1 * H, \quad \text{where } H = C_G(M_1).\ |H : Z(M)| = p\,.$$

If $H$ is abelian, then

$$G = M_1Z(G), \quad \text{where } Z(G) = H = C_G(M_1)\,, \quad M_1 < G\,.$$

This is the case (i) of our theorem.

One may assume that $H$ is nonabelian with $H' = G' = M_1'$ and $|H'| = p$. In addition, $Z(M)$ is an abelian maximal subgroup of $H$ and so Lemma 1.1 implies that $|H : Z(H)| = p^2$. Let $M_2$ be a minimal nonabelian subgroup of $H$ so that $H = M_2 Z(H)$; and hence

$$G = (M_1 * M_2)Z(H) \quad \text{with } Z(H) = Z(G) \text{ and } M_1' = M_2' = G' .$$

Note that $(M_1 * M_2)/Z(M_1 * M_2) \cong E_{p^4}$ and so $M_1 * M_2$ cannot be a proper subgroup of $G$. It follows that $G = M_1 * M_2$ with $M_1' = M_2'$ and so we have obtained the groups in part (ii) of our theorem.

It remains to be shown that such groups $G = M_1 * M_2$ with $M_1$ and $M_2$ being minimal nonabelian and $M_1' = M_2'$ have in fact the property $(*)$. We have (see Lemma 65.1)

$$Z(M_1) = \Phi(M_1), \quad Z(M_2) = \Phi(M_2);$$

hence

$$Z(M_1)Z(M_2) = \Phi(M_1)\Phi(M_2) = \Phi(G) = Z(G)$$

and

$$G/Z(G) \cong E_{p^4} \quad \text{with } G' = \langle z \rangle \le Z(G) \text{ and } |G'| = p .$$

Then Lemma 1.1 also implies that $G$ does not possess an abelian maximal subgroup. Let $X$ be any maximal subgroup of $G$ so that $X' = \langle z \rangle$. Therefore one may choose elements $x_1, x_2, x_3 \in X - Z(G)$ such that

$$X = \langle x_1, x_2, x_3 \rangle Z(G) \quad \text{and} \quad [x_1, x_2] = z .$$

Let us prove that $X$ possesses an abelian maximal subgroup. Indeed, if $[x_1, x_3] = 1$, then $\langle x_1, x_3 \rangle Z(G)$ is an abelian maximal subgroup of $X$. Assume that $[x_1, x_3] \neq 1$. In that case we replace $x_3$ with $x_3' = x_3^i$ for a suitable integer $i \not\equiv 0 \pmod p$ so that we may suppose that $[x_1, x_3'] = z^{-1}$. But then we get

$$[x_1, x_2 x_3'] = [x_1, x_2][x_1, x_3'] = zz^{-1} = 1 ,$$

and then $\langle x_1, x_2 x_3' \rangle Z(G)$ is an abelian maximal subgroup of $X$. By Lemma 1.1,

$$|X| = p|Z(X)|\,|X'| \Rightarrow |X : Z(X) = p^2 .$$

We have proved that our groups in part (ii) of our theorem have the property $(*)$.

(2) Suppose that $G$ is a group in Theorem 137.7 (b), i.e., $d(G) = 2$ and $|G'| = p^2$. It remains only to prove that in that case $G$ has the property $(*)$. Let $H$ be a $G$-invariant subgroup of order $p$ in $G'$. Then

$$d(G/H) = 2, \quad (G/H)' = G'/H \cong C_p ,$$

and so $G/H$ is minimal nonabelian (see Lemma 65.2 (a)). Let $M$ ($\ge G'$) be any maximal subgroup of $G$. Then $M/H$ is abelian and so either $M$ is abelian or $M' = H \cong C_p$.

Suppose that $M' = H$. By the structure of $G/H$ (see Lemma 65.1), we have $d(M/H) \leq 3$. But we have $H = M' \leq \Phi(M)$ and so we get also $d(M) \leq 3$. Let $K$ be a minimal nonabelian subgroup of $M$. Now

$$K' = M' = H \cong C_p \Rightarrow M = K * C_M(K) .$$

Suppose that $L = C_M(K)$ is nonabelian so that

$$L' = M' = K' \leq \Phi(L) , \quad |L/\Phi(L)| \geq p^2 .$$

But then $d(M) \geq 4$, a contradiction. Hence $L$ is abelian and so $L = Z(M)$, which implies that $M/Z(M) \cong K/Z(K) \cong E_{p^2}$ and we are done.

(3) Suppose that $G$ is a group in Theorem 137.7 (c), i.e.,

$$p > 2,\ d(G) = 2 , \quad cl(G) = 3 , \quad G' \cong E_{p^3} , \quad \mho_1(G) \leq Z(G) .$$

Again we have to prove only that $G$ has the property $(*)$. Indeed, since $d(G) = 2$, the quotient group $G'/K_3(G) \neq \{1\}$ must be cyclic, where $cl(G) = 3$ also implies that $K_3(G) = [G, G'] \leq Z(G)$. Hence we get $|G'/K_3(G)| = p$. Because of

$$\Phi(G) = G'\mho_1(G) , \quad \mho_1(G) \leq Z(G) , \quad G' \nleq Z(G) , \quad Z(G) \leq \Phi(G) ,$$

it follows that $Z(G) = \mho_1(G)K_3(G$ is of index $p$ in $\Phi(G)$. This implies $|G/Z(G)| = p^3$. Let $M$ be any maximal subgroup of $G$ so that

$$M > \Phi(G) , \quad |M : \Phi(G)| = p \Rightarrow |M : Z(G)| = p^2 .$$

Thus $M$ is either abelian or $Z(G) = Z(M)$ with $|M : Z(M)| = p^2$ and we are done.

(4) Finally, assume that $G$ is a group in Theorem 137.7 (d), i.e.,

$$d(G) = 3,\ cl(G) = 2,\ G' \cong E_{p^3} \text{ or } E_{p^2} , \quad \Phi(G) = Z(G) .$$

Let $M$ be any maximal subgroup of $G$. Then we have $M/Z(G) \cong E_{p^2}$ and so $M$ is either abelian or $Z(G) = Z(M)$, and so in this case $|M : Z(M)| = p^2$. Hence our group $G$ has the property $(*)$ and this completes the proof. □

**Problem.** Study the $p$-groups $G$, which are neither abelian nor minimal nonabelian, such that $|H : Z(H)| \leq p^3$ for all nonabelian $H < G$.

# § 152 $p$-central $p$-groups

A $p$-group $G$ is called $p$-central if either $p > 2$ and $\Omega_1(G) \leq Z(G)$ or $p = 2$ and $\Omega_2(G) \leq Z(G)$. If $G$ is a $p$-central $p$-group, then we know that $d(G) \leq d(Z(G))$ (see Theorem 15.1 and Theorem 15.3).

An interesting characterization of $p$-central $p$-groups is given in the next two theorems. For $p > 2$, the case was proved by A. Bianchi, A. Gillio Berta Mauri, and L. Verardi, and the case $p = 2$ was proved by A. Mann. However, our proofs are similar but they are somewhat simpler.

**Theorem 152.1.** *A $p$-group $G$, $p > 2$, is $p$-central if and only if the following condition holds:*

$(*)$ *Whenever $x, y \in G$ and $x^p = y^p$, then $x$ and $y$ commute.*

*Proof.* We use induction on $|G|$. Let $p > 2$ and let $G$ be a $p$-group satisfying $(*)$. Let $x \in G$ have order $p$ and let $y \in G$. Trying to show that $[x, y] = 1$ we may assume $[x, y] \neq 1$ and $G = \langle x, y\rangle$. Let $H$ be a maximal subgroup of $G$ containing $x$. By induction, $H$ is $p$-central and so $x \in Z(H)$. Set $E = \Omega_1(Z(H))$ so that $x \in E$, $E \lhd G$ and $G = E\langle y\rangle$. Set $E_0 = C_E(y)$ and $G_0 = E_0\langle y\rangle$ so that we have $E_0 < E$ and $G_0 < G$. Let $G_1 \leq G$ be such that $G_0 < G_1$ and $|G_1 : G_0| = p$. Set $E_1 = G_1 \cap E$, where we have

$$|E_1 : E_0| = p\,, \quad G_1 = E_1\langle y\rangle\,, \quad E_0 \leq Z(G_1)\,.$$

Take an element $e \in E_1 \setminus E_0$ so that we have $1 \neq [e, y] \in E \cap G_0 = E_0 \leq Z(G_1)$. It follows that $\langle e, y\rangle' = \langle [e, y]\rangle$ is of order $p$ and so $\langle e, y\rangle$ is minimal nonabelian (and so of class 2). We compute

$$(ye)^p = y^p e^p [e, y]^{\binom{p}{2}} = y^p\,,$$

and so, by the property $(*)$, $y$ commutes with $ye$ and so $y$ commutes with $e$, a contradiction. We have proved that $G$ is $p$-central.

Now, let $G$ be $p$-central and let $x^p = y^p$ for some $x, y \in G$. By induction, we may assume $G = \langle x, y\rangle$. Note that in that case $x^p = y^p \in Z(G)$ and so for each $g \in G$ we have $(x^g)^p = (x^p)^g = x^p$, which implies, by induction, $[x, x^g] = 1$ since the normal closure $A = \langle x\rangle^G$ is a proper subgroup of $G$. Hence $A$ is abelian and similarly, $B = \langle y\rangle^G$ is also abelian. Hence we get that $G = AB$ is of class $\leq 2$. It follows $[y^{-1}, x]^p = [y^{-1}, x^p] = 1$ and so we have

$$(xy^{-1})^p = x^p y^{-p} [y^{-1}, x]^{\binom{p}{2}} = 1\,,$$

which infers $xy^{-1} \in Z(G)$. Thus $x$ commutes with $xy^{-1}$ and with $y$ and we are done. □

**Theorem 152.2.** *A 2-group $G$ is 2-central if and only if the following condition holds:*

$(**)$ *Whenever $x, y \in G$ and $x^4 = y^4$, then $x$ and $y$ commute.*

*Proof.* We use induction on $|G|$. Let $G$ be a 2-group satisfying $(**)$. Let $x, y \in G$ with $x^4 = 1$. Trying to show that $[x, y] = 1$, we may assume $[x, y] \neq 1$ and $G = \langle x, y\rangle$. Let $H$ be a maximal subgroup of $G$ containing $x$. By induction, $H$ is 2-central and so $x \in Z(H)$. Set $E = \Omega_2(Z(H))$ so that $x \in E$, $E \lhd G$ and $G = E\langle y\rangle$.

Set $E_0 = C_E(y)$ and $G_0 = E_0\langle y\rangle$ so that we have $E_0 < E$ and $G_0 < G$. Let $G_1 \leq G$ be such that $G_0 < G_1$ and $|G_1 : G_0| = 2$. Set $E_1 = G_1 \cap E$, where we have $|E_1 : E_0| = 2$, $G_1 = E_1\langle y\rangle$ and $E_0 \leq Z(G_1)$. Note that

$$G_1' \leq G_0 \cap E_1 = E_0 \leq Z(G_1)$$

and so $G_1$ is of class 2.

Take an element $e \in E_1 - E_0$ so that $e^2 \in E_0 \leq Z(G_1)$ and therefore we get $[e, y]^2 = [e^2, y] = 1$, and so $[e, y]$ is of order 2. It follows that $\langle e, y\rangle' = \langle [e, y]\rangle$ is of order 2 and so $\langle e, y\rangle$ is minimal nonabelian. We get

$$(ye)^4 = y^4e^4[e, y]^6 = y^4 .$$

Hence, by the property $(**)$, $y$ commutes with $ye$ and so $y$ commutes with $e$, a contradiction.

Now, let $G$ be 2-central and let $x^4 = y^4$ for some $x, y \in G$. By induction, we may assume $G = \langle x, y\rangle$. As in the proof of Theorem 152.1, we show that $\langle x\rangle^G$ and $\langle y\rangle^G$ are abelian and so $G = \langle x\rangle^G\langle y\rangle^G$ is of class $\leq 2$. Indeed, $x^4 = y^4 \in Z(G)$ and so for each $g \in G$, one has $(x^g)^4 = (x^4)^g = x^4$, which implies (by induction) $[x, x^g] = 1$ because $\langle x\rangle^G$ is a proper subgroup of $G$. Hence $\langle x\rangle^G$ and similarly $\langle y\rangle^G$ are abelian. Since $x^4 \in Z(G)$ and so

$$[y^{-1}, x^2]^6 = [y^{-1}, x^{12}] = 1 ,$$

we get

$$(x^2y^{-1})^4 = x^8y^{-4}[y^{-1}, x^2]^6 = y^4 .$$

But $\langle x^2, y\rangle$ is a proper subgroup of $G$ and so induction implies that $x^2y^{-1}$ commutes with $y$ and so also $x^2$ commutes with $y$. Thus we have $[x, y]^2 = [x^2, y] = 1$. We compute

$$(xy^{-1})^4 = x^4y^{-4}[y^{-1}, x]^6 = [y^{-1}, x^6] = 1 .$$

But $G$ is 2-central and so $xy^{-1} \in Z(G)$. Hence $x$ commutes with $xy^{-1}$ and so also $x$ and $y$ commute. Our theorem is proved. □

**Definition 1.** Let $G$ be a $p$-group and let $N \lhd G$. Then $N$ is $p$-centrally embedded in $G$ if either $p > 2$ and $\Omega_1(N) \leq Z(G)$ or $p = 2$ and $\Omega_2(N) \leq Z(G)$.

**Definition 2.** A $p$-group $G$ has $p$-central class $\leq k$, if $G$ has a normal series $\{1\} = N_0 \leq N_1 \leq \cdots \leq N_k = G$ such that $N_i/N_{i-1}$ is $p$-centrally embedded in $G/N_{i-1}$, $i = 1, \ldots, k$. Such a series is termed a $p$-central series of $G$. If $k$ is the minimal length of such a normal series, then the $p$-central class of $G$ is exactly $k$, and we write $\mathrm{pcc}(G) = k$. A $p$-group $G$ has a small $p$-central class, if $\mathrm{pcc}(G) \leq p - 1$.

**Remark 1.** Obviously, the upper central series of a $p$-group $G$ is also a $p$-central series and so a $p$-central series of a $p$-group always exists.

**Theorem 152.3** (A. Mann). *Let $G$ be a $p$-group having a small $p$-central class. Then $\Omega_1(G)$ is of exponent $p$ and is contained in $Z_{p-1}(G)$.*

*Proof.* First, note that if $H \leq G$ and $\{N_i\}$ is a $p$-central series of $G$, then $\{N_i \cap H\}$ is a $p$-central series of $H$ and therefore $\mathrm{pcc}(H) \leq \mathrm{pcc}(G)$.

If $p = 2$, then $G$ is 2-central and there is nothing to prove. So we assume $p > 2$. Let $x, y \in G$ be of order $p$ and we want to prove that $o(xy) \leq p$. Therefore we assume that $o(xy) > p$ and so using induction on $|G|$, we get $G = \langle x, y\rangle$. Set $H = \langle x\rangle^G$, Then $H$ is generated by the conjugates of $x$ (elements of order $p$) and $H < G$ and so by induction $H$ is of exponent $p$. Taking intersections of $H$ with a $p$-central series $\{N_i\}$ of $G$ of length $\mathrm{pcc}(G)$, we see that $H \leq Z_{p-1}(G)$. But we have $G = H\langle y\rangle$ and so $G/H \cong C_p$, which shows that $\mathrm{cl}(G) \leq p - 1$. By Theorem 7.1 (b), $G$ is regular and so Theorem 7.2 (a) implies that $\Omega_1(G)$ is of exponent $p$. Hence we get $o(xy) \leq p$, a contradiction.

We have proved that $\Omega_1(G)$ is of exponent $p$. Taking intersections of $\Omega_1(G)$ with a $p$-central series $\{N_i\}$ of $G$ of length $\mathrm{pcc}(G) \leq p - 1$, we get that $\Omega_1(G) \leq Z_{p-1}(G)$. Our theorem is proved. □

# § 153 Some generalizations of 2-central 2-groups

We consider here some generalizations of 2-central 2-groups that have been considered in § 152. Recall that an element $x \neq 1$ in a 2-group $G$ is real if there is $y \in G$ such that $x^y = x^{-1}$ and a 2-group $G$ is 2-central if $\Omega_2(G) \le Z(G)$. All results of this section have been obtained by A. Mann.

**Definition 1.** Let $G$ be a 2-group. We say that $G$ satisfies $(*)$, $(**)$ or $(X)$ respectively, if the following holds in $G$:

$(*)$ Whenever $x^2 = y^2$ for some $x, y \in G$, then $[x, y] = 1$ (i.e., elements with the same square commute) or equivalently $(xy^{-1})^2 = 1$ (see Remark 1).

$(**)$ $\Omega_1(G) \le Z(G)$.

$(X)$ Both $(*)$ and $(**)$ hold.

**Remark 1.** If $x^2 = y^2$ implies $[x, y] = 1$, then we also have $(xy^{-1})^2 = x^2y^{-2} = 1$. If $x^2 = y^2$ implies $(xy^{-1})^2 = 1$, then

$$1 = (xy^{-1})^2 = xy^{-1}xy^{-1} = x^2(x^{-1}y^{-1}xy)y^{-2} = x^2[x, y]y^{-2} ,$$

and so $[x, y] = x^{-2}y^2 = 1$.

**Theorem 153.1.** *A 2-group $G$ satisfies $(**)$ $\iff$*

$$(xy^{-1})^2 = 1 \Rightarrow x^2 = y^2 \quad \textit{for any } x, y \in G.$$

*Proof.* Assume that $G$ satisfies $(**)$ and that $(xy^{-1})^2 = 1$. Then $xy^{-1} \in Z(G)$ and so $x$ commutes with $xy^{-1}$ and with $y^{-1}$. This gives

$$(xy^{-1})^2 = x^2y^{-2} = 1 , \quad x^2 = y^2 .$$

Conversely, assume that

$$(xy^{-1})^2 = 1 \Rightarrow x^2 = y^2 \quad \text{for any } x, y \in G.$$

Let $x$ be an involution and $y \in G$. Then

$$1 = x^2 = ((yx)^{-1}y)^2 \Rightarrow y^2 = (yx)^2 .$$

Thus, we have

$$1 = y^{-2}(yx)^2 = y^{-2}yxyx = y^{-1}xyx = y^{-1}x^{-1}yx = [y, x] .$$

Hence $x$ and $y$ commute and so $x \in Z(G)$ and we are done. □

**Theorem 153.2.** *A 2-group $G$ satisfies $(*)$ if and only if all real elements of $G$ are involutions.*

*Proof.* Assume that $G$ satisfies $(*)$. Let $1 \neq x \in G$ be a real element so that there is $y \in G$ such that $x^y = x^{-1}$. Then we have

$$(yx)^2 = yxyx = y^2x^yx = y^2$$

and so, by $(*)$, $yx$ and $y$ commute. But then $x$ and $y$ commute and so $x^y = x = x^{-1}$ and $x$ is an involution.

Suppose that all real elements of $G$ are involutions. Let $x, y \in G$ be such that $x^2 = y^2$. Set $z = xy^{-1}$ and compute

$$z^{y^{-1}} = y(xy^{-1})y^{-1} = yxy^{-2} = yxx^{-2} = yx^{-1} = z^{-1}\,.$$

If $z = 1$, then $x = y$ and so $[x, y] = 1$. If $z \neq 1$, then $z$ is real and so $z$ is an involution. We get

$$1 = z^2 = (xy^{-1})^2 = xy^{-1}xy^{-1} = xy^{-2}(yxy^{-1}) = xx^{-2}x^{y^{-1}} = x^{-1}x^{y^{-1}}\,,$$

and so $x^{y^{-1}} = x$ which implies again $[x, y] = 1$ and so $(*)$ holds. □

**Theorem 153.3.** *For a 2-group $G$, the following conditions are equivalent:*
(a) *$G$ satisfies $(X)$,*
(b) *if $x, y \in G$ and $x^2 = y^2$, then $x$ and $y$ belong to the same coset of $Z(G)$,*
(c) *every real element of $G$ is central.*

*Proof.* (a) is equivalent to $(*)$ and $(**)$ and so Theorem 153.2 implies that (a) is equivalent to (c).

If (a) holds and $x, y \in G$ with $x^2 = y^2$, then $[x, y] = 1$ (by $(X)$) and so $(xy^{-1})^2 = 1$ and (again by $(X)$) $xy^{-1} \in Z(G)$ and so (b) holds.

Conversely, if (b) holds and $x, y \in G$ with $x^2 = y^2$, then $x = yz$ for some $z \in Z(G)$, and so $[x, y] = 1$, which shows that $(*)$ holds. Also, taking $y = 1$, we get from $x^2 = 1 = 1^2$ that $x \in Z(G)$ and therefore $(**)$ holds. We have proved that (b) implies (a) and so (a) and (b) are equivalent. □

It is clear (by Theorem 153.2) that if we have $\Omega_2(G) \leq Z(G)$ in a 2-group $G$ (i.e., $G$ is 2-central), then $G$ is an $(X)$-group. But the converse is not true as the following example shows. Hence $(X)$-groups present a proper generalization of 2-central 2-groups.

**Example 1.** Consider the following nonmetacyclic minimal nonabelian group of order $2^{k+3}$, $k \geq 2$:

$$G = \langle x, z \mid x^4 = z^{2^k} = 1, \quad [x, z] = y, \quad y^2 = [y, x] = [y, z] = 1\rangle\,.$$

Then

$$\Omega_1(G) = \langle x^2, z^{2^{k-1}}, y\rangle \cong E_8 \quad \text{and} \quad \Omega_1(G) \leq Z(G)\,,$$

where $G' = \langle y\rangle$ is a maximal cyclic subgroup of $G$. Hence $G$ satisfies $(**)$. Let $h \neq 1$ be a real element in $G$. Then there is $g \in G$ so that $h^g = h^{-1}$. Hence $[h, g] = h^{-2}$, and so $h^{-2} \in \langle y \rangle$. Since $h^{-2} = y$ is not possible, we have $h^{-2} = 1$, and so $h$ is an involution. Thus all real elements in $G$ are involutions, and so Theorem 153.2 implies that $G$ satisfies $(*)$. It follows that $G$ is an $(X)$-group. But $G$ is not a 2-central group since $x \notin Z(G)$ and $o(x) = 4$.

**Theorem 153.4.** *Let G be a 2-group. Then:*
(a) *if G satisfies (X), then $G/\Omega_1(G)$ also satisfies (X),*
(b) *if G has commuting roots (i.e., G satisfies $(*)$) and H is a subgroup of G, then* $d(H) \le d(\Omega_1(G))$.

*Proof.* In both cases $G$ has commuting roots and so $\Omega_1(G)$ is elementary abelian and therefore $\Omega_1(G)$ consists of all real elements of $G$. Indeed, if $x$ and $y$ are involutions and $[x, y] \ne 1$, then $\langle x, y \rangle \cong D_{2^n}$, $n \ge 3$, and so $xy$ is a real element of order $2^{n-1} \ge 4$, contrary to Theorem 153.2.

(a) Assume that $G$ is an $(X)$-group and let $x\Omega_1(G)$ be an involution in $G/\Omega_1(G)$, where (by $(**)$) $\Omega_1(G) \le Z(G)$. Hence $x^2 \in Z(G)$, and so for any $y \in G$ we have $(x^y)^2 = (x^2)^y = x^2$, and so, by $(*)$, $x$ and $x^y$ commute. This gives

$$[x^{-1}, y]^2 = (x \cdot (y^{-1}x^{-1}y))^2 = (x(x^y)^{-1})^2 = x^2(x^y)^{-2} = 1 ,$$

and so

$$[x^{-1}, y] \in \Omega_1(G) \Rightarrow x^{-1}\Omega_1(G) = x\Omega_1(G) \in Z(G/\Omega_1(G)) .$$

So all involutions of $G/\Omega_1(G)$ are central and therefore $G/\Omega_1(G)$ satisfies $(**)$.

It remains to show that the quotient group $G/\Omega_1(G)$ satisfies $(*)$. By Theorem 153.2, it is enough to show that all real elements in $G/\Omega_1(G)$ are involutions. So, let $u\Omega_1(G)$ be a real element in $G/\Omega_1(G)$. Then there exists $y \in G$ such that $y^{-1}uy = u^{-1}z$ for some $z \in \Omega_1(G)$. It follows

$$y^{-1}u^2y = (y^{-1}uy)^2 = (u^{-1}z)^2 = u^{-2} ,$$

and so $u^2$ is real in $G$ and hence $u^2 \in \Omega_1(G)$. Thus $u\Omega_1(G)$ is an involution, as desired.

(b) Suppose that $G$ has commuting roots. We already know that $\Omega_1(G)$ is elementary abelian. Let $H$ be a subgroup of $G$. Since $\Omega_1(H) \le \Omega_1(G)$ and $H$ has also commuting roots, it suffices to show that $d(G) \le d(\Omega_1(G)$, because then $d(H) \le d(\Omega_1(H)) \le d(\Omega_1(G))$.

Now, $(*)$ says that elements from different cosets of $\Omega_1(G)$ have different squares. So, if $S$ is the set of all squares of the elements of $G$, then

$$|\mho_1(G)| \ge |S| \ge |G : \Omega_1(G)| \Rightarrow |G/\mho_1(G)| \le |\Omega_1(G)| \Rightarrow d(G) \le d(\Omega_1(G) ,$$

completing the proof. □

# § 154 Metacyclic $p$-groups covered by minimal nonabelian subgroups

One of the most difficult open problems in $p$-group theory is to classify $p$-groups that are covered by their minimal nonabelian subgroups (see Problem 860). We can solve this problem for metacyclic $p$-groups.

Recall (see § 26, Definition 7) that a metacyclic 2-group $G$ is said to be *ordinary metacyclic* (with respect to a subgroup $H$) if $G$ has a cyclic normal subgroup $H$ such that $G/H$ is cyclic and $G$ centralizes $H/\mho_2(H)$ (i.e., $[G,H] \le \mho_2(H)$) (clearly, $G$ centralizes $H/\mho_1(H)$ always).

Let $G$ be a nonabelian metacyclic $p$-group. If $p > 2$ or $p = 2$ and $G$ is ordinary metacyclic, then Theorem 149.3 implies that $G$ is covered by its minimal nonabelian subgroups (for $p > 2$ this also follows from Theorem 148.15). But there exist some nonabelian metacyclic 2-groups (for example, 2-groups of maximal class and order $> 2^3$; note that these groups are not ordinary metacyclic), which are not covered by their minimal nonabelian subgroups. Here we determine exactly such metacyclic 2-groups.

**Proposition 154.1.** *Let $G$ be a nonabelian metacyclic 2-group that is not covered by its minimal nonabelian subgroups. Then we have:*

$$G = \langle a, b \mid a^{2^m} = b^{2^n} = 1, z = a^{2^{m-1}}, a^b = a^{-1}z^{\epsilon}, b^{2^{n-1}} = z^{\eta}\rangle ,$$

*here $m \ge 3$, $n \ge 1$, $\epsilon$, $\eta \in \{0, 1\}$.*

*Proof.* Let $G$ be a nonabelian metacyclic 2-group that is not covered by its minimal nonabelian subgroups. In particular, $G$ is not minimal nonabelian and so $|G'| \ge 4$ (Lemma 65.2 (a)), and so $\exp(G) > 4$ since $G$ possesses a normal cyclic subgroup of order $> 4$ with cyclic quotient group. If $G$ is ordinary metacyclic, then (by what has been said above) $G$ is covered by its minimal nonabelian subgroups. Hence $G$ is not ordinary metacyclic. Therefore, $G = \langle a, b\rangle$, and one may assume that $a^b = a^{-1}v$ with $v \in \langle a^4\rangle$ and $o(a) = 2^m$, $m \ge 3$. Indeed, as $G$ is not ordinary metacyclic, the quotient group $\bar{G} = G/\langle a^4\rangle$ is nonabelian so $b$ inverts $\bar{a}$, and so $a^b = a^{-1}v$, where $v \in \langle a^4\rangle$. We get $v = a^{4i}$ so $v^{2^{m-2}} = (a^{4i})^{2^{m-2}}1$, and so

$$(a^{2^{m-2}})^b = (a^b)^{2^{m-2}} = (a^{-1}v)^{2^{m-2}} = a^{-2^{m-2}}v^{-2^{m-2}} = h^{-1} ,$$

where $h = a^{2^{m-2}}$ is of order 4. As $h^b = h^{-1} \ne h$, it follows that $|\langle a\rangle \cap \langle b\rangle| \le 2$ and $\langle h, b\rangle$ is minimal nonabelian (Lemma 65.2 (a), since $\langle h, b\rangle'$, being a proper subgroup of $\langle h\rangle$, has order 2). We may set

$$o(b) = 2^n, \; n \ge 1, \; b^{2^{n-1}} = z^{\eta}, \quad \text{where } z = a^{2^{m-1}}, \; \eta \in \{0, 1\} .$$

Moreover, we have $C_G(h) = \langle a, b^2\rangle$ and so each element $g \in G - \langle a, b^2\rangle$ inverts $h$, and therefore $\langle g, h\rangle$ is minimal nonabelian. Thus, the set in $G - \langle a, b^2\rangle$ is contained in the union of minimal nonabelian subgroups of $G$. We have

$$a^{b^2} = (a^{-1}v)^b = a(v^{-1}v^b), \quad \text{where } v^{-1}v^b \in \langle a^4 \rangle .$$

Since $b^2$ centralizes $\langle a \rangle / \langle a^4 \rangle$, the subgroup $\langle a, b^2 \rangle$ is either abelian or nonabelian and ordinary metacyclic. If $\langle a, b^2 \rangle$ is nonabelian and ordinary metacyclic, then (by the introduction above) each element in $\langle a, b^2 \rangle$ is contained in a minimal nonabelian subgroup, and so in this case $G$ is covered by its minimal nonabelian subgroups, a contradiction.

We have proved that $\langle a, b^2 \rangle$ must be abelian and so $b$ induces an involutory automorphism on $\langle a \rangle$. Thus, $a^b = a^{-1}z^{\epsilon}$, where $z$ is the involution in $\langle a \rangle$ and $\epsilon \in \{0, 1\}$. Then we see that the element $a$, being of order $2^m \geq 2^3$, is not contained in any minimal nonabelian subgroup of $G$. Indeed, assume that $a \in M < G$, where $M$ is minimal nonabelian. Then $M = \langle a \rangle (M \cap \langle b \rangle)$ so $M \cap \langle b \rangle \leq \langle b^2 \rangle$, as we have noticed, centralizes $a$, and we conclude that $M$ is abelian, a final contradiction. □

Note that if a nonabelian metacyclic 2-group $G$ of exponent $2^e$ has no minimal nonabelian subgroup of exponent $2^e$, then it is not covered by minimal nonabelian subgroups. Such $G$ are classified in Proposition 154.1.

**Problem 1.** Classify the metacyclic 2-groups $G$ with $|G'| > 2^k > 2$ covered by $\mathcal{A}_k$-subgroups (see Theorem 72.1).

**Problem 2.** Classify the nonabelian $p$-groups of exponent $p$ covered by minimal nonabelian subgroups (this is a partial case of #860).

# § 155 A new type of Thompson subgroup

We recall that the Thompson subgroup $\mathcal{J}(G)$ of a $p$-group $G$ is the characteristic subgroup generated by all abelian subgroups of $G$ of the maximal possible order. However, $\mathcal{J}(G)$ is not so important in the case $p = 2$. In the hope of finding a subgroup $\mathcal{K}(G)$ analogous to $\mathcal{J}(G)$ that will be valuable in studying 2-groups, J. G. Thompson proposed the following definition:

**Definition 1.** Let $G$ be a $p$-group. Then $\mathcal{B}_0(G)$ is the set of subgroups $H$ of $G$ that have the commutator quotient group $H/H'$ of the maximal possible order, and $\mathcal{B}(G)$ is the set of minimal elements in $\mathcal{B}_0(G)$ under inclusion $\leq$. Then the characteristic subgroup $\mathcal{K}(G)$ of $G$ is generated by all subgroups in $\mathcal{B}(G)$.

The first question is what is the structure of a subgroup $X$ in $\mathcal{B}(G)$? It turns out that $X$ is very close to abelian groups, namely, the class of $X$ is $\leq 2$ as the following result shows:

**Theorem 155.1** (J. G. Thompson). *If $G$ is a $p$-group and $|G : G'| > |H : H'|$ for every proper subgroup $H$ of $G$, then $G$ is of class $\leq 2$.*

*Proof.* Let $G$ be a minimal counterexample, so that $[G', G] \neq \{1\}$. Let $U$ be a normal subgroup of order $p$ that is contained in $[G', G]$. If $H/U$ is a proper subgroup of $G/U$, then we get

$$|G/U : (G/U)'| = |G : G'| > |H : H'| \geq |H/U : (H/U)'|,$$

so by the minimality of $G$, we see that $G/U$ is of class 2 and hence $[G', G] = U$ is of order $p$. This gives $[G, G', G] = [G', G, G] = \{1\}$ and then the Three Subgroups Lemma (Introduction, Exercise 13) implies $[G, G, G'] = [G', G'] = \{1\}$ and so $G'$ is abelian.

Set $C = C_G(G') \geq G'$. Then $G$ stabilizes the chain $G' > U > \{1\}$ and so $V_1 = G/C$ is elementary abelian. Indeed, for any $g, g_1 \in G$ and $h \in G'$, we have $h^g = hu$ and $h^{g_1} = hu_1$ for some $u, u_1 \in U$. This gives

$$h^{g^2} = (hu)^g = hu^2, \dots, h^{g^p} = hu^p = h, \quad h^{g^{-1}} = hu^{-1}, \quad h^{g_1^{-1}} = hu_1^{-1},$$

so that

$$h^{[g,g_1]} = h^{g^{-1}g_1^{-1}gg_1} = hu^{-1}u_1^{-1}uu_1 = h,$$

and we obtain $g^p \in C$ and $[g, g_1] \in C$.

For any $x \in G'$ and $g \in G$, we have $x^g = xu$ with some $u \in U$. This gives

$$(x^p)^g = (x^g)^p = (xu)^p = x^p \Rightarrow x^p \in Z(G).$$

We have proved that $V_2 = G'/(G' \cap Z(G))$ is also elementary abelian. Therefore, both $V_1$ and $V_2$ can be viewed as vector spaces over $GF(p)$, where $U \cong C_p$ can be identified with the Galois field $GF(p)$.

We define a map $\phi$ from $V_1 \times V_2$ into $U$ with

$$\phi(Cg, (G' \cap Z(G))x) = [g, x] \in U\,,$$

where $g \in G$ and $x \in G'$. This map is bilinear, i.e., it is homomorphic in each argument. Indeed, for any $g, g_1, g_2 \in G$ and $x, x_1, x_2 \in G'$, we have

$$\begin{aligned}[][g_1g_2, x] &= [g_1, x]^{g_2}[g_2, x] = [g_1, x][g_2, x]\\ [g, x_1x_2] &= [g, x_2][g, x_1]^{x_2} = [g, x_1][g, x_2]\,.\end{aligned}$$

Such a map $\phi$ is called a *pairing*. Also, by our construction, $\phi$ is nonsingular in both arguments. Indeed, if $x \in G' - Z(G)$, there is $g \in G - C$ such that $[g, x] \neq 1$ and for any $h \in G - C$, there is $y \in G' - Z(G)$ such that $[h, y] \neq 1$.

The pairing $\phi$ naturally gives rise to homomorphisms

$$\phi_1 : V_1 \to V_2^* \text{ and } \phi_2 : V_2 \to V_1^*\,,$$

where $V_1^*$ and $V_2^*$ are dual vector spaces of $V_1$ and $V_2$, respectively (noting that $V_1^*$ consists of all linear functionals from $V_1$ into GF($p$) and $V_2^*$ consists of all linear functionals from $V_2$ into GF($p$)) given by (for each $Cg \in V_1$ and $(G' \cap Z(G))h \in V_2$)

$$\begin{aligned}\phi_1(Cg)&\colon (G' \cap Z(G))x \to [g, x]\ (x \in G')\,,\\ \phi_2((G' \cap 6Z(G))h)&\colon Ck \to [k, h]\ (k \in G)\,.\end{aligned}$$

Since $\phi$ is nonsingular (in both arguments), $\phi_1$ is an isomorphism from $V_1$ into $V_2^*$ and $\phi_2$ is an isomorphism from $V_2$ into $V_1^*$. Also note that $V_1 \cong V_1^*$ and $V_2 \cong V_2^*$. This implies that the vector spaces $V_1$ and $V_2$ have the same dimension, and so $|G/C| = |G'/(G' \cap Z(G))|$.

Next, we see that $C$ stabilizes the chain $G > G' > \{1\}$, and so $C'$ centralizes $G$, that is, $C' \leq G' \cap Z(G)$. Indeed, let $c \in C$ and $g \in G$. Then we have

$$g^c = g(g^{-1}g^c) = g[g, c], \quad \text{where } [g, c] \in G'\,.$$

Also, for any $c_1, c_2 \in C$ and $g \in G$ we have

$$g^{c_1} = gh_1\,, \quad g^{c_2} = gh_2 \quad \text{for some } h_1, h_2 \in G'\,.$$

It follows that $g^{c_1^{-1}} = gh_1^{-1}, g^{c_2^{-1}} = gh_2^{-1}$ and so we have

$$g^{[c_1,c_2]} = g^{c_1^{-1}c_2^{-1}c_1c_2} = gh_1^{-1}h_2^{-1}h_1h_2 = g\,,$$

which implies $[c_1, c_2] \in Z(G)$ and so $C' \leq G' \cap Z(G)$.

Finally, we compute (using $|G/C| = |G'/(G' \cap Z(G))|$ and $C' \leq G' \cap Z(G)$):

$$\begin{aligned}|G : G'| &= |G : C|\,|C : G'| = |G'/(G' \cap Z(G))|\,|C : G'|\\ &= |C : (G' \cap Z(G))| = |C : C'| : |(G' \cap Z(G)) : C'| \leq |C : C'|\,.\end{aligned}$$

This contradicts our assumption, since $C < G$. The proof is complete. □

T. Laffey has proved [Laf1] that if a $p$-group $G$, $p > 2$, satisfies $\mathrm{d}(G) > \mathrm{d}(H)$ for all $H < G$, then $\mathrm{cl}(G) \leq 2$. Note that the minimal nonmetacyclic 2-groups $G$ satisfy the condition $\mathrm{d}(G) > \mathrm{d}(H)$ for all $H < G$ since $\mathrm{d}(G) = 3$ (Theorem 66.1).

**Problem 1.** Study the $p$-groups $G$ satisfying $\mathrm{d}(G) > \mathrm{d}(A)$ for all $A < G$ of class $\leq 2$.

**Problem 2.** Study the $p$-groups $G$ satisfying $\mathrm{d}(G) < \mathrm{d}(H)$ for all $H \in \Gamma_1$.

# § 156 Minimal number of generators of a $p$-group, $p > 2$

We present here some results of T. J. Laffey [Laf1]. To facilitate the proof of the main result (Theorem 156.2), we first prove an easy lemma.

**Lemma 156.1.** *Let $G$ be a $p$-group and $a, b \in G$ so that $[a, b, b] \in Z(G)$. Then:*
(a) $[a, b, b^i] = [a, b, b]^i$,
(b) $[a, b^i] = [a, b]^i [a, b, b]^{\binom{i}{2}}$.

*Proof.* We use induction on $i$.

(a) We have

$$[a, b, b^2] = [a, b, b][a, b, b]^b = [a, b, b]^2 .$$

Assuming that $[a, b, b^i] = [a, b, b]^i$ holds for some $i \geq 2$, we get

$$[a, b, b^{i+1}] = [a, b, bb^i] = [a, b, b^i][a, b, b]^{b^i} = [a, b, b]^i [a, b, b] = [a, b, b]^{i+1} .$$

(b) We have

$$\begin{aligned}[a, b^2] &= [a, b][a, b]^b = [a, b]^2([a, b]^{-1}[a, b]^b) = [a, b]^2[a, b, b] \\ &= [a, b]^2 [a, b, b]^{\binom{2}{1}} .\end{aligned}$$

Assuming that $[a, b^i] = [a, b]^i[a, b, b]^{\binom{i}{2}}$ holds for some $i \geq 2$, we get

$$\begin{aligned}[a, b^{i+1}] &= [a, b \cdot b^i] = [a, b^i][a, b]^{b^i} = [a, b]^i [a, b, b]^{\binom{i}{2}}[a, b]([a, b]^{-1}[a, b]^{b^i})\Box \\ &= [a, b]^{i+1}[a, b, b]^{\binom{i}{2}}[a, b, b^i] = (\text{by } (a)) \\ &= [a, b]^{i+1}[a, b, b]^{\binom{i}{2}+i} = [a, b]^{i+1}[a, b, b]^{\binom{i+1}{2}} .\end{aligned}$$

**Theorem 156.2** ([Laf1]). *Let $G$ be a $p$-group, $p > 2$. Then $G$ has a normal subgroup $K$ such that:*
(a) $d(G) = d(K)$,
(b) $\Phi(K) = K' \leq Z(K)$,
(c) *$K'$ is elementary abelian.*

*Proof.* We use induction on $|G|$. If $G$ is abelian, then we may set $K = \Omega_1(G)$. Thus we may assume that $G$ is nonabelian. Let $N$ be a subgroup of order $p$ in $G' \cap Z(G)$. By induction, $G/N$ has a normal subgroup $L/N$ satisfying $d(G) = d(G/N) = d(L/N)$ and $[L', L] \leq N$. Among all normal subgroups of $G$ contained in $L$ choose one, $K$, of minimal order subject to satisfying the condition $d(K) \geq d(G)$. Such a group $K$ exists, since $L$ itself satisfies this condition. We now show that $K$ satisfies the conclusions of our theorem.

(1) We have $K' = \Phi(K)$. Set $T/K' = \Omega_1(K/K')$. Then $T \trianglelefteq G$ and

$$d(T) \geq d(T/K') = d(K/K') = d(K) \geq d(G) .$$

By minimality of $|K|$, we get $T = K$, $K/K'$ is elementary abelian and so $K' = \Phi(K)$.

(2) We prove that $K' \leq Z(K)$. Assume that $K' \nleq Z(K)$. Since $K \leq L$ and $[L', L] \leq N$, we must have $[K', K] = N \cong C_p$. Because $[K, K', K] = [K', K, K] = \{1\}$, it follows (by the Three Subgroups lemma) $[K, K, K'] = [K', K'] = \{1\}$ and so $K'$ is abelian. If $h \in K'$ and $k \in K$, then $h^k = hn$ with $n \in N$. This gives

$$(h^p)^k = (h^k)^p = (hn)^p = h^p \Rightarrow h^p \in Z(K) .$$

It follows that $K'/(K' \cap Z(K))$ is elementary abelian. Set $C = C_K(K')$ so that $K/C$ is also elementary abelian (noting that $K' = \Phi(K)$, by (1)). Set $p^e = |K'/(K' \cap Z(K))|$ and $p^f = |K/C|$. Since $K/C$ and $K'/(K' \cap Z(K))$ are elementary abelian and each acts faithfully by conjugation on the other, so $|K/C| \leq p^e$ and $|K'/(K' \cap Z(K))| \leq p^f$. This gives

$$(*) \qquad |K/C| = |K'/(K' \cap Z(K))|.$$

(A detailed proof of this fact can be found in the proof of Theorem 155.2.)

Next,

$$[C, K, C] = [K, C, C] = \{1\} \Rightarrow [C, C, K] = \{1\} .$$

Thus $C' \leq K' \cap Z(K)$. Let $x \in K$ and $c \in C$. Since $K$ is of class 3 and $p > 2$, Lemma 156.1 gives

$$[c, x^p] = [c, x]^p [c, x, x]^{\binom{p}{2}} = [c, x]^p .$$

But $x^p \in \Phi(K) = K'$ and so $[c, x^p] = 1$, which gives $[c, x]^p = 1$. On the other hand, by Lemma 156.1,

$$[x, c^p] = [x, c]^p [x, c, c]^{\binom{p}{2}} = [x, c]^p$$

and so $[x, c^p] = 1$, which gives $c^p \in Z(K)$. Next,

$$c^p \in \Phi(K) = K' \Rightarrow \Phi(C) = \mho_1(C)C' \leq K' \cap Z(K) .$$

Using $(*)$, we get

$$|K|/|C| = |K'|/|K' \cap Z(K)|$$

and this gives

$$|K : K'| = |C : (K' \cap Z(K))| = |C : \Phi(C)|/|(K' \cap Z(K)) : \Phi(C)| .$$

Since $\Phi(K) = K'$, this gives $d(K) \leq d(C)$. But $C < K$, $C \trianglelefteq G$ and this contradicts the minimality of $|K|$.

(3) $K'$ is elementary abelian. Let $x, y \in K$. Since $y^p \in K'$, by (1), and $K' \leq Z(K)$, by (2), we have $[x, y]^p = [x, y^p] = 1$.

(4) Finally, we have $d(G) = d(K)$. Indeed, $K/K'$ has a subgroup $V/K' \trianglelefteq G/K'$ with $|V/K'| = p^{d(G)}$. Minimality of $|K|$ forces $V = K$ and thus $d(G) = d(K)$. This completes the proof. □

**Corollary 156.3.** *Let $G$ be a $p$-group, $p > 2$, such that $d(H) < d(G)$ for each proper subgroup $H$ of $G$. Then $G$ is of class $\leq 2$ and $\Phi(G) = G'$ is elementary abelian.*

*Proof.* We see that the normal subgroup $K$ of $G$ from Theorem 156.2 must be equal to $G$. □

The following result generalizes the well-known fact that if a $p$-group $G$ has only one subgroup of order $p$, $p > 2$, then $G$ is cyclic. Moreover, this result essentially generalizes Thompson's Theorem 15.1.

**Corollary 156.4.** *Let $G \neq \{1\}$ be a $p$-group, $p > 2$. Then* $\mathrm{d}(G) \leq \log_p|\Omega_1(G)|$.

*Proof.* Choose $K$ as in Theorem 156.2. Since $K$ is of class $\leq 2$ and $p > 2$, it follows that $K$ is regular (Theorem 7.1 (b)) and so $|K/\mho_1(K)| = |\Omega_1(K)|$ (Theorem 7.2 (d)). Then

$$\mathrm{d}(G) = \mathrm{d}(K) \leq \log_p|K/\mho_1(K)| = \log_p|\Omega_1(K)|\,,$$

completing the proof. □

In particular, if $p > 2$ and $|\Omega_1(G)| \leq p^2$, then $\mathrm{d}(G) \leq 2$. In fact, in that case, as we know (this follows from the results of §§ 7 and 12), $G$ is either metacyclic or a 3-group of maximal class. If $G$ is a 2-group with $|\Omega_1(G)| \leq 2^2$, then $\mathrm{d}(G) \leq 4$ (see § 50), and this estimate is best possible (for example: $G = \mathrm{Q}_{2^m} \times \mathrm{Q}_{2^n}$).

**Problem 1.** Does there exist a connection between $|\Omega_1(G)|$ and $|G : \Phi(G)|$ for 2-groups $G$ provided $\Omega_1(G)$ is elementary abelian? (It is easy to see that the equality $|G : \Phi(G)| = |\Omega_1(G)|^2$ is possible.) Does there exist a 2-group $G$ with elementary abelian $\Omega_1(G)$ satisfying $|G : \Phi(G)| > |\Omega_1(G)|^2$?

In view of Mann's extension of Theorem 15.1 on 2-groups it will be interesting to extend Corollary 156.4 on 2-groups $G$. We assume that, in the case under consideration, $|G : \Phi(\Phi(G))| \leq |\Omega_2(G)|$.

**Problem 2.** Classify the $p$-groups $G$ such that, whenever $A < B \leq G$, then $\mathrm{d}(B) \geq \mathrm{d}(A)$.

**Problem 3.** Classify the $p$-groups $G$ such that $|G : \mho_1(H)| \leq |\Omega_1(H)|$ for all $H \in \Gamma_1$.

# § 157 Some further properties of $p$-central $p$-groups

Recall that a $p$-group $G$ is called $p$-central if either $p > 2$ and $\Omega_1(G) \le Z(G)$ or $p = 2$ and $\Omega_2(G) \le Z(G)$. If $G$ is a $p$-central $p$-group, then we know that $d(G) \le d(Z(G))$ (see Theorem 15.1 and Theorem 15.3). An interesting characterization of $p$-central $p$-groups is given in Theorems 152.1 and 152.2. Here we show that if $G$ is a $p$-central $p$-group, then $\exp(\Omega_i(G)) \le p^i$ for all $i \ge 1$ (Corollary 157.2) and $\exp(G') = \exp(G/Z(G))$ (Theorem 157.6).

All results of this section are taken from [Laf3] and [Laf4]. To get further results concerning $p$-central $p$-groups we need the following very powerful lemma which will be also used in the next section:

**Lemma 157.1.** *Let $G$ be a $p$-group and set $O = \Omega_1(G')$ if $p > 2$ and $O = \Omega_2(G')$ if $p = 2$. We suppose that $O \le Z(G)$ (this is certainly satisfied if $G$ is $p$-central). Then for each subgroup $L$ with $O \le L \le Z(G)$, the quotient group $G/L$ is $p$-central.*

*Proof.* For any prime number $p$ we define the function $\delta$ with $\delta(2) = 2$ and $\delta(p) = 1$ for $p > 2$.

(i) We first show that if $x \in G$ with $x^{p^{\delta(p)}} \in Z(G)$, then $[G, x] \le O$.

Suppose that this does not hold and let $x$ be an element of minimal order $o(x)$ for which (i) fails. Let $n \ge 1$ be the greatest integer for which $[K_n(G), x] \nleq O$ and choose an element $s \in K_n(G)$ with $[s, x] \notin O$. The maximality of $n$ implies $[s, x, x] \in O \le Z(G)$ and so by Lemma 156.1 we get:

$$(*) \qquad 1 = [s, x^{p^{\delta(p)}}] = [s, x]^{p^{\delta(p)}} [s, x, x]^{\binom{p^{\delta(p)}}{2}} ,$$

where we note that in any case $p$ divides $\binom{p^{\delta(p)}}{2}$.

We claim that $[s, x, x]^p = 1$. Indeed, this is clear if $p > 2$, since in this case $O = \Omega_1(G') \le Z(G)$ is elementary abelian. So suppose that $p = 2$. In this case

$$[s, x^2] \in O = \Omega_2(G') \le Z(G)$$

because of minimality of $o(x)$ (for which (i) fails). Lemma 156.1 implies $[s, x^2] = [s, x]^2 [s, x, x]$ and so we compute

$$\begin{aligned} 1 &= [s, x^2, x] = [[s, x]^2 [s, x, x], x] = [[s, x]^2, x] \\ &= [s, x, x]^{[s,x]} [s, x, x] = [s, x, x]^2 . \end{aligned}$$

Hence from $(*)$ we get in any case $1 = [s, x]^{p^{\delta(p)}}$. But this means $[s, x] \in O$, which is a contradiction.

(ii) To prove the lemma, let $yL$ be an element in $G/L$ of order at most $p^{\delta(p)}$. Then $y^{p^{\delta(p)}} \in L \le Z(G)$, so, by (i), $[G, y] \in O \le L$. Hence $G/L$ is $p$-central and we are done. □

**Definition 1.** Let $X$ be a $p$-group. Then we set

$$\Omega(X) = \Omega_1(X) \text{ if } p > 2, \quad \text{and} \quad \Omega(X) = \Omega_2(X) \text{ if } p = 2 .$$

**Corollary 157.2.** *Let $G$ be a $p$-central $p$-group. Then* $\exp(\Omega_i(G)) \le p^i$ *for all* $i \ge 1$.

*Proof.* Set $L = \Omega(G) \le Z(G)$. Then Lemma 157.1 implies that $G/L$ is $p$-central, and so our result follows easily by induction. □

**Lemma 157.3.** *Let $G$ be a $p$-group and $x, y \in G$. Then for any $i \ge 1$,*

$$[y, x^i] = [y, x]^i [y, x, v_1] \dots [y, x, v_r]$$

*for some $v_1, \dots, v_r \in G$.*

*Proof.* We use induction on $i$. One has

$$[y, x^2] = [y, x][y, x]^x = [y, x]^2 \left([y, x]^{-1}[y, x]^x\right) = [y, x]^2 [y, x, x],$$

and we assume that the lemma holds for some $i \ge 2$. Then

$$\begin{aligned}[y, x^{i+1}] &= [y, xx^i] = [y, x^i][y, x]^{x^i} \\ &= [y, x^i][y, x]\left([y, x]^{-1}[y, x]^{x^i}\right) = [y, x^i][y, x][y, x, x^i] \\ &= [y, x]^i [y, x, v_1] \dots [y, x, v_r][y, x][y, x, x^i] \\ &= [y, x]^{i+1}\left([y, x]^{-1} \cdot [y, x, v_1] \dots [y, x, v_r] \cdot [y, x]\right)[y, x, x^i] \\ &= [y, x]^{i+1}[y, x, v_1]^{[y,x]} \dots [y, x, v_r]^{[y,x]}[y, x, x^i].\end{aligned}$$

Now, for any $j \in \{1, \dots, r\}$,

$$\begin{aligned}[y, x, v_j]^{[y,x]} &= \left([y, x, v_j]^{[y,x]}[y, x, v_j]^{-1}\right)[y, x, v_j] \\ &= [y, x, [y, x, v_j]^{-1}][y, x, v_j],\end{aligned}$$

completing the proof. □

**Proposition 157.4.** *Let $G$ be a $p$-central $p$-group. Let $x, y \in G$ and $Q = \langle x, y \rangle$. Then*

$$o([x, y]) \le o(xZ(Q)),$$

*where $o(xZ(Q))$ is the order of the element $xZ(Q)$ in $Q/Z(Q)$.*

*Proof.* Since $G$ is $p$-central, so $Q = \langle x, y \rangle$ is also $p$-central. We note that $o([x, y]) = o([y, x])$ since $[y, x] = [x, y]^{-1}$. Define characteristic subgroups $V_i$ of $Q$ inductively with

$$V_0 = Z(Q) \text{ and } V_{i+1} = \Omega(Q/V_i), \quad i \ge 0$$

(see Definition 1). Applications of Lemma 157.1 imply that all $Q/V_i$ are $p$-central, which gives $V_{i+1}/V_i \le Z(Q/V_i)$ and so the exponent of each $V_i/Q$ is at most $p^i$ if $p > 2$ and $4^i$ if $p = 2$ and that it has exactly this exponent if $V_i \ne Q$.

We now prove our proposition by induction on $o(xZ(Q)) = p^k$, $k \ge 1$. We want to prove that $o([y, x]) \le p^k$ and by induction we may assume that $o([y, x^p]) \le p^{k-1}$ and so $[y, x^p]^{p^{k-1}} = 1$.

By Lemma 157.3,

$$(*) \qquad [y, x^p] = [y, x]^p [y, x, v_1] \dots [y, x, v_r]$$

for some $v_1, \dots, v_r \in Q$. Now, if $s$ is the least positive integer for which $x \in V_s$, then $[y, x] \in V_{s-1}$ and thus, using the above considerations about the exponents, we get $o([y, x]Z(Q)) < p^k$ and so $[y, x]^{p^{k-1}} \in Z(Q)$. Since $o([y, x]Z(Q)) \le p^{k-1}$, it follows (by induction) that for any $v_j$ ($j \in \{1, 2, \dots, r\}$) we get $o([[y, x], v_j]) \le p^{k-1}$ and so $[y, x, v_j]^{p^{k-1}} = 1$. Then the relation $(*)$ implies that $[y, x]^p \in \Omega_{k-1}(Q)$. By Corollary 157.2,

$$\exp(\Omega_{k-1}(Q)) \le p^{k-1} \Rightarrow ([y, x]^p)^{p^{k-1}} = [y, x]^{p^k} = 1 ,$$

and we are done. □

**Proposition 157.5.** *Let $G$ be a $p$-central $p$-group. Let $x, y \in G$ and $Q = \langle x, y \rangle$. Then*

$$o([x, y]) = o(xZ(Q)) .$$

*Proof.* Let $o(xZ(Q)) = p^k$, $k \ge 1$, and we want to show that $o[y, x] = p^k$. If $k = 1$, then Proposition 157.4 implies that $o[y, x] \le p$. However, if $o[y, x] = 1$, then $Q = Z(Q)$ is abelian, which gives $o(xZ(Q)) = 1$, a contradiction. Hence we must have $o[y, x] = p$ and we are done in this case. Therefore, we may assume $k \ge 2$ and we use induction on $o(xZ(Q)) = p^k$. By induction, we may suppose $o([y, x^p]) = p^{k-1}$. By Lemma 157.3, the equality $(*)$ holds. Suppose that $o([y, x]) \le p^{k-1}$. Then

$$\Omega_1(\langle [y, x] \rangle) \le Z(Q) \Rightarrow o([y, x]Z(Q)) \le p^{k-2} .$$

By Proposition 157.4, $o([y, x, v_j]) \le p^{k-2}$ for all $j = 1, \dots r$. Since $o([y, x]^p) \le p^{k-2}$, the above relation $(*)$ gives that $[y, x^p] \in \Omega_{k-2}(Q)$. But then Corollary 157.2 implies $o([y, x^p]) \le p^{k-2}$, contrary to the above assumption. Hence we must have $o([y, x]) > p^{k-1}$ and then Proposition 157.4 implies $o([y, x]) = p^k$ and we are done. □

**Theorem 157.6.** *Let $G$ be a $p$-central $p$-group. Then* $\exp(G') = \exp(G/Z(G))$.

*Proof.* We may assume that $G$ is nonabelian. Set $\exp(G') = p^e$, $e \ge 1$, and also set $\exp(G/Z(G)) = p^f$, $f \ge 1$.

By Corollary 157.2, $\exp(G')$ is the maximum of the orders of simple commutators and so there are $x, y \in G$ such that $o([x, y]) = p^e$. By Proposition 157.5, $p^e = o(xZ(\langle x, y \rangle))$, where $xZ(\langle x, y \rangle)$ is an element in $\langle x, y \rangle / Z(\langle x, y \rangle)$. Hence $p^e$ is the least $p$-power of $x$ such that $x^{p^e}$ commutes with $y$. This gives

$$p^f = \exp(G/Z(G)) \ge p^e \Rightarrow f \ge e .$$

Conversely, let $x \in G - Z(G)$ be such that

$$o(xZ(G)) = p^f = \exp(G/Z(G)) .$$

Then $x^{p^f}$ is the least $p$-power of $x$ such that $x^{p^f} \in Z(G)$. Since $x^{p^{f-1}} \notin Z(G)$, there is $y \in G$ such that $[y, x^{p^{f-1}}] \neq 1$. Considering the subgroup $\langle x, y\rangle$ and the factor-group $\langle x, y\rangle/Z(\langle x, y\rangle)$, Proposition 157.5 implies that $o([x, y]) = o(xZ(\langle x, y\rangle))$. Since $x^{p^{f-1}}$ does not commute with $y$, $x^{p^f}$ is the smallest $p$-power of $x$ such that $x^{p^f}$ commutes with $y$, or equivalently, such that $x^{p^f} \in Z(\langle x, y\rangle)$ and so $o(xZ(\langle x, y\rangle)) = p^f$. Hence

$$o([x, y]) = p^f \Rightarrow p^e = \exp(G') \geq p^f \Rightarrow e \geq f\,.$$

We get $e = f$, completing the proof. □

# § 158 On extraspecial normal subgroups of $p$-groups

Here we give another application of the powerful Lemma 157.1. All results of this section are taken from [Laf4].

**Lemma 158.1.** *Let $G$ be a $p$-group and $E$ an extraspecial normal subgroup of $G$. Then* $C_G(E/Z(E)) = EC_G(E)$.

*Proof.* Set $|E/Z(E)| = p^k$ and let $x_1, \dots, x_k$ be a generating set of $E$. We also set $C = C_G(E/Z(E))$ and note that

$$EC_G(E) \le C\,, \quad |(EC_G(E))/C_G(E)| = p^k\,, \quad [E, C] \le Z(E)\,.$$

Define a map

$$\phi : C \to Z(E) \times Z(E) \times \cdots \times Z(E) \quad (k \text{ times })$$

with

$$\phi(c) = ([x_1, c], [x_2, c], \dots, [x_k, c])\,, \quad (c \in C)\,.$$

For $c_1, c_2 \in C$, we have

$$x_i, c_1c_2] = [x_i, c_2][x_i, c_1]^{c_2} = [x_i, c_1][x_i, c_2]\,, \quad i \in \{1, \dots, k\}\,,$$

and so

$$\phi(c_1c_2) = \phi(c_1)\phi(c_2)\,.$$

Thus, $\phi$ is a homomorphism whose kernel is equal to $C_G(E)$. Hence

$$|C/C_G(E)| \le p^k \Rightarrow C = EC_G(E)\,,$$

and the lemma is proved. □

**Theorem 158.2.** *Let $G$ be a noncyclic $p$-group with $Z(\Omega_1(G))$ cyclic and suppose that*

$$\Omega_1(G') \le Z(G) \text{ if } p > 2 \quad \text{and} \quad \Omega_2(G') \le Z(G) \text{ if } p = 2\,.$$

*Then $G$ is the central product of an extraspecial subgroup $E$ and the cyclic subgroup $Z(G)$, where $E \cap Z(G) = Z(E)$ (such $G$ are termed symplectic). Also, $\exp(E) = p$ if $p > 2$.*

*Proof.* (i) Suppose $p > 2$ so that $\Omega_1(G') \le Z(G)$. Note that $\Omega_1(Z(G)) \le Z(\Omega_1(G))$ and $Z(\Omega_1(G))$ is cyclic, so $Z(G)$ is cyclic. Since $G$ is noncyclic, so $G \ne Z(G)$ and $G$ is nonabelian. But $\Omega_1(G') \le Z(G)$ implies that $G'$ is also cyclic and $\Omega_1(G') \cong C_p$. By Lemma 157.1, $G/\Omega_1(G')$ is $p$-central and so $\Omega_1(G/\Omega_1(G')) \le Z(G/\Omega_1(G'))$, which implies that $\Omega_1(G)$ has class at most 2 and thus it has exponent $p$. If $\Omega_1(G) = \Omega_1(G') \cong C_p$, then $G$ is cyclic, a contradiction. Hence $|\Omega_1(G)| > p$ and since $Z(\Omega_1(G))$ is cyclic, we have $Z(\Omega_1(G)) = \Omega_1(G')$. Also,

$$[G, \Omega_1(G)] \le \Omega_1(G') \Rightarrow (\Omega_1(G))' = \Omega_1(G') \cong C_p$$

and therefore $\Omega_1(G)$ is extraspecial of exponent $p$.

On the other hand,

$$[G, \Omega_1(G)] = \Omega_1(G') = Z(\Omega_1(G))$$

also implies that

$$C_G(\Omega_1(G)/Z(\Omega_1(G))) = G .$$

By Lemma 158.1, $G = \Omega_1(G)C_G(\Omega_1(G))$. Now

$$\Omega_1(C_G(\Omega_1(G))) = Z(\Omega_1(G)) \cong C_p ,$$

so $C_G(\Omega_1(G))$ is cyclic and thus is equal to $Z(G)$, and we are done.

(ii) Suppose now that $p = 2$. We have here $\Omega_2(G') \leq Z(G)$. Note that $\Omega_1(Z(G)) \leq Z(\Omega_1(G))$ and $Z(\Omega_1(G))$ was supposed to be cyclic, so $Z(G)$ is cyclic. Since $G$ is noncyclic, so $G \neq Z(G)$ and therefore $G$ is nonabelian. But $\Omega_2(G') \leq Z(G)$ implies that $G'$ is also cyclic and $\{1\} \neq \Omega_2(G')$ is cyclic of order $\leq 4$.

By Lemma 157.1, $G/\Omega_2(G')$ is 2-central and so, in particular, $[G, \Omega_1(G)] \leq \Omega_2(G')$, and so $\Omega_1(G)$ is of class at most 2. If $\Omega_1(G)$ is abelian, then it is cyclic of order 2, so $G$ is generalized quaternion. But then $\Omega_2(G') \leq Z(G)$ implies $G \cong Q_8$ is quaternion and our theorem holds.

We may assume that $\Omega_1(G)$ is nonabelian. Since $[G, \Omega_1(G)] \leq Z(G)$ and $Z(G)$ is cyclic, we have $(\Omega_1(G))'$ has order 2 and this implies that $\Omega_1(G)$ is if exponent 4. Indeed, set

$$\langle z\rangle = \Omega_1(Z(G)) \leq \Omega_1(G)$$

and let $u, v$ be any two involutions in $\Omega_1(G)$. If $[u, v] \neq 1$, then $\langle u, v\rangle \cong D_8$ since $\Omega_1(G)$ is of class 2. Hence $[u, v]$ is an involution and by the above,

$$[u, v] \in [G, \Omega_1(G)] \leq Z(G) \Rightarrow [u, v] = z .$$

Since $\Omega_1(G)$ is generated by involutions, $\Omega_1(G)/\langle z\rangle$ is abelian and so $(\Omega_1(G))' = \langle z\rangle$ is of order 2. Let $x, y \in \Omega_1(G)$. In that case,

$$[x^2, y] = [x, y]^2 = 1 \Rightarrow x^2 \in Z(\Omega_1(G)) .$$

But $\Omega_1(G)$ is of exponent 4 and so $x^2$ is of order $\leq 2$ and so $x^2 \in \langle z\rangle$ since $Z(\Omega_1(G))$ is cyclic. We have proved that

$$\Phi(\Omega_1(G)) = (\Omega_1(G))' = \langle z\rangle$$

is of order 2. If $Z(\Omega_1(G))$ is of order 2, then $\Omega_1(G)$ is extraspecial and in that case we set $W = \Omega_1(G)$. In that case,

$$[G, \Omega_1(G)] \leq Z(G) \cap \Omega_1(G) = \langle z\rangle ,$$

and so we have $[G, W] = Z(W)$. If $Z(\Omega_1(G))$ is of order $> 2$, then $Z(\Omega_1(G)) \cong C_4$ since $Z(\Omega_1(G))$ is cyclic and $\exp(\Omega_1(G)) = 4$.

Assume that $Z(\Omega_1(G)) \cong C_4$. Then we set $Z(\Omega_1(G)) = \langle v\rangle$, where $v^2 = z$. Let $\{v, u_1, \ldots, u_t\}$, $t \geq 2$, be a minimal generating set for $\Omega_1(G)$, where we may assume that $u_1, \ldots, u_t$ are involutions. Indeed, if for some $i \in \{1, \ldots, t\}$ the generator $u_i$ is of order 4, then $u_i^2 = z$ and we replace $u_i$ with the involution $u_i v$. We set $W = \langle u_i, \ldots, u_t\rangle$ so that $W$ is nonabelian and therefore

$$W' = \langle z\rangle = \Phi(\Omega_1(G)) = \Phi(W)\,,$$

and so $W$ is a maximal subgroup of $\Omega_1(G)$. We get $\Omega_1(G) = W * \langle v\rangle$ with $W \cap \langle v\rangle = \langle z\rangle$ and so $Z(W) = \langle z\rangle$. Hence $W$ is extraspecial. We know that

$$[G, \Omega_1(G)] \leq Z(G) \cap \Omega_1(G) \leq \langle v\rangle\,,$$

and so for any $g \in G$ and $u_i \in W$ $(i \in \{1, \ldots, t\})$, we get $[g, u_i] = h \in \langle v\rangle \cap Z(G)$. Thus, $\langle g, u_i\rangle/\langle h\rangle$ is abelian and so $(\langle g, u_i\rangle)' = \langle h\rangle$ and $\langle g, u_i\rangle$ is of class $\leq 2$. It follows:

$$[g, u_i]^2 = [g, u_i^2] = [g, 1] = 1$$

and so $[g, u_i] \in \langle z\rangle$. We have proved that $[G, W] = Z(W) = \langle z\rangle$ and, in particular, $W$ is normal in $G$.

We have proved that in any case $\Omega_1(G)$ contains an extraspecial subgroup $W$ such that $|\Omega_1(G) : W| \leq 2$, $W$ is normal in $G$ and $[G, W] = Z(W)$. By Lemma 158.1, we get $G = WC_G(W)$. Since $C_G(W) \cap \Omega_1(G) = Z(\Omega_1(G))$, where $Z(\Omega_1(G))$ is cyclic (of order 2 or 4), it follows that $C_G(W)$ has only one involution. It follows that $C_G(W)$ is either cyclic or generalized quaternion. In the first case we are done. In the second case $C_G(W) = Q \cong Q_8$ since (by our assumption) $\Omega_2(G') \leq Z(G)$. But then $G = W * Q$ with $Q \cap W = Z(W)$, and therefore $G$ is extraspecial and so also in this case we are done. □

# § 159 2-groups all of whose cyclic subgroups $A, B$ with $A \cap B \neq \{1\}$ generate an abelian subgroup

This and the following section were written by the second author. These sections contain the solution to Problem 2916.

Heng Lv, Wei Zhou and Dapeng Yu (see [LZY]) have been studying $p$-groups $G$ all of whose cyclic subgroups are of index at most $p^2$ in their normal closure in $G$. The main obstacle in these investigations has been two-generator subgroups $\langle x, y\rangle$ $(x, y \in G)$ such that $\langle x\rangle \cap \langle y\rangle \neq \{1\}$ (see Lemma 144.4 and Lemma 144.5).

On the other hand, if $G$ is a noncyclic $p$-group such that any two distinct maximal cyclic subgroups of $G$ have intersection equal $\{1\}$, then we have either $\exp(G) = p$ or $G \cong D_{2^n}$, $n \geq 3$ (see Exercise 12 in Appendix 45). In fact, it is easy to prove this statement. Assume that $\exp(G) > p$ and let $L < G$ be a maximal cyclic subgroup of order $> p$. Let $L < M \leq G$ with $|M : L| = p$. By Theorem 1.2, we get $p = 2$ and $M$ is dihedral. Thus, all subgroups of $G$ containing $L$ and having order $2|L|$ are dihedral. Suppose that $M < G$. Then $N = N_G(M) > M$ and $N$ normalizes $L$. Let $L_0$ be the cyclic subgroup of order 4 in $L$ so that $C = C_N(L_0)$ covers $N/M$, which gives $C > L$. Let $K/L$ be a subgroup of order 2 in $C/L$. But then $K$ is not dihedral, a contradiction. Hence we have $M = G \cong D_{2^n}$, $n \geq 3$, and we are done.

Therefore, it is of interest to classify $p$-groups $G$ such that the following condition is satisfied:

(∗) Whenever $x, y \in G$ are such that $\langle x\rangle \cap \langle y\rangle \neq \{1\}$, then $\langle x, y\rangle' = \{1\}$.

In the first two Propositions 159.1 and 159.2 are determined $p$-groups $G$ satisfying the condition (∗), where $G$ is either minimal nonabelian or metacyclic and $p$ is any prime. It is clear that the condition (∗) says nothing in the case of $p$-groups of exponent $p$. But also in the case of 2-groups of exponent 4, the condition (∗) says very little, and we characterize such groups in Theorem 159.3.

We recall that a $p$-group $G$ is $p$-central if either $p > 2$ and $\Omega_1(G) \leq Z(G)$ or $p = 2$ and $\Omega_2(G) \leq Z(G)$. Also, we recall that a 2-group $G$ is quasidihedral if it has an abelian maximal subgroup A of exponent $> 2$ and an involution $t \in G - A$, which inverts each element in $A$.

In what follows $G$ will be a 2-group of exponent $> 4$ satisfying the condition (∗). In Theorem 159.4 we determine up to isomorphism all such groups $G$ which do not possess a normal elementary abelian subgroup of order 8. In Theorem 159.5 we classify our groups $G$ with the assumption that $\Omega_1(G) \leq Z(G)$. It turns out that these groups must be 2-central, i.e., we must also have $\Omega_2(G) \leq Z(G)$. In Proposition 159.6 it is shown that if $\Omega_2(G)$ is abelian, then $G$ is also 2-central. Modular 2-groups (i.e., 2-groups that are $D_8$-free) are classified in Theorem 159.7. Then we consider 2-groups $G$ which do not have $D_8$ as a subgroup or equivalently in which $\Omega_1(G)$ is elementary abelian but $\Omega_1(G) \nleq Z(G)$, and we show that such groups are of exponent 4 (Theorem 159.8). Finally, we classify in Theorem 159.9 our groups $G$ in which $\Omega_1(G)$ is nonabelian (i.e., $D_8$ is a subgroup in $G$). The starting point is the fact that in these groups $Z(G)$ must be elementary abelian (see Remark 1) and it turns out that $G$ has a quasidihedral normal subgroup of index at most 4.

**Proposition 159.1.** *Let $G$ be a $p$-group satisfying the condition $(*)$. If $G$ is minimal nonabelian, then $G$ is one of the following groups:*

(1) *If all generating elements (i.e., elements in $G - \Phi(G)$) are not of the same order, then $p = 2$ and either $G \cong D_8$ or $G \cong \mathcal{H}_{16}$ (= the nonmetacyclic minimal nonabelian group of order $2^4$) given with:*

$$\mathcal{H}_{16} = \langle a, t | a^4 = t^2 = 1, [a, t] = z, z^2 = [z, a] = [z, t] = 1 \rangle .$$

(2) *If all generating elements (i.e., elements in $G - \Phi(G)$) are of the same order, then $G$ is one of the groups:*
   (2a) *$G \cong S(p^3)$ (= the nonabelian group of order $p^3$ and exponent $p > 2$);*
   (2b) *$G = \langle a, b \mid a^{p^s} = b^{p^s} = 1, \quad [a, b] = a^{p^{s-1}} \rangle$, where either $p > 2$ and $s > 1$ or $p = 2$ and $s > 2$;*
   (2c) *$G = \langle a, b \mid a^{p^s} = b^{p^s} = 1, \quad [a, b] = c, \quad c^p = [c, a] = [c, b] = 1 \rangle$, where $s > 1$.*

*Proof.* Let $G = \langle a, b \rangle$ be a minimal nonabelian $p$-group satisfying the property $(*)$. Then we must have $\langle a \rangle \cap \langle b \rangle = \{1\}$. In particular, $G \ncong Q_8$. Set $o(a) = p^m$, $o(b) = p^n$ with $m \geq n$.

Suppose $m > n$. Then we have $(ab)^{p^n} = a^{p^n}[b, a]^{\binom{p^n}{2}}$ and so if $p > 2$ or $p = 2$ and $n > 1$, we get $(ab)^{p^n} = a^{p^n} \neq 1$ and $\langle a \rangle \cap \langle ab \rangle \neq \{1\}$. By the property $(*)$, we have $[a, ab] = [a, b] = 1$, a contradiction. Thus, if $m > n$, then we must have $p = 2$ and $n = 1$. If $m > 2$, then we get

$$(ab)^4 = a^4[b, a]^{\binom{4}{2}} = a^4 \neq 1$$

and so $\langle a \rangle \cap \langle ab \rangle \neq \{1\}$ implies again $[a, ab] = [a, b] = 1$, a contradiction. Hence we get $G = \langle a, b \rangle$ with $o(a) = 4$ and $o(b) = 2$. Set $[a, b] = z$. If $z \in \langle a \rangle$, then $\langle a \rangle \trianglelefteq G$ and so $G \equiv D_8$. If $z \notin \langle a \rangle$, then $(\langle a \rangle \times \langle z \rangle) \trianglelefteq G$ and $b \notin \langle a \rangle \times \langle z \rangle$, which implies that $G = (\langle a \rangle \times \langle z \rangle)\langle b \rangle$ is of order $2^4$. But we have $\langle a^2 \rangle \times \langle z \rangle \times \langle b \rangle \cong E_8$ and so $G \cong \mathcal{H}_{16}$ (= the nonmetacyclic minimal nonabelian group of order $2^4$). Conversely, $D_8$ and $\mathcal{H}_{16}$ obviously satisfy the assumption $(*)$.

We may assume that $o(a) = o(b) = p^s$ for each pair of elements $a, b$ that generate $G$. Then we have in the case $s > 1$ that $\exp(\Phi(G)) = p^{s-1}$, where $\Phi(G) = Z(G)$. It follows that all elements in $G - \Phi(G)$ are of the same order $p^s$. If $s = 1$, then we have $G \cong S(p^3)$ (= the nonabelian group of order $p^3$ and exponent $p > 2$). In what follows we also suppose that $s > 1$.

For an element $1 \neq x \in G$, we define the socle of $x$ (or of $\langle x \rangle$) to be the subgroup $\Omega_1(\langle x \rangle)$. Let $X_i/\Phi(G)$ be a subgroup of order $p$ in $G/\Phi(G) \cong E_{p^2}$, $i = 1, \dots p + 1$. Then obviously for a fixed $i$, all elements in $X_i - \Phi(G)$ have the same socle. Let $x_1, \dots, x_{p+1}$ be generating elements in $G - \Phi(G)$ such that $(\Phi(G)\langle x_i\rangle)/\Phi(G)$, $i = 1, \dots p+1$ are $p+1$ pairwise distinct subgroups of order $p$ in $G/\Phi(G)$. Then our group $G$ has the property $(*)$ if and only if $x_1, \dots, x_{p+1}$ have $p+1$ pairwise distinct socles.

Now we use the classification of minimal nonabelian $p$-groups from Exercise 1.8a (see also Lemma 65.1). We already know that $G \not\cong Q_8$. First suppose that $G$ is metacyclic and so in our case we have:

$$G = \langle a, b \mid a^{p^s} = b^{p^s} = 1, \quad a^b = a^{1+p^{s-1}}\rangle ,$$

where $s > 1$. We compute

$$(a^i b)^{p^{s-1}} = (a^{p^{s-1}})^i b^{p^{s-1}} [b, a^i]^{\binom{p^{s-1}}{2}} ,$$

where $i \in \{1, \dots, p-1\}$. Here

$$(\Phi(G)\langle a\rangle)/\Phi(G), \quad (\Phi(G)\langle b\rangle)/\Phi(G), \quad (\Phi(G)\langle a^i b\rangle)/\Phi(G), \quad i \in \{1, \dots, p-1\}$$

are $p+1$ pairwise distinct subgroups of order $p$ in $G/\Phi(G)$. If $p > 2$ or $p = 2$ and $s > 2$, then we have $[b, a^i]^{\binom{p^{s-1}}{2}} = 1$, and so the elements $a, b, a^i b$, $i \in \{1, \dots, p-1\}$, have $p+1$ pairwise distinct socles

$$\langle a^{p^{s-1}}\rangle, \quad \langle b^{p^{s-1}}\rangle, \quad (a^{p^{s-1}})^i \langle b^{p^{s-1}}\rangle, \quad i \in \{1, \dots, p-1\} .$$

Hence these groups have the property $(*)$. However, if $p = 2$ and $s = 2$, then we have

$$G = \left\langle a, b \,\middle|\, a^4 = b^4 = 1, a^b = a^{-1} \right\rangle \cong \mathcal{H}_2$$

and we get

$$(ab)^2 = a^2 b^2 [b, a] = a^2 b^2 a^2 = b^2 \neq 1 ,$$

which implies $[ab, b] = [a, b] = 1$, a contradiction. Hence the group $\mathcal{H}_2$ does not satisfy $(*)$. We have proved that all groups stated in (2b) satisfy $(*)$.

Finally, suppose that $G$ is nonmetacyclic and so here we have:

$$G = \left\langle a, b \,\middle|\, a^{p^s} = b^{p^s} = 1, [a, b] = c, c^p = [c, a] = [c, b] = 1 \right\rangle ,$$

where $s > 1$. We compute

$$(a^i b)^{p^{s-1}} = (a^{p^{s-1}})^i b^{p^{s-1}} [b, a^i]^{\binom{p^{s-1}}{2}} ,$$

where $i \in \{1, \dots, p-1\}$. If $p > 2$ or $p = 2$ with $s > 2$, then $[b, a^i]^{\binom{p^{s-1}}{2}} = 1$ and so the cyclic subgroups generated by elements $a, b, a^i b$, $i \in \{1, \dots, p-1\}$, have $p+1$ pairwise distinct socles

$$\langle a^{p^{s-1}}\rangle, \quad \langle b^{p^{s-1}}\rangle, \quad (a^{p^{s-1}})^i \langle b^{p^{s-1}}\rangle, \quad i \in \{1, \dots, p-1\} .$$

It remains to investigate the case $p = 2$ with $s = 2$. In this case the cyclic subgroups generated by elements $a$, $b$, $ab$ have pairwise distinct socles

$$\langle a^2 \rangle\,, \quad \langle b^2 \rangle\,, \quad \langle a^2 b^2 c \rangle\,,$$

since $(ab)^2 = a^2b^2[b,a] = a^2b^2c$ and $c \notin \langle a^2, b^2 \rangle$ because $\langle a^2, b^2, c \rangle \cong \mathrm{E}_8$. We have proved that all groups stated in (2c) satisfy $(*)$ and we are done. □

**Proposition 159.2.** *Let $G$ be a metacyclic $p$-group satisfying the condition $(*)$. If $G$ is neither abelian nor minimal nonabelian, then $G \cong \mathrm{D}_{2^n}$, $n > 3$.*

*Proof.* Let $G = \langle a, b \rangle$ with $\langle a \rangle \trianglelefteq G$ be a metacyclic $p$-group that is neither abelian nor minimal nonabelian and that has the property $(*)$. Then we have $\langle a \rangle \cap \langle b \rangle = \{1\}$ and $|G'| \geq p^2$ so that $\mathrm{o}(a) \geq p^3$. If $\mathrm{D}_8$ is a subgroup of $G$, then $G$ is a 2-group of maximal class. But $\mathrm{Q}_8$ is not a subgroup of $G$ and so we get $G \cong \mathrm{D}_{2^n}$, $n > 3$.

From now on we assume that $\mathrm{D}_8$ is not a subgroup of $G$. By Proposition 159.1, whenever $M$ is a minimal nonabelian subgroup of $G$, then all elements in $M - \Phi(M)$ are of the same order $p^s$, $s > 1$.

Set $\mathrm{o}(a) = p^m$, $m \geq 3$, and $\mathrm{o}(b) = p^n$, $n \geq 1$. First assume that either $p > 2$ or $p = 2$ and $G$ is ordinary metacyclic. Let $A$ be a maximal abelian subgroup of $G$ containing $\langle a \rangle$ and let $A < X \leq G$ be such that $|X : A| = p$. By Theorem 149.3, $X$ is minimal nonabelian. We have $X = (X \cap \langle b \rangle)\langle a \rangle$. By Proposition 159.1, $|X \cap \langle b \rangle| = \mathrm{o}(a) = p^m$ and since $X < G$, we get $p^n = \mathrm{o}(b) > p^m$ and $n > m$.

Let $B$ be a maximal abelian subgroup of $G$ containing $\langle b \rangle$ and let $B < Y \leq G$ be such that $|Y : B| = p$. By Theorem 149.3, $Y$ is minimal nonabelian. We have $Y = (Y \cap \langle a \rangle)\langle b \rangle$. By Proposition 159.1, $|Y \cap \langle a \rangle| = \mathrm{o}(b) = p^n$ and since $Y < G$, we get $p^m = \mathrm{o}(a) > p^n$ and $m > n$, a contradiction.

It remains to consider the case $p = 2$ and $G$ is not ordinary metacyclic. Then we have $a^b = a^{-1}h$, where $h \in \langle a^4 \rangle$. It follows that $v = a^{2^{m-2}}$ is of order 4 and we see that

$$v^b = \left(a^{2^{m-2}}\right)^b = \left(a^b\right)^{2^{m-2}} = \left(a^{-1}h\right)^{2^{m-2}} = a^{-2^{m-2}} = v^{-1}\,.$$

It follows that $\langle v, b \rangle$ is minimal nonabelian. But $\langle v, b \rangle \cong \mathrm{D}_8$ is excluded and so Proposition 159.1 gives a contradiction and we are done. □

Recall that $\mathcal{H}_2 = \langle x, y \mid x^4 = y^4 = 1\,, \quad x^y = x^{-1} \rangle$.

**Theorem 159.3.** *A 2-group $G$ of exponent 4 satisfies the condition $(*)$ if and only if $G$ does not possess elements $x, y$ of order 4 such that $x^y = x^{-1}$, or equivalently $G$ does not have subgroups isomorphic to $\mathrm{Q}_8$ or $\mathcal{H}_2$.*

*Proof.* Suppose that $G$ satisfies the condition $(*)$ and that there are elements $x, y \in G$ of order 4 such that $x^y = x^{-1}$. If $\langle x \rangle \cap \langle y \rangle \neq \{1\}$, then $\langle x, y \rangle \cong \mathrm{Q}_8$ and if $\langle x \rangle \cap \langle y \rangle = \{1\}$, then $\langle x, y \rangle \cong \mathcal{H}_2$. But both groups $\mathrm{Q}_8$ and $\mathcal{H}_2$ are minimal nonabelian and they do not satisfy the condition $(*)$ (see Proposition 159.1).

Conversely, assume that $Q_8$ and $\mathcal{H}_2$ are not subgroups of $G$. We want to show that $G$ satisfies $(*)$. If not, then there are elements $x, y \in G$ of order 4 such that $\langle x\rangle \cap \langle y\rangle = \langle z\rangle$, where $z$ is an involution and $H = \langle x, y\rangle$ is nonabelian. If $[x, y] = z$, then $\langle x, y\rangle \cong Q_8$, a contradiction. Hence $\langle x\rangle/\langle z\rangle$ and $\langle y\rangle/\langle z\rangle$ are two noncommuting subgroups of order 2 that generate the dihedral group $H/\langle z\rangle$. But $\exp(H/\langle z\rangle) \le 4$, and so $H/\langle z\rangle \cong D_8$ and $|H| = 2^4$. If $|H'| > 2$, then we have $|H'| = 4$ so that Proposition 1.6 (Taussky) implies that $H$ is of maximal class. Since $H$ is generated by two elements of order 4, it follows that $H \cong Q_{16}$. But then $H$ has a subgroup isomorphic to $Q_8$, a contradiction. Hence we have $|H'| = 2$ and so $H$ is minimal nonabelian of order $2^4$ and exponent 4 and $H$ is generated by two elements $x$ and $y$ of order 4 with $\langle x\rangle \cap \langle y\rangle \ne \{1\}$. This implies that $H \cong \mathcal{H}_2$, a contradiction. □

**Theorem 159.4.** *Let $G$ be a 2-group satisfying the condition $(*)$. Suppose that $G$ is neither metacyclic nor minimal nonabelian and that $G$ has no normal elementary abelian subgroup of order 8. Then $G$ is one of the following groups:*

(a) *$G$ is generalized dihedral (quasidihedral) and more precisely, $G$ has an abelian maximal subgroup $N$ of rank 2 and exponent $\ge 8$ and an involution $t \in G - N$ that inverts each element in $N$.*

(b)
$$G = \left\langle a, b, c \;\middle|\; a^{2^n} = b^{2^n} = c^4 = [a, b] = 1, a^c = b, b^c = b^{-1} \right\rangle,$$
*where $n \ge 3$. Here $G$ (of order $2^{2n+2}$) is a splitting and faithful extension of the abelian normal subgroup $\langle a, b\rangle \cong C_{2^n} \times C_{2^n}$ by a cyclic group $\langle c\rangle$ of order 4, where $c^2$ inverts each element in $\langle a, b\rangle$ and $|Z(G)| = 2$.*

(c)
$$G = \left\langle a, b, c \;\middle|\; a^{2^{n+1}} = b^{2^n} = c^4 = [a, b] = 1, a^c = ab, b^c = a^{-2}b^{-1} \right\rangle,$$
*where $n \ge 2$. Here $G$ (of order $2^{2n+3}$) is a splitting and faithful extension of the abelian normal subgroup $\langle a, b\rangle \cong C_{2^{n+1}} \times C_{2^n}$ by a cyclic group $\langle c\rangle$ of order 4, where $c^2$ inverts each element in $\langle a, b\rangle$ and $|Z(G)| = 2$.*

*Conversely, each of the 2-groups in (a), (b) and (c) satisfy the condition $(*)$.*

*Proof.* Since $G$ is not metacyclic, it follows that $G$ is neither abelian nor a 2-group of maximal class and so we may apply Theorem 50.1. Also, $G$ does not have subgroups isomorphic to $Q_8$ or $\mathcal{H}_2$ (Theorem 159.3), and so Theorem 66.1 implies that each minimal nonmetacyclic subgroup of $G$ is elementary abelian of order 8.

By Theorem 50.1, $G$ has a normal metacyclic subgroup $N$ such that $\Omega_2(N)$ is abelian of type $(4, 4)$ or $(4, 2)$, $C_G(\Omega_2(N)) \le N$, and $G/N \ne \{1\}$ is isomorphic to a subgroup of $D_8$.

(1) All elements in $G - N$ are of order $\le 4$.

Indeed, let $g$ be any element in $G - N$ and we consider the subgroup $G_0 = \Omega_2(N)\langle g\rangle$, where $\Omega_2(N) \trianglelefteq G_0$. Set $S = C_{\Omega_2(N)}(g)$ so that $S \le Z(G_0)$ and $S < \Omega_2(N)$. Let $S_1$ be a $G_0$-invariant subgroup in $\Omega_2(N)$ such that $S < S_1$ and $|S_1 : S| = 2$ and let

$x \in S_1 - S$. Then we have $1 \neq [x, g] \in S$ and so $x^2 \in S$ implies

$$1 = [x^2, g] = [x, g]^x [x, g] = [x, g]^2 .$$

Thus $[x, g]$ is an involution that commutes with $x$ and $g$ and so $\langle x, g\rangle' = \langle [x, g]\rangle$. Hence $\langle x, g\rangle$ is minimal nonabelian (see Lemma 65.2 (a)). By Proposition 159.1, the fact that $o(x) \leq 4$ implies that $o(g) \leq 4$ and we are done.

(2) If $X/N$ is any subgroup of order 2 in $G/N$, then $X$ is nonmetacyclic and so there is an involution $t \in X - N$ that centralizes $\Omega_1(N) \cong E_4$, and so $t$ stabilizes the chain $\Omega_2(N) > \Omega_1(N) > \{1\}$, which infers $[\Omega_2(N), t] \leq \Omega_1(N)$.

If this is false, then $X$ is nonabelian and metacyclic. Suppose, in addition, that $X$ is minimal nonabelian. Note that $X$ (of order $\geq 2^4$) is generated by elements in $X - N$ that are all of order $\leq 4$ (by (1)). But Proposition 159.1 implies that there is no such minimal nonabelian subgroup (satisfying $(*)$). Hence $X$ is neither abelian nor minimal nonabelian and then Proposition 159.2 gives that $X \cong D_{2^n}$, $n > 3$. This is not possible since $D_{2^n}$ does not have an abelian subgroup of type $(4, 2)$. We have proved that $X$ is nonmetacyclic. Let $M$ be a minimal nonmetacyclic subgroup of $X$ and then we know that $M \cong E_8$. But then $M \cap N = \Omega_1(N) \cong E_4$, and so $t \in M - N$ is the required involution.

(3) $G/N \neq \{1\}$ is of order $\leq 4$.

Suppose that $G/N \cong D_8$ and let $Nt_1$ and $Nt_2$ be involutory cosets that generate $G/N$, where $t_1$ and $t_2$ are involutions stabilizing the chain $\Omega_2(N) > \Omega_1(N) > \{1\}$ (by (2)). Then we have $\langle t_1, t_2\rangle = D \cong D_8$ so that $G = ND$ with $N \cap D = \{1\}$. In that case $t_1t_2$ is an element of order 4 that stabilizes the chain $\Omega_2(N) > \Omega_1(N) > \{1\}$. But then $(t_1t_2)^2$ is an involution in $G - N$ that centralizes $\Omega_2(N)$, a contradiction.

(4) We have $\exp(N) \geq 8$.

Assume that $\exp(N) = 4$ so that $N = \Omega_2(N)$ is abelian of type $(4, 4)$ or $(4, 2)$ and $\Omega_1(N) \cong E_4$. By (3), we have $G/N \neq \{1\}$ is of order $\leq 4$ and so (by (1)), $\exp(G) = 4$. Let $X/N$ be any subgroup of order 2 in $G/N$. By (2), there is an involution $t \in X - N$ that centralizes $\Omega_1(N)$ and $[N, t] \leq \Omega_1(N)$. Hence $X_0 = \langle t\rangle \times \Omega_1(N) \cong E_8$ and $X_0 \trianglelefteq X$. Now, $nt$ $(n \in N)$ is an involution in $X - N$ if and only if $(nt)^2 = 1$, or equivalently $n^t = n^{-1}$. Hence, if $t$ does not invert any element of order 4 in $N - \Omega_1(N)$, then we have $\Omega_1(X) = X_0$ and so $X_0 \trianglelefteq G$, contrary to our assumption that $G$ has no normal elementary abelian subgroups of order 8. Hence $t$ inverts at least one element $n_0 \in N - \Omega_1(N)$ (of order 4). Suppose that there is an element $v = n't \in X - N$ $(n' \in N)$ of order 4. Then we have

$$n_0^v = n_0^{n't} = n_0^t = n_0^{-1} .$$

But then Theorem 159.3 gives a contradiction. Hence all elements in $X - N$ must be involutions, which implies that $t$ inverts each element in $N$.

We have proved that whenever $X/N$ is a subgroup of order 2 in $G/N$, then all elements in $X - N$ are involutions that invert $N$. This shows that $G/N$ has only one subgroup $X/N$ of order 2 and so $G/N$ is cyclic. If $X = G$, then $X_0 \trianglelefteq G$, contrary to our assumption. Hence we must have $G/N \cong C_4$. Let $g \in G - X$ so that $o(g) = 4$ and $g^2 = t$ is an involution in $X - N$. But $X_0 = \Omega_1(N) \times \langle t \rangle \cong E_8$ is normal in $X$ and the element $g$ normalizes $\Omega_1(N)$ and centralizes $t$, and so $X_0 \trianglelefteq G$. This is a contradiction and we are done.

(5) If $X/N$ is any subgroup of order 2 in $G/N$, then all elements in $X-N$ are involutions that invert each element in $N$, and so $N$ is abelian of rank 2 and exponent $\geq 8$ and $G/N \neq \{1\}$ is cyclic of order $\leq 4$.

By (4), there is an element $a \in N$ of order 8. Then $a^4$ is an involution contained in $\Omega_1(N) \cong E_4$. If $a$ does not centralize $\Omega_1(N)$, then $\langle a, \Omega_1(N) \rangle \cong M_{2^4}$ and this is a minimal nonabelian group, which does not satisfy the condition $(*)$ (Proposition 159.1). Hence $a$ centralizes $\Omega_1(N)$. Suppose that there is an element $v \in X - N$ that is not an involution. By (1), we have $o(v) = 4$, and so $v^2 \in \Omega_1(N)$ and $(va)^2 \in \Omega_1(N)$. We get

$$(va)^2 = vava = v^2(v^{-1}av)a = v^2 a^v a \in \Omega_1(N)$$

and so $a^v = ua^{-1}$ for some $u \in \Omega_1(N)$. This gives

$$(a^2)^v = (a^v)^2 = (ua^{-1})^2 = u^2 a^{-2} = a^{-2}$$

and so the element $v$ of order 4 inverts the element $a^2$ of order 4, contrary to Theorem 159.3. We have proved that all elements in $X - N$ are involutions. This implies that $N$ is abelian of rank 2 and exponent $\geq 8$, $X$ is quasidihedral and $G/N \neq \{1\}$ is cyclic of order $\leq 4$.

(6) If $G/N \cong C_2$, then $G$ is quasidihedral and we have obtained the groups stated in part (a) of our theorem.

(7) If $G/N \cong C_4$, then we have either

$$N \cong C_{2^n} \times C_{2^n} \text{ with } n \geq 3 \quad \text{or} \quad N \cong C_{2^{n+1}} \times C_{2^n} \text{ with } n \geq 2\,.$$

Let $c \in G - N$ be such that $o(c) = 4$ and then $c^2 = t$ is an involution that inverts each element in $N$. We may set $N = \langle a \rangle \times \langle b \rangle$, where $o(a) = 2^m$, $o(b) = 2^n$, $m \geq n$ and $m \geq 3$. Suppose that $m \geq n + 2$. Then $\mho_n(N)$ is a $G$-invariant cyclic subgroup of order $\geq 4$ in $N$. But then $c$ normalizes $\mho_n(N)$ and $c^2 = t$ inverts $\mho_n(N)$, a contradiction. Hence we must have either $m = n$, $n \geq 3$ or $m = n + 1$, $n \geq 2$ and we are done.

From now on we assume that $G/N \cong C_4$. Then we have $G = N\langle c \rangle$, where $c$ is of order 4, $N \cap \langle c \rangle = \{1\}$ and the involution $c^2$ inverts each element in $N$ so that $C_N(c^2) = \Omega_1(N) = N_0 \cong E_4$.

(8) We have $C_G(N_0) = N\langle c^2\rangle$ so that $Z(G) = \langle z\rangle$ ($\leq N_0$) is of order 2 and if $u \in N_0 - \langle z\rangle$, then $u^c = uz$.

Indeed, if $N_0 \leq Z(G)$, then $G/N$ acts faithfully on $\Omega_2(N) \cong C_4 \times C_4$ and $G$ stabilizes the chain $\Omega_2(N) > N_0 > \{1\}$. But then $G/N$ would be elementary abelian, a contradiction.

(9) If $N \cong C_{2^n} \times C_{2^n}$, $n \geq 3$, then we obtain the groups stated in part (b) of our theorem.

Let $u, v$ be the involutions in $N_0 = \Omega_1(N)$ such that

$$u^c = v\,, \quad N_0 = \langle u, v\rangle\,, \quad z = uv \in Z(G)\,.$$

It follows that $u = u^{c^2} = v^c$. Let $a$ be an element of order $2^n \in N$ such that $\langle u\rangle = \Omega_1(\langle a\rangle)$ is the socle of $\langle a\rangle$, where $a^{2^{n-1}} = u$. Set $b = a^c$ so that

$$b^{2^{n-1}} = (a^c)^{2^{n-1}} = \left(a^{2^{n-1}}\right)^c = u^c = v\,,$$

and therefore $\langle v\rangle$ is the socle of $\langle b\rangle$. It follows that $N = \langle a\rangle \times \langle b\rangle$. Noting that $c^2$ inverts each element in $N$, we get $a^{-1} = a^{c^2} = (a^c)^c = b^c$. The structure of $G$ is determined.

(10) If $N \cong C_{2^{n+1}} \times C_{2^n}$, $n \geq 2$, then we obtain the groups stated in part (c) of our theorem.

We may set

$$N = \left\langle a, b' \;\middle|\; a^{2^{n+1}} = (b')^{2^n} = [a, b'] = 1\right\rangle\,,$$

where $n \geq 2$. Set $z = a^{2^n}$ and $u = (b')^{2^{n-1}}$ so that $(a^2b')^{2^{n-1}} = zu$. We have $\Phi(N) = \langle a^2\rangle \times \langle (b')^2\rangle$ and therefore $\exp(\Phi(N) = 2^n$. Since $\Phi(N)\langle b'\rangle = \Omega_n(N)$, it follows that $\Phi(N)\langle b'\rangle$ is $G$-invariant. Set

$$S = \Omega_{n-1}(\Phi(N)) = \langle a^4\rangle \times \langle (b')^2\rangle\,.$$

Since $\mho_n(N) = \langle z\rangle$, we have $\langle z\rangle = Z(G)$, and so $u^c = uz$. Thus $c$ maps all elements in $Sb'$ (that generate all cyclic subgroups of order $2^n$ in $N$ that have the socle $\langle u\rangle$) onto the elements in $Sa^2b'$ (that generate all cyclic subgroups of order $2^n$ in $N$ that have the socle $\langle uz\rangle$) and all element in $\Phi(N)b'$ are of order $2^n$.

There are three involutory cosets in the $N$ modulo $\Phi(N)$:

$$b'\Phi(N)\,, \quad a\Phi(N)\,, \quad ab'\Phi(N)\,,$$

and since all elements in $N - (\Phi(N)\langle b'\rangle)$ are of order $2^{n+1}$, it follows that these elements generate cyclic subgroups that all have the socle $\langle z\rangle$. Note also that (by the previous paragraph) we have $(b')^c = sa^2b'$ for some $s \in S$.

Assume (by way of contradiction) that $c$ fixes the cosets $a\Phi(N)$ and $ab'\Phi(N)$. Then we get $a^c = aa^{2i}(b')^{2j}$ for some integers $i, j$, where $a^{2i}(b')^{2j} \in \Phi(N)$. Since $c^2$ inverts each element in $N$, this gives

$$\begin{aligned} a^{-1} = a^{c^2} &= (aa^{2i}(b')^{2j})^c = a^c(a^c)^{2i}((b')^c)^{2j} \\ &= aa^{2i}(b')^{2j} \cdot a^{2i}a^{4i^2}(b')^{4ij} \cdot s^{2j}a^{4j}(b')^{2j}\,, \end{aligned}$$

and so we get

$$a^{-2} = a^{4i+4i^2+4j}(b')^{4j+4ij}s^{2j},$$

where the right hand side is an element in $S = \langle a^4 \rangle \times \langle (b')^2 \rangle$ but the left hand side is $a^{-2} \in \Phi(N) - S$, a contradiction.

We have proved that $c$ must interchange the cosets $a\Phi(N)$ and $ab'\Phi(N)$. Hence we get $a^c = ab'f$, where $f \in \Phi(N)$. We set $b = b'f \in b'\Phi(N)$ and so $o(b) = 2^n$ and the socle $\Omega_1(\langle b \rangle)$ is equal to $\langle u \rangle$ or $\langle uz \rangle$. Thus we have $\langle a \rangle \cap \langle b \rangle = \{1\}$, and so $N = \langle a \rangle \times \langle b \rangle$ and $a^c = ab$. This gives

$$a^{-1} = a^{c^2} = (a^c)^c = (ab)^c = a^c b^c = abb^c,$$

which gives finally $b^c = a^{-2}b^{-1}$ and the structure of $G$ is determined.

Conversely, we have only to check if the groups from parts (b) and (c) of our theorem satisfy the condition $(*)$. Set $N = \langle a, b \rangle$ and $N_0 = \Omega_1(N) \cong E_4$. If $(*)$ does not hold, then there are elements $c_1, c_2 \in G - N$ of order 4 such that $\langle c_1 \rangle \cap \langle c_2 \rangle = \langle t \rangle$, where $t \in G - N$ is an involution and $\langle c_1, c_2 \rangle$ is nonabelian. But $\langle c_1, c_2 \rangle$ is contained in $C_G(t) = N_0\langle c_1 \rangle$, where $N_0 \cap \langle c_1 \rangle = \{1\}$ and so $N_0\langle c_1 \rangle$ is minimal nonabelian (because $c_1$ does not centralize $N_0$) and is isomorphic to $\mathcal{H}_{16}$ from Proposition 159.1. Hence $\mathcal{H}_{16}$ satisfies the condition $(*)$, a contradiction. Thus all groups in our theorem satisfy the condition $(*)$ and our theorem is proved. □

**Theorem 159.5.** *Let $G$ be a nonabelian 2-group of exponent $2^e$, $e \geq 2$, satisfying the condition $(*)$. Assume that $\Omega_1(G) \leq Z(G)$. Then we have $\Omega_{e-1}(G) \leq Z(G)$, and so $G$ is of class 2 and $G'$ is elementary abelian. If $e > 2$, then $G$ is 2-central, i.e., $\Omega_2(G) \leq Z(G)$.*

*Conversely, if $G$ is a nonabelian 2-group of exponent $2^e$, $e \geq 3$, such that $\Omega_{e-1}(G) \leq Z(G)$, then $G$ satisfies the condition $(*)$.*

*Proof.* We may assume $e > 2$. Let $x, y \in G$ be such that $x^2 = y^2$. If $x^2 = 1$, then $x, y \in \Omega_1(G) \leq Z(G)$ and so $[x, y] = 1$. If $x^2 \neq 1$, then $\langle x \rangle \cap \langle y \rangle \neq \{1\}$ and our condition $(*)$ implies $[x, y] = 1$. Hence our 2-group $G$ has commuting roots, which together with $\Omega_1(G) \leq Z(G)$ shows that $G$ is an (X)-group according to Definition 1 in § 153.

By Theorem 153.4 (a) (A. Mann), $G/\Omega_1(G)$ is also an (X)-group. In particular, we have

$$\Omega_1(G/\Omega_1(G)) = L/\Omega_1(G) \leq Z(G/\Omega_1(G)).$$

Hence $L/\Omega_1(G)$ is elementary abelian and $[G, L] \leq \Omega_1(G)$. All elements in $L - \Omega_1(G)$ are of order 4 and if $h \in G$ is of order 4 then $h^2 \in \Omega_1(G)$, and so $h\Omega_1(G) \in \Omega_1(G/\Omega_1(G))$, which gives $h \in L$. We have proved that $L = \Omega_2(G)$ and so $\Omega_2(G)$ is of exponent 4 and $[G, \Omega_2(G)] \leq \Omega_1(G)$. In particular, we have $\Omega_2(G) < G$.

Suppose that $\Omega_2(G) \nleq Z(G)$. Then there are $h \in \Omega_2(G) - \Omega_1(G)$ and $g \in G - \Omega_2(G)$ such that $[g, h] = z \neq 1$. Since $[G, \Omega_2(G)] \leq \Omega_1(G) \leq Z(G)$, we see that $z$ is a central involution. Hence $\langle g, h \rangle' = \langle z \rangle$ is of order 2 and Lemma 65.2 (a) gives that $\langle g, h \rangle$ is minimal nonabelian. On the other hand, we have $o(h) = 4$ and $o(g) \geq 8$, which contradicts Proposition 159.1. We have proved that $\Omega_2(G) \le Z(G)$ and so $G$ is 2-central. By Corollary 157.2, we have $\exp(\Omega_i(G)) \le 2^i$ for all $i \ge 1$.

We want to prove that $\Omega_{e-1}(G) \le Z(G)$. Using induction on $i$, assume that $\Omega_i(G) \le Z(G)$, where $i \ge 2$ and $\Omega_{i+1}(G) < G$. By Lemma 157.1, $G/\Omega_i(G)$ is 2-central. In particular, we have

$$\Omega_{i+1}(G)/\Omega_i(G) \le Z(G/\Omega_i(G)) .$$

Assuming that $\Omega_{i+1}(G) \nleq Z(G)$, we see that there are elements $h \in \Omega_{i+1}(G) - \Omega_i(G)$ and $g \in G - \Omega_{i+1}(G)$ such that $[g, h] \ne 1$. We have $[g, h] \in \Omega_i(G) \le Z(G)$, and so $\langle g, h\rangle$ is of class 2. Next,

$$h^2 \in \Omega_i(G) \le Z(G) \Rightarrow 1 = [g, h^2] = [g, h]^2$$

and therefore $[g, h]$ is an involution. It follows that $\langle g, h\rangle' = \langle [g, h]\rangle$ is of order 2 and so Lemma 65.2 (a) implies that $\langle g, h\rangle$ is minimal nonabelian. But $o(h) = 2^{i+1}$ and $o(g) \ge 2^{i+2}$, where $i \ge 2$, and this contradicts Proposition 159.1. Hence we have $\Omega_{i+1}(G) \le Z(G)$ and therefore (by induction) we have proved that $\Omega_{e-1}(G) \le Z(G)$. In particular, $G$ is of class 2 and so for any $x, y \in G$ we get $[x, y]^2 = [x^2, y] = 1$ because $x^2 \in \Omega_{e-1}(G) \le Z(G)$, and so $G'$ is elementary abelian.

Conversely, let $G$ be a nonabelian 2-group of exponent $2^e$, $e \ge 3$, such that $\Omega_{e-1}(G) \le Z(G)$. In particular, we have $\Omega_2(G) \le Z(G)$ and so $G$ is 2-central. By Corollary 157.2, we have $\exp(\Omega_i(G)) \le 2^i$ for all $i \ge 1$. Therefore, all elements in $G - Z(G)$ are of order $2^e$ and $G' \le Z(G)$ so that $G$ is of class 2. Also, for any $x, y \in G$ we get $[x, y]^2 = [x^2, y] = 1$, and so $G'$ is elementary abelian. Let $x, y$ be any noncommuting elements in $G$. Then $o(x) = o(y) = 2^e$ and $\langle x\rangle$ and $\langle y\rangle$ have distinct socles $\Omega_1(\langle x\rangle)$ and $\Omega_1(\langle y\rangle)$ so that the condition $(*)$ holds. Indeed, assume that $x^{2^{e-1}} = y^{2^{e-1}}$. This gives

$$(xy)^{2^{e-1}} = x^{2^{e-1}} y^{2^{e-1}} [y, x]^{\binom{2^{e-1}}{2}} = x^{2^{e-1}} y^{2^{e-1}} = 1$$

and so $xy \in \Omega_{e-1}(G) \le Z(G)$ and $y = x^{-1}f$ for some $f \in Z(G)$. But then

$$[x, y] = [x, x^{-1}f] = 1 ,$$

a contradiction. Our theorem is proved. □

**Proposition 159.6.** *Let $G$ be a nonabelian 2-group satisfying the condition $(*)$. Assume that $\Omega_2(G)$ is abelian. Then $G$ is 2-central.*

*Proof.* Let $x, y \in G$ be such that $x^4 = y^4$. If $x^4 = 1$, then $x, y \in \Omega_2(G)$, and so $[x, y] = 1$. If $x^4 \ne 1$, then $\langle x\rangle \cap \langle y\rangle \ne \{1\}$ and our condition $(*)$ implies $[x, y] = 1$. Then, by Mann's Theorem 152.2, $G$ is 2-central and we are done. □

**Theorem 159.7.** *Let $G$ be a nonabelian modular 2-group satisfying the condition $(*)$. Then $\exp(G) = 2^e$, $e \ge 3$, $G$ is 2-central with an abelian maximal subgroup $A$ so that for each element $t \in G - A$, we have $a^t = a^{1+2^{e-1}}$ for all $a \in A$. We have $Z(G) = \Omega_{e-1}(G) < A$, $G' \le Z(G)$, $G' \cong A/Z(G)$ is elementary abelian (of order $\ge 2$) and each minimal nonabelian subgroup $M$ of $G$ is metacyclic of order $2^{2e}$ and is isomorphic to*

$$M \cong \left\langle a, b \;\middle|\; a^{2^e} = b^{2^e} = 1,\ a^b = a^{1+2^{e-1}} \right\rangle ,$$

*where $a \in A - Z(G)$ and $b \in G - A$.*

*Proof.* Let $G$ be a nonabelian 2-group with the property $(*)$ that is modular (i.e., $G$ is $D_8$-free). By Proposition 159.1, $Q_8$ is not a subgroup of $G$, and so $G$ is not Hamiltonian.

By Proposition 73.5, $\Omega_2(G)$ is abelian and so Proposition 159.6 implies that $G$ is 2-central of exponent $2^e$, $e \geq 3$. Hence Theorem 159.5 is applicable and we get $\Omega_{e-1}(G) \leq Z(G)$, $G/\Omega_{e-1}(G)$ is elementary abelian of order $\geq 4$ and $G'$ is elementary abelian.

Now we use Iwasawa's Theorem 73.15 characterizing modular $p$-groups that are not Hamiltonian. Hence our 2-group $G$ contains an abelian normal subgroup $N$ with a cyclic quotient group $G/N$ and there exists an element $t \in G$ such that $G = \langle N, t\rangle$ and a positive integer $s \geq 2$ such that $a^t = a^{1+2^s}$ for all $a \in N$. Since $G'$ is elementary abelian, we must have $s = e-1$ and $t \in G - \Omega_{e-1}(G)$ with $o(t) = 2^e$. But $t^2 \in \Omega_{e-1}(G) \leq Z(G)$, and so $A = N\langle t^2\rangle$ is an abelian maximal subgroup of $G$ and we see that for each $a \in A$ we also have $a^t = a^{1+2^{e-1}}$. For each $x \in A - \Omega_{e-1}(G)$, we have $x^t = x^{1+2^{e-1}} = xx^{2^{e-1}}$, where $x^{2^{e-1}}$ is an involution and so $x \notin Z(G)$. Hence $Z(G) = \Omega_{e-1}(G)$. By the proof of Lemma 1.1, we have $G' \cong A/Z(G)$ and our result follows. □

**Theorem 159.8.** *Let $G$ be a 2-group satisfying the condition $(*)$ and assume that $\Omega_1(G)$ is elementary abelian (i.e., $D_8$ is not a subgroup in $G$) but $\Omega_1(G) \nleq Z(G)$. Then $G$ is of exponent 4.*

*Proof.* Let $A$ be a maximal normal abelian subgroup of $G$ containing $E = \Omega_1(G)$. By our assumption we have $C = C_G(E) < G$.

First we show that all elements in $G - C$ are of order 4. Indeed, let $x \in G - C$ and then we consider the subgroup $G_0 = E\langle x\rangle$, where $E \trianglelefteq G_0$ and $E_0 = C_E(x) < E$. Note that $E_0 \leq Z(G_0)$ and let $E_1$ be a $G_0$-invariant subgroup such that $E_0 < E_1 \leq E$ and $|E_1 : E_0| = 2$. Take an element $e \in E_1 - E_0$ so that $1 \neq [e, x] \in E_0 \leq Z(G_0)$ and $e^2 \in E_0$ implies $1 = [e^2, x] = [e, x]^2$. Hence $\langle e, x\rangle' = \langle [e, x]\rangle$ is of order 2 and so, by Lemma 65.2 (a), $\langle e, x\rangle$ is minimal nonabelian. Then Proposition 159.1 gives $o(x) \leq 4$ and therefore we get here $o(x) = 4$.

Suppose by way of contradiction that $\exp(G) = 2^e$ with $e \geq 3$. By the previous paragraph, we have $\exp(C) = 2^e$. We claim that we also have $\exp(A) = 2^e$. This is clear if $A = C$, so we may assume $A < C$. In that case $C$ is nonabelian of exponent $2^e$, $e \geq 3$, and

$$E = \Omega_1(G) = \Omega_1(C) \leq Z(C)$$

so that we may use Theorem 195.5. It follows that $C$ is 2-central and $\Omega_{e-1}(C) \leq Z(C)$. Since $A$ is self centralizing, we have $\Omega_{e-1}(C) < A$ and so $A$ is of exponent $2^e$ and we are done.

Let $a$ be an element of order 8 in $A$ and let $x \in G - C$ so that $o(x) = 4$ and $x^2 \in E$. Since $ax \in G - C$, $ax$ is also of order 4 and $(ax)^2 \in E = \Omega_1(G)$. Then we have

$$(ax)^2 = axax = ax^2(x^{-1}ax) = ax^2a^x = x^2(aa^x) \in E\,,$$

and so $aa^x = e \in E$, which implies $a^x = a^{-1}e$ and

$$(a^2)^x = (a^x)^2 = (a^{-1}e)^2 = a^{-2},$$

and so the element $x$ of order 4 inverts the other element $a^2$ of order 4. But then we have $\langle a^2, x\rangle \cong Q_8$ or $\mathcal{H}_2$ (see Theorem 159.3), which is a contradiction. Hence $G$ is of exponent 4 and our theorem is proved. □

**Remark 1.** Let $G$ be a 2-group satisfying the condition $(*)$ and assume that $G$ possesses a subgroup

$$D = \langle v, t \mid v^4 = t^2 = 1, \quad v^t = v^{-1}\rangle \cong D_8 .$$

Then $C_G(D)$ is elementary abelian and so, in particular, $Z(G)$ is elementary abelian. Indeed, we set $z = v^2$ and $C = C_G(D)$ so that $C \cap D = \langle z\rangle$ and $Z(G) \le C$. Let $w$ be an element of order 4 in $C$. If $w^2 = z$, then $\langle v, tw\rangle \cong Q_8$. If $w^2 \neq z$, then $\langle v, tw\rangle \cong \mathcal{H}_2$. In both cases we have a contradiction (see Theorem 159.3). Hence $C$ is elementary abelian.

**Theorem 159.9.** *Let $G$ be a 2-group of exponent $\ge 8$ satisfying the condition $(*)$ and assume that $\Omega_1(G)$ is nonabelian (i.e., $D_8$ is a subgroup in $G$). Then $G$ has a maximal normal abelian subgroup $A$ of exponent $\ge 8$ such that $G/A \cong C_2$, $C_4$ or $Q_8$ and if $H/A$ is the subgroup of order 2 in $G/A$, then $H$ is quasidihedral (where all elements in $H - A$ are involutions) and we have $C_G(\Omega_1(A)) = H$.*

*Conversely, the above groups satisfy the condition $(*)$ unless $G/A \cong Q_8$ and $G$ has a subgroups $S$ isomorphic to $Q_8$ or $\mathcal{H}_2$ (see Theorem 159.3) so that $G = AS$.*

*Proof.* Since $D_8$ is a subgroup of $G$, Remark 1 implies that $Z(G)$ is elementary abelian. Let $E$ be a maximal normal elementary abelian subgroup of $G$ so that $Z(G) \le E$. If $E$ is also a maximal normal abelian subgroup of $G$, then $Z(G) < E$ and Lemma 57.1 implies that for each $x \in G - E$, there is $e \in E - Z(G)$ such that $\langle x, e\rangle$ is minimal nonabelian. By Proposition 159.1, we have $o(x) \le 4$. But then $G$ is of exponent 4, which was excluded by our assumption.

Let $A$ be a maximal normal abelian subgroup of $G$ that contains $E$. By the above, we have $E < A$ and $E = \Omega_1(A)$. Let $F$ be a $G$-invariant subgroup such that $E < F \le A$ and $|F : E| = 2$. Note that all elements in $F - E$ are of order 4. Set $C = C_G(F)$ so that $C \trianglelefteq G$ and $C < G$. By Remark 1, $D_8$ is not a subgroup of $C$ and so $\Omega_1(C) \trianglelefteq G$ is elementary abelian. The maximality of $E$ implies $E = \Omega_1(C)$.

Next, we show that all elements in $G - C$ are of order $\le 4$. Indeed, let $x \in G - C$ and we consider the subgroup $G_0 = F\langle x\rangle$, where $F \trianglelefteq G_0$. Set $F_0 = C_F(x) < F$ so that $F_0 \le Z(G_0)$ and let $F_1$ be a $G_0$-invariant subgroup such that $F_0 < F_1 \le F$ and $|F_1 : F_0| = 2$. Take an element $f \in F_1 - F_0$. Then we have

$$1 \neq [f, x] \in F_0 \le Z(G_0)$$

and $f^2 \in F_0$ implies $1 = [f^2, x] = [f, x]^2$ so that $\langle f, x\rangle' = \langle [f, x]\rangle$ is of order 2. Then Lemma 65.2 (a) gives that $\langle f, x\rangle$ is minimal nonabelian with $o(f) \le 4$. By Proposition 159.1, we get $o(x) \le 4$ and we are done.

Since $\exp(G) \ge 8$, we must have $\exp(C) \ge 8$. Suppose in addition that $\exp(A) = 4$. Then we have $A < C$, and so $C$ is nonabelian of exponent $\ge 8$ and $\Omega_1(C) = E \le Z(C)$. By Theorem 159.5, $C$ is 2-central, and so $A \le \Omega_2(C) \le Z(C)$, contrary to the fact that $A$ is self centralizing. Hence we have $\exp(A) \ge 8$. Let $H/C$ be any subgroup of order 2 in $G/C$. Suppose that there is an element $v$ of order 4 in $H - C$. Then we have $v^2 \in \Omega_1(C) = E$. Let $a$ be an element of order 8 in $A$. Then $va \in H - C$, and so $(va)^2 \in E$. We get

$$(va)^2 = vava = v^2(a^v a) \in E\,,$$

and so

$$a^v a = e \in E\,, \quad a^v = ea^{-1}\,, \quad (a^2)^v = (ea^{-1})^2 = a^{-2}\,,$$

which gives $\langle a^2, v\rangle \cong Q_8$ or $\mathcal{H}_2$ (see Theorem 159.3), a contradiction. Hence all elements in $H - C$ are involutions that invert each element in $C$. It follows that $A = C$ is abelian of exponent $\ge 8$ and so $H$ is quasidihedral. This also shows that $G/C$ has exactly one subgroup $H/A$ of order 2. If $G = H$, then $G/A \cong C_2$ and $G$ is quasidihedral.

Assume that $G > H$. Since $\exp(G/A) \le 4$, we must have either $G/A \cong C_4$ or $G/A \cong Q_8$. Let $g$ be an element of order 4 in $G - A$ and suppose that $g$ centralizes $E$. Then $\langle g\rangle$ stabilizes the chain $F > E > \{1\}$ and the involution $g^2 \in H - A$ would centralize $F$, a contradiction. This proves that $C_G(E) = H$ and we have obtained the groups stated in our theorem.

Conversely, we want to show that all obtained groups in our theorem satisfy the condition $(*)$ with only few exceptions. Indeed, let $K/A$ be a cyclic subgroup of order 4 in $G/A$. If $K$ does not satisfy $(*)$, then there are elements $v, w$ of order 4 in $K - A$ such that $v^2 = w^2 = i$ is an involution in $H - A$ and $\langle v, w\rangle$ is nonabelian. Our group $\langle v, w\rangle$ is contained in $C_K(i) = E\langle v\rangle$ with $E \cap \langle v\rangle = \{1\}$ and $E \times \langle i\rangle$ is an elementary abelian maximal subgroup of $E\langle v\rangle$ so that $\exp(E\langle v\rangle) = 4$. Since $E\langle v\rangle$ does not satisfy $(*)$, it follows by Theorem 159.3 that $Q_8$ or $\mathcal{H}_2$ must be subgroups of $E\langle v\rangle$. This is not possible since neither $Q_8$ nor $\mathcal{H}_2$ possess an elementary abelian maximal subgroup. Hence our group $K$ satisfies the condition $(*)$.

It remains to consider the case $G/A \cong Q_8$. Suppose that $G$ does not satisfy $(*)$. Then there are elements $v_1, v_2 \in G - A$ of order 4 such that $G = \langle v_1, v_2\rangle A$, $v_1^2 = v_2^2 = i$ is an involution in $H - A$, $\langle v_1, v_2\rangle$ is nonabelian and is contained in $C_G(i)$. We have $C_A(i) = E = \Omega_1(A)$ and $C_G(i)$ covers $G/A$. Thus

$$X = C_G(i) > E \quad \text{so that } X/E \cong Q_8 \text{ and } L = \Omega_1(X) = E \times \langle i\rangle\,.$$

Since $X$ is of exponent 4 and $X$ does not satisfy $(*)$, it follows by Theorem 159.3 that $X$ contains a subgroup $S$ that is isomorphic to $Q_8$ or $\mathcal{H}_2$. Since $S \cap L \le \Omega_1(S) = \Phi(S)$, it follows that $S$ covers $X/E$ and so $X = ES$ and $G = AS$. Our theorem is proved. □

# § 160 *p*-groups, *p* > 2, all of whose cyclic subgroups *A*, *B* with $A \cap B \neq \{1\}$ generate an abelian subgroup

We continue an investigation of $p$-groups $G$ for $p > 2$ that satisfy the following condition:

(∗) Whenever $x, y \in G$ are such that $\langle x\rangle \cap \langle y\rangle \neq \{1\}$, then $\langle x, y\rangle$ is abelian.

It is clear that the condition (∗) says nothing in case of $p$-groups of exponent $p$ and therefore we shall always assume that the exponent is $> p$.

In Theorem 160.1 we classify nonabelian $p$-groups $G$ with $p > 2$ and exponent $> p$ that satisfy the condition (∗) and $\Omega_1(G)$ is abelian. It turns out that in this case $G$ is $p$-central (i.e., $\Omega_1(G) \leq Z(G)$) of class 2 and $G'$ is elementary abelian.

In Theorem 160.5 we classify $p$-groups $G$ with $p > 2$ and exponent $> p$ that satisfy the condition (∗) and $\Omega_1(G)$ is nonabelian. It turns out that in this case $G$ has a proper subgroup $G_0$ such that $G_0 = H_p(G)$ and $G_0$ is either abelian or a group from Theorem 160.1. Hence in the second case, $G_0$ is $p$-central of class 2 and $G_0' \leq \Omega_1(G_0)$ is elementary abelian. We have as a rule $|G : G_0| = p$ and the case $|G : G_0| > p$ is possible only in the case when $\exp(G) = p^2$ and $\{1\} \neq \mho_1(G_0) \leq G_0'$.

The converse will be proved: that all $p$-groups described in Theorems 160.1 and 160.5 satisfy the condition (∗).

**Theorem 160.1.** *Let $G$ be a nonabelian $p$-group with $p > 2$ and exponent $p^e$, $e \geq 2$, which satisfies the condition (∗) and we assume that $\Omega_1(G)$ is abelian (and so $\Omega_1(G)$ is elementary abelian). Then $G$ is $p$-central, i.e.,*

$$\Omega_1(G) \leq Z(G), \quad \Omega_{e-1}(G) \leq Z(G),$$

*and $G' \leq \Omega_1(G)$ is elementary abelian.*

*Conversely, if $G$ is a nonabelian $p$-group, $p > 2$, of exponent $p^e$, $e \geq 2$, such that $\Omega_{e-1}(G) \leq Z(G)$, then $G$ satisfies the condition (∗).*

*Proof.* Let $x, y \in G$ be such that $x^p = y^p$. If $x^p = 1$, then $x, y \in \Omega_1(G)$ and therefore $[x, y] = 1$. If $x^p \neq 1$, then $\langle x\rangle \cap \langle y\rangle \neq \{1\}$, and so the condition (∗) implies again $[x, y] = 1$. By Theorem 152.1, $G$ is $p$-central, i.e., $\Omega_1(G) \leq Z(G)$. Then, by Corollary 157.2, $\exp(\Omega_i(G)) \leq p^i$ for all $i \geq 1$.

We want to show that, in fact, $\Omega_{e-1}(G) \leq Z(G)$. Proving this by induction, we assume that $\Omega_i(G) \leq Z(G)$ for an $i \geq 1$ and $\Omega_{i+1}(G) < G$ (i.e., $i < e-1$). By Lemma 157.1, $G/\Omega_i(G)$ is $p$-central. Hence

$$\Omega_{i+1}(G)/\Omega_i(G) \leq Z(G/\Omega_i(G)) \quad \text{and so} \quad [G, \Omega_{i+1}(G)] \leq \Omega_i(G).$$

Suppose that $\Omega_{i+1}(G) \nleq Z(G)$. Noting that $\langle G - \Omega_{i+1}(G)\rangle = G$, we see that there are elements

$$g \in G - \Omega_{i+1}(G)\,, \; h \in \Omega_{i+1}(G) - \Omega_i(G) \quad \text{such that } 1 \neq [g,h] \in \Omega_i(G) \leq Z(G)\,.$$

Thus,

$$\langle g,h\rangle' = \langle [g,h]\rangle \Rightarrow \mathrm{cl}(\langle g,h\rangle) = 2\,.$$

Next,

$$h^p \in \Omega_i(G) \leq Z(G) \Rightarrow 1 = [g,h^p] = [g,h]^p \Rightarrow |\langle g,h\rangle'| = p\,.$$

By Lemma 65.2 (a), $\langle g,h\rangle$ is minimal nonabelian with $o(g) \geq p^{i+2}$ and $o(h) = p^{i+1}$, where $i \geq 1$. By Proposition 159,1, this is a contradiction. We have proved that $\Omega_{i+1}(G) \leq Z(G)$, and so the induction gives $\Omega_{e-1}(G) \leq Z(G)$.

From the above, we also get $\exp(G/Z(G)) = p$. Theorem 157.6 implies that $\exp(G') = \exp(G/Z(G)) = p$. It follows $G' \leq \Omega_1(G) \leq Z(G)$ and so $G$ is of class 2 and $G'$ is elementary abelian. We have obtained the groups stated in our theorem.

Conversely, let $G$ be a nonabelian $p$-group with $p > 2$ and exponent $p^e$, $e \geq 2$, such that $\Omega_{e-1}(G) \leq Z(G)$. We want to show that the condition $(*)$ holds. Since $\Omega_1(G) \leq Z(G)$, our group $G$ is $p$-central. By Theorem 157.6,

$$\exp(G/Z(G)) = p = \exp(G') \Rightarrow G' \leq \Omega_1(G) \leq Z(G) \Rightarrow \mathrm{cl}(G) = 2$$

and $G'$ is elementary abelian. Let $x, y$ be any noncommuting elements in $G$ so that we must have

$$x, y \in G - \Omega_{e-1}(G) \Rightarrow o(x) = o(y) = p^e\,.$$

Since $1 \neq [x,y] \in \Omega_1(G) \leq Z(G)$, Lemma 65.2 (a) implies that $\langle x,y\rangle$ is minimal nonabelian. If $\Omega_1(\langle x\rangle) = \Omega_1(\langle y\rangle)$, then there is an integer $i \not\equiv 0 \pmod p$ such that

$$x^{ip^{e-1}} = y^{-p^{e-1}} \quad \text{and this gives } (x^i y)^{p^{e-1}} = x^{ip^{e-1}} y^{p^{e-1}} = 1\,,$$

which gives that $x^i y \in \Omega_{e-1}(G)$. Thus, $x^i = y^{-1}l$ for some $l \in \Omega_{e-1}(G) \leq Z(G)$ and therefore

$$[x^i, y] = [y^{-1}l, y] = 1 \Rightarrow [x,y] = 1\,,$$

a contradiction. Hence, $\Omega_1(\langle x\rangle) \neq \Omega_1(\langle y\rangle)$, and so condition $(*)$ holds. □

**Proposition 160.2.** *Let $G$ be a $p$-group with $p > 2$ that satisfies the condition $(*)$. Suppose that*

$$S = \langle a, b \mid a^p = b^p = 1, [a,b] = c, c^p = [c,a] = [c,b] = 1\rangle \cong S(p^3)$$

*is a subgroup in $G$. Then $C_G(S)$ is of exponent $p$. In particular, in this case $Z(G)$ is elementary abelian.*

*Proof.* Set $C = C_G(S)$ and let $d$ be an element of order $p^2$ in $C$. Then

$$[a, bd] = [a, d][a, b]^d = c^d = c \Rightarrow |\langle a, bd\rangle'| = |\langle c\rangle| = p\,.$$

Then Lemma 65.2 (a) implies that $\langle a, bd\rangle$ is minimal nonabelian. But $o(a) = p$ and $o(bd) = p^2$ and this contradicts Proposition 159.1. Hence $\exp(C) = p$. □

**Proposition 160.3.** *Let $X = \Omega_1(X)$ be a nonabelian p-group with $p > 2$ that satisfies the condition $(*)$. Then X possesses a minimal nonabelian subgroup isomorphic to $S(p^3)$ (= the nonabelian group of order $p^3$ and exponent p).*

*Proof.* Let $Y$ be a maximal normal abelian subgroup in $X$. Since $Y < X$ and $X$ is generated by elements of order $p$, there is an element $x \in X - Y$ of order $p$. By Lemma 57.1, there is an element $y \in Y$ such that $\langle x, y\rangle$ is minimal nonabelian. By Proposition 159.1, $o(x) = o(y) = p$ and so $\langle x, y\rangle \cong S(p^3)$. □

**Lemma 160.4.** *Let G be a p-group with an abelian normal subgroup A and an element g such that $G = \langle A, g\rangle$ and g does not centralize A. Then there is an element $a \in A$ such that $\langle a, g\rangle$ is minimal nonabelian.*

*Proof.* Set $A_0 = C_A(g)$ so that $A_0 \le Z(G)$ and $A_0 < A$. Let $A_1$ be a $G$-invariant subgroup such that $A_0 < A_1 \le A$ and $|A_1 : A_0| = p$ and let $a \in A_1 - A_0$. Then

$$1 \ne [a, g] \in A_0 \le Z(G) \text{ so that we have } \langle a, g\rangle' = \langle [a, g]\rangle\,,$$

and therefore $\langle a, g\rangle$ is of class 2. We have $a^p \in A_0$ and this gives $1 = [a^p, g] = [a, g]^p$, which shows that $\langle a, g\rangle'$ is of order $p$. By Lemma 65.2 (a), $\langle a, g\rangle$ is minimal nonabelian and we are done. □

**Theorem 160.5.** *Let G be a p-group with $p > 2$ and exponent $p^e$, $e \ge 2$, which satisfies the condition $(*)$ and we assume that $\Omega_1(G)$ is nonabelian. In this case G has a proper subgroup $G_0$ such that $G_0 = H_p(G) = C_G(\Omega_1(G_0))$, $G_0$ is of exponent $p^e$ and it is either abelian or a group from Theorem 160.1. Hence in the second case, $G_0$ is p-central of class 2, $\Omega_{e-1}(G_0) \le Z(G_0)$ and $G_0' \le \Omega_1(G_0)$ is elementary abelian. As a rule $|G : G_0| = p$, and the case $|G : G_0| > p$ is possible only if $\exp(G) = p^2$ and $\{1\} \ne \mho_1(G_0) \le G_0'$.*

*Conversely, all the above groups satisfy the condition $(*)$.*

*Proof.* Let $G$ be a $p$-group, $p > 2$, of exponent $p^e$, $e \ge 2$, satisfying the condition $(*)$ and assume that $\Omega_1(G)$ is nonabelian. By Proposition 160.3, $\Omega_1(G)$ possesses a minimal nonabelian subgroup $S \cong S(p^3)$. Then Proposition 160.2 implies that $Z(G)$ is elementary abelian.

Let $E$ be a maximal normal elementary abelian subgroup of $G$ so that $Z(G) \le E$ and let $A$ be a maximal normal abelian subgroup of $G$ which contains $E$. If $A = E$, then $Z(G) < E$ and if $g \in G - E$ is an element of order $p^2$ in $G$, Lemma 57.1 implies that there is an element $e \in E - Z(G)$ such that $\langle g, e\rangle$ is minimal nonabelian. Since $o(e) = p$, this contradicts Proposition 159.1. Hence $E < A$ and the maximality of $E$

gives $E = \Omega_1(A)$. Let $F$ be a $G$-invariant subgroup such that $E < F \le A$ and $|F : E| = p$. Note that all elements in $F - E$ are of order $p^2$.

Set $C = C_G(F)$ so that $A \le C$, $C \trianglelefteq G$, and $C < G$. Assume that $\Omega_1(C)$ is nonabelian. Then $\Omega_1(C)$ contains a subgroup $S \cong S(p^3)$ (Proposition 160.3) that centralizes the subgroup $F$ of exponent $p^2$, contrary to Proposition 160.2. Hence $\Omega_1(C) \trianglelefteq G$ is abelian, and so $\Omega_1(C)$ is a normal elementary abelian subgroup in $G$. Again the maximality of $E$ forces $E = \Omega_1(C)$.

Set $G_0 = C_G(E) \trianglelefteq G$ and assume that there is an element $i$ of order $p$ in $G_0 - C$. Considering the subgroup $F\langle i\rangle$, Lemma 160.4 implies that there is an element $f \in F - E$ (of order $p^2$) such that $\langle i, f\rangle$ is minimal nonabelian. But this contradicts Proposition 159.1 and therefore $E = \Omega_1(G_0)$. Since $\Omega_1(G)$ is nonabelian, we must have some elements of order $p$ in $G - G_0$ and, in particular, $G_0 < G$.

Assume that there is an element $g \in G - G_0$ of order $\ge p^2$. Then considering the subgroup $E\langle g\rangle$, Lemma 160.4 implies that there is an element $e \in E$ such that $\langle e, g\rangle$ is minimal nonabelian, contrary to Proposition 159.1. Hence all elements in $G - G_0$ are of order $p$ and so we have proved that $G_0 = H_p(G) = C_G(\Omega_1(G_0))$.

Since $E = \Omega_1(G_0) \le Z(G_0)$, it follows that $G_0$ is $p$-central of exponent $p^e$, and so $G_0$ is either abelian or a $p$-group of Theorem 160.1. In the second case, $G_0$ is of class 2, $\Omega_{e-1}(G_0) \le Z(G_0)$ and $G_0' \le E$ is elementary abelian.

Now we use the following result of G. T. Hogan and W. P. Kappe (see [HogK]): "In a metabelian $p$-group, $p > 2$, of exponent greater than $p$, all elements of order $> p$ generate a subgroup of index at most $p$."

Assume that $|G : G_0| > p$. Let $G_1/G_0 \cong E_{p^2}$ be a subgroup of order $p^2$ in $G/G_0$. If $G_0$ is abelian, then $G_1$ is metabelian. But then $H_p(G_1) \ne \{1\}$ and $|G_1 : H_p(G_1)| > p$, contrary to the above result of Hogan–Kappe. Hence $G_0$ is of class 2 and $G_1' \le G_0$ with $G_1' \ge G_0'$. Note that $\{1\} \ne G_0'$ is elementary abelian and so if $\exp(G_0/G_0') > p$, then the above result of Hogan–Kappe (applied on $G_1/G_0'$) gives a contradiction. Thus, $G_0/G_0'$ must be elementary abelian, and so $\exp(G) = p^2$ and $\{1\} \ne \mho_1(G_0) \le G_0'$. We have proved that as a rule $|G : G_0| = p$ and the case $|G : G_0| > p$ is possible only in the case when $\exp(G) = p^2$ and $\{1\} \ne \mho_1(G_0) \le G_0'$.

Conversely, since $G_0$ is either abelian or a group of Theorem 160.1, it is obvious that all groups from our theorem satisfy the condition $(*)$. Our theorem is proved. □

# § 161 $p$-groups where all subgroups not contained in the Frattini subgroup are quasinormal

This section was written by the second author.

We recall that a subgroup $H$ of a group $G$ is called *quasinormal* in $G$ if $HX = XH$ for all subgroups $X$ of $G$, i.e., $H$ is permutable with all subgroups of $G$. A $p$-group $G$ is modular if and only if each subgroup of $G$ is quasinormal. It is well known that a $p$-group $G$ is modular if and only if $G$ is $D_8$-free for $p = 2$ and $G$ is $S(p^3)$-free for $p > 2$, where $S(p^3)$ is the nonabelian group of order $p^3$ and exponent $p$.

**Definition 1.** A $p$-group $G$ is called an SP-group if every subgroup not contained in $\Phi(G)$ is quasinormal.

We present here a result of J. Kirtland showing that $p$-groups that are SP-groups are actually modular (see [Kirt1]). The proof is quite simple in the case $p > 2$ and more complicated in the case of 2-groups. In fact, Lemma 161.11 is crucial here, for which we give our own detailed proof.

**Theorem 161.1.** *If a $p$-group $G$ is an SP-group, then $G$ is modular.*

Before proving this theorem, we need to prove a series of lemmas concerning permutable subgroups and SP-groups.

**Lemma 161.2.** *If $H$ and $K$ are subgroups of a group $G$ such that $K = \langle K_i \mid i \in I\rangle$ (where $K_i$ are subgroups of $K$) and $HK_i = K_iH$ for all $i \in I$, then we have $HK = KH$.*

*Proof.* Let $h \in H$ and $k = k_1k_2 \dots k_n \in K$, where $k_i$ $(j = 1, \dots, n)$ are elements in some $K_i$, $i \in I$. Then we have

$$hk = h(k_1k_2 \dots k_n) = (hk_1)k_2 \dots k_n = (k_1'h_1)k_2 \dots k_n$$
$$= k_1'k_2'h_2 \dots k_n = k_1'k_2' \dots k_n'h_n\,,$$

where all $k_j'$ are contained in some $K_i$, $i \in I$, and all $h_r \in H$ for $r = 1, 2, \dots, n$. This shows that $HK \le KH$. Similarly, we see that $KH \le HK$ and we are done. □

**Lemma 161.3.** *If $L$ is a quasinormal cyclic subgroup in a $p$-group $G$, then any subgroup of $L$ is quasinormal in $G$.*

*Proof.* Let $H$ be any subgroup in $G$ so that $LH = HL = G_0$ is a subgroup in $G$. We show that each subgroup of $L$ is permutable with $H$. Set $L_0 = L \cap H$. It is clear that each subgroup of $L_0$ is permutable with $H$ and if $X$ is a subgroup of $L$ with $X \nleq L_0$, then we have $X > L_0$. Set $|X : L_0| = p^i$, $i \ge 1$, and let $G_i$ be a subgroup of $G_0$ such that $G_i > H$ and $|G_i : H| = p^i$. Then we have $|(G_i \cap L) : L_0| = p^i$ and so the uniqueness of subgroups of $L$ of each possible order implies $G_i \cap L = X$, and therefore $G_i = XH = HX$ and we are done. □

**Lemma 161.4.** *Let $H$ be a subgroup of a group $G$ that is permutable with all cyclic subgroups in $G$. Then $H$ is quasinormal.*

*Proof.* Let $K$ be any subgroup in $G$. Since $K$ is generated by some cyclic subgroups $K_i$ in $K$, the result follows from Lemma 161.2. □

**Lemma 161.5.** *Let $a_1, \dots, a_n$ be elements of a group $G$ such that*

$$\langle a_i\rangle\langle a_j\rangle = \langle a_j\rangle\langle a_i\rangle \; 1 \le i, j \le n .$$

*Then we have $\langle a_1\rangle \dots \langle a_n\rangle = \langle a_1, \dots, a_n\rangle$.*

*Proof.* The result is trivial for $n = 1$. Let $H = \langle a_1, \dots, a_n\rangle$ and we proceed by induction on $n$. Applying Lemma 161.2, it follows that $\langle a_1, \dots, a_{n-1}\rangle\langle a_n\rangle$ is a subgroup of $G$ with $\langle a_1, \dots, a_{n-1}\rangle\langle a_n\rangle \le H$. In addition, we obviously have $H \le \langle a_1, \dots, a_{n-1}\rangle\langle a_n\rangle$. Thus we have

$$\langle a_1, \dots, a_n\rangle = \langle a_1, \dots, a_{n-1}\rangle\langle a_n\rangle ,$$

and by induction,

$$H = \langle a_1, \dots, a_n\rangle = \langle a_1, \dots, a_{n-1}\rangle\langle a_n\rangle = \langle a_1\rangle \dots \langle a_n\rangle .$$

The proof is complete. □

**Lemma 161.6.** *Let $G$ be a $p$-group that is an SP-group. Then we have:*
(i) *if $N \trianglelefteq G$, then $G/N$ is an SP-group,*
(ii) *If $\{a_1, \dots, a_l\}$ is a generating set for $G$ with each $a_i \notin \Phi(G)$, $1 \le i \le l$, then $G = \langle a_1\rangle\langle a_2\rangle \dots \langle a_l\rangle$, where all $\langle a_i\rangle$ are quasinormal in $G$.*

*Proof.* (i) Let $H/N \le G/N$ such that

$$H/N \nleq \Phi(G/N) = (\Phi(G)N)/N .$$

Then $H \nleq \Phi(G)$. Let $K/N$ be an arbitrary subgroup of $G/N$. Since $HK = KH$, it follows that

$$(H/N)(K/N) = (K/N)(H/N) ,$$

and so $G/N$ is an SP-group.

(ii) Since each $\langle a_i\rangle$ is quasinormal, Lemma 161.5 implies that:

$$G = \langle a_1, \dots, a_l\rangle = \langle a_1\rangle \dots \langle a_l\rangle ,$$

completing the proof. □

**Lemma 161.7.** *Let $G$ be a $p$-group that is an SP-group. If $|G/\Phi(G)| = p^d$, then we have:*
(i) *$G = \langle a_1\rangle\langle a_2\rangle \dots \langle a_d\rangle$, where $a_i \notin \Phi(G)$ and each subgroup $\langle a_i\rangle$ is quasinormal in $G$, $1 \le i \le d$,*
(ii) *$\Phi(G) = \langle a_1^p\rangle\langle a_2^p\rangle \dots \langle a_d^p\rangle$ with each $\langle a_i^p\rangle$ quasinormal in $G$.*

*Proof.* (i) Note that $|G/\Phi(G)| = p^d$ implies that there is a generating set $\{a_1, \ldots, a_d\}$ for $G$ with each $a_i \notin \Phi(G)$ so that $\langle a_i \rangle$ is quasinormal. By Lemma 161.5, we get:

$$G = \langle a_1, a_2, \ldots, a_d \rangle = \langle a_1 \rangle \langle a_2 \rangle \ldots \langle a_d \rangle .$$

(ii) Since each $\langle a_i \rangle$ is quasinormal, Lemma 161.3 gives that each $\langle a_i^p \rangle$ is also quasinormal, $i = 1, 2, \ldots d$. Thus we have:

$$H = \langle a_1^p \rangle \ldots \langle a_d^p \rangle \le \mho_1(G) \le \Phi(G) .$$

Let $gH$ be a coset of $H$ in $G$. Since $G = \langle a_1 \rangle \langle a_2 \rangle \ldots \langle a_d \rangle$ and $\langle a_i \rangle$ are quasinormal in $G$ with $\langle a_i^p \rangle \le H$, it follows that

$$gH = a_1^{\alpha_1} \ldots a_d^{\alpha_d} , \qquad \text{where all } 0 \le \alpha_i \le p - 1 .$$

Thus there are at most $p^d$ cosets of $H$ in $G$. Since

$$H \le \Phi(G) , \quad |G : H| \le p^d , \quad |G/\Phi(G)| = p^d ,$$

it follows that $H = \Phi(G)$ and we are done. □

**Lemma 161.8.** *Let $G$ be a $p$-group that is an SP-group. Then we have:*
(i) *if $x$ is an element of order $p$ in $\Phi(G)$, then $x \in Z(\Phi(G))$,*
(ii) *$\Omega_1(G)$ is elementary abelian.*

*Proof.* (i) Let $x \in \Phi(G)$ with $o(x) = p$. By Lemma 161.7, we have

$$G = \langle a_1 \rangle \langle a_2 \rangle \ldots \langle a_d \rangle ,$$

where $|G/\Phi(G)| = p^d$ and

$$\Phi(G) = \langle a_1^p \rangle \langle a_2^p \rangle \ldots \langle a_d^p \rangle$$

and all $\langle a_i \rangle$ and $\langle a_i^p \rangle$ are quasinormal in $G$. We want to show that $[a_i^p, x] = 1$ for all $i = 1, \ldots, d$.

If $x \in \langle a_i^p \rangle$, then $[a_i^p, x] = 1$ and we are done. Hence we may assume that $x \notin \langle a_i^p \rangle$. Since $\langle a_i \rangle$ is permutable with $\langle x \rangle$, $\langle a_i \rangle$ is a cyclic subgroup of index $p$ in $\langle a_i, x \rangle$, and so $\langle a_i \rangle \trianglelefteq \langle a_i, x \rangle$. But $\langle a_i^p \rangle$ is also permutable with $\langle x \rangle$, and so if $o(a_i^p) \le p$ then $\langle a_i^p, x \rangle$ is of order $\le p^2$, and so abelian and $[a_i^p, x] = 1$. Thus we may assume that $o(a_i^p) \ge p^2$, and so we have $|\langle a_i, x \rangle| \ge p^4$. We may also assume that $\langle a_i, x \rangle$ is nonabelian and the structure of such a group is determined in Theorem 2.1.

If $\langle a_i, x \rangle \cong M_{p^s}$, $s \ge 4$, then $\langle a_i^p \rangle = Z(\langle a_i, x \rangle)$, which gives $[a_i^p, x] = 1$ and we are done. Hence we may assume that $p = 2$ and $\langle a_i, x \rangle$ is dihedral or semidihedral. In any case $o(a_i x) = 2$ or $4$. But $\langle a_i x \rangle \nleq \Phi(G)$, and so $\langle a_i x \rangle$ is permutable with $\langle x \rangle$, which gives that $\langle a_i x, x \rangle = \langle a_i x \rangle \langle x \rangle$ is of order $\le 2^3$. On the other hand, by the above, we know that $\langle a_i x, x \rangle = \langle a_i, x \rangle$ is of order $\ge 2^4$, a contradiction. We have proved that $[a_i^p, x] = 1$ for all $i = 1, \ldots, d$ and so we have $x \in Z(\Phi(G))$.

(ii) Let $x, y \in G$ be elements of order $p$. If $x$ or $y$ is not contained in $\Phi(G)$, then $\langle x\rangle$ is permutable with $\langle y\rangle$, and so $|\langle x, y\rangle| \leq p^2$ and $[x, y] = 1$. If both $x, y$ are contained in $\Phi(G)$, then by (i) $x, y \in Z(\Phi(G))$, and so again $[x, y] = 1$. Thus $\Omega_1(G)$ is elementary abelian. □

## Proof of Theorem 161.1 for $p > 2$

Let $G$ be a $p$-group, $p > 2$, which is an SP-group. By Lemma 161.6 (i), $G/\mho_1(G)$ is also an SP-group, and so Lemma 161.8(ii) implies that $G/\mho_1(G)$ is elementary abelian. In particular, we have $G' \leq \mho_1(G)$, and so $G$ is a powerful $p$-group. By Proposition 26.10, each element in $\Phi(G) = G'\mho_1(G) = \mho_1(G)$ is a $p$-th power of an element in $G$.

Let $\langle x\rangle$ be any cyclic subgroup in $\Phi(G)$ and let $\langle y\rangle$ be a maximal cyclic subgroup in $\Phi(G)$ that contains $\langle x\rangle$. By the above, there is an element $z \in G-\Phi(G)$ such that $z^p = y$. Hence $\langle z\rangle$ is quasinormal in $G$, and so Lemma 161.3 gives that $\langle x\rangle$ is also quasinormal. We have proved that each cyclic subgroup in $\Phi(G)$ is quasinormal in $G$.

Let $K = \langle x_1, \dots a_s\rangle$ be a subgroup in $\Phi(G)$ and let $H$ be any subgroup in $G$. Since all subgroups $\langle x_i\rangle$ $(i = 1, \dots s)$ are quasinormal in $G$ and they generate $K$, it follows by Lemma 161.2 that $HK = KH$ and we are done.

It remains to study those 2-groups that are SP-groups.

**Lemma 161.9.** *Let $G$ be a 2-group that is an SP-group. If $G$ is of exponent 4, then $G$ is a Dedekind group.*

*Proof.* By Lemma 161.8 (ii), $E = \Omega_1(G)$ is elementary abelian so that we have $\Phi(G) = \mho_1(G) \leq E$. We show that $E \leq Z(G)$. If this is false, then there is an involution $z \in E$ and an element $v \in G - E$ of order 4 such that $[v, z] \neq 1$. But $v \notin \Phi(G)$, and so $\langle v\rangle\langle z\rangle = \langle z\rangle\langle v\rangle$ is a subgroup in $G$ that is nonabelian of order 8. Hence we get $v^z = v^{-1}$, and so $\langle v, z\rangle \cong D_8$. But then $vz$ is an involution in $G - E$, and so we get

$$\langle v, z\rangle = \langle vz, z\rangle = \langle vz\rangle\langle z\rangle \cong E_4 ,$$

a contradiction. We have proved that $E \leq Z(G)$. Hence each cyclic subgroup in $\Phi(G)$ is quasinormal in $G$ and so $G$ is modular (Lemma 161.2). By Proposition 73.5, $G$ is either abelian or Hamiltonian and we are done. □

**Corollary 161.10.** *Let $G$ be a 2-group that is an SP-group. Then $G/\mho_2(G)$ is abelian or Hamiltonian.*

*Proof.* Since $G/\mho_2(G)$ is an SP-group (Lemma 161.6 (i)) of exponent 4, the result follows from Lemma 161.9. □

**Lemma 161.11.** *Let $G = \langle a\rangle\langle b\rangle$ be a 2-group, where $\langle a\rangle$ and $\langle b\rangle$ are quasinormal cyclic subgroups in $G$. Then $G$ is a modular metacyclic 2-group.*

*Proof.* For any normal subgroup $N$ of $G$, we have $\bar{G} = G/N = \langle\bar{a}\rangle\langle\bar{b}\rangle$ (bar convention), where $\langle\bar{a}\rangle$ and $\langle\bar{b}\rangle$ are quasinormal cyclic subgroups in $\bar{G}$.

Suppose that $G$ is nonmetacyclic. Since $d(G) = 2$, it follows that $G$ is nonabelian. Let $R$ be a $G$-invariant subgroup of index 2 in $G'$. By Lemma 65.2 (a), $X = G/R$ is minimal nonabelian and Theorem 36.1 implies that $X$ is nonmetacyclic. But a nonmetacyclic minimal nonabelian 2-group,

$$X = \langle x, y \mid x^{2^m} = y^{2^n} = 1, [x, y] = z, z^2 = [x, z] = [y, z] = 1\rangle$$

$m \geq 1$, $n \geq 1$, $m+n \geq 3$, possesses the factor-group $X/\langle x^2, y^2\rangle$ that is isomorphic to $D_8$. On the other hand, $D_8$ is not generated by two cyclic quasinormal subgroups, a contradiction.

We have proved that $G$ is metacyclic. Assume in addition that $G$ has a cyclic subgroup of index 2. Since abelian 2-groups, $Q_8$ and $M_{2^n}$, $n \geq 4$ are modular, we may assume that $G$ is of maximal class having a normal subgroup $N$ such that $G/N \cong D_8$. But this is a contradiction (as in the previous paragraph).

We may assume that $G$ has no cyclic subgroup of index 2. Let $\langle x\rangle$ be a normal cyclic subgroup, $o(x) \geq 4$, with the cyclic factor-group $G/\langle x\rangle$ and let $y \in G$ be such that $G = \langle x, y\rangle$. Assume in addition that $G$ is not ordinary metacyclic with respect to $\langle x\rangle$ (see Definition 7 in § 26). Then we have $x^y = x^{-1}v$, where $v \in \langle x^4\rangle$. Note that we have $\langle x\rangle \cap \langle y\rangle \leq \langle x^4\rangle$ since $G$ has no cyclic subgroup of index 2. We compute:

$$x^{y^2} = (x^{-1}v)^y = (x^{-1}v)^{-1}v^y = x(v^{-1}v^y),$$

so that

$$[x, y^2] = x^{-1}x^{y^2} = v^{-1}v^y \in \langle x^4\rangle,$$

and therefore $\langle x^4, y^2\rangle \trianglelefteq G$ and $G/\langle x^4, y^2\rangle \cong D_8$, a contradiction.

We have proved that $G$ is ordinary metacyclic with respect to $\langle x\rangle$, and so we have $x^y = x^{1+2^s}$ for some integer $s \geq 2$. This gives

$$(x^i)^y = (x^y)^i = (x^{1+2^s})^i = (x^i)^{1+2^s},$$

and so by Theorem 73.15 $G$ is modular and we are done. □

## Proof of Theorem 161.1 for $p = 2$

*Proof.* Let $G$ be a 2-group that is an SP-group. By Corollary 161.10, $G/\mho_2(G)$ is abelian or Hamiltonian. There are three cases to consider.

*Case 1.* If $G/\mho_2(G)$ is abelian, then $G$ is a powerful 2-group. By Proposition 26.23, a maximal cyclic subgroup in $\Phi(G) = \mho_1(G)$ is not a maximal cyclic subgroup in $G$. By Lemma 161.3, each cyclic subgroup in $\Phi(G)$ is quasinormal in $G$, and so Lemma 161.2 implies that each subgroup in $\Phi(G)$ is quasinormal in $G$ and we are done.

*Case 2.* Suppose that $G/\mho_2(G) \cong Q_8$. Since $\mho_2(G) \leq \Phi(G)$, the group $G$ is generated by two elements $a$ and $b$. By Lemma 161.7 (i), $G = \langle a\rangle\langle b\rangle$, where $\langle a\rangle$ and $\langle b\rangle$ are quasinormal in $G$. By Lemma 161.11, $G$ is a metacyclic modular 2-group and we are done.

*Case 3.* Assume that $G/\mho_2(G)$ is Hamiltonian but not quaternion. Hence we have:

$$\begin{aligned} G/\mho_2(G) &= \langle a\mho_2(G), b\mho_2(G)\rangle \times \langle c_1\mho_2(G)\rangle \times \cdots \times \langle c_n\mho_2(G)\rangle \\ &\cong Q_8 \times C_2 \times \cdots \times C_2\,, \end{aligned}$$

where

$$\langle a\mho_2(G), b\mho_2(G)\rangle \cong Q_8 \quad \text{and} \quad \langle c_i\mho_2(G)\rangle \cong C_2 \quad \text{for } 1 \leq i \leq n\,,\ n \geq 1\,.$$

We have $\Phi(G) = \mho_2(G)\langle a^2\rangle$, and so all $a, b, c_1, \ldots, c_n \notin \Phi(G)$. By Lemma 161.7,

$$G = \langle a\rangle\langle b\rangle\langle c_1\rangle \ldots \langle c_n\rangle \quad \text{and} \quad \Phi(G) = \langle a^2\rangle\langle b^2\rangle\langle c_1^2\rangle \ldots \langle c_n^2\rangle\,.$$

Hence, if all $c_i$, $i = 1, \ldots, n$ are involutions, then

$$\Phi(G) = \langle a^2\rangle\langle b^2\rangle = \Phi(\langle a\rangle\langle b\rangle)\,.$$

By Lemma 161.11, $\langle a\rangle\langle b\rangle$ is modular and so each subgroup in $\Phi(G) = \langle a^2\rangle\langle b^2\rangle$ is quasinormal in $\Phi(G)$. But then $G$ is modular and we are done.

Without loss of generality we may assume that (at least) $o(c_1) \geq 4$. First, we consider the elements $a, b$. Since $a, b \notin \Phi(G)$, the cyclic subgroups $\langle a\rangle$ and $\langle b\rangle$ are quasinormal in $G$. Thus, $H = \langle a\rangle\langle b\rangle$ is a subgroup in $G$ and Lemma 161.11 implies that $H$ is a modular metacyclic 2-group. But $H$ covers $(\langle a, b\rangle\mho_2(G))/\mho_2(G) \cong Q_8$ and so $H$ is not $Q_8$-free. By Corollary A.24.4, $H$ is Hamiltonian and since $H$ is two-generator, it follows that $H \cong Q_8$.

Set $C = \langle c_1, \ldots, c_n\rangle$. Then we have:

$$G = C\langle a, b\rangle \quad \text{with } \langle a, b\rangle \cong Q_8 \text{ and } C \cap \langle a, b\rangle = \{1\}\,,$$

and we note that all cyclic subgroups

$$\langle a\rangle\,, \quad \langle b\rangle\,, \quad \langle c_1\rangle, \ldots, \langle c_n\rangle$$

are quasinormal in $G$. Also note that $\mho_1(C) \neq \{1\}$ (since $o(c_1) \geq 4$) and that for all $c \in C$ we have $c^2 \in \mho_2(G)$.

It will be shown that for each $c \in C$ and $x \in \langle a, b\rangle$, we have $c^{x^{-1}} \in \langle c\rangle$. Indeed, let $x$ be any element of order 4 in $\langle a, b\rangle$. Since $x \notin \Phi(G)$, it follows that $\langle x\rangle$ is quasinormal in $G$. Hence we have

$$\langle c\rangle\langle x\rangle = \langle x\rangle\langle c\rangle \quad \text{with } \langle c\rangle \cap \langle x\rangle = \{1\}\,.$$

On the other hand,

$$\mho_2(G) \le C\,, \quad \langle x, c\rangle \cap C = \langle c\rangle\,, \quad c^2 \in \mho_2(G)\,,$$

and $x$ and $c$ commute modulo $\mho_2(G)$. Hence

$$[x, c] \in \mho_2(G) \cap \langle x, c\rangle \le C \cap \langle x, c\rangle = \langle c\rangle\,,$$

and therefore $\langle x\rangle$ normalizes $\langle c\rangle$ and we are done.

Let $g \in G$ and recall that $G = C\langle a, b\rangle$ with $C \cap \langle a, b\rangle = \{1\}$ so that $g = cx$, where $c \in C$ and $x \in \langle a, b\rangle$. We have $xc = c^\delta x$ since $c^{x^{-1}} \in \langle c\rangle$, where $\delta$ is odd and $1 \le \delta \le o(c) - 1$. In that case,

$$\begin{aligned} g^4 &= (cx)^4 = c(xc)xc(xc)x = c \cdot c^\delta x \cdot xc \cdot c^\delta x \cdot x = c^{\delta+1}x^2c^{\delta+1}x^2 \\ &= c^{\delta+1}(x^2c^{\delta+1}x^{-2})x^4 = c^{\delta+1}(c^{x^{-2}})^{\delta+1} = c^{\delta+1}(c^{\delta^2})^{\delta+1} = c^{\delta^3+\delta^2+\delta+1}\,. \end{aligned}$$

Since $\delta$ is odd, we may set $\delta = 2l + 1$, and this gives:

$$g^4 = c^{2^3l^3+2^4l^2+3(2^2l)+4} = c'^4 \quad \text{for some } c' \in C\,.$$

Thus for all $g \in G$, we get $g^4 = c'^4$ for some $c' \in C$, and so we have $\mho_2(G) \le \mho_2(C)$. But we have:

$$\mho_1(C) \le \mho_2(G) \le \mho_2(C) \quad \text{and so} \quad \mho_1(C) \le \mho_2(C)\,.$$

This gives $\mho_1(C) = \mho_2(C)$, which is a contradiction as $\mho_1(C) \ne \{1\}$. Thus all $c_i$, $i = 1, \ldots, n$ must be involutions and we are done. □

**Problem 1.** Classify the $p$-groups $G$ containing a maximal subgroup $H$ such that any two subgroups of $G$ not contained in $H$ are permutable.

**Problem 2.** Study the $p$-groups in which the set of quasinormal subgroups forms a lattice.

# § 162 The centralizer equality subgroup in a $p$-group

We recall that the Thompson subgroup $\mathrm{J}(G)$ in a $p$-group $G$ is the subgroup generated by all abelian subgroups of the maximal possible order in $G$, and $\mathrm{ZJ}(G)$ is its center, which is equal to the intersection of all of these abelian subgroups.

**Definition 1.** Let $G$ be a $p$-group. The centralizer equality subgroup $\mathrm{D}(G)$ of $G$ is the subgroup generated by all elements $x$ of $G$ satisfying $\mathrm{C}_G(x) = \mathrm{C}_G(x^p)$.

The centralizer equality subgroup $\mathrm{D}(G)$ is useful in some new investigations of A. Mann (see [Man25]). Also, the characteristic subgroup $\mathrm{D}(G)$ has very interesting properties (see Theorem 162.4).

**Proposition 162.1** (A. Mann). *Let $x$ be a noncentral element in a $p$-group $G$ and set $N = \langle x\rangle^G$. Then $\mathrm{cl}(N) \le b$, where $b = b(x)$ is the breadth of the element $x$, i.e., $|G : \mathrm{C}_G(x)| = p^b$.*

*Proof.* Set $C = \mathrm{C}_G(x)$.

If $b = 1$, then

$$|G : C| = p \Rightarrow C \lhd G \Rightarrow \mathrm{Z}(C) \lhd G \Rightarrow N = \langle x\rangle^G \le \mathrm{Z}(C)^G = \mathrm{Z}(C)$$

so $N$ is abelian, completing this case.

Now let $b > 1$ and $K = C_G$ be the core of $C$ in $G$. Then $x^g \in C^g$ for all $g \in G$ so that

$$\mathrm{C}_G(N) = \bigcap_{g\in G} \mathrm{C}_G(x^g) = \bigcap_{g\in G} C^g = K\,.$$

It follows from the displayed formula that $N \cap K = \mathrm{Z}(N)$.

If $x \in K$, then $N = \langle x\rangle^G \le K$ since $K \lhd G$ so $N = \mathrm{Z}(N)$ is abelian and $\mathrm{cl}(N) = 1 < b$.

Next we assume that $x \notin K$. Then $\mathrm{C}_{G/K}(xK) \ge \mathrm{Z}(G/K) > K/K$. As $C/K \le \mathrm{C}_{G/K}(xK)$ and $C/K$ has no nonidentity normal subgroup of $G/K$, it follows that $C/K < \mathrm{C}_{G/K}(xK)$, and we conclude that $b(xK) < b$. By induction on $b$,

$$\mathrm{cl}(N) = \mathrm{cl}(N/N \cap K) + 1 = \mathrm{cl}((NK)/K) + 1 \le (b-1) + 1 = b\,,$$

as was to be shown. □

**Proposition 162.2** (A. Mann). *Let $G$ be a metabelian $p$-group, and let $B$ be an abelian subgroup of the maximal possible order in $G$. Then $G$ contains a normal abelian subgroup $C \le B^G$ such that $|C| = |B|$.*

*Proof.* Let $B$ be an abelian subgroup of the maximal possible order in a metabelian $p$-group $G$. We choose $B$ in $B^G$ so that among the abelian subgroups of order $|B|$ in $B^G$ it has a maximal intersection with $G'$. Suppose that such a $B$ is not normal in $G$. Take an element $x \in \mathrm{N}_G(\mathrm{N}_G(B)) - \mathrm{N}_G(B)$. Then $B \ne B^x$ and $B$ and $B^x$ normalize each other so that

$$A = BB^x = B[B,x] = [B,x]B$$

is of class 2 and since both $B$ and $B^x$ are maximal abelian subgroups, we have $Z(A) = B \cap B^x$. We have $A > B$, and so $[B, x] \nleq B$. Set

$$E = Z(A)(B \cap G')[B, x] \leq A$$

so that $E \leq B^G$. Since $G'$ is abelian, so are $(B \cap G')[B, x]$ and $E$. Moreover,

$$EB = Z(A)(B \cap G')([B, x]B) = A \quad \text{and} \quad B \cap E \geq Z(A) = B \cap B^x ,$$

so $|E| \geq |B|$. The maximality implies $|E| = |B|$. But we have

$$E \cap G' \geq (B \cap G')[B, x] > B \cap G' \quad \text{since } [B, x] \nleq B ,$$

a contradiction. □

**Lemma 162.3** (A. Mann). *Let $G$ be a $p$-group of class $c \leq p$, and let $x, y \in G$. Then $[x^p, y] = 1$ if and only if $[x, y^p] = 1$.*

*Proof.* If $p = 2$, then the result is clear. Indeed, in that case $\mathrm{cl}(G) \leq 2$ and so we have $[x^2, y] = [x, y^2] = [x, y]^2$.

We will show that both equalities are equivalent to $[x, y]^p = 1$, where we may assume $p > 2$. Note that $\langle x, Z_{c-1}(G)\rangle$ is of class $\leq c - 1$ since $\langle x, Z_{c-1}(G)\rangle/Z_{c-2}(G)$ is abelian and because $G/Z_{c-1}(G) = Z_c(G)/Z_{c-1}(G)$ is abelian, we have $G' \leq Z_{c-1}(G)$. Thus

$$\mathrm{cl}(\langle x, G'\rangle) \leq c - 1 \leq p - 1 ,$$

and so $\langle x, G'\rangle$ is a regular $p$-group (see Theorem 7.1 (b)). In particular, $\langle x, x^y\rangle = \langle x, [x, y]\rangle$ is regular (noting that $[x, y] = x^{-1}x^y$). By Theorem 7.2 (a),

$$[x, y]^p = (x^{-1}x^y)^p = 1 \iff x^p = (x^y)^p = (x^p)^y .$$

Hence $[x^p, y] = 1$ is equivalent with $[x, y]^p = 1$. Similarly, $[x, y^p] = 1$ is equivalent with $[x, y]^p = 1$ and we are done. □

**Theorem 162.4** (A. Mann). *Let $G$ be a $p$-group. Then the centralizer equality subgroup $D(G)$ of $G$ has the following properties:*

(a) *$D(G)$ is abelian,*

(b) *if $D(G) \leq H \leq G$, and $\mathrm{cl}(H) \leq p$, then $D(G) \leq Z(H)$,*

(c) *$C_G(D(G))$ contains $Z_p(G)$ as well as all normal subgroups of $G$ of class less than $p$, and all all elements of breadth less than $p$,*

(d) *let $c < p$. If $N$ is maximal among the normal subgroups of $G$ of class $c$, then $D(G) \leq N$,*

(e) *$D(G) \leq ZJ(G)$.*

*Proof.* Write $D = D(G)$.

(a) Suppose that $D$ is nonabelian. Let $z \in Z_2(D) - Z(D)$ with $z^p \in Z(D)$. Let $x$ be one of the generating elements of $D$. Then $[x, z] \in Z(D)$ and so $\mathrm{cl}(\langle x, z\rangle) \leq 2$. Therefore

$[z, x^p] = [z^p, x] = 1$ and thus $z \in C_G(x^p) = C_G(x)$. Letting $x$ range over all generators of $D$, we obtain $z \in Z(D)$, a contradiction.

(b) Let $a$ be one of the generating elements of $D$ and let $x \in H$. We prove that $a$ commutes with $x$ by the induction on the order $o(x)$ of $x$. By the induction hypothesis, $a$ commutes with $x^p$. By Lemma 162.3, $x$ commutes with $a^p$ and so $x \in C_G(a^p) = C_G(a)$.

(c) and (d) Since $D$ is abelian, $\mathrm{cl}(DZ_p(G)) \leq p$. Similarly, if $N \trianglelefteq G$ and $\mathrm{cl}(N) < p$, by Fitting's lemma we have $\mathrm{cl}(DN) \leq p$. Thus the first two claims of (c) follow from (b).

Let $N$ be a maximal normal subgroup of $G$ of class $c$, where $c < p$. Then we know from the previous paragraph that $D$ centralizes $N$, and so $\mathrm{cl}(DN) = \mathrm{cl}(N)$. By the maximality of $N$, we have $DN = N$ and so $D \leq N$ and this proves (d).

For the last claim of (c) note that, by Proposition 162.1, an element of breadth $b$ is contained in a normal subgroup of class at most $b$.

(e) Let $B$ be an abelian subgroup of the maximal possible order in $G$. Then $K = DB$ is a metabelian group so, by Proposition 162.2, $K$ contains a normal abelian subgroup $C$ such that $|C| = |B|$ and $C \leq B^K$. Note that $D(K) \geq D = D(G)$. We may apply part (d) of our theorem to $K$ (with $c = 1$) to get $D \leq D(K) \leq C$. We get

$$K = DB \leq CB \leq B^K$$

(since $C \leq B^K$) and this implies $K = B$. Indeed, if $K > B$, then let $K_0$ be a maximal subgroup of $K$ containing $B$, in which case we would have $B^K \leq K_0 < K$, a contradiction. From $K = B$, we get $D \leq B$ and we are done. □

In the next result, a characterization of (the generators of) the subgroup $D(G)$ is given.

**Theorem 162.5** (A. Mann). *Let $G$ be a $p$-group. For an element $x \in G$ the following are equivalent:*

(1) $C_G(x) = C_G(x^p)$,

(2) *if $y \in G$ and $\mathrm{cl}(\langle x, y\rangle) \leq p$, then $xy = yx$.*

*Proof.* The equivalence of (1) and (2) will be proved by induction on $|G|$ in both directions.

First, let $x$ satisfy (2) and assume that $C = C_G(x) \neq C_G(x^p)$. Then we can find an element $y \in C_G(x^p) - C$ such that $y$ normalizes $C$ and $y^p \in C$. If $\langle x, y\rangle < G$, then the induction hypothesis gives that $y \in C$, a contradiction. Hence we have $G = \langle x, y\rangle$. Since $x \in C$, we also have $G = \langle C, y\rangle$. By the choice of $y$, $C \trianglelefteq G$ and $|G : C| = p$. Moreover, $x \in Z(C) \trianglelefteq G$, and therefore $W = \langle x\rangle^G$ is generated by $p$ elements $x, x^y, \dots, x^{y^{p-1}}$ and $G = \langle W, y\rangle$. Here $y^p$ commutes with $y$ and with all conjugates $x^{y^i}$, and therefore $y^p \in Z(G)$. Also, $x^p$ commutes with $y$ and $C$ and therefore $x^p \in Z(G)$. We see that

$$G/Z(G) = \langle yZ(G), WZ(G)\rangle/Z(G)\ ,$$

with $yZ(G)$ of order $p$ and $(WZ(G))/Z(G)$ elementary abelian of order at most $p^p$. If $|G/Z(G)| \leq p^p$, then $\mathrm{cl}(G) \leq p$, and so by the property (2), we have $xy = yx$, a contradiction. The only remaining possibility is that $|G/Z(G)| = p^{p+1}$, in which case $G/Z(G) \cong C_p \text{ wr } C_p$.

If $p > 2$, then Exercise 70 in Appendix 45 gives a contradiction. There remains the case that

$$p = 2 , \quad G/Z(G) \cong C_2 \text{ wr } C_2 \cong D_8 .$$

Then $\mathrm{cl}(G) \leq 3$ and $[x, y] \in Z_2(G)$, implying that $\mathrm{cl}(\langle x, [x, y]\rangle) \leq 2$. By the property (2), $x$ commutes with $[x, y]$. On the other hand, both $x^2$ and $y^2$ are central in $G$ and so we get:

$$1 = [y, x^2] = [y, x][y, x]^x = [y, x]^2[y, x, x] = [y, x]^2 ,$$
$$1 = [x, y^2] = [x, y][x, y]^y = [x, y]^2[x, y, y] = [x, y, y] .$$

Hence $G = \langle x, y\rangle$ is minimal nonabelian, and so $G$ is of class 2. By the property (2), $x$ commutes with $y$, another contradiction.

Conversely, let $x \in G$ satisfy $C_G(x) = C_G(x^p)$ and let $y \in G$ be such that $\mathrm{cl}(\langle x, y\rangle) \leq p$. If $\langle x, y\rangle < G$, then the induction implies that $xy = yx$. Hence we may assume $\langle x, y\rangle = G$. But $\langle x, y^p\rangle < G$ so that the induction implies $[y^p, x] = 1$. Then Lemma 162.3 gives $[y, x^p] = 1$, and so $y \in C_G(x^p) = C_G(x)$ and we are done. □

# § 163 Macdonald's theorem on $p$-groups all of whose proper subgroups are of class at most 2

In finite group theory the minimal non-$\theta$-groups are of fundamental importance (here $\theta$ is a group theoretical property inherited by subgroups). It suffices to cite applications of minimal nonabelian, minimal nonmetacyclic and minimal regular $p$-groups in finite $p$-group theory and minimal nonnilpotent groups in general finite group theory.

The purpose of this section is to prove a remarkable result of I. Macdonald. Note that the corresponding paper [Macd1] is difficult to read. Macdonald's proof was reworked (independently) by the second author and Mann; their proofs are essentially identical.

**Theorem 163.1.** *Let $G$ be a $p$-group all of whose proper subgroups are of class at most 2. Then* $\mathrm{cl}(G) \leq 3$. *In other words, minimal class-two $p$-groups have class* 3.

*Proof.* By Fitting's lemma, $G$ is of class at most 4, and so $\mathrm{K}_5(G) = \{1\}$ and $\mathrm{K}_4(G) \leq \mathrm{Z}(G)$.

Suppose that $\mathrm{d}(G) > 2$. Then $\langle x, y\rangle < G$ for any $x, y \in G$, and so $\mathrm{cl}(\langle x, y\rangle) \leq 2$, which yields

$$[x, y, y] = [x, y, x] = 1\,.$$

Thus, $G$ satisfies the second Engel condition, and so by [Lev1] (see also [Hup1, Satz III.6.9]), cl($G$($\leq$ 3, completing this case.

From now on we may assume $\mathrm{d}(G) = 2$. In that case, the quotient group $G'/\mathrm{K}_3(G)$ is cyclic. Then

$$[G, \mathrm{K}_3(G), G] = [\mathrm{K}_4(G), G] = \{1\} \quad \text{and} \quad [\mathrm{K}_3(G), G, G] = [\mathrm{K}_4(G), G] = \{1\}\,,$$

and so, by the Three Subgroups lemma,

$$[G, G, \mathrm{K}_3(G)] = [G', \mathrm{K}_3(G)] = \{1\}\,.$$

Since $\mathrm{K}_3(G) \leq \mathrm{Z}(G')$ and $G'/\mathrm{K}_3(G)$ is cyclic, it follows that $G'$ is abelian, and so $G$ is metabelian. For any $x, y \in G$, one has $\langle [x, y], y\rangle < G$, and so $\mathrm{cl}(\langle [x, y], y\rangle) \leq 2$, which yields

$$[x, y, y, y] = 1 \quad \text{and} \quad [y, x, y, y] = 1\,.$$

Thus, $G$ satisfies the third Engel condition. Since $G$ is metabelian, Exercise 86 in Appendix 45 implies that for all $a \in G'$ and $x, y \in G$, we have $[a, x, y] = [a, y, x]$.

Using the elementary commutator relations, one obtains for any $u, v, x, y \in G$:

$$(*) \qquad [uv, xy] = [uv, y][uv, x]^y = [u, y]^v[v, y][u, x]^{vy}[v, x]^y\,.$$

Since $G$ is two-generator, we may set $G = \langle a, b\rangle$. Suppose that we can show that $[a, b, a, b] = 1$. Then the fact that $G$ is metabelian also gives $[a, b, b, a] = 1$. On the other hand, $G$ satisfies the third Engel condition, and so we also get $[a, b, a, a] = 1$ and $[a, b, b, b] = 1$. Hence we get

$$[a, b, a] \in Z(G) \quad \text{and} \quad [a, b, b] \in Z(G) \Rightarrow K_3(G) \leq Z(G)\,.$$

But then $\mathrm{cl}(G) \leq 3$ and our theorem would be proved.

Using $(*)$ and the fact that $G$ is metabelian of class $\leq 4$ and that $G$ satisfies the third Engel condition, we get:

$$\begin{aligned}1 &= [a, ab, ab, ab] = [a, b, ab, ab] = [[[a, b], ab], ab] = [[a, b, b][a, b, a]^b, ab]\\ &= [a, b, b, b]^{[a,b,a]^b} \cdot [[a, b, a]^b, b] \cdot [a, b, b, a]^{[a,b,a]^b b} \cdot [[a, b, a]^b, a]^b\\ &= [a, b, a, b]^2\,,\end{aligned}$$

since $[a, b, b, b] = 1$,

$$\begin{aligned}[[a, b, a]^b, b] &= [[a, b, a]([a, b, a]^{-1}[a, b, a]^b), b]\\ &= [[a, b, a][a, b, a, b], b] = [a, b, a, b]\end{aligned}$$

(where we have used the fact that $[a, b, a, b] \in Z(G)$),

$$[a, b, b, a]^{[a,b,a]^b b} = [a, b, b, a] = [a, b, a, b]\,,$$

(where we have used the facts that

$$[a, b, b, a] \in Z(G) \quad \text{and} \quad [a, b, b, a] = [a, b, a, b])$$

and

$$\begin{aligned}[[a, b, a]^b, a]^b &= [[a, b, a]([a, b, a]^{-1}[a, b, a]^b), a]^b\\ &= [[a, b, a][a, b, a, b], a]^b = [a, b, a, a]^b = 1\end{aligned}$$

(where we have used again that $[a, b, a, b] \in Z(G)$ and that $G$ satisfies the third Engel condition).

We have obtained that $[a, b, a, b]^2 = 1$. Therefore if $p > 2$, then $[a, b, a, b] = 1$ and we are done. In what follows we assume that $p = 2$.

It follows from $\langle a^i b, \Phi(G)\rangle < G$ that $\mathrm{cl}(\langle a^i b, a^2\rangle) \leq 2$, and so we get for $i = 0, 1$:

$$\begin{aligned}1 &= [a^2, a^i b, a^i b] = [a^2, b, a^i b] = [[a, b]^a[a, b], a^i b] = [[a, b]^2[a, b, a], a^i b]\\ &= [[a, b]^2, b]^{[a,b,a]} \cdot [a, b, a, b] \cdot [[a, b]^2, a^i]^{[a,b,a]b} \cdot [a, b, a, a^i]^b\\ &= [[a, b]^2, b][a, b, a, b][[a, b]^2, a^i]^{[a,b,a]b}\,.\end{aligned}$$

For $i = 0$ this relation gives

(1) $$[[a, b]^2, b][a, b, a, b] = 1\,,$$

and for $i = 1$ we get:

(2) $$[[a, b]^2, b][a, b, a, b][[a, b]^2, a]^b = 1\,.$$

The relations (1) and (2) together give:

$$[[a, b]^2, a] = 1. \tag{3}$$

Since $\langle a, \Phi(G)\rangle < G$, we have $\mathrm{cl}(\langle a, b^2\rangle) \leq 2$, and so we get:

$$\begin{aligned} 1 &= [a, b^2, a] = [[a, b][a, b]^b, a] = [[a, b]^2([a, b]^{-1}[a, b]^b), a] = [[a, b]^2[a, b, b], a] \\ &= [[a, b]^2, a]^{[a,b,b]}[a, b, b, a] = [[a, b]^2, a][a, b, a, b] = [a, b, a, b]\,, \end{aligned}$$

where $[a, b, b, a] = [a, b, a, b]$ and relation (3) were used. Thus, $[a, b, a, b] = 1$, completing the proof. □

**Exercise 1.** Prove that the class of a $p$-group all of whose subgroups of index $p^2$ are abelian does not exceed 3.

**Problem.** (i) Study the minimal non-metabelian $p$-groups. (ii) Study the $p$-groups all of whose proper subgroups are of class ≤ 2 or metacyclic.

# § 164 Partitions and $H_p$-subgroups of a $p$-group

Let $G = A_1 \cup \dots \cup A_n$, where $n > 1$, $\{1\} < A_i < G$ and $A_i \cap A_j = \{1\}$ for all $i \neq j$. Then one says that $G$ admits a *nontrivial partition with components* $A_1, \dots, A_n$. The groups admitting a nontrivial partition were classified in papers of P. G. Kontorovich, R. Baer and M. Suzuki. A detailed account of groups admitting a nontrivial partition is contained in papers by R. Baer (see, for example, [Bae2]).

Below, the $p$-groups admitting a nontrivial partition are classified and some consequences from that result are deduced.

It appears that partitions of a $p$-group of exponent $> p$ are closely connected with its $H_p$-subgroup. Below, the subgroup $H_p(G) = \langle x \in G \mid o(x) > p \rangle$ is said to be the $H_p$-subgroup of a $p$-group $G$. It follows from Theorem 7.2 (b) that the $H_p$-subgroup of a regular $p$-group $G$ of exponent $> p$ coincides with $G$. If $G \cong \Sigma_{p^2} \in \mathrm{Syl}_p(S_{p^2})$, then it has two regular maximal subgroups of exponent $p^2$, and therefore $H_p(G) = G$. It is easy to deduce from this result, using induction on $n > 1$, that $H_p(\Sigma_{p^n}) = \Sigma_{p^n}$. If $G$ is an irregular $p$-group of maximal class of order $> p^{p+1}$, then $H_p(G) < G \iff H_p(G) \in \Gamma_1$ is absolutely regular (see Theorems 9.5 and 9.6).

O. Kegel [Keg2] has proved that if $H_p(G) < G$, then the $H_p(G)$-subgroup of an arbitrary finite group $G$ is nilpotent (see Appendix 49).

Suppose that a $p$-group $G$ is such that $\{1\} < H_p(G) < G$. Then any subgroup, say $M$, of $G$ of exponent $> p$ properly containing $H_p(G)$ is irregular. Indeed, in that case, $\Omega_1(M) = M$, and the assertion follows from Theorem 7.2 (b). It follows that if $L < H_p(G)$ is cyclic of order $> p$, then $C_G(L) \leq H_p(G)$. Indeed, if $x \in G - H_p(G)$ centralize $L$, then $\Omega_1(\langle x, L \rangle) \neq \langle x, L \rangle$, contrary to what has been said above.

**Remark 1.** Let a $p$-group $G$ be neither absolutely regular nor of maximal class and $\exp(G) > p$. Then $G$ contains a regular subgroup $H$ of order $p^{p+1}$ such that $H/\Omega_1(H)$ is of order $p$. This is the most important partial case of Theorem 97.1. Below we offer another proof. We proceed by induction on $|G|$. One may assume that $|G| > p^{p+1}$. If there is $H \in \Gamma_1$ of exponent $> p$ that is neither absolutely regular nor of maximal class, then $H$ possesses a subgroup with the required properties. Therefore, one may assume that any maximal subgroup of $G$ of exponent $> p$ is either absolutely regular or of maximal class. Assume that $E \in \Gamma_1$ is of exponent $p$; then $|E| \geq p^{p+1}$. In that case, the set $\Gamma_1$ has no absolutely regular members. Let $H \in \Gamma_1$ be of exponent $> p$; then, by induction, $H$ is of maximal class. Considering $H \cap E$, we see that $|H| = p^{p+1}$ (Theorem 9.6) so that $G = p^{p+2}$. Let $L < H$ be of index $p$ and exponent $> p$ (here we use Theorem 9.5). Then, by the above, any maximal subgroup of $G$ containing $L$ is of maximal class. By Exercise 10.10, $G$ is of maximal class, a contradiction. Thus, $E$ does not exist so that any member of the set $\Gamma_1$ is either absolutely regular or of maximal class. By Theorem 13.6, there is in $\Gamma_1$ an absolutely regular member $A$. By Theorem 12.1 (b), $G = AR$, where $R = \Omega_1(G)$ is of order $p^p$ and exponent $p$. Let $R < B \in \Gamma_1$. Then $B$ is of maximal class of order $\frac{1}{p}|G| = p^{p+1}$. Thus, all subgroups of $G$

containing $R$ as a subgroup of index $p$ are of maximal class. Then, by Exercise 10.10, $G$ is of maximal class, contrary to the hypothesis.

**Theorem 164.1.** *Suppose that a p-group G of exponent $p^e > p$ is such that* $H_p(G) < G$ *is of maximal class. Then G is of maximal class and order $p^{p+1}$, $p > 3$ (in that case,* $H_p(G)$ *is absolutely regular of order $p^p$).*

*Proof.* By Theorem 7.2, if $G$ is regular, then $H_p(G) = G$. Thus, $G$ is irregular so that $|G| \geq p^{p+1}$ (Theorem 7.1 (b)). As a proper $H_2$-subgroup of a 2-group $G$ is abelian, one has $p > 2$. Next we assume that $|G| > p^{p+1}$ (the irregular groups of order $p^{p+1}$ will be considered in the end of the proof). Then $H_p(G)$ is irregular (Theorem 9.5) so is $G$.

Assume that $G$ is of maximal class. Then $H_p(G) = G_1$ is the fundamental subgroup of $G$ (see Theorems 9.6 and 13.9). But it is known that $G_1$ of order $\geq p^4$ cannot be of maximal class (see Theorem 9.6 (e)). Thus, if $G$ is of maximal class, its order is $p^{p+1}$.

Next we assume that $G$ is not of maximal class. In that case, $|G| \geq p^{p+2}$.

By Remark 1, there is in $G$ a regular subgroup $F$ of order $p^{p+1}$ such that $|\Omega_1(F)| = p^p$. Then $F = H_p(F) \leq H_p(G)$. By Theorem 9.6, an irregular $p$-group of maximal class has no subgroup $\cong F$ (see Theorems 9.5 and 9.6).

Thus, the group $G$ satisfying the hypothesis, must be irregular of order $p^{p+1}$ (and so of maximal class, by Theorem 7.1 (b)). In that case, as we have noted, $p > 2$. As the fundamental subgroup of a 3-group of maximal class and order $3^4$ is abelian, we get $p > 3$. □

It follows easily from Theorem 12.1 (b) that if the $H_p$-subgroup of a $p$-group $G$ is proper absolutely regular, then $G$ is of maximal class

The following theorem classifies the $p$-groups admitting a nontrivial partition:

**Lemma 164.2.** *A p-group G of order $> p$ admits a nontrivial partition $\Sigma \iff$ either* $\exp(G) = p$ *or* $H_p(G) < G$.

*Proof.* If $H_p(G) < G$, then $G$ admits a nontrivial partition with the component $H_p(G)$, and all other components are subgroups of $G$ of order $p$ not contained in $H_p(G)$.

Assume that $\exp(G) > p$. Let $H \in \Sigma$ be of exponent $> p$ (such $H$ exists) and let $x \in Z(G)$ be of order $p$. Let $\langle c\rangle = C < H$ be cyclic of order $p^2$. Since $\langle cx\rangle \cap H > \{1\}$, we get $cx \in H \Rightarrow x \in H$. Thus, $\Omega_1(Z(G)) < H \Rightarrow Z(G) \leq H$. Assume that $F \in \Sigma - \{H\}$ is of exponent $> p$. Then $H \cap F \geq \Omega_1(Z(G)) > \{1\}$, a contradiction. Thus, $\Sigma$ has only one member, say $H$, of exponent $> p$; then $H \geq H_p(G)$. Thus, $H_p(G) \leq H < G$. □

*The second proof.* Let $L \triangleleft G$ be of order $p$ and $L \leq U < G$, where $U \in \Sigma$. Take a maximal cyclic $C < G$ of order $> p$. If $L < C$, then $C \leq U$. Assume that $L \nleq C$ and set $H = L \times C$. Let $C \leq V \in \Sigma$. If $C_1 < H$ is cyclic of index $p$ and $C_1 \neq C$, then $\{1\} < C \cap C_1$ so $L < CC_1 \leq V$, and we conclude that $V = U$. Thus, all cyclic subgroups of $G$ of order $> p$ are contained in $U$ so that $H_p(G) \leq U < G$. (This argument shows that if $W \in \Sigma - \{U\}$, then $U \cap Z(G) = \{1\}$.) □

If $G$ is a regular $p$-group admitting a nontrivial partition, then $\exp(G) = p$ (Theorem 164.2).

If $G$ is as in Theorem 164.2, then $Z(G) \le H_p(G)$ and $\exp(Z(G)) = p$. Indeed, if $x \in \Omega_1(Z(G))^\#$ and $C = \langle c\rangle$ is of order $p^2$, then $\langle cx\rangle \cap H_p(G) > \{1\}$ so that $cx \in H_p(G) \Rightarrow x \in H_p(G)$, a contradiction. Assume that $\exp(Z(G)) > p$. Let $D = \langle c\rangle \le Z(G)$ be cyclic of order $p^2$ and $x \in G$ be of order $p$. Then $c$, $cx$ are elements of $H_p(G)$ so $x \in H_p(G)$. Since $G = \Omega_1(G)$, we get $H_p(G) = G$, a contradiction.

**Remark 2.** If a $p$-group $G$ of exponent $> p$ admits a nontrivial partition all of whose components are regular, then $H_p(G)$ is a component of the partition. Indeed, let $A$ be a component of exponent $> p$. Then, by Theorem 164.2, $H_p(G) \le A = H_p(A) \le H_p(G)$, completing the proof.

Note that if a $p$-group $G$ is such that $\{1\} < H_p(G) < G$, then it admits a nontrivial partition with only one component of exponent $> p$. I do not know of any such $G$ of exponent $> p$ admitting a partition with two nonabelian components. In what follows we shall see that if a 2-group $G$ of exponent $> 2$ admits a nontrivial partition, then one of its components is $H_2(G)$ of index 2 so all remaining components have order 2.

**Exercise 1.** Classify the $p$-groups admitting a nontrivial partition all of whose members are cyclic (a cyclic partition).

**Solution.** Suppose that a $p$-group $G$ of exponent $> p$ admits a nontrivial cyclic partition. We claim that then $G$ is dihedral (any dihedral group admits a unique cyclic partition). In that case, $H_p(G)$ is cyclic so that $H_p(G)$ is the unique cyclic subgroup of its order in $G$. Now the result follows from Theorem 1.17 (b).

**Second solution.** Let $C < G$ be maximal cyclic of order $> p$ be a component of our partition and $C < B \le G$, where $|B : C| = p$. Then $B$ admits the induced partition, and so $C$ is the unique cyclic subgroup of index $p$ in $G$. It follows that $B$ is a 2-group of maximal class. Then $G$ is of maximal class (Exercise 10.10). Clearly, $G$ is dihedral.

**Exercise 2.** Suppose that among the components of a nontrivial partition of a $p$-group $G$ there is a cyclic component, say $C$, of order $> p$. Prove that then $G$ is a dihedral 2-group.

**Exercise 3.** Classify the metacyclic $p$-groups admitting a nontrivial partition.

**Solution.** Let $G$ be a metacyclic $p$-group admitting a nontrivial partition. It is easy to check that if $G$ is abelian, then its $\cong E_{p^2}$. Next we assume that $G$ is nonabelian. By Theorem 164.2, $H_p(G) < G$. It follows that $\Omega_1(G) = G$ so $G$ is a 2-group of maximal class (Proposition 10.19).

By Theorem 1.2, $G$ is dihedral.

**Second solution.** There is $C \lhd G$ such that $G/C$ is cyclic. Assume that $G$ is nonabelian; then $|C| > p$ and $C$ is maximal cyclic so it is a component of the partition. Let $U/C \le$

$G/C$ be of order $p$. Then $U$ admits a nontrivial partition (Exercise 5, below) so it is dihedral. Then $G$ is of maximal class (Exercise 10.10) so dihedral.

**Exercise 4.** Classify the minimal nonabelian $p$-groups $G$ admitting a nontrivial partition.

**Solution.** One has $|\Omega_1(G)| \leq p^3$ (Lemma 65.1). By Theorem 164.2, $\mathrm{H}_p(G) < G$. Then $\Omega_1(G) \geq \langle G - \mathrm{H}_p(G)\rangle = G$ so $|G| = p^3$; then $G \in \{S(p^3), \mathrm{D}_8\}$.

**Exercise 5.** Suppose that a group $G$ admits a nontrivial partition $\Sigma$. If $H < G$ and there is no $F \in \Sigma$ such that $H \leq F$, then $H$ admits a nontrivial partition.

**Exercise 6.** Suppose that a 3-group $G$ admits a nontrivial partition. Prove that all components of this partition are of class $\leq 2$.

**Exercise 7.** If a $p$-group $G$ of maximal class admits a nontrivial partition, then one of the following holds:
(a) $\exp(G) = p$,
(b) $|G| > p^p$, all components of the partition, except one which is of index $p$, have order $p$.

**Exercise 8.** Describe a $p$-group $G = A \operatorname{wr} B$ admitting a nontrivial partition.

**Exercise 9.** If a $p$-group $G$ of exponent $> p > 2$ admits a nontrivial partition and has a normal cyclic subgroup $L$ of order $> p$, then $|G : \mathrm{H}_p(G)| = p$. (**Hint.** Consider $\mathrm{C}_G(L)$.)

**Exercise 10.** Suppose that a nonabelian $p$-group $G$ of exponent $> p$ has an abelian subgroup $\mathrm{H}_p(G)$. Then $|G : \mathrm{H}_p(G)| = p$.

**Solution.** One may assume that $p > 2$. Let $A \leq G$ be minimal nonabelian; then $A \nleq \mathrm{H}_p(G)$. Therefore, all elements of the set $A - \mathrm{H}_p(G)$ have order $p$. It follows that $\Omega_1(A) = A$ so that $A \cong S(p^3)$. By the result of Mann (see footnote to #115) we have $|G : \mathrm{H}_2(G)| = p$.

**Exercise 11.** Suppose that $G$ is a $p$-group such that $\{1\} < \mathrm{H}_p(G) < G$.
(a) If $\mathrm{H}_p(G) < M \leq G$, then $M$ is irregular. In particular, $|\mathrm{H}_p(G)| \geq p^p$.
(b) If $|\mathrm{H}_p(G)| = p^p$, then $G$ is of maximal class and order $p^{p+1}$.

**Solution.** Since $\Omega_1(M) \geq \langle M - \mathrm{H}_p(G)\rangle = M$, it follows that $M$ is irregular (Theorem 7.2 (b)). Taking $|M : \mathrm{H}_p(G)| = p$ and using Theorem 7.1 (b), we conclude that $|\mathrm{H}_p(G)| \geq p^p$. Now let $|\mathrm{H}_p(G)| = p^p$ and let $M$ be as above. Then, by (a) and Theorem 7.1 (b), $\mathrm{cl}(M) = p$, i.e., $M$ is of maximal class. As $M$ is arbitrary containing $\mathrm{H}_p(G)$ as a subgroup of index $p$, it follows from Exercise 10.10 that $G$ is of maximal class. Assume that $|G| > p^{p+1}$. Then $G_1$, the fundamental subgroup of $G$, is absolutely regular of order $> p^p \geq |\mathrm{H}_p(G)|$. Since $\exp(G_1) > p$ and $\mathrm{H}_p(G_1) = G_1 \leq \mathrm{H}_p(G)$ (this follows from Theorem 7.2 (b)), we get a contradiction. Thus, $|G| = p^{p+1}$.

**Exercise 12.** If a $p$-group $G$ is such that $|\mathrm{H}_p(G)| = p^{p+1} < |G|$, then $\mathrm{H}_p(G)$ is regular.

**Exercise 13.** If $G$ is a $p$-group, then $\mho_1(\mathrm{H}_p(G)) = \mho_1(G)$.

**Solution.** Set $H = \mathrm{H}_p(G)$ and $\bar{G} = G/\mho_1(H)$. One may assume that $H < G$. All elements in $\bar{G} - \bar{H}$ have order $p$. Since $\exp(\bar{H}) = p$, it follows that $\exp(\bar{G}) = p$. Thus, $\mho_1(G) \le \mho_1(H)$.

**Exercise 14.** Classify the $p$-groups with a proper absolutely regular $\mathrm{H}_p$-subgroup. (**Hint.** Use Theorem 12.1 (b).)

**Exercise 15.** Let $W$ be a $p$-group of maximal class and order $> p^{p+1}$ and $G = W/\mathrm{K}_{p+1}(W)$. Then the $\mathrm{H}_p$-subgroup of $G$ is not of maximal class.

# § 165 $p$-groups $G$ all of whose subgroups containing $\Phi(G)$ as a subgroup of index $p$ are minimal nonabelian

We solve here the problem 2502 stated by the first author. This is another example of $p$-groups with 'many' minimal nonabelian subgroups (see § 149).

**Theorem 165.1.** *Let $G \neq \{1\}$ be a $p$-group all of whose subgroups containing $\Phi(G)$ as a subgroup of index $p$ are minimal nonabelian. Then $\mathrm{d}(G) = 2$, which implies that $G$ is an $\mathrm{A}_2$-group (see § 71) all of whose maximal subgroups are minimal nonabelian, and so the following two possibilities hold:*

(a) *$G$ is metacyclic of order $> p^4$ with $G' \cong \mathrm{C}_{p^2}$, where either $p > 2$ or $p = 2$ and $G$ is ordinary metacyclic (see Definition 7 in § 26),*

(b) *$G$ is a nonmetacyclic $p$-group of order $p^5$, $p > 2$, defined in Proposition 71.5 (b), where*

$$\Phi(G) = G' = \Omega_1(G) \cong \mathrm{E}_{p^3}\,, \quad \mathrm{Z}(G) = \mathrm{K}_3(G) = \mho_1(G) \cong \mathrm{E}_{p^2}\,.$$

*Proof.* Let $G$ be a $p$-group satisfying the assumptions of Theorem 165.1. Then we have $\mathrm{d}(G) \geq 2$ and $\Phi(G)$ is a self-centralizing abelian normal subgroup of $G$ so that $\Phi(G)$ is a maximal normal abelian subgroup of $G$.

First assume $\mathrm{d}(G) = 2$. Then $G$ is an $\mathrm{A}_2$-group all of whose maximal subgroups are minimal nonabelian. Note that we have here $|G| > p^4$ since $\Phi(G)$ is self centralizing in $G$, and so $|\Phi(G) > p^2$. Such groups are completely determined in § 71, and so we obtain the groups stated in parts (a) and (b) of our theorem.

From now on we assume $\mathrm{d}(G) \geq 3$. For any subgroup $B_i/\Phi(G)$ of order $p$ in $G/\Phi(G)$, $B_i$ is minimal nonabelian, and so $\Phi(B_i) = \mathrm{Z}(B_i)$ is a subgroup of index $p$ in $\Phi(G)$.

Suppose that $G/\Phi(G)$ does not possess a subgroup $B/\Phi(G)$ of order $p^2$ such that $\Phi(B) = \Phi(G)$. Then for all subgroups $B_i/\Phi(G)$ of order $p$ in $G/\Phi(G)$, we have that $\Phi(B_i) = U$ is a fixed subgroup with $|\Phi(G) : U| = p$. Indeed, if there are $B_i \neq B_j$ such that $\Phi(B_i) \neq \Phi(B_j)$, then $\Phi(B_i)\Phi(B_j) = \Phi(G)$, and so

$$(B_iB_j)/\Phi(G) \cong \mathrm{E}_{p^2} \quad \text{and} \quad \Phi(B_iB_j) = \Phi(G)\,,$$

contrary to our assumption. It follows $U = \mathrm{Z}(G)$ and $\mho_1(G) \leq U$. Hence $p > 2$ and there are elements $g_1, g_2 \in G$, where both $g_1$ and $g_2$ are not contained in any of the minimal nonabelian subgroups $B_i$, such that $[g_1, g_2] \in \Phi(G) - U$. But then we get

$$(\langle g_1, g_2\rangle\Phi(G))/\Phi(G) \cong \mathrm{E}_{p^2} \quad \text{and} \quad \Phi(\langle g_1, g_2\rangle\Phi(G)) = \Phi(G)\,,$$

a contradiction.

We have proved that $G/\Phi(G)$ has a subgroup $B/\Phi(G)$ of order $p^2$ such that $\Phi(B) = \Phi(G)$, and so $\mathrm{d}(B) = 2$. It follows that $B$ is an $\mathrm{A}_2$-group all of whose maximal subgroups are minimal nonabelian. Let $B < G_0 \le G$ be a subgroup of $G$ such that $|G_0 : B| = p$, and so $G_0/\Phi(G) \cong E_{p^3}$ and $d(G_0) = 3$. Suppose for a moment that $B$ is nonmetacyclic. If $|B| = p^4$, then $|\Phi(G)| = p^2$ and so $B$ would possess an abelian maximal subgroup, a contradiction (in fact, each $p$-group of order $p^4$ has an an abelian maximal subgroup). It follows that $B$ is isomorphic to a group of Proposition 71.5 (b), and so $p > 2$ and $\Phi(G) = \Phi(B) \cong E_{p^3}$. But $G_0/\Phi(G)$ acts faithfully on $\Phi(G)$, which is a contradiction since $\mathrm{Aut}(E_{p^3}) \cong S(p^3)$ (= the nonabelian group of order $p^3$ and exponent $p$). We have proved that $B$ is a metacyclic $A_2$-group and so $B' \cong C_{p^2}$.

In fact, we see that $G_0$ possesses maximal subgroups $X$ such that $\Phi(X) = \Phi(G) = \Phi(G_0)$ and whenever $X$ is such a subgroup, $X$ is metacyclic with $X' \cong C_{p^2}$. In particular, $\Phi(G)$ is metacyclic. However, if all maximal subgroups of $G_0$ were metacyclic, then $G_0$ is minimal nonmetacyclic and such groups are determined in Theorems 66.1 and 69.1. But when inspecting the structure of such groups, we see that $G_0$ would have an abelian maximal subgroup, a contradiction.

We have proved that $G_0$ must possess at least one maximal subgroup $C$ such that $\Phi(C) < \Phi(G)$. Then for each of $p + 1$ subgroups $C_i/\Phi(G)$ of order $p$ in $C/\Phi(G)$, $C_i$ is minimal nonabelian, and so

$$Z(C_i) = \Phi(C_i) = \Phi(C)\,, \quad |\Phi(G) : \Phi(C)| = p\,, \quad \Phi(C) = Z(C), \quad \text{and} \quad d(C) = 3\,.$$

If $Y$ is any nonabelian maximal subgroup of $C$, then

$$Z(C) = \Phi(C) < Y\,, \quad |Y : Z(C)| = p^2, \quad \text{and so} \quad Z(Y) = Z(C)$$

and therefore $Y$ has the largest possible center. We may apply Theorem 151.1 on our group $C$. Since $d(C) = 3$ and $C/Z(C) \cong E_{p^3}$, $C$ is a group of part (v) of that theorem. In particular, we get $C' \cong E_{p^2}$ or $C' \cong E_{p^3}$. But we have $\Phi(C) < \Phi(G)$ and we know that $\Phi(G)$ is metacyclic. Hence we must have $C' \cong E_{p^2}$.

We have proved that for each maximal subgroup $H$ of $G_0$, we have either $H' \cong C_{p^2}$ or $H' \cong E_{p^2}$ and both cases occur.

Assume that $G_0$ has exactly one maximal subgroup $C$ with $C' \cong E_{p^2}$. Then for all other maximal subgroups $X$ of $G_0$ we have $X' \cong C_{p^2}$. Such groups have been classified in § 146. It follows that $p = 2$, $|G_0' : C'| = 2$, and so $G_0'$ is abelian of type $(4, 2)$. Since $d(G_0) = 3$, Theorem 146.8 is applicable. It follows that $G_0$ has a maximal subgroup $A$ such that $A' = \Phi(G_0') \cong C_2$, a contradiction.

We have proved that $G_0$ has at least two maximal subgroups $C_1 \neq C_2$ such that $C_1' \cong C_2' \cong E_{p^2}$. Since $C_1'$ and $C_2'$ are contained in the abelian metacyclic subgroup $\Phi(G)$, we get $C_1' = C_2'$. By Exercise 1.69 (a), we have $|G_0' : C_1'| \le p$. On the other hand, $G_0$ has a maximal subgroup $B$ such that $B' \cong C_{p^2}$, and so $|G_0' : C_1'| = p$ and $G_0'$ is abelian of type $(p^2, p)$. We have

$$\Phi(G_0') = \mho_1(G_0') \cong C_p \quad \text{so that} \quad G_0'/\Phi(G_0') \cong E_{p^2}\,.$$

We consider the factor-group

$$\bar{G_0} = G_0/\Phi(G_0')\,, \quad \text{where } \bar{G_0}' \cong E_{p^2} \text{ and } d(\bar{G_0}) = 3\,.$$

Each maximal subgroup of $G_0$ has its derived subgroup isomorphic to $C_{p^2}$ or $E_{p^2}$, and so for each maximal subgroup $\bar{X}$ of $\bar{G_0}$, we have $\bar{X}' \cong C_p$. Since $d(\bar{G_0}) = 3$, Theorem 139.A implies

$$\mathrm{cl}(\bar{G_0}) = 2 \quad \text{and} \quad \Phi(\bar{G_0}) = Z(\bar{G_0}) \quad \text{and we know that } \bar{G_0}' \cong E_{p^2} .$$

By Lemma 146.7 applied on the group $\bar{G_0}$, we see that $\bar{G_0}$ has exactly one abelian maximal subgroup $\bar{A} = A/\Phi(G_0')$. But then $A$ is a maximal subgroup of $G_0$ with $A' \leq \Phi(G_0')$, and so $|A'| \leq p$, a final contradiction. We have proved that $d(G) \geq 3$ is not possible and so $d(G) = 2$, as was to be shown. □

# § 166 A characterization of $p$-groups of class $> 2$ all of whose proper subgroups are of class $\le 2$

Most of the results of this section were obtained by P. Li, H. Qu and J. Zeng [LQZ]. By a result of I. Macdonald (Theorem 163.1), the title groups are of class 3. Here we give a characterization of the title groups. P. Li has also given a complete classification of the title groups for $p > 3$ presenting the groups in terms of generators and relations.

**Theorem 166.1.** *Let $G$ be a $p$-group of class $> 2$ all of whose proper subgroups are of class $\le 2$. Then $G$ is metabelian of class 3 and $\mathrm{d}(G) = 2$ or 3. If $\mathrm{d}(G) = 2$, then $\mathrm{K}_3(G)$ is elementary abelian of order $p$ or $p^2$. If $\mathrm{d}(G) = 3$, then*

$$p = 3\,, \quad |\mathrm{K}_3(G)| = 3\,, \quad [x, y, y] = 1 \quad \textit{for all } x, y \in G$$

*(i.e., $G$ satisfies the second Engel condition).*

*Conversely, if $G$ is a $p$-group with*

$$\mathrm{d}(G) = 2\,, \quad \mathrm{cl}(G) = 3\,, \quad \exp(\mathrm{K}_3(G) = p\,,$$

*then each proper subgroup of $G$ is of class $\le 2$. Also, if $G$ is a $p$-group with $\mathrm{d}(G) = 3$, $\mathrm{K}_3(G)$ is of exponent $p$, and $G$ satisfies the second Engel condition, then each proper subgroup of $G$ is of class $\le 2$.*

In order to facilitate the proof of our theorem, we shall use two technical results from [Hup1].

**Lemma 166.2** ([Hup1], III, 1.11 Hilfssatz). *Suppose that a group $G$ is generated by the subset $X$. Then*

$$\mathrm{K}_i(G) = \langle [x_1, \dots, x_i],\ K_{i+1}(G) \mid x_j \in X \rangle\,.$$

**Lemma 166.3** ([Hup1], III, 6.8 Hilfssatz). *Let $G$ be a group, $k \ge 2$ and $g_i \in \mathrm{K}_{n_i}(G)$ ($i = 1, \dots, k$). Then*

$$[g_1^{a_1}, \dots, g_k^{a_k}] = [g_1, \dots, g_k]^{a_1 \dots a_k} t\,,$$

*with some $t \in \mathrm{K}_{n+1}(G)$, where $n = n_1 + \dots + n_k$.*

*Proof of Theorem 166.1.* Let $G$ be a $p$-group of class $> 2$ all of whose proper subgroups are of class $\le 2$. By Theorem 163.1, $G$ is of class 3.

(i) $G$ is metabelian. Indeed, we have

$$[G, G', G] = [G', G, G] = \{1\}\,,$$

and so, by the Three Subgroups lemma, we get $[G, G, G'] = [G', G'] = \{1\}$ and $G'$ is abelian.

(ii) We have $\mathrm{d}(G) = 2$ or 3. Suppose that $\mathrm{d}(G) \ge 4$. Then $\langle x, y, z\rangle < G$ for any $x, y, z \in G$, and so $\mathrm{cl}(\langle x, y, z\rangle) \le 2$. It follows that $[[x, y], z] = 1$, and so $\mathrm{cl}(G) \le 2$, a contradiction.

(iii) Suppose that $\mathrm{d}(G) = 3$. Then $p = 3$ and $[a, b, b] = 1$ for all $a, b \in G$. Indeed, for any $x, y \in G$, one has

$$\langle x, y\rangle < G \Rightarrow \mathrm{cl}(\langle x, y\rangle) \le 2 \Rightarrow [x, y, y] = 1 \,.$$

Hence, $G$ satisfies the second Engel condition. By [Hup1, III, 6.5 Satz], a $p$-group satisfying the second Engel condition is either of class $\le 2$ or $p = 3$ and $G$ is of class 3. Since in our case, $G$ is of class 3, we must have $p = 3$.

(iv) We prove that $\mho_1(G') \le Z(G)$. (iv1) First assume $\mathrm{d}(G) = 3$. Let $g \in G$. Then there exist maximal subgroups $M \neq N$ such that $g \in M \cap N$. We have $G' \le M \cap N$. Since $M$ and $N$ are of class $\le 2$, we get

$$M' \le Z(M)\,, \quad N' \le Z(N) \Rightarrow M'N' \le Z(M \cap N)\,,$$

which shows that $g$ centralizes $M'N'$. By Exercise 1.69 (a), $|G' : (M'N')| \le p$ and so, if $x \in G'$, then $x^p \in M'N'$, and so $x^p$ centralizes $g$ and therefore $\mho_1(G') \le Z(G)$. (iv2) We may assume that $\mathrm{d}(G) = 2$ and set $G = \langle a, b\rangle$. Since $G' = K_3(G)\langle [a, b]\rangle$, we see that $G'/K_3(G)$ is cyclic and $\mathrm{cl}(G) = 3$ implies that $K_3(G) \le Z(G)$.

Note that for any $x \in \Phi(G)$ and $y \in G$ we have $\langle x, y\rangle < G$, and so $\mathrm{cl}(\langle x, y\rangle) \le 2$, which implies $[x, y, y] = [y, x, y] = 1$. These facts will be used often in our following arguments.

We get

$$1 = [a^p, ab, ab] = [a^p, b, ab] = [a^p, b, b][a^p, b, a]^b = [a^p, b, a]^b \quad \text{and so } [a^p, b, a] = 1\,,$$

and this gives $[a^p, b] \in Z(G)$. Similarly, (interchanging $a$ and $b$) we get $[b^p, a] \in Z(G)$.

It follows from $[a^p, b] = z \in Z(G)$ that

$$(a^p)^b = za^p\,, \quad (a^p)^{b^{-1}} = z^{-1}a^p\,.$$

This implies $(a^p)^{[a,b]} = a^p$. Similarly,

$$[b^p, a] = u \in Z(G) \Rightarrow (b^p)^{[a,b]} = b^p\,.$$

We have obtained

$$[a, b, a^p] = 1 \quad \text{and} \quad [a, b, b^p] = 1\,.$$

But

$$\mathrm{cl}(\langle [a, b], a\rangle) \le 2\,, \quad \mathrm{cl}(\langle [a, b], b\rangle) \le 2\,,$$

and so we get finally:

$$1 = [a, b, a^p] = [[a, b]^p, a] \quad \text{and} \quad 1 = [a, b, b^p] = [[a, b]^p, b]\,,$$

and therefore $[a, b]^p \in Z(G)$. Since

$$G' = K_3(G)\langle [a, b]\rangle\,, \quad K_3(G) \le Z(G)\,,$$

we obtain again $\mho_1(G') \le Z(G)$ as required.

(v) We have $\exp(K_3(G)) = p$. We may assume

$$K_3(G) = \langle [x_1, y_1], \dots, [x_s, y_s]\rangle$$

with $x_i \in G'$ and $y_i \in G$, $1 \le i \le s$. Since $\mho_1(G') \le Z(G)$, by (iv), we get

$$[x_i, y_i]^p = [x_i^p, y_i] = 1\,, \quad 1 \le i \le s\,.$$

But $K_3(G)$ is abelian and so $\exp(K_3(G)) = p$.

(vi) If $d(G) = 2$, then $K_3(G)$ is elementary abelian of order $p$ or $p^2$. Indeed, we may set $G = \langle a, b\rangle$ and then

$$K_3(G) = \langle [a, b, a], [a, b, b]\rangle\,,$$

and so, by (v), $K_3(G)$ is elementary abelian of order $\le p^2$.

(vii) If $d(G) = 3$, then $K_3(G) \cong C_3$. Indeed, by (iii) we have $p = 3$ and by (v) we get $\exp(K_3(G)) = 3$. Also, for any $x, y, z \in G$ we have

$$\langle x, y\rangle < G\,, \quad \langle [x, y], z\rangle < G$$

and so we obtain:

$$[x, y, y] = [y, x, y] = 1\,, \quad [x, y^{-1}] = [x, y]^{-1} = [y, x]\,,$$
$$[y, x, z] = [[x, y]^{-1}, z] = [x, y, z]^{-1}\,.$$

Set $G = \langle a, b, c\rangle$. Then, by Lemma 166.2 and the above remarks,

$$K_3(G) = \langle [a, b, c], [b, c, a], [c, a, b]\rangle\,.$$

Since $G/K_3(G)$ is of class 2, we get

$$[ab, c] = [a, c][b, c]s, \quad \text{where } s \in K_3(G) \le Z(G)\,.$$

On the other hand, $\mathrm{cl}(\langle ab, c\rangle) \le 2$, and so

$$\begin{aligned} 1 &= [[ab, c], ab] = [[a, c][b, c]s, ab] = [[a, c][b, c], ab] \\ &= [a, c, ab]^{[b,c]}[b, c, ab] = [a, c, ab][b, c, ab] \\ &= [a, c, b][a, c, a]^b[b, c, b][b, c, a]^b = [a, c, b][b, c, a]\,. \end{aligned}$$

We have proved $[a, c, b][b, c, a] = 1$, which gives

$$[b, c, a] = [a, c, b]^{-1} = [[a, c]^{-1}, b] = [c, a, b]\,.$$

Hence, we have obtained the equality $[b, c, a] = [c, a, b]$. Similarly, we get (replacing in the last equality $a, b, c$ with $b, c, a$ in that order) $[c, a, b] = [a, b, c]$. Using the last two equalities, we get finally:

$$\mathrm{K}_3(G) = \langle [a, b, c], [b, c, a], [c, a, b] \rangle = \langle [a, b, c] \rangle ,$$

and we are done.

Conversely, let $G$ be a $p$-group with

$$\mathrm{d}(G) = 2 , \quad \mathrm{cl}(G) = 3 , \quad \exp(\mathrm{K}_3(G)) = p .$$

In that case, we show that each proper subgroup of $G$ is of class at most 2. Indeed, let $H$ be a maximal subgroup of $G$. Then there exist elements $a \in H - \Phi(G)$ and $b \in G - H$ so that $G = \langle a, b \rangle$. We have

$$H = \langle a, \Phi(G) \rangle = \langle a, G' \mho_1(G) \rangle = \langle a, G', S \rangle , \quad \text{where } S = \{g^p \mid g \in G\} .$$

By Lemma 166.2,

$$\mathrm{K}_3(H) = \langle [x, y, z] \mid x, y, z \in \{a\} \cup G' \cup S \rangle .$$

If an element in the set $\{x, y, z\}$ belongs to $S$, then Lemma 166.3 and the equality $\exp(\mathrm{K}_3(G)) = p$ imply $[x, y, z] = 1$. If an element in the set $\{x, y, z\}$ belongs to $G'$, then the fact that $G'$ is abelian (noting that each $p$-group of class 3 is metabelian, see (i)) and that $\mathrm{cl}(G) = 3$ imply also $[x, y, z] = 1$. Hence we get $\mathrm{K}_3(H) = \{1\}$.

Now assume that $G$ is a $p$-group such that

$$\mathrm{d}(G) = 3 , \quad \exp(\mathrm{K}_3(G)) = p , \quad \text{and } G \text{ satisfies the second Engel condition.}$$

Then we show that each proper subgroup of $G$ is of class at most 2. Since $G$ satisfies the second Engel condition, [Hup1, III, 6.5 Satz] implies that $\mathrm{cl}(G) = 3$ (noting that $\mathrm{K}_3(G)$ is of exponent $p$,and so $\mathrm{K}_3(G) \neq \{1\}$) and $p = 3$. Let $M$ be a maximal subgroup of $G$. There exist elements $a, b \in G$ such that $M = \langle a, b, \Phi(G) \rangle$. Then Lemma 166.2 and $\mathrm{cl}(M) \leq 3$ imply:

$$\mathrm{K}_3(M) = \langle [x, y, z] \mid x, y, z \in \{a, b\} \cup G' \cup S \rangle ,$$

where

$$S = \{g^p \mid g \in G\} , \quad \mho_1(G) = \langle S \rangle .$$

If an element in the set $\{x, y, z\}$ belongs to $S$, then Lemma 166.3 and $\exp(\mathrm{K}_3(G)) = p$ imply $[x, y, z] = 1$. If an element in the set $\{x, y, z\}$ belongs to $G'$, then the fact that $G'$ is abelian and that $\mathrm{cl}(G) = 3$ imply also $[x, y, z] = 1$. If $\{x, y, z\} \leq \{a, b\}$, then the fact that $G$ satisfies the second Engel condition gives again $[x, y, z] = 1$. Thus, $\mathrm{K}_3(M) = \{1\}$, completing the proof of the theorem. □

**Problem 1.** Classify the $p$-groups all of whose nonnormal subgroups are of class $\leq 2$.

**Problem 2.** Classify the irregular 3-groups all of whose regular subgroups are of class $\leq 2$.

# § 167 Nonabelian $p$-groups all of whose nonabelian subgroups contain the Frattini subgroup

In Theorem 165.1 we determined up to isomorphism the structure of a $p$-group $G \neq \{1\}$ all of whose subgroups containing $\Phi(G)$ as a subgroup of index $p$ are minimal nonabelian. It turns out that in such a case we have $\mathrm{d}(G) = 2$ and $G$ is an $\mathrm{A}_2$-group and so, in particular, each minimal nonabelian subgroup of $G$ contains $\Phi(G)$.

Conversely, the problem to classify the nonabelian $p$-groups $G$ such that each minimal nonabelian subgroup of $G$ contains $\Phi(G)$ is very difficult (Problem 2603). Here we classify such groups in the following result.

**Theorem 167.1.** *Let $G$ be a $p$-group that is neither abelian nor minimal nonabelian. Suppose that each minimal nonabelian subgroup of $G$ contains $\Phi(G)$. Then one of the following holds:*

(a) *$G$ is an $\mathrm{A}_2$-group (see § 71),*

(b) *$G$ is of class 2 with $\Phi(G) \leq \mathrm{Z}(G)$ and $G' \neq \{1\}$ is elementary abelian of order $\leq p^3$. Moreover if $|G'| = p$, then we have $G = H * S$, where $H$ is either extraspecial or minimal nonabelian of order $> p^3$ and $S$ is abelian with $H \cap S = H'$ and $\mho_1(S) \leq H'$.*

*Proof.* Let $G$ be a $p$-group satisfying the assumptions of our theorem. Then we have $\mathrm{d}(G) \geq 2$ and since $G$ is generated by its minimal nonabelian subgroups (Lemma 57.1), there is a minimal nonabelian subgroup $X \nleq \Phi(G)$. But then we have $X > \Phi(G)$, and so $\Phi(G)$ is abelian. Let $H/\Phi(G)$ be any subgroup of order $p$ in $G/\Phi(G)$ and assume that $H$ is nonabelian. Let $Y$ be a minimal nonabelian subgroup in $H$. Since $Y \geq \Phi(G)$ and $\Phi(G)$ is abelian, we have $Y = H$ is minimal nonabelian. If $\mathrm{d}(G) = 2$, then $G$ is an $\mathrm{A}_2$-group and this is a group in part (a) of our theorem. Hence we may always assume that $\mathrm{d}(G) \geq 3$ and that $G$ is not an $\mathcal{A}_2$-group. We have proved the following fact:

(i) $\Phi(G)$ is abelian and if $H/\Phi(G)$ is any subgroup of order $p$ in $G/\Phi(G)$, then $H$ is either abelian or minimal nonabelian. Also, we may assume that $\mathrm{d}(G) \geq 3$ and that $G$ is not an $\mathcal{A}_2$-group.

Since $G$ is nonabelian, there are $a, b \in G$ such that $[a, b] \neq 1$. Because

$$|(\Phi(G)\langle a, b\rangle)/\Phi(G)| \leq p^2 ,$$

there is a subgroup $H \geq \Phi(G)\langle a, b\rangle$ such that $|H/\Phi(G)| = p^2$ and $H$ is nonabelian. Suppose that $\mathrm{d}(H) > 2$. Then $H$ is not minimal nonabelian. But $H$ is nonabelian and so there is a minimal nonabelian subgroup $X$ in $H$ and we have $\Phi(G) < X < H$. This implies that $\Phi(X) < \Phi(G)$ and $|\Phi(G) : \Phi(X)| = p$. It follows that $\Phi(X) = \Phi(H)$ and $\mathrm{d}(H) = 3$. There are exactly $p^2$ maximal subgroups $H_i$ of $H$ that do not contain $\Phi(G)$, and so $H_i \cap \Phi(G) = \Phi(H)$ and $H_i$ covers $H/\Phi(G)$. But each minimal nonabelian subgroup of $G$ must contain $\Phi(G)$ and so all these $p^2$ maximal subgroups $H_i$ of $H$ are abelian. This contradicts Exercise 1.6 (a) which states that the number of abelian maximal subgroups in a nonabelian $p$-group is exactly $0$, $1$ or $p + 1$. We have proved that

$d(H) = 2$, and so $\Phi(H) = \Phi(G)$. By (i), each maximal subgroup of $H$ is abelian or minimal nonabelian. Hence $H$ is either minimal nonabelian or an $A_2$-group. We have proved:

(ii) Let $H/\Phi(G)$ be any subgroup of order $p^2$ in $G/\Phi(G)$. If $H$ is nonabelian, then $d(H) = 2$ and $H$ is either minimal nonabelian or an $\mathcal{A}_2$-group. Also, there exists a subgroup $H/\Phi(G)$ of order $p^2$ in $G/\Phi(G)$ such that $H$ is nonabelian.

Now it is easy to prove the following facts:

(iii) We set $G_1 = C_G(\Phi(G))$ and let $G_2$ be any complement of $G_1$ in $G$ modulo $\Phi(G)$, i.e., $G = G_1G_2$ with $G_1 \cap G_2 = \Phi(G)$. We assume $G_2 > \Phi(G)$. Then for each subgroup $K/\Phi(G)$ of order $p$ in $G_2/\Phi(G)$, $K$ is minimal nonabelian. If $|G_2/\Phi(G)| \geq p^2$, then

$$|G_2/\Phi(G)| = p^2\,, \quad \Phi(G_2) = \Phi(G)\,, \quad d(G_2) = 2\,,$$

and $G_2$ is an $\mathcal{A}_2$-group of order $> p^4$ given in Theorem 165.1, and so $G_2$ is either metacyclic with $G_2' \cong C_{p^2}$ (where either $p > 2$ or $p = 2$ and $G_2$ is ordinary metacyclic) or $G_2$ is a nonmetacyclic group of order $p^5$, $p > 2$, defined in Proposition 71.5 (b). Therefore

$$\Phi(G) = \Phi(G_2) = \Omega_1(G_2) = G_2' \cong E_{p^3} \quad \text{and} \quad Z(G_2) = K_3(G_2) = \mho_1(G_2) \cong E_{p^2}\,.$$

We have $G_1 > \Phi(G)$ and if $G_1$ is nonabelian, then $|G_1/\Phi(G)| \geq p^2$ and $G_1/\Phi(G)$ possesses subgroups $H/\Phi(G)$ of order $p^2$ such that $H$ is nonabelian. In fact, $H$ is then minimal nonabelian with $\Phi(H) = \Phi(G)$.

Indeed, since $\Phi(G)$ is self centralizing in $G_2$, we have (by (i)) for each subgroup $K/\Phi(G)$ of order $p$ in $G_2/\Phi(G)$ that $K$ is minimal nonabelian. If $|G_2/\Phi(G)| \geq p^2$, then whenever $Y/\Phi(G)$ is a subgroup of order $p^2$ in $G_2/\Phi(G)$, $Y$ is nonabelian, and so (by (ii)) $d(Y) = 2$ and $\Phi(Y) = \Phi(G) = \Phi(G_2)$. By Theorem 165.1, we get $|G_2/\Phi(G)| = p^2$, $d(G_2) = 2$, $G_2$ is an $\mathcal{A}_2$-group of order $> p^4$, and we get all properties of $G_2$ stated in (iii). By (i), $G_1 > \Phi(G)$ (otherwise, $G = G_2$ would be an $\mathcal{A}_2$-group). Suppose that $G_1$ is nonabelian so that $|G_1/\Phi(G)| \geq p^2$ and $G_1/\Phi(G)$ contains a subgroup $H/\Phi(G)$ of order $p^2$ such that $H$ is nonabelian (by (ii)). Again, by (ii),

$$d(H) = 2\,, \quad \Phi(H) = \Phi(G) = Z(H)\,,$$

and so $H$ is minimal nonabelian.

(iv) The case $|G_2/\Phi(G)| = p^2$ is not possible, where $G_2$ is defined in (iii).

Indeed, we assume that $|G_2/\Phi(G)| = p^2$ so that $G_2$ is an $\mathcal{A}_2$-group of order $> p^4$ with all properties stated in (iii). Since (by (iii)) $G_1 > \Phi(G)$, there is a subgroup $Y/\Phi(G)$ of order $p$ in $G_1/\Phi(G)$, where $Y$ is abelian. Set $K = YG_2$; then

$$\Phi(K) = \Phi(G_2) = \Phi(G) \quad \text{with } |K| \geq p^6\,,\ d(K) = 3\,.$$

Also, $K$ is not an $\mathcal{A}_2$-group since $G_2 < K$ and $K$ has no abelian maximal subgroup because $C_K(\Phi(G)) = Y$. There are exactly $p+1$ maximal subgroups $V_i$ of $K$ containing $Y$

and they are all $\mathcal{A}_2$-groups of order $> p^4$ and $d(V_i) = 2$ (by (ii)) with exactly one abelian maximal subgroup $Y$. There are exactly $p^2$ maximal subgroups $U_i$ of $K$ not containing $Y$ (one of them is equal $G_2$) and they are all $\mathcal{A}_2$-groups of order $> p^4$ of Theorem 165.1, where all maximal subgroups of each $U_i$ are minimal nonabelian. Using the results of § 71, we see that if one of the $V_i$'s is nonmetacyclic, then such a $V_i$ is an $\mathcal{A}_2$-group of Proposition 71.3, which implies that $Y$ is abelian of rank at least 3, and so in this case all $V_i$'s would be nonmetacyclic groups from Proposition 71.3, where $p > 2$ and for all $i$, $V_i' \cong E_{p^2}$.

First assume that $G_2$ is metacyclic. Then $\Phi(G)$ is metacyclic and so (by Theorem 165.1) all $p^2$ maximal subgroups $U_i$ of $K$ are metacyclic with $U_i' \cong C_{p^2}$. If one of the maximal subgroups $V_i$ is also metacyclic, then by the above all $p + 1$ maximal subgroups $V_i$ are metacyclic. But then all $p^2 + p + 1$ maximal subgroups of $K$ (noting that $d(K) = 3$) would be metacyclic, and so $K$ would be minimal nonmetacyclic of order $\geq p^6$. By Theorems 66.1 and 69.1 about minimal nonmetacyclic $p$-groups, we get a contradiction. Hence all $p + 1$ maximal subgroups $V_i$ are nonmetacyclic with $p > 2$ and $V_i' \cong E_{p^2}$. Since $V_i' \leq \Phi(G)$ and $\Phi(G)$ is metacyclic and abelian, we get

$$V_1' = V_2' = \cdots = V_{p+1}' \cong E_{p^2} .$$

By Exercise 1.69 (a), we have $|K' : V_1'| \leq p$ and since $U_i' \cong C_{p^2}$ and $U_i' \leq K'$, we get $|K' : V_1'| = p$, and so $K'$ is abelian of type $(p^2, p)$. We consider the factor-group

$$\bar{K} = K/\Phi(K') , \quad \text{where } \Phi(K') \cong C_p , \ \bar{K}' \cong E_{p^2} , \ d(\bar{K}) = 3 ,$$

and all maximal subgroups of $\bar{K}$ that have its derived subgroup of order $p$. By Theorem 139A, $\bar{K}$ is a group of part (c) of that theorem and so $cl(\bar{K}) = 2$ and $\Phi(\bar{K}) = Z(\bar{K})$. By Lemma 146.7, $\bar{K}$ has exactly one abelian maximal subgroup $\bar{A}$. But then $A$ is a maximal subgroup of $K$ with $A' \leq \Phi(K')$, and so $|A'| \leq p$, a contradiction.

Now assume that $G_2$ is nonmetacyclic. By (iii), we have in this case $p > 2$, $|G_2| = p^5$, $|K| = p^6$ and

$$\Phi(G) = \Phi(G_2) = \Omega_1(G_2) = G_2' \cong E_{p^3} \quad \text{and} \quad Z(G_2) = K_3(G_2) = \mho_1(G_2) \cong E_{p^2} .$$

All maximal subgroups $U_i$, $i = 1, 2, \ldots, p^2$ of $K$ are groups from Proposition 71.5 (b) with $Z(U_i) \cong E_{p^2}$ and all maximal subgroups $V_j$, $j = 1, 2, \ldots, p + 1$ of $K$ are nonmetacyclic groups from Proposition 71.3. Since all $V_j$ possess the abelian maximal subgroup $Y$, we get by Lemma 1.1

$$p^5 = |V_j| = p|V_j'|\,|Z(V_j)| \quad \text{and since } V_j' \cong E_{p^2} ,$$

we obtain

$$|Z(V_j)| = p^2 \quad \text{and} \quad Z(V_j) \leq \Phi(V_j) = \Phi(G) .$$

But

$$Z(G_2) \cong E_{p^2} \quad \text{and} \quad C_K(Z(G_2)) \geq \langle G_2, Y \rangle = K \quad \text{and so} \quad Z = Z(G_2) = Z(K) \cong E_{p^2} .$$

Also, from Proposition 71.3 we know that $K_3(V_j) \cong C_p$ and so $V_j' \nleq Z$ for all $j$. In addition, we have $U_i' = \Phi(G)$ for all $i$. Consider the factor-group $\bar{K} = K/Z$ of order $p^4$. We have $\bar{K}' = \Phi(G)/Z \cong C_p$ and for each maximal subgroup $\bar{U}_i$ and $\bar{V}_j$ of $\bar{K}$, we have

$$\bar{U}_i{}' \cong \bar{V}_j{}' \cong C_p \quad \text{since } U_i' = \Phi(G) \text{ and each } V_j' \text{ covers } \Phi(G)/Z\,.$$

On the other hand, it is very well known that each $p$-group of order $p^4$ has an abelian maximal subgroup, and so $\bar{K}$ has an abelian maximal subgroup. This is a final contradiction and so (iv) is proved.

(v) If $G_2 > \Phi(G)$, then we have $|G_2/\Phi(G)| = p$ and $G_2$ is minimal nonabelian. Also we have

$$|\Phi(G)| > p^2\,, \quad |G_1 : \Phi(G)| \geq p^2\,,$$

where $G_1$ and $G_2$ are defined in (iii).

Indeed by (iii) and (iv), the assumption $G_2 > \Phi(G)$ implies $|G_2/\Phi(G)| = p$ so that (i) and the fact that $G_2$ is nonabelian infer that $G_2$ is minimal nonabelian. Let $X/G_2$ be any subgroup of order $p$ in $G/G_2$. Then (ii) implies that $X$ is an $A_2$-group with $\Phi(X) = \Phi(G)$, $d(X) = 2$ and $X \cap G_1$ is a unique abelian maximal subgroup of $X$. By (i), $G$ is not an $A_2$-group, and so $G > X$, which gives $G_1 > X \cap G_1$, and so $|G_1 : \Phi(G)| \geq p^2$, as required.

Now assume that $|\Phi(G)| \leq p^2$ so that we must have $|\Phi(G)| = p^2$ and $G_2$ is minimal nonabelian of order $p^3$. If $|X'| = p$, then this together with the fact that $d(X) = 2$ gives (by Lemma 65.2 (a)) that $X$ is minimal nonabelian, a contradiction. Hence we have in our case $X' = \Phi(G)$ and $|X| = p^4$.

First we consider the case $p = 2$. Since $X$ is nonabelian and $|X/X'| = 4$, a well known result of O. Taussky implies that $X$ is of maximal class, and so $X \cong D_{16}$, $Q_{16}$ or $SD_{16}$ with $X \cap G_1 \cong C_8$. On the other hand, we have $X < G$ and let $Y$ be a subgroup of $G$ such that $Y > X$ and $|Y : X| = 2$. Noting that

$$\Phi(G_1) = \Phi(X \cap G_1) = \Phi(G) \cong C_4\,,$$

we see that $X \cap G_1 \cong C_8$ is a maximal cyclic subgroup in $Y \cap G_1$, where $|(Y \cap G_1) : (X \cap G_1)| = 2$. But $Y \cap G_1$ centralizes $\Phi(G)$, and so either $Y \cap G_1 \cong M_{16}$ or $Y \cap G_1$ is abelian of type $(8, 2)$. In any case, there is an involution $y \in (Y \cap G_1) - (X \cap G_1)$. Now we consider the subgroup $X_1 = \langle y \rangle G_2$ (instead of $X$) and we see (by the above argument) that $X_1$ is of maximal class. But $X_1 \cap G_1 = \langle y \rangle \times \Phi(G)$ is abelian of type $(4, 2)$, a contradiction.

It remains to consider the case $p > 2$. If $X' = \Phi(G) \cong C_{p^2}$, then Proposition 145.4 implies that each maximal subgroup of $X$ is nonabelian. This is a contradiction because $X \cap G_1$ is a unique abelian maximal subgroup of $X$. Hence we have $X' = \Phi(G) \cong E_{p^2}$. By Lemma 1.1 we have

$$p^4 = |X| = p|X'|\,|Z(X)| \quad \text{and so } |Z(X)| = p\,,\ Z(X) < \Phi(G)$$

and therefore $\text{cl}(X) = 3$. Let $L$ be a subgroup of $G$ such that $X < L$ and $|L : X| = p$. We have

$$|L| = p^5 \quad \text{and} \quad \Phi(L) = \Phi(X) = \Phi(G) = X' = L' \cong \text{E}_{p^2} .$$

Suppose for a moment that $L \cap G_1$ (of order $p^4$) is abelian. Then using again Lemma 1.1, we get

$$p^5 = |L| = p|L'|\,|\text{Z}(L)| \quad \text{and so } |\text{Z}(L)| = p^2 .$$

In this case $\text{Z}(L)$ (centralizing $\Phi(G)$) must be contained in $L \cap G_1$, which together with $|\text{Z}(X)| = p$ gives that there is an element $z \in (L \cap G_1) - X$ such that $z^p \in \Phi(G)$ and $z \in \text{Z}(L)$. But then $z$ centralizes $G_2$, and so considering the subgroup $X^* = \langle z\rangle G_2$ (instead of $X$), we get $|(X^*)'| = p$, a contradiction (by the above argument). Hence $L_1 = L \cap G_1$ is nonabelian and so (by (iii)) $L_1$ is minimal nonabelian with

$$\Phi(L_1) = \Phi(G) = \text{Z}(L_1) > \text{Z}(X) .$$

Since $\Phi(G) \nleq \text{Z}(L)$ and $\text{Z}(L) \leq \Phi(G)$, we have $\text{Z}(L) = \text{Z}(X)$ is of order $p$. If $L_1$ is non-metacyclic, then $L_1' \cong \text{C}_p$ is a maximal cyclic subgroup in $L_1$ and $L_1' = \text{Z}(L)$. But then

$$\Phi(G) = \Phi(L_1) = L_1' \times \mho_1(L_1) \quad \text{and} \quad \mho_1(L_1) \cong \text{C}_p$$

and $\mho_1(L_1)$ being characteristic in $L_1$ gives that $\mho_1(L_1) \leq \text{Z}(L)$, and so $\Phi(G) \leq \text{Z}(L)$, a contradiction. Hence $L_1$ is metacyclic, and so we may set:

$$L_1 = \langle a, b \mid |a^{p^2} = b^{p^2} = 1 , \quad a^b = a^{1+p}\rangle ,$$

where

$$L_1' = \langle a^p\rangle = \text{Z}(L) = \text{Z}(X) \quad \text{since } L_1' \trianglelefteq L .$$

Consider the maximal subgroup $X_1 = \langle b\rangle G_2$ of $L$, where $\langle b\rangle \times \langle a^p\rangle$ is a unique abelian maximal subgroup of $X_1$ (since $X_1' = \Phi(G)$). But then

$$\langle b^p\rangle = \mho_1(\langle b\rangle \times \langle a^p\rangle) \cong \text{C}_p$$

is a characteristic subgroup of $\langle b\rangle \times \langle a^p\rangle$, and so of $X_1$, which implies $\langle b^p\rangle \trianglelefteq L$, and so $b^p \in \text{Z}(L)$. It follows

$$\Phi(G) = \Phi(L_1) = \langle a^p\rangle \times \langle b^p\rangle \leq \text{Z}(L) ,$$

a contradiction. Statement (v) is completely proved.

(vi) Continuing with the assumption $G_2 > \Phi(G)$ in (v), we must have $p > 2$.

Indeed, assume $p = 2$. We have $|G_2/\Phi(G)| = 2$, $|\Phi(G)| > 4$, $G_2$ is minimal nonabelian and $|G_1 : \Phi(G)| \geq 4$. Let $X/G_2$ be any subgroup of order 2 in $G/G_2$. Then (ii) implies that $X$ is an $\mathcal{A}_2$-group with

$$\Phi(X) = \Phi(G) , \quad \text{d}(X) = 2 , \quad |X| > 2^4 ,$$

and $X \cap G_1$ is a unique abelian maximal subgroup of $X$. By § 71, $X$ is a metacyclic group from Proposition 71.2 (b) with $X' \cong C_4$. Let $S$ be a subgroup of $G$ such that $X < S$ and $|S : X| = 2$. Then $d(S) = 3$ and the maximal subgroup $S \cap G_1$ of $S$ is either abelian or minimal nonabelian. All other maximal subgroups of $S$ are metacyclic since they cover $S/S \cap G_1$, and so they are $\mathcal{A}_2$-groups from Proposition 71.2 (b). If $S \cap G_1$ were also metacyclic, then $S$ would be a minimal nonmetacyclic 2-group of order $\geq 2^6$, which contradicts Theorem 66.1. Hence $S$ has exactly one nonmetacyclic maximal subgroup $S \cap G_1$ (of order $\geq 2^5$). If $S \cap G_1$ is minimal nonabelian, then we have $\Omega_1(S \cap G_1) \cong E_8$. If $S \cap G_1$ is abelian, then the fact that $X \cap G_1$ is metacyclic and $|(S \cap G_1) : (X \cap G_1)| = 2$ also implies that $\Omega_1(S \cap G_1) \cong E_8$. Hence, in any case,

$$E = \Omega_1(S \cap G_1) \cong E_8 \quad \text{and} \quad E \trianglelefteq S .$$

But $S$ has exactly one nonmetacyclic maximal subgroup, and so $S/E$ is cyclic. It follows that $S' \leq E$, and so $S'$ is elementary abelian, contrary to the fact that $X < S$ and $X' \cong C_4$. Our claim is proved.

(vii) Continuing with the assumption $G_2 > \Phi(G)$ in (v), we have $p > 2$ and $G_1$ is abelian.

Indeed, assume that $G_1$ is nonabelian. We have (by (v) and (vi))

$$p > 2 , \quad |G_2/\Phi(G)| = p , \quad |\Phi(G)| > p^2 ,$$

$G_2$ is minimal nonabelian and $|G_1 : \Phi(G)| \geq p^2$. Let $X/G_2$ be any subgroup of order $p$ in $G/G_2$. Then (ii) implies that $X$ is an $A_2$-group with

$$\Phi(X) = \Phi(G) , \quad d(X) = 2 , \quad |X| > p^4 ,$$

and $X \cap G_1$ is a unique abelian maximal subgroup of $X$. By § 71, $X$ is a nonmetacyclic $p$-group from Proposition 71.3 with $X' \cong E_{p^2}$. By (iii), there is a subgroup $U$ in $G_1$ such that

$$\Phi(G) < U , \quad U/\Phi(G) \cong E_{p^2} \quad \text{and} \quad U \text{ is minimal nonabelian with } \Phi(U) = \Phi(G) .$$

Set $H = UG_2$ so that $d(H) = 3$ and each maximal subgroup $X_i$ of $H$ distinct from $U$ covers $H/U$, and so is an $A_2$-group isomorphic to a group from Proposition 71.3 and

$$\Phi(U) = \Phi(G) = \Phi(X_i) \quad \text{with } X_i' \cong E_{p^2} , \ \text{cl}(X_i) = 3 .$$

Let $x, y \in U - \Phi(G)$ be elements of minimal possible orders such that $\langle x, y\rangle$ covers $U/\Phi(G)$ and then we have $U = \langle x, y\rangle$. By Exercise P9 in Prerequisites from Volumes 1 and 2 (stated at the beginning of Volume 3), we have

$$\Phi(U) = \Phi(G) = \langle x^p, y^p, U'\rangle .$$

Set

$$X = \langle x\rangle G_2 , \quad Y = \langle y\rangle G_2 .$$

Then $X$ and $Y$ are $A_2$-groups isomorphic to some groups in Proposition 71.3. Note that $x \in X - \Phi(X)$ is a generating element in $X$ contained in the unique abelian maximal subgroup $\langle x\rangle\Phi(G)$ in $X$ and $y \in Y - \Phi(Y)$ is a generating element in $Y$ contained in the unique abelian maximal subgroup $\langle y\rangle\Phi(G)$ in $Y$. Looking at the groups in (i), (ii) and (iii) of Proposition 71.3 together with the minimality of orders of $x$ and $y$, we see that $o(x) \le p^2$, $o(y) \le p^2$ and only by groups in (iii) does that order equal exactly $p^2$. Since $|\Phi(G)| \ge p^3$, by the above, we must have $o(x) = o(y) = p^2$ so that $X$ and $Y$ must be groups from (iii) in Proposition 71.3 and we get

$$\Phi(U) = \Phi(G) = \langle x^p\rangle \times \langle y^p\rangle \times U' \cong E_{p^3} .$$

Next,

$$\mho_1(X \cap G_1) = \mho_1(\langle x\rangle \times \langle y^p\rangle \times U') = \langle x^p\rangle \le Z(X) \quad \text{and so } x^p \in Z(G) .$$

Similarly,

$$\mho_1(Y \cap G_1) = \mho_1(\langle x^p\rangle \times \langle y\rangle \times U') = \langle y^p\rangle \le Z(Y) \quad \text{and so } y^p \in Z(G) .$$

But $U' \le Z(H)$, and so also $U' \le Z(G)$. It follows that $\Phi(G) \le Z(G)$, a contradiction (because $cl(X_i) = 3$ and $\Phi(G) = \Phi(X_i)$). The statement (vii) is proved.

(viii) The case $G_2 > \Phi(G)$ is not possible. Hence we have $G = G_1$ is of class 2 with $\Phi(G) \le Z(G)$ and $G' \neq \{1\}$ is elementary abelian of order at most $p^3$.

Indeed, assume that $G_2 > \Phi(G)$. Then by (vii), $p > 2$ and $G_1$ is abelian. By (v),

$$|G_2/\Phi(G)| = p , \quad |\Phi(G)| > p^2 , \quad |G_1 : \Phi(G)| \ge p^2$$

and $G_2$ is minimal nonabelian. Let $X/G_2$ be any subgroup of order $p$ in $G/G_2$. Then (ii) implies that $X$ is an $A_2$-group with

$$\Phi(X) = \Phi(G) , \quad d(X) = 2 , \quad |X| > p^4 ,$$

and $X \cap G_1$ is a unique abelian maximal subgroup of $X$. By § 71, $X$ is a nonmetacyclic $p$-group from Proposition 71.3 with $X' \cong E_{p^2}$ and $cl(X) = 3$. For any $x, y \in G$ with $[x, y] \neq 1$, the subgroup $\langle x, y\rangle$ covers $G/G_1$ and since

$$|(\langle x, y\rangle\Phi(G))/\Phi(G)| \le p^2 ,$$

we get that $\langle x, y\rangle$ is either minimal nonabelian or an $A_2$-group (see (ii)) from Proposition 71.3, and so in any case we get $o([x, y]) = p$. It follows that $G'$ is elementary abelian of order $\ge p^2$. On the other hand, $\Phi(G)$ (being contained in a minimal nonabelian subgroup) is abelian of rank $\le 3$, and so $|G'| \le p^3$. Set $Z = Z(G_2)$ so that

$$Z = \Phi(G_2) , \quad |\Phi(G) : Z| = p \quad \text{with } Z \le Z(G) .$$

On the other hand, $Z(G) \le G_1$, and assume that there is an element $z \in G_1 - \Phi(G)$ such that $z^p \in \Phi(G)$ and $z \in Z(G)$. But then $X_1 = \langle z\rangle G_2$ is an $A_2$-group with $\Phi(X_1) = \Phi(G)$

and $z \in Z(X_1)$, a contradiction (because $z$ is a generating element of $X_1$ and $d(X_1) = 2$). Hence we have $Z = Z(G)$. By the proof of Lemma 1.1, we have $G_1/Z \cong G'$ and we know that $G_1/Z$ is of order $\geq p^3$ (since $|G_1 : \Phi(G)| \geq p^2$). But $|G'| \leq p^3$ and so we must have:

$$G' \cong E_{p^3}, \quad |G_1 : \Phi(G)| = p^2, \quad \text{and so } d(G) = 3 .$$

For any $g \in G - G_1$, we have $g^p \in Z$ and so $\mho_1(G) \leq Z$. Also, $X$ is a maximal subgroup of $G$ and $cl(X) = 3$ so that $X'$ covers $\Phi(G)/Z$ and we have $\Phi(G) = ZG'$ with $Z \cap G' \cong E_{p^2}$. For each maximal subgroup $X_i$ of $G$ distinct from $G_1$, we have that $X_i$ is an $A_2$-group with $d(X_i) = 2$ and $X_i' \cong E_{p^2}$. Consider the factor-group $L = G/Z$. Then we have

$$|L| = p^4, \quad d(L) = 3, \quad |L'| = p, \quad \text{and} \quad \exp(L) = p > 2 .$$

Also, $G_1/Z \cong E_{p^3}$ and all other $p^2 + p$ maximal subgroups $X_i/Z$ of $L$ have its derived subgroup of order $p$ since $cl(X_i) = 3$ and $X_i' \cong E_{p^2}$ covers $\Phi(G)/Z$. On the other hand, if $S \cong S(p^3)$ (the nonabelian group of order $p^3$ and exponent $p$) is a minimal nonabelian subgroup of $L$, then $L = S \times C$, where $C \cong C_p$. But then all $p + 1$ maximal subgroups of $L$ containing $C$ are abelian, a contradiction.

We have proved that $G = G_1$ is of class 2 with $\Phi(G) \leq Z(G)$. For any $x, y \in G$, we have

$$[x, y]^p = [x^p, y] = 1 ,$$

and so $G' \neq \{1\}$ is elementary abelian. On the other hand, $\Phi(G)$ is contained in a minimal nonabelian subgroup of $G$ and so $\Phi(G)$ is abelian of rank $\leq 3$. Hence $|G'| \leq p^3$ and (viii) is proved.

It remains to determine the structure of $G$ in the special case $|G'| = p$. Let $H$ be a subgroup in $G$ such that

$$G = HZ(G), \quad H \cap Z(G) = \Phi(G)$$

so that we have

$$\Phi(H) = \Phi(G) = Z(H) .$$

If $G' = \Phi(G)$, then $H$ is extraspecial. Setting $S = Z(G)$, we have $G = H * S$, $S$ is abelian, $H \cap S = H'$ and $\mho_1(S) \leq S$.

In what follows we assume in addition that $G' = H' < \Phi(G)$. Let $k_1, k_2 \in H - \Phi(G)$ be such that $\langle [k_1, k_2] \rangle = H'$. Then $K = \langle k_1, k_2 \rangle$ is minimal nonabelian and so

$$K > \Phi(G), \quad K/\Phi(G) \cong E_{p^2}, \quad \text{and} \quad \Phi(K) = \Phi(G) \quad \text{with } |K| > p^3 .$$

Set $M = C_H(K)$ so that we have

$$H = K * M \quad \text{with } K \cap M = \Phi(G) .$$

Assume that $K \neq H$ so that $M > \Phi(G)$. If $M$ is abelian, then $M \leq Z(H)$, a contradiction. Hence $M$ is nonabelian and let $\Phi(G) < M_0 \leq M$ be such that $M_0$ is minimal nonabelian, where

$$M_0/\Phi(G) \cong E_{p^2} \quad \text{and} \quad \Phi(M_0) = \Phi(G) .$$

Let $K_i/K$ be two distinct subgroups of order $p$ in $(KM_0)/K$, $i = 1, 2$ and set $L_i = K_i \cap M_0$. Then $K_i$ is an $A_2$-group (because each maximal subgroup of $K_i$ is abelian or minimal nonabelian) with $|K_i'| = p$, where $L_i = Z(K_i)$ and $|L_i : \Phi(G)| = p$. By Proposition 71.1 (noting that $|K_i| > p^4$), there is an element $l_i \in L_i - \Phi(G)$ such that $l_i^p \in K' = M_0'$, $i = 1, 2$. Since $M_0 = \langle l_1, l_2 \rangle$, Exercise P9 in Prerequisites from volumes 1 and 2 (contained at the beginning of volume 3) implies that

$$\Phi(M_0) = \langle l_1^p, l_2^p, M_0' \rangle = M_0'$$

is of order $p$, contrary to $\Phi(M_0) = \Phi(G) > G'$. We have proved that $K = H$ is minimal nonabelian of order $> p^3$ with

$$\Phi(H) = \Phi(G) = Z(H) .$$

Let $G_i/H$, $i = 1, 2, \ldots, r$, be a minimal generating set of subgroups of order $p$ in $G/H$ which generate $G/H$, where $|G/H| = p^r$ and $r > 1$ (noting that $G$ is neither minimal nonabelian nor an $A_2$-group [see (i)]). Let $Z_i = G_i \cap Z(G)$ so that $G_i$ is always (for all $i$) an $A_2$-group of order $> p^4$ with $|G_i'| = p$ and $Z_i = Z(G_i)$. By Proposition 71.1, there is an element $z_i \in Z_i - \Phi(G)$ such that $z_i^p \in H'$. Set

$$S = \langle z_1, z_2, \ldots, z_r \rangle H' \quad \text{so that } G = H * S,\ H \cap S = H',\ \mho_1(S) \le H'$$

and $S$ is abelian. The theorem is proved. □

**Problem.** Classify the $p$-groups $G$ all of whose maximal nonabelian subgroups with cyclic derived subgroups contain $\Phi(G)$ (contain $G'$, contain $\mho_1(G)$).

# § 168 $p$-groups with given intersections of certain subgroups

*1°*. The structure of a finite $p$-group depends essentially on intersections of some of its subgroups. For example, if any two distinct maximal subgroups of a noncyclic $p$-group $G$ of order $> p^3$ have a cyclic intersection, then $G$ has a cyclic subgroup of index $p$.

Let $k, n$ be integers, $1 < k < n$. The groups of order $p^n$, $p > 2$, in which the intersection of all their subgroups of order $p^k$ is $\cong C_{p^{k-1}}$, are classified in [ZW]. In this section we reprove the main results of [ZW] and consider some related questions. See also [Ber3, Golo1, Golo2].

**Proposition 168.1.** *Let $G$ be a noncyclic group of order $p^n$ and let $k$ be an integer such that $n > k > 2$. If any two distinct subgroups of $G$ of order $p^k$ have a cyclic intersection, then one of the following holds:*
(a) *$G$ has a cyclic subgroup of index $p$,*
(b) *$G$ is a 2-group of Lemma 42.1 (c) with $n - k = 3$.*

*Proof.* A group $G$ with a cyclic subgroup, say $C$, of index $p$ satisfies the hypothesis. Indeed, if $U$, $V$ are distinct subgroups of $G$ of order $p^k$, then $U \cap C = \Omega_{k-1}(C) = V \cap C$ so $U \cap V = \Omega_{k-1}(C)$ is cyclic.

In what follows we assume that $G$ has no cyclic subgroup of index $p$. Then it contains a normal subgroup $R \cong E_{p^2}$ (Lemma 1.4). By hypothesis, $G/R$ has only one subgroup of order $p^{k-2}$ so that it is either cyclic or $k - 2 = 1$ and $G/R$ is a generalized quaternion group (Proposition 1.3).

Let $G/R$ be cyclic. Then $R < \Omega_1(G) \cong E_{p^3}$ (since $G$ has no cyclic subgroup of index $p$ and $C_G(R) > R$), $G = C \cdot R$, where $C \cong C_{p^{n-2}}$ is cyclic. Write $A = \Omega_{k-2}(C) \cdot R$ and $B = \Omega_{k-1}(C) \times L$, where $L \le Z(G) \cap R$ is of order $p$. Then $A, B$ are distinct of order $p^k$ and $A \cap B = \Omega_{k-2}(C) \times L$ is noncyclic of order $p^{k-1}$, a contradiction. Thus, in that case $G$ has a cyclic subgroup of index $p$.

If $\Omega_1(G) = R$ and $G/R \cong Q_{2^{n-2}}$, then $|\Omega_2(G)| = 8$ so that $G$ is as in Lemma 42.1 (c); this $G$ satisfies the hypothesis. If $R < \Omega_1(G)$, $G/R \cong Q_{2^{n-2}}$, one has $k = 3$ and $\Omega_1(G) \cong E_8$. Let $T/R < G/R$ be cyclic of index 2. Then $T$ does not satisfy the hypothesis (see the previous paragraph), a final contradiction. □

Recall that two subgroups $A, B \le G$ are *incident* if $A \cap B \in \{A, B\}$.

**Lemma 168.2.** *Let $G$ be a noncyclic $p$-group of order $p^n > p^2$ and $1 \le k < n - 1$. If any two subgroups of $G$ of orders $p^k$ and $p^{k+1}$ are incident, then $p = 2$, $k = 1$ and $G$ is a generalized quaternion group.*

*Proof.* Let $H < G$ be of order $p^{k+1}$. By hypothesis, all subgroups of $G$ of order $p^k$ are contained in $H$. Since $G$ is noncyclic, there is in $G$ a subgroup $F \ne H$ of order $p^{k+1}$ (Proposition 1.3). Again, all subgroups of $G$ of order $p^k$ are contained in $F$. It follows that $H \cap F$ is the unique subgroup of order $p^k$ in $G$, and we conclude that $p = 2$, $k = 1$ and $G$ is a generalized quaternion group (Proposition 1.3). □

**Theorem 168.3.** *Let $n, k$ be integers such that $2 < k < n - 1$. Suppose that a noncyclic $p$-group $G$ has order $p^n$. If the intersection of any two nonincident subgroups of $G$ of orders $p^k$ and $p^{k+1}$ is cyclic, then either $G$ has a cyclic subgroup of index $p$ or $G$ is a (metacyclic) 2-group of Lemma 42.1 (c) with $k = 3$.*

*Proof.* Assume that $G$ has no cyclic subgroup of index $p$ (such $p$-groups satisfy the hypothesis). Then there is in $G$ a normal abelian subgroup $R \cong \mathrm{E}_{p^2}$ (Lemma 1.4). In that case, any two subgroups of $G/R$ of orders $p^{k-2}$ and $p^{k-1}$ are incident, and so $G/R$ is either cyclic or a generalized quaternion group and in the last case, $k = 3$ (Lemma 168.2).

(i) Assume that $G/R$ is cyclic. Then $H = \Omega_1(G) \cong \mathrm{E}_{p^3}$ since $G$ has no cyclic subgroup of index $p$. In that case, $G = C \cdot R$ (semidirect product), where $C \cong \mathrm{C}_{p^{n-2}}$. Let $S/R < G/R$ be of order $p^{k-2}$; then $|S| = p^k$ and $H = \Omega_1(G) < S$. Let $T = C \times L$, where $L \leq R \cap \mathrm{Z}(G)$ is of order $p$; then $|T| = p^{k+1}$. The subgroups $S$ and $T$ are nonincident and contain the subgroup $\Omega_1(C) \times L \cong \mathrm{E}_{p^2}$, a contradiction. Thus, in the case under consideration, $G$ has a cyclic subgroup of index $p$.

(ii) Now let $G/R$ be a generalized quaternion group and $k = 3$. Assume that $R < \Omega_1(G)$; then $\Omega_1(G) \cong \mathrm{E}_8$. Let $T/R < G/R$ be cyclic of index 2 in $G/R$. Applying (i) to $T$, we get a contradiction. Thus, $\Omega_1(G) = R$; then $|\Omega_2(G)| = 8$, hence $G$ is as in Lemma 42.1 (c). □

*2°.* In this subsection we replace cyclic intersections, as in the previous subsection, by absolutely regular ones. Recall that a $p$-group $G$ is absolutely regular provided $|G/\mho_1(G)| < p^p$. Since absolutely regular 2-groups are cyclic, we assume in what follows that $p > 2$ (see 1°).

Recall that a $p$-group $G$ is an $\mathrm{L}_m$-group provided $\Omega_1(G)$ is of order $p^m$ and exponent $p$ and $G/\Omega_1(G)$ is cyclic of order $> p$ (see §§ 17 and 18). For example, $\mathrm{M}_{p^n}$, $n > 3$, is an $\mathrm{L}_2$-group.

**Theorem 168.4.** *Suppose that $2 < p < k < n$ and a $p$-group $G$ of order $p^n$ is not absolutely regular. If any two distinct subgroups of $G$ of order $p^k$ have an absolutely regular intersection, then $G$ is either an $\mathrm{L}_p$-group or a $p$-group of maximal class.*

*Proof.* The irregular $p$-groups of maximal class satisfy the hypothesis (by Theorem 9.6, any proper non-absolutely regular subgroup, say $A$, of order $p^{k-1}$ of our group is contained in exactly one subgroup of order $p|A| = p^k$). In what follows we assume that $G$ is not of maximal class. Then $G$ contains a normal subgroup $R$ of order $p^p$ and exponent $p$ (Theorem 12.1 (a)). By hypothesis, $G/R$ has only one subgroup of order $p^{k-p} (\geq p)$. By Proposition 1.3, $G/R$ is cyclic (of order $p^{n-p} > p$) since $p > 2$. To prove that $G$ is an $\mathrm{L}_p$-group, it remains to show that $\Omega_1(G) = R$). Assume that this is false.

Write $H/R = \Omega_1(G/R)$; then $|H| = p^{p+1}$. Since $G/R$ is cyclic of order $> p$, $G$ is not of maximal class, and therefore, by Exercise 10.10, $H$ is not of maximal class. It follows that $H$ is regular (Theorem 7.1 (b)). By assumption, $\Omega_1(G) = H$. Then $\exp(H) = p$ (Theorem 7.2 (b)). In that case, $G = C \cdot R$ (semidirect product with kernel $R$), where $C$ is cyclic. Let $H \leq L < G$, where $|L| = p^k$ and $\Omega_{k-p}(C) < M \leq \Omega_{k-p}(C)R$, where $|M| = p^k$. Then $L \cap M$ contains a subgroup of order $p^p$ and exponent $p$, a contradiction. Thus, $\Omega_1(G) = R$ and $G/R$ is cyclic of order $p^{n-p} > p$ so that $G$ is an $\mathrm{L}_p$-group. □

**Theorem 168.5.** *Suppose that $p > 2, p < k < n-1$ and a $p$-group $G$ is not absolutely regular, $|G| = p^n$. If any two distinct subgroups of $G$ of orders $p^k$ and $p^{k+1}$ have absolutely regular intersection, then $G$ is either an $\mathrm{L}_p$-group or a $p$-group of maximal class.*

*Proof.* We present only the first part of the proof. Assume that $G$ is a counterexample. Then there is $R \lhd G$ of order $p^p$ and exponent $p$. By hypothesis, any subgroup of $\bar{G} = G/R$ of order $p^{k-p}$ is contained in a fixed subgroup of $\bar{G}$ of order $p^{k-p+1}$. Thus, the intersection of all subgroups of $\bar{G}$ of order $p^{k-p+1}$ has order $p^{k-p}$, moreover, it contains all such subgroups. It follows that $\bar{G}$ contains only one subgroup of order $p^{k-p}$, i.e., $\bar{G}$ is either cyclic or generalized quaternion group (Proposition 1.3); in the second case, $k - p = 1$. □

Thus, groups of Theorems 168.4 and 168.5 coincide.

Some related results are proved in Appendix 67 (see there Theorem A.67.1 and Proposition A.67.2).

**Exercise 1.** Let $k > 2$ be a fixed integer. Classify the nonabelian $p$-groups of order $p^n > p^{k+1}$, all of whose subgroups of order $p^k$, except one, are cyclic.

**Solution.** If $G$ has a cyclic subgroup of index $p$, it is $\cong \mathrm{M}_{p^n}$ (Theorem 1.2). Now let $G$ have no cyclic subgroup of index $p$. Let $R \lhd G$ be abelian of type $(p, p)$ (Lemma 1.4). In that case, $G/R$ has exactly one subgroup of order $p^{k-2}$. Then, by Proposition 1.3, $G/R$ is either cyclic or a generalized quaternion group. It is easy to see that in the first case, $\Omega_1(G) = R$, and then $G$ has a cyclic subgroup of index $p$. Let us consider the second case. Let $T/R < G/R$ be cyclic of index 2. By the first case, $\Omega_1(T) = R$ so that $\Omega_1(G) = R$. It follows that then $|\Omega_2(G)| = 8$ so that $G$ is a (metacyclic) group of Lemma 42.1 (c).

**Exercise 2.** Given $k > 1$, classify the $p$-groups of order $p^n > p^{k+2}$, all of whose subgroups of order $p^k$ generate the subgroup of order $p^{k+2}$.

**Exercise 3.** Given $k > 1$, classify the $p$-groups of order $p^n > p^{k+2}$, all of whose subgroups of orders $p^k$ and $p^{k+2}$ are incident.

**Exercise 4.** Given $k > 1$, classify the $p$-groups of order $p^n > p^{k+2}$, all of whose nonincident subgroups of orders $p^k$ and $p^{k+2}$ have cyclic intersection.

**Problem 1.** Given $k > 1$, classify the $p$-groups in which all subgroups of order $p^k$ are contained in all subgroups of order $p^{k+3}$. Using this, study the $p$-groups in which the intersection of any two nonincident subgroups of orders $p^k$ and $p^{k+3}$ is cyclic (absolutely regular provided $k > p$).

**Problem 2.** Study the $p$-groups in which the intersection of any two distinct maximal subgroups is either abelian or minimal nonabelian.

**Problem 3.** Study the 2-groups in which the intersection of any two distinct maximal subgroups is metacyclic.

# § 169 Nonabelian $p$-groups $G$ with $\langle A, B\rangle$ minimal nonabelian for any two distinct maximal cyclic subgroups $A, B$ of $G$

The purpose of this section is to solve the problem 2897 of the first author. As a result, a new characterization of the ordinary quaternion group is obtained.

**Theorem 169.1** (Janko). *Let $G$ be a nonabelian $p$-group such that for any two distinct maximal cyclic subgroups $A \neq B$ of $G$, the subgroup $\langle A, B\rangle$ is minimal nonabelian. Then $G$ is isomorphic to the ordinary quaternion group $Q_8$.*

*Proof.* Let $G$ be a nonabelian $p$-group satisfying the assumption of the theorem and let $M = \langle a, b\rangle$ be a minimal nonabelian subgroup in $G$, where $a, b$ are any two generating elements for $M$. Then $\langle a\rangle \neq \langle b\rangle$ are nonincident maximal cyclic subgroups in $M$. Let $A \geq \langle a\rangle$ and $B \geq \langle b\rangle$ be maximal cyclic subgroups of $G$ containing $\langle a\rangle$ and $\langle b\rangle$ respectively, so that we have $A \neq B$. By our assumption, $\langle A, B\rangle$ is minimal nonabelian. Since $\langle A, B\rangle \geq \langle a, b\rangle = M$, we have $\langle A, B\rangle = M$. This implies that $A = \langle a\rangle$ and $B = \langle b\rangle$, and so $\langle a\rangle$ and $\langle b\rangle$ are two distinct maximal cyclic subgroups in $G$.

Suppose that $\langle a\rangle$ is not normal in $M$. Then there is $x \in M$ such that

$$\langle a\rangle^x \neq \langle a\rangle ,$$

and so $\langle a\rangle$ and $\langle a\rangle^x$ are two distinct maximal cyclic subgroups in $G$. On the one hand, $\langle a, a^x\rangle$ is a proper subgroup of $M$ (indeed, the normal closure in $M$ of any proper subgroup of $M$ is a proper subgroup of $M$), and so $\langle a, a^x\rangle$ is abelian. On the other hand, by hypothesis, $\langle a, a^x\rangle$ is minimal nonabelian, and this is a contradiction.

We have proved that the subgroup $\langle a\rangle$ is normal in $M$, and similarly the subgroup $\langle b\rangle$ is also normal in $M$. In particular, $M$ is metacyclic (see Lemma 65.1) and we have

$$\langle a\rangle \cap \langle b\rangle \geq M' \cong C_p .$$

It follows that $M$ is a metacyclic minimal nonabelian group and that $M$ is not a splitting extension for any two cyclic subgroups that generate $M$. By the classification of minimal nonabelian groups (see Lemma 65.1), we have $M \cong Q_8$.

We have proved that each minimal nonabelian subgroup of $G$ is isomorphic to $Q_8$. By Corollary A.17.3, we have

$$G = Q \times V, \quad \text{where } Q \cong Q_{2^n},\ n \geq 3, \text{ and } \exp(V) \leq 2 .$$

However, if $n > 3$, then $Q$ is generated by some two distinct maximal cyclic subgroups (that are also maximal cyclic in $G$) and $Q$ is not minimal nonabelian, contrary to our assumption. Hence we have $n = 3$ and so $G$ is nonabelian Dedekindian.

Assume that $V \neq \{1\}$ and let $z$ be an involution in $V$. Then $\langle z\rangle$ is a maximal cyclic subgroup in $G$. Let $\langle v\rangle$ be a cyclic subgroup of order 4 in $Q \cong Q_8$ so that $\langle v\rangle$ is also a maximal cyclic subgroup in $G$. But $\langle v, z\rangle$ is abelian, contrary to our assumption. We have proved that $G = Q \cong Q_8$. □

**Exercise 1.** Suppose that a $p$-group $G$ contains a maximal cyclic subgroup $A = \langle a\rangle$ such that $M = \langle A, B\rangle$ is minimal nonabelian for any maximal cyclic subgroup $B \neq A$ of $G$. Then $A \triangleleft M$.

**Solution.** Assume that $A$ is not normal in $M$. Then there is $x \in M$ such that $A^x \neq A$ so that $\langle A, A^x\rangle = M$. This is a contradiction since $A^M < M$. Thus, $A \triangleleft M$.

**Exercise 2.** Suppose that any two distinct maximal cyclic subgroups of a $p$-group $G$ generate the maximal metacyclic subgroup of $G$. Is it true that $G$ is Dedekindian?

**Problem 1.** Classify the nonabelian $p$-groups $G$ such that, whenever $A, B < G$ are distinct maximal cyclic, then $\langle A, B\rangle$ is either abelian or minimal nonabelian.

**Problem 2.** Classify the nonabelian $p$-groups $G$ such that, whenever $A, B < G$ are distinct maximal cyclic and $A \cap B > \{1\}$ $(A \cap B = \{1\})$, then $\langle A, B\rangle$ is (i) minimal nonabelian, (ii) either abelian or minimal nonabelian.

**Problem 3.** Classify the nonabelian $p$-groups $G$ such that, whenever $M \leq G$ is minimal nonabelian and $C < M$ is maximal cyclic not contained in $\Phi(M)$, then $C$ is a maximal cyclic subgroup of $G$.

# § 170 $p$-groups with many minimal nonabelian subgroups, 2

Let $G$ be a nonabelian $p$-group and let $A$ be a maximal normal abelian subgroup of $G$. Then Lemma 57.1 implies that for any $x \in G - A$, there is $a \in A - Z(G)$ so that $\langle a, x\rangle$ is minimal nonabelian. This is a general lemma and this statement is true for an arbitrary $p$-group $G$. The first author came to the interesting idea to ask what can be said about the structure of a $p$-group $G$ with a maximal normal abelian subgroup $A$ such that for any $x \in G - A$ and any $a \in A - Z(G)$, the subgroup $\langle a, x\rangle$ is minimal nonabelian (see Problem 1980(ii)). We find an answer in the following theorem.

**Theorem 170.1** (Janko). *Let $G$ be a nonabelian $p$-group with a maximal normal abelian subgroup $A$ so that for all $x \in G - A$ and $a \in A - Z(G)$, the subgroup $\langle a, x\rangle$ is minimal nonabelian. Then we have:*

(a) $\mho_1(G) \le Z(G)$, $A \le Z_2(G)$, $[A, G] \le \Omega_1(Z(G))$, $G' \le A$,

(b) $K_4(G) = \{1\}$, *i.e., the class of $G$ is $\le 3$,*

(c) *if $p = 2$, then the class of $G$ is equal to 2.*

*Proof.* Let $G$ be a nonabelian $p$-group satisfying the assumption of the theorem. We have $Z(G) < A < G$ and let $x \in G-A$ and $a \in A-Z(G)$ so that we have by our assumption that $\langle a, x\rangle$ is minimal nonabelian. Then we have $x^p \in Z(\langle a, x\rangle)$, and so $x^p$ centralizes $a$, which implies that $x^p \in C_G(A)$, and we conclude that $x^p \in A$. But if $x^p \in A - Z(G)$, then by our assumption $\langle x, x^p\rangle$ must be minimal nonabelian, a contradiction. Hence we have $x^p \in Z(G)$. Also we have $a^p \in Z(\langle a, x\rangle)$, and so $a^p$ centralizes $x$. This implies $a^p \in Z(G)$ in view of the arbitrariness of $x \in G - A$. We have proved that $\mho_1(G) \le Z(G)$.

Further we have:

$$[a, x] \in [A, G] \le A \quad \text{and} \quad C_p \cong \langle [a, x]\rangle = \langle a, x\rangle' \le Z(\langle a, x\rangle)\,,$$

so that $[a, x]$ centralizes $x$, which forces

$$[a, x] \in C_G(\langle G - A\rangle) = Z(G)\,.$$

We have proved that $[A, G]$ is elementary abelian and is contained in $Z(G)$. In particular, we get $A \le Z_2(G)$.

Let $a \in A - Z(G)$ and $x, y \in G$. Since $a \in Z_2(G)$, we have

$$a^x = az\,, \quad a^y = au\,, \quad \text{where } z, u \in Z(G)$$

and this also gives

$$a^{x^{-1}} = az^{-1} \quad \text{and} \quad a^{y^{-1}} = au^{-1}\,.$$

Thus we get

$$a^{[x,y]} = a^{x^{-1}y^{-1}xy} = az^{-1}u^{-1}zu = a\,.$$

It follows that $[x, y]$ centralizes $a$, and so $[x, y] \in A$. We have proved that $G' \leq A$.

Since $[A, G] \leq Z(G)$ and $G' \leq A$, we have

$$K_3(G) = [G', G] \leq [A, G] \leq Z(G) ,$$

and so $G$ is of class at most 3. If $p = 2$, then

$$\Phi(G) = \mho_1(G) \leq Z(G)$$

and so in this case $G$ is of class 2. Our theorem is proved. □

**Remark 1.** The Blackburn $p$-groups, $p > 2$, of order $p^5$ and class 3, defined in Proposition 71.5 (b), satisfy the hypothesis of Theorem 170.1.

**Problem.** Suppose that $A$ is a fixed maximal abelian subgroup of a nonabelian $p$-group $G$, $x \in G - A$ and $y \in A - Z(G)$. Set $H = \langle x, y \rangle$. Study the structure of $G$ provided $H$ (i) is minimal nonabelian, (ii) is of class 2, (iii) is metacyclic, (iv) is a group with cyclic derived subgroup (Frattini subgroup), (v) is of maximal class, (vi) has an abelian subgroup of index $p$.

# § 171 Characterizations of Dedekindian 2-groups

The intersection $\mathcal{N}(G)$ of normalizers of all subgroups of a group $G$ is said to be the *norm* of $G$ (see § 143). The subgroup $\mathcal{N}(G)$ is characteristic in $G$ and $Z(G) \leq \mathcal{N}(G)$. If $H \leq G$, then $H \cap \mathcal{N}(G) \leq \mathcal{N}(H)$. Clearly, the norm is Dedekindian, i.e., all its subgroups are normal. If $L \lhd G$ and $L < \mathcal{N}(G)$, then $\mathcal{N}(G)/L \leq \mathcal{N}(G/L)$. Moreover, if $A \leq \mathcal{N}(G)$, $L \leq A$ is normal in $G$, then $A/L \leq \mathcal{N}(G/L)$. The last fact we use freely in what follows.

By Corollary 140.8, $\mathcal{N}(G) \leq Z_2(G)$, the second member of the upper central series of $G$ (this result is due to E. Shenkman). If $\Omega_1(G) = G$, then $\mathcal{N}(G) = Z(G)$.

A subgroup $H$ of a group $G$ is said to be *quasinormal* if $H$ is permutable with all subgroups of $G$. All nonnormal subgroups of the group $M_{p^n}$ are quasinormal. All quasinormal subgroups of a 2-group of maximal class are normal (see Remark 2). A $p$-group is said to be *modular* if all of its subgroups are quasinormal (for nonnilpotent groups the definition is another). The property to be modular is inherited by subgroups and epimorphic images. The modular $p$-groups were classified by K. Iwasawa [Iwa] (see also § 73 and [Schm5, §§ 2.3 and 2.4]); this is a fairly deep result. This section is independent of Iwasawa's classification.

R. Baer [Bae3] has proved that if the norm $\mathcal{N}(G)$ of a 2-group is nonabelian, then $G$ is Dedekindian (see also Theorem 143.1 and Appendix 24). In this section we offer alternate proof of this theorem. Some arguments from Baer's proof are used essentially in the proof of our main result, Theorem 171.4.

We begin with the following two easy lemmas.

**Lemma 171.1.** *Suppose that a $p$-group $G$ contains three distinct elementary abelian subgroups of index $p$, say $A, B, C$.*

(a) *If $p = 2$, then $G$ is elementary abelian.*

(b) *If $p > 2$, then either $G$ is elementary abelian or $G = S(p^3) \times E$, where $S(p^3)$ is nonabelian of order $p^3$ and exponent $p$ and $E$ is elementary abelian.*

*Proof.* Since $\Omega_1(G) = G$, it follows that if $G$ is abelian, it is elementary. Next we assume that $G$ is nonabelian. Then $A \cap B = Z(G) = B \cap C = C \cap A$ has index $p^2$ in $G$ so $\mathrm{cl}(G) = 2$ and $\exp(G) \leq p^2$. Therefore, by Theorems 7.1 (b) and 7.2 (b), if $p > 2$, then $\exp(G) = p$. If $|G| = p^3$, then $p > 2$ and $G \cong S(p^3)$. Next we assume that $|G| > p^3$.

(i) Suppose that $p > 2$. Let $M \leq G$ be minimal nonabelian; then $M \cong S(p^3)$ (Lemma 65.1) so $M < G$ and $MZ(G) = G$, by the product formula. Then $Z(G) = Z(M) \times E$ for some $E < Z(G)$ of index $p$. In that case, $G = M \times E$, completing the proof of (a).

(ii) Let $p = 2$. It follows from $G = A \cup B \cup C$ that $G$ is elementary abelian. □

**Lemma 171.2.** *If $G$ is a 2-group of maximal class and order $> 2^3$, then $\mathcal{N}(G) = Z(G)$, unless $G \cong Q_{2^n}$, in which case $\mathcal{N}(G) \cong C_4$.*

*Proof.* Let $L, M < G$ be nonnormal subgroups of order 4 lying in distinct maximal subgroups of $G$. Then $LM \neq ML$ so $M, L \nleq \mathcal{N}(G)$. The cyclic subgroup of $G$ of order 8 does not normalizes a nonnormal subgroup of order 4. The normal (cyclic) subgroup of order 4 normalizes all subgroups of $G \iff G$ is generalized quaternion group. □

**Theorem 171.3.** *If a 2-group $G$ contains a subgroup $Q \cong Q_8$ that normalizes all subgroups of $G$ of order $\leq 8$, then $G$ is Dedekindian.*

*Proof.* Assume that $G$ is non-Dedekindian; then $|G| > |Q| = 2^3$.

If $G$ has only one subgroup of order 2, it is a generalized quaternion group. Let $Q < H \leq G$, where $|H : Q| = 2$. Then $Q$ does not normalize any subgroup of $H$ of order 4 not contained in $Q$. Thus, $|\Omega_1(G)| > 2$. Obviously:

(i) If $x \in G - Q$ is an involution, then $x$ centralizes $Q$.

(ii) We claim that if $X < G$ is cyclic of order 4, then $Q \cap X > \{1\}$. Assume that this is false. First assume that $H = QX = Q \times X$; then $Z(H) = \Omega_1(Q) \times \Omega_1(X) \cong E_4$. If $L < H$ is of order 2 such that $L \nleq Q$ and $L \nleq X$, then $H/L$ (of order 16) is the central product of $Q_1 \cong Q$ and $X_1 \cong X$ with amalgamated subgroup of order 2. But $H/L$ contains a subgroup, say $K/L$, of order 2 that is not normalized by $QL/L$ (Appendix 16); then the subgroup $K$ of order 4 is not normalized by $Q$, a contradiction. Let $Q = AB$, where $A, B$ are cyclic of order 4. One may assume, without loss of generality, that $F = A \cdot X$ is a nonabelian semidirect product with kernel $X$ (otherwise, $Q \times X$, which is impossible, by the above). Then $F/\Omega_1(A) \cong D_8$ so $A/\Omega_1(A)$ does not normalize some subgroup, say $M/\Omega_1(A) < F/\Omega_1(A)$, of order 2. Then $A$ does not normalize the subgroup $M$ of order 4, a contradiction. Thus, $Q \cap X > \{1\}$.

(iii) We claim that $\exp(C_G(Q)) = 2$. Indeed, if a cyclic $X < G$ of order 4 is contained in $C_G(Q)$, then $H = Q * X$ is of order 16, by (ii). In that case, by Appendix 16, the non-Dedekindian group $H$ contains a nonnormal subgroup, say $L$, of order 2, a contradiction since $Q$ does not normalize $L$.

(iv) If $|G'| = 2$, then $G$ is Dedekindian. Indeed, by Lemma 4.2, $G = Q * U$ (central product), where $U = C_G(Q)$. Since $\exp(U) = 2$, by (iii), we get $U = Z(Q) \times E$, where $\exp(E) = 2$. Then $G = Q \times E$ is Dedekindian.

Next we use induction on $|G|$. By induction, all maximal subgroups of $G$ containing $Q$ are (nonabelian) Dedekindian and, if $H$ is one of such maximal subgroups, then $Q' = H' \trianglelefteq G$ since $H'$ is characteristic in $H \trianglelefteq G$, and the quotient group $H/H' = H/Q'$ is elementary abelian, by Theorem 1.20. The number $\mu(Q)$ of maximal subgroups of $G$ containing $Q$, is odd. Write $\bar{G} = G/Q'$.

(v) Assume that $\mu(Q) > 1$; then $\mu(Q) \geq 3$. In that case the quotient group $\bar{G}$ contains three distinct elementary abelian subgroups of index 2 (see (iii)) so it is elementary abelian (Lemma 171.1), and we conclude that $G' = Q'$ has order 2. It follows from (iv) that $G$ is Dedekindian.

(vi) Thus, there is a unique maximal subgroup, say $A$, containing $Q$. If $A = Q$ then, as in (i), $G$ is Dedekindian. Next we assume that $|G| > 16$. By induction, $A$ is (nonabelian) Dedekindian. Let $y \in G - A$ and $Y = \langle y \rangle$; then $o(y) \leq \exp(G) \leq 2\exp(A) = 8$ so that

$Q$ normalizes $Y$, by hypothesis. It follows from $\mu(Q) = 1$ and $QY \nleq A$ that $QY = G$. Since $|G : Q| > 2$, we get $|Y| > 2$ and, by (ii), $|Y| = 8$ (indeed, $Q \cap Y > \{1\}$). Since $A \cap Y$ is cyclic of order 4, it follows that $Q \cap Y > \{1\}$ again, and, by the product formula, $|G| = 32$. Let $Q = UV$, where $U, V < Q$ are of order 4. Set $C = UY$, $D = VY$ (recall that $Q$ normalized $Y$). Then $C$ and $D$ are not of maximal class since $U \leq \mathcal{N}(C)$ and $V \leq \mathcal{N}(D)$ and $U, V \nleq Y$ (see Lemma 171.2). Besides, $|C : Y| = 2 = |D : Y|$. In that case, by Theorem 1.2, $\mho_1(Y)$ centralizes $U$ and $V$ (indeed, $\mho_1(Y) \leq \mathrm{Z}(C) \cap \mathrm{Z}(D)$) so it also centralizes $Q = UV$. It follows that $\mathrm{C}_G(Q)$ contains a cyclic subgroup $\mho_1(Y)$ of order $4 > 2$, contrary to (iii). □

In particular, (Baer) if the norm of a 2-group $G$ is nonabelian, then $G$ is Dedekindian.

**Remark 1.** (I) A minimal nonabelian $p$-group $G$ is modular $\iff$ one of the following holds: (a) $G \cong \mathrm{Q}_8$, (b) $G = \langle a, b \mid a^{2^m} = b^{2^n} = 1,\ m > 2,\ a^b = a^{1+2^{m-1}} \rangle$, or (c) $p > 2$, $G$ is metacyclic. This follows immediately from the following fact: A $p$-group $G$ is modular $\iff$ it has no sections isomorphic to $S(p^3)$ and $\mathrm{D}_8$ (see Appendices 24 and 59). (II) If a nonabelian modular group has exponent 4, it is Dedekindian [Schm5, Lemma 2.3.7]. Indeed, let $M \leq G$ be minimal nonabelian. Then, by (I), $M \cong \mathrm{Q}_8$. Therefore, by Corollary A.17.3, $G \cong Q \times E$, where $Q \cong \mathrm{Q}_8$ and $\exp(E) \leq 2$, i.e., $G$ is Dedekindian. (III) It follows that a $\mathrm{Q}_8$-free modular 2-group $G$ is powerful. Indeed, by (II), $G/\mho_2(G)$ is abelian so $G' \leq \mho_2(G)$ (see § 26).

**Remark 2.** Suppose that a 2-group $G$ is of maximal class and order $> 2^3$. We claim that all quasinormal subgroups of $G$ are normal. Assume that there is a quasinormal but nonnormal $H < G$. Let $H_G = \bigcap_{x \in G} H^x$; then $H_G = H \cap T$, where $T < G$ is cyclic of index 2. Without loss of generality one may assume that $H_G = \{1\}$. Then $|H| = 2$ and $G$ is either dihedral or semidihedral. In that case, there is in $G$ a subgroup $F$ of order 2 such that $HF \neq FH$, a contradiction.[1]

The following theorem improves Theorem 171.3.

**Theorem 171.4.** *Suppose that $Q \cong \mathrm{Q}_8$ is a proper subgroup of a 2-group $G$. Then the following conditions are equivalent:*

(a) *$G$ is Dedekindian, and*

(b) *whenever a cyclic $A < G$ is of order $\leq 8$, then the subgroup $\langle A, Q \rangle$ is modular (then $QA = AQ$).*

*Proof.* (a) ⇒ (b) is obvious so it remains to prove that (b) ⇒ (a).

We proceed by induction on $|G|$. Then all $\mu(Q)$ maximal subgroups of $G$ containing $Q$ are Dedekindian. The number $\mu(Q)$ is odd.

1 A slight modification of this argument shows that if $p > 2$ and $G$ is a $p$-group of maximal class and order $> p^3$ with abelian subgroup $G_1$ of index $p$, then all quasinormal subgroups of $G$ are normal.

(i) Suppose that $|G| = 16$. By hypothesis, $G$ is modular so that it is not of maximal class (Remark 2), hence $\exp(G) = 4$ (Proposition 1.6 and Theorem 1.2). By part (II) of Remark 1, $G$ is Dedekindian. In what follows we assume that $|G| > 16$.

(ii) Assume that $\exp(C_G(Q)) > 2$. Let $X < C_G(Q)$ be cyclic of order 4. Then $H = Q * X$ (of order $\leq 32$) is modular and $\exp(H) = 4$ (if $Q \cap X > \{1\}$, then $H/Q'$ is elementary abelian). As $\exp(Z(H)) = 4$, $H$ is non-Dedekindian (Theorem 1.20). However, by part (II) of Remark 1, $H$ is Dedekindian, and this is a contradiction. Thus, $\exp(C_G(Q)) = 2$.

(iii) Assume that $\mu(Q) > 1$; then $\mu(Q) \geq 3$. If $Q < M \in \Gamma_1$, then $Q' = M'$, $Q' \triangleleft G$ and $M/Q'$ is elementary abelian. Thus, $G/Q'$ has 3 distinct elementary abelian maximal subgroups so $G/Q'$ is elementary abelian (Lemma 171.1). It follows that $G' = Q'$. Then $G = QC_G(Q)$ (Lemma 4.2). By (ii), $\exp(C_G(Q)) = 2$ so that $C_G(Q) = Z(G)$. It follows that $Z(G) = Z(Q) \times E$, where $E < G$ is of exponent 2. Then $G = Q \times E$ is Dedekindian.

Next we assume that $\mu(Q) = 1$. Let $H$ be the unique maximal subgroup of $G$ containing $Q$. By induction, $H$ is Dedekindian.

Let $Y < G$ be cyclic not contained in $H$ such that $|Y|$ is as large as possible. Since $\exp(G) \leq 2 \cdot \exp(H) = 8$, we get $|Y| \leq 8$ so that, by hypothesis, $QY = YQ$ since $\langle Q, Y \rangle$ is modular. Since $\mu(Q) = 1$ and $QY \nleq H \in \Gamma_1$, we get $QY = G$. It follows from $|G| > 16$ that $|Y| > 2$. If $|Y| = 4$, then $|G| = 32$ and $\exp(G) = 4$, by the choice of $Y$, hence $G$ is modular, by hypothesis. Therefore, $G$ is Dedekindian, by part (II) of Remark 1.

It remains to consider the case when $|Y| = 8$. Then $Y \cap H$ is cyclic of order 4 so $Q' = \mho_1(H) < Y \cap H$, i.e., $Y \cap Q > \{1\}$. By the product formula, $|G| = 32$ and $G$ is modular, by hypothesis. Let $Q = AB$, where $A, B < Q$ are of order 4 and such that the subgroups $S = AY$ and $T = BY$ are nonabelian (such $A$ and $B$ exist since $Y$ does not centralize $Q$, by (ii)). By Remark 2, the subgroups $S, T$ are not of maximal class so that $S, T \cong M_{2^4} \cong T$ (Theorem 1.2). Then $Z(S) = \mho_1(Y) = Z(T)$ is of order 4. In that case, $C_G(\mho_1(Y)) \geq AB = Q$, contrary to (ii). □

**Lemma 171.5.** *If a modular 2-group $G$ has a section $\cong Q_8$, then it has a subgroup $\cong Q_8$.*

*Proof.* Suppose that the lemma holds for all modular 2-groups of order $< |G|$. Let $M/L \cong Q_8$ be a section of a modular 2-group $G$. One may assume that $M = G$ (otherwise, the result follows by induction applied to $M$). Let $R \triangleleft G$ be a subgroup of order 2 in $L$. Then $G/R$ has a subgroup $H/R \cong Q_8$, by induction, and one may assume that $H = G$. Then $G/L \cong Q_8$ so that $|G| = 16$. By Theorem 1.2, $G$ is not of maximal class so $\exp(G) = 4$ (Theorem 1.2). Then $G$ is Dedekindian, by part (II) of Remark 1, and we are done, by Theorem 1.20. □

**Corollary 171.6** (Iwasawa [Iwa]). *If a modular 2-group has a section $\cong Q_8$, it is Dedekindian.*

This follows from Theorem 171.4 and Lemma 171.5 (see also Theorem 44.13, Corollary A.24.2 and [Schm3, Section 2.3]).

It follows that if a nonabelian modular 2-group is non-Dedekindian, it has a characteristic subgroup of index 2. Indeed, this group, by Corollary 171.6, is $Q_8$-free, and so the assertion follows from Ward's Theorem 56.1. Next, it is proved in [Waa3] that every nonabelian modular $p$-group $G$, $p > 2$, contains a characteristic subgroup of index $p$.

**Remark 3.** Best and Taussky [BT] have proved that for a $p$-group $G$ the following conditions are equivalent: (a) $A \triangleleft B \triangleleft G \Rightarrow A \triangleleft G$ (i.e., normality is transitive), and (b) $G$ is Dedekindian. Indeed, all subnormal subgroups of $G$ are normal. As all subgroups of $G$ are subnormal, we are done.

**Exercise 1.** (a) Study the 2-groups $G$ containing a proper nonabelian metacyclic subgroup $\mathcal{H}_{2.2}$ of order 16 and exponent 4 such that all subgroups of $\mathcal{H}_{2,2}$ are permutable with all subgroups of $G$ not contained in $\mathcal{H}_{2,2}$.
(b) The same question for $M_{2^n}$ instead of $\mathcal{H}_{2,2}$.

**Exercise 2.** Suppose that $G$ is a $p$-group with $|G'| = p^k > p$. Prove that there is a nonnormal subgroup $H$ of order $p^k$ such that $|H_G| = p^{k-1}$. (**Hint.** The group $G$ is non-Dedekindian [Theorem 1.20]. Use Theorem 1.23.)

**Exercise 3.** Classify the nonabelian $p$-groups containing exactly two elementary abelian subgroups, say $A$ and $B$, of index $p$.

**Solution.** The subgroup $A \cap B = Z(G)$ is elementary abelian of index $p^2$ in $G$. Let $M \leq G$ be minimal nonabelian; then $G = MZ(G)$. As $M \cap A$ and $M \cap B$ are distinct elementary abelian of index $p$ in $M$, one has $M \cong D_8$ (check!). As in Lemma 171.1, one has $G \times E$, where $\exp(E) \leq 2$.

**Exercise 4.** Let $D \cong D_{2^n}$ be a proper subgroup of a 2-group and suppose that $G$ is not of maximal class. Then there is $L < G$ of order 2, $L \nleq D$ such that $D$ does not normalize $L$.

**Solution.** Assume that the assertion is false. By Theorem 1.17 (a), there is in $G$ a subgroup $L = \langle x \rangle$ of order 2 not contained in $D$ (indeed, if $L$ does not exist, then $G$ is of maximal class, by that theorem). In that case, $H = DL = D \times L$, by assumption. If $a$ is a noncentral involution of $D$, then $D$ does not normalize the subgroup $\langle ax \rangle$ of order 2, contrary to the assumption.

**Problem 1.** Let a noncentral subgroup $H$ of a $p$-group $G$ normalize all those cyclic subgroups of $G$ that are not incident with $H$. Describe the structures of groups $H$ and $G$.

**Problem 2.** Classify the $p$-groups all of whose nonquasinormal subgroups (i) are cyclic, and (ii) have equal order.

**Problem 3.** Study the $p$-groups $G$ with $|G'| = p^k$ without nonnormal subgroups of order $> p^k$.

**Problem 4.** Study the non-Dedekindian $p$-groups all of whose maximal nonnormal subgroups are quasinormal.

**Problem 5.** Study the intersection of normalizers of all nonquasinormal subgroups of $G$.

**Problem 6.** Classify the $p$-groups $G$ with transitive quasinormality.

**Problem 7.** Classify the nonabelian $p$-groups in which all subgroups of $\mathcal{A}_1$-subgroups of $G$ are quasinormal in $G$.

# § 172 On 2-groups with small centralizers of elements

The following three results were proved by the second author.

**Theorem 172.1.** *If $G$ is a 2-group containing a minimal nonmetacyclic subgroup $H$ of order $2^5$ such that $C_G(x) \le H$ for all elements $x \in H$ of order 4, then $G = H$.*

*Proof.* By Theorem 66.1, $H$ is a unique special 2-group with

$$\Omega_1(H) = H' = Z(H) = \Phi(H) \cong E_4 ,$$

which possesses exactly one abelian maximal subgroup $A$ (of type $(4, 4)$) and all other six maximal subgroups of $H$ are isomorphic to the metacyclic minimal nonabelian group of order $2^4$ and exponent 4 given by:

$$\mathcal{H}_2 = \langle x, y \mid x^4 = y^4 = 1, x^y = x^{-1} \rangle .$$

Assume $H < G$ and set $K = N_G(H)$ so that $K > H$ and the abelian subgroup $A$ is normal in $K$. Note that $A$ has exactly 12 elements of order 4 and for each such element $x$ we have $C_G(x) = A$. This implies

$$|K : H| = 2 , \quad |K| = 2^6 , \quad N_G(A) = K .$$

Suppose that $A$ is not characteristic in $K$. Then there is $\phi \in \mathrm{Aut}(K)$ such that $A \neq A^{\phi}$. Since $A^{\phi} \nleq H$, it follows that $A^{\phi}$ covers $K/H$ and $|A^{\phi} \cap H| = 2^3$. But then $A^{\phi} \cap H$ contains elements of order 4, contrary to our assumption. We have proved that $A$ is characteristic in $K$ and this implies that $G = K$ is of order $2^6$.

Our assumption implies that for each $g \in G - H$, we have $g^2 \in Z(H)$, and so we have

$$\mho_1(G) = \Phi(G) = G' = Z(H) \cong E_4 .$$

For an element $h \in H - A$ we have that $C_H(h) = \langle h \rangle Z(H)$ is of order 8 because $A$ is the unique abelian maximal subgroup of $H$. But then $C_G(h) = C_H(h)$ is of order 8 and so $h$ has 8 conjugates in $G$, contrary to the fact that $|G'| = 4$. Our theorem is proved. □

**Theorem 172.2.** *If $G$ is a 2-group containing a minimal nonmetacyclic subgroup $H$ of order $2^5$ such that $C_G(Z(H)) = H$, then either $G = H$ or $|G : H| = 2$ and $G/H$ acts faithfully on $Z(H)$.*

*Proof.* Suppose that $G > H$. Since $Z(H)$ is characteristic in $H$, one has $N_G(H) \le N_G(Z(H))$. By N/C-Theorem, we have

$$2 = |N_G(Z(H)) : C_G(Z(H))| = |N_G(Z(H)) : H| .$$

Since $H < N_G(H)$, it follows that $N_G(H) = N_G(Z(H))$. Write $K = N_G(H)$ and $Z = Z(H)$. Assume, by way of contradiction, that $K < G$. Let $K < L \le G$ be such that $|L : K| = 2$.

Then $Z^x \neq Z$ for any $x \in L - K$ since $K = N_G(Z)$. It follows from $\Omega_1(H) = Z$ that $Z^x \nleq H$, and so $K = HZ^x$. Note that $Z^x \lhd K^x = K$ so that

$$2 = |K : H| = |K^x : C_G(Z)^x| = |K : C_G(Z^x)|\,,$$

and we conclude that $\Phi(K) < C_G(Z^x)$. It follows from $Z = \Phi(H) \leq \Phi(K)$ that $Z \leq C_G(Z^x)$, and therefore $Z^x \leq C_G(Z) = H$, contrary to what has been noted above. □

**Proposition 172.3.** *Let a nonabelian p-group G be neither abelian nor minimal nonabelian. Then there is a nonabelian $H < G$ such that $C_G(H)$ is abelian.*

*Proof.* Suppose that $G$ has no such subgroup as $H$. Let $A$ be any maximal abelian subgroup of $G$ so that we have $A < G$. Let $B > A$ be such that $|B : A| = p$; in that case $B$ is nonabelian. But then $C_G(B) \leq C_G(A) = A$ is abelian and so (by our assumption) we get $B = G$. We have proved that each maximal abelian subgroup of $G$ is a maximal subgroup of $G$. In particular, $G$ has more than one abelian maximal subgroup and this infers $|G'| = p$. By Lemma 1.1, we get $|G : Z(G)| = p^2$. Let $M < G$ be a minimal nonabelian subgroup of $G$ so that we have $G = MZ(G)$. But then $C_G(M) = Z(G)$ is abelian, contrary to our assumption. □

In fact, in Proposition 172.3 the following fact is proved: A nonabelian $p$-group $G$ is either minimal nonabelian or possesses a proper nonabelian subgroup $H$ with abelian subgroup of index $p$ such that $C_G(H)$ is abelian.

# § 173 Nonabelian $p$-groups with exactly one noncyclic maximal abelian subgroup

The first author has noted that the modular $p$-groups $M_{p^{n+1}}$ of order $p^{n+1}$, $n \geq 2$, (where in the case $p = 2$ we have $n \geq 3$) possess exactly one noncyclic maximal abelian subgroup and he proposed (Problem 3104) to classify all nonabelian $p$-groups with this property. We classify such groups in the following result.

**Theorem 173.1** (Janko). *Let $G$ be a nonabelian $p$-group with exactly one noncyclic maximal abelian subgroup $A$. Then $|G : A| = p$ and for each element $x \in G - A$, we have $o(x) = p^n$, $n \geq 2$ is fixed (in the case $p = 2$ we have $n \geq 3$) and $Z(G) = C_A(x) = \langle x^p \rangle$ is cyclic. Each minimal nonabelian subgroup of $G$ is isomorphic to $M_{p^{n+1}}$ and $G' \cong A/Z(G)$.*

*If $p = 2$, then either $G \cong M_{2^{k+1}}$, $k \geq 3$, or*

$$G = \langle g, h \mid g^{2^n} = h^{2^m} = 1, m \geq 3, n \geq 3, g^{2^{n-1}} = h^{2^{m-1}}, h^g = h^{-1} \rangle ,$$

*where $G$ is metacyclic of order $2^{m+n-1}$,*

$$Z(G) = \langle g^2 \rangle \cong C_{2^{n-1}} , \quad G' = \langle h^2 \rangle \cong C_{2^{m-1}} ,$$

*and $A = \langle h, g^2 \rangle$ is the unique abelian maximal subgroup of $G$.*

*Proof.* Let $A$ be the unique noncyclic maximal abelian subgroup in $G$. We have $A < G$ and $A \triangleleft G$. Let $E$ be a $G$-invariant abelian subgroup of type $(p, p)$ contained in $A$. Suppose that $|G : A| > p$ and let $x$ be an element in $C_G(E) - A$. Let $B$ be a maximal abelian subgroup in $G$ containing $\langle E, x \rangle$. It follows that $B \neq A$ and $B$ is noncyclic, a contradiction.

We have proved that $|G : A| = p$. Let $x$ be an arbitrary element in $G - A$. Set $Z = C_A(x)$ so that $\{1\} \neq Z = Z(G)$ and $x^p \in Z < A$. Let $C$ be a maximal abelian subgroup of $G$ containing $\langle Z, x \rangle$. Since $C \neq A$, it follows by our assumption that $C$ is cyclic. This implies that $Z$ is cyclic and $Z = \langle x^p \rangle$. In particular, each element $x \in G - A$ has the same order $p^n$, $n \geq 2$, where $Z \cong C_{p^{n-1}}$ and $Z = \langle x^p \rangle$. In addition $\langle x \rangle$ is a maximal abelian subgroup in $G$.

Let $M$ be any minimal nonabelian subgroup in $G$. Since $M \nleq A$, there is an element $x \in M - A$. But $\langle x \rangle \cong C_{p^n}$ $(n \geq 2)$ is a maximal abelian subgroup in $M$ and so $|M : \langle x \rangle| = p$ and $|M| = p^{n+1}$. We have proved that each minimal nonabelian subgroup $M$ in $G$ has the same order $p^{n+1}$ $(n \geq 2$ is fixed) and $M$ possesses a cyclic subgroup of index $p$. Hence we have either $M \cong M_{p^{n+1}}$, $n \geq 2$, (and in the case $p = 2$ we have $n \geq 3$) or $M \cong Q_8$. Note that $M \cong D_8$ is not possible since in that case we would have in $M - A$ elements of order 2 or 4, contrary to the fact that all elements in $G - A$ have the same order.

Suppose that $M \cong Q_8$ so that each minimal nonabelian subgroup in $G$ is isomorphic to the ordinary quaternion group. By Corollary A.17.3, we have $G = Q \times V$, where $Q \cong Q_{2^m}$, $m \geq 3$, and $\exp(V) \leq 2$. Since $Z(G)$ is cyclic, we get $V = \{1\}$, and so $G = Q$.

But each maximal abelian subgroup in $Q_{2^m}$ is cyclic, a contradiction. We have proved that each minimal nonabelian subgroup of $G$ is isomorphic to $M_{p^{n+1}}$ ($n$ fixed). Also, by the proof of Lemma 1.1, we have $G' \cong A/Z(G)$.

Suppose that $p = 2$. Since any two maximal abelian subgroups in $G$ have cyclic intersection, we may use Theorem 91.1, where all such 2-groups are determined up to isomorphism. It follows that either $G \cong M_{2^{k+1}}$, $k \geq 3$, (the case (b) of Theorem 91.1) or $G$ is a group of part (e) of that theorem. Indeed, in case (a) of Theorem 91.1, $G$ has either more than one noncyclic maximal abelian subgroup or each maximal abelian subgroup of $G$ is cyclic. In cases (c) and (d) of Theorem 91.1, $G$ has more than one noncyclic maximal abelian subgroup. Our theorem is proved. □

**Exercise 1.** Give another proof that, in Theorem 173.1, $|G : A| = p$.

**Solution.** One may assume that $|G| > p^3$. Since $G$ has a cyclic maximal abelian subgroup, it follows that $Z(G)$ is cyclic. Since $G$ has only one noncyclic maximal abelian subgroup, it is not a 2-group of maximal class. Therefore $G$ has a normal abelian subgroup $R$ of type $(p, p)$ and $C_G(R)$ is maximal in $G$ since $Z(G)$ is cyclic. Assume that $C_G(R)$ is nonabelian. Let $A$ be a maximal abelian subgroup of $C_G(R)$. Then $R < A$ and $A$ is maximal abelian subgroup of $G$. Thus, $A$ is a unique maximal abelian subgroup of the nonabelian group $C_G(R)$, which is impossible. Thus, $A = C_G(R)$ is a noncyclic abelian maximal subgroup of $G$.

**Exercise 2.** Let $A < G$ be such as in Exercise 1. Prove that if $x \in G - A$, then $Z(G) < \langle x \rangle$. In particular, $\Omega_1(G) \leq A$.

**Solution.** One may assume that $|G| > p^3$. Assume that $x \in G - A$. Then $C_G(x) = \langle x \rangle Z(G)$ is maximal abelian $\neq A$, and the result follows.

It follows from Exercise 2 that, whenever $x \in G - A$, then $C_G(x) = \langle x \rangle$.

**Exercise 3.** If $G$ is as in Theorem 173.1.

(a) If $G$ is minimal nonabelian, then $G \cong M_{p^n}$.

(b) If $M < G$ is minimal nonabelian, then $M \cong M_{p^n}$.

**Solution.** (a) Since $G$ has only one noncyclic maximal subgroup, it follows by Theorem 1.2 that $G \cong M_{p^n}$.

(b) Assume that $X, Y$ are two distinct noncyclic maximal subgroups of $M$ and let $X < U < G$, $Y < V < G$, where $U, V$ are maximal abelian in $G$. Then, by the hypothesis, $U = V$. Since $XY = M$, we get $M \leq U$, which is a contradiction. Thus, $M$ has at most one noncyclic maximal subgroup so $M \in \{Q_8, M_{p^n}\}$. Assume that $M \cong Q_8$. Since $G$ is not of maximal class, we get $C_G(M) \nleq M$ (Proposition 10.17). Let $M < R \leq MC_G(M)$ be of order 16. Then $R$ has two distinct noncyclic maximal abelian subgroups, say $X, Y$. Then maximal abelian subgroups of $G$, containing $X, Y$ respectively, are distinct and noncyclic, which is a contradiction. Thus, $M \cong M_{p^n}$, and the proof of (b) is complete.

**Exercise 4.** Any nonabelian subgroup of the $p$-group $G$ of Theorem 173.1 has exactly one noncyclic maximal abelian subgroup.

Let a $p$-group $G$ be such as in Theorem 173.1. Suppose that $M \not\cong Q_8$ is a nonabelian metacyclic subgroup of $G$. Set $R = \Omega_1(M)$ and let $U/R$ be a maximal abelian subgroup of $M/R$. Since $M' < R$ (indeed, $M'$ is cyclic so $|M'| < |R| = p^2$), it follows from Lemma 65.2(a) that $U$ is either abelian or minimal nonabelian. In the second case, $U \cong \mathrm{M}_{p^n}$ so that $U/R$ is cyclic. In the first case $U = M \cap A$.

**Problem 1.** Classify the $p$-groups all of whose noncyclic maximal abelian subgroups are (i) conjugate, (ii) isomorphic.

**Problem 2.** Classify the $p$-groups all of whose nonabelian maximal subgroups are metacyclic.

# § 174 Classification of $p$-groups all of whose nonnormal subgroups are cyclic or abelian of type $(p, p)$

This section was written by the second author.

The ultimate goal is to classify $p$-groups in which any two distinct conjugate subgroups have a cyclic intersection (Problem 1572). Note that the title groups satisfy this condition, and so the classification of these groups can be considered as a first step in solving Problem 1572.

In view of Theorem 16.2, where $p$-groups all of whose nonnormal subgroups are cyclic have been classified, it follows that in determining the title groups $G$ we may assume that $G$ possesses a nonnormal abelian subgroup of type $(p, p)$. We prove the following result.

**Theorem 174.1** (Janko). *Let $G$ be a $p$-group all of whose nonnormal subgroups are cyclic or abelian of type $(p, p)$. Assume in addition that $G$ possesses a nonnormal abelian subgroup of type $(p, p)$. Then $G$ is one of the following groups (where $\mathrm{S}(p^3)$, $p > 2$, denotes the nonabelian group of order $p^3$ and exponent $p$):*

(a) *$G \cong \mathrm{D}_{16}$ or $\mathrm{SD}_{16}$,*

(b) *$G = LZ$, where $L \cong \mathrm{S}(p^3)$, $p > 2$, is normal in $G$, $Z \cong \mathrm{C}_{p^2}$, $L \cap Z = \mathrm{Z}(L) = \mathrm{Z}(G)$,*

(c) *$G$ is any nonabelian group of order $p^4$ with an elementary abelian subgroup of index $p$,*

(d) *$p = 2$ and $G \cong (\mathrm{D}_8 * \mathrm{Q}_8) \times \mathrm{C}_2$, where $\mathrm{D}_8 \cap \mathrm{Q}_8 = (\mathrm{D}_8)'$ or $G \cong \mathrm{H}_{16} * \mathrm{Q}_8$ with $\mathrm{H}_{16} \cap \mathrm{Q}_8 = (\mathrm{H}_{16})'$, where $\mathrm{H}_{16}$ is the nonmetacyclic minimal nonabelian group of order 16,*

(e) *$G \cong \mathrm{M}_{p^{s+1}} \times \mathrm{C}_p$, $s \geq 3$,*

(f) *$G = (Z * S) \times \mathrm{C}_p$, where $Z \cong \mathrm{C}_{p^{s+1}}$, $s \geq 1$, $Z \cap S = S'$, and either $p = 2$ and $S \cong \mathrm{D}_8$ or $p > 2$ and $S \cong \mathrm{S}(p^3)$ or $G = Z * S$, where $Z \cong \mathrm{C}_{p^{s+1}}$, $s \geq 1$, $Z \cap S = S'$, and $S$ is the nonmetacyclic minimal nonabelian group of order $p^4$,*

(g) *$G$ is an $\mathrm{A}_2$-group of order $p^5$ from Proposition 71.4 (b2) for $\alpha = 1$,*

(h) *$G \cong \mathrm{Q}_8 * \mathrm{Q}_8 * \mathrm{Q}_8$, an extraspecial group of order $2^7$ and type "$-$",*

(i) *$G = (A_1 * A_2)\mathrm{Z}(G)$, where $A_1$ and $A_2$ are minimal nonabelian $p$-groups and $\mathrm{Z}(G)$ is cyclic. In the case $p = 2$, $A_1$ and $A_2$ are isomorphic to one of $\mathrm{D}_8$, $\mathrm{Q}_8$ and $\mathrm{M}_{2^n}$, $n \geq 4$, where in the case $A_1 \cong \mathrm{Q}_8$ and $A_2 \cong \mathrm{D}_8$ we must have $|\mathrm{Z}(G)| > 2$. In the case $p > 2$, $A_1$ and $A_2$ are isomorphic to one of $\mathrm{S}(p^3)$ or $\mathrm{M}_{p^n}$, $n \geq 3$.*

*Conversely, all the above groups satisfy the assumptions of the theorem.*

*Proof.* Let $G$ be a $p$-group all of whose nonnormal subgroups are cyclic or abelian of type $(p, p)$ and we assume that $G$ possesses a nonnormal abelian subgroup $H$ of type $(p, p)$. We set $K = \mathrm{N}_G(H)$ so that we have $H < K < G$ and $K \trianglelefteq G$. Since each subgroup $X$ of $G$ with $X > H$ is normal in $G$, it follows that $K/H$ is Dedekindian and $K/H$ has exactly one subgroup of order $p$. This implies that $K/H \neq \{1\}$ is either cyclic or $p = 2$ and $K/H \cong Q_8$. Let $L/H$ be a unique subgroup of order $p$ in $K/H$ so that $L \trianglelefteq G$ and $\Omega_1(K) \leq L$. If $g \in G - K$, then $L = \langle H, H^g \rangle$ and so we have $\Omega_1(K) = L$.

Suppose that $K$ does not possess a $G$-invariant abelian subgroup of type $(p, p)$. By Lemma 1.4, we get $p = 2$ and $K$ is of maximal class. But $H$ is a normal 4-subgroup in $K$, and so $K \cong D_8$. Since $C_G(H) = C_K(H) = H$, it follows by Proposition 1.8 that $G$ is also a 2-group of maximal class. In this case $H$ has exactly two conjugates in $K = L \cong D_8$, and so $|G : K| = 2$ and $|G| = 2^4$. It follows that $G \cong D_{16}$ or $SD_{16}$ and we have obtained the groups stated in part (a) of our theorem.

In what follows, we may assume that $K$ possesses a $G$-invariant abelian subgroup $U$ of type $(p, p)$. Since $\Omega_1(K) = L$, we have $U \leq L$, and so $L = HU$ with $|H \cap U| = p$. If $L$ is abelian, then $L \cong E_{p^3}$. If $L$ is nonabelian, then in the case $p > 2$ we have $L \cong S(p^3)$ and in the case $p = 2$ we must have $L \cong D_8$. But the last case cannot happen since $U \trianglelefteq G$ and $L$ has exactly two 4-subgroups, which would imply that also $H \trianglelefteq G$, a contradiction. Hence we have either $L \cong E_{p^3}$ or $p > 2$ and $L \cong S(p^3)$.

Suppose that $p > 2$ and $L \cong S(p^3)$. In that case we have

$$\langle z \rangle = H \cap U = L' = Z(L) \leq Z(G) .$$

If $C_G(L) > \langle z \rangle$, then take an element $x \in C_G(L) - \langle z \rangle$ such that $x^p \in \langle z \rangle$ and consider the abelian subgroup $S = \langle h, z, x \rangle$ of order $p^3$, where $h$ is any element in $H - \langle z \rangle$. By our assumptions, we have $S \trianglelefteq G$. But $L \cap S = H = \langle h, z \rangle$, and so $H \trianglelefteq G$, a contradiction. We have proved that $C_G(L) = \langle z \rangle$. Since an $S_p$-subgroup of $\mathrm{Aut}(L)$ is isomorphic to $S(p^3)$, it follows that $|G : L| = p$ and $K = L$ so that $|G| = p^4$. Also note that $G/\langle z \rangle \cong S(p^3)$ and $G/K$ acting on $p+1$ subgroups of order $p^2$ (containing $\langle z \rangle$) fixes $U$ and acts transitively on $p$ other ones. Hence $U$ is the unique $G$-invariant subgroup of order $p^2$ in $L$. Set $V = C_G(U)$ so that $V$ is an abelian normal subgroup of order $p^3$ in $G$ and we have $G = LV$ with $L \cap V = U$. If $V \cong E_{p^3}$, then we get a group stated in part (c) of our theorem. Hence we may assume that there is an element $t$ of order $p^2$ in $V - U$ such that $t^p = z$. We have obtained a group from part (b) of our theorem.

From now on we may assume that $L \cong E_{p^3}$. If $|G/L| = p$, then $K = L$ is elementary abelian of order $p^3$ and index $p$ and again we have obtained the groups from part (c) of our theorem. Thus we may assume in what follows that $|G/L| > p$.

In the rest of the proof we fix our notation for

$$E_{p^2} \cong H , \quad K = N_G(H) \neq G , \quad \Omega_1(K) = L , \quad E_{p^2} \cong U \trianglelefteq G ,$$

where $L = HU$, $\quad H \cap U \cong C_p$, and $\{1\} \neq K/H$ is either cyclic or $p = 2$ and $K/H \cong Q_8$. We also fix our assumptions that $L \cong E_{p^3}$ and $|G/L| > p$.

(i) First assume that there is a central element $z$ in $G$ of order $p$ that is contained in $H$.

In that case we have $|G : K| = p$ so that $K > L$, and therefore there is an element $v \in K - L$ of order $p^2$ with $v^p \in L - H$. We may choose a $G$-invariant subgroup $U \leq L$ of order $p^2$ so that $U \leq Z(G)$. The socle $\Omega_1(X)$ of any cyclic subgroup $X$ in $G$ of composite order is contained in $U$.

Indeed, acting with $G/K$ on $p+1$ subgroups of order $p^2$ in $L$ that contain $\langle z \rangle$, we see that $|G : K| = p$. Since $|G/L| > p$, we have $K > L$, and so there is an element $v \in K - L$ of order $p^2$, where $v^p \in L - H$. Considering $\langle v, z \rangle \cong C_{p^2} \times C_p$, we obtain

$$\langle v, z \rangle \trianglelefteq G \quad \text{and so } \mho_1(\langle v, z \rangle) = \langle v^p \rangle \trianglelefteq G\,.$$

Then we may set $E_{p^2} \cong U = \langle z, v^p \rangle \leq Z(G)$. Let $X$ be any cyclic subgroup of composite order in $G$ and assume that $\Omega_1(X) \nleq U$. But then $\Omega_1(X) \leq K$, and so $\Omega_1(X) \leq L$. Take an element $1 \neq u \in U \leq Z(G)$ and consider the subgroup $X \times \langle u \rangle \trianglelefteq G$ so that we get $\Omega_1(X) \trianglelefteq G$. Since $\Omega_1(X) \nleq U$, we get $L \leq Z(G)$, and so $H \trianglelefteq G$, a contradiction.

(i1) Suppose that $K/L$ is noncyclic. Then we have $p = 2$, $K/H \cong Q_8$, $|G| = 2^6$ and $K/L \cong E_4$. Since $\mho_1(K) \leq U \leq Z(G)$, $K/U$ is elementary abelian. Considering the Dedekindian group $G/U$ of order $2^4$ that possesses an elementary abelian subgroup $K/U$ of index 2, it follows that $G/U$ is abelian, and so $G' \leq U$. Any two noncommuting elements in $G$ generate here a minimal nonabelian subgroup (see Lemma 65.2). For any $g, h \in G$ we have $[g^2, h] = [g, h]^2 = 1$, and so $\mho_1(G) \leq Z(G)$. In particular, for any $g \in G - K$, $g^2 \in K - L$ is not possible and so $g^2 \in L$ and this implies $g^2 \in U$. Hence $\mho_1(G) \leq U$ and $\exp(G) = 4$. Since $Z(G) \leq K$, we get $Z(G) = U$. Because $G/L \cong E_8$, we have $C_G(L) > L$, and so $C_G(L) \leq K$ implies $C_K(L) > L$. Thus there is $v \in C_K(L) - L$ such that $v^2 \in U - H$. Let $h \in H - U$ and consider the subgroup $\langle h, v \rangle \cong C_2 \times C_4$ so that $\langle h, v \rangle \trianglelefteq G$ and

$$\Omega_1(\langle h, v \rangle) = \langle h, v^2 \rangle \trianglelefteq G\,.$$

If $\langle h, v^2 \rangle \nleq Z(K)$, then there is $g \in G - K$ centralizing $\langle h, v^2 \rangle$, a contradiction. We have proved that $H \leq Z(K)$, and so $C_G(L) = K$.

We have $Z(K) = L$, and so $|K'| = 2$ and $U = K' \times (H \cap U)$. Suppose that $\mho_1(K) = U$. Then there are elements $v_1, v_2 \in K - L$ such that $z_1 = v_1^2 \neq z_2 = v_2^2$, where $z_1, z_2 \in U - H$. Let $h \in H - U$ and $g \in G - K$. Since

$$\langle h, v_1 \rangle \cong C_2 \times C_4 \quad \text{and} \quad \langle h, v_2 \rangle \cong C_2 \times C_4\,,$$

we have

$$\langle h, v_1 \rangle \trianglelefteq G \text{ and } \langle h, v_2 \rangle \trianglelefteq G \quad \text{and so } \langle h, z_1 \rangle \trianglelefteq G \text{ and } \langle h, z_2 \rangle \trianglelefteq G\,.$$

But this gives $h^g = hz_1 = hz_2$ and $z_1 = z_2$, a contradiction.

We have proved that $\mho_1(K) = \langle u \rangle$ is of order 2, where $u \in U - H$. It follows that $K/\langle u \rangle$ is elementary abelian, and so $\mho_1(K) = K' = \langle u \rangle$. Let $k_1, k_2 \in K - L$ be such that $\langle k_1, k_2 \rangle$ covers $K/L$. Since $k_1^2 = k_2^2 = u$ and $[k_1, k_2] = u$, we get

$$Q = \langle k_1, k_2 \rangle \cong Q_8\,, \quad K = H \times Q,\,, \; L = H \times \langle u \rangle\,,$$

where $Q \trianglelefteq G$.

Since $G' \leq U$ is elementary abelian, it follows that $G$ induces on $Q$ only inner automorphisms of $Q$, and so we have $G = Q * C$, where $C = C_G(Q)$ and $Q \cap C = \langle u \rangle$, $K \cap C = L$. We also have $Z(C) = Z(G) = U$. By Lemma 1.1 we get $|C'| = 2$. On the other hand, let $h \in H - U$, $g \in C - L$ and $v \in Q$ with $v^2 = u$. Since

$$C_2 \times C_4 \cong \langle h, v \rangle \trianglelefteq G, \quad \text{it follows that } \Omega_1(\langle h, v \rangle) = \langle h, u \rangle \trianglelefteq G.$$

Thus we get $h^g = hu$ and so $u \in C'$. We have proved that $C' = Q' = \langle u \rangle = G'$.

Let $g$ be an element in $C - L$ and $h \in H - U$. If $g^2 \in U - \langle u \rangle$, then $C = \langle g, h \rangle \cong H_{16}$, where $H_{16}$ denotes the nonmetacyclic minimal nonabelian group of order 16. If $g^2 \in \langle u \rangle$, then we have $\langle g, h \rangle \cong D_8$, and so in this case $C = \langle g, h \rangle \times \langle z \rangle$, where $\langle z \rangle = H \cap U$. We have obtained the groups stated in part (d) of our theorem.

(i2) Suppose that $\{1\} \neq K/L$ is cyclic so that $K/H$ is cyclic of order $\geq p^2$. In this case we show that $G/L$ is abelian.

Indeed, assume that $G/L$ is nonabelian. Since $G/L$ is Dedekindian, it follows that $p = 2$ and $G/L \cong Q_8$. We also have $\Omega_1(G) = L$. Since $C_G(L) > L$ and $C_G(L) \leq K$, we get $C_K(L) > L$. Let $v \in C_K(L) - L$ with $o(v) = 4$ so that $v^2 \in U - H$ and let $h \in H - U$. Then

$$C_2 \times C_4 \cong \langle h, v \rangle \trianglelefteq G, \quad \text{and} \quad \langle h, v^2 \rangle \trianglelefteq G.$$

If $\langle h, v^2 \rangle \not\leq Z(K)$, then there is $g \in G - K$, which centralizes $h$, a contradiction. Hence $\langle h, v^2 \rangle \leq Z(K)$, and so $H \leq Z(K)$, which implies that $K$ is abelian.

Since $G/U$ is Dedekindian and nonabelian, it follows that $G/U$ is Hamiltonian. Let $Q/U$ be a subgroup in $G/U$ that is isomorphic to $Q_8$ and set

$$Q_0/U = Z(Q/U) = (Q/U)'.$$

Let $Q_1/U$ and $Q_2/U$ be two distinct cyclic subgroups of order 4 in $Q/U$ so that $Q_1$ and $Q_2$ are abelian and $Q_1 \cap Q_2 = Q_0$. It follows that $Q_0 \leq Z(Q)$, and so $Q_0 = Z(Q)$. By Lemma 1.1, $|Q'| = 2$ and since $Q'$ covers $Q_0/U$, it follows that $Q_0 = U \times Q' \cong E_8$. But then $Q_0 = \Omega_1(G) = L$, and so $K = C_G(L) \geq Q$ is nonabelian, a contradiction. We have proved that $G/L$ is abelian, and so $G/L$ is either cyclic of order $\geq p^2$ or $G/L$ is abelian of type $(p^s, p)$, $s \geq 1$.

(i2a) Assume that $G/L$ is cyclic. Let $g \in G - K$ so that $\langle g \rangle$ covers $G/L$ and let $\langle t \rangle = \Omega_1(\langle g \rangle)$ be the socle of $\langle g \rangle$, where $t \in U - H$ and $o(g) = p^s$, $s \geq 3$. We may set $t = g^{p^{s-1}}$, and so $\langle g^p \rangle$ covers $K/H \cong C_{p^{s-1}}$. Also set $v = g^{p^{s-2}}$ so that $\langle v \rangle \cong C_{p^2}$ and $v^p = t$.

Since $\langle g \rangle$ stabilizes the chain $L > U > \{1\}$, it follows that $\langle g^p \rangle$ centralizes $L$ and so $K$ is abelian. Consider the abelian subgroup $\langle h, v \rangle \cong C_p \times C_{p^2}$, where $h$ is any element in $H - U$. Since $\langle h, v \rangle \trianglelefteq G$, we get

$$\Omega_1(\langle h, v \rangle) = \langle h, t \rangle \trianglelefteq G.$$

Thus we get $h^g = ht^i$ for some $i \not\equiv 0 \pmod p$ and so $G' \geq \langle t \rangle$. On the other hand,

$$Z(G) = C_K(g) = \langle g^p, U \rangle \quad \text{and so } |G : Z(G)| = p^2.$$

By Lemma 1.1, we get

$$|G| = p|Z(G)|\,|G'| \quad \text{and so } |G'| = p \text{ and } G' = \langle t\rangle\,.$$

We have $\langle g, h\rangle \cong M_{p^{s+1}}$ and if we set $\langle z\rangle = H \cap U$, then

$$G = \langle z\rangle \times \langle g, h\rangle \cong C_p \times M_{p^{s+1}}\,.$$

We have obtained the groups stated in part (e) of our theorem.

(i2b) Assume that $G/L$ is abelian of type $(p^s, p)$, $s \geq 1$, and $K$ is abelian. Let $v \in K - L$ be such that $\langle v\rangle$ covers $K/L \cong C_{p^s}$, $s \geq 1$. Then $t = v^{p^s} \in U - H$ so that

$$K/H \cong C_{p^{s+1}} \quad \text{and} \quad K = H \times \langle v\rangle \cong E_{p^2} \times C_{p^{s+1}}\,.$$

Since $G/L$ is abelian of type $(p^s, p)$, there is an element $w \in G - K$ such that $w^p \in L$, and so $w^p \in U$. Let $h \in H - U$ and consider the abelian subgroup

$$\langle h, v\rangle \cong C_p \times C_{p^{s+1}}\,, \quad s \geq 1\,.$$

Since $\langle h, v\rangle \trianglelefteq G$, we get $\langle h, t\rangle \trianglelefteq G$, and so $h^w = ht$ (where we replace $h$ with a suitable power $h^j$, $j \not\equiv 0$ [mod $p$], if necessary). In particular, we get $G' \geq \langle t\rangle$.

Suppose that $G/U$ is nonabelian so that $p = 2$ and $G/U$ is Hamiltonian. But $G/L$ is abelian, and so

$$(G/U)' = \mho_1(G/U) = L/U\,.$$

Hence there is an element $m \in G$ such that $m^2 \in L - U$, a contradiction. We have proved that $G/U$ is abelian, and so $\langle t\rangle \leq G' \leq U \leq Z(G)$, and therefore $G$ is of class 2 with an elementary abelian commutator subgroup.

Note that

$$C_p \times C_{p^{s+1}} \cong \langle h, v\rangle \trianglelefteq G \quad \text{and so } [h, w] \in \langle h, v\rangle \cap U = \langle t\rangle\,,$$

which implies that $\langle v\rangle \trianglelefteq G$, and therefore $p - 1$ other cyclic maximal subgroups of $\langle h, v\rangle$ are also normal in $G$.

In the case $\langle v\rangle \not\leq Z(G)$ we get $v^w = vt^j$ for some integer $j \not\equiv 0 \pmod p$. Solve the congruence $ij \equiv -1 \pmod p$, where $i \not\equiv 0 \pmod p$. Then we compute:

$$(v^i h)^w = (v^w)^i h^w = (vt^j)^i ht = v^i t^{-1} ht = v^i h\,,$$

where $\langle v^i h\rangle \cong C_{p^{s+1}}$ is also a cyclic maximal subgroup in $\langle h, v\rangle$ and $\langle v^i h\rangle \leq Z(G)$. Thus replacing $\langle v\rangle$ with $\langle v^i h\rangle$, we may assume from the start that $\langle v\rangle \leq Z(G)$. We get

$$Z(G) = C_K(w) = \langle v\rangle U \quad \text{and so } |G : Z(G)| = p^2\,.$$

By Lemma 1.1 we get

$$|G| = p|Z(G)||G'| \quad \text{and so } |G'| = p \text{ and } G' = \langle t\rangle\,.$$

First suppose that $w^p \in U - \langle t\rangle$. Then $S = \langle h, w\rangle$ is the nonmetacyclic minimal nonabelian group of order $p^4$. If we set $Z = \langle v\rangle$, then we get

$$G = Z * S\,, \quad \text{where } Z \cong C_{p^{s+1}} \text{ and } Z \cap S = S'\,.$$

Assume that $w^p \in \langle t\rangle$ and set $\langle z\rangle = U \cap H$. Then $S = \langle h, w\rangle$ is isomorphic to $D_8$ in the case $p = 2$ and to $S(p^3)$ or $M_{p^3}$ in the case $p > 2$. Setting again $Z = \langle v\rangle \cong C_{p^{s+1}}$ we have $Z \le Z(G)$, $S \cap Z = S'$ and $G = \langle z\rangle \times (S * Z)$. However, in the case $p > 2$ and $S \cong M_{p^3}$, we have $S * Z = S_1 * Z$, where $S_1 \cong S(p^3)$ for a suitable subgroup $S_1$ in $S * Z$. We have obtained all groups stated in part (f) of our theorem.

(i2c) Assume that $G/L$ is abelian of type $(p^s, p)$, $s \ge 1$, and $K$ is nonabelian. We have $K/L \cong C_{p^s}$, $s \ge 1$. Let $v \in K - L$ be such that $\langle v\rangle$ covers $K/L$. Then $1 \ne t = v^{p^s} \in U - H$ so that $K/H \cong C_{p^{s+1}}$. Acting with $K$ on $L$, we see that $K$ stabilizes the chain $L > U > \{1\}$. Hence if $s > 1$, then there is an element $v_0$ of order $p^2$ in $K$ that centralizes $L$ and $v_0^p \in U - H$. For an element $h \in H - U$ we consider

$$C_p \times C_{p^2} \cong \langle h, v_0\rangle \trianglelefteq G \quad \text{and so } E_{p^2} \cong \langle h, v_0^p\rangle \trianglelefteq G\,.$$

If $\langle h, v_0^p\rangle \nleq Z(K)$, then there is an element $g \in G - K$ that centralizes $h$, a contradiction. Thus we must have $\langle h, v_0^p\rangle \le Z(K)$ and this implies that $K$ is abelian, a contradiction. We have proved that $s = 1$, and so $t = v^p$ and $|G| = p^5$. Since $C_L(v) = U = Z(K)$, Lemma 1.1 gives that $|K'| = p$. On the other hand, $K' \le H$ and since $K' \le Z(G)$, we get $K' = H \cap U$. For any $h \in H - U$, we have $\langle [h, v]\rangle = K'$, and so $K$ is the nonmetacyclic minimal nonabelian group of order $p^4$ and $\Phi(K) = U$. Because $G/L \cong E_{p^2}$, we have $\exp(G) = p^2$, and so for any $x \in G - L$, we have $x^p \in U$ and $\mho_1(G) \le U$. For $p = 2$, $G/U$ is elementary abelian. For $p > 2$, the fact that $G/U$ is Dedekindian implies that $G/U$ is abelian, and so again $G/U$ is elementary abelian. We have proved that $\Phi(G) = U$, and so $G' \le U$ and $d(G) = 3$. Since $Z(G) \le K$, we also get $Z(G) = U$. If $G' = K'$, then $H \trianglelefteq G$, a contradiction. Thus $G' = U$, and so $G$ is special.

By Lemma 146.7, $G$ has exactly one abelian maximal subgroup $A$ and for each subgroup $X_i$ of order $p$ in $G'$ $(i = 1, 2, \dots, p + 1)$ there are exactly $p$ pairwise distinct maximal subgroups $L_{ij}$ $(j = 1, 2, \dots, p)$ of $G$ such that $L'_{ij} = X_i$.

Suppose that $G$ possesses a nonabelian subgroup $S$ of order $p^3$ so that $S$ is minimal nonabelian and $S \trianglelefteq G$. But then $E_{p^2} \cong G' \le S$ and since $G' = Z(G)$, we get that $S$ is abelian, a contradiction. Hence $G$ is an $A_2$-group since each subgroup of index $p^2$ in $G$ is abelian and $K$ is a minimal nonabelian maximal subgroup in $G$. If there is an element $g \in G - K$ of order $p$, then $\langle g, h\rangle$ (with $h \in H - U$) is minimal nonabelian of order $p^3$, a contradiction. We have proved that $E_{p^3} \cong L = \Omega_1(G)$, and so a unique abelian maximal subgroup $A$ of $G$ is of type $(p^2, p^2)$. Indeed, $A$ contains $U = \Phi(G)$ and $|K \cap A| = p^3$. If $L \le A$, then there is an element $g \in G - K$ that centralizes $L$, a contradiction. Hence we have $A \cap L = U = \Omega_1(A)$, which shows that $A \cong C_{p^2} \times C_{p^2}$.

By the results of § 71, it follows that $G$ is one of the $A_2$-groups from Theorem 71.4 (b2) with $\alpha = 1$. We have obtained the groups from part (g) of our theorem.

(ii) We assume that whenever $H$ is a nonnormal abelian subgroup of type $(p, p)$ in $G$, then $H \cap Z(G) = \{1\}$. Let $z$ be a central element of $G$ that is contained in $L - H$ so that we have $L = \Omega_1(K) = \langle z\rangle \times H \cong E_{p^3}$ and $L \cap Z(G) = \langle z\rangle$. For any $1 \neq h \in H$, we have $\langle h, z\rangle \trianglelefteq G$, and therefore $H \cap \langle h, z\rangle = \langle h\rangle \trianglelefteq K$. Thus $H \leq Z(K)$ and $C_G(L) = K$. It follows that $G/K$ acts faithfully on $L$ and stabilizes the chain $L > \langle z\rangle > \{1\}$ and $[H, G] = \langle z\rangle$. Thus $\{1\} \neq G/K$ is elementary abelian of order $\leq p^2$. However, if $|G/K| = p$, then there is an element $g \in G - K$ centralizing an element $1 \neq h \in H$, and so $h \in Z(G)$, a contradiction. We have proved that we have $G/K \cong E_{p^2}$.

Let $X$ be any cyclic subgroup of composite order in $G$. Since $\Omega_1(X) \leq K$, we have $\Omega_1(X) \leq L = \Omega_1(K)$. Suppose that $\Omega_1(X) \neq \langle z\rangle$. In this case we have

$$X \times \langle z\rangle \trianglelefteq G \quad \text{and so } \Omega_1(X) \trianglelefteq G\,.$$

This is a contradiction since $L \cap Z(G) = \langle z\rangle$. We have proved that the socle of each cyclic subgroup of composite order in $G$ is equal $\langle z\rangle \leq G'$.

We have $Z(G) \leq K$, and so we have

$$Z(G) \cap L = Z(G) \cap \Omega_1(K) = \langle z\rangle\,.$$

This implies that $Z(G)$ is cyclic and we also have $|G : Z(G)| \geq p^4$.

(ii1) First assume that $K/H \cong Q_8$. In this case we have $|G| = 2^7$. Let $K_i$ be any of the three maximal subgroups of $K$ containing $H$ so that $K_i/H \cong C_4$, and therefore each $K_i$ is abelian. Hence $|K'| = 2$, and so $K' \trianglelefteq G$ and $K' \leq L$ implies that $K' = \langle z\rangle$. Let $v_1, v_2 \in K - L$ be such that $\langle v_1, v_2\rangle$ covers $K/L$. Because $v_1^2 = v_2^2 = z$ and $[v_1, v_2] = z$, we get $Q = \langle v_1, v_2\rangle \cong Q_8$ so that $K = H \times Q$ and $Q \trianglelefteq G$. For each $K_i$ $(i = 1, 2, 3)$ we have $K_i \trianglelefteq G$, and so $K_i \cap Q \trianglelefteq G$. Thus $G$ induces on $Q$ only inner automorphisms of $Q$, which gives $G = Q * M$ with $Q \cap M = \langle z\rangle = Q'$ and $M \cap K = L$, where $M = C_G(Q)$ covers $G/K$. We have $\mho_1(M) \leq \langle z\rangle$, and so $Q/\langle z\rangle$ is elementary abelian. We get $G' = \Phi(G) = Z(G) = \langle z\rangle$, and so $G$ is extraspecial of order $2^7$. Since $M' = \Phi(M) = Z(M) = \langle z\rangle$, it follows that $M$ is extraspecial of order $2^5$ containing an elementary abelian subgroup $L$ of order 8, and so $M \cong Q_8 \times Q_8$ and $G \cong Q_8 \times Q_8 \times Q_8$. We have obtained the group stated in part (h) of our theorem.

(ii2) Assume that $K/H$ is cyclic. Then $K = H \times \langle v\rangle$ is abelian, where $\langle v\rangle \cong C_{p^s}$, $s \geq 1$, and $\langle v\rangle \geq \langle z\rangle \leq G' \cap Z(G)$.

(ii2a) First suppose that $G' = \langle z\rangle$. Then each cyclic subgroup of composite order is normal in $G$. Let $x, y \in G$ so that we have $[x^p, y] = [x, y]^p = 1$, and therefore $\mho_1(G) \leq Z(G)$. Hence we have $\Phi(G) = G'\mho_1(G) \leq Z(G)$ and we know that $Z(G)$ is cyclic. Hence $\Phi(G)$ is also cyclic and $G' = \Omega_1(\Phi(G))$. Since $v^p \in Z(G)$, we have $|G : Z(G)| = p^4$ or $p^5$. If $M$ is any minimal nonabelian subgroup in $G$, then either $M \cong S(p^3)$ or $Z(M) = \Phi(M) = \mho_1(M)$, and so, in this case, $M$ has a cyclic subgroup of index $p$. This gives:

$$\text{If } p = 2 \text{ then } M \in \{D_8, Q_8, M_{2^n}\,,\quad n \geq 4\}\,.$$
$$\text{If } p > 2 \text{ then } M \in \{S(p^3), M_{p^n}\,,\quad n \geq 3\}\,.$$

Let $A_1$ be any minimal nonabelian subgroup in $G$. Then we have $G = A_1 * C$, where $C = C_G(A_1)$ with $A_1 \cap C = Z(A_1)$. If $C$ is abelian, then $C = Z(G)$ and $|G : Z(G)| = p^2$, a contradiction. Thus, $C$ is nonabelian and $Z(C) = Z(G)$, where $|C : Z(C)| = p^2$ or $p^3$. Let $A_2$ be a minimal nonabelian subgroup in $C$. Then we have $C = A_2 * C^*$, where $C^* = C_C(A_2)$ and $A_2 \cap C^* = Z(A_2)$. Note that $Z(C^*) = Z(C)$, and so if $C^*$ are nonabelian, then we get $|C^* : Z(C^*)| \geq p^2$, and so $|C : Z(C)| \geq p^4$, a contradiction. Hence $C^*$ is abelian, and so $C^* = Z(C) = Z(G)$. We have proved that $G = A_1 * A_2Z(G)$, where $Z(G)$ is cyclic. Finally, if $p = 2$ and $A_1 \cong Q_8$ and $A_2 \cong D_8$, then we must have $|Z(G)| > 2$. Indeed, if we have in this case $|Z(G)| = 2$, then $G \cong Q_8 * D_8$ and this group does not possess an elementary abelian subgroup of order 8. We have obtained the groups in part (i) of our theorem.

(ii2b) Finally assume that $G' > \langle z \rangle$. Set $H = \langle h_1, h_2 \rangle$ and we know that $\langle h_1, z \rangle \trianglelefteq G$, $\langle h_2, z \rangle \trianglelefteq G$ and both $G/\langle h_1, z \rangle$ and $G/\langle h_2, z \rangle$ are Dedekindian. If both $G/\langle h_1, z \rangle$ and $G/\langle h_2, z \rangle$ are abelian, then we get $G' \leq \langle h_1, z \rangle \cap \langle h_2, z \rangle = \langle z \rangle$, contrary to our assumption. Hence we must have $p = 2$ and we may assume that $G/\langle h_1, z \rangle$ is Hamiltonian.

Let $Q/\langle h_1, z \rangle$ be an ordinary quaternion subgroup in $G/\langle h_1, z \rangle$ and set

$$C/\langle h_1, z \rangle = (Q/\langle h_1, z \rangle)'$$

so that $Q'$ covers $C/\langle h_1, z \rangle$. Since $G/K \cong E_4$, we have $G' \leq K$ and we know that $K$ is abelian. It follows that $C = \langle h_1, z \rangle Q' \leq K$, and so $C$ is abelian of order 8. For each $x \in Q - C$ we have $x^2 \in C - \langle h_1, z \rangle$. On the other hand, the socle of each cyclic subgroup of composite order in $G$ is equal $\langle z \rangle$, and so $o(x^2) = 4$, and therefore $C$ is abelian of type $(4, 2)$. We get $\Omega_1(Q) = \langle h_1, z \rangle$, $\Omega_2(Q) = C$, and all elements in $Q - C$ are of order 8. Also we have $Q \cap L = \langle h_1, z \rangle$. If $Q' = C$, then $|Q : Q'| = 4$ and Lemma 1.6 (O. Taussky) would imply that $Q$ is of maximal class (and order $2^5$), contrary to the fact that $\Omega_1(Q) = \langle h_1, z \rangle \cong E_4$. On the other hand, $Q'$ must cover $C/\langle h_1, z \rangle$, and so we have $Q' \cong C_4$.

By Lemma 42.1, we have

$$Q = \langle a, b \mid a^8 = b^8 = 1, a^4 = b^4 = z, a^b = a^{-1} \rangle ,$$

where $Q' = \langle a^2 \rangle$, $Z(Q) = \langle b^2 \rangle$, $\Omega_2(Q) = \langle a^2, b^2 \rangle$, and $\Omega_1(Q) = \langle z, a^2b^2 \rangle$. Since $Z(Q) = \langle b^2 \rangle$, we have $C_Q(b) = \langle b \rangle$, and so $C_{\langle h_1 z \rangle}(b) = \langle z \rangle$. On the other hand, $b^2 \in K > L$, and therefore $b^2$ centralizes $L$, and so $b$ induces an involutory automorphism on $L \cong E_8$. Hence $C_L(b) \cong E_4$, and so there exists an involution $e \in H - \langle h_1 \rangle$ such that $[e, b] = 1$.

We have

$$C_2 \times C_8 \cong \langle e, b \rangle \trianglelefteq G, \quad \text{where } \Omega_1(\langle e, b \rangle) = \langle e, z \rangle .$$

On the other hand,

$$b^a = a^{-1}ba = b(b^{-1}a^{-1}b)a = ba^2 ,$$

which shows that $a^2 \in \langle e, b \rangle$. But then $\langle e, b \rangle$ contains $\langle e, z, a^2b^2 \rangle \cong E_8$, contrary to

$$\Omega_1(\langle e, b \rangle) = \langle e, z \rangle \cong E_4 .$$

We have proved that the case $G' > \langle z\rangle$ cannot occur.
It remains to be proved the converse: that all groups $G$ stated in our theorem satisfy the assumptions of that theorem. In fact, we have to prove that each noncyclic subgroup of order $\geq p^3$ is normal in $G$ and that $G$ has a nonnormal abelian subgroup of type $(p, p)$.

If $G \cong \mathrm{D}_{16}$ or $G \cong \mathrm{SD}_{16}$, a 4-subgroup in $G$ is not normal in $G$.

Let $G$ be a $p$-group in part (b) of our theorem. Then we have $L' < G' < L$, where $G' \cong \mathrm{E}_{p^2}$. For an element $l \in L - G'$, set $H = \langle L', l\rangle \cong \mathrm{E}_{p^2}$. If $H \trianglelefteq G$, then $|G/H| = p^2$ implies that $G' \leq H$, a contradiction. Hence $H$ is not normal in $G$.

Let $E$ be an elementary abelian maximal subgroup in a nonabelian $p$-group $G$ of order $p^4$ (from part (c) of our theorem). Then we have $1 \neq G' < E$. Let $\mathrm{E}_{p^2} \cong H$ be any subgroup of order $p^2$ in $E$ that does not contain $G'$. If $H \trianglelefteq G$, then $|G/H| = p^2$ implies that $G' \leq H$, a contradiction. Hence $H$ is not normal in $G$.

Let $G$ be a 2-group of order $2^6$ from part (d) of our theorem. Note that $\mathrm{Z}(G) \cong \mathrm{E}_4$ implies that $G$ has no abelian maximal subgroup. Indeed, if $G$ would have an abelian maximal subgroup, then we may use Lemma 1.1 and we get

$$|G| = 2^6 = 2|G'|\,|\mathrm{Z}(G)| = 2^3|G'| \quad \text{and} \quad |G'| = 2^3\,,$$

which contradicts the fact that $|G'| = 2$. Let $S$ be a noncyclic subgroup of order $\geq 2^3$ and assume that $S$ is not normal in $G$. Then $G' \nleq S$, and so $S$ is noncyclic abelian. If $|S| = 2^4$, then $S \times G'$ would be an abelian maximal subgroup of $G$, a contradiction. Assume that $|S| = 2^3$. Since $G$ has no elementary abelian subgroups of order $2^4$, we get that $S$ is abelian of type $(4, 2)$. In the case $G \cong (\mathrm{D}_8 * \mathrm{Q}_8) \times \mathrm{C}_2$, we have $\mho_1(G) = G'$, and so (since $G' \nleq S$) we must be in the case

$$\mathrm{H}_{16} * \mathrm{Q}_8 \cong G = D * Q\,, \quad \text{where } D \cong \mathrm{H}_{16}\,, \quad Q \cong \mathrm{Q}_8 \text{ and } D \cap Q = D' = \langle z\rangle = Q'\,,$$

and $z$ is not a square of any element in $D$. Since all elements in $G - D$ are of order 4, we have $\Omega_1(S) \leq D$, and so

$$\mathrm{E}_8 \cong \Omega_1(D) = \Omega_1(S) \times D' = \Omega_1(S) \times \langle z\rangle\,.$$

We have

$$\mathrm{C}_D(\Omega_1(S)) = \Omega_1(S) \times \langle z\rangle = \Omega_1(D) \quad \text{and} \quad \mathrm{C}_G(\Omega_1(S)) = \Omega_1(D) * Q\,,$$
$$\text{where } \mho_1(\mathrm{C}_G(\Omega_1(S))) = \langle z\rangle\,.$$

but $S \leq \mathrm{C}_G(\Omega_1(S))$, and so $G' = \langle z\rangle \leq S$, a contradiction. It is easy to see that $G$ possesses a nonnormal abelian subgroup $H \cong \mathrm{E}_4$. Set $H = \langle t, u\rangle$, where $t$ is a noncentral involution in $G$ and $u$ is a central involution in $G$ such that $\langle u\rangle \neq G'$. Then we have $G' \nleq H$. If $H \trianglelefteq G$, then there is $g \in G$ such that $[g, t] \neq 1$, and so $G' = \langle [g, t]\rangle \leq H$, a contradiction. Hence $H = \langle t, u\rangle$ is not normal in $G$.

Let $G = M \times \langle t\rangle$, where $M \cong \mathrm{M}_{p^{s+1}}$, $s \geq 3$, and $\langle t\rangle \cong \mathrm{C}_p$ (which are groups of part (e) of our theorem). We have $\Omega_1(G) \cong \mathrm{E}_{p^3}$, $\Omega_2(G)$ is abelian of type $(p^2, p, p)$ with $\mho_1(\Omega_2(G)) = G' = \mathrm{C}_p$. Thus any subgroup of order $\geq p^3$ is normal in $G$. Let $H$ be a complement of $G'$ in $\Omega_1(G)$ so that $H \nleq Z(G)$, and so $H$ is not normal in $G$. Indeed, if in this case $H \trianglelefteq G$, then $[G, H] \neq \{1\}$ and $[G, H] \leq H$, and so $G' \leq H$, a contradiction.

Let $G$ be a group of part (f) of our theorem. Let $X$ be any subgroup of $G$ of order $\geq p^3$ that is not normal in $G$. Then we have $G' = S' \nleq X$, and so $X$ is abelian of order $\geq p^3$ with $X \cap Z = \{1\}$. But $|G/Z| = p^3$, and so $|X| = p^3$ and $G = Z \times X$ is abelian, a contradiction. Let $H = \langle t, u\rangle \cong E_{p^2}$, where $t$ is a noncentral element of order $p$ in $S$ and $u$ is a central element of order $p$ in $G$ with $\langle u\rangle \neq G'$. Then we have $G' \nleq H$, and so $H$ is not normal in $G$.

Let $G$ be a group of order $p^5$ given in part (g) of our theorem. Then $G$ is special with $G' \cong E_{p^2}$ and $G$ is an $A_2$-group. Let $Y$ be any subgroup of $G$ of order $p^3$ that does not contain $G'$. Since $|G : Y| = p^2$ and $G$ is an $A_2$-group, it follows that $Y$ is abelian of type $(p^2, p)$. Then $A = G'Y$ is a unique abelian maximal subgroup of $G$ and we know that $A \cong C_{p^2} \times C_{p^2}$. But then $E_{p^2} \cong \Omega_1(A) = \Phi(A) = G'$, a contradiction. Let $H$ be an abelian subgroup of order $p^2$ contained in $\Omega_1(G) \cong E_{p^3}$ distinct from $G'$. If $H \trianglelefteq G$, then $G = HA$ and $G/H$ is abelian so that $G' \leq H$, a contradiction. Hence $H$ is not normal in $G$.

Let $G \cong Q_8 * Q_8 * Q_8$ be the extraspecial group of order $2^7$ given in part (h) of our theorem. Let $X$ be any subgroup of order $\geq 2^3$ and assume that $X$ is not normal in $G$. Then $X \cap G' = \{1\}$, and so $X$ is elementary abelian. But then $X \times G'$ is an elementary abelian subgroup of order $\geq 2^4$ in $G$. Since $G$ is extraspecial of order $2^7$ and type "$-$", there are no such elementary abelian subgroups in $G$. Hence $X \trianglelefteq G$. Let $H$ be a 4-subgroup in $G$ with $H \cap G' = \{1\}$. If $H \trianglelefteq G$, then $H \cap Z(G) \neq \{1\}$, a contradiction.

Finally, let $G$ be a group stated in part (i) of our theorem. Then we have

$$\Omega_1(Z(G)) = G', \quad \text{where } Z(G) \text{ is cyclic} .$$

Also note that $|G : Z(G)| = p^4$, and so $G$ does not possess an abelian maximal subgroup. Indeed, if $G$ would have an abelian maximal subgroup, then Lemma 1.1 implies that

$$|G| = p|G'|\,|Z(G)|, \quad \text{where } |G'| = p ,$$

a contradiction. Let $X$ be any subgroup of order $\geq p^3$ in $G$. Then we claim that $X \trianglelefteq G$. Indeed, assume that $X$ is not normal in $G$. Then we have $G' \nleq X$, and so $X \cap Z(G) = \{1\}$, and therefore $X$ is abelian of order $\geq p^3$. But then $Z(G) \times X$ is an abelian subgroup of index $\leq p$ in $G$, a contradiction. It remains to be shown that $G = (A_1 * A_2)Z(G)$ possesses an abelian subgroup of type $(p, p)$ that is not normal in $G$. If $A_1$ and $A_2$ possess noncentral elements $a_1 \in A_1$ and $a_2 \in A_2$ of order $p$, then $H = \langle a_1, a_2\rangle \cong E_{p^2}$ and $H$ is not normal in $G$ since $H \cap Z(G) = \{1\}$. If $p > 2$, then

$$A_1, A_2 \in \{S(p^3), M_{p^n}, n \geq 3\}$$

and in this case there are such elements $a_1$ and $a_2$. If $p = 2$, then we have

$$A_1, A_2 \in \{D_8, Q_8, M_{2^n}, n \geq 4\}$$

and we may replace $A_1$ and $A_2$ with suitable other minimal nonabelian subgroups of $G$ so that again we find noncentral involutions $a_1 \in A_1$ and $a_2 \in A_2$. Indeed we have:

$$Q_8 * Q_8 = D_8 * D_8 , Q_8 * M_{2^n} = D_8 * M_{2^n} , \quad n \geq 4 ,$$

and

$$(D_8 * Q_8)Z(G) = (D_8 * D_8)Z(G) , \quad \text{where } |Z(G)| > 2 .$$

Theorem 174.1 is completely proved. □

**Exercise 1.** Classify the noncyclic $p$-groups such that the intersection of any two distinct conjugate subgroups is trivial (all groups of Theorem 1.25 satisfy this condition).

**Exercise 2.** Classify the noncyclic $p$-groups such that the intersection of any two distinct conjugate subgroups is of order $\leq p$.

**Exercise 3.** Classify the noncyclic $p$-groups such that the intersection of any two distinct conjugate subgroups is of order $\leq p^2$.

**Hint.** Let $R \triangleleft G$ be of order $p^2$. Then $G/R$ is either Dedekindian or a group of Exercise 1.

**Exercise 4.** Classify the noncyclic $p$-groups such that the intersection of any two distinct conjugate subgroups is either cyclic or a 2-group of maximal class.

**Hint.** Assume that $G$ is not a 2-group of maximal class. Then it contains a normal subgroup $R \cong E_{p^2}$. Let $A/R < G/R$ be nonnormal and $x \in G - N_G(A)$. Then $R \leq D = A \cap A^x$. It follows that $D \cong D_8$, and hence $p = 2$.

**Problem 1.** Classify the $p$-groups all of whose nonnormal abelian subgroups are cyclic.

**Problem 2.** Classify the $p$-groups all of whose nonnormal abelian subgroups are either cyclic or noncyclic of order $p^2$.

**Problem 3.** Classify the 2-groups $G$ all of whose nonnormal subgroups are either cyclic or a 2-group of maximal class.[1]

**Problem 4.** Study the $p$-groups all of whose nonnormal subgroups are (i) metacyclic, (ii) absolutely regular, (iii) of exponent $p$, (iv) groups with cyclic derived subgroups.

**Problem 5.** Classify the $p$-groups all of whose nonquasinormal subgroups are (i) cyclic, (ii) either cyclic or 2-groups of maximal class, (iii) metacyclic, (iv) absolutely regular, (v) of exponent $p$.

---

1 Assume that $G$ is not a 2-group of maximal class. Let $R \triangleleft G$ be abelian of type $(2, 2)$ and let $A/R$ be nonnormal in $G/R$. Then $A \cong D_8$. It follows that either $G/R$ is abelian or all nonnormal subgroups of $G/R$ have order 2. In the second case $G/R$ is a group of Theorem 1.25.

**Problem 6.** Classify the $p$-groups in which the normal closure of any nonnormal cyclic subgroup is abelian.

**Problem 7.** Classify the $p$-groups in which the normal closure of any nonnormal metacyclic subgroup is metacyclic.

**Problem 8.** Study the $p$-groups $G$ such that, whenever $H < G$, there is in $H$ a subgroup $F$ of index $\leq p$ that is normal (quasinormal) in $G$.

**Problem 9.** Study the $p$-groups $G$ in which the intersection of their two quasinormal subgroups is quasinormal in $G$. (In that case, the set of quasinormal subgroups of $G$ is a lattice.)

# § 175 Classification of $p$-groups all of whose nonnormal subgroups are cyclic, abelian of type $(p, p)$ or ordinary quaternion

The ultimate goal is to classify $p$-groups in which any two distinct conjugate subgroups have cyclic intersection (Problem 1572). Note that the title groups satisfy this condition, and so the classification of these groups can be considered as one of the first steps in solving problem 1572.

In view of Theorems 16.2 and 174.1, where $p$-groups all of whose nonnormal subgroups are cyclic or abelian of type $(p, p)$ have been classified, it follows that in determining the title groups $G$ we may assume that $p = 2$ and that $G$ possesses a nonnormal subgroup that is ordinary quaternion. In fact we shall prove the following result:

**Theorem 175.A** (Janko). *Let $G$ be a 2-group all of whose nonnormal subgroups are either cyclic or abelian of type $(2, 2)$ or ordinary quaternion. Assume in addition that $G$ possesses a nonnormal subgroup $H$ that is isomorphic to $Q_8$. Then $G$ is isomorphic to one of the following groups:*

(a) *$G \cong Q_{32}$ (a generalized quaternion group of order 32).*

(b) *$G$ is a unique 2-group of order $> 2^4$ with the property that $\Omega_2(G) \cong Q_8 \times C_2$ and we have $|G| = 2^5$, where this group (of class 3) is defined in part A2 (a) of Theorem 49.1.*

(c) *$G$ is a splitting extension of a cyclic noncentral normal subgroup of order 4 by $Q_8$.*

(d) *$G = H_1 \times H_2$, where $H_1 \cong H_2 \cong Q_8$.*

(e) *$G = \langle h_0, h_1\rangle\langle g\rangle$, where*

$$\langle h_0, h_1\rangle \cong Q_8\,, \quad Z(\langle h_0, h_1\rangle) = \langle z\rangle\,,$$
$$\langle g\rangle \cong C_{2^n}\,, \quad n \geq 3\,, \quad \langle h_0, h_1\rangle \cap \langle g\rangle = \{1\}\,,$$
$$\Omega_1(\langle g\rangle) = \langle z'\rangle\,, \quad g^2 \in Z(G)\,, \quad [g, h_0] = 1\,, \quad [g, h_1] = z^{\epsilon}z'\,, \quad \epsilon = 0, 1\,.$$

*Here we have*

$$|G| = 2^{n+3}\,, \quad n \geq 3\,, \quad G' = \Omega_1(G) = \langle z, z'\rangle \cong E_4\,, \quad cl(G) = 2,$$
$$Z(G) = \langle g^2\rangle \times \langle z\rangle \cong C_{2^{n-1}} \times C_2\,.$$

(f) *$G = C * Q$, where*

$$C \cong \mathcal{H}_2 = \langle a, b \mid a^4 = b^4 = 1\,, \quad a^b = a^{-1}\rangle\,, \quad Q \cong Q_8\,, \quad C \cap Q = \langle a^2b^2\rangle = Q'\,.$$

*Conversely, all the above groups satisfy the assumptions of the theorem.*

First we shall prove a series of lemmas concerning 2-groups $G$ that satisfy the assumptions of Theorem 175.A, where $H$ always denotes a nonnormal subgroup in $G$ that is isomorphic to $Q_8$. Set $K = N_G(H)$ so that $H < K < G$ and $K \trianglelefteq G$. Let $L$ be a unique subgroup in $G$ that contains $H$ as a subgroup of index 2. We fix this notation in the whole section.

**Lemma 175.1.** *The factor-group $K/H \neq \{1\}$ is either cyclic or isomorphic to $Q_8$ and $G/L \neq \{1\}$ is Dedekindian. We have $\Omega_1(K) \leq L$ and if $K$ does not possess a $G$-invariant 4-subgroup, then $G \cong Q_{2^5}$ (the case (a) of Theorem 175.A). From now on we shall assume that $K$ possesses a $G$-invariant 4-subgroup $U$. We have in that case $L = HU$ with $U_0 = H \cap U = Z(H) \leq Z(G)$ and $G/U$ is also Dedekindian.*

*Proof.* Since $K/H$ is Dedekindian and $L/H$ is a unique subgroup of order 2 in $K/H$, it follows that $K/H \neq \{1\}$ is either cyclic or isomorphic to $Q_8$, which also implies that $\Omega_1(K) \leq L$.

Assume that $K$ has no $G$-invariant 4-subgroup. By Lemma 1.4, $K$ is a 2-group of maximal class and then $K = L$ is of order $2^4$. We have $C_G(H) = C_K(H) < H$ and then Proposition 10.17 implies that $G$ is also of maximal class. Since $K \trianglelefteq G$, we must have $|G/K| = 2$, and so $|G| = 2^5$. The only possibility is $G \cong Q_{2^5}$ and this group obviously satisfies the assumptions of Theorem 175.A.

From now on we shall assume that $K$ has a $G$-invariant 4-subgroup $U$. Since $\Omega_1(K) \leq L$, we have $U \leq L$ and so $L = HU$ with $U_0 = H \cap U = Z(H)$. But $L' \leq H \cap U$, and so we have $L' = U_0 \leq Z(G)$. Also, $G/U$ is Dedekindian. □

**Lemma 175.2.** *We have*

$$U = Z(L) \leq G' , \ K = H * C_G(H) \quad \text{with } U \leq C_G(H) , \ H \cap C_G(H) = U_0 .$$

*Also, $G/K$ is elementary abelian of order 2 or 4 and $\Omega_1(K) = U$.*

*Proof.* Since $L' = H' = U_0$, we get

$$L = H * Z , \quad \text{where } Z \cong C_4 \text{ or } E_4 \text{ and } H \cap Z = U_0 .$$

However, if $Z \cong C_4$, then $H$ would be a unique subgroup in $L$ that is isomorphic to $Q_8$ and this gives $H \trianglelefteq G$, a contradiction. Hence we have $Z \cong E_4$, and so

$$U = \Omega_1(L) = \Omega_1(K) = Z(L) .$$

Let $H_1$ be any cyclic subgroup of order 4 in $H$. Then

$$H_1 U \trianglelefteq G \quad \text{and so } H_1 = (H_1 U) \cap H \trianglelefteq K .$$

Thus each element in $K$ induces on $H$ an inner automorphism of $H$, and so we get

$$K = H * C_G(H) \quad \text{with } U \leq C_G(H) \text{ and } H \cap C_G(H) = U_0 .$$

For an element $x \in G - K$, there is an element $h \in H$ of order 4 such that $h^x \in L - H$. But $\langle h \rangle U \trianglelefteq G$ with $h^2 \in U_0$, and so $h^x = hu$ for some $u \in U - U_0$. Then we have $[h, x] = u$, and so we get $U \leq G'$.

There are exactly three maximal subgroups of $L$ that contain $U$ and they all are abelian of type $(4, 2)$. The other four maximal subgroups of $L$ that do not contain $U$ are isomorphic to $Q_8$. This gives $1 \neq |G/K| \leq 4$.

For any element $y \in H - U_0$ and any $g \in G - K$, we have

$$y^2 \in U_0\,,\ U\langle y\rangle \trianglelefteq G \text{ and } y^g = yu\,, \quad \text{where } u \in U\,.$$

This gives

$$y^{g^2} = (yu)^g = (yu)u^g = (yu)uu_0 = yu_0 \text{ with some } u_0 \in U_0\,.$$

Hence $g^2 \in K$, and so $G/L$ is elementary abelian of order $\le 4$. □

**Lemma 175.3.** *If $U \nleq Z(G)$, then $G$ is the group of order $2^5$ and class 3 from part (b) of Theorem 175.A and this group satisfies the assumptions of that theorem. From now on we shall always suppose that $U \le Z(G)$.*

*Proof.* Assume that $U \nleq Z(G)$. Note that $K/H \cong C_G(H)/U_0$ is either cyclic or isomorphic to $Q_8$. Hence if $K > L$, then $C_G(H) = C_K(H) > U$, and so there is an element $k$ of order 4 in $C_K(H) - U$ such that $k^2 \in U - U_0$. In that case we have

$$U\langle k\rangle = U_0 \times \langle k\rangle \cong C_2 \times C_4 \trianglelefteq G\,.$$

But then we get $\langle k^2\rangle \trianglelefteq G$, and so $U \le Z(G)$, a contradiction.

We have proved that $K = L$. Suppose that $G - K$ contains an element $y$ of order $\le 4$ that does not centralize $U$. Since $y^2 \in U$, we get $D = U\langle y\rangle \cong D_8 \trianglelefteq G$. Let $V$ be a 4-subgroup in $D$ that is distinct from $U$. Because $U \trianglelefteq G$, we also get $V \trianglelefteq G$ and $V \cap K = U_0 = Z(D)$. But then we have

$$[H, V] \le K \cap V = U_0 < H$$

and so $V$ normalizes $H$, a contradiction. Hence each element in $G - K$ of order $\le 4$ centralizes $U$ and since $U \nleq Z(G)$, there is an element $x$ of order 8 in $G - K$ so that we have $x^2 \in L - U$ and $\langle x^4\rangle = U_0$. Note that $\langle x\rangle U \trianglelefteq G$ and we have either $\langle x\rangle U \cong C_8 \times C_2$ or $\langle x\rangle U \cong M_{16}$. In any case, $\langle x^2\rangle$ is characteristic in $\langle x\rangle U$, and so $\langle x^2\rangle \trianglelefteq G$. Then there are exactly three maximal subgroups of $K = L$ that contain $\langle x^2\rangle$, where two of them are isomorphic to $Q_8$ and $\langle x^2\rangle U \cong C_4 \times C_2$. Thus acting with $G/K$ on four maximal subgroups of $L$ which are isomorphic to $Q_8$, we get $|G : K| = 2$, and so $|G| = 2^5$. Since $U \le Z(K)$ (noting that $K = L$), each element in $G - K$ does not centralize $U$, and so (by the above argument) all elements in $G - K$ are of order 8.

We have proved that $\Omega_2(G) = K = L \cong C_2 \times Q_8$, and so by Theorem 52.1, $G$ is isomorphic to the group defined in part A2 (a) of Theorem 49.1. Since $\Omega_1(G) = G' = U$, this group obviously satisfies the assumptions of Theorem 175.A and we are done. □

**Lemma 175.4.** *The factor-group $G/U$ is abelian, and so we have $G' = U \le Z(G)$. Since for all $x, y \in G$ we get $[x^2, y] = [x, y]^2 = 1$, it follows that $\Phi(G) \le Z(G)$.*

*Proof.* Assume that $G/U$ is nonabelian so that $G/U$ is Hamiltonian. Let $Q/U$ be an ordinary quaternion subgroup in $G/U$, where by our assumption we have $U \le Z(G)$ (see Lemma 175.3). Set

$$Q_0/U = (Q/U)' = Z(Q/U)\,, \quad \text{where } |Q_0 : U| = 2\,.$$

Let $Q_1/U$ and $Q_2/U$ be two distinct cyclic subgroups of order 4 in $Q/U$ so that $Q_1$ and $Q_2$ are two distinct abelian maximal subgroups in $Q$. This implies that $|Q'| = 2$. On the other hand, $Q'$ covers $Q_0/U = (Q/U)'$ and so $Q_0 = U \times Q' \cong E_8$. For each $l \in Q - Q_0$, we have $l^2 \in Q_0 - U$ and $l^2 \in K$ (since $G/K$ is elementary abelian of order $\leq 4$). But then $Q_0 \leq K$, which contradicts Lemma 175.2 that states that $\Omega_1(K) = U$. □

**Lemma 175.5.** *There are no involutions in $G - K$, and so we have $U = G' = \Omega_1(G) \leq Z(G)$.*

*Proof.* Set $Z(H) = H' = \langle z \rangle$ and suppose that there is an involution $i$ in $G - K$. Then $H \neq H^i$ and $i$ normalizes $H_0 = H \cap H^i \cong C_4$. It follows that $H_0\langle i \rangle \cong C_4 \times C_2$ or $D_8$ and $H_0\langle i \rangle \trianglelefteq G$. If $\langle z, i \rangle$ is not normal in $G$, then $H_0\langle i \rangle \cong D_8$ and there is $g \in G$ that induces on $H_0\langle i \rangle$ an outer automorphism (which permutes two 4-subgroups in $H_0\langle i \rangle$). But in that case we have $[(H_0\langle i \rangle), \langle g \rangle] = H_0 \cong C_4$, contrary to the fact that $G' = U \cong E_4$. It follows that we have $E = \langle z, i \rangle \trianglelefteq G$. But then we have $[H, E] \leq K \cap E = \langle z \rangle$, and so $i$ normalizes $H$, a contradiction. □

**Lemma 175.6.** *The factor-group $K/H$ is cyclic.*

*Proof.* Assume that $K/H$ is noncyclic so that setting $Z(H) = H' = \langle z \rangle$ we get

$$Q_8 \cong K/H \cong C_G(H)/\langle z \rangle \quad \text{and therefore } Z(C_G(H)) = U \text{ and} Z(K) = U = Z(G)\,.$$

By Lemma 175.4, we have $\Phi(G) \leq Z(G)$, and so $\Phi(G) = U$. On the other hand, $|K| = 2^6$, and so $|G| \geq 2^7$ and $d(G) \geq 5$. By Lemma 175.5, $G$ has no normal elementary abelian subgroup of order 8, and so by the four-generator theorem (see Theorem 50.3), we must have $d(G) \leq 4$, a contradiction. □

*Proof of Theorem 175.A.* We continue with the situation that we have reached after Lemma 175.6. Hence we have

$$U = G' = \Omega_1(G) \leq Z(G)\,, \quad \Phi(G) \leq Z(G)\,,$$
$$K = H \times \langle a \rangle \quad \text{with } \langle a \rangle \cong C_{2^n}\,,\ n \geq 1\,,\ L = H \times \Omega_1(\langle a \rangle)\,,$$

and $G/K \neq \{1\}$ is elementary abelian of order $\leq 4$.

(i) First assume $K = L$. In this case $G$ is a special group of order $2^5$ or $2^6$ with

$$\Omega_1(G) = \Phi(G) = Z(G) = G' = U \cong E_4 \quad \text{and we set } Z(H) = \langle z \rangle\,.$$

Let $G_0/K$ be any fixed subgroup of order 2 in $G/K$ and let $x \in G_0 - K$. Then $x$ normalizes

$$H_0 = \langle h_0 \rangle = H \cap H^x \cong C_4\,.$$

If $x$ inverts $h_0$, then for an element $h \in H - H_0$, we have $hx \in G_0 - K$ and $hx$ centralizes $H_0$. Hence there is an element $v \in G_0 - K$ such that $v$ centralizes an element $h_0 \in H$ of order 4. If $v^2 = z$, then $h_0 v$ is an involution in $G - K$, a contradiction. Hence we have $v^2 = z' \in U - \langle z\rangle$. Since $H$ is not normal in $G_0$, we have for any $h_1 \in H - \langle h_0\rangle$, $[h_1, v] \in \{z', zz'\}$. However, if $[h_1, v] = zz'$, then we get

$$(h_1v)^2 = h_1^2v^2[h_1, v] = zz'(zz') = 1 ,$$

and so $h_1v$ is an involution in $G - K$, a contradiction. Thus we get $[h_1, v] = z' = v^2$, and so $\langle v\rangle \trianglelefteq G_0$. It follows that $G_0$ is a splitting extension of the cyclic noncentral normal subgroup $\langle v\rangle$ of order 4 (with $v^2 = z'$) by $H \cong \mathrm{Q}_8$. We have obtained the group stated in part (c) of Theorem 175.A. Note that $(h_0v)^2 = zz'$, $\langle h_0v\rangle$ centralizes $\langle h_0\rangle$ and $[h_1, h_0v] = zz'$, and so $G_0$ is also a splitting extension of the cyclic noncentral normal subgroup $\langle h_0v\rangle$ of order 4 (with $(h_0v)^2 = zz'$) by $H \cong \mathrm{Q}_8$.

Suppose now in addition that we have $G/K \cong \mathrm{E}_4$. If a cyclic subgroup $\langle h\rangle$ of order 4 in $H$ is normal in $G$, then acting with $G/K$ on four quaternion subgroups in $K = L$, we see that $G$ interchanges two quaternion subgroups that contain $\langle h\rangle$, and so $G$ interchanges also the other two quaternion subgroups in $K$. But this implies that $|G/K| = 2$, a contradiction. Hence if $G_i/K$ are three subgroups of order 2 in $G/K$, $i = 1, 2, 3$, then each $G_i$ normalizes exactly one of the three cyclic subgroups of order 4 in $H$. This implies that there is an element $w \in G - G_0$ such that $w$ centralizes $h_1$ (from the previous paragraph), $w^2 = z'$ and $[h_0, w] = z'$ so that $K\langle w\rangle$ is a splitting extension of the cyclic noncentral normal subgroup $\langle w\rangle$ of order 4 (with $w^2 = z'$) by $H \cong \mathrm{Q}_8$. We have

$$[h_0, vw] = z', [h_1, vw] = z', [h_0h_1, vw] = 1 ,$$

and so $H$ normalizes $\langle vw\rangle$ with $H \cap \langle vw\rangle = \{1\}$. By the above, we must have $(vw)^2 = z'$, and so we have

$$z' = (vw)^2 = v^2w^2[v, w] = z'z'[v, w] = [v, w] ,$$

which implies that

$$\langle v, w\rangle \cong \mathrm{Q}_8 \quad \text{with } \mathrm{Z}(\langle v, w\rangle) = \langle z'\rangle .$$

But $H$ normalizes both $\langle v\rangle$ and $\langle w\rangle$, and so $H_1 = \langle v, w\rangle \trianglelefteq G$. The structure of $G$ is uniquely determined. We verify that we also have $H_2 = \langle h_1w, h_0v\rangle \cong \mathrm{Q}_8$ with $\mathrm{Z}(\langle h_1w, h_0v\rangle) = \langle zz'\rangle$ and $[H_1, H_2] = \{1\}$. Since $H_1 \cap H_2 = \{1\}$, we have obtained the group $G = H_1 \times H_2$ from part (d) of Theorem 175.A.

Finally, in both cases of groups $G$ in parts (c) and (d) of Theorem 175.A, we have $\Omega_1(G) = G' \cong \mathrm{E}_4$. If $X$ is any subgroup in $G$ of order $\geq 2^3$ and if $X$ contains only one involution, then $X \cong \mathrm{Q}_8$, and if $X$ contains more than one involution, then $X \geq G'$, and so $X \trianglelefteq G$. Thus in both cases the assumptions of Theorem 175.A are satisfied.

(ii) Now assume that $K > L$, and so $|C_G(H) : U| \geq 2$. Since $G/L$ is abelian, $G/K$ is elementary abelian of order 2 or 4, and $K/L$ is cyclic of order $\geq 2$, we have to consider two subcases.

(ii1) $G/K$ has a subgroup $G_0/K$ of order 2 such that $G_0/L$ is cyclic of order $\geq 4$ and either $G = G_0$ or $G = G_0G_1$ with $G_0 \cap G_1 = L$ and $|G_1 : L| = 2$. We set $\mathrm{Z}(H) = \langle z\rangle$. Let

$g$ be an arbitrary element in $G_0 - K$ so that $\langle g\rangle$ covers $G_0/L$. Since $g^2 \in Z(G)$, we have $g^2 \in C_G(H)$. Because $K/H$ is cyclic but $U \leq C_G(H)$ is noncyclic and $C_G(H)/\langle z\rangle \cong K/H$, we get

$$C_G(H) = \langle z\rangle \times \langle g^2\rangle \quad \text{with } o(g^2) \geq 4 \Rightarrow o(g) \geq 8\,.$$

Let $\langle z'\rangle = \Omega_1(\langle g\rangle)$ be the socle of $\langle g\rangle$, where $U = \langle z, z'\rangle$. We have

$$H_0 = \langle h_0\rangle = H \cap H^g \cong C_4$$

is $\langle g\rangle$-invariant, and so $H_0 \trianglelefteq G_0$. But $h_1 \in H - H_0$ inverts $\langle h_0\rangle$, and so $C_G(h_0)$ covers $G_0/K$. Therefore we may choose $g \in C_G(h_0) - K$ so that we may assume $[g, h_0] = 1$. But $H$ is not normal in $G_0$, and so $[h_1, g] \in \{z', zz'\}$ and we may set $[h_1, g] = z^{\epsilon}z'$, where $\epsilon = 0, 1$. We have obtained the groups from part (e) of Theorem 175.A that obviously satisfy the assumptions of that theorem.

Continuing with this case, we assume that

$$G = G_0G_1 \quad \text{with } G_0 \cap G_1 = L = HU\,,\ |G_1 : L| = 2\,.$$

The group $G_1$ is isomorphic to a group in part (c) of Theorem 175.A, and so there is an element $v \in G_1 - L$ of order 4 such that $v^2 = z'$ and $H$ normalizes but does not centralize $\langle v\rangle$ (see arguments in (i)). On the other hand, $g^2 \in Z(G)$ and $o(g^2) \geq 4$, and so there is an element $w$ of order 4 in $\langle g^2\rangle$. But then $vw$ is an involution in $G - K$, contrary to Lemma 175.5.

(ii2) $G = KG^*$, where $K \cap G^* = L$ and $G^*/L$ is elementary abelian of order 2 or 4. Also we have $K = H \times \langle a\rangle$, where $o(a) \geq 4$. Also we set $Z(H) = \langle z\rangle$ and $\Omega_1(\langle a\rangle) = \langle z'\rangle$ so that $U = \langle z, z'\rangle$. In any case, we have in $G^* - L$ an element $v$ of order 4 such that $v^2 = z'$ and $H$ normalizes but does not centralize $\langle v\rangle$. We have $Z(G) \leq C_G(H) = U\langle a\rangle$. If $Z(G) > U$, then there is an element $w$ of order 4 in $\langle a\rangle$ with $w^2 = z'$ and $[v, w] = 1$. But then $vw$ is an involution in $G - K$, contrary to Lemma 175.5.

We have proved that $\Omega_1(G) = Z(G) = U$ and so, in particular, $o(a) = 4$ and $a \notin Z(G)$. This also gives that $\exp(G) = 4$ (because $\mho_1(G) \leq Z(G)$). Hence $G$ is a special group of order $2^6$ or $2^7$. But $G$ has no normal elementary abelian subgroup of order 8, and so by the four-generator theorem we must have $d(G) \leq 4$. Since $\Phi(G) = U$, we must have $|G| = 2^6$ and $|G^* : L| = 2$. We may set $H = \langle h_0, h_1\rangle$ so that $[h_0, v] = 1$ and $[h_1, v] = z'$. Set $[a, v] = u$, where $1 \neq u \in U$. We compute:

$$(va)^2 = v^2a^2u = z'z'u = u \neq 1\,,$$
$$(v(ah_0))^2 = z'(zz')u = uz \qquad \text{and so } u \neq z\,,$$
$$(v(ah_1))^2 = z'(zz')uz' = u(zz') \qquad \text{and so } u \neq zz'\,.$$

It follows that $u = z'$, and so $[a, v] = z'$ and $Q = \langle a, v\rangle \cong Q_8$, which is normalized but not centralized by $H$ and $Q \cap H = \{1\}$. The structure of $G$ is uniquely determined.

Set $C = \langle h_0, h_1a\rangle$. Since

$$h_0^2 = z\,,\quad (h_1a)^2 = zz'\,,\quad [h_0, h_1a] = z\,,$$

we have that $C \cong \mathcal{H}_2$ and $C \cap Q = \langle z' \rangle$, where $z'$ is not a square in $C$. Also we have $[C, Q] = \{1\}$, and therefore we have obtained the group in part (f) of Theorem 175.A, which obviously satisfies the assumptions of that theorem. □

# § 176 Classification of $p$-groups with a cyclic intersection of any two distinct conjugate subgroups

The purpose of this section is to give a complete classification of $p$-groups in which any two distinct conjugate subgroups have a cyclic intersection (problem 1572).

In Theorems 16.2, 174.1 and 175.A we completely determined non-Dedekindian $p$-groups all of whose nonnormal subgroups are either cyclic, abelian of type $(p, p)$ or ordinary quaternion. Since in these groups any two distinct conjugate subgroups have a cyclic intersection, so these results can be considered as a good start in solving problem 1572. Therefore, in handling the title groups $G$ we may always assume that there is in $G$ a nonnormal subgroup that is neither cyclic nor abelian of type $(p, p)$ nor an ordinary quaternion group. The following result will be proved:

**Theorem 176.A.** *Let $G$ be a $p$-group with a cyclic intersection of any two distinct conjugate subgroups. Assume in addition that $G$ has a nonnormal subgroup that is neither cyclic nor abelian of type $(p, p)$ nor an ordinary quaternion group. Then $G$ is metabelian and $G$ is either a 2-group of maximal class and order $\geq 2^5$ (if $|G| = 2^5$, then $G \cong \mathrm{D}_{32}$ or $\mathrm{SD}_{32}$) or $G$ is a $p$-group of class at most 3 with $G' \neq \{1\}$ elementary abelian of order at most $p^2$ and $G$ is isomorphic to one of the groups defined in Propositions 176.3 (b2), 176.5, 176.7, 176.8, 176.9, 175.10, 176.11 and 176.12.*

*Conversely, all these groups satisfy the assumptions of our theorem.*

This theorem will be proved with a series of Propositions: 176.1 to 176.12.

The following exercises are considered as an introduction in the theme of this section. Below we consider intersections, generally speaking, of non-conjugate subgroups.

**Exercise 1.** If $R$ is a subgroup of order $\geq p^4$ and exponent $p$ of a title group $G$, then $R$ is normal in $G$.

**Solution.** As any two distinct maximal subgroups of $R$, say $K, L$ are such that $K \cap L$ is noncyclic, it follows that $K, L \trianglelefteq G$; then $R = KL \trianglelefteq G$.

In particular, $G$ has no nonnormal subgroup of order $\geq p^4$ and exponent $p$.

**Exercise 2.** Let $E \cong \mathrm{E}_{p^k}$ be a subgroup of a title group $G$. Then (a) $k \leq 3$, (b) if, in addition, $E \leq \mathrm{Z}(G)$, then $k \leq 2$, (c) if $H < G$ is nonnormal, then $\mathrm{Z}(G) \cap H$ is of order $\leq p$. In particular, $|\Omega_1(\mathrm{Z}(G))| \leq p^2$.

**Solution.** (a) Assume that $k > 3$. By Exercise 1, all maximal subgroups of $E$ are $G$-invariant and $E \trianglelefteq G$. As any subgroup, say $L$, of order $p$ from $E$ is an intersection

of some maximal subgroups of $E$, it follows that $L \leq Z(G) \Rightarrow E \leq Z(G)$. Let $H < G$ be nonnormal. Then $Z(G) \cap H$ has order at most $p$.
One may assume that $H \cap E = \{1\}$. Let $A, B$ be subgroups of $E$ of order $p^2$, $A \cap B = \{1\}$. Then $H = (H \times A) \cap (H \times B) \lhd G$, a contradiction.

(b) Assume that $E_{p^3} \cong E \leq Z(G)$. As above, one may assume that $H \cap E = \{1\}$. Let $A, B < E$ be of order $p^2$ such that $A \cap B = L$ is of order $p$. Then $H \times L = HA \cap HB \lhd G$. It is easily to deduce from this that $H \lhd G$, a contradiction.

**Exercise 3.** Let $G$ be a title group, $p > 3$. Then $G$ has no nonnormal irregular subgroup of maximal class.

**Solution.** Assume that $H < G$ is a nonnormal irregular of maximal class; then $|G : H| > p$. One has $|\Omega_1(\Phi(H))| = p^{p-1} \geq p^4$ and $\exp(\Omega_1(\Phi(H))) = p$. By Exercise 1, $\Omega_1(\Phi(H)) \leq H_G$ is noncyclic, contrary to Exercise 2 (a).

**Exercise 4.** Let $G$ be a title group, $p > 3$. Is it true that $G$ has no nonnormal irregular subgroup? (**Hint.** By Theorem 12.1 (a), there is in $G$ a subgroup $R$ of order $p^{p-1}$ and exponent $p$. Mimic the solution of the previous exercise. Answer is 'yes'.)

**Exercise 5.** Let $G$ be a title group and $A < G$ be a nonnormal nonmetacyclic minimal nonabelian subgroup. Study the structure of $G$.

**Exercise 6.** The following conditions for a non-Dedekindian $p$-group $G$ are equivalent: (a) $H_G = \{1\}$ for any nonnormal $H < G$, and (b) $G'$ is the unique minimal normal subgroup of $G$.

**Hint.** By Theorem 1.23, there is a $G$-invariant $K < G'$ of index $p$ such that $G/K$ is non-Dedekindian. Let $H/K < G/K$ be nonnormal. Then $K \leq H_G = \{1\}$ so that $K = \{1\}$. It follows that $|G'| = p$, a prime. Assume that $L \neq G'$ is $G$-invariant of order $p$. Then $R = G' \times L$ is an abelian subgroup of type $(p, p)$ in $Z(G)$. It follows that $H \cap R = \{1\}$. The subgroups $HG'$ and $HL$ are normal in $G$ (for example, $(HL)_G = L > \{1\}$) and $HG' \cap HL = H$ so $H \lhd G$, a contradiction. Thus, $G'$ is the unique minimal normal subgroup of $G$. If $T < G$ is nonnormal, then $G' \nleq T$ so that $T_G = \{1\}$.

**Proposition 176.1.** *Let $G$ be a $p$-group with a cyclic intersection of any two distinct conjugate subgroups. Then each nonnormal subgroup $X$ in $G$ possesses a cyclic subgroup of index $p$.*

*Proof.* Let $H$ be a maximal nonnormal subgroup of $G$ containing $X$. Let $L > H$ be such that $|L : H| = p$ so that we have $L \trianglelefteq G$. Since $H$ is not normal in $G$, there is $g \in G - L$ such that $H^g \neq H$. Hence we have $L = HH^g$ and $|H : (H \cap H^g)| = p$. By our assumption, $H \cap H^g$ is cyclic, and so $H$ has a cyclic subgroup of index $p$. Since $X \leq H$, it follows that $X$ also has a cyclic subgroup of index $p$. □

Throughout this section we assume:

(∗) *G is a p-group with a cyclic intersection of any two distinct conjugate subgroups. Assume in addition that G has a maximal nonnormal subgroup H that is neither cyclic nor abelian of type* $(p, p)$ *nor an ordinary quaternion group. We set* $K = N_G(H)$ *so that* $H < K < G$ *and* $K \trianglelefteq G$ *and let* $L/H$ *be a unique subgroup of order p in* $K/H$*, where* $L \trianglelefteq G$*. This notation will be fixed in the whole section.*

**Proposition 176.2.** *We have that* $K/H \neq \{1\}$ *is either cyclic or* $p = 2$ *and* $K/H \cong Q_8$*. Also we have* $\Omega_1(K) \leq L$*.*

*If K does not possess a G-invariant subgroup isomorphic to* $E_{p^2}$*, then G is a 2-group of maximal class and order* $\geq 2^5$ *and if* $|G| = 2^5$*, then* $G \cong D_{32}$ *or* $SD_{32}$ *and all these groups satisfy our assumption* (∗).

*From now on we always assume that K has a G-invariant subgroup U isomorphic to* $E_{p^2}$ *and then we have* $L = HU$ *with* $U_0 = H \cap U \cong C_p$ *and* $G/U$ *is Dedekindian.*

*Proof.* Suppose that $K/H$ has two distinct subgroups $K_1/H$ and $K_2/H$ of order $p$. Then $K_1 \trianglelefteq G$, $K_2 \trianglelefteq G$, and so $K_1 \cap K_2 = H \trianglelefteq G$, a contradiction. Hence $L/H$ is a unique subgroup of order $p$ in $K/H$, and so $K/H$ is either cyclic or generalized quaternion. On the other hand, $K/H$ is Dedekindian, and so $K/H \neq \{1\}$ is either cyclic or $p = 2$ and $K/H \cong Q_8$. In any case, we have $\Omega_1(K) \leq L$.

Assume that $K$ does not have a $G$-invariant abelian subgroup of type $(p, p)$. By Lemma 1.1, we have $p = 2$ and $K$ is a 2-group of maximal class and order $\geq 2^4$. In that case $K/H \cong Q_8$ cannot happen, and so $K/H$ is cyclic. It follows that $K' \leq H$ and $K/K' \cong E_4$, and so $K = L$ and $K'$ is a cyclic subgroup of index 2 in $H$ and $K' \trianglelefteq G$. Since $H$ has only two conjugates in $G$, we have $|G : K| = 2$, and so $|G| \geq 2^5$. Since $H$ is not normal in $G$, we have $G' > K'$, and so $|G : G'| = 4$. By Proposition 1.7 (Taussky), $G$ is a 2-group of maximal class and order $\geq 2^5$. However, $Q_{32}$ does not satisfy (∗), and so if $|G| = 2^5$, then $G \cong D_{32}$ or $SD_{32}$.

Conversely, let $G$ be a 2-group of maximal class and order $\geq 2^5$. Let $Z$ be a unique cyclic subgroup of index 2 in $G$. Let $H$ be any nonnormal subgroup in $G$ so that we have $H \nleq Z$ and set $H_0 = H \cap Z \trianglelefteq G$ with $|H : H_0| = 2$. Hence if $g \in G$ is such that $H^g \neq H$, then we have $H \cap H^g = H_0$ is cyclic.

In what follows we shall always assume that $K$ possesses a $G$-invariant abelian subgroup $U$ of type $(p, p)$. Since $\Omega_1(K) \leq L$, we have $U \leq L$. On the other hand, $G/U$ is Dedekindian, and so $U \nleq H$. We get $L = HU$ with $U_0 = H \cap U \cong C_p$. □

**Proposition 176.3.** *Assuming that G is not a 2-group of maximal class, then it follows that* $|G : K| = p$ *and we may choose a G-invariant abelian subgroup U of type* $(p, p)$ *in L so that* $C_p \cong U_0 = H \cap U \leq Z(G)$*. Also,* $G'$ *covers* $U/U_0$ *and we have one of the following possibilities:*

(a) *One has*

$$p = 2\,, \quad H \cong \mathrm{D}_8\,, \quad \mathrm{Z}(L) = U \le G' \quad \text{and} \quad K = H * \mathrm{C}_G(H) \text{ with}$$
$$U \le \mathrm{C}_G(H) \quad \text{and} \quad H \cap \mathrm{C}_G(H) = U_0\,.$$

*Also, the unique cyclic subgroup of order 4 in $H$ is normal in $G$.*

(b) *The subgroup $H \cong \mathrm{M}_{p^n}$, $n \ge 3$, (if $p = 2$, then $n \ge 4$) or $H$ is abelian of type $(p^s, p)$, $s \ge 2$. Set $H_0 = \Omega_1(H)$ and then $H_0 \cong \mathrm{E}_{p^2}$, $\mathrm{N}_G(H_0) = K$ and $K/H_0$ is Dedekindian. There are two subcases:*

(b1) *If $S = H_0U$ is abelian, then $S \trianglelefteq G$ is elementary abelian of order $p^3$ and either $H \cong \mathrm{M}_{p^n}$, $n \ge 3$, (if $p = 2$. then $n \ge 4$) and in this case we have either $U = \Omega_1(\mathrm{Z}(L))$, $L' = U_0$, and $U \le G'$, or $H$ is abelian of type $(p^s, p)$, $s \ge 2$, and in this case $L$ is abelian of type $(p^s, p, p)$ with $\mho_1(L) = \mho_1(H) \ge U_0$.*

(b2) *If $S = H_0U$ is nonabelian, then $p > 2$, $S \cong \mathrm{S}(p^3) \trianglelefteq G$ (the nonabelian group of order $p^3$ and exponent $p$) with $\mathrm{Z}(S) = U_0$. We have*

$$G = (Z * S)\langle e\rangle\,, \text{ where } \mathrm{C}_{p^m} \cong Z = \mathrm{C}_G(S) \trianglelefteq G, m \ge 2, S \cong \mathrm{S}(p^3) \trianglelefteq G\,,$$
$$Z \cap S = \mathrm{Z}(S) = U_0, Z\langle e\rangle = \langle e\rangle \cong \mathrm{C}_{p^{m+1}} \text{ or } o(e) = p \text{ and } Z\langle e\rangle$$

*is either abelian of type $(p^m, p)$ or $Z\langle e\rangle \cong \mathrm{M}_{p^{m+1}}$, where in any case $e$ induces on $S$ an outer automorphism of order $p$ (normalizing $U$ and fusing the other $p$ maximal subgroups of $S$). One has $\mathrm{E}_{p^2} \cong G' = U < S$ and $G$ is a group of class 3. Next, $\Omega_1(Z * S) = S$ and if $Z\langle e\rangle = \langle e\rangle \cong \mathrm{C}_{p^{m+1}}$, then $\Omega_1(G) = S$. Conversely, groups $G$ defined in (b2) satisfy our assumption $(*)$.*

*Proof.* By Proposition 176.1, $H$ possesses a cyclic subgroup of index $p$.

(i) First assume that $H$ is a 2-group of maximal class. In that case $U_0 = U \cap H = \mathrm{Z}(H)$. If $|H| > 2^3$, then we have $H/U_0 \cong L/U \cong \mathrm{D}_{2^n}$, $n \ge 3$, contrary to the fact that $G/U$ is Dedekindian. It follows that $H \cong \mathrm{D}_8$ and because $|L/U| = 4$, we get $L' \le H \cap U = U_0$ and so $L' = U_0 \le \mathrm{Z}(G)$. Then we have $L = H * Z$, where $Z = \mathrm{C}_L(H)$, $Z \cap H = U_0$ and $Z \cong \mathrm{C}_4$ or $\mathrm{E}_4$.

Let $\langle h\rangle$ be a unique cyclic subgroup of order 4 in $H$ and let $x \in G - K$ so that $H^x \ne H$. Since $H \cap H^x$ is cyclic, we get $H \cap H^x = \langle h\rangle$ for all $x \in G - K$. This gives $\langle h\rangle \trianglelefteq G$. But $L/\langle h\rangle \cong \mathrm{E}_4$, and so $L$ contains exactly two distinct conjugates of $H$ in $G$ and this implies $|G : K| = 2$. Let $t$ be an involution in $H - \langle h\rangle$. Because $U\langle t\rangle \trianglelefteq G$ and $H$ is not normal in $G$, we get for an $x \in G - K$, $t^x \notin H$, and therefore one obtains $t^x = tu$ with some $u \in U - U_0$. Hence $[t, x] = u \in G'$, which implies that $G'$ covers $U/U_0$, and so in this case $U \le G'$.

Assume for a moment that $Z \cong \mathrm{C}_4$. In this case it is well known that $L \cong \mathrm{D}_8 * \mathrm{C}_4$ (central product) contains a unique subgroup $Q$ isomorphic to $\mathrm{Q}_8$, and so $Q \trianglelefteq G$. For any cyclic subgroup $\langle v\rangle$ of order 4 in $Q$ we have $U_0 < \langle v\rangle$ and $U\langle v\rangle \trianglelefteq G$. But then $\langle v\rangle = (U\langle v\rangle) \cap Q \trianglelefteq G$, and so $G$ induces on $Q$ only inner automorphisms of $Q$. We get $G = Q * C$, where $C = \mathrm{C}_G(Q)$ and $Q \cap C = U_0$. Since $Q$ does not centralize $U$, we have

$U \nleq C$, and so $U \cap C = U_0 = Q'$. On the other hand, we get $G' = Q'C' = U_0C' \leq C$, contrary to $U \leq G'$. We have proved that $Z \cong E_4$ and $Z = Z(L) \trianglelefteq G$.

Suppose that $U \neq Z$ so that $U \cap Z = U_0$, $S = UZ \cong E_8$ and $S \trianglelefteq G$. Acting with an element $x \in G - K$ on three subgroups of order 4 in $S$ that contain $U_0 \leq Z(G)$, we see that $Z \trianglelefteq G$, $U \trianglelefteq G$ and so also we have $E_4 \cong S \cap H \trianglelefteq G$. But we know that a cyclic subgroup of order 4 in $H$ is normal in $G$ and so we get $H \trianglelefteq G$, a contradiction. We have proved that $U = Z = Z(L)$.

Let $t$ be any involution in $H$. Since $U\langle t\rangle \trianglelefteq G$ and $H \trianglelefteq K$, it follows that $(U\langle t\rangle) \cap H = \langle t, U_0\rangle \trianglelefteq K$. Thus, each element in $K$ induces on $H$ only inner automorphisms of $H$. It follows:

$$K = H * C_G(H) \quad \text{with } U \leq C_G(H) = C_K(H) \text{ and } H \cap C_G(H) = U_0\,.$$

(ii) Now suppose that $H \cong M_{p^n}$, $n \geq 3$, (where in case $p = 2$ we have $n \geq 4$) or $H$ is abelian of type $(p^s, p)$, $s \geq 2$. Set $H_0 = \Omega_1(H) \cong E_{p^2}$ so that $H_0 \trianglelefteq K$. It follows that $N_G(H_0) = K$ and $K/H_0$ is Dedekindian. Set $S = H_0U \trianglelefteq G$. We have

$$L/U \cong H/U_0\,, \quad \text{where } H' \leq U_0 \leq Z(H), \text{ and so } L' \leq H \cap U = U_0\,.$$

If $L$ is nonabelian, then $L' = U_0 \leq Z(G)$. In that case we act with $G/K$ on $p + 1$ subgroups of order $p^2$ in $S$ that contain $U_0 \leq Z(G)$, where $U$ is the only one of them that is normal in $G$ and all $p$ other ones are fused with $G/K$, and so we get $|G : K| = p$. Also, if $h_0 \in H_0 - U_0$ and $x \in G - K$, then $h_0^x = h_0u$ with $u \in U - U_0$. Hence $G'$ covers $U/U_0$, and so we have in this case $U \leq G'$.

Now assume that $L$ is abelian so that $L$ is of type $(p^s, p, p)$. If $U_0 \leq Z(G)$, then with the same arguments as above, we get $|G : K| = p$ and $G'$ covers $U/U_0$. Now suppose that $U_0 \nleq Z(G)$. Then there is a subgroup $U_1$ of order $p$ in $U$ such that $U = U_0 \times U_1$ and $U_1 \leq Z(G)$. We have $\mho_1(L) = \mho_1(H) \trianglelefteq G$ and $\mho_1(H) \neq \{1\}$ is cyclic. Let $H_1$ be the subgroup of order $p$ in $\mho_1(H)$ so that we get $H_1 \leq Z(G)$. Then we replace $U$ with

$$E_{p^2} \cong U^* = U_1 \times H_1 \leq Z(G)\,, \quad \text{where } U_0^* = U^* \cap H = H_1 \leq Z(G) \text{ and set } S^* = H_0U^*\,.$$

Now, working with $U^*$, $U_0^* \leq Z(G)$ and $S^* = H_0U^*$ (instead of $U$, $U_0$ and $S$), we get with the same arguments as above that $|G : K| = p$ and that $G'$ covers $U^*/U_0^*$. We write again $U$ and $U_0$ instead of $U^*$ and $U_0^*$, respectively, so that we may always assume that $U_0 = U \cap H \leq Z(G)$.

(ii1) Assume that $S = H_0U$ is abelian so that $S \cong E_{p^3}$ and $S \trianglelefteq G$. Suppose in addition that $H \cong M_{p^n}$, $n \geq 3$, (where in case $p = 2$ we have $n \geq 4$). Then we have $L' = H' = U_0 \leq Z(G)$ and $U \leq G'$. Let $\langle a\rangle$ be a cyclic subgroup of index $p$ in $H$ so that $\langle a\rangle$ covers $H/H_0$ (and $L/S$) and $\langle a\rangle \cap H_0 = U_0 = \langle z\rangle$. Let $t \in H_0 - U_0$ so that we may set $[a, t] = z$. Suppose, by way of contradiction, that $U \nleq Z(L)$. In that case, $|L : C_L(U)| = p$, and so $C_L(U) = \langle a^p\rangle S$. We may choose an element $u \in U - U_0$ so that $[a, u] = z^{-1}$. Then we get $[a, ut] = z^{-1}z = 1$ so that we have $Z(L) = \langle a^p\rangle \times \langle ut\rangle$ and $E_{p^2} \cong \Omega_1(Z(L)) = \langle ut, z\rangle \trianglelefteq G$.

But we know that $C_p \cong G/K$ acts transitively on $p$ maximal subgroups of $S$ that contain $U_0 \leq Z(G)$ and that are distinct from $U$. Since $\langle ut, z\rangle \neq U$, we have a contradiction. Thus we have proved that $U \leq Z(L)$, and so $U = \Omega_1(Z(L))$.

Now assume that $H$ is abelian of type $(p^s, p)$, $s \geq 2$. Suppose, by way of contradiction, that $L$ is nonabelian. In that case we have $L' = U_0 \leq Z(G)$ and $C_L(H) = H$. By Lemma 1.1, we get $|L| = p|Z(L)|\,|L'|$, and so $|L : Z(L)| = p^2$. Since $Z(L) < H$, it follows that $Z(L)$ is a maximal subgroup of $H$. If $Z(L) \geq H_0$, then $H_0 = \Omega_1(Z(L))$, which implies that $H_0 \trianglelefteq G$, a contradiction. It follows that $Z(L)$ is a cyclic subgroup of index $p$ in $H$, and so $Z(L)$ covers $H/H_0$ and $L/S$. Hence we get that $L = Z(L)S$ is abelian, a contradiction. We have proved that $L$ is abelian of type $(p^s, p, p)$. Then we get $\mho_1(L) = \mho_1(H)$ and $\mho_1(H)$ is cyclic of order $\geq p$. Let $H_1$ be the subgroup of order $p$ in $\mho_1(H)$ so that $H_1 \leq Z(G)$ and $H_1 \leq H_0$. If $H_1 \neq U_0$, then $H_0 = H_1 \times U_0 \leq Z(G)$, contrary to $N_G(H_0) = K$. Hence we have $H_1 = U_0$, and so $\mho_1(L) = \mho_1(H) \geq U_0$.

(ii2) Assume that $S = H_0U$ is nonabelian. If $p = 2$, then $S \cong D_8$. But $U$ and $H_0$ are the only two 4-subgroups in $S$ and since $U \trianglelefteq G$, it follows that $H_0 \trianglelefteq G$, a contradiction. Hence we have $p > 2$ and $S \cong S(p^3)$ (the nonabelian group of order $p^3$ and exponent $p$) with $S' = Z(S) = U_0$. We know that $U \leq G'$. On the other hand, $G/U$ is Dedekindian and so abelian, which implies that $G' \leq U$, and therefore we have $G' = U < S \trianglelefteq G$. Since $U = G' \nleq Z(S)$, it follows that $G$ is of class 3. Also, $U$ is a unique normal abelian subgroup of type $(p, p)$ in $G$. Indeed, if $V \cong E_{p^2}$, $V \trianglelefteq G$ and $V \neq U$, then the fact that $G/V$ is abelian Dedekindian implies that $G' \leq V \cap U < U$, a contradiction. Set $Z = C_G(S)$ so that $Z \trianglelefteq G$ and $Z \cap S = U_0$. We know that $Z$ does not have a $G$-invariant abelian subgroup of type $(p, p)$, and so Lemma 1.4 implies that $Z \cong C_{p^m}$, $m \geq 1$, is cyclic and so $\Omega_1(Z * S) = S$. If $Z * S = G$, then $G' = U_0 \cong C_p$, a contradiction. Hence we have $Z * S < G$. On the other hand, a Sylow $p$-subgroup of $\mathrm{Aut}(S)$ is isomorphic to $S(p^3)$, and so $G/Z \cong S(p^3)$ and $|G : (Z * S)| = p$. We know that $|G| \geq p^5$ because $|H| \geq p^3$, and so $L = HU(< G)$ is of order $\geq p^4$. This implies that we have $m \geq 2$. Let $e$ be an element in $G - (Z * S)$ so that $e$ fixes $U$ and fuses the other $p$ maximal subgroups of $S$. Since $G/Z \cong S(p^3)$ is of exponent $p$, we have $e^p \in Z$. If $Z\langle e\rangle$ is cyclic, then $Z\langle e\rangle = \langle e\rangle \cong C_{p^{m+1}}$. In this case, $G/S$ is cyclic of order $\geq p^2$ and $\Omega_1(Z * S) = S$ together with $|Z| \geq p^2$ implies $\Omega_1(G) = S$. If $Z\langle e\rangle$ is noncyclic, then $Z\langle e\rangle$ splits over $Z$ and we may assume that $o(e) = p$. In this case $Z\langle e\rangle$ is either abelian of type $(p^m, p)$ or $Z\langle e\rangle \cong M_{p^{m+1}}$. We have obtained the groups stated in part (b2) of our proposition.

It remains to be proved that these groups $G$ satisfy our condition $(*)$. Let $X$ be any noncyclic and nonnormal subgroup of order $\geq p^3$ in $G$. First assume that $|X \cap S| = p^2$ so that we have $X \cap S = S_i$ for some $i \in \{1, 2, \ldots, p\}$, where $\{S_1, S_2, \ldots, S_p\}$ is the set of maximal subgroups of $S$ distinct from $U$ that are acted upon transitively by $G/(Z * S)$. Since $\Omega_1(Z * S) = S$, we have $\Omega_1(X \cap Z * S) = S_i$ and this implies that $X \leq Z * S$. Since $X \geq S_i > U_0 = (Z * S)'$, it follows that $N_G(X) = N_G(S_i) = Z * S$ and then for each $g \in G - (Z * S)$, the intersection $X \cap X^g$ is cyclic.

Now assume that $|X \cap S| = p$. (If $|X \cap S| = 1$, then $X \cap (Z * S) = \{1\}$ and then $|X| \le p$, a contradiction.) In this case, $X_0 = X \cap (Z * S)$ is cyclic of order $\ge p^2$, $X \nleq Z * S$, and so $|X : X_0| = p$. On the other hand, $\mho_1(Z * S) = \mho_1(Z) \ge U_0$, and so $X_0 \ge U_0$. We get $N_G(X_0) \ge \langle Z * S, X\rangle = G$. Hence for each $g \in G$ with $X^g \ne X$, we see that $X \cap X^g = X_0$ is cyclic.

Finally, $ZS_i \cong C_{p^m} \times C_p$, $m \ge 2$, is not normal in $G$ but $Z \trianglelefteq G$, and so our condition $(*)$ is satisfied. Proposition 176.3 is completely proved. □

**Proposition 176.4.** *If $U \cong E_{p^2}$ is a $G$-invariant subgroup contained in $K = N_G(H)$ such that $U_0 = H \cap U \le Z(G)$, then we have $G' \le U$. Hence $G'$ is elementary abelian of order $\le p^2$, and so $G$ is of class at most 3.*

*Proof.* Assume that $G/U$ is nonabelian so that we have $p = 2$ and $G/U$ is Hamiltonian. Let $Q/U$ be any ordinary quaternion subgroup in $G/U$ and we set $Q_0/U = (Q/U)' = Z(Q/U) = (G/U)'$. We have $|Q : C_Q(U)| \le 2$, and so $Q_0 < C_Q(U)$ and let $y \in C_Q(U) - Q_0$ so that $y^2 \in Q_0 - U$. Hence $U\langle y\rangle$ is an abelian maximal subgroup in $Q$. By lemma 1.1,

$$2^5 = |Q| = 2|Q'|\,|Z(Q)|\,, \quad \text{where } Z(Q) \le Q_0 \text{ and } Q_0 \cong E_8 \text{ or } Q_0 \cong C_4 \times C_2\,.$$

If $Q' = Q_0$, then $|Q : Q'| = 4$, and so by Proposition 1.6 (Taussky), $Q$ is of maximal class and order $2^5$, contrary to $U \trianglelefteq Q$. Thus, we have $Q' < Q_0$ and $Q'$ covers $Q_0/Q$.

(i) First suppose that $Q_0 \cong E_8$. We know that $G/Q_0$ is elementary abelian, and so in this case $\exp(G) = 4$. In particular, we must have (according to Proposition 176.3) $H \cong D_8$ or $C_4 \times C_2$. Consider again an abelian maximal subgroup $U \times \langle y\rangle$ of $Q$, where $\langle y\rangle \cong C_4$ and $y^2 \in Q_0 - U$. Since $U \times \langle y\rangle \trianglelefteq G$, we get $y^2 \in Z(G)$. Hence $y^2$ is an involution in $K$ and since $\Omega_1(K) \le L$ (see Propositions 176.2 and 176.3), we get $Q_0 = \langle y^2\rangle \times U \le L$. Set $H_0 = Q_0 \cap H \cong E_4$, where $H_0 > U_0$ and $N_G(H_0) = K$. Now act with $G/K$ on three subgroups of order 4 in $Q_0$ that contain $U_0 \le Z(G)$. We see that only $U$ is normal in $G$ and $H_0 \ne H_0^g$ with some $g \in G - K$. But $y^2 \in Q_0 - U$ and $y^2 \in Z(G)$, and so $\langle y^2, U_0\rangle \trianglelefteq G$, a contradiction.

(ii) We have proved that $Q_0 \cong C_4 \times C_2$ so that all elements in $Q_0 - U$ are of order 4 and all elements in $Q - Q_0$ are of order 8. Since $Q'$ covers $Q_0/U$ and $Q' < Q_0$, we get $Q' \cong C_4$. On the other hand, $\Omega_2(Q) = Q_0 \cong C_4 \times C_2$ and so Lemma 42.1 implies that $Q$ can be defined with $Q = \langle a, b \mid a^8 = b^8 = 1\,,\quad a^4 = b^4 = z\,,\quad a^b = a^{-1}\rangle$, where

$$Q' = \langle a^2\rangle \cong C_4\,,\quad Z(Q) = \langle b^2\rangle \cong C_4\,,\quad \Omega_2(Q) = \langle a^2, b^2\rangle = Q_0 \cong C_4 \times C_2\,,$$
$$\Omega_1(Q) = U = \langle z, a^2b^2\rangle \cong E_4\,,\quad U_0 = \langle z\rangle\,,$$

and $A = \langle a, b^2\rangle \cong C_8 \times C_2$ is a unique abelian maximal subgroup of $Q$. Also, it is easy to see that $\langle a\rangle$ is a characteristic subgroup in $Q$. Indeed, if $\theta \in \mathrm{Aut}(Q)$, then $A^\theta = A$, and so $b^\theta \in Q - A$. Suppose that $\langle a\rangle^\theta \ne \langle a\rangle$. Then $\langle a\rangle^\theta = \langle ab^2\rangle$ and we get $(ab^2)^{b^\theta} = a^{-1}b^{-2} = a^{b^\theta}(b^2)^{b^\theta} = a^{-1}b^2$, and so $b^4 = 1$, a contradiction.

(iii) We know from Proposition 176.3 that $G'$ covers $U/U_0$ and since $G/Q_0$ is elementary abelian (and so $\exp(G) = 8$), we have $G' \le Q_0$. But $Q' = \langle a^2 \rangle$ with $\langle a^4 \rangle = \langle z \rangle = U_0$ and so we get $G' = Q_0$. In particular, we have $G > Q$ and $|G| \ge 2^6$.

Since $C_Q(U) = A = \langle a, b^2 \rangle$ and $|Q : A| = 2$, we see that $C = C_G(U)$ covers $G/Q$, where $C \cap Q = A$ and $C > A$. On the other hand, $C/U$ does not possess an ordinary quaternion subgroup, and so $C/U$ is abelian, and therefore $C$ is of class $\le 2$ with $C' \le U \le Z(C)$. Indeed, if $Q_1/U \cong Q_8$ and $Q_1 \le C$, then by (ii) (since $Q/U$ was an arbitrary ordinary quaternion subgroup in $G/U$), we have $U \nleq Z(Q_1)$, which is not the case. For any $x, y \in C$, we have $[x^2, y] = [x, y]^2 = 1$, and so $\mho_1(C) \le Z(C)$. Since $a \in C$ and $a^2 \in Q_0 - U$, it follows that $Q_0 \le Z(C)$, and so $C = C_G(U) = C_G(Q_0)$. In particular, we get $C_G(b^2) \ge \langle Q, C \rangle = G$, which shows that $b^2 \in Z(G)$.

(iv) Now we show that $C_G(Q) = Z(Q) = \langle b^2 \rangle = Z(G)$. Indeed, set $R = C_G(Q)$, where $R \cap Q = Z(Q) = \langle b^2 \rangle \le Z(G)$ and $b^4 = z$ with $\langle z \rangle = U_0$. First suppose that $R$ has a $G$-invariant 4-subgroup $U_1$. If $U_1 > \langle z \rangle$, then set $U_1 = U^*$ and if $U_1 \ngtr \langle z \rangle$, then considering $E_8 \cong U_1 \times \langle z \rangle$, we may choose in $U_1 \times \langle z \rangle$ a $G$-invariant 4-subgroup $U^*$ such that $U^* > \langle z \rangle$ and we have in any case $U^* \cap U = \langle z \rangle = U^* \cap Q$. Since $U^* \cap H = \langle z \rangle = U_0 \le Z(G)$ and $|(HU^*) : H| = 2$, we have $HU^* \le K = N_G(H)$, and so $L = HU^*$. By Proposition 176.3 (using $U^*$ instead of $U$), we get that $G'$ covers $U^*/U_0$, contrary to to the fact that $G' = Q_0$. Hence $R$ does not have a $G$-invariant 4-subgroup. By Lemma 1.4, $R$ is either cyclic or $R$ is of maximal class. But $\langle b^2 \rangle \cong C_4$ and $\langle b^2 \rangle \le Z(R)$, and so $R$ must be cyclic. Assume that $R > \langle b^2 \rangle$, which together with $\exp(G) = 8$ gives $R \cong C_8$. We may choose a generator $r$ of $R$ so that $r^2 = b^{-2}$ and then $i = rb$ is an involution in $G - Q$ since $i^2 = (rb)^2 = r^2 b^2 = b^{-2} b^2 = 1$. We have $a^i = a^{rb} = a^b = a^{-1}$ and so $[a, i] = a^{-2} \notin U$, contrary to the fact that $G/U$ is Hamiltonian, where for each $x \in G$ with $x^2 \in U$ we must have $[G, x] \le U$.

(v) We study the automorphisms of $Q$ induced on $Q$ by elements of $C$, where $C \cap Q = A$. Now, $A$ induces on $Q$ the inner automorphisms given by:

$$b^a = a^{-1}ba = b(b^{-1}a^{-1}b)a = ba^2\,, \quad b^{a^2} = (ba^2)^a = ba^4 = bz\,.$$

Let $x \in C - A$ so that $x$ centralizes $Q_0 = \langle a^2, b^2 \rangle$ and $x$ normalizes $\langle a \rangle$ (because $\langle a \rangle$ is characteristic in $Q$), which gives $a^x = az^\epsilon$, where $\epsilon \in \{0, 1\}$. Note that $b^x = by$ with some $y \in A = \langle a, b^2 \rangle$. But $x$ normalizes (centralizes) $Q_0 = \langle a^2, b^2 \rangle \cong C_4 \times C_2$, and so $x$ must also normalize $\langle a^2, b \rangle \cong M_{16}$, and so $y \in \langle a^2, b^2 \rangle$. Then we get (noting that $b^2 \in Z(G)$):

$$b^2 = (b^2)^x = (b^x)^2 = (by)^2 = byby = b^2(b^{-1}yb)y = b^2 y^b y\,,$$

and so we have $y^b = y^{-1}$ and this implies $y \in \langle a^2 \rangle$.

(vi) We have proved that each element $x \in C - A$ induces on $Q$ an automorphism given by $b^x = by$, where $y \in \langle a^2 \rangle$ and $a^x = az$. Indeed, if $\epsilon = 0$, i.e., $a^x = a$, then $x$ would induce on $Q$ an inner automorphism, contrary to $C_G(Q) = Z(Q)$. Since $b^{x^2} = by^2$

and $a^{x^2} = a$, we have $x^2 \in Q$. Setting $G_0 = \langle x \rangle Q$, where $|G_0 : Q| = 2$, we see that $G_0 = G$, and so $G' = Q' = \langle a^2 \rangle \cong C_4$ because $[b, x] = y \in \langle a^2 \rangle$ and $[a, x] = z = a^4$, and so $G/\langle a^2 \rangle$ is abelian. On the other hand, we know that $G' = Q_0$. This is a final contradiction and our proposition is proved. □

**Proposition 176.5.** *Suppose that we have the case* (a) *of Proposition 176.3, where* $H \cong D_8$. *Then* $K/H$ *is cyclic and we have the following possibilities:*

(a) $G = (\langle a \rangle \times \langle b \rangle)\langle i \rangle$, *where* $\langle a \rangle \cong \langle b \rangle \cong C_4$ *and* $i$ *is an involution with* $a^i = a^{-1}$ *and* $b^i = b^{-1}$ *or* $b^i = ba^2b^2$.

(b) *$G$ is a unique group of order* $2^5$ *and class* 3 *with* $\Omega_2(G) \cong C_2 \times D_8$, *which is defined in Theorem 52.2 (a) for* $n = 2$.

(c) $G = (\langle h \rangle \times \langle g \rangle)\langle i \rangle$, *where* $\langle h \rangle \cong C_4$, $\langle g \rangle \cong C_{2^m}$, $m \geq 3$, *and* $i$ *is an involution with* $h^i = h^{-1}$ *and* $g^i = g^{1+2^{m-1}}$. *Here*

$$|G| = 2^{m+3}, \quad G' = \langle h^2, g^{2^{m-1}} \rangle \cong E_4, \quad G' \leq Z(G),$$
$$Z(G) = \langle h^2 \rangle \times \langle g^2 \rangle \cong C_2 \times C_{2^{m-1}}.$$

*Finally,* $\langle h, i \rangle \cong D_8$ *and* $\langle g, i \rangle \cong M_{2^{m+1}}$ *are not normal in* $G$.

(d) *$G$ is a special group of order* $2^6$ *given with:*

$$G = (H \times \langle a \rangle)\langle g \rangle, \quad \text{where } H = \langle h, i \mid h^4 = i^2 = 1, \quad h^i = h^{-1}, \quad h^2 = z \rangle \cong D_8,$$
$$\langle a \rangle \cong C_4, \quad a^2 = z', \quad g^2 = zz', \quad [g, h] = 1, \quad [g, i] = [g, a] = z'.$$

*Here* $G' = \langle z, z' \rangle \cong E_4$, $\langle h, i \rangle \cong D_8$ *is not normal in $G$ but* $\langle h \rangle \trianglelefteq G$, *and* $\langle i, a \rangle \cong C_2 \times C_4$ *is not normal in $G$ but* $\langle a \rangle \trianglelefteq G$.

*Conversely, all the above groups satisfy our assumption* $(*)$.

*Proof.* By Proposition 176.4, we have $G' = U \cong E_4$.

(i) First assume $K/H \cong Q_8$ so that we have $|G| = 2^7$. We set $C = C_G(H) = C_K(H)$ so that we have $K = H * C$ with $U \leq C$, $H \cap C = U_0$ and $C/U_0 \cong Q_8$. Let $C_1/U_0$ and $C_2/U_0$ be two distinct cyclic subgroups of order 4 in $C/U_0$ so that $C_1$ and $C_2$ are abelian and $C_1 \cap C_2 = U$. It follows that $U \leq Z(C)$, and so we get $U = Z(K)$ and $|C'| = 2$, and therefore we have $U = U_0 \times C'$, where we set $U_0 = \langle z \rangle$ and $C' = \langle z' \rangle$. Also we have $C = C_G(L)$ and $C \trianglelefteq G$, $C' \trianglelefteq G$, which implies $U \leq Z(G)$. Thus we get $U = Z(G) = G'$ and for any $x, y \in G$ we have $[x^2, y] = [x, y]^2 = 1$, and therefore $\mho_1(G) \leq Z(G)$, and so $U = \Phi(G)$, which shows that $G$ is special. Set

$$H = \langle h, t \mid h^4 = t^2 = 1, \quad h^t = h^{-1} \rangle \cong D_8;$$

then $\langle h \rangle \trianglelefteq G$ (Proposition 176.3 (a)).

(i1) Suppose that $C$ splits over $U_0$, and so, in this case, $C = \langle z\rangle \times C_0$, where $C_0 = \langle c_1, c_2\rangle \cong Q_8$, $C_0' = \langle z'\rangle$. Since $\langle t\rangle \times C_0$ has no cyclic subgroup of index 2, Proposition 176.1 implies that $\langle t\rangle \times C_0 \trianglelefteq G$. But then $C_0 = C \cap (\langle t\rangle \times C_0) \trianglelefteq G$ and each element in $G$ induces on $C_0$ an inner automorphism (otherwise, a cyclic subgroup of order 4 in $C_0$ would be contained in $G'$, contrary to Proposition 176.4). This implies $G = C_0 * G_0$, where

$$G_0 = C_G(C_0), \quad C_0 \cap G_0 = \langle z'\rangle = Z(C_0), \quad G_0 \cap K = L, \quad K = H \times C_0,$$

and $G_0$ is special of order $2^5$ with $Z(G_0) = U$. Since $\langle h\rangle \trianglelefteq G$ and $h^t = h^{-1}$, there is $g \in G_0 - L$ such that $[g, h] = 1$. But $\langle t\rangle U \trianglelefteq G$ and $H$ is not normal in $G$, and so $t^g = tu$ with $u \in \{z', zz'\}$. However, if $t^g = tzz'$, then we replace $g$ with $g' = gh$ (noting that $g' \in G_0 - L$ and $g'$ also centralizes $h$) and get $t^{g'} = (tzz')^h = (tz)zz' = tz'$. Hence, writing again $g$ instead of $g'$, one may assume from the start that $t^g = tz'$, and so $[t, g] = z'$. We have $g^2 \in U$, and so we have $g^2 \in \{1, z', zz', z\}$.

If $g^2 = 1$, then $[g, t] = z'$ implies $\langle g, t\rangle \cong D_8$ with $\langle g, t\rangle' = \langle z'\rangle$, where the unique cyclic subgroup $\langle gt\rangle$ of order 4 in $\langle g, t\rangle$ must be normal in $G$. Indeed, if $\langle g, t\rangle \trianglelefteq G$, then $\langle gt\rangle \trianglelefteq G$, and if $\langle g, t\rangle$ is not normal in $G$, then Proposition 176.3 (a) implies that $\langle gt\rangle \trianglelefteq G$. However, $[gt, h] = z$ but $(gt)^2 = [g, t] = z' \neq z$, and so $\langle gt\rangle$ is not normal in $G$, a contradiction. We shall use this kind of argument here several times.

If $g^2 = z'$, then $c_1^2 = z'$ together with $[g, c_1] = 1$ implies that $gc_1$ is an involution. In that case, $[t, gc_1] = z'$ shows that $\langle t, gc_1\rangle \cong D_8$ with $\langle t, gc_1\rangle' = \langle z'\rangle$. But then $C_4 \cong \langle tgc_1\rangle$ is not normal in $G$ since $[tgc_1, h] = z$, a contradiction.

If $g^2 = zz'$, then $(gh)^2 = z' = c_1^2$ together with $[gh, c_1] = 1$ implies that $ghc_1$ is an involution. In that case, $[t, ghc_1] = z'z$ shows that $\langle t, ghc_1\rangle \cong D_8$ with $\langle t, ghc_1\rangle' = \langle z'z\rangle$. But then $C_4 \cong \langle tghc_1\rangle$ is not normal in $G$ since $[tghc_1, g] = z'$, a contradiction.

If $g^2 = z$, then $gh$ is an involution. In this case, $[t, gh] = z'z$ shows that $\langle t, gh\rangle \cong D_8$ with $\langle t, gh\rangle' = \langle z'z\rangle$. But then $C_4 \cong \langle tgh\rangle$ is not normal in $G$ since $[tgh, g] = z'$, a contradiction.

(i2) We have proved that $C$ does not split over $U_0$. Since $C$ is two-generator with $C' = \langle z'\rangle$, it follows that $C$ is minimal nonabelian. We have $\Omega_1(C) = U \cong E_4$, and so $C$ is metacyclic. Hence we may choose generators $c_1, c_2$ of $C$ so that

$$\mathcal{H}_2 \cong C = \langle c_1, c_2 \mid c_1^4 = c_2^4 = 1, c_1^{c_2} = c_1^{-1}\rangle,$$

where $c_1^2 = z'$, $c_2^2 = zz'$, $z$ is not a square in $C$.

Since $\langle h\rangle \trianglelefteq G$ and $h^t = h^{-1}$, it follows that $C_G(h)$ covers $G/K$. Let $g \in C_G(h) - K$ so that $[h, g] = 1$ and $g^2 \in \langle z, z'\rangle$. Because $\langle t\rangle U \trianglelefteq G$, $\langle h\rangle \trianglelefteq G$ and $H$ is not normal in $G$, it follows that $t^g = tu$ with $u \in U - U_0$. Replacing $g$ with $gh$, if necessary, we may assume from the start that $t^g = tz'$, and so we have $[g, t] = z'$.

If $g$ normalizes $\langle c_1\rangle$, then replacing $g$ with $g' = gc_2$ (if necessary), we may assume that $g'$ centralizes $\langle c_1\rangle$ (and we note that $g'$ acts the same way on $H$ as $g$ does). In this case we write again $g$ instead of $g'$ and we have $[g, c_1] = z^\epsilon$ with $\epsilon = 0$. If $g$ does not normalize $\langle c_1\rangle$, then we have $[g, c_1] = zz'$ or $[g, c_1] = z$. If in this case $[g, c_1] = zz'$, then again replacing $g$ with $g' = gc_2$, we get

$$[g', c_1] = [gc_2, c_1] = (zz')z' = z\,.$$

Hence writing again $g$ instead of $g'$, we may assume from the start that $[g, c_1] = z^{\epsilon}$ with $\epsilon = 1$. Hence we have in any case $[g, c_1] = z^{\epsilon}$, where $\epsilon \in \{0, 1\}$.

If $g^2 = 1$, then $[g, t] = z'$ shows that $\langle g, t\rangle \cong D_8$ with $\langle g, t\rangle' = \langle z'\rangle$. But then $C_4 \cong \langle gt\rangle$ is not normal in $G$ since $[gt, h] = z$, a contradiction.

Assume that $g^2 = z'$. If $\epsilon = 0$, then we have $[g, c_1] = 1$, and so $gc_1$ is an involution. Then $[t, gc_1] = z'$ shows that $\langle t, gc_1\rangle \cong D_8$ with $\langle t, gc_1\rangle' = \langle z'\rangle$. But then $C_4 \cong \langle tgc_1\rangle$ is not normal in $G$ since $[tgc_1, h] = z$, a contradiction. Thus we must have $\epsilon = 1$, and so $[g, c_1] = z$. We compute:

$$(ghc_1)^2 = z'z \cdot z' \cdot [c_1, gh] = zz = 1\,,$$

and so $ghc_1$ is an involution. Then $[t, ghc_1] = z'z$ shows that $\langle t, ghc_1\rangle \cong D_8$ with $\langle t, ghc_1\rangle' = \langle z'z\rangle$. But then $C_4 \cong \langle tghc_1\rangle$ is not normal in $G$ since $[tghc_1, h] = z$, a contradiction.

If $g^2 = z$, then $gh$ is an involution. Then $[t, gh] = z'z$ shows that $\langle t, gh\rangle \cong D_8$ with $\langle t, gh\rangle' = \langle z'z\rangle$. But then $C_4 \cong \langle tgh\rangle$ is not normal in $G$ since $[tgh, g] = z'$, a contradiction.

Suppose that $g^2 = zz'$. Assume in addition that $\epsilon = 0$, and so $[g, c_1] = 1$. In this case $(ghc_1)^2 = zz' \cdot z \cdot z' = 1$, and so $ghc_1$ is an involution. Then $[t, ghc_1] = z'z$ shows that $\langle t, ghc_1\rangle \cong D_8$ with $\langle t, ghc_1\rangle' = \langle z'z\rangle$. But then $C_4 \cong \langle tghc_1\rangle$ is not normal in $G$ since $[tghc_1, g] = z'$, a contradiction. Hence we must have $\epsilon = 1$, and so $[g, c_1] = z$. In this case, $gc_1$ is an involution since $(gc_1)^2 = zz' \cdot z' \cdot z = 1$. Then $[t, gc_1] = z'$ shows that $\langle t, gc_1\rangle \cong D_8$ with $\langle t, gc_1\rangle' = \langle z'\rangle$. But then $C_4 \cong \langle tgc_1\rangle$ is not normal in $G$ since $[tgc_1, g] = z'z$, a contradiction. We have finally proved that here $K/H \cong Q_8$ is not possible.

(ii) Now assume that $K/H \neq \{1\}$ is cyclic. Here we have $K = H \times \langle a\rangle$ with $o(a) = 2^n$, $n \geq 1$, where we set

$$\Omega_1(\langle a\rangle) = \langle z'\rangle\,, \quad U_0 = \langle z\rangle = Z(H)\,,$$
$$\langle h, h' \mid h^4 = (h')^2 = 1\,, \quad [h, h'] = z\,, \quad z^2 = 1\rangle \cong D_8\,, \quad U = \langle z, z'\rangle = G'\,.$$

Since $\langle h\rangle \trianglelefteq G$ (Proposition 176.3 (a)) and $h^{h'} = h^{-1}$, it follows that $C_G(h)$ covers $G/H \cong C_2$. Let $g \in C_G(h) - K$ so that we have $(h')^g = h'u$ for some $u \in U - U_0$ (noting that $\langle h\rangle \trianglelefteq G$, $\langle U\langle h'\rangle\rangle \trianglelefteq G$ but $H$ is not normal in $G$), and so replacing $g$ with $gh$ (if necessary), we may assume from the start that $(h')^g = h'z'$, and so we have $[g, h'] = z'$.

(ii1) Assume that $K = L$ and $z' \in Z(G)$. In this case we have $Z(K) = Z(L) = U = Z(G)$ and $\mho_1(G) \leq Z(G)$. Hence $G$ is a special group of order $2^5$. In particular, all elements in $G - K$ are of order $\leq 4$. Suppose that there is an involution $t \in C_G(h) - K$. Then

$[h', t] = u \in U - \langle z\rangle$, and therefore $\langle h', t\rangle \cong D_8$ with $\langle h', t\rangle' = \langle u\rangle$. Then we must have $C_4 \cong \langle h't\rangle \trianglelefteq G$. On the other hand, $[h't, h] = z$, a contradiction. Hence there is no involution in $C_G(h) - K$. If $g^2 = z$, then $hg$ is an involution in $C_G(h) - K$, a contradiction. Hence we have $g^2 \in \{z', zz'\}$ and $\langle h, g\rangle = \langle h\rangle \times \langle g\rangle \cong C_4 \times C_4$. Set $h' = i$ so that $G = (\langle h\rangle \times \langle g\rangle)\langle i\rangle$ with $h^i = h^{-1}, \quad g^i = gz'$. We have obtained two groups of order $2^5$ stated in part (a) of our proposition, which obviously satisfy our assumption $(*)$.

(ii2) Assume that $K = L$ and $z' \notin Z(G)$. Then we have $[g, z'] = z$. Suppose that there is an element $y \in G - K$ of order $\le 4$. We claim that in this case we have $y^2 \in U$. Indeed, if $y^2$ is a noncentral involution in $K = L$, then $y^2$ inverts $\langle h\rangle$ and $y$ normalizes $\langle h\rangle$ (since $\langle h\rangle \trianglelefteq G$), a contradiction. Hence we have $y^2 \in U$, and so $y^2 \in \langle z\rangle$ since $[y, z'] = z$. We get $D = \langle y, U\rangle \cong D_8$ and $D \trianglelefteq G$ with $Z(D) = \langle z\rangle = D'$. Since $G' = U$ is elementary abelian, each element in $G$ induces an inner automorphism on $D$. Hence, $G = D * C$, where $C = C_G(D)$ and $D \cap C = \langle z\rangle$. Since $|C| = 2^3$ and $z \in Z(C)$, one has $C' \le \langle z\rangle$. This implies $G' = \langle z\rangle$, contrary to Proposition 176.3 (a). Thus, all elements in $G - K$ are of order 8, and so $\Omega_2(G) \cong C_2 \times D_8$. As $g$ centralizes $\langle h\rangle$, we must have $\langle g^2\rangle = \langle h\rangle$, and so one may assume that $g^2 = h$. Indeed, if $\langle g^2\rangle = \langle hz'\rangle$, then $g$ would centralize $h$ and $hz'$, and so $g$ would centralize $z'$, a contradiction. We have obtained a unique group $G$ of order $2^5$ and class 3 with $\Omega_2(G) \cong C_2 \times D_8$, which is defined in Theorem 52.2 (a) for $n = 2$ (stated in part (b) of our proposition). This group obviously satisfies our assumption $(*)$.

(ii3) Assume that $K > L$, i.e., $o(a) = 2^n$, $n \ge 2$. Then there is an element $w \in \langle a\rangle$ of order 4 so that $w^2 = z'$. One has $\langle z, w\rangle = \langle z\rangle \times \langle w\rangle \trianglelefteq G$, and so $\mho_1(\langle z\rangle \times \langle w\rangle) = \langle z'\rangle \trianglelefteq G$, which implies that $G' = U \le Z(G)$. One also has $\mho_1(G) \le Z(G)$. Since $G/L$ is abelian and $K/L \ne \{1\}$ is cyclic, we have here two subcases:

(ii3a) Suppose that $G/L$ is cyclic, and so if $g \in C_G(h) - K$, then

$$\langle g\rangle \text{ covers } G/L, \quad [h', g] = z' \qquad \text{with } \langle z'\rangle = \Omega_1(\langle g^2\rangle) \text{ and } o(g) = 2^m, \quad m \ge 3\,.$$

Hence, $\langle g, h'\rangle \cong M_{2^{m+1}}$. Setting $h' = i$, one obtains $G = (\langle h\rangle \times \langle g\rangle)\langle i\rangle$, where

$$\langle h\rangle \cong C_4, \quad \langle g\rangle \cong C_{2^m}, \quad m \ge 3, \quad h^i = h^{-1}, \quad g^i = g^{1+2^{m-1}}.$$

We have obtained the groups stated in part (c) of our proposition. Conversely, let $X$ be a nonnormal and noncyclic subgroup of order $\ge 2^3$ in $G$. We see that $A = \langle h\rangle \times \langle g\rangle$ is an abelian maximal subgroup in $G$. If $X \cap A$ is noncyclic, then $X \cap A \ge \langle z, z'\rangle = G'$, and so $X \trianglelefteq G$, a contradiction. Hence $X \cap A$ is cyclic and then $X \nleq A$ so that $|X : (X \cap A)| = 2$. It follows that $N_G(X \cap A) \ge \langle A, X\rangle = G$, and so $X \cap A \trianglelefteq G$. Thus, if $g \in G$ is such that $X^g \ne X$, then $X \cap X^g = X \cap A$ is cyclic. Finally, $\langle h, i\rangle \cong D_8$ and $[i, g] = z' \notin \langle h, i\rangle$, and so $\langle h, i\rangle$ is not normal in $G$. Hence our groups satisfy the assumption $(*)$.

(ii3b) $G/L$ is noncyclic abelian so that $G/L$ splits over $K/L$, where $K = H \times \langle a\rangle$ with $o(a) = 2^n$, $n \ge 2$, and $\Omega_1(\langle a\rangle) = \langle z'\rangle$. We have $G = KG_0$, where $K \cap G_0 = L$ and

$|G_0 : L| = 2$. It follows from $G' = U = \langle z, z' \rangle \le Z(G)$ and $\mho_1(G) \le Z(G)$) that $G_0$ is one of two groups defined in part (a) of this proposition, where there is $g \in G_0 - L$ such that

$$\langle g, h \rangle = \langle g \rangle \times \langle h \rangle\,, \quad [h', g] = z'\,, \quad g^2 = z^{\epsilon} z' \quad \text{with } \epsilon = 0, 1\,.$$

Suppose that $\epsilon = 0$ so that $g^2 = z'$, and so $h'$ inverts each element in $\langle g, h \rangle$. Consider the subgroup $H_1 = \langle h', g \rangle \cong D_8$ with $Z(\langle h', g \rangle) = \langle z' \rangle$. If $H_1 \trianglelefteq G$, then $\langle g \rangle \trianglelefteq G$ and if $H_1$ is not normal in $G$, then Proposition 176.3 (a) also shows that $\langle g \rangle \trianglelefteq G$. Hence, in any case, $\langle g \rangle \trianglelefteq G$. Since $\langle a \rangle$ centralizes $h'$, it follows that $\langle a \rangle \times \langle z \rangle$ normalizes $H_1$. On the other hand, $[h, h'] = z$, and so $\langle h \rangle$ does not normalize $H_1$, and so $N_G(H_1) = H_1(\langle a \rangle \times \langle z \rangle)$. If $w$ is an element of order 4 in $\langle a \rangle$, then we have $w^2 = z'$, and so $(H_1\langle w \rangle)/H_1$ and $(H_1\langle z \rangle)/H_1$ are two distinct subgroups of order 2 in $N_G(H_1)/H_1$, contrary to Proposition 176.2. Thus, $\epsilon = 1$, and so $g^2 = zz'$.

Assume that there is an element $w \in \langle a \rangle$ of order 4 such that $w^2 = z'$ and $[w, g] = 1$. Then we have

$$(wg)^2 = w^2 g^2 = z' \cdot zz' = z,\ [wg, h] = 1\,,$$

and so $hwg$ is an involution. From $[h', hwg] = zz'$ it follows that $\langle h', hwg \rangle \cong D_8$ with $Z(\langle h', hwg \rangle) = \langle zz' \rangle$. But then $C_4 \cong \langle h'hwg \rangle$ is not normal in $G$ since $[h'hwg, h] = z$, a contradiction. Thus, there is no such an element $w \in \langle a \rangle$. This implies

$$n = 2\,, \quad o(a) = 4\,, \quad \exp(G) = 4\,, \quad a^2 = z'\,,$$
$$[a, g] \ne 1\,, \quad Z(G) = U = G' = \Phi(G)\,,$$

and so $G$ is special of order $2^6$. It remains to determine $[a, g] \ne 1$.

Suppose that $[a, g] = z$. Then $(ag)^2 = z' \cdot zz' \cdot z = 1$, and so $ag$ is an involution. Next, $[h', ag] = z'$ implies $\langle h', ag \rangle \cong D_8$ with $Z(\langle h', ag \rangle) = \langle z' \rangle$. But then $C_4 \cong \langle h'ag \rangle$ is not normal in $G$ since $[h'ag, h] = z$, a contradiction.

Suppose that $[a, g] = zz'$. Then

$$(gah')^2 = zz' \cdot z' \cdot zz' \cdot z' = 1 \Rightarrow gah' \text{ is an involution}\,.$$

Since $[gah', h'] = z'$, one obtains $\langle gah', h' \rangle \cong D_8$ with $Z(\langle gah', h' \rangle) = \langle z' \rangle$. But then $C_4 \cong \langle gah'h' \rangle = \langle ga \rangle$ is not normal in $G$ since $[ga, g] = zz'$, a contradiction.

Hence we must have $[a, g] = z'$, and so the structure of $G$ is uniquely determined. We set $h' = i$, and so one obtains a special group $G$ of order $2^6$ given with: $G = (H \times \langle a \rangle)\langle g \rangle$, where

$$H = \langle h, i \mid h^4 = i^2 = 1, h^i = h^{-1}, h^2 = z \rangle \cong D_8\,, \quad \langle a \rangle \cong C_4\,,$$
$$a^2 = z'\,, \quad g^2 = zz'\,, \quad [g, h] = 1\,, \quad [g, i] = [g, a] = z'\,.$$

Then $G' = \langle z, z' \rangle \cong E_4$, $\langle h, i \rangle \cong D_8$ is not normal in $G$ but $\langle h \rangle \trianglelefteq G$, and $\langle i, a \rangle \cong C_2 \times C_4$ is not normal in $G$ but $\langle a \rangle \trianglelefteq G$. Thus, the group stated in part (d) of the proposition is obtained.

It remains to prove that this group $G$ satisfies our assumption $(*)$. We first show that there are no involutions in $G-K$, where $K = H \times \langle a \rangle$. Indeed, suppose that $gh^{\alpha} i^{\beta} a^{\gamma}$ with $\alpha, \beta, \gamma \in \{0, 1\}$ is an involution. Then

$$1 = (gh^{\alpha} i^{\beta} a^{\gamma})^2 = zz' \cdot z^{\alpha} \cdot (z')^{\gamma} \cdot (z')^{\beta} \cdot (z')^{\gamma} \cdot z^{\alpha\beta} = z^{1+\alpha+\alpha\beta} (z')^{1+\beta} ,$$

which implies $\beta = 1$ and then we get $z = 1$, a contradiction. Thus, $\Omega_1(G) = L = HU$, where $U = \langle z, z' \rangle$. There are exactly two conjugate classes of noncentral involutions in $G$ with representatives $i$ (4 conjugates) and $hi$ (4 conjugates) and

$$\mathrm{C}_G(i) = \langle i, z \rangle \times \langle a \rangle \cong \mathrm{E}_4 \times \mathrm{C}_4 \quad \text{and}$$
$$\mathrm{C}_G(hi) = \langle hi, z \rangle \times \langle a \rangle \cong \mathrm{E}_4 \times \mathrm{C}_4 .$$

Let $X$ be a noncyclic nonnormal subgroup of order $\geq 2^3$ that contains more than one involution (so that $X \cong \mathrm{Q}_8$ is excluded). Then $G' = U = \langle z, z' \rangle \nleq X$ and $|X| \in \{2^3, 2^4\}$ (noting that all subgroups of order $\geq 2^5$ are normal in $G$).

First assume that $|X| = 2^4$. In this case $X \nleq K$ since $\Phi(K) = \langle z, z' \rangle$ and $|K| = 2^5$. Next, $|X : (X \cap K)| = 2$ and $|X \cap K| = 2^3$. All elements in $X - K$ are of order 4, and so $\mho_1(X) \neq \{1\}$ and this implies that there is exactly one central involution $z_0$ in $G$ that is contained in $X \cap K$ and therefore we have $\mho_1(X) = \langle z_0 \rangle$ and $\mathrm{d}(X) = 3$. But $X \cap K$ must contain another involution $i' \neq z_0$ that is noncentral in $G$ and we know (by the above) that $\mathrm{C}_G(i') = \mathrm{C}_K(i')$ is abelian. In particular, $X$ is nonabelian and $X' = \langle z_0 \rangle$. Because $\mathrm{d}(X) = 3$, $X$ is not minimal nonabelian. Let $X_0$ be any minimal nonabelian subgroup in $X$. If $X_0 \cong \mathrm{D}_8$, then (since there are no involutions in $X - K$) $X_0 = X \cap K$. Since $G' \cong \mathrm{E}_4$, it follows that $X$ induces on $X_0$ only inner automorphisms of $X_0$, which implies that $\mathrm{C}_X(i') \nleq K$, a contradiction. Hence each minimal nonabelian subgroup of $X$ is isomorphic to $\mathrm{Q}_8$. By Corollary A.17.3, $X = \langle t \rangle \times Q$, where $t$ is an involution and $Q \cong \mathrm{Q}_8$ with $\mathrm{Z}(Q) = X' = \langle z_0 \rangle$. Thus $t$ is a noncentral involution in $G$, contrary to the fact that $\mathrm{C}_G(t)$ must be abelian.

Thus, $|X| = 2^3$. First assume that $X \nleq K$. Since $X$ contains more than one involution, it follows that $X \cap K$ contains a noncentral involution $i'$ of $G$. We know that $\mathrm{C}_G(i') \leq K$, and so $X$ is nonabelian. But then $X \cong \mathrm{D}_8$, which is not possible since there are no involutions in $X - K$. Thus, $X \leq K$.

If $X \cong \mathrm{E}_8$, then $X \leq L$, where $L = H \times \langle z' \rangle$. But then $X \geq \langle z, z' \rangle = G'$, a contradiction. It follows that either $X \cong \mathrm{D}_8$ or $X \cong \mathrm{C}_4 \times \mathrm{C}_2$. First assume that $X \cong \mathrm{D}_8$. Because, in this case, $\Omega_1(X) = X$ and $\Omega_1(K) = L$, it follows that $X \leq L$. But then $X$ is conjugate in $G$ to $H = \langle h, i \rangle$ or to $H^* = \langle hz', i \rangle$, where both $\langle h \rangle$ and $\langle hz' \rangle$ are normal in $G$.

Finally, suppose that $X \cong \mathrm{C}_4 \times \mathrm{C}_2$. Because in this case $\{1\} \neq \mho_1(X) \leq \langle z, z' \rangle$, it follows that $X$ contains exactly one central involution of $G$ and two noncentral involutions of $G$. Then $X$ is conjugate in $G$ to $X_1 = \langle i \rangle \times \langle v \rangle$ or to $X_2 = \langle hi \rangle \times \langle w \rangle$, where $\langle v \rangle \cong \langle w \rangle \cong \mathrm{C}_4$. It follows from $X_1 \leq \mathrm{C}_G(i) = \mathrm{C}_K(i) = \langle i, z \rangle \times \langle a \rangle$ that $X_1 = \langle i \rangle \times \langle a \rangle$ or $X_1 = \langle i \rangle \times \langle az \rangle$. Similarly,

$$X_2 \leq \mathrm{C}_G(hi) = \mathrm{C}_K(hi) = \langle hi, z \rangle \times \langle a \rangle$$

implies $X_2 = \langle hi\rangle \times \langle a\rangle$ or $X_2 = \langle hi\rangle \times \langle az\rangle$. On the other hand, $\langle a\rangle \trianglelefteq G$ and $\langle az\rangle \trianglelefteq G$, completing the proof. □

**Proposition 176.6.** *Suppose that the case* (b1) *of Proposition 176.3 holds. Then $H$ possesses exactly one $G$-invariant cyclic subgroup of index $p$.*

*Proof.* We have $H \cong \mathrm{M}_{p^n}$, $n \geq 3$, (if $p = 2$, then $n \geq 4$) or $H$ is abelian of type $(p^s, p)$, $s \geq 2$. Set $H_0 = \Omega_1(H)$ and then

$$H_0 \cong \mathrm{E}_{p^2}\,, \quad \mathrm{N}_G(H_0) = \mathrm{N}_G(H) = K\,, \quad |G/K| = p\,, \quad U_0 = U \cap H = \langle z\rangle \leq \mathrm{Z}(G)\,,$$

and let $g \in G - K$. Note that $H$ has exactly $p$ cyclic subgroups of index $p$. By Proposition 176.4, $G' \leq U$, and so $[K, H] \leq H \cap U = U_0 = \langle z\rangle$. This implies that each cyclic subgroup of index $p$ in $H$ is $K$-invariant. Assume, by way of contradiction, that $H$ has no $G$-invariant cyclic subgroup of index $p$. Since $H \cap H^g$ is a cyclic subgroup of index $p$ in $H$, there is a cyclic subgroup $\langle h\rangle$ of index $p$ in $H$ such that $\langle h\rangle^g = \langle ht\rangle$ for some $t \in H_0 - \langle z\rangle$. Then $h^g = htv$ with some $v \in \langle (ht)^p\rangle = \langle h^p\rangle$. In that case, $h^{-1}h^g = [h, g] = tv \in U \cap H = \langle z\rangle$. Since $v \in \langle h^p\rangle$ and (by Proposition 176.3 (b1)) $\langle h^p\rangle \geq \langle z\rangle$, it follows that $t \in \langle h^p\rangle$, a contradiction. Since $H$ is not normal in $G$, then $H$ possesses exactly one $G$-invariant cyclic subgroup of index $p$ and we are done. □

**Proposition 176.7.** *Suppose that case* (b1) *of Proposition 176.3 holds and assume in addition that $K/H_0$ is Hamiltonian (and so $p = 2$), where $H_0 = \Omega_1(H) \cong \mathrm{E}_4$, and that $G$ has no nonnormal subgroup $\cong \mathrm{D}_8$. Then $G$ is of order $2^7$ and class 2 that has a normal subgroup $K$ of index 2, where $K = (\langle h\rangle \times Q)\langle t\rangle$ with*

$$\langle h\rangle \cong \mathrm{C}_4\,, \quad h^2 = z\,, \quad Q = \langle a, b\rangle \cong \mathrm{Q}_8\,, \quad Q' = \langle u\rangle\,,$$

*$t$ is an involution commuting with $h$ and $a$ and $[b, t] = z$. There is $g \in G - K$ such that one of the following holds:*

(a) *$g^2 = uz$, $g$ centralizes $Q$, $[g, h] = z$, $[g, t] = u$ (and here $G$ is a special group with $G' = \langle u, z\rangle \cong \mathrm{E}_4$ and $\Omega_1(G) = G' \times \langle t\rangle \cong \mathrm{E}_8$),*

(b) *$g^2 = h$, $g$ centralizes $Q$, $[g, t] = uz$ (and here $G$ is of exponent 8 with $G' = \langle u, z\rangle \cong \mathrm{E}_4$, $\mathrm{Z}(G) = G'\langle h\rangle \cong \mathrm{C}_4 \times \mathrm{C}_2$, $\Omega_1(G) = G' \times \langle t\rangle \cong \mathrm{E}_8$ and $\Omega_2(G) = K$).*

*Conversely, the above two groups satisfy our assumption* $(*)$.

*Proof.* One has

$$\begin{aligned} H_0 = \Omega_1(H) \cong \mathrm{E}_4\,, \quad \mathrm{N}_G(H_0) = \mathrm{N}_G(H) = K\,, \quad |G : K| = 2\,, \\ \mathrm{E}_8 \cong S = H_0U \trianglelefteq G\,, \quad U \cap H_0 = U \cap H = U_0 = \langle z\rangle \leq \mathrm{Z}(G)\,, \end{aligned}$$

and $K/H_0$ is Hamiltonian. By Proposition 176.4, we have $G' \leq U$ and this gives $(K/H_0)' = S/H_0 = \mho_1(K/H_0)$, and so $\exp(K) = 4$ and $H \cong \mathrm{C}_4 \times \mathrm{C}_2$. By Proposition 176.3 (b1), $L = HU$ is abelian of type $(4, 2, 2)$, $\mho_1(L) = \mho_1(H) = U_0 = \langle z\rangle$, and so we have $S = \Omega_1(L) = \Omega_1(K)$.

Let $Q/H_0$ be an ordinary quaternion subgroup of $K/H_0$. Since $(Q/H_0)' = (K/H_0)' = S/H_0$, it follows that $S < Q$. Also, $S/H_0$ is a unique subgroup of order 2 in $Q/H_0$, and so we have $Q \cap H = H_0$ and $Q \cap L = S$. Since $Q/H_0 \cong \mathrm{Q}_8$ is isomorphic to a subgroup of $K/H$, Proposition 176.2 implies that $K/H \cong \mathrm{Q}_8$, and so we get $K = HQ$ with $H \cap Q = H_0$.

We have $|Q : \mathrm{C}_Q(H_0)| \le 2$, and so, if $a \in \mathrm{C}_Q(H_0) - S$, then $a^2 \in S - H_0$, and so $A = \langle a \rangle \times H_0 \cong \mathrm{C}_4 \times \mathrm{E}_4$ (containing $U$) is an abelian maximal subgroup of $Q$, $A \trianglelefteq G$ and

$$\mho_1(A) = \langle a^2 \rangle \le \mathrm{Z}(G), \mathrm{E}_4 \cong \langle a^2, z \rangle \trianglelefteq G .$$

On the other hand, $G/K$ acts on the three maximal subgroups of $S$ that contain $\langle z \rangle \le \mathrm{Z}(G)$ fixing $U$ and fusing the other two (since $\mathrm{N}_G(H_0) = K$), and so $\langle a^2, z \rangle = U$ and $U \le \mathrm{Z}(G)$. In particular, $G$ is of class 2 with an elementary abelian commutator subgroup of order $\le 4$ (contained in $U$) and this implies that $\mho_1(G) \le \mathrm{Z}(G)$. Indeed, if $x, y \in G$, then $[x^2, y] = [x, y]^2 = 1$. Next, $\mho_1(K) \le S$ and, since $S \cap \mathrm{Z}(G) = U$, one obtains $\mho_1(K) \le U$, and so $\Phi(K) = U$. For each element $k \in K - L$, we have $k^2 \in U - \langle z \rangle$.

By Proposition 176.6, $H$ possesses exactly one cyclic subgroup $\langle h \rangle$ of index 2 that is normal in $G$ and we have $h^2 = z$. Note that for an element $u \in U - \langle z \rangle$, the cyclic subgroup $\langle hu \rangle \cong \mathrm{C}_4$ is also $G$-invariant. But the abelian normal subgroup $L$ possesses exactly four cyclic subgroups of order 4 and so the other two cyclic subgroups of order 4 in $L$ (which are distinct from $\langle h \rangle$ and $\langle hu \rangle$) must be fused in $G$. Indeed, if $t \in H_0 - \langle z \rangle$ and $g \in G - K$, then we have $t^g = tu$ for some $u \in U - \langle z \rangle$, and so $\langle ht \rangle^g = \langle htu \rangle$.

By Proposition 176.4, $G/U$ is abelian, and so $G/L$ is abelian and $K/L \cong \mathrm{E}_4$. Assume that $G/L$ is not elementary abelian. Then there is $x \in G - K$ such that $x^2 \in K - L$. But then $x^2 \in \mathrm{Z}(G)$, contrary to the fact that $K/H \cong \mathrm{Q}_8$. Hence, $G/L \cong \mathrm{E}_8$. For any $g \in G - K$, one has $g^2 \in L \cap \mathrm{Z}(G)$, and so either $g^2 \in U$ or $g^2 \in L - S$ and in the second case we have either $g^2 \in \langle h \rangle$ or $g^2 \in \langle hu \rangle$ with $u \in U - \langle z \rangle$. Note that $H_1 = \langle hu, t \rangle$ is also a maximal nonnormal subgroup in $G$ with $\Omega_1(H_1) = \Omega_1(H) = \langle z, t \rangle$. Indeed, if $H_1$ is not maximal nonnormal, then let $H_1^*$ containing $H_1$ be a maximal nonnormal subgroup in $G$. Since $\exp(G) \le 8$ and $\exp(K) = 4$, it follows that either $H_1^* \cong \mathrm{C}_8 \times \mathrm{C}_2$ or $H_1^* \cong \mathrm{M}_{16}$, and so $H_1^* \not\le K$. But $\Omega_1(H_1^*) = \Omega_1(H_1) = H_0 = \langle z, t \rangle$, and so $H_0 \trianglelefteq G$, a contradiction. Thus, in the case where there is $g \in G - K$ with $g^2 \in \langle hu \rangle$, one may replace $H$ with $H_1$ (and write again $H$ instead of $H_1$) so that one may assume from the start that $g^2 \in \langle h \rangle$ and then (by a suitable choice of a generator of $\langle g \rangle$), $g^2 = h$.

Let $k$ be any element in $K - L$ that commutes with $t \in H_0 - \langle z \rangle$. Then we have $k^2 \in U - \langle z \rangle$ so that $\langle k, t \rangle \cong \mathrm{C}_4 \times \mathrm{C}_2$. Let us prove that in this case at least one of cyclic subgroups $\langle k \rangle$, $\langle kt \rangle$ is $G$-invariant. If $\langle k, t \rangle \trianglelefteq G$, then both $\langle k \rangle$ and $\langle kt \rangle$ are $G$-invariant because $G' \le U$. (If there is $x \in G$ such that $k^x = kt$ or $k^x = k^{-1}t$, then we have either $t \in G'$ or $k^2 t \in G'$, and so $t \in U$, a contradiction.) If $\langle k, t \rangle$ is not normal in $G$, then it is easy to see that $\langle k, t \rangle$ is a maximal nonnormal subgroup in $G$. Indeed, if $H^* > \langle k, t \rangle$ is a maximal nonnormal subgroup in $G$, then by Proposition 176.3, either $H^* \cong \mathrm{C}_8 \times \mathrm{C}_2$ or $H^* \cong \mathrm{M}_{16}$ (noting that $\exp(G) \le 8$), and so $k$ or $kt$ is a square in $H^*$, and therefore $k$ or $kt$ is contained in $\mathrm{Z}(G)$, contrary to the fact that $K/H \cong \mathrm{Q}_8$. Hence $\langle k, t \rangle$

is a maximal nonnormal subgroup in $G$, and so, by Proposition 176.6, one of $\langle k\rangle$ or $\langle kt\rangle$ is $G$-invariant. Since $\langle k, t\rangle \cap \langle h\rangle = \{1\}$ and $\langle h\rangle \trianglelefteq G$, it follows that $k$ or $kt$ commutes with $h$. But $t$ commutes with $h$, and so in any case $k$ commutes with $h$. Thus, whenever $k \in K - L$ commutes with $t \in H_0 - \langle z\rangle$, then $k$ also commutes with $h$.

Suppose, by way of contradiction, that $t \in Z(Q)$. Let $a, b \in Q - S$ be such that $\langle a, b\rangle$ covers $Q/S$ and set $a^2 = u \in U - \langle z\rangle$. By the above, both $a$ and $b$ commute with $h$. One has $[a, b] \in U - \langle z\rangle$, and so $[a, b] \in \{u, uz\}$. Suppose at the moment that $[a, b] = uz$. By the previous paragraph, we know that $\langle a\rangle$ or $\langle at\rangle$ is normal in $G$. On the other hand,

$$a^b = a(uz)\,, \quad (at)^b = (at)(uz) \quad \text{with } a^2 = (at)^2 = u\,,$$

and so both $\langle a\rangle$ and $\langle at\rangle$ are non-$G$-invariant, a contradiction. Thus, $[a, b] = u$. Consideration of the subgroup $\langle ah\rangle \times \langle t\rangle$ shows that one of the subgroups $\langle ah\rangle$ or $\langle aht\rangle$ must be $G$-invariant. But

$$(ah)^2 = (aht)^2 = uz\,, \quad (ah)^b = (ah)u\,, \quad (aht)^b = (aht)u\,,$$

and so both $\langle ah\rangle$ and $\langle aht\rangle$ are not $G$-invariant, a contradiction.

Thus, $t \notin Z(Q)$. Then $|Q : C_Q(t)| = 2$. Let $a \in C_Q(t) - S$ and $b \in Q - C_Q(t)$ so that

$$\langle a, b\rangle \text{ covers } Q/S\,, \quad [a, b] \in U - \langle z\rangle\,, \quad [a, h] = 1\,, \quad \text{and } [b, t] = z\,.$$

In particular, $Q' = G' = U$. Set $a^2 = u \in U - \langle z\rangle$. If $[a, b] = uz$, then replace $a$ with $a' = at$ (noting that $[a', h] = 1$ and $(a')^2 = u$) and then $[a', b] = [at, b] = uz \cdot z = u$. Let us write $a$ instead of $a'$ so that one may assume from the start that $[a, b] = u$. If $b^2 = uz$, then replace $b$ with $b' = bt$ (noting that $[a, b'] = [a, bt] = u$ and $[b', t] = [bt, t] = z$) and one obtains

$$(b')^2 = (bt)^2 = b^2t^2[t, b] = uz \cdot z = u\,.$$

Hence writing $b$ instead of $b'$, we may assume from the start that $b^2 = u$. We have obtained that $Q^* = \langle a, b\rangle \cong Q_8$. It follows from $(at)^b = (at)(uz)$ and $(at)^2 = u$ that $\langle at\rangle$ is not normal in $G$. This implies that $\langle a\rangle \triangleleft G$. Also note that $b$ has four conjugates in $Q$ and $Q \trianglelefteq G$. Since $|G'| = 4$, $b$ has exactly four conjugates in $G$, and so $C_G(b)$ must cover $G/Q$. Let $g \in C_G(b) - K$; then $g$ normalizes $\langle a\rangle$. If $a^g = a^{-1} = au$, then replace $g$ with $g' = gb \in G - K$ so that

$$a^{g'} = a^{gb} = (au)^b = (au)u = a\,.$$

Noting that $g'$ also commutes with $b$, one may write $g$ instead of $g'$ so that we may assume from the start that $g \in G - K$ centralizes $Q^* = \langle a, b\rangle$. Since $t^b = tz$ and $t^g = tu'$ with some $u' \in U - \langle z\rangle$, it follows that the conjugate class of $t$ in $G$ contains four elements (and they all lie in $S - U$).

Now it is easy to see that there are no involutions contained in $G - K$, and so $\Omega_1(G) = S = G' \times \langle t\rangle \cong E_8$. Indeed, assume that there is an involution $i \in G - K$.

Then $D = \langle i, t\rangle \cong D_8$ and by the assumption, $D \trianglelefteq G$. Since $G' \cong E_4$ is elementary abelian, each element in $G$ induces on $D$ an inner automorphism. In particular, both 4-subgroups in $D$ are $G$-invariant. But then $t$ would have only two conjugates in $G$, a contradiction.

It remains to determine:

$$g^2\,, \quad h^g = hz^\epsilon\,, \quad h^b = hz^\eta\,, \quad \text{and} \quad t^g = tuz^\zeta\,, \quad \text{where } \epsilon, \eta, \zeta \in \{0, 1\}\,.$$

Consideration of the subgroup $\langle ah\rangle \times \langle t\rangle$ shows (by the above) that at least one of the cyclic subgroups $\langle ah\rangle$ or $\langle aht\rangle$ must be normal in $G$. Since

$$\langle h, a, t\rangle = \langle h\rangle \times \langle a\rangle \times \langle t\rangle \cong C_4 \times C_4 \times C_2$$

is abelian, it is enough to consider the action of elements $b$ and $g$ on these cyclic subgroups. One has

$$(ah)^2 = (aht)^2 = uz\,, \quad \text{and} \quad (ah)^b = (ah)uz^\eta\,, \quad (ah)^g = (ah)z^\epsilon\,,$$
$$(aht)^b = (aht)uz^{\eta+1}\,, \quad (aht)^g = (aht)uz^{\epsilon+\zeta}\,.$$

If $\eta = 1$, then $(aht)^b = (aht)u$, and so $\langle aht\rangle$ is not normal in $G$. Then $\langle ah\rangle \trianglelefteq G$, and so $\epsilon = 0$.

If $\eta = 0$, then $(ah)^b = (ah)u$, and so $\langle ah\rangle$ is not normal in $G$. Then $\langle aht\rangle \trianglelefteq G \Rightarrow \epsilon + \zeta = 1$.

(i) First assume that $g^2 \in \{u, z, uz\}$. If $\epsilon = 0$, then $h^g = h$, and so $g$ centralizes $\langle h\rangle \times \langle a\rangle \cong C_4 \times C_4$ and then there is an involution in the coset $g\langle h, a\rangle$, a contradiction. Hence we must have $\epsilon = 1$. By the above, $\eta = 0$ and $\zeta = 0$. Hence, in this case, $h^g = hz$, $h^b = h$, and $t^g = tu$. If $g^2 = u$, then $[g, a] = 1$ implies that $ga$ is an involution, a contradiction. If $g^2 = z$, then $(tb)^2 = uz$ and

$$(gtb)^2 = z \cdot uz \cdot [tb, g] = u \cdot u = 1$$

so that $gtb$ is an involution, a contradiction. Hence, $g^2 = uz$. The structure of $G$ is determined as given in part (a) of the proposition. We check that there are no involutions in $G - K$. Indeed, assume that $gh^\alpha t^\beta a^\gamma b^\delta u'$ with $u' \in U = Z(G)$ and $\alpha, \beta, \gamma, \delta \in \{0, 1\}$, is an involution. Then

$$1 = (gh^\alpha t^\beta a^\gamma b^\delta u')^2 = u^{1+\beta+\gamma+\delta+\gamma\delta} z^{1+\beta\delta}\,,$$

and so $\beta = \delta = 1 \Rightarrow u = 1$, a contradiction.

It remains to prove that this special group $G$ of order $2^7$ satisfies our condition $(*)$. Let $X$ be a noncyclic and nonnormal subgroup of order $\geq 2^3$ that has more than one involution. Then $|X \cap S| = 4$ and $X \cap U = \langle u'\rangle$, where $u'$ is a central involution and $S = \Omega_1(G) = U \times \langle t\rangle$. But all four involutions in $S - U$ are conjugate in $G$ noting that $C_G(t) = \langle h\rangle \times \langle a\rangle \times \langle t\rangle$. Therefore, one may assume that $t \in X$, and so $\Omega_1(X) = \langle t, u'\rangle =$

$X \cap S$. Then $X \le N_G(\langle t, u' \rangle)$ and, since $\Omega_1(X)$ contains at most two conjugates $t$ and $tu'$ of $t$, it follows that $X$ cannot cover $G/C_G(t)$. Therefore either $X \le C_G(t)$ or $X \nleq C_G(t)$. in which case there are one of the following three possibilities:

$$X \le C_G(t)\langle b \rangle \quad \text{or} \quad X \le C_G(t)\langle g \rangle \quad \text{or} \quad X \le C_G(t)\langle bg \rangle .$$

First assume that $X \nleq C_G(t)$, in which case there are three subcases:

(1) If $X \le C_G(t)\langle b \rangle$, then $t^b = tz$, and so $u' = z$. If $x \in X - C_G(t)$, then $x^2 \in U - \langle z \rangle$, which gives $X \ge U = G'$, a contradiction.

(2) Assume that $X \le C_G(t)\langle g \rangle$ and then we have $t^g = tu$, and so $u' = u$. If in this case $x \in X - C_G(t)$, then

$$x = ga^{\alpha} t^{\beta} h^{\gamma} u'' \ (u'' \in U, \quad \alpha, \quad \beta, \quad \gamma \in \{0, 1\}) \quad \text{and then } x^2 = u^{1+\alpha+\beta} z ,$$

which implies $X \ge U = G'$, a contradiction.

(3) Suppose that $X \le C_G(t)\langle bg \rangle$; then $t^{bg} = tuz$, and so $u' = uz$. If in this case $x \in X - C_G(t)$, then

$$x = bga^{\alpha} t^{\beta} h^{\gamma} u'' \ (u'' \in U, \quad \alpha, \quad \beta, \quad \gamma \in \{0, 1\}) \quad \text{and then } x^2 = u^{\beta} z^{1+\beta} .$$

If $\beta = 0$, then $x^2 = z$. If $\beta = 1$, then $x^2 = u$. In any case, $X \ge U = G'$, a contradiction.

Now assume $X \le C_G(t) = (\langle h \rangle \times \langle a \rangle) \times \langle t \rangle$. Since $X \ngeq G' = U$, it follows

$$X \in \{\langle hu^{\mu} \rangle \times \langle t \rangle , \quad \langle az^{\nu} \rangle \times \langle t \rangle , \quad \langle ahz^{\sigma} \rangle \times \langle t \rangle , \quad \text{where } \mu , \quad \nu , \quad \sigma \in \{0, 1\}.\}$$

If $X = \langle hu^{\mu} \rangle \times \langle t \rangle$, then $\langle hu^{\mu} \rangle \trianglelefteq G$.
If $X = \langle az^{\nu} \rangle \times \langle t \rangle$, then $\langle az^{\nu} \rangle \trianglelefteq G$.
If $X = \langle ahz^{\sigma} \rangle \times \langle t \rangle$, then $\langle ahz^{\sigma} t \rangle \trianglelefteq G$ since

$$(ahz^{\sigma} t)^2 = uz , \quad [ahz^{\sigma} t, b] = uz , \quad \text{and} \quad [ahz^{\sigma} t, g] = uz .$$

Thus, condition $(*)$ is satisfied because, for example, $\langle h \rangle \times \langle t \rangle$ is not normal in $G$ (noting that $t^g = tu$).

(ii) Assume that $g^2 = h$. In this case, $h \in Z(G)$ and hence $\epsilon = 0, \eta = 0$. It follows (from the above) that $\zeta = 1$, and so $t^g = tuz$. The structure of $G$ is determined as given in part (b) of the proposition. For each $k \in K$ one has $(gk)^4 = g^4 = z$. Thus, all elements in $G - K$ are of order 8, and so $\Omega_1(G) = S = G' \times \langle t \rangle \cong E_8$.

Conversely, let $X$ be a noncyclic and nonnormal subgroup of order $\ge 2^3$ in $G$ that has more than one involution. Since four noncentral involutions in $S - U$ form a single conjugate class in $G$, one may assume $t \in X$. In addition, $X$ contains exactly one central involution $u' \in U$ so that we have $\Omega_1(X) = \langle t, u' \rangle$.

First suppose that $X \nleq K$ so that $X$ contains elements of order 8, which implies that $z \in X$, and so we have $\Omega_1(X) = \langle t, z \rangle = H_0$. But then $H_0 \trianglelefteq G$, contrary to $t^g = tuz$.

Thus, $X \leq K$. Suppose that

$$X \nleq C_G(t) = (\langle h\rangle \times \langle a\rangle) \times \langle t\rangle \quad \text{and let } x \in X - C_G(t)\,.$$

Then $t^x = tz$ and $x^2 \in U - \langle z\rangle$, and so $X \geq U = \langle u, z\rangle = G'$, a contradiction.

Thus, $X \leq C_G(t)$, and since $\langle u, z\rangle \nleq X$, we get $X \cong C_4 \times C_2$. There are three subcases:
If $X = \langle hu^\mu\rangle \times \langle t\rangle$ ($\mu \in \{0, 1\}$), then $\langle hu^\mu\rangle \trianglelefteq G$.
If $X = \langle az^\nu\rangle \times \langle t\rangle$ ($\nu \in \{0, 1\}$), then $\langle az^\nu\rangle \trianglelefteq G$.
If $X = \langle ahz^\sigma\rangle \times \langle t\rangle$ ($\sigma \in \{0, 1\}$), then $\langle ahz^\sigma t\rangle \trianglelefteq G$ since

$$(ahz^\sigma t)^2 = uz\,, \quad [ahz^\sigma t, b] = uz\,, \quad \text{and} \quad [ahz^\sigma t, g] = uz\,.$$

Thus, the condition $(*)$ is satisfied because, for example, $\langle h\rangle \times \langle t\rangle$ is not normal in $G$ (noting that $t^g = tuz$). The proposition is proved. □

**Proposition 176.8.** *Suppose that our group $G$ has the commutator group $G'$ of order $p$. Then we have $|G : Z(G)| = p^2$, $Z(G)$ is of rank 2, $\Omega_1(G) \nleq Z(G)$ and $Z(G)$ possesses cyclic subgroups of order $\geq p^2$ that do not contain $G'$.*

*Conversely, all these groups satisfy our condition $(*)$.*

*Proof.* By Propositions 176.2 and 176.3, we must be in case (b1) of Proposition 176.3, where $H$ is abelian of type $(p^s, p)$, $s \geq 2$, $L = HU$ is abelian of type $(p^s, p, p)$ with

$$\mho_1(L) = \mho_1(H) \geq U_0 = H \cap U = \langle z\rangle \leq Z(G)\,.$$

By Proposition 176.3, $G'$ covers $U/\langle z\rangle$, and so one may set $G' = \langle u\rangle$, where $u \in U - \langle z\rangle$ so that $U \leq Z(G)$. Then $N_G(H_0) = N_G(H) = K$, where

$$H_0 = \Omega_1(H) \cong E_{p^2}\,, \quad S = H_0U \cong E_{p^3}\,, \quad S = \Omega_1(K)\,.$$

Note that $G/K \cong C_p$ acts transitively on $p$ subgroups of order $p^2$ in $S$ that contain $\langle z\rangle$ and that are distinct from $U$, and so $Z(G) \cap S = U$. Since $Z(G) \leq K$, it follows that $Z(G)$ is of rank 2 and $\Omega_1(G) \nleq Z(G)$. By Proposition 176.3, $G$ has no nonnormal subgroup isomorphic to $D_8$, and so by Proposition 176.7, $K/H_0$ is abelian. This implies that $K$ is abelian and so Lemma 1.1 gives at once that $|G : Z(G)| = p^2$. By Proposition 176.6, $H$ has exactly one $G$-invariant cyclic subgroup $\langle h\rangle \cong C_{p^s}$, $s \geq 2$, where $\langle h\rangle \cap U = \langle z\rangle$, and so $G' \nleq \langle h\rangle$. But $[G, \langle h\rangle] \leq \langle h\rangle \cap G' = \{1\}$, and so $\langle h\rangle \leq Z(G)$. Thus, $Z(G)$ contains cyclic subgroups of order $\geq p^2$ that do not contain $G'$. The groups stated in the proposition are obtained.

Conversely, let $X$ be any noncyclic and nonnormal subgroup of order $\geq p^3$ in a group $G$ described in our proposition. Since $G' \nleq X$, it follows that $X$ is abelian, and so $X$ does not cover $G/Z(G)$ and $X \nleq Z(G)$. Then $|X : (X \cap Z(G))| = p$ and $X_0 = X \cap Z(G)$ is cyclic (since $E_{p^2} \cong \Omega_1(Z(G))$ contains $G'$). If $g \in G$ with $X^g \neq X$, then $X \cap X^g = X_0$ is cyclic. Let $\langle k\rangle$ be a maximal cyclic subgroup of order $\geq p^2$ in $Z(G)$ that does not contain $G'$ and let $i$ be an element of order $p$ in $\Omega_1(G) - Z(G)$. Then $\langle k\rangle \times \langle i\rangle$ does not contain $G'$, and so $\langle k\rangle \times \langle i\rangle$ is a maximal nonnormal subgroup of $G$ of type $(p^r, p)$, $r \geq 2$. Indeed, if $\langle k\rangle \times \langle i\rangle \trianglelefteq G$, then

$$[G, (\langle k\rangle \times \langle i\rangle)] \leq (\langle k\rangle \times \langle i\rangle) \cap G' = \{1\},$$

and so $i \in Z(G)$, a contradiction. The maximality of the cyclic subgroup $\langle k\rangle$ in $Z(G)$ also shows that $\langle k\rangle \times \langle i\rangle$ is a maximal nonnormal subgroup in $G$. □

**Proposition 176.9.** *Suppose that the case* (b1) *of Proposition 176.3 holds, where* $H \cong M_{p^n}$, $n \geq 3$ *(if* $p = 2$*, then* $n \geq 4$*),* $G$ *is of class 3 and* $G$ *does not have nonnormal subgroups isomorphic to* $D_8$ *or such one that leads to the case* (b2) *of Proposition 176.3. Then we have* $p = 2$*,* $G$ *has the following subgroup of index 2:*

$$M_{2^{n+1}} \cong \langle g, u \mid g^{2^n} = u^2 = 1, \quad [g, u] = z = g^{2^{n-1}}\rangle, n \geq 4,$$

*and* $G = \langle g, u\rangle\langle t\rangle$*, where* $t$ *is an involution with* $[g, t] = u$ *and* $[u, t] = 1$*.*

*Next,*

$$|G| = 2^{n+2}, \quad n \geq 4, \quad \textit{with } G' = \langle u, z\rangle \cong E_4,\ [G, G'] = \langle z\rangle,\ \Omega_1(G) = \langle u, z, t\rangle \cong E_8,$$
$$Z(G) = \langle g^4\rangle \cong C_{2^{n-2}} \textit{ and } \langle g^2, t\rangle \cong M_{2^n}$$

*is a nonnormal subgroup in* $G$ *with* $\langle g^2\rangle \trianglelefteq G$*.*

*Conversely, these groups satisfy the condition* $(*)$*.*

*Proof.* By Proposition 176.4, $G' \leq U$, and so we have $G' = U \nleq Z(G)$. Also, Proposition 176.7 implies that $K/\Omega_1(H)$ is abelian, where $\Omega_1(H) \cong E_{p^2}$, and so $K' = H' = \langle z\rangle \leq Z(G)$. By Proposition 176.2, $K/H$ is cyclic of order $\geq p$. Finally, Proposition 176.3 also implies that $U = \Omega_1(Z(L))$, where $L = HU \trianglelefteq G$. By Proposition 176.6, $H$ possesses a $G$-invariant cyclic subgroup $\langle h\rangle$ of index $p$ and there is an element $t$ of order $p$ in $H - \langle h\rangle$ so that $\langle [h, t]\rangle = \langle z\rangle$. For any $g \in G - K$, we have $t^g = tu'$ for some $u' \in U - \langle z\rangle$, where $G/K \cong C_p$, $S = \langle t\rangle U \cong E_{p^3}$ is normal in $G$ and $S = \Omega_1(K)$. It follows that all $p^2$ subgroups of order $p$ contained in $(S - U) \cup \{1\}$ form a single conjugate class in $G$.

Since $K' = H'$, we get $\mho_1(K) \leq Z(K)$ and $K = H * C$, where $C = C_K(H)$ and $H \cap C = \langle h^p\rangle \geq \langle z\rangle$. On the other hand, $K/H \cong C/\langle h^p\rangle$ is cyclic, and so $C$ is abelian of rank 2 (because $\Omega_1(C) = U$), $C = Z(K)$ and $K_1 = \langle h\rangle C$ is an abelian subgroup of index 2 in $K$ with $\Omega_1(K_1) = U$.

No element in $U - \langle z\rangle$ is a $p$-th power of an element in $G$. Indeed, if there is $x \in G$ such that $x^p \in U - \langle z\rangle$, then we consider the subgroup $U\langle x\rangle \trianglelefteq G$ of order $p^3$. Since $\langle z\rangle \leq Z(G)$ and $x$ commutes with $x^p$, it follows that $U\langle x\rangle$ is abelian of type $(p^2, p)$. But then $\mho_1(U\langle x\rangle) = \langle x^p\rangle \trianglelefteq G$, and so $U \leq Z(G)$, a contradiction.

Since $\Omega_1(K_1) = U$ and no element in $U - \langle z\rangle$ is a $p$-th power of an element in $K_1$, it follows that:

$$K_1 = \langle k\rangle \times \langle u\rangle \quad \text{with } u \in U - \langle z\rangle,\ o(k) \geq p^{n-1},\ \langle k\rangle \geq \langle z\rangle.$$

Note that $\mho_1(K_1) = \langle k^p \rangle \le Z(K)$, and so $\langle k^p \rangle \times \langle u \rangle \le Z(K)$. Suppose that $K > L$, in which case $o(k) \ge p^n$. But then $\Omega_{n-1}(K_1) \le \langle k^p \rangle \times \langle u \rangle \le Z(K)$ and $h \in \Omega_{n-1}(K_1)$ implies $h \in Z(K)$, a contradiction.

Thus, $K = L$. Since $\langle h \rangle \trianglelefteq G$, one obtains:

$$[G, \langle h \rangle] \le \langle h \rangle \cap G' = \langle h \rangle \cap U = \langle z \rangle \quad \text{and so } [G, \langle h \rangle] = \langle z \rangle .$$

It follows that $C_G(h)$ covers $G/K$ and $C_K(h) = \langle h \rangle U$. Hence, if $g \in C_G(h) - K$, then we have $g^p \in \langle h \rangle U$ and note that $|C_G(h) : \langle h \rangle| = p^2$. Thus, if $g^p \in (\langle h \rangle U) - \langle h \rangle$, then $C_G(h)$ would be abelian and $C_G(U) \ge \langle C_G(h), t \rangle = G$, a contradiction. Thus, $g^p \in \langle h \rangle$ and this implies that either $o(g) = p^n$, in which case we may set $g^p = h$, or one may assume that $o(g) = p$.

First assume that $p > 2$. Assume in addition that $g^p = h$. We have $[g, t] = u$ with some $u \in U - \langle z \rangle$ and $u^g = uz$, where $\langle g^{p^{n-1}} \rangle = \langle z \rangle \le Z(G)$. It follows that:

$$[g^2, t] = [g, t]^g [g, t] = (uz)u = u^2 z$$

and we prove that $[g^i, t] = u^i z^{\binom{i}{2}}$ for all $i \ge 2$. Indeed, by induction:

$$\begin{aligned} [g^{i+1}, t] &= [g^i g, t] = [g^i, t]^g [g, t] = (u^i z^{\binom{i}{2}})^g u \\ &= (uz)^i z^{\binom{i}{2}} u = u^{i+1} (z^{i+\binom{i}{2}}) = u^{i+1} z^{\binom{i+1}{2}} . \end{aligned}$$

It follows that $[h, t] = [g^p, t] = u^p z^{\binom{p}{2}} = 1$, which is a contradiction.

One may assume in the case $p > 2$ that $o(g) = p$, where $[g, h] = 1$, $h^{p^{n-2}} = z$, $n \ge 3$, and $z \in Z(G)$. It is possible to choose a suitable power $t^j$ in $\langle t \rangle$, $j \not\equiv 0 \pmod p$, so that one can set from the start that $[h, t] = z$. Then $[g, t] = u$ for some $u \in U - \langle z \rangle$; then $[g, u] = z^i$ with some $i \not\equiv 0 \pmod p$. Note that $H^* = \langle g \rangle \times \langle h \rangle \cong C_p \times C_{p^{n-1}}$, $\quad n \ge 3$ is a maximal nonnormal subgroup in $G$ since $|G : H^*| = p^2$ and $[g, t] = u \notin H^*$. Since $\Omega_1(H^*)U = \langle g, z \rangle U \cong S(p^3)$, case (b2) of Proposition 176.3 with respect to $H^*$ holds. But this was excluded by our assumptions.

Thus, $p = 2$. Assume, in addition, that $o(g) = 2$. Then $\langle t, g \rangle \cong D_8$ and $[h, t] = z \notin \langle t, g \rangle$, and so $\langle t, g \rangle$ is a nonnormal subgroup isomorphic to $D_8$, contrary to our assumptions. Thus we have in this case $g^2 = h$. Also

$$o(g) = 2^n , \quad n \ge 4 , \quad [g, t] = u \in U - \langle z \rangle , \quad z = g^{2^{n-1}} , \quad [g, u] = z ,$$

so that $\langle g, u \rangle \cong M_{2^{n+1}}$ is of index 2 in $G$. Next, $\langle h, t \rangle = \langle g^2, t \rangle \cong M_{2^n}$ and $\langle h, t \rangle$ is not normal in $G$ since $[g, t] = u$. We have obtained the groups $G$ stated in our proposition.

We check that there are no involutions in $G - K$, where $K = L = \langle g^2, t \rangle \times \langle u \rangle$, and so $\Omega_1(G) = \langle u, z, t \rangle \cong E_8$. Indeed, suppose that $gh^i u^j t^k$ is an element in $G - K$, where $g^2 = h$, $i$ is any integer and $j, k \in \{0, 1\}$. Then

$$x = (gh^i u^j t^k)^2 = h^{2i+1} u^k z^{j+ik} ,$$

and so $\langle x \rangle \ge \langle z \rangle$. If $x = 1$ then $k = 0$, and so $h^{2i+1} z^j = 1$, a contradiction.

Conversely, let $X$ be any noncyclic and nonnormal subgroup in $G$ of order $\geq 2^3$ containing more than one involution. Then one may assume (up to conjugacy in $G$) that $t \in X$, and so $\Omega_1(X) = \langle t, u' \rangle$ with some involution $u' \in U$. If $X \nleq K$, then by the above calculation we see that $X$ contains $z$, and so we have $\Omega_1(X) = \langle t, z \rangle$. But then for an element $x \in X - K$, we have $[x, t] \in U - \langle z \rangle$, and so in this case $X \geq G' = \langle u, z \rangle$, a contradiction. Hence we have $X \leq K$. Note that $\langle h \rangle \trianglelefteq G$ and $\langle hu \rangle \trianglelefteq G$. Since $|X| \geq 2^3$, it follows that $X \cap \langle h \rangle \neq \{1\}$, and so $z \in X$ and $\Omega_1(X) = \langle t, z \rangle$. Hence, $X = \langle t \rangle (X \cap \langle h \rangle)$ or $X = \langle t \rangle (X \cap \langle hu \rangle)$. But both $X \cap \langle h \rangle$ and $X \cap \langle hu \rangle$ are normal in $G$ and we are done. Our group $G$ satisfies the condition $(*)$. □

**Proposition 176.10.** *Suppose that case* (b1) *of Proposition 176.3 holds, where $H \cong \mathrm{M}_{p^3}$, $p > 2$, and $G$ is of class 2. Then we have the following possibilities:*

(a) *$G$ is a splitting extension of a cyclic normal subgroup $\langle g \rangle \cong \mathrm{C}_{p^m}$, $m \geq 3$, by*

$$\mathrm{M}_{p^3} \cong \langle h, t \mid h^{p^2} = t^p = 1, [h, t] = h^p = z \rangle ,$$

*where $[g, h] = 1$ and $[g, t] = u$ with $\langle u \rangle = \Omega_1(\langle g \rangle)$. Then*

$$|G| = p^{m+3}, \quad m \geq 3, \quad \mathrm{E}_{p^2} \cong G' = \langle u, z \rangle, \quad \mathrm{Z}(G) = \langle g^p \rangle \times \langle z \rangle \cong \mathrm{C}_{p^{m-1}} \times \mathrm{C}_p ,$$

*$\langle g, h \rangle \cong \mathrm{C}_{p^m} \times \mathrm{C}_{p^2}$ is a unique abelian maximal subgroup of $G$,*

$$\Omega_1(G) = \langle u, z, t \rangle \cong \mathrm{E}_{p^3}$$

*and*

$$\langle h, t \rangle \cong \mathrm{M}_{p^3} \quad \text{and} \quad \langle g, t \rangle \cong \mathrm{M}_{p^{m+1}}$$

*are nonnormal subgroups in $G$ with $\langle h \rangle \trianglelefteq G$ and $\langle g \rangle \trianglelefteq G$.*

(b) *$G = (\langle g \rangle \times \langle h \rangle)\langle t \rangle$, where*

$$\langle g \rangle \cong \langle h \rangle \cong \mathrm{C}_{p^2}, \quad g^p = u, \quad h^p = z, \quad t \in \mathrm{C}_g((\langle u, z \rangle)) ,$$
$$[h, t] = z, \quad [g, t] = u^i z^j, \quad i \not\equiv 0 \pmod{p} .$$

*Here $G$ is a special group of order $p^5$ with*

$$\mathrm{E}_{p^2} \cong G' = \langle u, z \rangle, \quad \Omega_1(G) = \langle u, z, t \rangle \cong \mathrm{E}_{p^3}$$

*and $\langle h, t \rangle \cong \mathrm{M}_{p^3}$ is not normal in $G$ with $\langle h \rangle \trianglelefteq G$.*

*Conversely, all groups in (a) and (b) satisfy our assumption $(*)$.*

*Proof.* By Proposition 176.6, $H$ possesses a $G$-invariant cyclic subgroup $\langle h \rangle \cong \mathrm{C}_{p^2}$; and then one may set:

$$H = \langle h, t \mid h^{p^2} = t^p = 1 , \quad [h, t] = h^p = z \rangle .$$

Since $K/\langle t, z \rangle$ is abelian, one has $K' = H' = \langle z \rangle$, and so $K = H * C$ with $H \cap C = \langle z \rangle$, where $C = \mathrm{C}_K(H)$. Also, $K/H \cong C/\langle z \rangle$ is cyclic of order $\geq p$, and so $C$ and $C_1 = \langle h \rangle C$ are abelian, where $\Omega_1(C_1) = U = G' \leq \mathrm{Z}(G)$ and $\mho_1(G) \leq \mathrm{Z}(G)$.

Since $[G, \langle h\rangle] = \langle z\rangle$, we have $G = \langle t\rangle C_G(h)$. Set $S = U \times \langle t\rangle \cong E_{p^3}$ and because $|G : C_G(t)| = p^2$, all $p^2$ subgroups of order $p$ in $(S-U)\cup\{1\}$ form a single conjugate class in $G$. We have $\Omega_1(K) = S$ and, in fact, $\Omega_1(G) = S$. Indeed, if $g \in G-K$ is of order $p$, then $\langle g, t\rangle \cong S(p^3)$ with $u' = [g, t] \in U - \langle z\rangle$. It follows from $\langle g, t\rangle \cap K = \langle t, u'\rangle \cong E_{p^2}$, that $z \notin \langle g, t\rangle$. But $[h, t] = z$, and so $\langle g, t\rangle$ is not normal in $G$, contrary to Proposition 176.1.

(i) First assume that $G/L$ is cyclic of order $\geq p^2$, where $L = HU$. Let $g \in C_G(h) - K$ so that $\langle g\rangle$ covers $G/L$ and $\langle g^p\rangle \leq Z(G)$ covers $K/H$ (which is cyclic of order $\geq p^2$). Hence we have $\Omega_1\langle g\rangle = \langle u\rangle$, where $o(g) = p^m$, $m \geq 3$, $u \in U - \langle z\rangle$ and $[g, t] = uz^i$ for some integer $i$ (mod $p$). We replace $g$ with $g' = h^{-i}g \in C_G(h) - K$ so that

$$[g', t] = [h^{-i}g, t] = z^{-i}(uz^i) = u\,, \quad \text{where } (g')^{p^{m-1}} = (h^{-i}g)^{p^{m-1}} = g^{p^{m-1}}$$

with $\langle g^{p^{m-1}}\rangle = \langle u\rangle$. Thus, one may assume from the start that $[g, t] = u$, and so $\langle g, t\rangle \cong M_{p^{m+1}}$ with $\langle g\rangle \trianglelefteq G$. But $[h, t] = z = h^p$, and so $z \notin \langle g, t\rangle$ and therefore $\langle g, t\rangle$ is a maximal nonnormal subgroup in $G$. Our group $G$ is a splitting extension of $\langle g\rangle$ by $\langle h, t\rangle$, and so we have obtained the groups stated in part (a) of our proposition. Let us check that $\Omega_1(G) = S = \langle u, z, t\rangle \cong E_{p^3}$. Indeed, let $1 \neq t' \in \langle t\rangle$ and suppose that $x = t'g^rh^s$ ($r, s$ are any integers) is an element of order $p$ in $G - \langle g, h\rangle$. Then

$$1 = (t'(g^rh^s))^p = (t')^p g^{pr}h^{ps}[g^rh^s, t']^{\binom{p}{2}} = g^{pr}h^{ps}\,.$$

Hence $r \equiv 0 \pmod{p^{m-1}}$, $s \equiv 0 \pmod p$, and so $x \in S$.

Conversely, let $X$ be a noncyclic and nonnormal subgroup of order $\geq p^3$ in $G$. We may assume (up to conjugacy in $G$) that $t \in X$, and so $\Omega_1(X) = \langle t, u'\rangle \cong E_{p^2}$, where $u'$ is an element of order $p$ in $U$. Set $X_0 = X \cap \langle g, h\rangle$ so that $X_0$ is cyclic and $N_G(X_0) \geq \langle g, h\rangle\langle t\rangle = G$. The condition $(*)$ is satisfied.

(ii) Assume that either $K = L$ or $K > L$ but $G/L$ is noncyclic so that $G/K$ splits over $K/L$. In any case we have $G = KG_0$ with $K \cap G_0 = L$ and $|G_0 : L| = p$. We have $C_{G_0}(h) = (\langle h\rangle U)\langle g\rangle$ for some $g \in G_0-K$. Since there are no elements of order $p$ in $G_0-K$, we have $o(g) \geq p^2$, and so $g^p \in Z(G)\cap L$ implies that $1 \neq g^p \in U$. If $g^p \in \langle z\rangle$, then $\langle g, h\rangle$ would contain elements of order $p$ in $G_0 - K$, a contradiction. Hence, $g^p = u \in U - \langle z\rangle$.

Suppose that $K > L$. Then there is $a \in C - U$ of order $p^2$ so that $a^p = u' \in U - \langle z\rangle$. Consider the subgroup $\langle h\rangle \times \langle g\rangle \cong C_{p^2} \times C_{p^2}$; each element in $\mho_1(\langle g, h\rangle) = \langle u, z\rangle$ is a $p$-th power of an element in $\langle g, h\rangle$. Thus, there is $y \in \langle g, h\rangle - K$ such that $y^p = (u')^{-1}$. But then

$$(ay)^p = a^py^p[y, a]^{\binom{p}{2}} = u'(u')^{-1} = 1\,,$$

and so $ay$ is an element of order $p$ in $G-K$, a contradiction. Hence, $K = L$. In this case, $[g, t] = u^iz^j$ with $i \not\equiv 0 \pmod p$, and so one obtains a special group of order $p^5$ stated in part (b) of the proposition. Let us check that $\Omega_1(G) = S = \langle u, z, t\rangle \cong E_{p^3}$. Indeed, let $1 \neq t' \in \langle t\rangle$ and suppose that $x = t'g^rh^s$ ($r, s$ are any integers) is an element of order $p$ in $G - \langle g, h\rangle$. Then

$$1 = (t'(g^rh^s))^p = (t')^p g^{pr}h^{ps}[g^rh^s, t']^{\binom{p}{2}} = g^{pr}h^{ps}\,.$$

Hence $r \equiv 0 \pmod p$, $s \equiv 0 \pmod p$, and so $x \in S$.

Conversely, let $X$ be a noncyclic and nonnormal subgroup of order $p^3$ in $G$. One may assume (up to conjugacy in $G$) that $t \in X$, and so $\Omega_1(X) = \langle t, u' \rangle \cong \mathrm{E}_{p^2}$, where $u' \in U$ is of order $p$. Set $X_0 = X \cap \langle g, h\rangle$ so that $X_0$ is cyclic of order $p^2$ and $\mathrm{N}_G(X_0) \geq \langle g, h\rangle\langle t\rangle = G$. The condition $(*)$ is satisfied. □

**Proposition 176.11.** *Suppose that the case* (b1) *of Proposition 176.3 holds, where $H \cong \mathrm{M}_{p^n}$, $n \geq 4$, is a nonnormal subgroup of maximal possible order in $G$ (which is isomorphic to some $\mathrm{M}_{p^m}$, $m \geq 4$), $\mathrm{cl}(G) = 2$ and assume that $G$ has no nonnormal subgroups isomorphic to $\mathrm{M}_{p^3}$ with $p > 2$ or $\mathrm{D}_8$. Then the following possibilities hold:*

(a) *$G = (\langle h\rangle \times \langle g\rangle)\langle t\rangle$, where*

$$\langle h\rangle \cong \mathrm{C}_{p^{n-1}}, \quad n \geq 4, \quad \langle g\rangle \cong \mathrm{C}_{p^m}, \quad m \geq 3, \quad \langle t\rangle \cong \mathrm{C}_p,$$
$$[h, t] = z \quad \text{with } \langle z\rangle = \Omega_1(\langle h\rangle), \quad [g, t] = z^i u \quad \text{with } \langle u\rangle = \Omega_1(\langle g\rangle), \quad i \text{ integer},$$

*and $t$ centralizes $\langle u, z\rangle$. Here $|G| = p^{m+n}$, $m \geq 3$, $n \geq 4$,*

$$\mathrm{E}_{p^2} \cong G' = \langle u, z\rangle \leq \mathrm{Z}(G), \quad \Omega_1(G) = \langle u, z, t\rangle \cong \mathrm{E}_{p^3},$$

*$\langle g, h\rangle \cong \mathrm{C}_{p^m} \times \mathrm{C}_{p^{n-1}}$ is a unique abelian maximal subgroup of $G$ and $\langle h, t\rangle \cong \mathrm{M}_{p^n}$ is nonnormal in $G$ with $\langle h\rangle \trianglelefteq G$.*

(b) *$G = (\langle k\rangle \times \langle g\rangle)\langle t\rangle$, where*

$$\langle g\rangle \cong \mathrm{C}_{p^n}, \quad n \geq 4, \quad \langle k\rangle \cong \mathrm{C}_{p^m}, \quad 2 \leq m \leq n-2, \quad \langle t\rangle \cong \mathrm{C}_p,$$
$$[k, t] = z \quad \text{with } \langle z\rangle = \Omega_1(\langle g\rangle), \quad [g, t] = u \quad \text{with } \langle u\rangle = \Omega_1(\langle k\rangle),$$

*and $t$ centralizes $\langle u, z\rangle$. Here we have $|G| = p^{m+n+1}$, $n \geq 4$, $2 \leq m \leq n-2$,*

$$\mathrm{E}_{p^2} \cong G' = \langle u, z\rangle \leq \mathrm{Z}(G), \quad \Omega_1(G) = \langle u, z, t\rangle \cong \mathrm{E}_{p^3},$$

*$\langle g, k\rangle \cong \mathrm{C}_{p^n} \times \mathrm{C}_{p^m}$ is a unique abelian maximal subgroup of $G$ and $\langle kg^p, t\rangle \cong \mathrm{M}_{p^n}$ is nonnormal in $G$ with $\langle kg^p\rangle \trianglelefteq G$.*

*Conversely, all groups in (a) and (b) satisfy the condition $(*)$.*

*Proof.* By Proposition 176.4, $G' \leq U$, and so $G' = U \leq \mathrm{Z}(G)$ and $\mho_1(G) \leq \mathrm{Z}(G)$. Also, Proposition 176.7 implies that $K/\Omega_1(H)$ is abelian, and so $K/H$ is cyclic (by Proposition 176.2), where $\Omega_1(H) \cong \mathrm{E}_{p^2}$ and therefore we have $K' = H' = \langle z\rangle \leq \mathrm{Z}(G)$. By Proposition 176.2, $K/H$ is cyclic of order $\geq p$. Finally, Proposition 176.3 also implies that $U = \Omega_1(\mathrm{Z}(L))$, where $L = HU \trianglelefteq G$. By Proposition 176.6, $H$ possesses a $G$-invariant cyclic subgroup $\langle h\rangle$ of index $p$ and there is an element $t$ of order $p$ in $H - \langle h\rangle$ so that $\langle [h, t]\rangle = \langle z\rangle$. For any $g \in G - K$, we have $t^g = tu'$ for some $u' \in U - \langle z\rangle$, where $G/K \cong \mathrm{C}_p$, $S = \langle t\rangle U \cong \mathrm{E}_{p^3}$ is normal in $G$ and $S = \Omega_1(K)$. It follows that all $p^2$ subgroups of order $p$ contained in $(S - U) \cup \{1\}$ form a single conjugate class in $G$.

Since $K' = H'$, we get $K = H * C$, where $C = \mathrm{C}_K(H)$ and $H \cap C = \langle h^p\rangle \geq \langle z\rangle$. On the other hand, $K/H \cong C/\langle h^p\rangle$ is cyclic, and so $C$ is abelian of rank 2 (because $\Omega_1(C) = U$),

$C = Z(K)$ and $K_1 = \langle h\rangle C$ is an abelian subgroup of index 2 in $K$ with $\Omega_1(K_1) = U$. Since $\langle h\rangle \trianglelefteq G$, we get

$$[G, \langle h\rangle] \le \langle h\rangle \cap G' = \langle h\rangle \cap U = \langle z\rangle ,$$

and so $[G, \langle h\rangle] = \langle z\rangle$. It follows that $G = \langle t\rangle C_G(h)$.

It is easy to see that there are no elements of order $p$ in $G-K$. Indeed, suppose that there is an element $i$ of order $p$ in $G-K$. Since $[i, t] = u \in U - \langle z\rangle$, we get that $D = \langle i, t\rangle$ is isomorphic to $D_8$ in the case $p = 2$ and $D$ is isomorphic to $S(p^3)$ in the case $p > 2$. On the other hand, $D \cap K = \langle t, u\rangle \cong E_{p^2}$ and we have $[h, t] = z$, where $\langle z\rangle = \Omega_1(\langle h\rangle)$. Hence $D$ is not normal in $G$. But the case $D \cong D_8$ is excluded by our assumptions and the case $D \cong S(p^3)$ is not possible by Proposition 176.1.

First we consider the case where $G/L$ (being abelian as a factor-group of the abelian group $G/U$) is not cyclic of order $\ge p^2$. Hence we have either $G/L \cong C_p$ (i.e., $K = L$) or $G/L$ is abelian of type $(p^r, p)$, $r \ge 1$ (noting that $K/H$ is cyclic, and so $K/L$ is cyclic). In any case, $G/L$ splits over $K/L$, and so $G$ has a normal subgroup $G_0$ such that $G = KG_0$ with $K \cap G_0 = L$ and $|G_0 : L| = p$. Since $[G_0, \langle h\rangle] = \langle z\rangle$, it follows that $C_{G_0}(h)$ covers $G_0/L$, where $C_L(h) = \langle h\rangle U$, and so $C_{G_0}(h)$ is abelian of rank 2 with $\Omega_1(C_{G_0}(h)) = U$ (noting that there are no elements of order $p$ in $G_0 - L$). If $C_{G_0}(h)$ is abelian of type $(p^n, p)$, then there is $g_1 \in C_{G_0}(h) - (\langle h\rangle U)$ such that $(g_1)^p = hu^i$ $(0 \le i \le p-1)$, where $u \in U - \langle z\rangle$. But then $(g_1)^p = hu^i \in Z(G)$, and so $h \in Z(G)$, a contradiction. Hence $C_{G_0}(h)$ is of type $(p^{n-1}, p^2)$ and therefore there is $g \in C_{G_0}(h) - K$ such that $g^p = u \in U - \langle z\rangle$. One may assume that $[t, g] = uz^i$ $(0 \le i \le p-1)$ (by replacing $t$ with a suitable power $\ne 1$ of $t$, if necessary) and then we choose an element $h' \in \langle h^p\rangle$ such that $(h')^p = z^i$ (noting that $o(h) = p^{n-1} \ge p^3$). Then we take the element $g' = h'g \in G_0 - K$ and compute:

$$(g')^p = (h')^p g^p = uz^i \quad \text{and} \quad [t, g'] = [t, h'g] = [t, g] = uz^i .$$

Hence, in the case $p = 2$ we have $\langle g', t\rangle \cong D_8$ and then $g't$ is an involution in $G_0 - K$, a contradiction. If $p > 2$, then $\langle g', t\rangle \cong M_{p^3}$. But

$$\langle g', t\rangle \cap K = \langle (g')^p, t\rangle \cong E_{p^2} \quad \text{and} \quad 1 \ne [h, t] \in \langle z\rangle \notin \langle g', t\rangle .$$

Thus, $\langle g', t\rangle$ is a nonnormal subgroup in $G$ isomorphic to $M_{p^3}$, $p > 2$, which was excluded by our assumptions.

Thus, $G/L$ is cyclic of order $\ge p^2$. Let $g \in C_G(h) - K$ so that $\langle g\rangle$ covers $G/L$ and we have $g^p \in Z(G)$. But $K/H$ is cyclic of order $\ge p^2$, and so $\langle g^p\rangle$ (covering $K/L$) covers $K/H$. Hence $\langle g\rangle$ covers $C_G(h)/\langle h\rangle$ and so $A = C_G(h)$ is abelian of rank 2 because $\Omega_1(A) = U$. We also have $|A/\langle h\rangle| \ge p^3$.

(i) First assume that $A$ splits over $\langle h\rangle$. Then we may set

$$A = \langle h\rangle \times \langle g\rangle \quad \text{with } o(g) = p^m , \ m \ge 3 , \ \Omega_1(\langle g\rangle) = \langle u\rangle .$$

Then $[h, t] = z$ with $\Omega_1(\langle h\rangle) = \langle z\rangle$ and $[g, t] = z^i u$, where $i$ is an integer modulo $p$.

Thus, the groups stated in part (a) of the proposition are obtained. Let us check that $\Omega_1(G) = S = \langle u, z, t\rangle \cong E_{p^3}$. Indeed, let $1 \neq t' \in \langle t\rangle$ and let $x = t'h^r g^s$ ($r, s$ are any integers) be an element of order $p$. Then one has in the case $p > 2$:

$$1 = (t'(h^r g^s))^p = (t')^p h^{rp} g^{sp} [h^r g^s, t']^{\binom{p}{2}} = h^{rp} g^{sp} .$$

This implies

$$r \equiv 0 \pmod{p^{n-2}} \quad \text{and} \quad s \equiv 0 \pmod{p^{m-1}} \quad \text{and so } x \in S .$$

Suppose that $p = 2$. Then

$$1 = (t(h^r g^s))^2 = t^2 h^{2r} g^{2s} [h^r g^s, t] = h^{2r} g^{2s} z^r z^{is} u^s = (h^{2r} z^{r+is})(g^{2s} u^s) .$$

This implies $r \equiv 0 \pmod{2^{n-3}}$ and $s \equiv 0 \pmod{2^{m-2}}$. Since $n \geq 4$ and $m \geq 3$, this gives $z^{r+is} = u^s = 1$ and then we get $h^{2r} g^{2s} = 1$, and therefore $r \equiv 0 \pmod{2^{n-2}}$, $s \equiv 0 \pmod{2^{m-1}}$ and $x \in S$.

(ii) Assume that $A$ does not split over $\langle h\rangle$. Then we have for an element $g \in A - K$ the following facts:

$$A = \langle h\rangle\langle g\rangle , \quad \langle h\rangle \cap \langle g\rangle \geq \langle z\rangle \quad \text{and} \quad o(h) = p^{n-1} < o(g) .$$

Suppose that $o(g) > p^n$. Then $o(g^p) \geq p^n$ and $g^p \in Z(G)$. In this case,

$$(hg^p)^{p^{n-1}} = g^{p^n} \geq \langle z\rangle , \quad [t, hg^p] = [t, h] , \quad \langle [t, h]\rangle = \langle z\rangle , \quad [t, g] = u' \in U - \langle z\rangle ,$$

and this shows that $\langle t, hg^p\rangle \cong M_{p^r}$, $r \geq n + 1$, is nonnormal in $G$, contrary to our maximality assumption.

Thus, $o(g) = p^n$. Also one has:

$$|A : \langle g\rangle| = |\langle h\rangle : (\langle h\rangle \cap \langle g\rangle)| = p^m \quad \text{with } m \leq n-2 \text{ since } \langle h\rangle \cap \langle g\rangle \geq \langle z\rangle .$$

If $m \leq 1$, then $A = \langle g\rangle U$, and so $\langle g^p\rangle U = A \cap K \leq Z(G)$, contrary to $h \notin Z(G)$. Hence, $m \geq 2$. Since $\langle g^p\rangle$ (of order $p^{n-1}$) splits in $A \cap K$, we get $A \cap K = \langle k\rangle \times \langle g^p\rangle$, and so

$$A = \langle k\rangle \times \langle g\rangle \quad \text{with } o(k) = p^m , \ 2 \leq m \leq n-2 .$$

Because $[A \cap K, \langle t\rangle] = \langle z\rangle$, one obtains $[k, t] = z$, where $\langle z\rangle = \Omega_1(\langle g\rangle$.

Further we have $[g, t] = uz^i$ ($i$ some integer) with $\langle u\rangle = \Omega_1(\langle k\rangle)$. We may replace $g$ with $g' = k^{-i}g$ so that we have:

$$(g')^{p^{n-1}} = (k^{-i}g)^{p^{n-1}} = g^{p^{n-1}} , \quad \langle g^{p^{n-1}}\rangle = \langle z\rangle , \quad [g', t] = [k^{-i}g, t] = z^{-i}(uz^i) = u ,$$

and so writing again $g$ instead of $g'$, one can assume from the start that $[g, t] = u$. Also

$$1 \neq (kg^p)^{p^{n-2}} = g^{p^{n-1}} \geq \langle z\rangle , \quad [kg^p, t] = z , \quad [g, t] = u ,$$

and so $\langle kg^p, t\rangle \cong \mathrm{M}_{p^n}$ is nonnormal in $G$ with $\langle kg^p\rangle \trianglelefteq G$. The groups stated in part (b) of our proposition have been obtained.

Let us check that $\Omega_1(G) = S = \langle u, z, t\rangle \cong \mathrm{E}_{p^3}$. Indeed, let $1 \neq t' \in \langle t\rangle$ and let $x = t'k^r g^s$ ($r, s$ are any integers) be an element of order $p$. Then one has in the case $p > 2$:

$$1 = (t'(k^r g^s))^p = (t')^p k^{rp} g^{sp} [k^r g^s, t']^{\binom{p}{2}} = k^{rp} g^{sp} .$$

This implies

$$r \equiv 0 \pmod{p^{m-1}} \quad \text{and} \quad s \equiv 0 \pmod{p^{n-1}} \quad \text{and so } x \in S .$$

Suppose that $p = 2$. Then

$$1 = (t(k^r g^s))^2 = t^2 k^{2r} g^{2s} [k^r g^s, t] = k^{2r} g^{2s} z^r u^s = (k^{2r} u^s)(g^{2s} z^r) .$$

This implies $s \equiv 0 \pmod{2^{n-2}}$ and so $1 = k^{2r}(g^{2s}z^r)$ and $r \equiv 0 \pmod{2^{m-1}}$, which gives $g^{2s} = 1$ and $s \equiv 0 \pmod{2^{n-1}}$. Hence one again obtains $x \in S$.

It remains to prove in the case of both groups in parts (a) and (b) of our proposition that the assumption $(*)$ is satisfied. Indeed, let $A$ be a unique abelian maximal subgroup of $G$, where $t \in G - A$ (since $\Omega_1(A) = U = G'$). Let $X$ be a noncyclic and nonnormal subgroup of order $\geq p^3$ in $G$, which in the case $p = 2$ has more than one involution. Since $X \not\leq G'$ and all noncentral subgroups of order $p$ form a single conjugate class in $G$ (with a representative $\langle t\rangle$), we may assume that $t \in X$. We set $X_0 = X \cap A$, where $X_0$ is cyclic since $\Omega_1(X) = \langle t, u'\rangle$ for some $1 \neq u' \in G' = \Omega_1(A)$. But then $\mathrm{N}_G(X_0) \geq \langle A, t\rangle = G$, completing the proof. □

In the next proposition we collect all the remaining $p$-groups satisfying the condition $(*)$.

**Proposition 176.12.** *Suppose that $G$ is a $p$-group satisfying $(*)$ that is not a 2-group of maximal class, $G$ has no nonnormal subgroups isomorphic to $\mathrm{D}_8$ or $\mathrm{M}_{p^n}$, $|G'| = p^2$, $K/\Omega_1(H)$ is abelian for each abelian noncyclic maximal nonnormal subgroup $H$ of order $\geq p^3$ in $G$, and $G$ has no nonnormal abelian subgroups that lead to the case* (b2) *of Proposition 176.3. Then the following possibilities hold:*

(a) *$G$ has a maximal subgroup*

$$\mathrm{M}_{p^{s+2}} \cong \langle g, u \mid g^{p^{s+1}} = u^p = 1, \quad [u, g] = z, \quad \langle z\rangle = \Omega_1(\langle g\rangle)\rangle, p > 2, \quad s \geq 2,$$
$$G = \langle g, u\rangle\langle t\rangle, \quad \text{where } o(t) = p,\ [g, t] = u,\ [u, t] = 1 .$$

*These groups are actually $\mathrm{A}_2$-groups defined in Proposition 71.3 (i), where $\langle g^p, t\rangle \cong \mathrm{C}_{p^s} \times \mathrm{C}_p$ is not normal in $G$ with $\langle g^p\rangle \trianglelefteq G$.*

(b) *$G$ is a special group of order $2^5$ with a unique abelian maximal subgroup*

$$K = \langle h\rangle \times \langle u\rangle \times \langle t\rangle, \quad \langle h\rangle \cong \mathrm{C}_4, \quad h^2 = z, \quad \langle u\rangle \cong \langle t\rangle \cong \mathrm{C}_2,$$

*and*

$$G = K\langle g\rangle, \quad \text{where } g^2 = z,\ [g, h] = z,\ [g, u] = 1,\ [g, t] = u .$$

*Here*

$$G' = \langle u, z\rangle \cong E_4\,, \quad \Omega_1(G) = \langle u, z, t\rangle \cong E_8\,, \,,$$

*and* $\langle h, t\rangle \cong C_4 \times C_2$ *is a nonnormal subgroup in* $G$ *with* $\langle h\rangle \trianglelefteq G$.

(c) *G has a maximal subgroup*

$$\langle h, g \mid h^{p^s} = g^{p^r} = 1\,, \quad h^{p^{s-1}} = z\,, \quad [g, h] = z\,\rangle, s \geq 4\,, \quad 3 \leq r < s$$

*and*

$$G = \langle h, g\rangle\langle t\rangle \text{ with } t^p = 1\,, \quad [h, t] = 1\,, \quad [g, t] = uz^i\,, \quad i \not\equiv 0 (\text{mod } p)\,,$$
$$\langle u\rangle = \Omega_1(\langle g\rangle)\,, \quad [u, t] = 1\,.$$

*Here*

$$|G| = p^{r+s+1}\,, \quad E_{p^2} \cong G' = \langle u, z\rangle \leq Z(G)\,, \quad \Omega_1(G) = \langle u, z, t\rangle \cong E_{p^3}\,,$$

*and*

$$K = \langle t, h, g^p\rangle \cong C_p \times C_{p^s} \times C_{p^{r-1}}$$

*is a unique abelian maximal subgroup in* $G$ *and* $\langle h, t\rangle \cong C_{p^s} \times C_p$ *is an abelian maximal nonnormal subgroup in* $G$ *with* $\langle h\rangle \trianglelefteq G$.

(d) *G is a 2-group that possesses a normal subgroup* $G_0 = L\langle g\rangle$, *where*

$$L = \langle h\rangle \times \langle u\rangle \times \langle t\rangle\,, \quad \langle h\rangle \cong C_4\,, \quad h^2 = z\,, \quad \langle u\rangle \cong \langle t\rangle \cong C_2\,,$$
$$g^2 = z\,, \quad [g, h] = z\,, \quad [g, u] = 1\,, \quad [g, t] = u\,,$$

*which is a special group of order* $2^5$ *with* $G_0' = \langle u, z\rangle \cong E_4$. *Then the following possibilities for* $G = G_0\langle k\rangle$ *hold:*

(d1) $k^4 = u$, $[k, g] = 1$, $[k, t] = z$, $[k, h] = z$, *and here*

$$|G| = 2^7\,, \quad \exp(G) = 8\,, \quad Z(G) = G'\langle k^2\rangle \cong C_4 \times C_2\,.$$

(d2) $k^2 = u$, $[k, g] = [k, t] = [k, h] = 1$, *and here we have*

$$|G| = 2^6\,, \quad \exp(G) = 4\,, \quad Z(G) = G'\langle k\rangle \cong C_4 \times C_2\,.$$

(d3) $k^2 = uz$, $[k, g] = [k, h] = 1$, $[k, t] = z$ *and here G is a special group of order* $2^6$ *with* $Z(G) = \langle u, z\rangle \cong E_4$.

*In all three cases*

$$E_4 \cong G' = \langle u, z\rangle \leq Z(G)\,, \quad \Omega_1(G) = G' \times \langle t\rangle \cong E_8\,,$$

*and* $\langle h, t\rangle \cong C_4 \times C_2$ *is an abelian maximal nonnormal subgroup in* $G$ *with* $\langle h\rangle \trianglelefteq G$.

(e) *We have* $G = (\langle a\rangle \times \langle b\rangle)\langle t\rangle$, *where*

$$\langle a\rangle \cong C_{p^{s+1}}\,, \quad \langle b\rangle \cong C_{p^r}\,, \quad \langle t\rangle \cong C_p\,, \quad s \geq 2\,, \quad 2 \leq r \leq s+1\,,$$
$$z = a^{p^s}\,, \quad u = b^{p^{r-1}}\,, \quad [b,t] = z\,, \quad [a,t] = u^i z^j\,,$$
$$i \not\equiv 0 (\mathrm{mod}\ p)\,, \quad [z,t] = [u,t] = 1\,.$$

*If* $r = s+1$, *then* $j \not\equiv \xi - i\xi^{-1} (\mathrm{mod}\ p)$ *for all integers* $\xi \not\equiv 0\ (\mathrm{mod}\ p)$. *Here*

$$|G| = p^{r+s+2}\,, \quad G' = \langle u, z\rangle \cong E_{p^2}\,, \quad \Omega_1(G) = G' \times \langle t\rangle \cong E_{p^3}\,, \quad \mathrm{cl}(G) = 2$$

*with*

$$\Phi(G) = \mho_1(G) = Z(G) = \langle a^p\rangle \times \langle b^p\rangle \cong C_{p^s} \times C_{p^{r-1}}\,.$$

*Finally,* $\langle a^p\rangle \times \langle t\rangle \cong C_{p^s} \times C_p$ *is a maximal nonnormal subgroup of* $G$ *with* $\langle a^p\rangle \trianglelefteq G$. *Conversely, all the above groups from (a) to (e) satisfy condition* $(*)$.

*Proof.* Let $G$ be a $p$-group satisfying all assumptions of this proposition. Let $H$ be a maximal nonnormal subgroup of a maximal possible order in $G$ that is abelian of type $(p^s, p)$, $s \geq 2$.

Set $U_0 = U \cap H = \langle z\rangle \leq Z(G)$ and $H_0 = \Omega_1(H) = \langle t, z\rangle$ so that $S = H_0 U \cong E_{p^3}$, $S = \Omega_1(K) = \Omega_1(L)$ and $L$ is abelian with $\mho_1(L) = \mho_1(H) \geq U_0$. Also, $K/H_0$ is abelian and since $G' \leq U$ (Proposition 176.4), we have here $G' = U$ (see Proposition 176.3 (b1)) because by our assumption $|G'| = p^2$, and so $K' \leq \langle z\rangle$ and $G/L$ is abelian. By Proposition 176.6, $H$ possesses a $G$-invariant cyclic subgroup $\langle h\rangle \cong C_{p^s}$ that contains $z$, and so we have $H = \langle h\rangle \times \langle t\rangle$. Also, $N_G(H_0) = K$ and, by Proposition 176.2, $K/H$ is cyclic of order $\geq p$. By Proposition 176.3, for each $g \in G - K$, one has $[g,t] = u \in U - \langle z\rangle$, where $|G/K| = p$. All these facts will be used in the proof of this proposition.

First we prove that there are no elements of order $p$ in $G - K$ and so $\Omega_1(G) = S = G' \times \langle t\rangle \cong E_{p^3}$. Indeed, let $i \in G - K$ be of order $p$. Then $[i,t] = u' \in U - \langle z\rangle$, and so $\langle h, i\rangle$ is not normal in $G$ because $\langle h, i\rangle \cap K = \langle h\rangle$. It follows that $H^* = \langle h, i\rangle = \langle h\rangle \times \langle i\rangle$ is abelian and the fact that $|H^*| = |H|$ together with the maximality of $|H|$ implies that $H^*$ is another maximal nonnormal subgroup in $G$ of type $(p^s, p)$. Since $H^* \cap U = \langle z\rangle \leq Z(G)$, it follows that $H^*U$ is the unique normal subgroup of $G$ that contains $H^*$ with $|(H^*U) : H^*| = p$. By our assumptions, we have that $\Omega_1(H^*) = \langle z, i\rangle$ centralizes $U$. Thus $C_G(U) \geq L\langle i\rangle$ and since $u'$ commutes with $i$ and $t$, one obtains together with $[i,t] = u'$ that $D = \langle i, t\rangle \cong D_8$ if $p = 2$ and $D = \langle i, t\rangle \cong S(p^3)$ if $p > 2$ and, in any case, $D' = Z(D) = \langle u'\rangle$.

If $D \cong D_8$, then our assumptions imply $D \trianglelefteq G$ and if $D \cong S(p^3)$, then Proposition 176.1 gives that $D \trianglelefteq G$. Hence in any case we have $D \trianglelefteq G$, and so $D' = \langle u'\rangle \leq Z(G)$. This gives that $G' = U = \langle z\rangle \times \langle u'\rangle \leq Z(G)$ and, therefore, $\mathrm{cl}(G) = 2$ with $\mho_1(G) \leq Z(G)$. Since $D \cap G' = \langle u'\rangle$, it follows that no element in $G$ induces an outer automorphism on $D$. We get $G = D * C$, where $C = C_G(D)$ and $C \cap D = \langle u'\rangle$.

Note that $\langle h\rangle U \leq C$ and $C_G(t) = C \times \langle t\rangle$, which together with the fact that no element in $G - K$ centralizes $t$ implies that $C_G(t) = K$. Also, we have $|G : C_G(i)| = p$, and so if $K$ would be abelian, then $C = C_K(i)$ is abelian and then $G' = D' = \langle u' \rangle$ is of order $p$, a contradiction. Hence $K$ is nonabelian, and so $K' = \langle z \rangle = C'$ since $K = C \times \langle t \rangle$. If $\langle h \rangle \le Z(K)$, then $L \le Z(K)$, and so the fact that $K/L$ is cyclic gives that $K$ is abelian, a contradiction. Hence we get $\langle h \rangle \nleq Z(K)$, and so, in particular, $K > L$.

We have $K = C_K(i) \times \langle t \rangle$ and since $K/H$ is cyclic of order $\ge p^2$ and

$$K/H \cong C_K(i)/C_H(i) = C_K(i)/\langle h \rangle\,,$$

there is $k \in C_K(i) = C$ so that $\langle k \rangle$ covers $C_K(i)/\langle h \rangle$ and $[h, k] = z$. Since

$$C = C_K(i) = \langle h, k \rangle \quad \text{with } [h, k] = z\,, \quad \langle z \rangle = \langle h \rangle \cap U \text{ and } U = \Omega_1(C) \le Z(G)$$

and noting that $\Omega_1(K) = U \times \langle t \rangle \cong E_{p^3}$, it follows that $C$ is metacyclic minimal nonabelian without a cyclic subgroup of index $p$. Hence we may set

$$C = \langle a, b \mid a^{p^\alpha} = b^{p^\beta} = 1\,, \quad [a, b] = z = a^{p^{\alpha-1}} \rangle\,,$$

where $\alpha \ge 2$, $\beta \ge 2$ and $b^{p^{\beta-1}} = u \in U - \langle z \rangle$. Also

$$G = C * \langle i, t \rangle \quad \text{with } C \cap \langle i, t \rangle = \langle u' \rangle\,,\ u' \in U - \langle z \rangle \text{ and } D = \langle i, t \rangle \cong D_8 \text{ or } S(p^3)\,.$$

Let us consider the subgroup $H_1 = \langle b \rangle \times \langle i \rangle \cong C_{p^\beta} \times C_p\,, \quad \beta \ge 2$. Since $H_1 \cap C = \langle b \rangle$ and $[a, b] = z \notin H_1$, it follows that $H_1$ is nonnormal in $G$. Suppose that $H_1$ is not a maximal nonnormal subgroup in $G$. Then there is an element $b' \in G$ such that $b = i^\gamma (b')^p$, where $\gamma$ is an integer mod $p$ and $(b')^p \in \mho_1(G) \le Z(G)$. Then

$$[a, b] = [a, i^\gamma (b')^p] = [a, i]^\gamma = 1\,,$$

a contradiction. Hence $H_1$ is a maximal nonnormal subgroup in $G$. By Proposition 176.6, $H_1$ possesses a $G$-invariant subgroup $\langle bi^\delta \rangle$ of index $p$, where $\delta$ is an integer mod $p$ and $\Omega_1(\langle bi^\delta \rangle) = \langle u \rangle$. On the other hand, we have $[a, bi^\delta] = [a, b] = z$, a contradiction. Thus, there are no elements of order $p$ in $G - K$.

Now assume that $G$ is of class 3. In that case no element in $U - \langle z \rangle$ is a $p$-th power of an element in $G$. Indeed, if there is $x \in G$ such that $x^p \in U - \langle z \rangle$, then we consider the subgroup $U\langle x \rangle \trianglelefteq G$ of order $p^3$. Since $\langle z \rangle \le Z(G)$ and $x$ commutes with $x^p$, it follows that $U\langle x \rangle$ is abelian of type $(p^2, p)$. But then $\mho_1(U\langle x \rangle) = \langle x^p \rangle \triangleleft G$, and so $G' = U \le Z(G)$, a contradiction.

Note that $G/K \cong C_p$ acts transitively on $p$ subgroups of order $p^2$ in $S = U \times \langle t \rangle$ that contain $\langle z \rangle$ and that are distinct from $U$. Assume for a moment that $t \notin Z(K)$. Then we have $K' = \langle z \rangle$ and $K > L$. Let $k \in K - C_K(t)$ so that $\langle k \rangle$ covers $K/H$. Suppose that $\langle k' \rangle = \Omega_1(\langle k \rangle) \nleq U$. Then $k' \in Z(K)$ and, if $U \nleq Z(K)$, then $\Omega_1(Z(K)) = \langle z, k' \rangle \trianglelefteq G$, a contradiction. Hence $U \le Z(K)$, and so $S \le Z(K)$, which implies that $t \in Z(K)$, a contradiction. Thus, $\Omega_1(\langle k \rangle) \le U$, and so $\Omega_1(\langle k \rangle) = \langle z \rangle$ and $o(k) \ge p^3$. Since $\langle [k, t] \rangle = \langle z \rangle$, one obtains $\langle k, t \rangle \cong M_{p^m}\,, \quad m \ge 4$. On the other hand, $[g, t] = u' \in U - \langle z \rangle$ for $g \in G - K$, and so $\langle k, t \rangle$ is not normal in $G$, contrary to our assumptions. Thus, $t \in Z(K)$, and so $C_G(t) = K$.

If $U \nleq Z(K)$, then $H_0 = \Omega_1(Z(K)) \trianglelefteq G$, a contradiction. Hence, $U \leq Z(K)$, and so $S = \Omega_1(Z(K)) = \Omega_1(G)$. Given $x \in G - K$, one has $C_U(x) = \langle z\rangle$, and therefore, by the above, $C_S(x) = \langle z\rangle$. In particular, $p > 2$ and $\Omega_1(\langle x\rangle) = \langle z\rangle$.

Suppose that for some $y \in K$ we have $y^p \in S - U$. Then $\langle y\rangle S \trianglelefteq G$ and $\mho_1(\langle y\rangle S) = \langle y^p\rangle \leq Z(G)$, a contradiction. Hence for each $x \in G$ of composite order, the socle $\Omega_1(\langle x\rangle)$ is equal to $\langle z\rangle$.

Assume that $\langle h\rangle \nleq Z(K)$ so that $K > L$. Let $k \in K$ be such that $\langle k\rangle$ covers $K/H$; then $\Omega_1(\langle k\rangle) = \langle z\rangle$ implies $o(k) \geq p^3$. It follows that $\langle h, k\rangle$ is a splitting metacyclic minimal nonabelian subgroup with $\langle [h, k]\rangle = \langle z\rangle$. We may set

$$\langle h, k\rangle = \langle a, b \mid a^{p^\alpha} = b^{p^\beta} = 1\,,\ [a, b] = z = a^{p^{\alpha-1}}\rangle\,,$$

where $\alpha \geq 3$ and $\beta \geq 1$. By the previous paragraph, $\beta = 1$, and then $b \in Z(K)$, a contradiction.

Thus, $h \in Z(K)$, and so $L \leq Z(K)$, which together with the fact that $K/L$ is cyclic implies that $K$ is abelian. Hence $K$ is abelian of rank 3, and therefore one may set

$$K = \langle a\rangle \times \langle u\rangle \times \langle t\rangle \quad \text{with } \Omega_1(\langle a\rangle) = \langle z\rangle\,,\ o(a) \geq p^s\,,\ \text{and } \langle z, u\rangle = U\,.$$

Since $[t, g] \in U - \langle z\rangle$ for each element $g \in G - K$, it follows that $\langle a\rangle \times \langle t\rangle$ is nonnormal in $G$, which together with the maximality of $|H|$ gives $o(a) = p^s$, and so we have $K = L$.

Let $g \in G - K$. Since $C_S(g) = \langle z\rangle$, it follows that $C_K(g)$ is cyclic. By Lemma 1.1, $C_K(g) = \langle h'\rangle$ covers $K/S$, and so $\langle h'\rangle \cong C_{p^s}$ and $\langle h'\rangle = Z(G)$ so that $g^p \in \langle h'\rangle$. But $G - K$ has no element of order $p$, and so $\langle g, h'\rangle = \langle g\rangle$ is cyclic of order $p^{s+1}$. One may assume without loss of generality that $g^p = h$. Then we may set $[g, t] = u \in U - \langle z\rangle$ and $[u, g] = z$, where $\langle z\rangle = \Omega_1(\langle g\rangle)$. The group $G$ has a maximal subgroup

$$M_{p^{s+2}} \cong \langle g, u \mid g^{p^{s+1}} = u^p = 1\,,\ [u, g] = z\,,\ \langle z\rangle = \Omega_1(\langle g\rangle)\rangle\,,$$

where $p > 2$, $s \geq 2$ and $G = \langle g, u\rangle\langle t\rangle$ with $o(t) = p$, $[g, t] = u$ and $[u, t] = 1$. Thus, the groups of (a) of our proposition are obtained. It turns out that these groups are actually $A_2$-groups that are defined in Proposition 71.3 (i). Conversely, it is easy to check that these groups satisfy our condition $(*)$.

From now on one may assume that $G$ is of class 2. Since $G' = U \cong E_{p^2}$, we also have $\mho_1(G) \leq Z(G)$. Also $\Omega_1(Z(G)) = U$, and so no element in $S - U$ is a $p$-th power of any element in $G$.

(i) Assume that $K = L$. In this case $|G/Z(G)| = p^3$, by Lemma 1.1. Then $\langle h\rangle \trianglelefteq G$ but $\langle h\rangle \nleq Z(G)$, and so $Z(G) = U\langle h^p\rangle$. Hence, $1 \neq g^p \in U\langle h^p\rangle$ for each $g \in G - K$.

(i1) First suppose that $1 \neq g^p \in \langle h^p\rangle \geq \langle z\rangle$. Since there are no elements of order $p$ in $\langle g, h\rangle - \langle h\rangle$ and $\langle g, h\rangle$ is nonabelian (because $\langle h\rangle \nleq Z(G)$) with $\Omega_1(\langle g, h\rangle) = \langle z\rangle$, it follows that $p = 2$ and $\langle g, h\rangle \cong Q_8$. Hence

$$\langle h\rangle \cong C_4\,,\quad g^2 = z\,,\quad [g, h] = z\,,\quad [g, t] = u \in U - \langle z\rangle\,.$$

Thus, the special group of order $2^5$ of part (b) is obtained, and this group satisfies the condition $(*)$.

(i2) Now assume that $g^p \in (U\langle h^p\rangle) - \langle h^p\rangle$ so one may set

$$g^p = uh', \quad \text{where } u \in U - \langle z\rangle\,,\ \langle z\rangle = \Omega_1(\langle h\rangle)\,,\ h' \in \langle h^p\rangle\,.$$

Let $h_0 \in \langle h\rangle$ be such that $h_0^p = (h')^{-1}$. Then replace $g$ with $gh_0 \in G - K$ and compute:

$$(gh_0)^p = g^p h_0^p [h_0, g]^{\binom{p}{2}} = (uh')(h')^{-1}z' = uz' \in U - \langle z\rangle\,,$$
$$\text{where } [h_0, g]^{\binom{p}{2}} = z' \in \langle z\rangle\,.$$

In this case we may choose from the start $g \in G - K$ so that $g^p = u \in U - \langle z\rangle$. Then $[g, t] = uz^i$ for some integer $i$ modulo $p$ (where $t$ is replaced with a suitable power $t^j$ ($j \not\equiv 0 \bmod p$). Let $h^* \in \langle h\rangle$ be such that $(h^*)^p = z^i$.

Assume that either $p > 2$ or $p = 2$ and $s \geq 3$ (where in the last case $[h^*, g] = 1$). Let us consider the subgroup $\langle g', t\rangle$, where $g' = gh^* \in G - K$. Then

$$(g')^p = g^p(h^*)^p[h^*, g]^{\binom{p}{2}} = uz^i = [g, t] = [gh^*, t] = [g', t]\,,$$

and so

$$\langle g', t\rangle \cong \mathrm{D}_8 \text{ if } p = 2 \quad\text{and}\quad \langle g', t\rangle \cong \mathrm{M}_{p^3} \text{ if } p > 2\,.$$

On the other hand, $1 \neq [h, g'] \in \langle z\rangle$, and so $\langle g', t\rangle$ is nonnormal in $G$, contrary to the assumptions.

Thus, $p = 2$ and $s = 2$ so that $\langle h\rangle \cong \mathrm{C}_4$ and $G$ is a special group of order $2^5$ with $g^2 = u \in U - \langle z\rangle$, $h^2 = z$, $[g, h] = z$ and $[g, t] = uz^i$, $i = 0, 1$. However, if $i = 0$, then $\langle g, t\rangle \cong \mathrm{D}_8$ is not normal in $G$, a contradiction. Thus $i = 1$, and so $[g, t] = uz$. The structure of $G$ is uniquely determined.

Let us prove that the special 2-group obtained in the previous paragraph is in fact isomorphic to the special group of order $2^5$ from part (i1) of our proof. Indeed, set $g' = gt$ and $u' = uz$. Then

$$(g')^2 = (gt)^2 = u(uz) = z = h^2\,, \quad [g', h] = [gt, h] = z$$
$$\text{and} \quad [g', t] = [gt, t] = uz = u'\,.$$

In addition, $[g', u'] = [h, t] = 1$, and so, writing again $g, u$ instead of $g', u'$, respectively, we see that the relations for the special group of order $2^5$ defined in ($i$1) are obtained.

In what follows let $K > L$.

(ii) Suppose that $G/L$ is cyclic of order $\geq p^2$. Let $g \in G - K$ so that $\langle g\rangle$ covers $G/L$. But $g^p \in \mathrm{Z}(G)$ and $\langle g^p\rangle$ covers $K/L \neq \{1\}$. Since $K/H$ is cyclic of order $\geq p^2$, it follows that $\langle g^p\rangle$ covers $K/H$, and so $K = H\langle g^p\rangle$ is abelian. Since $G' = U \cong \mathrm{E}_{p^2}$, Lemma 1.1 implies that $|G : Z(G)| = p^3$. On the other hand, $\langle h^p, g^p\rangle \le Z(G)$ and $|K_1 : \langle h^p, g^p\rangle| = p$, where $K_1 = \langle h, g^p\rangle$, $K = \langle t\rangle \times K_1$ is of rank 3. It follows that $Z(G) = \langle h^p, g^p\rangle$. In particular, $U \le \langle h^p, g^p\rangle$ since $U \le Z(G)$, so that $\Omega_1(K_1) = U$ and $h \notin Z(G)$. One may set $[g, h] = z$. There are exactly $p$ conjugate classes of noncentral subgroups of order $p$ in $G$ with the representatives $\langle tz^i\rangle$, $0 \le i \le p-1$. It follows (using also Proposition 176.6) that any abelian maximal nonnormal subgroup in $G$ of type $(p^r, p)$, $r \ge 2$, is contained in $C_G(tz^i) = K$.

Suppose that $K_1$ is of exponent $p^r$, where $r > s$. Let $k$ be an element of order $p^r$ in $K_1$ and consider the subgroup $\langle t\rangle \times \langle k\rangle$. If $\langle t\rangle \times \langle k\rangle$ is not normal in $G$, then $\langle t\rangle \times \langle k\rangle$ is maximal nonnormal in $G$ of order $> |H| = p^s$, contrary to the assumptions. Hence $\langle t\rangle \times \langle k\rangle \trianglelefteq G$. Since $[g, t] \in U - \langle z\rangle$, it follows that $\Omega_1(\langle k\rangle) = \langle u\rangle$ with $u \in U - \langle z\rangle$. Then $[g, K_1] = \langle z\rangle$ since $[g, h] = z$, and so $k \in K_1 \Rightarrow [g, k] \in \langle z\rangle$. But $\Omega_1(\langle t, k\rangle) = \langle t, u\rangle$, and so $[g, k] = 1$, and therefore $k \in Z(G)$. Now consider the subgroup

$$\langle t\rangle \times \langle hk\rangle, \quad \text{where } hk \in K_1\,,\ o(hk) = p^r\,,\ \Omega_1(\langle hk\rangle) = \langle u\rangle\,.$$

If $\langle t\rangle \times \langle hk\rangle$ is not normal in $G$, then $\langle t\rangle \times \langle hk\rangle$ is maximal nonnormal in $G$ of order $> |H|$, a contradiction. Hence $\langle t\rangle \times \langle hk\rangle \trianglelefteq G$. But $[g, hk] = [g, h][g, k] = z$ and $z \notin \Omega_1(\langle t\rangle \times \langle hk\rangle) = \langle t, u\rangle$, a contradiction. Thus, $\exp(K) = \exp(K_1) = p^s$, and therefore $o(g) \le p^{s+1}$ and all elements in $G - K$ are of order $\le p^{s+1}$.

There are elements of order $p^s$ or $p^{s+1}$ in $G - K$. Indeed, assume that $o(g) \le p^{s-1}$ for some $g \in G - K$. In that case, $s \ge 3$ since $\Omega_1(G) = U \times \langle t\rangle$. Compute:

$$(gh)^{p^{s-1}} = g^{p^{s-1}} h^{p^{s-1}} [h, g]^{\binom{p^{s-1}}{2}} = h^{p^{s-1}} = z\,,$$

where $\langle z\rangle = \Omega_1(\langle h\rangle)$, and so $o(gh) = p^s$.

If there is $g \in G - K$ of order $p^{s+1}$, then all elements in $G - K$ are of order $p^{s+1}$. Indeed, for any $x \in K$ and any integer $i \not\equiv 0 \pmod p$ one has:

$$(g^i x)^{p^s} = (g^i)^{p^s} x^{p^s} [x, g^i]^{\binom{p^s}{2}} = (g^i)^{p^s} \ne 1\,.$$

(ii1) Suppose that $G - K$ contains elements of order $p^s$. Let $g$ be an element of the minimal possible order $p^r$ in $G - K$. Then $3 \le r \le s$. Indeed, $\langle g\rangle$ covers $G/L$ (which is cyclic of order $\ge p^2$) and there are no elements of order $p$ in $G - L$, and so $o(g) \ge p^3$.

The element $g^{p^{r-1}} \in U$ has order $p$. Write $g^{p^{r-1}} = z$, where $\langle z\rangle = \Omega_1(\langle h\rangle)$. Let $h'$ be an element in $\langle h\rangle$ such that $(h')^{p^{r-1}} = z^{-1}$. Then (noting that $r \ge 3$), one has

$$(h'g)^{p^{r-1}} = (h')^{p^{r-1}} g^{p^{r-1}} [g, h']^{\binom{p^{r-1}}{2}} = z^{-1}z = 1\,,$$

and so $o(h'g) \le p^{r-1}$, a contradiction. Thus, $\langle g\rangle$ splits over $\langle h\rangle$, and so $\Omega_1(\langle g\rangle) = \langle u\rangle$ with $u \in U - \langle z\rangle$.

Set $h^{p^{s-1}} = z$, $s \ge 3$, and then replacing $g$ with $g^j$ for some integer $j \not\equiv 0 \pmod p$, one may set $[g, h] = z$. Replacing $t$ with $t^l$ for some suitable integer $l \not\equiv 0 \pmod p$, one may assume that $[g, t] = uz^i$ for some integer $i \pmod p$. If $[g, t] = u$ (i.e., $i \equiv 0 \pmod p$), then $\langle g, t\rangle \cong \mathrm{M}_{p^{r+1}}$, $r \geq 3$. But $[g, h] = z \notin \langle g, t\rangle$, and so $\langle g, t\rangle$ is not normal in $G$, contrary to our assumptions. Hence, $i \not\equiv 0 \pmod p$.

Assume that $r = s$, and so $o(g) = p^s$. Write $g^{p^{s-1}} = u$ and then, changing $t$ with a suitable power $t^j$, $j \not\equiv 0 \pmod p$, one may set $[g, t] = uz^i$ with $i \not\equiv 0 \pmod p$. Let $h' \in \langle h\rangle$ be such that $(h')^{p^{s-1}} = z^i$. Then (noting that $s \geq 3$):

$$(gh')^{p^{s-1}} = uz^i[h', g]^{\binom{p^{s-1}}{2}} = uz^i ,$$

and so $\langle gh', t\rangle \cong \mathrm{M}_{p^{s+1}}$ since $[gh', t] = [g, t] = uz^i$. On the other hand, $1 \neq [gh', h] \in \langle z\rangle$, and so $\langle gh', t\rangle$ is not normal in $G$, a contradiction. Thus, $o(g) = p^r$ with $3 \leq r < s$ so that $s \geq 4$. The groups stated in part (c) (which obviously satisfy condition $(*)$) are obtained.

(ii2) Suppose that all elements in $G - K$ are of order $p^{s+1}$.

(ii2a) First assume that there is $g \in G - K$ such that $\langle g\rangle$ splits over $\langle h\rangle$. It is possible to choose a generator $g$ in $\langle g\rangle$ so that $[g, h] = z = h^{p^{s-1}}$, $s \geq 2$. Write $u = g^{p^s} \in U - \langle z\rangle$ and choose a generator $t \in \langle z\rangle$ so that $[g, t] = uz^i$, where $i$ is an integer modulo $p$. Suppose that $i \equiv 0 \pmod p$. Then $\langle g, t\rangle \cong \mathrm{M}_{p^{s+2}}$. But $[g, h] = z \notin \langle g, t\rangle$, and so $\langle g, t\rangle$ is not normal in $G$, contrary to our assumptions. Hence $i \not\equiv 0 \pmod p$. Note that the socle $\Omega_1(\langle x\rangle)$ is equal to $\langle u\rangle$ for each $x \in G - K$.

Consider the subgroup $X = \langle t, h^\alpha g^p\rangle \cong \mathrm{C}_p \times \mathrm{C}_{p^s}$, where $g^p \in \mathrm{Z}(G)$ and $\alpha$ is any fixed integer with $\alpha \not\equiv 0 \pmod p$. In this case, for every integer $j \pmod p$,

$$(t^j h^\alpha g^p)^{p^{s-1}} = (h^{p^{s-1}})^\alpha g^{p^s} = z^\alpha u ,$$

and so $\langle t^j h^\alpha g^p\rangle \cong \mathrm{C}_{p^s}$ is a maximal cyclic subgroup in $G$ since its socle is $\langle z^\alpha u\rangle$. Then $\Omega_1(X) = \langle t, z^\alpha u\rangle$ and

$$[g, h^\alpha g^p] = [g, h^\alpha] = z^\alpha \notin X$$

implies that $X$ is not normal in $G$. This gives

$$\mathrm{N}_G(X) = \mathrm{N}_G(\Omega_1(X)) = K .$$

Then $[g, t] = uz^i$, and so $z^i u \notin \Omega_1(X) = \langle t, z^\alpha u\rangle$. In particular, $i \not\equiv \alpha \pmod p$ for any integer $\alpha \not\equiv 0 \pmod p$. But this implies $i \equiv 0 \pmod p$, a contradiction.

(ii2b) By what has been proved already, $\langle g\rangle$ does not split over $\langle h\rangle$ for each $g \in G - K$. Hence

$$\langle g\rangle \cap \langle h\rangle \geq \langle z\rangle , \quad \langle g, h\rangle' = \langle z\rangle = \Omega_1(\langle h\rangle) \quad \text{and therefore}$$
$$\langle g\rangle \trianglelefteq \langle g, h\rangle \quad \text{with } p \leq |\langle g, h\rangle : \langle g\rangle| \leq p^{s-1} .$$

Since $\langle g^p\rangle$ is of order $p^s = \exp(\langle g^p, h\rangle)$, it follows that $\langle g^p\rangle$ splits in $\langle g^p, h\rangle$, and so

$$\langle g^p, h\rangle = \langle k\rangle \times \langle g^p\rangle \quad \text{with } K = \langle t\rangle \times (\langle k\rangle \times \langle g^p\rangle) \quad \text{and}$$
$$\langle k\rangle\langle g\rangle = \langle g, h\rangle \quad \text{with } \langle k\rangle \cap \langle g\rangle = \{1\} .$$

Now $\langle[k,g]\rangle = \langle z\rangle$ and $\Omega_1(\langle k\rangle\langle g\rangle) = U \le \mathrm{Z}(G)$ imply $o(k) = p^r$, $2 \le r \le s-1$, and so $s \ge 3$. It is possible to set $u = k^{p^{r-1}} \in U - \langle z\rangle$ and $[g,k] = z = g^{p^s}$. Also note that the socle $\Omega_1(\langle x\rangle)$ for each $x \in G - K$ is equal to $\langle z\rangle$.

It is possible to choose a generator $t$ in $\langle t\rangle$ so that $[g,t] = uz^i$ for some integer $i$ mod $p$. Consider the subgroup

$$Y = \langle k\rangle \times \langle t\rangle \cong \mathrm{C}_p \times \mathrm{C}_{p^r}\,, \quad 2 \le r \le s-1\,,$$

which is not normal in $G$ since $[g,k] = z \notin Y$. Then $\mathrm{N}_G(Y) = K$, and so $\mathrm{N}_G(\langle t,u\rangle) = K$, where $\langle t,u\rangle = \Omega_1(Y)$. In that case, $[g,t] = uz^i \notin \Omega_1(Y)$, and so $i \not\equiv 0 \pmod p$.

Choose $g' \in \langle g^p\rangle$ such that $o(g') = p^r$ and $(g')^{p^{r-1}} = z$ and note that $g' \in \mathrm{Z}(G)$. Let us consider for each $\alpha \not\equiv 0 \pmod p$ the subgroup

$$V = \langle k^\alpha g'\rangle \times \langle t\rangle \cong \mathrm{C}_{p^r} \times \mathrm{C}_p$$
$$\text{with } (k^\alpha g')^{p^{r-1}} = u^\alpha z \text{ so that } \Omega_1(V) = \langle t, u^\alpha z\rangle\,.$$

Since $[g, k^\alpha g'] = z^\alpha \notin \Omega_1(V)$, we have $\mathrm{N}_G(V) = K$, and so also $\mathrm{N}_G(\langle t, u^\alpha z\rangle) = K$. Because $[g,t] = uz^i$, it follows that $uz^i \notin \langle u^\alpha z\rangle$ for each $\alpha \not\equiv 0 \pmod p$. Then there is an integer $j \not\equiv 0 \pmod p$ such that $ij \equiv 1 \pmod p$. Then

$$(uz^i)^j = u^j z^{ij} = u^j z \notin \langle u^\alpha z\rangle$$

for each $\alpha \not\equiv 0 \pmod p$, a contradiction.

(iii) It remains to consider the remaining case, where $G/L$ is noncyclic. Since $G/L$ is abelian and $K/L \ne \{1\}$ is cyclic, it follows that $G/L$ splits over $K/L$, and so $G = KG_0$ with $K \cap G_0 = L$ and $|G_0 : L| = p$. Also, $K/H$ is cyclic of order $\ge p^2$ and

$$H = \langle h\rangle \times \langle t\rangle \cong \mathrm{C}_{p^s} \times \mathrm{C}_p\,, \quad s \ge 2\,, \quad \text{where } \mathrm{C}_{p^s} \cong \langle h\rangle \trianglelefteq G\,, \quad \langle t\rangle \cong \mathrm{C}_p\,,$$
$$\Omega_1(H) = \langle z\rangle\,, \quad G' = U \cong \mathrm{E}_{p^2}\,, \quad L = UH \text{ is abelian and } U \le \mathrm{Z}(G)\,.$$

(iii1) Suppose first that $\langle h\rangle \not\le \mathrm{Z}(G_0)$ so that $U = G_0' \cong \mathrm{E}_{p^2}$, and therefore, by (i), $p = 2$ and $G_0$ is the uniquely determined special 2-group of order $2^5$ (stated in part (b),

$$L = \langle h\rangle \times \langle u\rangle \times \langle t\rangle \cong \mathrm{C}_4 \times \mathrm{C}_2 \times \mathrm{C}_2\,, \quad \langle h\rangle \cong \mathrm{C}_4\,, \quad h^2 = z\,, \quad \langle u\rangle \cong \langle t\rangle \cong \mathrm{C}_2\,,$$
$$G_0 \cong L\langle g\rangle \quad \text{with } g^2 = z\,, \quad [g,h] = z\,, \quad [g,u] = 1\,, \quad \text{and } [g,t] = u\,.$$

Since $\mathrm{Z}(G_0) = U$, it follows that $1 \ne x^2 \in U$ for each $x \in K - L$ such that $x^2 \in L$. Let $k \in K - L$ be such that $\langle k\rangle$ covers the cyclic group $K/H$ of order $\ge 4$. Thus, $\Omega_1(\langle k\rangle) = \langle u\rangle$ or $\langle uz\rangle$, and so $K$ splits over $H$.

Because $\mathrm{C}_{G_0}(g) = U\langle g\rangle$, and so $|G_0 : \mathrm{C}_{G_0}(g)| = 4$, one has together with $|G'| = 4$ that $|G : \mathrm{C}_G(g)| = 4$. But $G = K\langle g\rangle$, and so $\mathrm{C}_G(g) = \mathrm{C}_K(g)\langle g\rangle$, which implies that $|K : \mathrm{C}_K(g)| = 4$. On the other hand, $|H : \mathrm{C}_H(g)| = 4$, and therefore $\mathrm{C}_K(g)$ covers $K/H$.

It follows that it is possible to choose $k \in C_K(g)$ such that $\langle k\rangle$ covers $K/H$. Hence, one may assume that $[g, k] = 1$.

Case (1). Suppose that $|K : L| > 2$ so that $o(k) \geq 8$. Then there is an element $k'$ of order 4 in $\langle k\rangle$ such that $k' \in Z(G)$. Note that $(tg)^2 = uz$, and so, if $(k')^2 = uz$, then $k'(tg)$ is an involution in $G - K$, a contradiction. Hence, in this case $(k')^2 = u$. Set $o(k) = 2^n$, $n \geq 3$; then $k^{2^{n-1}} = u$. Assume for a moment that $[k, h] = [k, t] = 1$, which together with $[k, g] = 1$ (from the previous paragraph) implies that $k \in Z(G)$. In that case, $(gk)^{2^{n-1}} = u$ and $[gk, t] = u$ so that $\langle gk, t\rangle \cong M_{2^{n+1}}$ with $n \geq 3$. But $[h, gk] = z \notin \langle gk, t\rangle$, and so $\langle gk, t\rangle$ is not normal in $G$, contrary to our assumptions. Thus, $k \notin Z(G)$.

Assume that $[k, t] = 1$. Then $[k, h] = z$. Consider in this case the subgroup

$$\langle t\rangle \times \langle k\rangle\,, \quad \text{where } o(k) = 2^n = \exp(G)\,, \quad n \geq 3\,.$$

Since $[h, k] = z \notin \langle t, k\rangle$, it follows that $\langle t, k\rangle$ is a maximal nonnormal subgroup in $G$ of order $> |H|$, contrary to our assumptions. Thus, $[k, t] = z$ (noting that $K' \leq \langle z\rangle$).

Let us consider the subgroup $\langle t\rangle \times \langle hk'\rangle$, where $k' \in \langle k\rangle$ is of order 4 and $k' \in Z(G)$. Here $\Omega_1(\langle t, hk'\rangle) = \langle t, uz\rangle$. Because $[g, t] = u$, it follows that $\langle t, hk'\rangle \cong C_2 \times C_4$ is abelian non-$G$-invariant. By the maximality of $|H|$, $\langle t, hk'\rangle$ is a maximal nonnormal subgroup of $G$. Then Proposition 176.6 implies that either $\langle hk'\rangle \trianglelefteq G$ or $\langle thk'\rangle \trianglelefteq G$. But $[hk', g] = z$, and so $\langle hk'\rangle$ is not normal in $G$. Hence, $\langle thk'\rangle \trianglelefteq G$. From $[thk', k] = z[h, k]$ follows that $[h, k] = z$.

Finally assume that $n > 3$ so that the subgroup $\langle t\rangle \times \langle k^2\rangle \cong C_2 \times C_{2^{n-1}}$ is not normal in $G$ (since $[t, k] = z$), contrary to the maximality of $|H|$. Hence we get $n = 3$, $o(k) = 8$ and $|G| = 2^7$. The group of (d1) of our proposition is obtained.

Case (2). Suppose that $|K : L| = 2$ and $k \in Z(G)$. Here $o(k) = 4$ and $k^2 \in \{u, uz\}$. If $k^2 = uz$, then $(gt)^2 = uz$ together with $[k, gt] = 1$ implies that $gtk$ is an involution in $G - K$, a contradiction. Hence in this case $k^2 = u$, and the group of order $2^6$ stated in part (d2) is obtained.

Case (3). Assume that $|K : L| = 2$ and $k \notin Z(G)$. Then

$$k^2 = uz^{\epsilon}\,, \quad \epsilon \in \{0, 1\}\,, \quad [k, t] = z^{\eta}\,,$$
$$[k, h] = z^{\delta}\,, \quad \eta, \delta \in \{0, 1\}\,,$$

and $\eta = \delta = 0$ is not possible.

As the set $G - K$ has no involution, there is a unique solution

$$\epsilon = 1\,, \quad \eta = 1\,, \quad \delta = 0\,,$$

and so the special group of order $2^6$ of part (d3) is obtained.

Conversely, all groups from part (d) satisfy the condition $(*)$.

(iii2) Suppose that $\langle h\rangle \leq Z(G_0)$. Then, for each $g \in G_0 - L$, one has

$$G_0' = \langle [g, t]\rangle \quad \text{with } [g, t] = u \in U - \langle z\rangle\,, \quad \langle z\rangle = \Omega_1(\langle h\rangle)\,.$$

In that case,

$$Z(G_0) = \langle h\rangle \times \langle u\rangle \cong C_{p^s} \times C_p\,, \quad s \geq 2\,.$$

Since $1 \neq g^p \in Z(G_0)$ and there are no elements of order $p$ in $G - K$, it follows that $A = Z(G_0)\langle g\rangle$ is abelian of rank 2. Hence $A$ is either of type $(p^s, p^2)$ or $(p^{s+1}, p)$.

Suppose, by way of contradiction, that $A$ is of type $(p^s, p^2)$. In that case, there is $g_0 \in A - Z(G_0)$ such that $g_0^2 = u$, where $\langle u\rangle = G_0'$. If $p = 2$, then $\langle g_0, t\rangle \cong D_8$, and so $g_0 t$ is an involution in $G_0 - K$, a contradiction. Hence $p > 2$ and $M = \langle g_0, t\rangle \cong M_{p^3}$. By our assumptions, $M \trianglelefteq G$. Note that $G' \cap M = U \cap M = \langle u\rangle$, Set $C = C_G(M)$ so that $C \cap M = \langle u\rangle$. If $C * M < G$, then $G/C \cong S(p^3)$ (which is an $S_p$-subgroup of Aut$(M)$), contrary to $U = G' \leq C$. Hence $G = M * C$. It follows from $\langle h\rangle \leq C$, $\langle h\rangle \trianglelefteq G$ and $[t, C] = \{1\}$ that $C \leq K$, and so $K = C \times \langle t\rangle$. Now $C < K$ and $K' \leq \langle z\rangle$ implies $C' \leq \langle z\rangle$. If $C' = \{1\}$, then $G' = C'M' = \langle u\rangle$, a contradiction. Hence, $C' = \langle z\rangle$. Note that $\{1\} \neq K/L$ is cyclic, where $L = (\langle h\rangle U) \times \langle t\rangle$ and $K = CL$ with $C \cap L = \langle h\rangle U$. Thus $\{1\} \neq C/(\langle h\rangle \times \langle u\rangle)$ is cyclic. Let $c \in C$ be such that $\langle c\rangle$ covers $C/(\langle h\rangle \times \langle u\rangle)$; and so $\langle [h, c]\rangle = \langle z\rangle$. Since $K/H$ is cyclic of order $\geq p^2$, $\langle c\rangle$ also covers $K/H$, and so $\langle c\rangle$ covers $C/(H \cap C) = C/\langle h\rangle$. It follows that $C$ is metacyclic minimal nonabelian without a cyclic subgroup of index $p$ (noting that $E_{p^2} \cong \Omega_1(C) = U \leq Z(G)$). Hence, one may set

$$C = \langle a\rangle\langle b\rangle \quad \text{with } \langle a\rangle > \langle z\rangle = C'\,, \quad \langle a\rangle \cap \langle b\rangle = \{1\}\,, \quad \langle b\rangle \cong C_{p^r}\,, \quad r \geq 2\,,$$

and $\Omega_1(\langle b\rangle) = \langle uz^i\rangle$, where $i$ is an integer modulo $p$. Consider the subgroup

$$\langle b\rangle \times \langle t\rangle \cong C_{p^r} \times C_p\,,$$

which is non-$G$-invariant since $\langle [a, b]\rangle = \langle z\rangle$ and $z \notin \langle b, t\rangle$. Let us prove that $\langle b, t\rangle$ is a maximal nonnormal subgroup in $G$. Indeed, let $X > \langle b, t\rangle$ be a maximal nonnormal subgroup in $G$. If $X \cap C > \langle b\rangle$, then $\langle z\rangle \leq X$, and so $\langle z, uz^i\rangle = G' \leq X$, a contradiction. Hence, $X \cap C = \langle b\rangle$. Because $G/C \cong E_{p^2}$, it follows that $X$ must contain $x \in G - (C \times \langle t\rangle) = G - K$. On the other hand, $C_G(t) = C \times \langle t\rangle = K$, and so $[x, t] \neq 1$ and $X$ is nonabelian, contrary to our assumptions. Finally, by Proposition 176.6, $\langle bt^j\rangle \trianglelefteq G$ for some integer $j$ modulo $p$, where $\Omega_1(\langle bt^j\rangle) = \langle uz^i\rangle$. On the other hand,

$$[a, bt^j] = [a, b]\,, \quad \text{where } \langle [a, b]\rangle = \langle z\rangle \neq \langle uz^i\rangle\,,$$

a final contradiction.

Thus, $A = Z(G_0)\langle g\rangle$ is abelian of type $(p^{s+1}, p)$. It follows that all elements of order $p^s$ in $\langle h\rangle U$ are central in $G$ (noting that $U \leq Z(G)$). Replacing $H$ with $H^* = \langle t\rangle \times \langle hu^i\rangle$ for some integer $i$ mod $p$ (which is also a maximal nonnormal abelian subgroup of type $(p^s, p)$) so that $\langle g^p\rangle = \langle hu^i\rangle$, and then working with $H^*$ instead of $H$, one may assume from the start that there is $g \in G - K$ such that $g^p = h$, where $\langle h\rangle \leq Z(G)$ and we set $g^{p^s} = z$. If $t \in Z(K)$, then $L \leq Z(K)$, and since $K/L$ is cyclic, $K$ would in that case be abelian.

(iii2a) First assume that $K$ is nonabelian, i.e., $t \notin Z(K)$. Then $K' = \langle z\rangle$, and so, if $k \in K - L$ is such that $\langle k\rangle$ covers $K/H$ (which is cyclic of order $\geq p^2$), then it is possible to set (by choosing a suitable generator $t$ of $\langle t\rangle$) $[k, t] = z$.

It is easy to see that $\langle k\rangle$ splits over $H$. Indeed, if $\langle k\rangle$ does not split over $H$, then $\langle k\rangle \cap H = \langle k\rangle \cap \langle h\rangle$ since $Z(G) \cap L = \langle h\rangle U$ and so $\langle k\rangle > \langle z\rangle$. It follows that $\langle k, t\rangle \cong M_{p^{n+1}}$ with $n \geq 3$ since $[k, t] = z$. On the other hand, $[g, t] \in U - \langle z\rangle$, and so $[g, t] \notin \langle k, t\rangle$, which implies that $\langle k, t\rangle$ is not normal in $G$, contrary to our assumptions. Hence $\langle k\rangle$ splits over $H$ and one may set $o(k) = p^r$, $r \geq 2$, and $k^{p^{r-1}} = u \in U - \langle z\rangle$.

If $o(k^p) > p^s$, then $\langle t\rangle \times \langle k^p\rangle \cong C_p \times C_{p^{r-1}}$ is non-$G$-invariant (since $[k, t] = z \notin \langle t, k^p\rangle$), contrary to the maximality of $|H| = p^{s+1}$. Hence, $r \leq s + 1$. Set $[g, t] = u^i z^j$ with $i \not\equiv 0 \pmod p$. Here

$$\Phi(G) = \mho_1(G) = Z(G) = \langle g^p\rangle \times \langle k^p\rangle \quad \text{and so } |G : \Phi(G)| = p^3 .$$

By Lemma 146.7, $G$ possesses a unique abelian maximal subgroup $A^*$. Because $|G : C_G(t)| = p^2$, it follows that $t \in G - A^*$ and

$$C_{A^*}(t) = Z(G) = \langle h\rangle \times \langle k^p\rangle , \quad A^*/Z(G) \cong G' = U = \Omega_1(A^*) ,$$

so that $A^*$ is of type $(p^{s+1}, p^r)$, where $s \geq 2$ and $2 \leq r \leq s + 1$. Indeed, the map $a \to [a, t]$ $(a \in A^*)$ is a homomorphism from $A^*$ onto $G'$, and so $A^*/Z(G) \cong G'$.

Case (a). $r < s + 1$. In this case, one may set

$$A^* = \langle a\rangle \times \langle b\rangle , \quad \text{where } \langle a\rangle \cong C_{p^{s+1}} , \ \langle b\rangle \cong C_{p^r} , \ z = a^{p^s} , \ u = b^{p^{r-1}} .$$

Take an element $a' \in \langle a^p\rangle \leq Z(G)$ such that $o(a') = p^r$ and $(a')^{p^{r-1}} = z$. Suppose that $[b, t] \notin \langle z\rangle$. Then $[b, t] = z^i u$ ($i$ is an integer modulo $p$) for a suitable choice of a generator $t$ of $\langle t\rangle$. We get

$$((a')^i b)^{p^{r-1}} = z^i u \quad \text{and} \quad [(a')^i b, t] = [b, t] = z^i u ,$$

and therefore either $p = 2$, $r = 2$ and $\langle (a')^i b, t\rangle \cong D_8$ or else $\langle (a')^i b, t\rangle \cong M_{p^{r+1}}$. But $|G : C_G(t)| = p^2$, and so for some $g \in G$ we get $\langle [g, t]\rangle \neq \langle z^i u\rangle$, and so $\langle (a')^i b, t\rangle$ is not normal in $G$, contrary to our assumptions. Hence choosing a suitable generator $t$ of $\langle t\rangle$, one obtains $[b, t] = z$. Then $[a, t] = u^i z^j$ with $i \not\equiv 0 \pmod p$.

Case (b). $r = s + 1$. Let $b \in A^* - \Phi(G)$ be such that $[b, t] = z$ and set $b^{p^s} = u$, where $\langle u\rangle \neq \langle z\rangle$. Let $a \in A^* - \Phi(G)$ be such that $a^{p^s} = z$; then

$$A^* = \langle a\rangle \times \langle b\rangle \cong C_{p^{s+1}} \times C_{p^{s+1}} \quad \text{and}$$
$$[a, t] = u^i z^j , \quad i \not\equiv 0 \pmod p .$$

In this critical case, $j \not\equiv \xi - i\xi^{-1} \pmod p$ for all integers $\xi \not\equiv 0 \pmod p$. Indeed, assume that for some $\xi \not\equiv 0 \pmod p$; the $j \equiv \xi - i\xi^{-1} \pmod p$. In that case we solve the congruence $i\mu \equiv \xi \pmod p$ with some $\mu \not\equiv 0 \pmod p$. We compute (noting that $s \geq 2$):

$$(a^\mu b)^{p^s} = (a^{p^s})^\mu b^{p^s} [b, a^\mu]^{\binom{p^s}{2}} = z^\mu u$$

and

$$[a^\mu b, t] = (u^i z^j)^\mu z = z^{1+j\mu} u^{i\mu} = z^{1+(\xi - i\xi^{-1})\mu} u^\xi$$
$$= z^{1+\xi\mu-\xi^{-1}i\mu} u^\xi = z^{1+\xi\mu-1} u^\xi = z^{\xi\mu} u^\xi = (z^\mu u)^\xi .$$

It follows that $\langle a^\mu b, t\rangle \cong \mathrm{M}_{p^{s+2}}$ and, since $[b, t] = z \notin \langle a^\mu b, t\rangle$, it follows that $\langle a^\mu b, t\rangle$ is not normal in $G$, contrary to our assumptions. The groups of part (e) are obtained.

Conversely, in any group $G$ from part (e), for each $x \in A^* - \mathrm{Z}(G)$, $\langle x\rangle$ is not normal in $G$, and so $\mathrm{D}_8$ or $\mathrm{M}_{p^n}$ cannot be subgroups of $G$, where $A^*$ is the unique abelian maximal subgroup of $G$. Furthermore, let $X$ be any maximal nonnormal abelian subgroup of $G$ of order $\geq p^3$ that has more than one subgroup of order $p$. Since $G$ has exactly one conjugacy class of noncentral subgroups of order $p$ with the representative $\langle t\rangle$, one may assume that $t \in X$. It follows that $X = \langle t\rangle \times X_0$, where $X_0$ is any maximal cyclic subgroup in $\mathrm{Z}(G)$. Hence, condition $(*)$ holds.

(iii2b) It remains to consider the case $t \in \mathrm{Z}(K)$ so that $K$ is abelian and $K > L$. Since $K/\mathrm{C}_K(g) \cong G'$ (Lemma 1.1), there is $k \in K - L$ such that $\langle k\rangle$ covers $K/H$ and $[g, k] = z = g^{p^s}$, $s \geq 2$, where $[g, t] \in U - \langle z\rangle$ with $p^r = o(k) \geq p^2$. Since $K = \langle t\rangle \times \langle h, k\rangle$ and $\mathrm{Z}(G) = \langle h, k^p\rangle$, it follows that $U \leq \langle h, k^p\rangle$ because $U \leq \mathrm{Z}(G)$. Hence, $\Omega_1(\langle h, k\rangle) = U = G'$. Consider the subgroup

$$\langle t\rangle \times \langle k\rangle \cong \mathrm{C}_p \times \mathrm{C}_{p^r} , \quad r \geq 2 .$$

If $\Omega_1(\langle k\rangle) = \langle z\rangle$, then $[g, t] \in U - \langle z\rangle$ shows that $\langle t, k\rangle$ is not normal in $G$. If $\Omega_1(\langle k\rangle) = \langle u\rangle$ with $u \in U - \langle z\rangle$, then $[g, k] = z$ shows that again $\langle t, k\rangle$ is not normal in $G$. The maximality of $|H|$ shows that $r \leq s$, and so $\exp(K) = p^s$. It follows that $\langle h\rangle$ splits in $\langle h, k\rangle$, and so

$$\langle h, k\rangle = \langle h\rangle \times \langle k'\rangle$$
$$\text{with } \Omega_1(\langle k'\rangle) = \langle u\rangle , \ u \in U - \langle z\rangle \text{ and } o(k') \geq p^2 .$$

Since $[g, t] \in U - \langle z\rangle$, there is an integer $j$ modulo $p$ so that $[g, t^j k'] = z$. Because $\Omega_1(\langle t^j k'\rangle) = \langle u\rangle$, one may assume from the start that (replacing $k$ with $t^j k'$ and writing $k$ again):

$$K = \langle t\rangle \times \langle h\rangle \times \langle k\rangle , \quad o(k) = p^r , \quad 2 \leq r \leq s ,$$
$$k^{p^{r-1}} = u \in U - \langle z\rangle \quad \text{and } [g, k] = z = g^{p^s} .$$

Replacing $t$ with some other generator of $\langle t\rangle$ (if necessary), one may assume from the start that $[g, t] = uz^i$ for some integer $i$ modulo $p$.

For any integer $\alpha \not\equiv 0 \pmod p$ and any $x \in K$, one obtains (noting that $s \geq 2$)

$$(g^\alpha x)^{p^s} = z^\alpha [x, g^\alpha]^{\binom{p^s}{2}} = z^\alpha ,$$

and so $\Omega_1(G) = \langle t\rangle \times U \cong \mathrm{E}_{p^3}$ and the socle of each cyclic subgroup of $G$ that is not contained in $K$ is equal to $\langle z\rangle$.

Let $h'$ be an element of order $p^r$ in $\langle h\rangle$ such that $(h')^{p^{r-1}} = z$. For any fixed $\alpha \not\equiv 0 \pmod p$ let us consider the subgroup

$$\langle t\rangle \times \langle (h')^{\alpha}k\rangle \cong C_p \times C_{p^r}\,, \quad \text{where } ((h')^{\alpha}k)^{p^{r-1}} = z^{\alpha}u$$

and note that $\langle (h')^{\alpha}k\rangle \cong C_{p^r}$, $r \geq 2$, is a maximal cyclic subgroup in $G$ with the socle $\langle z^{\alpha}u\rangle$. In that case, $[g, (h')^{\alpha}k] = z \notin \langle t, (h')^{\alpha}k\rangle$, so that $\langle t, (h')^{\alpha}k\rangle$ is a maximal nonnormal subgroup in $G$. By Proposition 176.6, there is a unique integer $j \pmod p$ such that $\langle t^j(h')^{\alpha}k\rangle \trianglelefteq G$. Hence

$$[g, t^j(h')^{\alpha}k] = (uz^i)^j z = z^{1+ij}u^j \in \langle z^{\alpha}u\rangle\,,$$

which shows that $j \not\equiv 0 \pmod p$, and therefore

$$z^{1+ij}u^j = z^{\alpha j}u^j \quad \text{so that } 1 + ij \equiv \alpha j \text{ or } j(\alpha - i) \equiv 1 \pmod p\,.$$

Hence, for any fixed $\alpha \not\equiv 0 \pmod p$, there is $j \not\equiv 0 \pmod p$ such that $j(\alpha - i) \equiv 1 \pmod p$, and so $i \equiv 0 \pmod p$). Thus $[g, t] = u$.

Because $[g, k] = z$ and $\langle k\rangle \cong C_{p^r}$, $r \geq 2$, is a maximal cyclic subgroup in $G$ with the socle $\langle u\rangle$, it follows that $\langle t\rangle \times \langle k\rangle \cong C_p \times C_{p^r}$ is a maximal nonnormal subgroup in $G$. By Proposition 176.6, there is a unique integer $m \pmod p$ such that $\langle t^m k\rangle \trianglelefteq G$. But

$$[g, t^m k] = [g, t]^m [g, k] = u^m z\,,$$

a final contradiction (since $\Omega_1(\langle t^m k\rangle) = \langle u\rangle$). □

*Proof of Theorem 176.A.* It is easy to see, by inspection of Propositions 176.1 to 176.12, that all possible cases have been investigated, and so the theorem is proved. □

Theorem 176.A is a deep generalization of Theorem 16.2.

**Problem 1.** Classify the $p$-groups $G$ such that $H_G$ is cyclic for any nonnormal $H < G$. (See Theorem 176.A.) In particular, classify the $p$-groups $G$ such that $|H_G| \leq p$ for any nonnormal $H < G$ (see Exercise 6).

**Problem 2.** Classify the 2-groups $G$ such that $H \cap H^x$ is either cyclic or a generalized quaternion group for any nonnormal $H < G$ and any $x \in G - N_G(H)$.

**Problem 3.** Classify the 2-groups $G$ such that $H \cap H^x$ is either cyclic or a group of maximal class for any nonnormal $H < G$ and any $x \in G - N_G(H)$.

**Problem 4.** Study the $p$-groups $G$ such that $H_G$ is abelian for any nonnormal $H < G$ (see Theorem 176.A).

**Problem 5.** Classify the $p$-groups $G$ such that (i) $H_G = \{1\}$ for any nonnormal metacyclic $H < G$, (ii) all maximal metacyclic subgroups of $G$ are normal, (iii) $H_G$ is metacyclic for any nonnormal $H < G$.

**Problem 6.** Classify the $p$-groups $G$ such that $H_G$ is absolutely regular for any nonnormal $H < G$ (all $p$-groups of maximal class satisfy this condition).

**Problem 7.** Classify the $p$-groups all of whose nonnormal subgroups are either (i) abelian or minimal nonabelian, (ii) abelian or metacyclic. (See §§ 174 and 175. See also [FA], where the $p$-groups all of whose nonnormal subgroups are abelian are classified.)

# § 177 On the norm of a $p$-group

The norm $\mathcal{N}(G)$ of a group $G$ is the intersection of normalizers of all subgroups of $G$. The characteristic subgroup $\mathcal{N}(G)$ is Dedekindian and contains the center $Z(G)$ of $G$. By Baer's theorem (see §§ 143 and 171), if the norm $\mathcal{N}(G)$ is nonabelian, then a 2-group is Dedekindian. Therefore, in all interesting cases the subgroup $\mathcal{N}(G)$ is abelian. Only for few groups the norm is known. For example, if $G$ is a $p$-group with $\Omega_1(G) = G$, then $\mathcal{N}(G) = Z(G)$ (indeed, in this case, $\mathcal{N}(G)$ is centralized by all elements of $G$ of order $p$). In particular, if $\Sigma_{p^n} \in \mathrm{Syl}_p(\mathrm{S}_{p^n})$, then $\mathcal{N}(\Sigma_{p^n}) = Z(\Sigma_{p^n})$ since $\Omega_1(\Sigma_{p^n}) = \Sigma_{p^n}$. If $G$ is a 2-group of maximal class, then $\mathcal{N}(G) = Z(G)$, unless $G$ is a generalized quaternion group (see Lemma 171.2). As the proof of Theorem 177.1 shows, finding the norm even for so well-known groups as minimal nonabelian $p$-groups may be a fairly nontrivial task.

In this section we find all minimal nonabelian $p$-groups $G$ of order $> p^3$ satisfying $Z(G) < \mathcal{N}(G)$. Below we prove the following:

**Theorem 177.1.** *Suppose that $G$ is a minimal nonabelian $p$-group satisfying $Z(G) < \mathcal{N}(G) < G$. Then $G = \langle a, b\rangle$ is metacyclic, where*

$$o(a) = p^m\,, \quad o(b) = p^n\,, \quad \langle a\rangle \cap \langle b\rangle = \{1\}\,, \quad a^b = a^{1+p^{m-1}}\,, \quad m > n,\ m+n > 3\,.$$

The minimal nonabelian $p$-groups are described in Exercise 1.8a and Lemma 65.1, and we assume that this description is known.

If $G$ is a metacyclic $p$-group, we set $w(G) = w = \max\{k \mid |\Omega_k(G)| = p^{2k}\}$. Write $R(G) = \Omega_w(G)$ (see § 124). Then $G/R(G)$ is either cyclic or a 2-group of maximal class (Lemma 1.4).

Theorem 177.1 is an immediate consequence of the following four lemmas.

**Lemma 177.2.** (a) *If $G$ is a two-generator $p$-group such that $\mathcal{N}(G) \in \Gamma_1$, then $G$ is metacyclic. In particular, if $G$ is minimal nonabelian and $Z(G) < \mathcal{N}(G)$, then $G$ is metacyclic.*

(b) *If $G$ is a non-Dedekindian two-generator $p$-group such that $\mathcal{N}(G) \nleq \Phi(G)$, then $G$ is metacyclic.*

*Proof.* (a) If $x \in G - \mathcal{N}(G)$ and $X = \langle x\rangle$, then $X \lhd G$ since $N_G(X) \ge X\mathcal{N}(G) = G$. We claim that $G/X$ is cyclic. Indeed, otherwise, since $d(G) = 2$ so that $d(G/X) = 2 = d(G)$, we have $X \le \Phi(G) < \mathcal{N}(G)$, a contradiction. The last assertion follows since $\Phi(G) = Z(G) < \mathcal{N}(G)$ and all minimal nonabelian subgroups are two-generator.

(b) One has $\mathcal{N}(G)\Phi(G) \in \Gamma_1$ since $G/\Phi(G) \cong \mathrm{E}_{p^2}$. Let $R < G'$ be $G$-invariant of index $p$ such that $G/R$ is non-Dedekindian (Proposition 1.23); note that $R < \Phi(G)$. Then $G/R$ is minimal nonabelian (Lemma 65.2 (a)). In this case, $\mathcal{N}(G/R)$ is maximal in $G/R$. Assume that this is false. Then $\mathcal{N}(G/R) = Z(G/R) = \Phi(G/R) = \Phi(G)/R$. It follows that $\mathcal{N}(G)R \le \Phi(G)$ so $\mathcal{N}(G) \le \Phi(G)$, contrary to the hypothesis. As $\mathcal{N}(G/R)$ is maximal in the minimal nonabelian group $G/R$ that is two-generator, (a) implies that $G/R$ is metacyclic. Then, by Theorem 36.1, $G$ is metacyclic. □

**Lemma 177.3** (see Theorem A.80.1). *Suppose that $G = R(G)$ is a metacyclic minimal nonabelian p-group of order $p^{2w}$ and exponent $p^w$. Then all G-invariant cyclic subgroups of order $p^w$ generate a maximal subgroup, say M, of G and all cyclic subgroups of M of order $p^w$ are G-invariant. If L is one of such subgroups, then $G/L$ is cyclic so that $G = C \cdot L$ (semidirect product with kernel L).*

**Lemma 177.4.** *Let $G = R(G)$ be a metacyclic minimal nonabelian p-group of order $p^{2w}$. Then $\mathcal{N}(G) = Z(G)$.*

*Proof.* Let $A \triangleleft G$ be cyclic such that $G/A$ is cyclic; then $|A| = p^w$ and $G = B \cdot A$ for any nonnormal cyclic $B < G$ of order $p^w$. As $B$ is nonnormal in $G$, we get $A \nleq \mathcal{N}(G)$. Assume that $B < \mathcal{N}(G)$ and let $L < G$ be nonnormal cyclic of order $p^w$. Then $B$ normalizes $L$ so that $BL$, being a proper subgroup of $G$, is abelian; hence $L$ centralizes $B$. Thus, $C_G(B)$ contains all non-$G$-invariant cyclic subgroups of order $p^w$ of $G$. Let $A < H \in \Gamma_1$; then all cyclic subgroups of order $p^w$ in $H$ are $G$-invariant and $H$ contains all such subgroups (Lemma A.177.3). One has

$$c_w(G) = \frac{p^{2w} - p^{2w-2}}{(p-1)p^{w-1}} = (p+1)p^{w-1}, \quad c_w(H) = \frac{p^{2w-1} - p^{2w-2}}{(p-1)p^{w-1}} = p^{w-1}.$$

It follows that there are in $G$ exactly $c_w(G) - c_w(H) = p^w$ nonnormal cyclic subgroups of order $p^w$. If $F \in \Gamma_1$, then $c_w(F) = c_w(H) = p^{w-1}$. It follows that the nonnormal cyclic subgroups of order $p^w$ generate $G$, and we conclude that $C_G(B) = G$, a contradiction. Thus, $\exp(\mathcal{N}(G)) < p^w$ so that $\mathcal{N}(G) = Z(G)$. □

**Lemma 177.5.** *Let $G > R(G)$ be a metacyclic minimal nonabelian p-group of order $> p^3$ and*

$$G = \langle a, b \mid o(a) = p^m, \quad o(b) = p^n, \quad \langle a\rangle \cap \langle b\rangle = \{1\}, \quad a^b = a^{1+p^{m-1}}\rangle.$$

(a) *If $m < n$, then $\mathcal{N}(G) = Z(G)$.*
(b) *If $m > n$, then $\mathcal{N}(G) > Z(G)$.*

*Proof.* By hypothesis, $n \neq m > 1$, $G' = \Omega_1(A)$, where $A = \langle a\rangle$. Write $B = \langle b\rangle$.

(a) Let $m < n$. Since $\bar{G} = G/\mho_m(B) = R(G/\mho_m(B))$ of order $p^{2m}$ and exponent $p^m$ is nonabelian, in view of $G' \nleq B$, then $\mathcal{N}(\bar{G}) = Z(\bar{G}) = \Phi(\bar{G})$ (Lemma 177.4), so that $\mathcal{N}(G) \notin \Gamma_1$. In that case, $\mathcal{N}(G) = Z(G)$, completing the proof of (a).

(b) Now let $m > n$; then $\Omega_{n-1}(G) < Z(G)$. If $C < G$ is cyclic of order $> p^n$, then $G' \leq A \cap C \Rightarrow C \triangleleft G$, i.e., all cyclic subgroups of order $> p^n$ are $G$-invariant. It follows that all nonnormal cyclic subgroups of $G$ have order $p^n$. Clearly, $\exp(\Phi(G)) < p^m$. Let $|C| = p^m$; then $C \nleq \Phi(G) \Rightarrow G/C$ is cyclic so that $G' < C$. Therefore, if $D < G$ is nonnormal cyclic of order $p^n$, then $C \cap D = \{1\} \Rightarrow G = D \cdot C$ (semidirect product with kernel $C$). It follows that $C$ does not normalize $D$. Thus, $\exp(\mathcal{N}(G)) < p^m$. As $R(G) = \Omega_n(G)$ is abelian, any cyclic subgroup of $G$ of order $p^n$ centralizes all subgroups of orders $\leq p^n$. As any cyclic subgroup of $G$ of order $> p^n$ is $G$-invariant, one has $\Omega_n(G) \leq \mathcal{N}(G)$, and hence $F = \Omega_n(G)Z(G) \leq \mathcal{N}(G)$. Since $F \in \Gamma_1$, (b) is proved. □

Now Theorem 177.1 follows immediately from Lemmas 177.2–177.5.

**Exercise 1.** Study the $p$-groups $G$ containing a proper subgroup $H \cong \mathcal{H}_{2,2} = \langle a, b \mid a^4 = b^4 = 1, \quad a^b = a^3 \rangle$ normalizing all subgroups of $G$ nonincident with $H$.

**Exercise 2.** Let $N \leq \Phi(G)$, where $G$ is a $p$-group. If $N \leq \mathcal{N}(H)$ for all $H \in \Gamma_1$, then $N \leq \mathcal{N}(G)$.

**Solution.** Let $B < G$ and $B \leq H \in \Gamma_1$. Then $N \leq N_H(B)$ since $N \leq \mathcal{N}(H)$.

By Schenkman's theorem (see Corollary 140.8), $\mathcal{N}(G) \leq Z_2(G)$ so $|\mathcal{N}(G)| \leq p^2$ if $G$ is of maximal class of order $> p^3$. In the proof of Theorem 177.6 and in the solution of the following exercise we do not use this fairly deep result.

**Exercise 3.** Suppose that $G$ is a $p$-group of maximal class and order $> p^3$. Then $|\mathcal{N}(G)| \leq p^2$.

**Solution.** One may assume that $p > 2$. By Theorem 9.6 (f), there is in $G$ a subgroup $R$ of order $p^2$ such that $N_G(R) = N$ is nonabelian of order $p^3$. It follows that $\mathcal{N}(G) < N$. Thus, $|\mathcal{N}(G)| \leq p^2$. In particular, $\mathcal{N}(G) \leq Z_2(G)$.

**Exercise 4.** Is it true that if all minimal nonabelian subgroups of a $p$-group $G$ are nonmetacyclic, then $Z(G) = \mathcal{N}(G)$?

**Theorem 177.6.** *Suppose that $G$ is a $p$-group of maximal class and order $> p^3$. Let $N \lhd G$ be of order $p^2$ and $A = C_G(N)$. Then one of the following holds:*
(a) *if $\Omega_1(A) < \Omega_1(G)$, then $\mathcal{N}(G) = Z(G)$,*
(b) *if $\Omega_1(A) = \Omega_1(G)$, then $\mathcal{N}(G) = N$.*

*Proof.* In view of Lemma 171.2, one may assume that $p > 2$; then $N \cong E_{p^2}$ (Lemma 1.4 and Exercise 9.1 (b)).

(a) Suppose that there is $x \in G - A$ of order $p$; then $x$ does not centralize $N$ so $N$ does not normalize $\langle x \rangle$. It follows that

$$N \nleq \mathcal{N}(G) \Rightarrow |\mathcal{N}(G)| = p \Rightarrow \mathcal{N}(G) = Z(G),$$

proving (a).

(b) Now, assuming that $\Omega_1(G) = \Omega_1(A)$, we have to prove that $\mathcal{N}(G) = N$. In that case, all elements of the set $G - A$ have order $p^2$ (Theorem 13.19 (a)). If $C < G$ is cyclic of order $p^2$ and $C \nleq A$, then $CN \cong M_{p^3}$ (Theorem 13.19) so $N \leq N_G(C)$. Thus, $N$ normalizes all cyclic subgroups of $G$ of order $p^2$ not contained in $A$. If $U < G$ is such that $U \nleq A$, then $U$ is generated by elements of order $p^2$ since $U = \langle U - A \rangle$ and all elements of the set $U - A$, as a subset of the set $G - A$, have order $p^2$. Therefore, $N$ normalizes $U$. Hence, $N$

normalizes all subgroups of $G$ not contained in $A$. As $N \leq Z(A)$, then $N$ normalizes all subgroups of $A$. Thus, $N$ normalizes all subgroups of $G$; hence $N \leq \mathcal{N}(G)$. By Exercise 3, $N = \mathcal{N}(G)$. □

Let $G_1$ be the fundamental subgroup of a $p$-group of maximal class $G$ of order $> p^{p+1}$. It follows from Theorem 177.6 that $\mathcal{N}(G) = Z_2(G) \iff \Omega_1(G) = \Omega_1(G_1)$.

**Remark 1.** Suppose that a group $G$ contains a proper nonabelian normal $p$-subgroup $P$ such that $P/\Phi(P)$ is minimal normal in $G/\Phi(P)$. Then either $\mathcal{N}(P) \leq \Phi(P)$ or $P$ is Dedekindian, $p = 2$. Indeed, it follows from $|P/\Phi(P)| > p$ that $G/P$ is not a $p$-group. Assume that $\mathcal{N}(P) \nleq \Phi(P)$. Note that $\mathcal{N}(P) \triangleleft G$. Then $\mathcal{N}(P)\Phi(P)$, being characteristic in $P$, is normal in $G$. As $P/\Phi(P) > \{1\}$ is minimal normal in $G/\Phi(P)$, one has $\mathcal{N}(P)\Phi(P) = P \Rightarrow \mathcal{N}(P) = P$. Then $P$ is nonabelian Dedekindian, which implies $p = 2$ (Theorem 1.20).

**Problem 1.** Find the norm of (i) a metacyclic $p$-group, (ii) a two-generator $p$-group with a cyclic derived subgroup, (iii) a special $p$-group, (iv) a $p$-group with an abelian subgroup of index $p$.

**Problem 2.** Study the $p$-groups $G$ with cyclic $G'$ such that $G' \nleq \mathcal{N}(G)$.

**Problem 3.** Study the $p$-groups $G$ satisfying $|G : \mathcal{N}(G)| \leq p^2$.

**Problem 4.** Classify the $p$-groups such that $|H : \mathcal{N}(H)| \leq p$ for all $H \in \Gamma_1$.

**Problem 5.** Find $\mathcal{N}(A \times B)$, where $A$ and $B$ are $p$-groups.

**Problem 6.** Study the $p$-groups $G$ such that $H \cap \mathcal{N}(G) = \mathcal{N}(H)$ for all nonabelian $H \leq G$.

**Problem 7.** Find the norms of prime power $\mathcal{A}_k$-groups, $k = 2, 3$ (see § 71 and [ZZLS]).

**Problem 8.** Study the $p$-groups $G$ containing a nontrivial subgroup $H$ normalizing all subgroups of $G$ nonincident with $H$.

**Problem 9.** Denote by $\mathcal{QN}(G)$ the intersection of normalizers of all nonquasinormal subgroups of a group $G$. We call $\mathcal{QN}(G)$ the *quasinorm* of $G$. Classify the $p$-groups with the nonabelian quasinorm. Replace in all previous propositions and problems, where this is possible, the word ‘norm’ with ‘quasinorm’.

**Problem 10.** Study the $p$-groups $G$ satisfying (i) $|\mathcal{N}(G)| = p$, (ii) $|\mathcal{QN}(G)| = p$.

# § 178 *p*-groups whose character tables are strongly equivalent to character tables of metacyclic *p*-groups, and some related topics

The first author is indebted to Martin Isaacs for constructive discussion and essential help.

*$1^o$ p-groups with identical character tables.* Let $p$ be a prime divisor of the order $|G|$ of a group $G$ and $\chi \in \mathrm{Irr}(G)$. Recall (see [BZ, §§ 4.5 and 4.6]) that the generalized Frobenius–Schur indicator (for a fixed prime $p$) of the character $\chi$ is equal to

(1) $$\nu_p(\chi) = |G|^{-1} \sum_{g\in G} \chi(g^p).$$

Set $\chi^{(p)}(g) = \chi(g^p)$ $(g \in G)$. It is known that the class function $\chi^{(p)}$ is a generalized character of $G$ (see [BZ, §§ 4.5 and 4.6]). Therefore, $\nu_p(\chi) = \langle\chi^{(p)}, 1_G\rangle$ is a rational integer.[1]

**Definition 1.** Let $G, G_0$ be groups, let $\mathrm{X}(G), \mathrm{X}(G_0)$ be their character tables and let $\mathrm{CL}(G), \mathrm{CL}(G_0)$ be the sets of conjugacy classes of $G, G_0$, respectively. We say that $\mathrm{X}(G), \mathrm{X}(G_0)$ are equal (or equivalent), if there are bijections of $\mathrm{Irr}(G)$ onto $\mathrm{Irr}(G_0)$ and $\mathrm{CL}(G)$ onto $\mathrm{CL}(G_0)$ such that, provided $\chi \in \mathrm{Irr}(G)$, $\chi_0 \in \mathrm{Irr}(G_0)$ are corresponding characters and $K \in \mathrm{CL}(G)$, $K_0 \in \mathrm{CL}(G_0)$ are corresponding classes, then $\chi(x) = \chi_0(x_0)$ for any $x \in K$, $x_0 \in K_0$. In this case, we write $\mathrm{X}(G) = \mathrm{X}(G_0)$.

Suppose that groups $G, G_0$ have equal character tables. It is known that the structures of $G$ and $G_0$ are connected closely but are not identical in the general case (for example, $\mathrm{X}(D_8) = \mathrm{X}(Q_8)$). If, in addition, for any two corresponding characters $\chi \in \mathrm{Irr}(G)$ and $\chi_0 \in \mathrm{Irr}(G_0)$ one has $\nu_p(\chi) = \nu_p(\chi_0)$, then the character tables $\mathrm{X}(G)$ and $\mathrm{X}(G_0)$ (or groups $G$ and $G_0$) are said to be *strongly equivalent* with respect to $p$ (in what follows we will omit the string 'with respect to $p$' since $p$ will be clear from the context).

Let

(2) $$\vartheta_p(G) = |\{x \in G \mid x^p = 1\}|$$

be the number of solutions of the equation $x^p = 1$ in a group $G$. Then, as we know (see equality (14) in Chapter 5 of [BZ]),

(3) $$\vartheta_p(G) = \sum_{\chi\in\mathrm{Irr}(G)} \nu_p(\chi)\chi(1).$$

Clearly, $\vartheta_p(G)$ uniquely determines $c_1(G)$, the number of cyclic subgroups of order $p$ in $G$: $c_1(G) = \frac{1}{p-1}(\vartheta_p(G) - 1)$.

---

1 Frobenius and Schur have considered only $\nu_2(\chi)$.

Let $p$-groups $G$ and $G_0$ have equal character tables and $\chi \in \mathrm{Irr}(G)$ and $\chi_0 \in \mathrm{Irr}(G_0)$ are corresponding characters. Then it is not necessarily true that $\nu_p(\chi) = \nu_p(\chi_0)$. Indeed, $\mathrm{X}(\mathrm{D}_8) = \mathrm{X}(\mathrm{Q}_8)$. Let $\chi$ and $\chi_0$ be the (corresponding) nonlinear irreducible characters of $G = \mathrm{D}_8$ and $G_0 = \mathrm{Q}_8$. One has

$$6 = \vartheta_2(G) = |G/G'| + \chi(1)\nu_2(\chi) = 4 + 2\nu_2(\chi)$$

so that $\nu_2(\chi) = 1$. Next,

$$2 = \vartheta_2(G_0) = |G_0/G_0'| + \chi_1(1)\nu_2(\chi_0) = 4 + 2\nu_2(\chi_0)$$

so that $\nu_2(\chi_0) = -1 \neq 1 = \nu_2(\chi)$.

The character table $\mathrm{X}(G)$ of $G$ determines the lattice of normal subgroups of $G$. Let $H \triangleleft G$ and let $\mathcal{M} = \{\chi \in \mathrm{Irr}(G) \mid H \le \ker(\chi)\}$; then $H = \bigcap_{\chi \in \mathcal{M}} \ker(\chi)$. Suppose that groups $G$, $G_0$ have equal character tables. Let $\mathcal{M}_0$ be the set of characters corresponding to characters belonging to the set $\mathcal{M}$. Write $H_0 = \bigcap_{\chi_0 \in \mathcal{M}_0} \ker(\chi_0)$. Then

$$\mathrm{X}(G_0/H_0) = \mathrm{X}(G/H). \tag{4}$$

If, in addition, the tables $\mathrm{X}(G)$ and $\mathrm{X}(G_0)$ are strongly equivalent, then the character tables from (4) are also strongly equivalent and, by (3),

$$\vartheta_p(G) = \vartheta_p(G_0), \quad \vartheta_p(G/H) = \vartheta_p(G_0/H_0). \tag{5}$$

Let $\mathrm{X}(G) = \mathrm{X}(G_0)$; then $|G| = |G_0|$. If $G_0$ is abelian, then $G$ is also abelian and, moreover, $G \cong G_0$. It follows, in the general case, that $G/G' \cong G_0/G_0'$. In particular, if $G_0$ is a $p$-group, then $\mathrm{d}(G) = \mathrm{d}(G_0)$.

**Remark 1.** Let $n > 2$ if $p > 2$ and $n > 3$ if $p = 2$. Let $\mu_n(G)$ be the number of normal subgroups $D$ of a $p$-group $G$ such that $G/D$ is metacyclic of order $p^n$. Then, if $G$ in nonmetacyclic, $p \mid \mu_n(G)$ (Theorem 47.1).

Now we are ready to prove the following theorem:

**Theorem 178.1.** *Suppose that $G_0$ is a metacyclic $p$-group. If a group $G$ is such that $\mathrm{X}(G)$ and $\mathrm{X}(G_0)$ are strongly equivalent, then $G$ is also a metacyclic $p$-group.*

*Proof.* It follows from the hypothesis that $|G| = |G_0|$, so that $G$ is a $p$-group. Then, as we have noted, $\mathrm{d}(G) = \mathrm{d}(G_0) = 2$. Assume that $G$ is a counterexample of minimal order. In that case, $G_0$ is nonabelian; then $G$ is also nonabelian and $G/G' \cong G_0/G_0'$.

Let $|G_0| = p^{n+1}$; then $|G| = p^{n+1}$. Let $N \triangleleft G$ be of order $p$ and let $N_0$ be a corresponding normal subgroup of $G_0$. Since $\mathrm{X}(G/N)$ and $\mathrm{X}(G_0/N_0)$ are strongly equivalent and as $G_0/N_0$ is metacyclic, then the quotient group $G/N$ is also metacyclic, by induction. It follows that $\mu_n(G) \equiv 1 \pmod p$ (see Remark 1). Since $G$ is nonmetacyclic, one has $|G| = p^3$ for $p > 2$ and $|G| = 2^4$ for $p = 2$, by Theorem 47.1.

(i) Suppose that $p > 2$; then $n = 2$ and $G$ is nonmetacyclic of order $p^3$ so that $G$ is nonabelian of exponent $p > 2$. Since $\mathrm{c}_1(G) = 1 + p + p^2 \neq 1 + p = \mathrm{c}_1(G_0)$, the groups $G$, $G_0$ are not strongly equivalent (see formula (3)). Thus, the theorem holds for $p > 2$.

(ii) Let $p = 2$; then $n = 3$ and $G$ is a nonmetacyclic group of order $2^4$.
If $G_0$ is of maximal class, then $|G/G'| = |G_0/G_0'| = 4$, so $G$ is also of maximal class (Taussky) hence metacyclic, contrary to the assumption.

Since $G$ is not of maximal class, one has $|G'| = |G_0'| = 2$ and $\mathrm{d}(G) = 2$ so that $G/G' \cong G_0/G_0'$ is abelian of type $(4, 2)$, and we conclude that $G$ and $G_0$ are minimal nonabelian (Lemma 65.2 (a)) and $|\Omega_1(G)| = 8$ (Lemma 65.1). In that case, $\vartheta_2(G) = 8 > 4 = \vartheta_2(G_0)$ so that $\mathrm{X}(G)$ and $\mathrm{X}(G_0)$ are not strongly equivalent (see formula (3)). Thus, $G$ is not a counterexample, i.e., $G$ is metacyclic. □

In the two following proofs of Theorem 178.1 for $p > 2$ we assume that $G$ is a counterexample of minimal order.

*Second proof for $p > 2$; a small change also allows us to consider the case $p = 2$.* Let $R$ be a $G$-invariant subgroup of index $p$ in $G'$. Since $\mathrm{d}(G) = 2$, the subgroup $R$ is defined uniquely (this follows easily from Lemma 36.5 (a)). Let $R_0$ be the corresponding normal subgroup of $G_0$; then $R_0 < G_0'$ has index $p$ too. Both quotient groups $G/R$ and $G_0/R_0$ are minimal nonabelian (Lemma 65.2 (a)) and, as $G$ is nonmetacyclic, $G/R$ is also nonmetacyclic (Theorem 36.1), and hence $c_1(G/R) = 1 + p + p^2$ (Lemma 65.1). Since the metacyclic $p$-group $G_0/R_0$ has $< 1 + p + p^2$ cyclic subgroups of order $p$, we get a contradiction. □

*Third proof for $p > 2$.* Setting $H = \mho_1(G)$, we get $|G/H| > p^2$ (Theorems 9.11 or 36.8) so that $\vartheta_p(G/H) \geq p^3$. Let $H_0$ be the corresponding normal subgroup of $G_0$. Since $G_0/H_0$ is metacyclic, one has $\vartheta_p(G_0/H_0) < p^3$. It follows that $\vartheta_p(G/H) \neq \vartheta_p(G_0/H_0)$ so the quotient groups $G/H$, $G_0/H_0$ are not strongly equivalent, a contradiction. □

Theorem 178.1 is not trivial since the presented proofs are based on the fairly deep Theorems 47.1, 36.1 and 36.8.

In his letter dated July 22, 2012, Isaacs noted that if $\mathrm{X}(G) = \mathrm{X}(G_0)$, where $G_0$ is the unique metacyclic group of order 16 and exponent 4, then $G$ is not necessarily metacyclic (in that case, $G = \langle a, b \mid a^4 = b^2 = 1, \quad c = [a, b], \quad [a, c] = [b, c] = 1\rangle$).

**Proposition 178.2.** *Suppose that character tables of p-groups $G$ and $G_0$ are strongly equivalent. If $G_0$ is absolutely regular, so is $G$.*

*Proof.* Assume that $G$ is not absolutely regular so that $\vartheta_p(G/H) = p^p$ for some normal subgroup $H$ of $G$ containing $\mho_1(G)$ (Theorem 9.8 (a)). Let $H_0 \lhd G_0$ be the subgroup corresponding to $H$. Then $G_0/H_0$ of order $p^p$ is not of exponent $p$ since $G_0$ is absolutely regular. It follows that $\vartheta_p(G/H) > \vartheta_p(G_0/H_0)$, a contradiction. □

*$2^o$ p-groups whose lattices of normal subgroups are isomorphic and related topics.* Let $\mathcal{L}_N(W)$ be the lattice of normal subgroups of a group $W$. If $G$, $G_0$ are $p$-groups such that $\mathrm{X}(G) = \mathrm{X}(G_0)$, then $\mathcal{L}_N(G) \cong \mathcal{L}_N(G_0)$. But the converse assertion is not true. Therefore, it is interesting to study the pairs of $p$-groups $G$, $G_0$ satisfying the last condition. Clearly, $\mathcal{L}_N(G) \cong \mathcal{L}_N(G_0) \Rightarrow \mathrm{d}(G) = \mathrm{d}(G_0)$ since $|G/\Phi(G)| = |G_0/\Phi(G_0)|$.

**Remark 2.** Let $G, G_0$ be $p$-groups such that $G_0$ is abelian and $\mathcal{L}_N(G) \cong \mathcal{L}_N(G_0)$. We claim that $G$ is also abelian. To prove this we proceed by induction on $|G|$. One may assume that $G_0$ is noncyclic. Then it contains two distinct subgroups $K_0, L_0$ of order $p$. Let $K, L$ be corresponding members of the set $\mathcal{L}_N(G)$. In that case, by induction, $G/K$ and $G/L$ are abelian so $G$ is also abelian. It is easy to prove that $G \cong G_0$. Indeed, $c_1(G) = c_1(G_0)$ so that $\Omega_1(G) \cong \Omega_1(G_0)$. Proceeding by induction, one has $G/\Omega_1(G) \cong G_0/\Omega_1(G_0)$, and this implies $G \cong G_0$, as desired. In the case under consideration, $\mathcal{L}_N(G) \cong \mathcal{L}(G_0)$, where $\mathcal{L}(G_0)$ is the lattice of subgroups of $G_0$. If $\mathcal{L}(G) \cong \mathcal{L}(G_0)$, where a $p$-group $G_0$ is abelian, then $G$ may be nonabelian (if $G_0$ is abelian of type $(p^n, p)$, $n > 2$, and $G \cong \mathrm{M}_{p^{n+1}}$, then $\mathcal{L}(G) \cong \mathcal{L}(G_0)$).

It is trivial that if a $p$-group $G$ has the same character table as a $p$-group $G_0$ of maximal class, then $G$ is also of maximal class. If $p$-groups $G$ and $G_0$ of equal order are of maximal class, then $\mathcal{L}_N(G) \cong \mathcal{L}_N(G_0)$. As the following proposition shows, the converse is also true:

**Proposition 178.3.** *Suppose that $p$-groups $G$ and $G_0$ are such that $\mathcal{L}_N(G) \cong \mathcal{L}_N(G_0)$. If $G_0$ is of maximal class so is $G$.*

*Proof.* It follows from $d(G) = 2$ that $G$ is noncyclic. By Exercise 9.1 (b), there is in $G_0$ exactly one normal subgroup of every index $> p$ so the same holds in $G$. Now the result follows from Lemma 9.1. □

If the lattices of subgroups of $p$-groups $G$ and $G_0$ are isomorphic and $G_0$ is of maximal class, then $G$ may be not as Theorem A.42.2 shows (this is possible only if $|G| < p^{p+2}$; see Theorem 13.2 (b)).

**Exercise 1.** Suppose that $G, G_0$ are $p$-groups such that $\mathcal{L}_N(G) \cong \mathcal{L}_N(G_0)$ and $G_0$ is minimal nonabelian. Then $G$ is also minimal nonabelian.

**Exercise 2.** Suppose that $G, G_0$ are $p$-groups such that $\mathrm{X}(G) = \mathrm{X}(G_0)$ and $G_0$ is special. Is it true that $G$ is special?

**Definition 2.** A pair of $p$-groups $G$ and $G_0$ satisfy the condition $(\mathcal{C}_1)$ if the following holds: (i) $\mathcal{L}_N(G) \cong \mathcal{L}_N(G_0)$, and (ii) if $N \lhd G$ and $N_0 \lhd G_0$ are corresponding normal subgroups, then $c_1(G/N) = c_1(G_0/N_0)(\iff \vartheta_p(G/N) = \vartheta_p(G_0/N_0))$.

Obviously, pairs $G, G_0$ of $p$-groups from Theorems 178.1 and Proposition 178.2 satisfy the condition $(\mathcal{C}_1)$.

**Exercise 3.** Suppose that a pair $G$, $G_0$ of $p$-groups satisfy the condition $(\mathcal{C}_1)$.
(a) If $G_0$ is metacyclic, so is $G$.
(b) If $G_0$ is absolutely regular, so is $G$.

**Solution.** (a) Let $G_0$ be noncyclic. One has $G/\mho_1(G) \cong G_0/\mho_1(G_0) \cong \mathrm{E}_{p^2}$. If $p > 2$, the result follows from Theorem 9.11. Now let $p = 2$. One may assume that $G_0$ is not of maximal class. Let $R \in \mathcal{L}_N(G)$ be of index $p$ in $G'$. Then $G/R$ is minimal nonabelian (Lemma 65.2 (a)). If $R_0$ is the corresponding normal subgroup of $G_0$, then $G_0/R_0$ is also minimal nonabelian and $\mathrm{c}_1(G_0/R_0) = 1 + 2 \Rightarrow \mathrm{c}_1(G/R) = 1 + 2$, and we conclude that $G/R$ is metacyclic (Lemma 65.1). Then $G$ is metacyclic, by Theorem 36.1.

**Exercise 4.** Let a $p$-group $G$ be strongly equivalent to a minimal nonabelian $p$-group $G_0$. Is it true that $G \cong G_0$?

**Exercise 5.** Let $G$, $G_0$ be minimal nonabelian groups of order $p^4$ and exponent $p^2$, $p > 2$, $G \not\cong G_0$. Is it true that $G$, $G_0$ have identical character tables?

**Theorem 178.4** (Isaacs, personal communication). *Suppose that $G_0$ is a $p$-group with abelian subgroup of index $p$. If a $p$-group $G$ is such that $\mathrm{X}(G) = \mathrm{X}(G_0)$, then $G$ also possesses an abelian subgroup of index $p$.*

*Proof.* Say that $G_0$ and $G$ are the groups satisfying the hypothesis, and $G_0$ has an abelian subgroup of index $p$ but $G$ does not. By Ito's Theorem on degrees [BZ, Theorem 7.2.3], $\mathrm{cd}(G_0) = \{1, p\}$ so that $\mathrm{cd}(G) = \{1, p\}$. Let $\mathrm{Z}(G)$ be the center of $G$. We know that $|G : \mathrm{Z}(G)| = p^3$ ([BZ, Theorem 14.9]), unless $G$ has an abelian subgroup of index $p$, which we may assume. Let $A_0$ be abelian of index $p$ in $G_0$ and let $x_0 \in A_0 - \mathrm{Z}(G_0)$ (since $G$ is nonabelian so is $G_0$; then $\mathrm{Z}(G_0) < A_0$). In that case, the class of $x_0$ has size $p$. Then $G$ has a class of size $p$, and let $y$ be an element of that class. Let $C = \mathrm{C}_G(y)$ so $|G : C| = p$; clearly, $\mathrm{Z}(G) \leq C$. We claim that $C$ is abelian. We know that $y \notin \mathrm{Z}(G)$ so the subgroup $K = \langle \mathrm{Z}(G), y\rangle$ has index $\leq p^2$ in $G$, and thus has index $\leq p$ in $C$. But $K$ is central in $C$, and thus $C$ is abelian. □

**Exercise 6.** Let $G$ be a $p$-group such that $|G : \mathrm{Z}(G)| = p^3$.
(a) If $x \in G$ is such that $|G : \mathrm{C}_G(x)| = p$, then $\mathrm{C}_G(x)$ is abelian.
(b) If $G$ has no abelian subgroup of index $p$, then $|G : \mathrm{C}_G(x)| = p^2$ for all $x \in G - \mathrm{Z}(G)$.

**Exercise 7.** If the center of a minimal nonabelian $p$-group $G$ is cyclic, then either $|G| = p^3$ or $G \cong \mathrm{M}_{p^n}$.

**Exercise 8.** Classify the $p$-groups $G$ such that $\mathrm{X}(G) \cong \mathrm{X}(G_0)$, where $G_0 = \Sigma_{p^2}$ is a Sylow $p$-subgroup of the symmetric group of degree $p^2$. (**Hint.** The group $G$ is of maximal class with an abelian subgroup of index $p$, by Theorem 178.4. If $p = 2$, then $G$ need not necessarily be isomorphic with $G_0$.)

**Exercise 9.** If a $p$-group $G$ is such that $\mathcal{L}_N(G) = \mathcal{L}_N(G_0)$, where $G_0 \cong \mathrm{M}_{p^n}$, then $G \cong G_0$, unless $n = 3$ (one has $\mathcal{L}_N(\mathrm{M}_{p^3}) \cong \mathcal{L}_N(S(p^3))$).

**Solution.** Let $R_0 < G_0$ be the unique normal subgroup of order $p$ and let $R$ be a corresponding subgroup of $G$. By Exercise 1, $G$ is minimal nonabelian. Since $R$ is the unique subgroup of order $p$ in $Z(G)$, we conclude that $Z(G)$ is cyclic. Now the result follows from Exercise 7.

**Exercise 10.** Suppose that a metacyclic minimal nonabelian $p$-group

$$G_0 = \langle x, y \mid o(x) = p^m > p^2, o(y) = p^n, x^y = x^{1+p^{m-1}} \rangle .$$

Let a $p$-group $G$ be such that $\mathcal{L}_N(G) \cong \mathcal{L}_N(G_0)$. Is it true that then $G$ is also a metacyclic minimal nonabelian $p$-group?

**Exercise 11.** Let $G$ be a noncyclic $p$-group of order $p^n > p^3$ having exactly $p+1$ normal subgroups of orders $p^2, \dots, p^{n-1}$. Is it true that $G \cong M_{p^n}$?

**Proposition 178.5.** *Suppose that $G_0$ is a metacyclic $p$-group such that $G_0/G_0'$ has no cyclic subgroup of index $p$. If $G$ is a $p$-group such that $\mathcal{L}_N(G) \cong \mathcal{L}_N(G_0)$, then $G$ is also metacyclic.*

*Proof.* By hypothesis, $|G_0/G_0'| \geq p^4$. There is $T_0 \triangleleft G_0$ such that $G_0/T_0$ is abelian of type $(p^2, p^2)$. As $G_0$ is metacyclic, one has $T_0 = \mho_2(G_0)$. Let $T$ be the corresponding member of the set $\mathcal{L}_N(G)$. Then $G/T$ is also abelian of type $(p^2, p^2)$, by Remark 2. Let $R \in \mathcal{L}_N(G) - \{T\}$ be of index $p^4$ in $G$ and let $R_0$ be the corresponding member of the set $\mathcal{L}_N(G)$ (see Theorem 47.1). In that case, $G_0/R_0$ has a cyclic subgroup of index $p$ since $R_0 \neq \mho_2(G_0)$, hence, in view of $\mathcal{L}_N(G/R) \cong \mathcal{L}_N(G_0/R_0)$, the quotient group $G/R$ has a cyclic subgroup of index $p$, so is metacyclic (Remark 2 and Exercise 7 since $G/R$ is not a 2-group of maximal class). Thus, any epimorphic image of $G$ of order $p^4$ is metacyclic. Therefore, by Theorem 47.1, the group $G$ is metacyclic, completing the proof. □

*3° Problems.*

**Problem 1.** Let $G$, $G_0$ be $p$-groups such that $\mathcal{L}_N(G) \cong \mathcal{L}_N(G_0)$. Suppose that $G_0$ contains an abelian subgroup of index $p$. Is is true that $G$ has an abelian subgroup of index $p$? (See Theorem 178.4.)

**Problem 2.** (Isaacs). Is it true that the character table of a $p$-group determines the minimal index of its abelian (or normal abelian) subgroup?

**Problem 3.** Let $p$-groups $G$ and $G_0$ be such that $G$ is regular and $G_0$ is irregular, $p > 2$. Is it is possible that $X(G)$ and $X(G_0)$ are strongly equivalent?

**Problem 4.** Let regular $p$-groups $G$ and $G_0$ be such that $|\Omega_1(G)| \geq |\Omega_1(G_0)|$ and $X(G) = X(G_0)$. Is it possible to estimate $\max\{\frac{|\Omega_1(G)|}{|\Omega_1(G_0)|}\}$?

**Problem 5.** Do there exist two nonisomorphic metacyclic $p$-groups whose character tables are strongly equivalent?

**Problem 6.** (Isaacs). Two $p$-groups $G, G_0$ are said to be *Brauer connected* (we also write: $G \operatorname{brc} G_0$), if the following holds: (a) $\mathrm{X}(G) = \mathrm{X}(G_0)$, and (b) if $K, K_0$ are corresponding $G$- and $G_0$-classes, $x \in K$, $x_0 \in K_0$, then elements $x^p$ and $x_0^p$ lie in corresponding classes of $G$, $G_0$, respectively. Is it true that two $p$-groups $G, G_0$ are Brauer connected $\iff$ they are strongly equivalent?

**Problem 7.** Suppose that $G$ and $G_0$ are $p$-groups and $G_0$ is metacyclic. Describe the structure of $G$ provided $\mathcal{L}_N(G) \cong \mathcal{L}_N(G_0)$.

# § 179 $p$-groups with the same numbers of subgroups of small indices and orders as in a metacyclic $p$-group

Given $n$ and a $p$-group $G$, let $s_n(G)$ ($c_n(G)$) be the number of subgroups (cyclic subgroups) of order $p^n$ in $G$. These numbers were computed in the case where $G$ is a metacyclic $p$-group (see § 124 and Appendix 98; the first result of this type was proved by Mann in Theorem 124.1). It is known that if $H$ is the elementary abelian $p$-group of order $p^n$ and a $p$-group $G$ satisfies

$$(*) \qquad s_k(G) = s_k(H)$$

for some fixed $k \in \{2, \ldots, n-1\}$, then $G \cong H$ (Theorem 5.17). However, if $H$ is an abelian $p$-group and $(*)$ holds for all $k$, then it need not necessarily $G \cong H$ (for example, if $H$ is abelian of type $(p^n, p)$, $n > 1$ and $G \cong M_{p^{n+1}}$, then $(*)$ holds for all $k$). Another example: $G$ is a nonabelian metacyclic group of order $p^4$ and exponent $p^2$ and $H$ is abelian of type $(p^2, p^2)$. Note that if $p$-groups $G$ and $H$ are lattice isomorphic, then they satisfy the condition $(*)$ for all $k$. Therefore, lattice isomorphism is a more restrictive condition than the condition $(*)$ for all $k$.

It follows from Theorem 1.17 (a,b) that if 2-groups $G$ and $H$ satisfy the condition $(*)$ for $k = 1, 2$ and $H$ is a 2-group of maximal class, then $G \cong H$.

This section was inspired by [Zha2].[1]

In this section we study the $p$-groups $G$ such that $(*)$ holds for some $k$, where $H$ is a metacyclic $p$-group. Below, we repeat some results from § 124.6$^o$. Next, we state a number of open problems, some of which are challenging.

We begin with the following theorem:

**Theorem 179.1** ([Man33]). *Let $H$ be a metacyclic $p$-group, $p > 2$. If $G$ is a $p$-group such that $G$ and $H$ satisfy the condition $(*)$ for $k = 1, 2$, then one of the following holds:*
(a) *$G$ is a 3-group of maximal class and order $3^4$ with $|\Omega_1(G)| = 3^2$.*
(b) *$G$ is metacyclic.*

*Proof.* One may assume that $|H| = p^m > p^3$ and $H$ has no cyclic subgroup of index $p$. Then $s_1(G) = s_1(H) = 1 + p$ so $G$ is either metacyclic or a 3-group of maximal class (Theorem 13.7). Assume that $G$ is a 3-group of maximal class; then $|\Omega_1(G)| = 3^2$. One has (note that $H$ is regular):

$$s_2(H) = 1 + \frac{3^4 - 3^2}{3(3-1)} = 1 + 3 + 3^2 = s_2(G)\,.$$

1 It is proved in that paper that if $p$-groups $G$, $H$, $p > 2$, satisfy the condition $(*)$ and, in addition, $\nu_k(G) = \nu_k(H)$ for all $k$ and $H$ is metacyclic, then $G \cong H$ (here $\nu_k(H)$ is the number of normal subgroups of order $p^k$ in a $p$-group $H$).

If $|G| = 3^4$, then $c_2(G) = \frac{3^4-3^2}{3(3-1)} = 3^2+3$ so that $s_2(G) = 1+3+3^2 = s_2(H)$. If $|G| > 3^4$ and $G_1$ is the fundamental subgroup of $G$, then $s_2(G_1) = s_2(H)$; and we get a contradiction since not all subgroups of $G$ of order $3^2$ are contained in $G_1$. □

**Remark 1.** The case $p = 2$ of Theorem 179.1 is a fairly difficult problem that can be solved only for partial classes of metacyclic 2-groups $H$. Let $H$ be a nonabelian metacyclic 2-group. If $H$ is of maximal class so is $G$, by Theorem 1.17. Now assume that $H$ is not of maximal class. Then $E_4 \cong \Omega_1(H) \cong \Omega_1(G)$. If $|\Omega_2(H)| = 8$, then $|\Omega_2(G)| = 8$ so $G \cong M_{2^n}$ or a group from Lemma 42.1 (c). If $|\Omega_2(H)| = 2^4$, then it is possible that $\Omega_2(G) \cong Q_8 \times C_2$, i.e., $G$ is nonmetacyclic. It is interesting to find the 2-groups $G$ such that the condition $(*)$ is fulfilled for $k = 1, 2, 3$, where $H$ is a metacyclic 2-group.

**Exercise 1.** Let $H$ be a minimal nonmetacyclic 2-group (see Theorem 66.1) and 2-groups $G$ and $H$ satisfy $(*)$ for all $k$. Describe all possible structures of $G$.

**Exercise 2.** Let $H \cong \Sigma_{p^2} \in \mathrm{Syl}_p(S_{p^2})$, $p > 3$.
(a) Classify the $p$-groups $G$ satisfying $s_p(G) = s_p(H)$ and $s_{p-1}(G) = s_{p-1}(H)$.
(b) Classify the $p$-groups $G$ satisfying $(*)$ for all $k$. Is it true that $G \cong H$?

**Hint.** (a) We have $d(G) = d(H) = 2$. There is in $\Gamma_1(H)$ a subgroup $E \cong E_{p^p}$, and all other maximal subgroups of $H$ are of maximal class (Fitting's lemma), so two-generator. By Hall's enumeration principle, one has: $s_{p-1}(H) = \frac{p^p-1}{p-1} + (p+1)p - p = 1 + p + 2p^2 + p^3 + \cdots + p^{p-1}$. Hence, if $p = 3$, then $s_2(H) = 1 + 3 + 3 \cdot 3^2$.

**Theorem 179.2.** *Let $H$ be a metacyclic $p$-group of order $p^n$. If $G$ is a $p$-group satisfying $(*)$ for $k \in \{n-1, n-2\}$, then one and only one of the following holds:*
(a) *$G$ is metacyclic,*
(b) *$G$ is a minimal nonmetacyclic group of order $3^4$ (see Theorem 69.1),*
(c) *$p > 2$, $G$ is neither metacyclic nor minimal nonmetacyclic, $G/\mho_1(G) \cong S(p^3)$, $K_3(G) = \mho_1(G)$, and all maximal subgroups of $G$ are two-generator.*

We need the following lemma:

**Lemma 179.3.** *Suppose that $H$ is a noncyclic metacyclic $p$-group of order $p^n > p^3$.*
(a) *If $H$ is a 2-group of maximal class, then $s_{n-2}(H) = 5$.*
(b) *If $H$ has a cyclic subgroup of index $p$, then $s_{n-2}(H) = p + 1$, unless $H$ is as in (a).*
(c) *If $H$ has no cyclic subgroup of index $p$, then $s_{n-2}(G) = 1 + p + p^2$.*

To prove this, use Hall's enumeration principle (see § 5).

*Proof of Theorem 179.2.* If $n \leq 3$, then $G$ is metacyclic. Therefore, in what follows we assume that $n > 3$. We also may assume that $H$ is noncyclic.

Let $\Gamma_1(X)$ be the set of maximal subgroups of a $p$-group $X$. Instead of $\Gamma_1(G)$ we write, as usually in this book, $\Gamma_1$. Then

$$|\Gamma_1| = s_{n-1}(G) = s_{n-1}(H) = |\Gamma_1(H)| = p + 1, \quad \text{i.e., } d(G) = 2\,.$$

(i) Suppose that $H$ possesses a cyclic subgroup of index $p$.

(i1) Let $H$ be a 2-group of maximal class. Then $5 = \mathrm{s}_{n-2}(H) = \mathrm{s}_{n-2}(G)$ so that $G$ is a 2-group of maximal class (Lemma 179.3 and Theorem 5.5), so metacyclic.

(i2) Let $H$ be either abelian or $\cong \mathrm{M}_{p^{n+1}}$. In that case, $p + 1 = \mathrm{s}_{n-2}(H) = \mathrm{s}_{n-2}(G)$. By Hall's enumeration principle,

$$p + 1 = \mathrm{s}_{n-2}(G) = \sum_{K \in \Gamma_1} \mathrm{s}_{n-2}(K) - p\,. \tag{1}$$

It follows from (1) that there is $K \in \Gamma_1$ such that $\mathrm{s}_{n-2}(K) = 1$; then $K$ is cyclic. In that case, by Lemma 179.2, $G$ is either abelian of type $(p^{n-1}, p)$ or $\cong \mathrm{M}_{p^n}$.
Next we assume that $H$ has no cyclic subgroup of index $p$. In that case, by Lemma 179.2, one has:

$$\mathrm{s}_{n-2}(H) = 1 + p + p^2 = \mathrm{s}_{n-2}(G)\,. \tag{2}$$

(ii) Using Hall's enumeration principle, we deduce from (2) that all maximal subgroups of $G$ are two-generator.

(ii1) Let $p = 2$. Then, by the above, $G$ and all maximal subgroups of $G$ are two-generator. In that case, by Corollary 36.6, $G$ is metacyclic, proving the theorem in this case.

(ii2) Next we assume that that $p > 2$. Then $G$ is as in (c), by Corollary 36.6. □

See also Appendix 98 where the number $\mathrm{c}_n(G)$ is computed for all metacyclic 2-groups $G$ (if $p > 2$, then a metacyclic $p$-group is regular, so $\mathrm{c}_n(G) = \frac{|\Omega_n(G) - \Omega_{n-1}(G)|}{(p-1)p^{n-1}}$).

**Exercise 3.** Let $H$ be a minimal nonabelian $p$-group. Classify the $p$-groups $G$ satisfying $(*)$ for all $k$.

**Exercise 4.** Let $H$ be a nonabelian Dedekindian 2-group. Classify the 2-groups $G$ satisfying $(*)$ for all $k$.

**Exercise 5.** Let $\nu_n(X)$ be the number of normal subgroups of order $p^n$ in a $p$-group $X$. Let $H$ be a minimal nonabelian $p$-group. Classify the $p$-groups $G$ such that $\nu_k(G) = \nu_k(H)$ for all $k$.

If $\nu_k(G) = \nu_k(H)$ for all $k$, where $G$ is a $p$-group and $H$ is a $p$-group of maximal class, then $G$ is also of maximal class. This follows from Lemma 9.1.

**Exercise 6.** Let $H = D \times A$, where $D$ is nonabelian of order $p^3$ and $A$ is an elementary abelian $p$-group. Classify the $p$-groups $G$ satisfying $(*)$ for all $k$.

**Exercise 7.** Classify the 2-groups $G$ such that $G$ and $H \cong \mathrm{Q}_{2^m} \times \mathrm{Q}_{2^n}$ satisfy $(*)$ for all $k$.

**Exercise 8.** Classify the $p$-groups $G$ such that $G$ and an extraspecial $p$-group $H$ satisfy the condition (a) $(*)$ for all $k$, (b) $\nu_k(G) = \nu_k(H)$ for all $k$.

**Exercise 9.** Let $H$ be a 3-group of maximal class. Classify the 3-groups $G$ such that $G$ satisfies $(*)$ for all $k$.

**Exercise 10.** Classify the $p$-groups $G$ such that $G$ and the group $H$, a Sylow $p$-subgroup of the holomorph of a cyclic $p$-group, satisfy $(*)$ for all $k$.

**Exercise 11.** Classify the $p$-groups $G$ such that $G$ and $H$ satisfy $(*)$ for all $k$, where (a) $H = C_{p^m} \operatorname{wr} C_p$, (b) $H = C_p \operatorname{wr} C_{p^m}$.

**Exercise 12.** Let $H \cong \Sigma_9 \in \mathrm{Syl}_3(S_9)$ and $G$ be a nonmetacyclic minimal nonabelian group of order $3^4$. Is it true that $G$ and $H$ satisfy $(*)$ for $k = 2, 3$?

**Exercise 13.** Classify the $p$-groups $G$ such that $G$ and $H = \langle a, b \mid o(a) = o(b) = p^2,\ \ a^b = a^{1+p}\}$ satisfy $\nu_k(G) = \nu_k(H)$ for all $k$.

**Hint.** Consider the group $G = \langle a, b \mid o(a) = p^2,\ \ o(b) = p,\ \ c = [a, b],\ \ c^p = [a, c] = [b, c] = 1\rangle$.

**Exercise 14.** Classify the 2-groups $G$ such that $G$ and a minimal nonmetacyclic group $H$ of order $2^5$ satisfy (a) $(*)$ for all $k$, (b) $\nu_k(G) = \nu_k(H)$ for all $k$.

**Exercise 15.** Let $H$ be a minimal nonmetacyclic group of order $2^4$. Classify the 2-groups $G$ satisfying (a) $(*)$ for all $k$, (b) $\nu_k(G) = \nu_k(H)$ for all $k$.

**Exercise 16.** Let a central product $H = \mathcal{H}_{2,2} * C_{2^n}$, $n > 1$, be of order $2^{n+3}$. Classify the 2-groups $G$ satisfying (a) $(*)$ for all $k$, (b) $\nu_k(G) = \nu_k(H)$ for all $k$.

**Exercise 17.** Let a central product $H = \mathcal{H}_{2,2} * Q_8$ ($H = \mathcal{H}_{2,2} * D_8$), $n > 1$, be of order $2^6$. Classify the 2-groups $G$ satisfying (a) $(*)$ for all $k$, (b) $\nu_k(G) = \nu_k(H)$ for all $k$.

**Exercise 18.** Let $H = \mathcal{H}_{2,2} * \mathcal{H}_{2,2}$ be a central product of order $2^6$. Classify the 2-groups $G$ satisfying (a) $(*)$ for all $k$, (b) $\nu_k(G) = \nu_k(H)$ for all $k$.

**Exercise 19.** Study the $p$-groups $G$ of order $p^n$ satisfying $s_{n-2}(G) = 1 + p + p^2$.

**Exercise 20.** Let $H$ be a $p$-group of maximal class and order $p^n > p^{p+1}$. Study the $p$-groups $G$ such that $G$ and $H$ satisfy $s_{n-2}(G) = s_{n-2}(H)$.

**Exercise 21.** Try to prove Theorem 179.1 using the classification of minimal nonmetacyclic $p$-groups.

**Exercise 22.** In Exercises 1-21, where this is possible, replace condition $(*)$ with:

$$(**) \qquad c_k(G) = c_k(H) \quad \text{for all } k.$$

**Problem 1.** Let $H = D_1 \times \cdots \times D_n$, where all $D_i$ are nonabelian of order $p^3$. Classify the $p$-groups $G$ satisfying $(*)$ for all $k$.

**Problem 2.** Let $H$ be a minimal nonabelian $p$-group. Classify the $p$-groups $G$ satisfying $(*)$ for $k = 1, 2$.

**Problem 3.** Study the $p$-groups $G$ such that $v_k(G) = v_k(H)$ for all $k$, where $H$ is special.

**Problem 4.** Let $H$ be a group of exponent $p$ and let a $p$-group $G$ satisfy $(*)$ for all $k > 1$. Study the structure of $G$.

**Problem 5.** Classify the $p$-groups $G$ satisfying $(*)$ for all $k$, where $H$ is an $\mathcal{A}_2$-group.

**Problem 6.** Let $H \cong \Sigma_{p^n} \in \mathrm{Syl}_p(\mathrm{S}_{p^n})$. Study the $p$-groups $G$ satisfying $(*)$ for all $k$. The same problem for $H \cong \mathrm{UT}(n, p) \in \mathrm{Syl}_p(\mathrm{GL}(n, p))$.

**Problem 7.** Classify the regular $p$-groups $G$ such that $G$ and $H$ satisfy $(*)$ for all $k > 1$, where $H$ is irregular of order $p^{p+1}$.

# § 180 *p*-groups all of whose noncyclic abelian subgroups are normal

The purpose of this section is to classify $p$-groups all of whose noncyclic abelian subgroups are normal (problem 2315). Since $p$-groups all of whose nonnormal subgroups are cyclic have been classified in § 16, we may assume that $G$ possesses noncyclic nonnormal subgroups. In fact, we shall prove the following result:

**Theorem 180.A.** *Let $G$ be a $p$-group all of whose noncyclic abelian subgroups are normal and assume that $G$ possesses noncyclic nonnormal subgroups. Then $G$ is a 2-group and we have the following possibilities:*
(a) *$G$ is a generalized quaternion group of order $\geq 2^5$,*
(b) *$\Omega_1(G) = G' \cong E_4$ and $G$ possesses a maximal nonnormal subgroup that is isomorphic to $Q_8$,*
(c) $G \cong B_{2^{m+2}} = \langle b, c \mid b^{2^m} = c^8 = 1, \quad c^4 = b^{2^{m-1}}, \quad b^c = b^{-1}\rangle, m \geq 4.$
*Conversely, all the above groups satisfy the assumptions of the theorem.*

This theorem will be proved with the series of Propositions 180.1–180.4.

**Proposition 180.1.** *Let $G$ be a title $p$-group that possesses a noncyclic nonnormal subgroup $H$. Then $H$ is either generalized quaternion or $A = C_G(\Omega_1(H))$ is an abelian maximal subgroup in $G$ of type $(p^n, p)$, $n \geq 3$, and $H$ is isomorphic either to $M_{p^m}$, $m \geq 3$ (if $p = 2$, then $m \geq 4$) or to $B_{2^{m+2}}$, $m \geq 3$, where*

$$B_{2^{m+2}} = \langle b, c \mid b^{2^m} = c^8 = 1, \ c^4 = b^{2^{m-1}}, \ b^c = b^{-1}\rangle .$$

*Proof.* If $H$ has no normal abelian subgroup of type $(p, p)$, then $H$ is a 2-group of maximal class. It follows that $H$ is in fact generalized quaternion.

Assume that $H$ possesses a normal abelian subgroup $U$ of type $(p, p)$. Then $U \trianglelefteq G$, and so $|G : C_G(U)| \leq p$. If $g \in G_1 = C_G(U)$, then $\langle g, U\rangle \trianglelefteq G$. Hence each cyclic subgroup of $G_1/U$ is normal in $G/U$, and so each subgroup of $G_1/U$ is normal in $G/U$. In particular, $G_1/U$ is Dedekindian. Setting $H_1 = C_H(U)$, we get $|H : H_1| = p$, $H_1 \trianglelefteq G$, $G = HG_1$ and $H \cap G_1 = H_1$.

Let $g_1 \in G_1 - U$ and assume that $o(g_1) = p$. Then for each $1 \neq u \in U$, $\langle u, g_1\rangle \trianglelefteq G$, and so $\langle u, g_1\rangle \cap U = \langle u\rangle \leq Z(G)$ and $U \leq Z(G)$ (giving $H = H_1 \trianglelefteq G$), a contradiction. We have proved that $\Omega_1(G_1) = U$ and we set $\langle z\rangle = U \cap Z(G) \cong C_p$. Let $x \in G_1 - U$ be such that $1 \neq x^p \in U$. Then $\langle x, U\rangle \trianglelefteq G$ implies $\langle x^p\rangle \trianglelefteq G$, and so $\langle x^p\rangle = \langle z\rangle$. Hence the socle of each cyclic subgroup in $G_1$ of composite order is equal $\langle z\rangle$. Note that $|G : H| \geq p^2$, and so $|G_1 : H_1| \geq p^2$.

We prove that $G_1/U$ is abelian. Indeed, suppose that $p = 2$ and $G_1/U$ is Hamiltonian. Let $Q/U$ be an ordinary quaternion subgroup in $G_1/U$. Let $Q_1/U$ and $Q_2/U$ be two distinct cyclic subgroups of order 4 in $Q/U$ so that $Q_1$ and $Q_2$ are two distinct abelian maximal subgroups in $Q$. This implies that $|Q'| = 2$. But $Q' \nleq U$, and so $Q'$ is a subgroup of order 2 in $G_1$ that is not contained in $U$, a contradiction. We have proved that $G_1' \le U$, and so $G_1$ is of class $\le 2$ with $\mho_1(G_1) \le Z(G_1)$.

Assume, by way of contradiction, that $G_1/U$ is noncyclic. Let $E/U \cong E_{p^2}$, where $E \le G_1$ and either $p > 2$ or $p = 2$ and $E$ contains an element $x$ of order 4 with $x \in Z(G_1)$. Let $x, y \in E - U$ be such that $\langle x, y\rangle$ covers $E/U$, $x^p = z$, $y^p = z^{-1}$ and in the case $p = 2$, $x \in Z(G_1)$. Then $xy \in E - U$ and

$$(xy)^p = x^p y^p [y, x]^{\binom{p}{2}} = zz^{-1} = 1\,,$$

a contradiction. Hence we must have in this case $p = 2$ and $G_1$ does not posses central elements of order 4. Since $\mho_1(G_1) \le Z(G_1)$, we have $\exp(G_1/U) = 2$. Hence $G_1/\langle z\rangle$ is elementary abelian, and so $G_1' \le \langle z\rangle$. Let $E/U \cong E_4$ with $E \le G_1$ and let $x, y \in E - U$ be such that $\langle x, y\rangle$ covers $E/U$. Then we have $x^2 = y^2 = z$ and $[x, y] = z$ so that $\langle x, y\rangle = Q \cong Q_4$, $E = Q \times \langle u\rangle$ with some $u \in U - \langle z\rangle$ and $G_1' = Q' = \langle z\rangle$. We get $G_1 = Q * C_{G_1}(Q)$, where $Q \cap C_{G_1}(Q) = \langle z\rangle$ and $U = \langle u, z\rangle \le C_{G_1}(Q)$. If $U < C_{G_1}(Q)$, then there is an element $v$ of order 4 in $C_{G_1}(Q)$. But then $vx \in G_1 - U$ is an involution, a contradiction. Hence $U = C_{G_1}(Q)$, and so $G_1 = E = Q \times \langle u\rangle$. We have $|G_1| = 2^4$, and so the fact that $|G_1/H_1| \ge 4$ gives $U = H_1$, and therefore $H \cong D_8$. But $D_8$ possesses two distinct 4-subgroups (which are both normal in $G$), and so $H \trianglelefteq G$, a contradiction.

We have proved that $G_1/U$ is cyclic (of order $\ge p^2$), and so $G_1$ is abelian of type $(p^n, p)$, $n \ge 3$. We set $G_1 = \langle a\rangle \times \langle u\rangle$, where $o(a) = p^n$, $a^{p^{n-1}} = z \in Z(G)$, $o(u) = p$, and $C_G(u) = G_1$. If $y \in H - H_1$ and $o(y) = p$, then $E_{p^2} \cong \langle y, z\rangle \trianglelefteq G$, and so $H \trianglelefteq G$, a contradiction. Hence we have $\Omega_1(H) = U = \langle u, z\rangle$.

Suppose that $H/U$ is cyclic of order $\ge p$, where in the case $p = 2$, $H/U$ is cyclic of order $\ge 4$ because $H \cong D_8$ is not possible. Let $h \in H - H_1$ so that $\langle h\rangle$ covers $H/U$, $\Omega_1(\langle h\rangle) = \langle z\rangle$, $|H : \langle h\rangle| = p$, $\langle [h, u]\rangle = \langle z\rangle$, and so $H \cong M_{p^m}$, $m \ge 3$, where in the case $p = 2$, $m \ge 4$ and $n \ge 4$.

Assume that $H/U$ is noncyclic and $p > 2$. In particular, $|H_1/U| \ge p$ so that $n \ge 4$. Then $\langle a^p\rangle = \mho_1(G_1) \trianglelefteq G$, and so each $x \in G - G_1$ induces an automorphism of order $\le p$ on $\langle a^p\rangle$. Hence there is an element $h_1 \in H_1 \cap \langle a^p\rangle$ of order $p^2$ such that $h_1 \in Z(G)$ and $h_1^p = z$. Since $H_1/U$ is cyclic and $H/U$ is noncyclic with $p > 2$, it follows that there is $h \in H - H_1$ such that $1 \ne h^p \in U$ and $h^p = z^{-1}$. But then $(hh_1)^p = 1$ and $hh_1 \in H - H_1$, a contradiction.

Suppose that $H/U$ is noncyclic and $p = 2$. Again, $|H_1/U| \ge 2$. Assume that there is $y \in H - H_1$ and $o(y) = 4$. Then $y^2 = z$ and $\langle y, u\rangle \cong D_8$ so that $\langle y, u\rangle \trianglelefteq G$, which forces $H \trianglelefteq G$, a contradiction. Hence $\Omega_2(H) = \Omega_2(H_1)$ is abelian of type $(4, 2)$. If $|H_1/U| = 2$, then $H_1 \cong C_4 \times C_2$ and for an element $y \in H - H_1$ of order 8, $\langle y\rangle$ would cover $H/U$, contrary to our assumption that $H/U$ is noncyclic. Thus, $|H_1 : U| \ge 4$, and so $|H| \ge 2^5$. Then Lemma 42.1 implies that

$$H \cong B_{2^{m+2}} = \langle b, c \mid b^{2^m} = c^8 = 1\,,\ c^4 = b^{2^{m-1}}\,,\ b^c = b^{-1}\rangle\,,$$

where $m \ge 3$, and in this case we have $n \ge 5$. □

**Proposition 180.2.** *Let $G$ be a title $p$-group and let $H$ be any noncyclic maximal nonnormal subgroup in $G$. Set $K = N_G(H)$ so that $H < K < G$ and $K \trianglelefteq G$. Let $L/H$ be a unique subgroup of order $p$ in $K/H$. Then $\{1\} \neq K/H$ is either cyclic or $p = 2$ and $K/H \cong Q_8$, $G/L$ is Dedekindian and $\Omega_1(K) = \Omega_1(L)$. We fix the notation introduced in this proposition in the rest of the section.*

*Proof.* Since $L/H$ is a unique subgroup of order $p$ in $K/H$, it follows that $K/H$ is either cyclic or $p = 2$ and $K/H$ is generalized quaternion. But $K/H$ is Dedekindian, and so in the second case $K/H \cong Q_8$. This also implies that $\Omega_1(K) = \Omega_1(L)$. Finally, each subgroup $X$ of $G$ containing $H$ properly is normal in $G$, and therefore $G/L$ is Dedekindian. □

**Proposition 180.3.** *Let $G$ be a title $p$-group that is not generalized quaternion. Assume that a noncyclic maximal nonnormal subgroup $H$ in $G$ is generalized quaternion. Then $H \cong Q_8$, $L \cong Q_8 \times C_2$, $\Omega_1(G) = Z(L) = G' \cong E_4$. Conversely, in all such groups any noncyclic abelian subgroup is normal in $G$.*

*Proof.* Assume that $H \cong Q_{2^m}$, $m \geq 3$, and set $\langle z \rangle = Z(H)$ so that $\langle z \rangle \leq Z(K)$.

(i) There are involutions in $K - H$. Indeed, if there are no involutions in $K - H$, then $K \cong Q_{2^{m+1}}$ and $K = L$ with $\langle z \rangle = Z(K)$ so that $\langle z \rangle \leq Z(G)$. Because $G$ is not generalized quaternion, there is an involution $t \in G - K$. We have $U = \langle z, t \rangle \trianglelefteq G$. Let $\langle v \rangle$ be a nonnormal cyclic subgroup of order 4 in $K$ so that $v^2 = z$ and $U\langle v \rangle \cong D_8$ or $C_4 \times C_2$, which implies that $U\langle v \rangle \trianglelefteq G$. But then $\langle v \rangle = K \cap (U\langle v \rangle) \trianglelefteq G$, a contradiction.

(ii) We have

$$H \cong Q_8 , \quad L = HU \quad \text{with } U \cong E_4 , \ U \trianglelefteq G , \ H \cap U = \langle z \rangle \text{ and } L' = \langle z \rangle \leq Z(G) .$$

Indeed, let $t$ be an involution in $K - H$ so that $t \in L - H$ and $U = \langle t, z \rangle \trianglelefteq G$. Suppose that $m > 3$ and let $\langle v \rangle$ be a nonnormal cyclic subgroup of order 4 in $H$ so that $v^2 = z$ and $U\langle v \rangle \cong D_8$ or $C_4 \times C_2$, which forces that $U\langle v \rangle \trianglelefteq G$. But then $\langle v \rangle = H \cap (U\langle v \rangle) \trianglelefteq K$, a contradiction.

(iii) We have $U = Z(L)$ so that $L \cong Q_8 \times C_2$.
Since $L' = H' = \langle z \rangle$, $L = H * Z$ with $Z = C_L(H) \cong C_4$ or $E_4$ and $H \cap Z = \langle z \rangle$. But if $Z \cong C_4$, then $H$ is a unique ordinary quaternion subgroup in $L$, and so $H \trianglelefteq G$, a contradiction.

(iv) We have $K = H * C_G(H)$, where $U \leq C_G(H)$ and $H \cap C_G(H) = \langle z \rangle$.

Indeed, for any element $v$ of order 4 in $H$ and $u \in U - \langle z \rangle$, we have $\langle v \rangle \times \langle u \rangle \trianglelefteq G$, and so $\langle v \rangle \trianglelefteq K$. Thus, each element in $K$ induces on $H$ an inner automorphism of $H$, and so $K = H * C_K(H)$.

(v) $\{1\} \neq G/K$ is elementary abelian of order $\leq 4$.
There are exactly three maximal subgroups of $L$ that contain $U$ and they are all abelian of type $(4, 2)$. The other four maximal subgroups of $L$ (which do not contain

$U$) are isomorphic to $Q_8$. This gives already that $|G/K| \leq 4$. For any $y \in H - \langle z \rangle$ and $g \in G - K$ we have $y^2 = z$ and $U\langle y \rangle \trianglelefteq G$ so that $y^g = yu$ with some $u \in U$. Hence

$$y^{g^2} = (yu)^g = (yu)u^g = (yu)(uz^{\epsilon}) = yz^{\epsilon}\,, \quad \epsilon = 0, 1\,.$$

Thus, we have obtained $g^2 \in K$.

(vi) There are no involutions in $G - K$.

Indeed, let $t \in G - K$ be an involution so that $E_4 \cong \langle z, t \rangle \trianglelefteq G$. Let $v$ be any element of order 4 in $H$ so that $v^2 = z$ and $\langle v, t \rangle \cong D_8$ or $C_4 \times C_2$. In any case, $\langle v, t \rangle \trianglelefteq G$. But then $\langle v \rangle = K \cap \langle v, t \rangle \trianglelefteq G$, and so $H \trianglelefteq G$, a contradiction.

(vii) If $U \nleq Z(G)$, then $G$ is a group of order $2^5$ with $U = \Omega_1(G) = G' \cong E_4$, and so $G$ is a group from part (b) of Theorem 180.A.

First suppose $K > L$. Then $C_G(H) = C_K(H) > U$, where $C_K(H)/\langle z \rangle$ is cyclic or ordinary quaternion. Hence there is an element $k$ of order 4 in $C_K(H) - U$ such that $k^2 \in U - \langle z \rangle$. In that case we get $U\langle k \rangle = \langle z \rangle \times \langle k \rangle \trianglelefteq G$ and this forces $\langle k^2 \rangle \in Z(G)$, a contradiction.

We have proved that $K = L$. Suppose that $G - K$ contains an element $y$ of order 4 that does not centralize $U$. Since $y^2 \in K$, it follows that $y^2 \in U$, and so $D = U\langle y \rangle \cong D_8$. But then there are involutions in $D - K$, a contradiction.

Suppose that all elements in $G - K$ are of order 8. Then $\Omega_2(G) = K \cong Q_8 \times C_2$ and Theorem 52.1 implies that $G$ is isomorphic to a group of order $2^5$ and class 3, which is defined in part A2 (a) of Theorem 49.1. In particular, $\Omega_1(G) = G' \cong E_4$ and $G$ is a group from part (b) of Theorem 180.A.

It follows that we may assume in the sequel that $G - K$ contains an element $v$ of order 4 such that $1 \neq v^2 \in U$ and (by the above) $v$ centralizes $U$. Set $G_0 = K\langle v \rangle$ so that $|G_0| = 2^5$ and $U = Z(G_0)$. Since $H$ is not normal in $G_0$, there is $w \in H - \langle z \rangle$ such that $[w, v] \notin H$ and $w^2 = z$. On the other hand, $U\langle w \rangle \cong C_4 \times C_2$, and so $U\langle w \rangle \trianglelefteq G$, where $|(U\langle w \rangle) : U| = 2$. Hence we get $[w, v] \in U - \langle z \rangle$, and so $U \leq G_0'$. If $G_0' > U$, then $|G_0 : G_0'| = 4$ and a result of O. Taussky implies that $G_0$ is of maximal class, a contradiction. We have proved that $G_0' = Z(G_0) = U \cong E_4$, and so $\mho_1(G_0) \leq Z(G_0)$. This gives that all elements in $G_0 - K$ are of order 4. Let $x \in G_0 - K$ so that $1 \neq x^2 \in U$. If $x^2 \in U - \langle z \rangle$, then $U\langle x \rangle \cong C_4 \times C_2 \trianglelefteq G$, and so $x^2 \in Z(G)$, a contradiction. It follows that for each $x \in G_0 - K$, $x^2 = z$, and so $\mho_1(G_0) = \langle z \rangle$. But then $G_0/\langle z \rangle$ is elementary abelian, a contradiction.

(viii) We may assume in what follows $U \leq Z(G)$ and then $U \leq G'$.

Indeed, for $g \in G - K$, there is $w \in H - \langle z \rangle$ such that $U\langle w \rangle \trianglelefteq G$ and $[g, w] \in U - \langle z \rangle$ and this implies $U \leq G'$.

(ix) $G/U$ is Dedekindian.

Let $X/U$ be any subgroup in $G/U$. If $x \in X$, then $U\langle x \rangle \trianglelefteq G$ (being abelian of rank 2). Hence $X \trianglelefteq G$ and we are done.

(x) $G/U$ is abelian, and so $U = G' = \Omega_1(G)$. We have obtained groups from part (b) of Theorem 180.A.

Suppose that $G/U$ is Hamiltonian and let $Q/U$ be an ordinary quaternion subgroup in $G/U$. Let $Q_1/U$ and $Q_2/U$ be two distinct cyclic subgroups of order 4 in $Q/U$ so that $Q_1$ and $Q_2$ are two distinct abelian maximal subgroups in $Q$. This implies that $|Q'| = 2$. But $Q' \nleq U$, contrary to $\Omega_1(G) = U$. Our proposition is proved. □

**Proposition 180.4.** *Let $G$ be a title $p$-group that possesses a noncyclic maximal nonnormal subgroup $H$ that is not generalized quaternion. Then $p = 2$ and*

$$G \cong \mathrm{B}_{2^{m+2}} = \langle b, c \mid b^{2^m} = c^8 = 1\,,\; c^4 = b^{2^{m-1}}\,,\; b^c = b^{-1}\rangle\,, \quad m \geq 4\,.$$

*Proof.* By Proposition 180.1, $H$ is isomorphic either to $\mathrm{M}_{p^s}$, $s \geq 3$ (if $p = 2$, then $s \geq 4$) or $\mathrm{B}_{2^{s+2}}$, $s \geq 3$. We set $|G| = p^m$ and since $|G : H| \geq p^2$, it follows $m \geq 5$ (and in the case $p = 2$, $m \geq 6$). We set $H_0 = \Omega_1(H)$ and $A = \mathrm{C}_G(H_0)$ so that Proposition 180.1 gives that $A$ is abelian of type $(p^n, p)$, $n \geq 3$, and $|G : A| = p$. Set $\Omega_1(H') = \langle z\rangle$ so that $\langle z\rangle < H_0$, $\langle z\rangle \leq \mathrm{Z}(H)$, and therefore $\langle z\rangle \leq \mathrm{Z}(G)$. If $p = 2$, then $\mathrm{Z}(H)$ is cyclic of order $\geq 4$ and $\mathrm{Z}(H) \leq \mathrm{Z}(G)$.

A cyclic subgroup of order $p^n$, $n \geq 3$, in $A$ is of index $p^2$ in $G$. Suppose that $G$ has an element $g \in G - A$ of order $p$. Then $\mathrm{E}_{p^2} \cong \langle g, z\rangle \trianglelefteq G$ and so for each $a \in A$, $[g, a] \in A \cap \langle g, z\rangle = \langle z\rangle$. This implies $G' = \langle z\rangle$, and so $H \trianglelefteq G$, a contradiction. We have proved that $\Omega_1(G) = \Omega_1(A) = H_0 \cong \mathrm{E}_{p^2}$.

Assume that $G$ has a cyclic subgroup of index $p$. Because $G$ is nonabelian, $|G| \geq p^5$, and $\Omega_1(G) \cong \mathrm{E}_{p^2}$, we get $G \cong \mathrm{M}_{p^l}$, $l \geq 5$. But then $|G'| = p$ and so $H \trianglelefteq G$, a contradiction. We have proved that $\exp(G) = p^n = p^{m-2}$.

First suppose that $p > 2$. Since $|G| = p^m$, $m \geq 5$, and $\exp(G) = p^{m-2}$, we may use Theorem 74.1. Because $\Omega_1(G) \cong \mathrm{E}_{p^2}$ and $m > 4$, we get that $G$ is metacyclic (case (a) of Theorem 74.1). However, $G$ has an abelian maximal subgroup $A$, and so Proposition 149.1 implies that $G$ is minimal nonabelian. But then $G' = H'$ is of order $p$, and so $H \trianglelefteq G$, a contradiction.

We have proved that $p = 2$. Since $|G| = 2^m$, $m \geq 6$, and $\exp(G) = 2^{m-2}$, we may use Theorem 74.2. Because $\Omega_1(G) \cong \mathrm{E}_4$, $|G| \geq 2^6$ and $\mathrm{C}_G(\Omega_1(G))$ is abelian of type $(2^n, 2)$, $n \geq 3$, it follows that only case (c) of Theorem 74.2 is possible, in which case $G$ is a $\mathrm{U}_2$-group. Such groups are completely determined in § 67.

Not one of the cases (a, b, c) of Theorem 67.1 could occur since $\Omega_1(G) \cong \mathrm{E}_4$ and $\mathrm{C}_G(\Omega_1(G))$ is abelian.

Since $\mathrm{C}_G(\Omega_1(G))$ is abelian and $\mathrm{Z}(H)$ is cyclic of order $\geq 4$ with $\mathrm{Z}(H) \leq \mathrm{Z}(G)$, the cases (a) and (c) of Theorem 67.2 cannot occur. It follows that we must have case (b) of Theorem 67.2, where

$$G \cong \mathrm{B}_{2^{m+2}} = \langle b, c \mid b^{2^m} = c^8 = 1\,,\; c^4 = b^{2^{m-1}} = z\,,\; b^c = b^{-1}\,,\; b^{2^{m-2}}c^2 = u\rangle\,, \quad m \geq 4\,,$$

and here we have

$$\mathrm{Z}(G) = \langle c^2\rangle \cong \mathrm{C}_4\,, \quad G' = \langle b^2\rangle \cong \mathrm{C}_{2^{m-1}}\,, \quad \Omega_1(G) = \langle u, z\rangle \cong \mathrm{E}_4\,,$$
$$\Omega_2(G) = \langle c^2, u\rangle \cong \mathrm{C}_4 \times \mathrm{C}_2\,, \quad \text{and} \quad A = \mathrm{C}_G(\langle c^2, a\rangle) \cong \mathrm{C}_{2^m} \times \mathrm{C}_2$$

is a unique abelian maximal subgroup in $G$. Also, $H = \langle b^{2^{m-2}}, c\rangle$ is a maximal nonnormal subgroup of $G$ and we have $H \cong M_{16}$ if $m = 4$ and $H \cong B_{2^m}$ if $m > 4$.

Conversely, let $X$ be any noncyclic abelian subgroup in $G \cong B_{2^{m+2}}$, $m \geq 4$. Then

$$\Omega_1(X) = \Omega_1(G) = \langle u, z\rangle \quad \text{and so } X \leq C_G(\langle u, z\rangle) = A\,.$$

But $A$ is characteristic in $G$ and each subgroup $X$ in $A$ containing $\Omega_1(A) = \langle u, z\rangle$ is characteristic in $A$, and so $X \trianglelefteq G$. Hence each noncyclic abelian subgroup of $G$ is normal in $G$.

Finally, assume that $G$ is a $U_2$-group of Theorem 67.3. But not one of the cases (a–d) of that theorem can occur since $\Omega_1(G) \cong E_4$, $|G| \geq 2^6$, $C_G(\Omega_1(G))$ is abelian and $Z(H) \leq Z(G)$, where $Z(H)$ is cyclic of order $\geq 4$. Our proposition is proved. □

# § 181 $p$-groups all of whose nonnormal abelian subgroups lie in the center of their normalizers

The purpose of this section is to initiate a study of the $p$-groups satisfying the following:

**Problem** (Berkovich). Study the $p$-groups all of whose nonnormal abelian subgroups are contained in the centers of their normalizers. In other words, if $A < G$ is such an abelian subgroup, then $\mathrm{N}_G(A) = \mathrm{C}_G(A)$.

Our aim in this section is to prove the following result:

**Theorem 181.1.** *Let $G$ be a non-Dedekindian $p$-group all of whose nonnormal abelian subgroups lie in the center of their normalizers. Then $G$ is of class 2.*

*Proof.* Since every maximal abelian subgroup of a nonabelian $p$-group is not contained in the center of its normalizer, it follows that each maximal abelian subgroup of $G$ is normal.

Let $A$ be a maximal abelian subgroup in $G$ so that $A \trianglelefteq G$. Let $X/A \neq \{1\}$ be any cyclic subgroup in $G/A$ and let $x \in X - A$ be such that $X = A\langle x\rangle$; then $X$ is nonabelian. Set $A_1 = \mathrm{C}_A(x)$ so that

$$A_1 \geq \langle x\rangle \cap A \quad \text{and} \quad A > A_1$$

since $A_1$ is abelian and $A$ is nonabelian. The centralizer

$$X_1 = \mathrm{C}_X(x) = A_1\langle x\rangle$$

is a maximal abelian subgroup of the nonabelian subgroup $X$. Let $B$ be a maximal abelian subgroup in $G$ containing $X_1$; then

$$A \cap B = A_1\,, \quad X \cap B = X_1\,, \quad \text{and } B \trianglelefteq G\,.$$

The last relation follows from the result of the first paragraph. Let $a \in A$ and $x_1 \in X_1$. Then

$$[a, x_1] \in A \cap B = A_1 < X_1\,,$$

and so $A$ normalizes $X_1$, and therefore $X_1 \trianglelefteq X = \langle x\rangle A$. In particular, $A$ normalizes $X_1$. If $X_1$ is not normal in $G$, then, by hypothesis, $A$ as a subgroup of $\mathrm{N}_G(X_1)$, centralizes $X_1$, a contradiction since $x \in X_1$ does not centralizes $A$. Thus,

$$X_1 \trianglelefteq G \Rightarrow X = AX_1 \trianglelefteq G\,.$$

It follows that $G/A$ is a Dedekindian group. As $\mathrm{C}_G(a) \geq A$ for each $a \in A$, it follows that $\mathrm{C}_G(a) \trianglelefteq G$.

Let $g \in G$ and let $B$ be a maximal abelian subgroup in $G$ containing $g$. Then $B \trianglelefteq G$ and, by the previous paragraph, $\mathrm{C}_G(g) \trianglelefteq G$. Thus, the centralizer of each element in $G$ is $G$-invariant. By Theorem 27.1, $G$ is either of class 2 or $G$ is a 3-group of class 3.

Suppose that $G$ is a 3-group of class 3. Let $A$ be a maximal abelian subgroup in $G$ so that

$$A \trianglelefteq G \quad \text{and} \quad G/A \text{ is Dedekindian}$$

and in this case, since $p = 3 > 2$, $G/A$ is abelian, by Theorem 1.20. Hence $G' \leq A$, and, obviously, $Z(G) < A$. Thus, $G'Z(G)$ is contained in the intersection, say $D$, of all maximal abelian subgroups of $G$. It follows that $G/D$ is abelian. Note that $G$ is covered by their maximal abelian subgroups so that any element of $D$ is contained in $Z(G)$. Thus, $D = Z(G)$, and hence $cl(G) = 2$. □

**Exercise 1.** If $G$ is a $p$-group of Theorem 181.1, then all nonabelian subgroups of $G$ are normal.

**Solution.** Let $U < S \leq G$, where $S$ is minimal nonabelian and $|S : U| = p$. As $S \leq N_G(U)$, it follows that $U \triangleleft G$. Thus, all maximal subgroups of $S$ are normal in $G$ so that $S \triangleleft G$. As all minimal nonabelian subgroups are normal in $G$, any nonabelian subgroup of $G$, being generated by minimal nonabelian subgroups, is $G$-invariant.

**Exercise 2.** If a $p$-group $G$ is as in Theorem 181.1, then $G/Z(G)$ is a subgroup of a direct product of Dedekindian groups.

**Solution.** Let $A < G$ be maximal abelian, then $A \triangleleft G$. Let $A < X \leq G$. Then $X$ is $G$-invariant, by Exercise 1, since $X$ is nonabelian. It follows that $G/A$ is Dedekindian. As the intersection of all maximal abelian subgroups of $G$ coincides with $Z(G)$, it follows that $G/Z(G)$ is a subgroup of the direct product of Dedekindian groups.

By Theorem 27.1, $G/Z(G)$, where $G$ is as in Theorem 181.1, is abelian so that $cl(G) = 2$. This yields another proof of Theorem 181.1.

**Exercise 3.** If $G$ is a $p$-group of Theorem 181.1, then normalizers of all abelian subgroups of $G$ are $G$-invariant.

**Solution.** Let $A < G$ be abelian. If $N_G(A)$ is abelian, then it is maximal abelian in $G$, and so $G$-invariant, by Theorem 181.1. If $N_G(A)$ is nonabelian, the result follows from Exercise 1.

**Problem 1.** Study the non-Dedekindian $p$-groups all of whose nonnormal cyclic subgroups are contained in centers of their normalizers.

**Problem 2.** Study the non-Dedekindian $p$-groups such that $N_G(H) = HC_G(H)$ for any nonnormal nonabelian $H < G$. (The $p$-groups all of whose nonnormal subgroups are nonabelian are classified in [FA].)

**Problem 3.** Study the non-Dedekindian $p$-groups such that $N_G(H) = HC_G(H)$ for any nonnormal metacyclic $H < G$.

**Problem 4.** Study the non-Dedekindian $p$-groups such that $N_G(H) = HC_G(H)$ for minimal nonabelian $H < G$.

**Problem 5.** Classify the non-Dedekindian $p$-groups all of whose nonnormal abelian subgroups of order $> p^2$ are contained in centers of their normalizers.

**Problem 6.** Classify the non-Dedekindian $p$-groups all of whose maximal abelian subgroups and nonabelian subgroups are normal.

# § 182 $p$-groups with a special maximal cyclic subgroup

The purpose of this section is to generalize Theorem 169.1. This solves problem 2899; we also classify the groups from Problem 3 below.

**Theorem 182.1** (Janko). *Let $G$ be a $p$-group containing a maximal cyclic subgroup $A < G$ such that $\langle A, B\rangle$ is minimal nonabelian for any maximal cyclic subgroup $B$ distinct from $A$. Then $|G : A| = p$ and $G$ is isomorphic to*

$$M_{p^n}\,, \quad n \geq 3 \text{ (if } p = 2, \text{ then } n \geq 4)\,, \quad D_8 \quad \text{or} \quad Q_8\,.$$

**Remark 1.** Below we freely use the following obvious fact: If in the notation of Theorem 182.1, $x \in G - C_G(A)$, the subgroup $\langle x, A\rangle$ is minimal nonabelian.

*Proof of Theorem 182.1.* Let $B$ be any maximal cyclic subgroup distinct from $A$. If $A$ is not normal in $\langle A, B\rangle$, then there is $x \in \langle A, B\rangle$ such that $A \neq A^x$. Here $A^x$ is also a maximal cyclic subgroup in $G$, and so $\langle A, A^x\rangle$ is minimal nonabelian, by hypothesis. But $\langle A, A^x\rangle < \langle A, B\rangle$, a contradiction since $\langle A, B\rangle$ is minimal nonabelian. Thus, each maximal cyclic subgroup of $G$ normalizes $A$, and so $A \lhd G$.

Set $Z = \Omega_1(A)$ so that for each $g \in G$ either $[A, g] = \{1\}$ or else $\langle A, g\rangle$ is metacyclic minimal nonabelian and then $[A, g] = Z$. Hence $[A, G] = Z$ since $A \nleq Z(G)$, by hypothesis. Write $A = \langle a\rangle$ and $M = C_G(A)$. Let us consider the map

$$g \mapsto [a, g] \quad (g \in G)$$

from $G$ onto $Z$. Since $Z \leq Z(G)$, this map is a homomorphism. Indeed, for any $g_1, g_2 \in G$,

$$[a, g_1g_2] = [a, g_2][a, g_1]^{g_2} = [a, g_2][a, g_1] = [a, g_1][a, g_2]\,.$$

It follows that $G/M \cong Z$ because $M$ is the kernel of this homomorphism, and so $|G : M| = p$.

Suppose that $M > A$ (in that case, $M$ is noncyclic) and let $A \leq H < M$ be such that $|M : H| = p$. Then each $x \in M - H$ centralizes $A$, and so $\langle x\rangle$ is not a maximal cyclic subgroup in $G$, and therefore there is $y \in G$ such that $y^p = x$. Hence each element in $M - H$ is $p$-th power of an element in $G$ and all elements in $M - H$ generate $M$. It follows that $\mho_1(G) = \langle M - H\rangle = M$ has index $p$ in $G$, and so $G$ is cyclic, a contradiction.

Thus, $M = A$ is a cyclic subgroup of index $p$ in $G$, and so $G$ is minimal nonabelian since it is generated by $A$ and another maximal cyclic subgroup distinct from $A$ (see the Remark). Therefore, by Theorem 1.2, $G$ is isomorphic to one of the following groups:

$$M_{p^n}\,, \quad n \geq 3 \text{ (if } p = 2, \text{ then } n \geq 4)\,, \quad D_8 \quad \text{or} \quad Q_8$$

(note that if a 2-group of maximal class is minimal nonabelian, its order is equal to 8). The proof is complete. □

Obviously, any group from the conclusion of Theorem 182.1 satisfies its hypothesis.

It follows from Theorem 182.1 that if a nonabelian $p$-group $G$ is generated by any two distinct maximal cyclic subgroups, then $G \cong Q_8$.

**Theorem 182.2** (Janko). *Let $G$ be a nonabelian $p$-group containing a maximal cyclic subgroup $A$ such that, whenever $S \le G$ is minimal nonabelian, then $\langle A, S\rangle = G$. It follows that the normal closure $H = A^G$ is abelian and $G$ has an abelian maximal subgroup $M$ containing $H$. We have the following possibilities:*

(a) *if $A \nleq Z(G)$, then $G/H$ is cyclic,*

(b) *if $A \le Z(G)$, then $G = S * A$ for some minimal nonabelian subgroup $S$.*

*Conversely, all the above groups satisfy the assumption of the theorem.*

*Proof.* First we show that the normal closure $H = A^G$ is abelian. Indeed, assume that there is $g \in G$ such that $\langle A, A^g\rangle$ is nonabelian. Let $X$ be a minimal nonabelian subgroup in $\langle A, A^g\rangle$. Then

$$\langle A, X\rangle \le \langle A, A^g\rangle < G ,$$

a contradiction.

Let $M$ be a maximal normal abelian subgroup in $G$ that contains $H = A^G$. Let $x \in G - M$ be such that $x^p \in M$. By Lemma 57.1, there is $y \in M$ so that $\langle x, y\rangle$ is minimal nonabelian. By our assumption,

$$\langle x, y, A\rangle = G \quad \text{and so } G = \langle M, x\rangle .$$

We have proved that $|G : M| = p$, and so $M$ is an abelian maximal subgroup of $G$ containing $H$.

(i) First assume $A \nleq Z(G)$. Let $g \in G - M$ so that $g^p \in Z(G)$ and $g$ does not centralize $H$ because $A \le H \le M$. Set $H_0 = C_H(g)$, where $H_0 < H$ and $H_0 \le Z(G)$. Let $H_1$ be a $G$-invariant subgroup with

$$H_0 < H_1 \le H , \quad |H_1 : H_0| = p \quad \text{and} \quad \langle g\rangle \cap H \le H_0 .$$

Consider the subgroup $G_1 = H_1\langle g\rangle$ and note that

$$Z(G_1) = H_0\langle g^p\rangle \quad \text{so that } |G_1 : Z(G_1)| = p^2 .$$

This implies that $\mathrm{cl}(G_1) = 2$ and with some $h \in H_1 - H_0$,

$$[g, h] \ne 1 , \quad \text{and} \quad [g, h]^p = [g^p, h] = 1 .$$

By Lemma 65.2 (a), $\langle g, h\rangle$ is minimal nonabelian. By our assumption, $\langle A, g, h\rangle = G$, and therefore $G = H\langle g\rangle$ so that $G/H$ is cyclic. We have obtained the groups stated in part (a) of our theorem.

Conversely, let $X$ be any minimal nonabelian subgroup in a group of part (a) of our theorem. Then $X \nleq M$ and let $y \in X - M$ so that $\langle y\rangle$ covers $G/H$. But $H = A^G = A^X$, and so $H \le \langle A, X\rangle$, which implies $\langle A, X\rangle = G$.

(ii) Suppose $A \leq Z(G)$. Let $S$ be a minimal nonabelian subgroup in $G$ so that, by our assumption, $G = A * S$. Set $S_0 = Z(S) = \Phi(S)$ and then

$$\Phi(G) = S_0\Phi(A) \quad \text{and} \quad Z(G) = S_0 * A\,.$$

We have obtained the groups stated in part (b) of our theorem.

Conversely, let $G = S * A$ for some minimal nonabelian subgroup $S$ in $G$. Let $X$ be any other minimal nonabelian subgroup in $G$. Then $X$ covers $G/Z(G)$ and $X * A$ covers $G/\Phi(G)$. But then $X * A = G$, and so $G$ satisfies the assumption of our theorem. □

**Problem 1.** Classify the nonabelian $p$-groups $G$ such that, whenever $A, B < G$ are maximal cyclic, then either $[A, B] = \{1\}$ or $\langle A, B\rangle$ is minimal nonabelian. Consider in detail the case when $\exp(G) = p$.

**Problem 2.** Study the nonabelian $p$-groups $G$ such that, whenever $x, y \in G$ generate a minimal nonabelian subgroup, then $\langle x\rangle$ and $\langle y\rangle$ are maximal cyclic subgroups of $G$.

**Problem 3.** Classify the nonabelian $p$-groups $G$ containing a maximal cyclic subgroup $A$ such that, whenever $S < G$ is minimal nonabelian, then $\langle A, S\rangle = G$.

**Problem 4.** Classify the nonmetacyclic $p$-groups $G$ containing a maximal metacyclic subgroup $A$ such that $G = \langle A, B\rangle$, where $B < G$ is cyclic.

**Problem 5.** Study the $p$-groups $G$ that are generated by any two minimal nonabelian subgroups (maximal abelian subgroups) not contained in the Frattini subgroup $\Phi(G)$ (the derived subgroup $G'$).

**Exercise 1.** Let $A$ and $G$ be as in Problem 1 and $A \nleq Z(G)$. Prove that $A \triangleleft G$. Moreover, $G$ is one of the groups in conclusion of Theorem 182.1.

**Hint.** See the proof of Theorem 182.1.

# § 183 *p*-groups generated by any two distinct maximal abelian subgroups

The purpose of this section is to prove the following result which solves problem 2929:

**Theorem 183.1** (Janko). *Let $G$ be a nonabelian $p$-group that is generated by any two distinct maximal abelian subgroups. Then for any two distinct maximal abelian subgroups $A$ and $B$ in $G$ one has*

$$G = AB\,, \quad A \cap B = Z(G)\,, \quad \mathrm{cl}(G) = 2\,,$$
$$\mho_1(G) \le Z(G) \quad \text{and} \quad G' \text{ is elementary abelian.}$$

*If $g \in G - Z(G)$, then $C_G(g)$ is maximal abelian. Moreover, maximal abelian subgroups of $G$ partition $G$ mod $Z(G)$. If $A < G$ is maximal abelian, then $|G : A| = |A : Z(G)|$.*

*Proof.* Suppose that $G$ possesses a nonnormal maximal abelian subgroup $X$. Set $M = N_G(X)$ so that $M < G$. Let $M < H \le G$ be such that $|H : M| = p$ and let $h \in H - M$. Then $X^h \ne X$ and $X^h \le M$. On the other hand, $X^h$ is also a maximal abelian subgroup in $G$ but $\langle X, X^h\rangle \le M < G$, a contradiction. Thus, each maximal abelian subgroup is normal in $G$.

Let $A \ne B$ be any two distinct maximal abelian subgroups in $G$; then

$$\text{(1)} \qquad A \lhd G\,, \quad B \lhd G\,, \quad G = AB\,, \quad A \cap B = Z(G)\,.$$

Since $G/Z(G) = (A/Z(G)) \times (B/Z(G))$ is abelian, it follows:

$$G' \le A \cap B = Z(G) \Rightarrow \mathrm{cl}(G) = 2\,.$$

By Theorem 91.2, the centralizer of each noncentral element is a maximal abelian subgroup of $G$. Hence maximal abelian subgroups of $G$ form a partition of $G$ modulo $Z(G)$. Also, all maximal abelian subgroups of $G$ have equal order, by the product formula, and so $|G : A| = |A : Z(G)|$ for any maximal abelian $A < G$. Indeed, if $B < G$ is another maximal abelian, then, as $|B : Z(G)| = |G : A|$, it follows that all maximal abelian subgroups of $G$ have equal order and

$$\text{(2)} \qquad |G| = |AB| = \frac{|A|\,|B|}{|Z(G)|} = |A| \cdot |A : Z(G)|\,,$$

and the desired equality is established.

Let $g \in G - Z(G)$. Let us prove that $A = C_G(g)$ is a maximal abelian subgroup in $G$. Indeed, assume that $A$ is nonabelian. Then it has two distinct maximal abelian subgroups, say $U_1$ and $V_1$. Let $U_1 \le U < G$ and $V_1 \le V < G$, where $U$, $V$ are maximal abelian. As $\langle U_1, V_1\rangle$ is a nonabelian subgroup of $\langle U, V\rangle$, it follows that $U \ne V$. In that case, $x \in (U \cap V) - Z(G)$, contrary to the result of the previous paragraph. Thus, $A < G$ is maximal abelian hence the centralizer of any noncentral element of $G$ is abelian. Let

$B \neq A$ be another maximal abelian subgroup in $G$; then (1) holds. Let $Z(G) < B_1 \leq B$ be such that $|B_1 : Z(G)| = p$; then $B_1 \triangleleft G$ since $cl(G) = 2$, and let $b \in B_1 - Z(G)$. Then $g \notin A = C_G(b)$, hence

$$1 \neq [b, g] \in Z(G) \text{ and } [b, g]^p = [b^p, g] = 1 \quad \text{since } b^p \in Z(G) \text{ and } cl(G) = 2\,.$$

It follows from Lemma 65.2 (a) that the subgroup $\langle g, b\rangle$ is minimal nonabelian and this implies

$$[g^p, b] = 1 \quad \text{and so } C_G(g^p) \geq \langle A, b\rangle$$

is nonabelian, and therefore $g^p \leq Z(G)$. Thus, $\mho_1(G) \leq Z(G)$, and so for any $x, y \in G$,

$$[x, y]^p = [x^p, y] = 1\,,$$

which implies that $G'$ is elementary abelian, completing the proof. □

Extraspecial $p$-groups and minimal nonabelian $p$-groups satisfy the hypothesis of Theorem 183.1.

**Exercise 1.** The following conditions for a nonabelian $p$-group $G$ are equivalent:
(a) $A \cap B = Z(G)$ and $|G : A| = |A : Z(G)|$ for any two distinct maximal abelian subgroups $A, B < G$,
(b) any two distinct maximal abelian subgroups generate $G$.

**Hint.** See the proof of Theorem 183.1.

It follows from Exercise 1 that all $p$-groups for which conclusion of Theorem 183.1 is fulfilled satisfy its hypothesis.

**Exercise 2.** If any two distinct minimal nonabelian subgroups generate a $p$-group $G$, then $G$ is an $\mathcal{A}_2$-group. In particular, all minimal nonabelian subgroups of $G$ have equal order.

**Solution.** Let $U \leq G$ be an $\mathcal{A}_2$-subgroup of minimal order. By Remark 76.1, there are in $U$ two distinct minimal nonabelian subgroups, say $A$ and $B$. Then $G = AB \leq U$ so that $G = U$ is an $\mathcal{A}_2$-group. Next, $A$ and $B$, as maximal subgroups of $G$, have equal order.

**Exercise 3.** If $G$ is as in Theorem 183.1 and $S \leq G$ is minimal nonabelian, then $Z(S) \leq Z(G)$.

**Solution.** If $g \in Z(S)$, then $C_G(g)$ is nonabelian. It follows from Theorem 183.1 that $C_G(g) = G$.

**Problem 1.** Classify the nonabelian $p$-groups all of whose maximal abelian subgroups are normal and have the same order.

**Problem 2.** Study the nonabelian $p$-groups all of whose maximal abelian subgroups are isomorphic.

**Problem 3.** Classify the nonabelian $p$-groups $G$ containing a minimal nonabelian subgroup $S$ such that $\langle S, A \rangle = G$ for any maximal abelian $A < G$ such that $A \nleq S$.

**Problem 4.** Study the nonmetacyclic $p$-groups generated by any two different maximal metacyclic subgroups.

**Problem 5.** Study the $p$-groups $G$ such that for any nonincident minimal nonabelian $A < G$ and maximal metacyclic $M < G$ one has $\langle A, M \rangle = G$.

**Problem 6.** Classify the special $p$-groups satisfying the hypothesis of Theorem 183.1.

# § 184 $p$-groups in which the intersection of any two distinct conjugate subgroups is cyclic or generalized quaternion

The purpose of this section is to classify the title groups (problem 2584). More precisely, we shall reduce this problem to the classifications from §§ 16, 174, 175 and 176. To facilitate the proof of the main result (Theorem 184.3), we shall first present two simple auxiliary results.

**Lemma 184.1** (see also Lemma 1.22). *Let $G$ be an arbitrary non-Dedekindian $p$-group. Let $H$ be a maximal nonnormal subgroup in $G$. Set $K = N_G(H)$ so that $H < K < G$, $K \trianglelefteq G$ and then $K/H$ is cyclic or $p = 2$ and $K/H$ is ordinary quaternion. There is a unique subgroup $L$ in $G$ such that $L > H$ and $|L : H| = p$ so that $L$ is normal in $G$, $G/L$ is Dedekindian and $\Omega_1(K) \leq L$.*

*Proof.* Since $K/H$ has exactly one subgroup $L/H$ of order $p$, it follows that $K/H$ is either cyclic or generalized quaternion. On the other hand, $K/H$ is Dedekindian, and so $K/H$ is either cyclic or $K/H \cong Q_8$. This also gives $\Omega_1(K) \leq L$ and $G/L$ is Dedekindian. □

**Lemma 184.2.** *Let $G$ be a non-Dedekindian 2-group all of whose subgroups of composite order are normal in $G$. Then we have the following possibilities:*

(a) *$G = D * Z$, where $D \cong D_8$, $Z \cong C_{2^m}$, $m \geq 1$, and $D \cap Z = Z(D)$,*

(b) *$G \cong M_{2^n}$, $n \geq 4$,*

(c) *$G = D * Q$ with $D \cong D_8$, $Q \cong Q_8$ and $D \cap Q = Z(D) = Z(Q)$. Here $G$ is the extraspecial group of order $2^5$ and type "$-$".*

*Conversely, all the above groups satisfy the assumption of the lemma.*

*Proof.* This result is actually the content of Theorem 1.25 for $p = 2$. □

**Theorem 184.3** (Janko). *Let $G$ be a non-Dedekindian $p$-group in which the intersection of any two distinct conjugate subgroups is either cyclic or generalized quaternion. Suppose that $p = 2$ and at least two distinct conjugate subgroups intersect in $Q_{2^n}$, $n \geq 3$. Then $n = 3$, $G' \cong E_4$ and we have the following two possibilities:*

(a) *$G = (H_0 \times H_0^*) * S$, where*

$$H_0 \cong H_0^* \cong Q_8\,, \quad \langle z \rangle = Z(H_0)\,, \quad \langle z^* \rangle = Z(H_0^*)\,,$$
$$S' = \langle zz^* \rangle\,, \quad S \cap (H_0 \times H_0^*) = \langle z, z^* \rangle\,,$$

*and either $S \cong C_2 \times D_8$ or $S$ is the nonmetacyclic minimal nonabelian group of order $2^4$. In this case both groups are special of order $2^8$.*

(b)

$$G = H_0 * C\,, \quad \text{where } H_0 \cong Q_8\,, \quad H_0 \cap C = Z(H_0) = \langle z \rangle\,,$$

*and $C$ is a non-Dedekindian group in which any two distinct conjugate subgroups have cyclic intersection. Moreover, $C/\langle z\rangle$ is isomorphic to one of the groups stated in Lemma 184.2 and $C$ has a maximal nonnormal subgroup*

$$Z \cong \mathrm{E}_4 \text{ or } \mathrm{C}_4 \quad \text{with } Z > \langle z\rangle \text{ and } |C : \mathrm{N}_C(Z)| = 2\,.$$

*Proof.* Let $G$ be a title $p$-group. In view of § 176, we may assume $p = 2$ and $G$ possesses a nonnormal subgroup $S$ such that for some $x \in G$,

$$S \neq S^x \quad \text{and} \quad S_0 = S \cap S^x \cong \mathrm{Q}_{2^n}\,, \quad n \geq 3\,.$$

Let $H$ be a maximal nonnormal subgroup in $G$ containing $S$. Then we use the notation for $H, K, L$ together with the results stated in Lemma 184.1. Let $g \in G - K$ so that $H^g \neq H$, $L = HH^g$ and $H_0 = H \cap H^g$ is a subgroup of index 2 in $H$. Note that $H$ contains $S_0 \cong \mathrm{Q}_{2^n}$, $n \geq 3$, with $S_0 < H$.

If $L$ has no $G$-invariant 4-subgroup, then Lemma 1.4 implies that $L$ is of maximal class and order $\geq 2^5$ (because $S_0 < H$). Let $L_0$ be a unique cyclic subgroup of index 2 in $L$ so that $L_0$ is characteristic in $L \trianglelefteq G$. But then $|S : (S \cap L_0)| = 2$ and $S \cap L_0 \trianglelefteq G$, contrary to the above fact that $S_0 = S \cap S^x \cong \mathrm{Q}_{2^n}$, $n \geq 3$. Hence $L$ has a $G$-invariant 4-subgroup $U$. Because $G/U$ is Dedekindian, we have $U \not\leq H$ so that $L = HU$ and $\mathrm{C}_2 \cong U_0 = H \cap U \leq \mathrm{Z}(K)$. Assume that $H_0 = H \cap H^g$ is cyclic. Since $|H : H_0| = 2$ and $\mathrm{Q}_{2^n} \cong S_0 < H$, Theorem 1.2 implies that $H$ is of maximal class and order $\geq 2^4$ with $U_0 = \mathrm{Z}(H)$. But then $L/U \cong H/U_0$ is dihedral of order $\geq 2^3$, contrary to the fact that $G/U$ is Dedekindian. We have proved that $H_0 \cong \mathrm{Q}_{2^m}$, $m \geq 3$. Suppose that $m > 3$. If $U_0 \leq H_0$, then $U_0 = \mathrm{Z}(H_0)$, and so $H_0/U_0$ is dihedral of order $\geq 2^3$, contrary to the fact that $H/U_0 \cong L/U$ is Dedekindian. If $U_0 \not\leq H_0$, then $H = U_0 \times H_0$ and $H/U_0 \cong H_0 \cong \mathrm{Q}_{2^m}$, $m > 3$, which again contradicts the fact that $H/U_0 \cong L/U$ is Dedekindian. It follows that $H_0 \cong \mathrm{Q}_8$, and so $|H| = 2^4$ and $S = H$. We have proved the following partial result, which was also obtained independently by the first author:

(i) Whenever $S$ is a nonnormal subgroup in $G$ such that for some $x \in G$,

$$S \neq S^x \quad \text{and} \quad S \cap S^x \cong \mathrm{Q}_{2^n}\,, \quad n \geq 3\,,$$
$$\text{then } S \cap S^x \cong \mathrm{Q}_8\,, \quad |S| = 2^4 \text{ and } S = H$$

is a maximal nonnormal subgroup in $G$. Setting

$$K = \mathrm{N}_G(H) \quad \text{and} \quad K \geq L > H \quad \text{with } |L : H| = 2$$

(as in Lemma 184.1), we have $L = HU$, where $U$ is a $G$-invariant 4-subgroup and

$$\mathrm{C}_2 \cong U_0 = H \cap U \leq \mathrm{Z}(K)\,.$$

For any fixed $g \in G - K$,

$$L = HH^g \quad \text{and} \quad H_0 = H \cap H^g \cong \mathrm{Q}_8\,.$$

Now, we want to determine the structure of $H$ and $L$. If $C_H(H_0) < H_0$, then Proposition 10.17 implies that $H$ is of maximal class and order $2^4$. In this case,

$$Z(H) = Z(H_0) = U_0 \quad \text{and} \quad H/U_0 \cong D_8 ,$$

contrary to the fact that $H/U_0 \cong L/U$ is Dedekindian. Hence we have $C_H(H_0) \nleq H_0$, and so

$$H = H_0 * Z , \quad \text{where } Z \cong E_4 \text{ or } C_4 \text{ and } H_0 \cap Z = Z(H_0) = \langle z \rangle .$$

Similarly, if $C_{H^g}(H_0) < H_0$, then $H^g$ would be of maximal class, a contradiction. Hence $C_{H^g}(H_0) \nleq H_0$, and so $C_L(H_0)$ covers $L/H_0 \cong E_4$. This gives

$$L = H_0 * C_L(H_0) \quad \text{with } H_0 \cap C_L(H_0) = \langle z \rangle = Z(H_0) \text{ and } |C_L(H_0)| = 8 .$$

But $(C_L(H_0))' \leq \langle z \rangle$, and so we get $L' = \langle z \rangle \leq Z(G)$. Also, we may assume that $U > \langle z \rangle$, and so $U \cap H = \langle z \rangle$. Indeed, if $U \ngeq \langle z \rangle$, then let $1 \neq u_0 \in U \cap Z(G)$ so that $U^* = \langle z, u_0 \rangle \cong E_4$ is $G$-invariant and $U^* \leq L$. Thus, replacing $U$ with $U^*$ and writing $U$ again (if necessary), we may always assume that $U > \langle z \rangle$. We have proved:

(ii) We have

$$H = H_0 * Z \quad \text{with } Z \cong E_4 \text{ or } C_4 \text{ and } H_0 \cap Z = Z(H_0) = \langle z \rangle \text{ and}$$
$$L = H_0 * C_L(H_0) , \quad \text{where } H_0 \cap C_L(H_0) = \langle z \rangle = L' \leq Z(G) \text{ and } |C_L(H_0)| = 8 .$$

Also, we may assume that $U \cap H = \langle z \rangle$.

Since $H$ has no cyclic subgroup of index 2, it follows that for each $x \in G - K$, $H \cap H^x \cong Q_8$. Moreover, if $x^2 \in K$, then

$$(H \cap H^x)^x = H^x \cap H^{x^2} = H^x \cap H = H \cap H^x ,$$

and so $H \cap H^x$ is $\langle x \rangle$-invariant.

(iii) $G/K \neq \{1\}$ is elementary abelian, and so for each $g \in G - K$, $H_0 = H \cap H^g$ is $\langle g \rangle$-invariant.

Indeed, assume that $\exp(G/K) > 2$. Then we may choose elements $y, g \in G - K$ so that $y^2 = g$ and $g^2 \in K$. Let $h \in Z(H) - L' = Z(H) - \langle z \rangle$ so that $h^2 \in \langle z \rangle$. We have $U\langle h \rangle \trianglelefteq G$ and $|(U\langle h \rangle) : U| = 2$ so that $h^y = hu$ with some $u \in U$, where $u^y = uz^\epsilon$, $\epsilon = 0, 1$, since $U \cap H = \langle z \rangle \leq Z(G)$. Then

$$h^g = h^{y^2} = (hu)^y = (hu)(uz^\epsilon) = hz^\epsilon ,$$

and so $h^g \in H$. But $Q_8 \cong H \cap H^g$ is $\langle g \rangle$-invariant and so $H = \langle h, H \cap H^g \rangle$ is $\langle g \rangle$-invariant, a contradiction.

(iv) We prove that $U \leq Z(L)$ so that

$$C_L(H_0) = Z(L) \cong C_4 \times C_2 \quad \text{or} \quad E_8 \quad \text{and} \quad U \leq G' .$$

Indeed, let $g \in G - K$ so that $H_0 = H \cap H^g \cong Q_8$ is $\langle g \rangle$-invariant and let

$$h \in Z(H) - \langle z \rangle\,, \quad \text{where } \langle z \rangle = L' = Z(H_0) \text{ and } h^2 \in \langle z \rangle\,.$$

Since $U > \langle z \rangle$, $U\langle h \rangle \trianglelefteq G$ and $|(U\langle h \rangle) : U| = 2$, we must have $h^g = hu$ for some $u \in U$. If $u \in \langle z \rangle$, then $g$ normalizes $H = H_0\langle h \rangle$, a contradiction. Thus, $u \in U - \langle z \rangle$ and this already proves that $U \leq G'$. But $h$ centralizes $H_0$, and so $h^g = hu$ centralizes $H_0$, and so also $u$ centralizes $H_0$. Hence $U$ centralizes $H_0$. Suppose that $h$ does not centralize $U$. Then $D = U\langle h \rangle \trianglelefteq G$ and $D \cong D_8$. But $U \trianglelefteq G$ and the unique cyclic subgroup of order 4 in $D$ is also normal in $G$. It follows that all three maximal subgroups of $D$ are normal in $G$. In particular, $\langle h, z \rangle$ (which is isomorphic to $E_4$ or $C_4$) is normal in $G$ and then $H = \langle h, H_0 \rangle$ is $\langle g \rangle$-invariant, a contradiction. We have proved that $U \leq Z(L)$, and so

$$C_L(H_0) = Z(L) \cong C_4 \times C_2 \text{ or } E_8\,.$$

(v) We have $|G/K| = 2$ and then we set for a fixed $g \in G - K$, $Q_8 \cong H \cap H^g = H_0$, where $g$ normalizes $H_0$.

Assume that $|G/K| > 2$ and let $G_0/K$ be a 4-subgroup in $G/K$. Let $g_1, g_2$ be elements in $G_0 - K$ so that $K\langle g_1, g_2 \rangle = G_0$. Let $h \in Z(H) - \langle z \rangle$ with $h^2 \in \langle z \rangle$ and we know that with a fixed $u \in U - \langle z \rangle$, we may set:

$$h^{g_1} = huz^{\epsilon_1}\,, \quad h^{g_2} = huz^{\epsilon_2}\,, \quad \text{where } \epsilon_1, \epsilon_2 \in \{0, 1\}\,.$$

Noting that $u^{g_2} = uz^{\epsilon_3}$ with $\epsilon_3 \in \{0, 1\}$, we get:

$$h^{g_1 g_2} = (huz^{\epsilon_1})^{g_2} = (huz^{\epsilon_2})(uz^{\epsilon_3})z^{\epsilon_1} = hz^{\epsilon_1+\epsilon_2+\epsilon_3} \in H\,.$$

On the other hand, $H \cap H^{g_1 g_2} \cong Q_8$ is $\langle g_1 g_2 \rangle$-invariant, and so $H = \langle h, H \cap H^{g_1 g_2} \rangle$ is $\langle g_1 g_2 \rangle$-invariant, a contradiction. We have proved that $|G/K| = 2$.

(vi) We have $H_0 = H \cap H^g$ $(g \in G - K)$ is normal in $G$.

Indeed, $H_0$ is $\langle g \rangle$-invariant, and so for any $kg \in G - K$ with $k \in K$, we have $kg = gk'$ for some $k' \in K$ and this gives:

$$H_0^{kg} = (H \cap H^g)^{kg} = H^{kg} \cap H^{gkg} = H^g \cap H^{g^2 k'} = H^g \cap H = H \cap H^g = H_0\,.$$

It follows that each element in $G - K$ normalizes $H_0$ and since all elements in $G - K$ generate $G$, we obtain $H_0 \trianglelefteq G$.

Set $C = C_G(H_0)$ so that $|G/(H_0C)| \leq 2$. If $|G/(H_0C)| = 2$, then $G/C \cong D_8$. Here we have used the fact that a Sylow 2-subgroup of $\mathrm{Aut}(Q_8)$ is isomorphic to $D_8$. But $U \leq C$ and $G/U$ is Dedekindian, a contradiction. We have proved:

(vii) Setting $C = C_G(H_0)$, we have $G = H_0 * C$ with

$$H_0 \cap C = \langle z \rangle \quad \text{and} \quad U < Z(L) \leq C\,.$$

Now, we want to determine the structure of $C/\langle z\rangle$. Set $Z = \mathrm{Z}(H)$ so that $Z > \langle z\rangle$, $Z \cong E_4$ or $C_4$, $K = H_0\mathrm{N}_C(Z)$, where $|C : \mathrm{N}_C(Z)| = 2$ and $Z$ is a maximal nonnormal subgroup in $C$. In particular, $C$ is non-Dedekindian. Let $X/\langle z\rangle$ be any subgroup of order $\geq 4$ in $C/\langle z\rangle$ and assume that $X$ is not normal in $C$. Then $T = XH_0$ is nonnormal in $G$, contrary to our results in (i). We have proved that each subgroup of composite order in $C/\langle z\rangle$ is normal in $C/\langle z\rangle$. We may use Lemma 184.2, and so we get:

(viii) $C/\langle z\rangle$ is non-Dedekindian and is isomorphic to one of the groups stated in Lemma 184.2. In particular, $U/\langle z\rangle = (C/\langle z\rangle)'$, and so $G' = U \cong E_4$. Moreover, setting $Z = \mathrm{Z}(H)$, we have $Z > \langle z\rangle$, $Z \cong E_4$ or $C_4$ and $Z$ is a maximal nonnormal subgroup in $C$ with $|C : \mathrm{N}_C(Z)| = 2$.

If the intersection of any two distinct conjugate subgroups in $C$ is cyclic, then we have obtained the groups from part (b) of our theorem and such groups have been determined in §§ 16, 174, 175 and 176.

It remains to study the case where two distinct conjugate subgroups in $C$ intersect in $Q_8$. We may use all results (i)–(viii) obtained so far. Let $H^*$ be a maximal nonnormal subgroup in $C$ such that

$$H^* = H_0^* * Z^*$$

$$\text{with } H_0^* \cong Q_8\,,\ H_0^* \trianglelefteq C\,,\ Z^* \cong C_4 \text{ or } E_4 \text{ and } H_0^* \cap Z^* = \mathrm{Z}(H_0^*) = \langle z*\rangle\,.$$

Because each subgroup of order $\geq 4$ in $C/\langle z\rangle$ is normal in $C/\langle z\rangle$, it follows that $H^* \cap \langle z\rangle = \{1\}$. Since

$$E_4 \cong \langle z, z*\rangle = H_0'(H_0^*)' = G'\,,$$

we have

$$U = \langle z, z*\rangle = G' \leq \mathrm{Z}(G) \quad (\text{noting that } H_0^* \trianglelefteq G)\,,$$

and therefore $G$ is of class 2. Also,

$$H^* \times \langle z\rangle = H^*U = L^* \trianglelefteq G\,,$$

where $L^*$ is the unique subgroup of $C$ containing $H^*$ as a subgroup of index 2, and so we have $L^* < C$ (see Lemma 184.1). It follows that $|C/\langle z\rangle| \geq 2^5$ and $C/\langle z\rangle$ cannot be isomorphic to a group in case (b) of Lemma 184.2 (because $M_{2^n}$, $n \geq 4$, is minimal nonabelian and so cannot possess the proper subgroup $(H_0^*\langle z\rangle)/\langle z\rangle \cong Q_8$).

We show that in fact $C/\langle z\rangle$ must be isomorphic to the extraspecial group $D_8 * Q_8$ of order $2^5$ from part (c) of Lemma 184.2. Indeed, since $H_0 \cong Q_8$ normalizes $H^*$ and $H_0 \cap H^* = \{1\}$, Lemma 184.1 implies that $\mathrm{N}_G(H^*) = K^* = H_0 \times H^*$. On the other hand, by (v), $|G : K^*| = |C : L^*| = 2$, giving $|C/\langle z\rangle| = 2^5$. We note that $E_4 \cong G' \leq \mathrm{Z}(G)$ also implies $\mho_1(G) \leq \mathrm{Z}(G)$. Suppose that there is an element $c$ of order 8 in $C - L^*$. Then $c^2 \in \mathrm{Z}(G) \cap L^*$ and $o(c^2) = 4$. But no element of order 4 in $\langle z\rangle \times H_0^*$ lies in $\mathrm{Z}(G)$, and so $c^2 \in L^* - (\langle z\rangle \times H_0^*)$ implying that $\mathrm{Z}(L^*)$ (of type $(4, 2)$) lies in $\mathrm{Z}(G)$. Then

$$\mathrm{Z}(H^*) \leq \mathrm{Z}(L^*) \leq \mathrm{Z}(G) \quad \text{gives that } H^* \trianglelefteq G\,,$$

a contradiction. We have proved that $\exp(C) = \exp(G) = 4$. In particular, $C/\langle z\rangle$ (of order $2^5$) cannot be isomorphic to the case (a) of Lemma 184.2 with $m = 3$. We have proved that $C/\langle z\rangle \cong D_8 * Q_8$ is extraspecial of order $2^5$. Since $U/\langle z\rangle = (C/\langle z\rangle)'$, we have $C/U \cong E_{2^4}$, and so $G/U \cong E_{2^6}$.

We have $G = (H_0 \times H_0^*) * S$, where

$$H_0 \cong H_0^* \cong Q_8\,, \quad S \cap (H_0 \times H_0^*) = \langle z, z^*\rangle = G' = \mho_1(G) \quad \text{and} \quad |S| = 2^4\,.$$

If $S' = G'$, then by a well-known result of O. Taussky, $S$ would be of maximal class, contrary to $\mathrm{cl}(G) = 2$. Hence we have $|S'| \le 2$. If $S' \le \langle z\rangle$, then each subgroup of $H_0S$ containing $H_0$ as a subgroup of index 2 is normalized by $S$. In particular, $H$ would be normal in $G$, a contradiction. Similarly, $S' \le \langle z^*\rangle$ gives that $H^* \trianglelefteq G$, a contradiction. Hence we must have $S' = \langle zz^*\rangle$. Note that $C = H_0^* * S$. If $S$ is not minimal nonabelian, then $S \cong C_2 \times D_8$ or $S \cong C_2 \times Q_8$. But if $S \cong C_2 \times Q_8$, then $C/\langle z\rangle \cong Q_8 * Q_8$, a contradiction. Hence in this case $S \cong C_2 \times D_8$.

Now assume that $S$ is minimal nonabelian so that

$$\Phi(S) = \mho_1(S) = Z(S) = \langle z, z^*\rangle = U\,.$$

If $S$ is metacyclic, then it is well known that for each $s \in S - U$, we have either $s^2 = zz^*$ or $s^2 = z^*$ (say) and in this case $z$ is not a square in $S$, and so there is no subgroup $H \le H_0S$ containing $H_0$ as a subgroup of index 2 that is different from $H_0U \trianglelefteq G$. Hence $S$ is nonmetacyclic and we have obtained two special 2-groups of order $2^8$ stated in part (a) of our theorem. □

# § 185 2-groups in which the intersection of any two distinct conjugate subgroups is either cyclic or of maximal class

We study the groups in which the intersection of any two distinct conjugate subgroups is either cyclic or of maximal class, and solve a special case of a problem, posed by the first author and solved by the second author. In order to facilitate the proof of the main result (Theorem 185.2) that is quite surprising, we shall first prove the following useful lemma:

**Lemma 185.1.** *Let $G$ be a 2-group in which the intersection of any two distinct conjugate subgroups is either cyclic or of maximal class. Then for each normal 4-subgroup $U$ in $G$, $G/U$ is Dedekindian.*

*Proof.* Assume that $G$ is a title 2-group that has a normal 4-subgroup $U$ such that $G/U$ is non-Dedekindian. Then each subgroup $X/U$ of order 2 is normal in $G/U$. Indeed, if $X$ is not normal in $G$, then there is $g \in G$ such that $X \cap X^g = U$, a contradiction.

Let $H/U$ be a nonnormal subgroup in $G/U$ so that there is $x \in G$ such that $H \neq H^x$ and set $D = H \cap H^x$. Since $D$ is not cyclic, it follows that $D$ is of maximal class, and so $D \cong D_8$. But $|D : U| = 2$, and so $D \trianglelefteq G$. Let $H_0/D$ be a subgroup of order 2 in $H/D$. Note that $H_0$ (being of order $2^4$ and having a normal 4-subgroup) is not of maximal class and so Proposition 10.17 implies $C_{H_0}(D) \nleq D$. We get

$$H_0 = D * Z, \quad \text{where } |Z| = 4 \text{ and } Z \cap D = Z(D) < U .$$

Since $|(ZU) : U| = 2$, we have $ZU \trianglelefteq G$. It follows that $H_0 \trianglelefteq G$, and so $H_0^x = H_0$. But then $H_0 \leq H \cap H^x$, a contradiction. Our lemma is proved. □

**Theorem 185.2** (Janko). *Let $G$ be a 2-group in which the intersection of any two distinct conjugate subgroups is either cyclic or of maximal class. Then the intersection of any two distinct conjugate subgroups is either cyclic or ordinary quaternion and such groups are classified in §§ 176 and 184.*

*Proof.* Let $G$ be a title group and assume (in view of Theorem 184.3) that $G$ has a non-normal subgroup $S$ such that for some $x \in G$, $S \neq S^x$ and $S_0 = S \cap S^x$ is dihedral or semidihedral.

Let $H$ be a maximal nonnormal subgroup of $G$ that contains $S$. Then we use the notation for the subgroups $H, K, L$ together with the results stated in Lemma 184.1. Let $g \in G - K$ so that

$$H \neq H^g , \quad L = HH^g \text{ and } H_0 = H \cap H_0 \text{ is of index 2 in } H .$$

Suppose that $L$ has no $G$-invariant 4-subgroup. Then Lemma 1.4 implies that $L$ is of maximal class and order $\geq 2^5$ (noting that $S_0 < H$). Let $L_0$ be a unique cyclic subgroup of index 2 in $L$ so that $L_0$ is characteristic in $L \trianglelefteq G$ and therefore $L_0 \trianglelefteq G$. But then $|S : (S \cap L_0)| = 2$ and $S \cap L_0 \trianglelefteq G$, contrary to our assumption that $S_0 = S \cap S^x$ is dihedral or semidihedral.

We have proved that $L$ possesses a $G$-invariant subgroup $U$. By Lemma 185.1, $G/U$ is Dedekindian, and so

$$U \nleq H \quad \text{so that } L = HU \text{ and } \mathrm{C}_2 \cong U_0 = H \cap U \leq \mathrm{Z}(K)\,.$$

Assume that $H_0$ is cyclic. Since $|H : H_0| = 2$ and $S_0 < H$, Theorem 1.2 implies that $H$ is of maximal class, and so $H$ is dihedral or semidihedral of order $\geq 2^4$ with $\mathrm{Z}(H) = U_0$. Then $L/U \cong H/U_0$ is dihedral of order $\geq 2^3$, contrary to the fact that $G/U$ is Dedekindian. Hence $H_0$ is of maximal class.

Suppose that $|H_0| \geq 2^4$. If $U_0 \leq H_0$, then $U_0 = \mathrm{Z}(H_0)$, and so $H_0/U_0$ is dihedral of order $\geq 2^3$, contrary to the fact that $L/U \cong H/U_0$ is Dedekindian. If $U_0 \nleq H_0$, then $H = U_0 \times H_0$ and $H/U_0 \cong H_0$, which again contradicts the fact that $L/U \cong H/U_0$ is Dedekindian. It follows that $|H_0| = 2^3$, and so $H_0 \cong \mathrm{D}_8$ or $\mathrm{Q}_8$. In particular,

$$|H| = 2^4\,, \quad S = H\,, \quad x \in G - K\,, \quad S_0 = H \cap H^x = S \cap S^x \cong \mathrm{D}_8\,.$$

Also, $L = HH^x$. If $U_0 = U \cap H \nleq S_0$, then

$$H = U_0 \times S_0 \quad \text{and} \quad L/U \cong H/U_0 \cong S_0 \cong \mathrm{D}_8\,,$$

contrary to the fact that $G/U$ is Dedekindian. Hence we have

$$U_0 = U \cap H = \langle z \rangle = \mathrm{Z}(S_0)\,.$$

If $\mathrm{C}_H(S_0) < S_0$, then Proposition 10.17 implies that $H$ is of maximal class and order $2^4$. In this case,

$$\mathrm{Z}(H) = \mathrm{Z}(S_0) = U_0 \quad \text{and} \quad H/U_0 \cong \mathrm{D}_8\,,$$

contrary to the fact that $H/U_0 \cong L/U$ is Dedekindian. Hence we have $\mathrm{C}_H(S_0) \nleq S_0$, and so

$$H = S_0 * Z\,, \quad \text{where } Z = \mathrm{Z}(H) \cong \mathrm{E}_4 \quad \text{or } \mathrm{C}_4 \text{ and } S_0 \cap Z = \mathrm{Z}(S_0) = \langle z \rangle\,.$$

Thus, $H' = S_0' = \langle z \rangle$, and so $L/U \cong H/\langle z \rangle$ is abelian, which implies

$$L' \leq U \cap H = \langle z \rangle \quad \text{and} \quad L' = S_0' = \langle z \rangle \leq \mathrm{Z}(G)\,.$$

This also gives

$$L = S_0 * \mathrm{C}_L(S_0) \quad \text{with } |\mathrm{C}_L(S_0)| = 8 \text{ and } S_0 \cap \mathrm{C}_L(S_0) = \langle z \rangle\,.$$

Since $H$ has no cyclic subgroup of index 2, it follows that for each $l \in G - K$, $H \cap H^l \cong \mathrm{Q}_8$ or $\mathrm{D}_8$. Moreover, if $l^2 \in K$, then

$$(H \cap H^l)^l = H^l \cap H^{l^2} = H^l \cap H = H \cap H^l\,,$$

and so $H \cap H^l$ is $\langle l \rangle$-invariant.

It is easy to see that $G/K \neq \{1\}$ is elementary abelian. Indeed, assume that $\exp(G/K) > 2$. Then we may choose elements $y, g \in G - K$ so that $y^2 = g$ and $g^2 \in K$. Let $h \in Z(H) - \langle z \rangle$ so that $h^2 \in \langle z \rangle$. We have $U\langle h \rangle \trianglelefteq G$ and $|(U\langle h \rangle) : U| = 2$ so that $h^y = hu$ for some $u \in U$, where $u^y = uz^\epsilon$, $\epsilon = 0, 1$, since $U \cap H = \langle z \rangle \leq Z(G)$. Then

$$h^g = h^{y^2} = (hu)^y = (hu)(uz^\epsilon) = hz^\epsilon \in H .$$

But $H \cap H^g \cong Q_8$ or $D_8$ is $\langle g \rangle$-invariant, and so $H = \langle h, H \cap H^g \rangle$ is also $\langle g \rangle$-invariant, a contradiction.

We show now that $U \leq Z(L)$ so that

$$C_L(S_0) = Z(L) \cong C_4 \times C_2 \text{ or } E_8 .$$

Indeed, note that $S_0 = H \cap H^x \cong D_8$ is $\langle x \rangle$-invariant and let

$$h \in Z(H) - \langle z \rangle , \quad \text{where } \langle z \rangle = L' = Z(S_0) \text{ and } h^2 \in \langle z \rangle .$$

Since $U > \langle z \rangle$, $U\langle h \rangle \trianglelefteq G$ and $|(U\langle h \rangle) : U| = 2$, we must have $h^x = hu$ for some $u \in U$. If $u \in \langle z \rangle$, then $x$ normalizes $H = S_0\langle h \rangle$, a contradiction. Thus, $u \in U - \langle z \rangle$. But $h$ centralizes $S_0$, and so $h^x = hu$ centralizes $S_0$, and so also $u$ centralizes $S_0$. Hence $U$ centralizes $S_0$. Suppose that $h$ does not centralize $U$. Then $D = U\langle h \rangle \trianglelefteq G$ and $D \cong D_8$. But $U \trianglelefteq G$ and the unique cyclic subgroup of order 4 in $D$ is also normal in $G$. It follows that all three maximal subgroups of $D$ are normal in $G$. In particular, $\langle h, z \rangle$ is normal in $G$ and then $H = \langle h, S_0 \rangle$ is $\langle x \rangle$-invariant, a contradiction.

We have $|G/K| = 2$. Indeed, assume that $|G/K| > 2$ and let $G_0/K$ be a 4-subgroup in $G/K$. Let $g_1, g_2$ be elements in $G_0 - K$ so that $K\langle g_1, g_2 \rangle = G_0$. Let $h \in Z(H) - \langle z \rangle$ with $h^2 \in \langle z \rangle$ and we know that with a fixed $u \in U - \langle z \rangle$, we may set

$$h^{g_1} = huz^{\epsilon_1} , \quad h^{g_2} = huz^{\epsilon_2} , \quad \text{where } \epsilon_1, \epsilon_2 \in \{0, 1\} .$$

Noting that $u^{g_2} = uz^{\epsilon_3}$ with $\epsilon_3 \in \{0, 1\}$, we get

$$h^{g_1 g_2} = (huz^{\epsilon_1})^{g_2} = (huz^{\epsilon_2})(uz^{\epsilon_3})z^{\epsilon_1} = hz^{\epsilon_1+\epsilon_2+\epsilon_3} \in H .$$

On the other hand, $H \cap H^{g_1 g_2}$ is $\langle g_1 g_2 \rangle$-invariant, and so $H = \langle h, H \cap H^{g_1 g_2} \rangle$ is $\langle g_1 g_2 \rangle$-invariant, a contradiction.

It follows that we may set $G = K\langle x \rangle$, where $S_0 = H \cap H^x$ is $\langle x \rangle$-invariant. For any $kx \in G - K$ with $k \in K$, we have $kx = xk'$ for some $k' \in K$ and this gives:

$$S_0^{kx} = (H \cap H^x)^{kx} = H^{kx} \cap H^{xkx} = H^x \cap H^{x^2 k'} = H^x \cap H = H \cap H^x = S_0 .$$

It follows that each element in $G - K$ normalizes $S_0$ and since all elements in $G - K$ generate $G$, we obtain $D_8 \cong S_0 \trianglelefteq G$.

Set $C = C_G(S_0)$ and note that $\mathrm{Aut}(S_0) \cong D_8$, $U \leq C$ and $S_0 \cap C = \langle z \rangle$. If $S_0 * C < G$, then $G/C \cong D_8$, which contradicts the fact that $G/U$ is Dedekindian. Hence we have $G = S_0 * C$.

Let $V$ be a 4-subgroup in $S_0$ so that $V \trianglelefteq G$. By Lemma 185.1, $G/V$ is Dedekindian. But then $H$ (containing $S_0$) is normal in $G$, a final contradiction. Our theorem is proved. □

The groups from the conclusion of Theorem 185.2 are classified in §§ 176 and 184.

**Exercise 1.** Suppose that $G$ is a $p$-group in which the intersection of any two distinct conjugate subgroups is metacyclic. Then:
(a) if $p > 2$, then $G$ is metabelian,
(b) if $p = 2$, then the derived length of $G$ is at most 3.

**Solution.** (a) Let $p > 2$. If $G$ has no normal elementary abelian subgroup of order $p^3$, then $G$ is one of the groups of Theorem 13.7. It is easy to see that all of these groups are metabelian. Now suppose that $G$ has a normal subgroup $E \cong \mathrm{E}_{p^3}$. Then $G/E$ is abelian so $G$ is metabelian.
(b) Now let $p = 2$. Suppose that $G$ has no normal elementary abelian subgroup of order 8. Then, by Theorem 50.1, the derived length of $G$ is at most 3. Now assume that there is $E \cong \mathrm{E}_8$ normal in $G$. Then $G/E$ is Dedekindian. In this case the derived length of $G$ is at most 3.

**Problem 0.** Study the irregular $p$-groups in which the intersection of any two distinct subgroups of order $p^p$ and exponent $p$ has order $p^{p-1}$.

**Problem 1.** Classify the 2-groups in which the intersection of any two distinct conjugate subgroups is either cyclic or abelian of type $(2, 2)$ or of maximal class.

**Problem 2.** Classify the $p$-groups in which the intersection of any two distinct conjugate cyclic subgroups is equal to $\{1\}$.

In particular, classify the $p$-groups $G$ satisfying $C_G = \{1\}$ for any nonnormal cyclic $C < G$.

**Problem 3.** Study the 2-groups $G$ such that $G/\mathrm{Z}(G)$ is Dedekindian. Study in detail the case where $|\mathrm{Z}(G)| = 2$.

**Problem 4.** Classify the $p$-groups, $p > 2$, in which the intersection of any two distinct conjugate subgroups is (i) absolutely regular, (ii) either absolutely regular or of maximal class.

If $G$ is neither absolutely regular nor of maximal class, it contains a normal subgroup $R$ of order $p^p$ and exponent $p$. If, in addition, $G$ satisfies Problem 4 (i) (Problem 4(ii)) and $|G| > p^{p+1}$ (and $|G| > p^{p+2}$), then $G/R$ is abelian.

**Problem 5.** Classify the $p$-groups in which the intersection of any two distinct conjugate nonquasinormal cyclic subgroups is equal to $\{1\}$.

This is an essentially more difficult analog of Problem 2. Similarly, one can state analogs of other problems.

# § 186 *p*-groups in which the intersection of any two distinct conjugate subgroups is either cyclic or abelian of type $(p, p)$

In this section we present the solution of a problem posed by the first author and solved by the second author.

The class of title groups is already very large. Therefore we start with a special case, where a title $p$-group $G$ has an abelian normal subgroup $U$ of type $(p, p)$ such that $G/U$ is non-Dedekindian (Theorem 186.1). We notice that in this special case each nonnormal subgroup $X$ of $G$ containing $U$ is of order $p^3$. It is surprising that it is so also true in the general case! More precisely, whenever $H$ is a nonnormal subgroup in a title group $G$ such that for some $g \in G$, $H^g \neq H$ and $H \cap H^g \cong \mathrm{E}_{p^2}$, then $H$ is of order $p^3$ (Theorem 186.2). Finally, in Theorem 186.3 we classify the title groups $G$ that have a nonnormal subgroup $H$ such that the conjugate class of $H$ contains distinct conjugates that intersect in a cyclic group, as well as there being distinct conjugates that intersect in a subgroup isomorphic to $\mathrm{E}_{p^2}$.

**Theorem 186.1.** *Let $G$ be a $p$-group in which the intersection of any two distinct conjugate subgroups is either cyclic or abelian of type $(p, p)$. If $G$ has an abelian normal subgroup $U$ of type $(p, p)$ such that $G/U$ is non-Dedekindian, then we have one of the following possibilities:*

(a) *$G/U \cong \mathrm{M}_{p^n}$, $n \geq 3$, and if $p = 2$, then $n \geq 4$,*

(b) *$G/U = Z * G_0$ with $Z$ cyclic, $Z \cap G_0 = G_0'$, where in case $p = 2$, $G_0 \cong \mathrm{D}_8$ and if $p > 2$, then $G_0 \cong \mathrm{S}(p^3)$ (the nonabelian group of order $p^3$ and exponent $p$),*

(c) *$p = 2$ and $G/U \cong \mathrm{D}_8 * \mathrm{Q}_8$ is extraspecial of order $2^5$.*

*In particular, each nonnormal subgroup $X$ of $G$ containing $U$ is of order $p^3$ and also $|G : \mathrm{N}_G(X)| = p$. In addition, $G/U$ has a unique normal subgroup $Y/U$ of order $p$ and we have $(G/U)' = Y/U$.*

*Proof.* Let $H/U$ be any maximal nonnormal subgroup of $G/U$. Let L be a unique subgroup of $G$ such that $H < L$ and $|L : H| = p$. Then $L \trianglelefteq G$ and $G/L$ is Dedekindian. Let $g \in G - \mathrm{N}_G(H)$ so that $H^g \neq H$ and $H^g < L$. It follows that $|H : (H \cap H^g)| = p$, and so $U \leq H \cap H^g$ gives $H \cap H^g = U$ and $|H/U| = p$.

We have proved that each nonnormal subgroup of $G/U$ is of order $p$. By Theorem 1.25, $G/U$ is isomorphic to one of the groups (a), (b) or (c) stated in our theorem. In particular, we see that each nonnormal subgroup $X$ of $G$ containing $U$ is of order $p^3$ and $|G : \mathrm{N}_G(X)| = p$. Also, $(G/U)' = Y/U$ is a unique normal subgroup of order $p$ in $G/U$ since $\mathrm{Z}(G/U)$ is cyclic and $(G/U)' \cong \mathrm{C}_p$. □

**Theorem 186.2.** *Let $G$ be a $p$-group in which the intersection of any two distinct conjugate subgroups is either cyclic or abelian of type $(p, p)$. Whenever $H$ is a nonnormal subgroup in $G$ such that for some $x \in G$, $H \neq H^x$ and $H \cap H^x \cong \mathrm{E}_{p^2}$, then $H$ is a maximal* *nonnormal subgroup in $G$ and $H$ is of order $p^3$. Supposing that $G$ has such a subgroup $H$, then $|G'| \le p^3$, and so $G$ is of class $\le 4$.*

*Proof.* Let $H_0$ be a nonnormal subgroup of the maximal order such that there is $x \in G$ so that $H_0^x \ne H_0$ and $H_0 \cap H_0^x \cong E_{p^2}$. Assume in addition that $H_0$ is not a maximal nonnormal subgroup in $G$. Let $H > H_0$ be a maximal nonnormal subgroup containing $H_0$. Since $|H_0| \ge p^3$, we have $|H| \ge p^4$.

Let $H$, $K = N_G(H)$ and $L \trianglelefteq G$ with $|L : H| = p$ be according to the notation of Lemma 184.1, where we also use the results of that general lemma. Let $g \in G - K$ so that $H^g \ne H$, $L = HH^g$ and $H_1 = H \cap H^g$ is a cyclic subgroup of index $p$ in $H$ with $|H_1| \ge p^3$ and $H_0 \nleq H_1$. We have $\Phi(H) = \mho_1(H_1)$ and $|H_1 : \Phi(H)| = p$ so that $\Phi(H)$ is cyclic of order $\ge p^2$.

Assume for a moment that the above element $x$ is contained in $G - K$. But then $H \cap H^x$ is cyclic and then also $H_0 \cap H_0^x$ (as a subgroup of $H \cap H^x$) is cyclic, a contradiction. Thus, $x \in K$, and so $x$ normalizes $H$ and $\Phi(H)$.

If $|H_0 \cap H_1| \le p$, then $|H_0| \le p^2$, a contradiction. Hence $|H_0 \cap H_1| \ge p^2$, and so $|H_0 \cap \Phi(H)| \ge p^2$ (since $H_1$ is cyclic). But $H_0 \cap \Phi(H)$ is characteristic in $H$, and so $x$ normalizes the cyclic subgroup $H_0 \cap \Phi(H)$ of order $\ge p^2$, contrary to our assumption that $H_0 \cap H_0^x \cong E_{p^2}$.

We have proved that $H_0 = H$ is a maximal nonnormal subgroup in $G$, and so $x \in G - K$ with $H \cap H^x \cong E_{p^2}$ and $|H| = p^3$.

Since $H \trianglelefteq K$ and $H^x \trianglelefteq K$, we get $E_{p^2} \cong H \cap H^x \trianglelefteq K$. If $L \trianglelefteq G$ does not contain an abelian $G$-invariant subgroup of type $(p, p)$, then Lemma 1.4 implies that $L$ is either cyclic or a 2-group of maximal class. This is a contradiction since $|L| = p^4$ and $E_{p^2} \cong H \cap H^x \trianglelefteq L$. Hence $L$ possesses a $G$-invariant abelian subgroup $U$ of type $(p, p)$. If $G/U$ is Dedekindian, then $|(G/U)'| \le p$. If $G/U$ is non-Dedekindian, then Theorem 186.1 implies $|(G/U)'| = p$. In any case, $|G'| \le p^3$, and so $\mathrm{cl}(G) \le 4$. □

**Theorem 186.3.** *Let $G$ be a $p$-group in which the intersection of any two distinct conjugate subgroups is either cyclic or abelian of type $(p, p)$. Suppose that $G$ has a nonnormal subgroup $H$ such that the conjugate class of $H$ contains distinct conjugates that intersect in a cyclic group, as well as there being distinct conjugates that intersect in a subgroup isomorphic to $E_{p^2}$. Then $H$ is a maximal nonnormal subgroup in $G$ and $H$ is isomorphic to $C_{p^2} \times C_p$, $D_8$ or $M_{p^3}$, $p > 2$. Setting $K = N_G(H) \trianglelefteq G$, we have $|G/K| = p^2$, $K/H$ is cyclic or ordinary quaternion, and if $L$ is a unique subgroup in $G$ such that $H < L \le K$ with $|L : H| = p$, then $L \trianglelefteq G$, $G/L$ is Dedekindian and $L = H \times \langle u \rangle$, where $\langle u \rangle \cong C_p$. Setting*

$$\mho_1(L) = \mho_1(H) = \langle z \rangle \le Z(G),$$

*the subgroup $U = \langle u, z \rangle \cong E_{p^2}$ is $G$-invariant.*

*If $G/K \cong C_{p^2}$, then $p = 2$, $G/U$ is non-Dedekindian and*

$$G/U \cong M_{2^n}, \quad n \ge 4, \quad \text{or } G/U = Z * G_0$$
$$\text{with } Z \cong C_{2^{n-1}}, \quad G_0 \cong D_8, \quad \text{and } Z \cap G_0 = G_0'.$$

*If $G/U$ is Dedekindian, then $G/K \cong \mathrm{E}_{p^2}$ and if $H_i$, $i = 1, \dots, p+1$ is the set of maximal subgroups (of order $p^2$) of $H$, then $\mathrm{N}_G(H_i)$ is the set of $p+1$ maximal subgroups of $G$ containing $K$.*

*Proof.* By Theorem 186.2, $|H| = p^3$ and $H$ is a maximal nonnormal subgroup in $G$. We use Lemma 184.1, where $K = \mathrm{N}_G(H)$ and $L$ is a unique subgroup in $G$ with $H < L \le K$, $|L : H| = p$ and $L \trianglelefteq G$. It follows that $K/H \neq \{1\}$ is either cyclic or ordinary quaternion and $G/L$ is Dedekindian.

By our assumptions, $H$ has an abelian subgroup of type $(p, p)$ and $\exp(H) = p^2$. This gives one of the following possibilities:

$$H \cong \mathrm{C}_{p^2} \times \mathrm{C}_p\,, \quad H \cong \mathrm{D}_8 \quad \text{or } H \cong \mathrm{M}_{p^3}\,, \quad p > 2\,.$$

There is $x \in G - K$ so that $H \neq H^x$, $L = HH^x$ and $V = H \cap H^x \cong \mathrm{E}_{p^2}$. Note that $V \trianglelefteq K$ and $L/V \cong \mathrm{E}_{p^2}$. Suppose that $L$ does not possess a $G$-invariant abelian subgroup of type $(p, p)$. Then $p = 2$ and $L$ is of maximal class, contrary to the fact that $\mathrm{E}_4 \cong V \trianglelefteq L$ and that $|L| = 2^4$. Thus, $L$ has a $G$-invariant abelian subgroup $U$ of type $(p, p)$. By our assumptions, $U \nleq H$, and so $L = HU$ and $H \cap U = U_0 \cong \mathrm{C}_p$. If $U \cap V = \{1\}$, then $L = V \times U \cong \mathrm{E}_{p^4}$, contrary to $\exp(H) = p^2$. Hence $H \cap U = V \cap U = U_0$.

There is $y \in G - K$ so that $H \neq H^y$, $L = HH^y$ and $W = H \cap H^y \cong \mathrm{C}_{p^2}$ with $W \trianglelefteq K$. Suppose that $W \ngeq U_0$ so that

$$L = W \times U \quad \text{and} \quad \mho_1(W) = \Phi(L) \cong \mathrm{C}_p \quad \text{with } V = U_0 \times \mho_1(W)\,.$$

Note that in this case $\Phi(L) \le \mathrm{Z}(G)$. Let $z \neq 1$ be an element in $U$ such that $z \in \mathrm{Z}(G)$. If $\langle z \rangle = U_0$, then $V \le \mathrm{Z}(G)$, contrary to $H \cap H^y \cong \mathrm{C}_{p^2}$. Thus, $z \in U - U_0$ and

$$U^* = \Phi(L) \times \langle z \rangle \le \mathrm{Z}(G)\,, \quad U^* \cong \mathrm{E}_{p^2} \quad \text{and} \quad U^* \cap H = U^* \cap V = \Phi(L) = \mho_1(H)\,.$$

Hence replacing $U$ with $U^*$ (if necessary), we may assume from the start that

$$\mho_1(H) = U_0 = U \cap H = \Phi(H)\,.$$

Since $L/V \cong \mathrm{E}_{p^2}$, it follows $U_0 \le \mho_1(L) \le V$. If $\mho_1(L) = V$, then $V \trianglelefteq G$, contrary to our assumptions. Hence

$$\mho_1(L) = \mho_1(H) = U_0 = \langle z \rangle \le \mathrm{Z}(G)\,.$$

Suppose that $L$ is nonabelian. Then $L' \le U$ (since $L/U \cong H/\langle z \rangle \cong \mathrm{E}_{p^2}$) and $L' \le H$, and so $L' = H \cap U = \langle z \rangle$. If $H$ is abelian, then $H^x$ is also abelian, and so $V = H \cap H^x = \mathrm{Z}(L) \trianglelefteq G$, a contradiction. Thus, if $L$ is nonabelian, then $L' = H' = \langle z \rangle$.

We have proved that in any case

$$\Phi(L) = \Phi(H) = \langle z \rangle \cong \mathrm{C}_p\,.$$

Since $H$ is a maximal subgroup of $L$ and $L$ has exactly $p^2 + p + 1$ maximal subgroups, it follows that $|G : K| \le p^2$.

Suppose, by way of contradiction, that $|G : K| = p$. If $p = 2$, then for an $x \in G - K$, $x^2 \in K$, $H \trianglelefteq K$, $H^x \trianglelefteq K$, and so $H \cap H^x \trianglelefteq K$ and

$$(H \cap H^x)^x = H^x \cap H^{x^2} = H^x \cap H = H \cap H^x ,$$

and so $H \cap H^x \trianglelefteq G$, contrary to our assumptions. Hence we have $p > 2$. In that case, $H \cong C_{p^2} \times C_p$ or $H \cong M_{p^3}$. In both cases, $\Omega_1(H) = V \cong E_{p^2}$. By our assumption, there is $g \in G - K$ such that $H \cap H^g \cong E_{p^2}$ and also $H \cap H^g = V \trianglelefteq K$. In this case, $g$ sends (by conjugation) each element in $H - V$ (of order $p^2$) onto some element in $L - H$, and so $g$ must send each element in $V$ onto an element in $V$, and therefore $g$ normalizes $V$. But then $V \trianglelefteq G$ and then any two distinct conjugates of $H$ intersect in $V \cong E_{p^2}$, a contradiction. We have proved that $|G : K| = p^2$.

Suppose again that $L$ is nonabelian so that

$$L' = H' = \langle z \rangle , \quad H \cong D_8 \text{ or } H \cong M_{p^3} , \quad p > 2 , \quad \text{and } L = H * Z \quad \text{with}$$
$$H \cap Z = \langle z \rangle \text{ and } Z \cong E_{p^2} \text{ or } Z \cong C_{p^2} , \quad \text{where } Z = Z(L) \trianglelefteq G .$$

Assume, by way of contradiction, $Z \cong C_{p^2}$. All $p + 1$ maximal subgroups of $L$ that contain $Z$ are abelian, and so all other $p^2$ maximal subgroups $H_i$ of $L$ are nonabelian of order $p^3$ and $H_i \cap Z = \langle z \rangle$, $i = 1, 2, \ldots, p^2$. If $p = 2$, then

$$L = H * Z \cong Q_8 * C_4$$

and $L$ has exactly one subgroup isomorphic to $Q_8$ and three subgroups isomorphic to $D_8$. If $p > 2$, then

$$L = H * Z \cong S(p^3) * C_{p^2}$$

and $L$ has exactly one subgroup isomorphic to $S(p^3)$ (the nonabelian group of order $p^3$ and exponent $p$) and $p^2 - 1$ subgroups isomorphic to $M_{p^3}$. In both cases we have a contradiction with the fact that $L$ has exactly $p^2$ conjugates of $H$. We have proved that if $L$ is nonabelian, then $Z \cong E_{p^2}$, where $Z \trianglelefteq G$, and so we may set $Z = U$. Thus, in any case (where $L$ is abelian or nonabelian) we may set $L = H \times \langle u \rangle$ with a suitable $\langle u \rangle \cong C_p$ so that $U = \langle u, z \rangle \cong E_{p^2}$ is $G$-invariant.

Assume that $G/K \cong C_{p^2}$. If $p > 2$, then a Sylow $p$-subgroup of $\mathrm{Aut}(L/\langle z \rangle)$ is isomorphic to $S(p^3)$, and so acting with $G/K$ on $L/\langle z \rangle$, we see that an element $x \in G - K$ centralizes $L/\langle z \rangle$, and so, in particular, $x$ normalizes $H$ (where $H > \langle z \rangle$), a contradiction. Hence in this case,

$$p = 2 \quad \text{and} \quad H \cong C_4 \times C_2 \text{ or } H \cong D_8 .$$

Let $g \in G - K$ be such that $\langle g \rangle$ covers $G/K$. Obviously, $\langle g \rangle$ does not normalize any subgroup of order 4 in $H$ (which is maximal in $H$). Since there are exactly three abelian maximal subgroups of $L$ containing $U$, there is (at least) one of them denoted with $L_0$ that is $\langle g \rangle$-invariant. If $h_0 \in (L_0 \cap H) - \langle z \rangle$, then

$$h_0^2 \in \langle z \rangle \quad \text{and} \quad h_0^g = h_0 u \quad \text{for some } u \in U - \langle z \rangle , \text{ where } u^g = uz^\epsilon , \ \epsilon = 0, 1 .$$

If $\langle g\rangle$ centralizes $L/U \cong H/\langle z\rangle \cong E_4$, then for $h_1 \in H - L_0$, $h_1^g = h_1uz^\delta$, $\delta = 0, 1$. But then

$$h_0^{g^2} = (h_0u)^g = (h_0u)(uz^\epsilon) = h_0z^\epsilon\ ,$$
$$h_1^{g^2} = (h_1uz^\delta)(uz^\epsilon)z^\delta = h_1z^\epsilon\ ,$$

and so $g^2 \in G - K$ normalizes $H = \langle h_0, h_1\rangle$, a contradiction. Hence $\langle g\rangle$ acts nontrivially on $L/U \cong E_4$, which shows that $(L\langle g\rangle)/U$ is two-generator ($\langle h_1, g\rangle = L\langle g\rangle$) and non-Dedekindian. Also, $|(L\langle g\rangle)/U| \geq 2^4$, and so Theorem 186.1 (applied to $L\langle g\rangle$) gives $(L\langle g\rangle)/U \cong M_{2^n}$, $n \geq 4$. But then $G/U$ is also non-Dedekindian and since $L_0/U$ is a normal subgroup of order 2 in $G/U$, Theorem 186.1 also implies

$$L_0/U = (G/U)' \quad \text{and so } \langle g\rangle \text{ covers } L_0/U\ .$$

It follows that either

$$G/U \cong M_{2^n}\ , \quad n \geq 4\ , \quad \text{or} \quad G/U = Z * G_0 \quad \text{with } Z \cong C_{2^m}\ ,\ G_0 \cong D_8 \text{ and } Z \cap G_0 = G_0'\ ,$$

where in the second case we must have $m = n - 1$ because $M_{2^n}$, $n \geq 4$, is in any case a subgroup of $G/U$.

Finally, assume that $G/U$ is Dedekindian. Then by the above, $G/K \cong E_{p^2}$. Let $H_i$, $i = 1, \ldots, p+1$, be the set of maximal subgroups of $H$ so that $H_i > \langle z\rangle$ and $|H_i| = p^2$. Then $L_i = H_iU \trianglelefteq G$ and we act with $G/K$ on $p+1$ maximal subgroups of $L_i$ containing $\langle z\rangle \leq Z(G)$. Then only $U$ is $G$-invariant and the other $p$ maximal subgroups of $L_i$ containing $\langle z\rangle$ are acted upon transitively by $G$ because $H_i = H \cap L_i \trianglelefteq K$ but $H_i$ is not normal in $G$. Since $G/K \cong E_{p^2}$, there is $x \in G-K$ such that $x$ normalizes $H_i$, which together with $H_i \trianglelefteq K$ gives that $N_G(H_i) = K\langle x\rangle$ is a maximal subgroup of $G$ containing $K$. Because $N_G(H) = K$, it follows that for $j \neq i$, $N_G(H_j)$ must be a maximal subgroup of $G$ containing $K$ distinct from $N_G(H_i)$. The proof is complete. □

# § 187 *p*-groups in which the intersection of any two distinct conjugate cyclic subgroups is trivial

We shall prove here a surprising result: that in the title groups all cyclic subgroups of composite order are normal, and such $p$-groups have been characterized in Theorem 63.4. This solves Problem 2555.

Our method of proof is different from the methods of solving the intersection problems that have been developed in §§ 16, 174, 175, 176, 184, 185 and 186. In fact, the method of proof in the case $p = 2$ (Theorem 187.1) is distinct from that one in case $p > 2$ (Theorem 187.2).

Note that here we actually classify $p$-groups all of whose cyclic subgroups are TI-subgroups. See also Corollary 187.3, where we determine completely $p$-groups all of whose metacyclic subgroups are TI-subgroups.

**Theorem 187.1.** *Let $G$ be a non-Dedekindian 2-group in which the intersection of any two distinct conjugate cyclic subgroups is trivial. Then each cyclic subgroup of composite order is normal in $G$, and so $G$ is either quasidihedral (i.e., $|G : \mathrm{H}_2(G)| = 2$) or $|G'| = 2$ and $\Phi(G)$ is cyclic.*

*Proof.* Assume that $G$ possesses a cyclic subgroup $X$ of composite order that is not normal in $G$. Set $N = \mathrm{N}_G(X)$ and let $N < G_0 \le G$ be such that $|G_0 : N| = 2$. Take $g \in G_0 - N$, where $g^2 \in N$. Then $X \cap X^g = \{1\}$ and $\mathrm{N}_G(X^g) = N$ giving that $\langle X, X^g\rangle = X \times X^g$ is abelian. Note that for each subgroup $\{1\} \ne X_0 \le X$, $\mathrm{N}_G(X_0) = \mathrm{N}_G(X_0^g) = N$.

Choose an element $a$ of order 4 in $X$ and set $b = a^g$. Then

$$\langle a, b\rangle = \langle a\rangle \times \langle b\rangle \cong \mathrm{C}_4 \times \mathrm{C}_4$$

is $\langle g\rangle$-invariant since $b^g = a^{g^2} \in \langle a\rangle$. We have two possibilities: $b^g = a$ or $b^g = a^{-1}$.

(i) First assume $b^g = a$. Then

$$\mathrm{C}_{\langle a,b\rangle}(g) = \langle ab\rangle \cong \mathrm{C}_4 \quad \text{and therefore } \langle g\rangle \cap \langle a, b\rangle \le \langle ab\rangle\,.$$

Set $H = \langle a, g\rangle$ so that $\mathrm{Z}(H) = \langle ab\rangle\langle g^2\rangle$.

If $g^2 \ne 1$, then $\langle g\rangle \cap \mathrm{Z}(H) \ne \{1\}$ and this gives $\langle g\rangle \trianglelefteq H$. But

$$[a, g] = a^{-1}(g^{-1}ag) = a^{-1}b \notin \langle g\rangle \quad \text{since } \langle g\rangle \cap \langle a, b\rangle \le \langle ab\rangle\,,$$

a contradiction. Hence $g$ is an involution and $\langle g, ab\rangle \cong \mathrm{C}_2 \times \mathrm{C}_4$.

We have $\langle ab\rangle \le \mathrm{Z}(H)$, and so $\langle ab\rangle \trianglelefteq H$. Also,

$$\langle abg\rangle \cong \mathrm{C}_4 \quad \text{with } (abg)^2 = a^2b^2 \in \mathrm{Z}(H) \text{ and so } \langle abg\rangle \trianglelefteq H\,.$$

Thus,

$$\langle g, ab\rangle \trianglelefteq H \quad \text{and so } \Omega_1(\langle g, ab\rangle) = \langle g, a^2b^2\rangle \trianglelefteq H\,.$$

But $[a,g] = a^{-1}b \notin \langle g, a^2b^2\rangle$, a contradiction.

(ii) Suppose $b^g = a^{-1}$. In this case

$$[a,g] = a^{-1}b\,, \quad a^{g^2} = b^g = a^{-1}\,, \quad b^{g^2} = (a^{-1})^g = b^{-1}\,,$$

and so $g^2$ inverts each element in $\langle a,b\rangle$, and therefore $\langle g\rangle \cap \langle a,b\rangle \le \langle a^2b^2\rangle$. We set again $H = \langle a,g\rangle$, and so

$$H' = \langle a^2, b^2, a^{-1}b\rangle \cong C_4 \times C_2 \quad \text{and} \quad Z(H) = \langle g^4, a^2b^2\rangle\,.$$

If $g^4 \neq 1$, then $\langle g\rangle \cap Z(H) \neq \{1\}$, and so $\langle g\rangle \trianglelefteq H$. But $[a,g] = a^{-1}b \notin \langle g\rangle$ since

$$\langle g\rangle \cap \langle a,b\rangle \le \langle a^2b^2\rangle\,.$$

Hence $g^4 = 1$, and so $H = \langle a,b,g\rangle$ is a semidirect product of $\langle a,b\rangle \cong C_4 \times C_4$ by $\langle g\rangle \cong C_4$.

Since $a^2$ centralizes $\langle g^2\rangle$, it follows that $a^2$ normalizes $\langle g\rangle \cong C_4$. But $a^2b^2 \in Z(H)$, and so $a^2b^2$ also normalizes $\langle g\rangle$, and so $E_4 \cong \langle a^2, b^2\rangle$ normalizes $\langle g\rangle$. Since $\langle g\rangle \cap \langle a^2, b^2\rangle = \{1\}$ and $g$ also normalizes $\langle a^2, b^2\rangle$, it follows that $g$ centralizes $\langle a^2, b^2\rangle$, a contradiction.

We have proved that in $G$ each cyclic subgroup of composite order is normal in $G$. Hence, by Theorem 63.4, $G$ is either quasidihedral (i.e., $|G : H_2(G)| = 2$) or $|G'| = 2$ and $\Phi(G)$ is cyclic. The proof is complete. □

**Theorem 187.2.** *Let $G$ be a $p$-group, $p > 2$, in which the intersection of any two distinct conjugate cyclic subgroups is trivial. Then each cyclic subgroup of composite order is normal in $G$ and $G$ is either abelian or of exponent $p$ or $|G'| = p$ and $\Phi(G)$ is cyclic.*

*Proof.* Since the assumption of our theorem is hereditary for subgroups, we may use induction on $|G|$. Assume that $G$ is a minimal counterexample to our theorem. Then $G$ possesses a nonnormal cyclic subgroup $X$ of order $\ge p^2$. Set $N = N_G(X)$ so that $N < G$ and let $N < G_0 \le G$ be such that $|G_0 : N| = p$. Take $g \in G_0 - N$, where $g^p \in N$. Let $a \in X$ be of order $p^2$. Then $N_G(\langle a\rangle) = N$.

Set

$$A = \langle a, a^g, a^{g^2}, \ldots, a^{g^{p-1}}\rangle = \langle a\rangle^{G_0}$$

and we note that each $\langle a^{g^i}\rangle$ is normal in $N$ and for any two distinct conjugates $\langle a^{g^i}\rangle$ and $\langle a^{g^j}\rangle$, we have $\langle a^{g^i}\rangle \cap \langle a^{g^j}\rangle = \{1\}$. Thus $A$ is abelian of exponent $p^2$ and type $(p^2, p^2, \ldots)$. Also, the rank of $A$ is at least 2 and $A$ is $\langle g\rangle$-invariant.

By the minimality of our counterexample, we have $G = A\langle g\rangle$. Also, we have

$$\Phi(A) = \Omega_1(A) \quad \text{and} \quad |A : \Omega_1(A)| \ge p^2\,,$$

and so each maximal subgroup of $A$ is of exponent $p^2$. Therefore each maximal subgroup of $G$ is of exponent $> p$. Hence, by induction, each maximal subgroup $H$ of $G$ is either abelian or $|H'| = p$ and $\Phi(H)$ is cyclic, and so our group $G$ is a title group from § 137.

Suppose that $g^p$ does not centralize $\langle a \rangle$. Then we may set $[a, g^p] = a^p$ and acting with $g$ on this relation, we also get $[a^g, g^p] = (a^g)^p$. This implies that the proper subgroup $A\langle g^p \rangle$ has a commutator group containing $\langle a^p, (a^g)^p \rangle \cong E_{p^2}$, a contradiction. It follows that $g^p$ centralizes $\langle a \rangle$.

On the other hand,

$$[a, g] = a^{-1}g^{-1}ag = a^{-1}a^g \text{ is of order } p^2 \quad \text{since } \langle a, a^g \rangle \cong C_{p^2} \times C_{p^2} .$$

But, by Proposition 137.5, $[a, g]^p = [a, g^p] = 1$, a contradiction.

We have proved that in $G$ each cyclic subgroup of composite order is normal in $G$. Hence, by Theorem 63.4, $G$ is either abelian or of exponent $p$ or $|G'| = p$ and $\Phi(G)$ is cyclic. Our theorem is proved. □

**Corollary 187.3.** *Let $G$ be a non-Dedekindian $p$-group in which the intersection of any two distinct conjugate metacyclic subgroups is trivial (i.e., all metacyclic subgroups are TI-subgroups). Then each subgroup of composite order is normal in $G$, and so $G$ is isomorphic to one of the $p$-groups defined in Theorem 1.25.*

*Conversely, all $p$-groups from Theorem 1.25 satisfy the assumption of the corollary.*

*Proof.* By Theorems 187.1 and 187.2, each cyclic subgroup of composite order is normal in $G$.

(i) First we show that $|G'| = p$ and $\Phi(G)$ is cyclic.

In case $p = 2$, Theorem 187.1 shows that we have to consider only the case that $G$ is quasidihedral, where $A = H_2(G)$ is abelian of index 2 and exponent $> 2$. Let $i$ be an involution in $G - A$ so that $i$ inverts each element in $A$. Let $z$ be any involution in $A$ and obviously, $z \in Z(G)$. Any two conjugates of the metacyclic subgroup $\langle z, i \rangle \cong E_4$ contain $z$, and so, by our assumption, $\langle z, i \rangle \trianglelefteq G$. It follows

$$G' = [A, i] \le A \cap \langle z, i \rangle = \langle z \rangle .$$

In particular, $A$ has only one involution $z$, and so $A$ is cyclic and $A \cong C_4$. It follows $G \cong D_8$. We have proved that $|G'| = 2$ and $\Phi(G)$ is cyclic.

Assume $p > 2$. By Theorem 187.2, we have to consider only the case where $G$ is a nonabelian group of exponent $p$. Let $1 \neq z \in Z(G)$ and let $X/\langle z \rangle$ be any subgroup of order $p$ in $G/\langle z \rangle$. Then $X \cong E_{p^2}$ is metacyclic but any two conjugates of $X$ contain $\langle z \rangle$, and so $X \trianglelefteq G$. It follows that $G/\langle z \rangle$ is elementary abelian, and so $G' = \langle z \rangle \cong C_p$ and $\Phi(G) = \langle z \rangle$ is cyclic.

(ii) Let $M$ be any nonnormal subgroup in $G$. Then $G' \not\le M$ and since $|G'| = p$, we get $M \cap G' = \{1\}$. We know that $\Phi(G)$ is cyclic, and so $\Omega_1(\Phi(G)) = G'$, which implies $M \cap \Phi(G) = \{1\}$. It follows that $M$ is elementary abelian, and therefore $L = M \times G'$ is an elementary abelian normal subgroup with $|L : M| = p$.

Suppose, by way of contradiction, that $|M| > p$. Because $M$ is generated by its elementary abelian subgroups of order $p^2$ and $M$ is not normal in $G$, it follows that $M$ contains a subgroup $M_0 \cong \mathrm{E}_{p^2}$ that is not normal in $G$. On the other hand, $L_0 = M_0 \times G' \cong \mathrm{E}_{p^3}$ is normal in $G$. Let $g \in G$ be such that $M_0^g \neq M$. Since $M_0^g \leq L_0$, we have $M_0 \cap M_0^g \neq \{1\}$, a contradiction.

We have proved that each nonnormal subgroup in $G$ is of order $p$. By Theorem 1.25, $G$ is isomorphic to one of the $p$-groups defined in that theorem.

Conversely, it is obvious that each metacyclic subgroup in any $p$-group defined in Theorem 1.25 is a TI-subgroup. □

# § 188 *p*-groups with small subgroups generated by two conjugate elements

We present here some results of H. Lv, W. Z. Zhou and H. Xu [LZX] concerning so-called $K_1$-groups. A $p$-group $G$ in which

$$|\langle x, x^y\rangle : \langle x\rangle| \le p \quad \text{for all } x, y \in G$$

is called a $K_1$-group. This is actually a generalization of $p$-groups considered in Problem 1762. Note that the $K_1$-property is hereditary to subgroups and factor-groups. We prove the following result:

**Theorem 188.1** (Lv, Zhou, Xu [LZX]). *Let $G$ be a $p$-group that is a $K_1$-group. Then the following holds:*

(a) *If $p > 2$, then $\mho_1(G) \le Z(G)$, $\exp(G') \le p$ and $\mathrm{cl}(G) \le 4$. Moreover, if $p > 3$, then $\mathrm{cl}(G) \le 3$.*

(b) *If $p = 2$, then $\mho_2(G) \le Z(G)$, $\exp(G') \le 4$ and $\mathrm{cl}(G) \le 3$.*

We first prove some auxiliary lemmas. The 'left normed' commutator $[x_1, \dots, x_n]$ for $n > 2$ is defined inductively as

$$[[x_1, \dots, x_{n-1}], x_n] .$$

We write

$$[x_1, (n-1)y] \quad \text{for } [x_1, \dots, x_n], \quad \text{where } x_2 = \dots = x_n = y .$$

**Lemma 188.2.** *In metabelian groups for $n \ge 2$ the following identities hold:*

(1) $$[x^n, y] = \prod_{i=1}^{n} [x, y, (i-1)x]^{\binom{n}{i}} ,$$

(2) $$[x, y^n] = \prod_{i=1}^{n} [x, iy]^{\binom{n}{i}} .$$

*Proof.* We prove (1) by induction on $n$. For $n = 2$ we get

$$[x^2, y] = [x, y]^x [x, y] = [x, y]([x, y]^{-1}[x, y]^x)[x, y] = [x, y]^2 [x, y, x]$$
$$= \prod_{i=1}^{2} [x, y, (i-1)x]^{\binom{2}{i}} .$$

Supposing that (1) holds for $n \geq 2$, we get for $n+1$:

$$\begin{aligned}
[x^{n+1}, y] &= [x^n x, y] = [x^n, y]^x [x, y] = [x^n, y][x^n, y, x][x, y] \\
&= [x, y] \prod_{i=1}^{n} [x, y, (i-1)x]^{\binom{n}{i}} [\prod_{i=1}^{n} [x, y, (i-1)x]^{\binom{n}{i}}, x] \\
&= [x, y] \prod_{i=1}^{n} [x, y, (i-1)x]^{\binom{n}{i}} \prod_{i=1}^{n} [x, y, ix]^{\binom{n}{i}} \\
&= [x, y]^{\binom{n+1}{1}} [x, y, x]^{\binom{n2}{+}\binom{n}{1}} [x, y, 2x]^{\binom{n3}{+}\binom{n}{2}} \dots [x, y, (n-1)x]^{\binom{nn}{+}\binom{n}{n-1}} [x, y, nx]^{\binom{n}{n}} \\
&= [x, y]^{\binom{n+1}{1}} [x, y, x]^{\binom{n+1}{2}} [x, y, 2x]^{\binom{n+1}{3}} \dots [x, y, (n-1)x]^{\binom{n+1}{n}} [x, y, nx]^{\binom{n+1}{n+1}} \\
&= \prod_{i=1}^{n+1} [x, y, (i-1)x]^{\binom{n+1}{i}} ,
\end{aligned}$$

where we use the obvious equality $\binom{n}{j} + \binom{n}{j-1} = \binom{n+1}{j}$.

To prove (2), we have for $n = 2$,

$$[x, y^2] = [x, y][x, y]^y = [x, y][x, y]([x, y]^{-1}[x, y]^y) = [x, y]^2[x, y, y] = \prod_{i=1}^{2} [x, iy]^{\binom{2}{i}} .$$

Assuming that (2) holds for $n \geq 2$, we get for $n+1$:

$$\begin{aligned}
[x, y^{n+1}] &= [x, y^n y] = [x, y][x, y^n]^y = [x, y][x, y^n][x, y^n, y] \\
&= [x, y] \prod_{i=1}^{n} [x, iy]^{\binom{n}{i}} [\prod_{i=1}^{n} [x, iy]^{\binom{n}{i}}, y] \\
&= [x, y] \prod_{i=1}^{n} [x, iy]^{\binom{n}{i}} \prod_{i=1}^{n} [x, (i+1)y]^{\binom{n}{i}} \\
&= [x, y]^{\binom{n+1}{1}} [x, 2y]^{\binom{n2}{+}\binom{n}{1}} [x, 3y]^{\binom{n3}{+}\binom{n}{2}} \dots [x, ny]^{\binom{nn}{+}\binom{n}{n-1}} [x, (n+1)y]^{\binom{n}{n}} \\
&= [x, y]^{\binom{n+1}{1}} [x, 2y]^{\binom{n+1}{2}} [x, 3y]^{\binom{n+1}{3}} \dots [x, ny]^{\binom{n+1}{n}} [x, (n+1)y]^{\binom{n+1}{n+1}} \\
&= \prod_{i=1}^{n+1} [x, iy]^{\binom{n+1}{i}} ,
\end{aligned}$$

completing the proof. □

**Lemma 188.3.** *Suppose that $G$ is a $p$-group that is a $K_1$-group. Then for any element $x$ of order $p$, $\langle x \rangle^G$ is elementary abelian. If $x, y \in G$ are both of order $p$ with $\langle x \rangle \neq \langle y \rangle$, then*

$$\langle x, y \rangle \cong E_{p^2} , \quad D_8 \text{ or } S(p^3) ,$$

*where $S(p^3)$ is the nonabelian group of order $p^3$ and exponent $p > 2$.*

*Proof.* For any $g \in G$, one has $|\langle x, x^g \rangle| \leq p^2$ since $o(x) = p$ and $\langle x, x^g \rangle$ is abelian. Hence the subgroup $\langle x \rangle^G$ is elementary abelian. Let $x, y \in G$ be both of order $p$ with

$\langle x\rangle \neq \langle y\rangle$. Then $H = \langle x\rangle^G \langle y\rangle^G$ is of class $\leq 2$ with $H'$ elementary abelian. Hence either $[x, y] = 1$ and $\langle x, y\rangle \cong \mathrm{E}_{p^2}$ or $o([x, y]) = p$ with $[x, y] \in \mathrm{Z}(\langle x, y\rangle)$ so that $\langle x, y\rangle$ is minimal nonabelian of order $p^3$ (Lemma 65.2 (a)) generated by two elements of order $p$, and therefore in this case $\langle x, y\rangle \cong \mathrm{D}_8$ or $\mathrm{S}(p^3)$. □

**Lemma 188.4.** *Suppose that $G = \langle x, y\rangle$ is a $p$-group that is a $\mathrm{K}_1$-group with $\langle x\rangle \cap \langle y\rangle = \{1\}$. Then $\mathrm{cl}(G) \leq 3$ and $G'$ is elementary abelian of order $\leq p^3$. Moreover, if $p > 2$, then $\mho_1(G) \leq \mathrm{Z}(G)$ and if $p = 2$, then $\mho_2(G) < \mathrm{Z}(G)$.*

*Proof.* One may assume that $G$ is nonabelian. Since $G$ is a $\mathrm{K}_1$-group, it follows that:

$$|\langle x, x^g\rangle : \langle x\rangle| \leq p \text{ for each } g \in G \quad \text{and so } \langle x^p\rangle^g \leq \langle x\rangle\,,$$

which gives

$$\langle x^p\rangle^g = \langle x^p\rangle \quad \text{and} \quad \langle x^p\rangle \trianglelefteq G\,.$$

Similarly, $\langle y^p\rangle \trianglelefteq G$. Set $N = \langle x^p, y^p\rangle = \langle x^p\rangle \times \langle y^p\rangle$.

By Lemma 188.3, either $G/N \cong \mathrm{E}_{p^2}$ or $G/N$ is minimal nonabelian of order $p^3$ generated by two elements of order $p$, and so in this case, $G/N \cong \mathrm{D}_8$ or $S(p^3)$. It follows, working in $G/N$:

$$|\langle xN\rangle^{G/N}| \leq p^2 \quad \text{and so } |(\langle x\rangle^G N) : N| \leq p^2\,.$$

By the product formula,

$$|\langle x\rangle^G N| = (|\langle x\rangle^G|\,|N|) : |\langle x\rangle^G \cap N| \quad \text{and so } |\langle x\rangle^G| : |\langle x\rangle^G \cap N| = |\langle x\rangle^G N| : |N| \leq p^2\,,$$

which implies

$$|\langle x\rangle^G| \leq p^2 |\langle x\rangle^G \cap N| = p^2 |\langle x^p\rangle(\langle x\rangle^G \cap \langle y^p\rangle)|\,.$$

Let $\bar{G} = G/\langle x^p\rangle$ and $\bar{x} = x\langle x^p\rangle$. By Lemma 188.3, $\langle \bar{x}\rangle^{\bar{G}}$ is elementary abelian, and so $\langle x\rangle^G/\langle x^p\rangle$ is elementary abelian. Since $\langle x\rangle \cap \langle y\rangle = \{1\}$, we obtain $|\langle x\rangle^G \cap \langle y\rangle| \leq p$. Because

$$|\langle x\rangle^G| \leq p^2 |\langle x^p\rangle(\langle x\rangle^G \cap \langle y^p\rangle)| \leq p^3 |\langle x^p\rangle|\,,$$

we get

$$|\langle x\rangle^G : \langle x\rangle| \leq p^2 \text{ and if } |\langle x\rangle^G : \langle x\rangle| = p^2\,, \text{ then } |\langle x\rangle^G \cap \langle y\rangle| = p\,.$$

Similarly,

$$|\langle y\rangle^G : \langle y\rangle| \leq p^2 \text{ and if } |\langle y\rangle^G : \langle y\rangle| = p^2\,, \text{ then } |\langle y\rangle^G \cap \langle x\rangle| = p\,.$$

Set $M = \langle x\rangle^G \cap \langle y\rangle^G$ so that $|M| \leq p^3$ and if $|M| = p^3$, then $\Omega_1(\langle x\rangle) \leq M$ and $\Omega_1(\langle y\rangle) \leq M$, which implies that $o(x) \geq p^2$, $o(y) \geq p^2$ and $M$ is abelian. Since $G = \langle x, y\rangle$, it follows that $G/\langle x\rangle^G$ and $G/\langle y\rangle^G$ are cyclic, and so $G' \leq M$, which gives that $G'$ is abelian of order $\leq p^3$. If $|G'| \leq p^2$, then $\mathrm{cl}(G) \leq 3$. If $|G'| = p^3$, then

$$G' = M\,, \quad \Omega_1(\langle x\rangle) \leq M\,, \quad \Omega_1(\langle y\rangle) \leq M\,,$$

which together with $o(x) \geq p^2$, $o(y) \geq p^2$ implies that $\Omega_1(\langle x\rangle) \times \Omega_1(\langle y\rangle) \leq Z(G)$, and so again $\mathrm{cl}(G) \leq 3$.

Now we prove that $G'$ is elementary abelian. Since $G'$ is abelian, it suffices to show that $a^p = 1$ for any $a \in G'$. Because $\langle x\rangle \cap \langle y\rangle = \{1\}$, then either $\langle a\rangle \cap \langle x\rangle = \{1\}$ or $\langle a\rangle \cap \langle y\rangle = \{1\}$. Without loss of generality, we may assume $\langle a\rangle \cap \langle x\rangle = \{1\}$. But $a \in \langle x\rangle^G$ and we know that $\langle x\rangle^G/\langle x^p\rangle$ is elementary abelian. This gives $a^p \in \langle x^p\rangle$, and so $a^p = 1$.

First suppose that $p > 2$. By Lemma 188.2,

$$[x^p, y] = \prod_{i=1}^{p}[x, y, (i-1)x]^{\binom{p}{i}} = 1 \quad \text{and so } x^p \in Z(G)$$

and similarly,

$$[x, y^p] = \prod_{i=1}^{p}[x, iy]^{\binom{p}{i}} = 1 \quad \text{and so } y^p \in Z(G)\,.$$

It follows that $N \leq Z(G)$ and since $G/N$ is of exponent $p$, we get $\mho_1(G) \leq Z(G)$.

Now suppose $p = 2$. By Lemma 188.2, for any $g, h \in G$ we get

$$[g, h^4] = \prod_{i=1}^{4}[g, ih]^{\binom{4}{i}} = 1 \quad \text{and so } h^4 \in Z(G)$$

and this implies $\mho_2(G) \leq Z(G)$. The proof is complete. □

*Proof of Theorem 188.1.* Let $x, y \in G$ and $H = \langle x, y\rangle$. Without loss of generality, suppose

$$o(x) = p^m \geq o(y) = p^n\,.$$

Since

$$|\langle x, x^g\rangle : \langle x\rangle| \leq p \text{ for any } g \in G\,, \text{ we have } \langle x^p\rangle^g = \langle x^p\rangle\,.$$

Thus, $\langle x^p\rangle \trianglelefteq G$ and similarly, $\langle y^p\rangle \trianglelefteq G$.

(i) Assume $p > 2$. We shall prove in this case that $\mho_1(H) \leq Z(H)$ and $H'$ is elementary abelian. This then implies

$$[x^p, y] = [x, y^p] = 1 \quad \text{and} \quad o([x, y]) \leq p\,,$$

and therefore $\mho_1(G) \leq Z(G)$ and by Lemma 188.3, $\exp(G') \leq p$. If $\langle x\rangle \cap \langle y\rangle = \{1\}$, then Lemma 188.4 gives the desired conclusions. Therefore we suppose $\langle x\rangle \cap \langle y\rangle \neq \{1\}$.

If $m \leq 2$, then $n \leq 2$ and since $\langle x\rangle \cap \langle y\rangle \neq \{1\}$, we need only to consider the case $m = n = 2$. Here we have $\langle x^p\rangle = \langle y^p\rangle \leq Z(G)$. By Lemma 188.3,

$$\mho_1(H) \leq \langle x^p\rangle \leq Z(H) \quad \text{and} \quad |H'| \leq p^2$$

so that $H'$ is abelian. By Lemma 188.2 (2),

$$1 = [x, y^p] = [x, y]^p[x, y, y]^{\binom{p}{2}} = [x, y]^p\,,$$

and so $[x, y]^p = 1$, which implies that $H'$ is elementary abelian, as required.

From now on we consider the possibility $m \geq 3$. If there is an element $y_1 \in H$ such that $H = \langle x, y_1\rangle$ and $\langle x\rangle \cap \langle y_1\rangle = \{1\}$, then Lemma 188.4 implies $\mho_1(H) \leq Z(H)$ and $H'$ is elementary abelian. We shall prove this by induction on $n$.

First suppose $m > n \geq 2$. Set $H_1 = \langle x^{p^{m-n}}, y\rangle$ and note that $C_{p^n} \cong \langle x^{p^{m-n}}\rangle \trianglelefteq H$. By Theorem 7.1 (c), $H_1$ is regular since $H_1' \leq \langle x^{p^{m-n}}\rangle$ is cyclic. Since $\langle x\rangle \cap \langle y\rangle \neq \{1\}$, there is an integer $s \not\equiv 0 \pmod p$ such that $(x^{sp^{m-n}})^{p^{n-1}} = y^{p^{n-1}}$. By Theorem 7.2 (a), $(x^{-sp^{m-n}}y)^{p^{n-1}} = 1$. Because

$$H = \langle x, x^{-sp^{m-n}}y\rangle \quad \text{and} \quad o(x^{-sp^{m-n}}y) < p^n = o(y)\,,$$

we get by induction that there is $y_1 \in H$ satisfying the above requirements.

Next, suppose $m = n$. Let $\langle x\rangle \cap \langle y\rangle = \langle y^{p^k}\rangle$, where $k \geq 1$. If $k = 1$, then Lemma 188.3 implies that

$$H/\langle y^p\rangle \cong E_{p^2} \quad \text{or} \quad S(p^3)\,.$$

We have $\mho_1(H) \leq \langle y^p\rangle \leq Z(H)$. Let $H^*$ be any maximal subgroup of $H$ so that $H^* \geq \langle y^p\rangle$ and $H^*/\langle y^p\rangle$ is elementary abelian of order $\leq p^2$. This gives $|(H^*)'| \leq p$. By Proposition 137.5, for any $h_1, h_2 \in H$, one has $1 = [h_1^p, h_2] = [h_1, h_2]^p$, and so $H'$ is elementary abelian of order $\leq p^2$. In this case, $H$ has the required properties.

If $k \geq 2$, then we consider $\bar{H} = H/\langle y^{p^{k+1}}\rangle$ and use the bar convention. By Lemma 188.4, $(\bar{H}/\langle \bar{y}^{p^k}\rangle)'$ is an elementary abelian group of order $\leq p^3$ and $\mathrm{cl}(\bar{H}/\langle \bar{y}^{p^k}\rangle) \leq 3$. It follows that:

$$\mathrm{cl}(\bar{H}) \leq 4 \quad \text{and} \quad |\bar{H}'| \leq p^4 \quad \text{with } |\bar{H}' : \Omega_1(\bar{H}')| \leq p\,.$$

Indeed, by Theorem 7.1 (c), $\bar{H}'$ is regular, and so Theorem 7.2 (d) implies $|\Omega_1(\bar{H}')| = |\bar{H}' : \mho_1(\bar{H}')|$.

We have $\bar{y}^{p^k} = \bar{x}^{-sp^k}$ for some integer $s$. By Theorem A.1.3 (Hall–Petrescu formula),

$$(\bar{x}^s\bar{y})^{p^k} = \bar{x}^{sp^k}\bar{y}^{p^k}c_2^{\binom{p^k}{2}}c_3^{\binom{p^k}{3}}c_4^{\binom{p^k}{4}} = \bar{x}^{sp^k}\bar{y}^{p^k} = 1\,,$$

where $c_i = K_i(\bar{H})$, $i = 2, 3, 4$. Then $o(\bar{x}^s\bar{y}) \leq p^k$. Hence $o(x^s y) < p^n$ and $H = \langle x, x^s y\rangle$. By induction, there is $y_1 \in H$ such that $H = \langle x, y_1\rangle$ and $\langle x\rangle \cap \langle y_1\rangle = \{1\}$, as required.

Let $\bar{G} = G/\mho_1(G)$, where $\mho_1(G) \leq Z(G)$. By Lemma 188.3, for any $\bar{g}, \bar{h} \in \bar{G}$, the subgroup $\langle \bar{g}, \bar{h}\rangle$ is either elementary abelian of order $\leq p^2$ or $\langle \bar{g}, \bar{h}\rangle \cong S(p^3)$. In any case, $[\bar{g}, \bar{h}, \bar{h}] = 1$, which shows that $\bar{G}$ is 2-Engel group. By Satz 6.5 in [Hup1], if $p \geq 5$, then $\mathrm{cl}(\bar{G}) \leq 2$, and so $\mathrm{cl}(G) \leq 3$ and if $p = 3$, then $\mathrm{cl}(\bar{G}) \leq 3$, and so $\mathrm{cl}(G) \leq 4$.

(ii) Suppose $p = 2$. We shall show $\mho_2(H) \leq Z(H)$ and this implies $\mho_2(G) \leq Z(G)$. If $\langle x\rangle \cap \langle y\rangle = \{1\}$, then Lemma 188.4 implies $\mho_2(H) \leq Z(H)$ and $H'$ is elementary abelian. From now on we suppose $\langle x\rangle \cap \langle y\rangle \neq \{1\}$.

If $m = n$ and $\langle x\rangle \cap \langle y\rangle = \langle x^2\rangle$, then Lemma 188.3 implies that $H/\langle x^2\rangle$ is either elementary abelian of order $\leq 4$ or $H/\langle x^2\rangle \cong D_8$, and so $\mho_2(H) \leq Z(H)$.

If $m = n$ and $\langle x\rangle \cap \langle y\rangle = \langle x^4\rangle$, then we consider

$$\bar{H} = H/\langle x^4\rangle = \langle \bar{x}, \bar{y}\rangle, \quad \text{where } \langle \bar{x}\rangle \cong \langle \bar{y}\rangle \cong C_4 \text{ and } \langle \bar{x}\rangle \cap \langle \bar{y}\rangle = \{1\}.$$

We set $E_4 \cong \bar{N} = \langle \bar{x}^2\rangle \times \langle \bar{y}^2\rangle \le Z(\bar{H})$, where by Lemma 188.3, $\bar{H}/\bar{N} \cong E_4$ or $D_8$. By Lemma 188.4, $\mathrm{cl}(\bar{H}) \le 3$ and $\bar{H}'$ is elementary abelian. By Theorem A.1.3.,

$$(\bar{x}\bar{y})^4 = \bar{x}^4\bar{y}^4 c_2^{\binom{4}{2}} c_3^{\binom{4}{3}} = 1, \quad \text{where } c_i \in K_i(\bar{H}), \quad i = 2, 3,$$

and so $\exp(\bar{H}) \le 4$ and $\mho_2(H) \le \langle x^4\rangle \le Z(H)$.

Suppose $m = n$ and $\langle x\rangle \cap \langle y\rangle = \langle x^{2^k}\rangle$, where $k \ge 3$. By Lemma 188.4, $(H/\langle x^{2^k}\rangle)'$ is elementary abelian of order $\le 2^3$ and $\mathrm{cl}(H/\langle x^{2^k}\rangle) \le 3$. Set $\bar{H} = H/\langle x^{2^{k+1}}\rangle$. It follows that $|\bar{H}'| \le 2^4$, $\exp(\bar{H}') \le 4$ and $\mathrm{cl}(\bar{H}) \le 4$. By Theorem A.1.3. (Hall–Petrescu formula),

$$(\bar{x}\bar{y})^{2^k} = \bar{x}^{2^k}\bar{y}^{2^k} c_2^{\binom{2^k}{2}} c_3^{\binom{2^k}{3}} c_4^{\binom{2^k}{4}} = \bar{x}^{2^k}\bar{y}^{2^k} = 1, \quad \text{where } c_i \in K_i(\bar{H}), \quad i = 2, 3, 4.$$

Then $o(\bar{x}\bar{y}) \le 2^k$, and so $o(xy) < 2^m$. Also note that $H = \langle x, xy\rangle$.

It follows that we need only to consider the case $H = \langle x, y\rangle$, where

$$o(x) = 2^m > o(y) = 2^n \quad \text{and} \quad \langle x\rangle \cap \langle y\rangle \ne \{1\}.$$

If $m \le 3$, then it is easy to see that $\mho_2(H) \le \langle x\rangle \cap \langle y\rangle \le Z(H)$. Indeed, in this case we need only to consider the case $m = 3$, $n = 2$ and $\langle x\rangle \cap \langle y\rangle = \langle x^4\rangle = \langle y^2\rangle$. Set $\bar{H} = H/\langle x^4\rangle$ so that Lemma 188.4 implies that $\bar{H}'$ is elementary abelian and $\mathrm{cl}(\bar{H}) \le 3$. By the Hall–Petrescu formula,

$$(\bar{x}\bar{y})^4 = \bar{x}^4\bar{y}^4 c_2^{\binom{4}{2}} c_3^{\binom{4}{3}} = 1, \quad \text{where } c_i \in K_i(\bar{H}), \quad i = 2, 3.$$

By Lemma 188.3, $H/\langle x^2\rangle \cong E_4$ or $D_8$, and so $\exp(\bar{H}) \le 4$, which finally implies $\mho_2(H) \le \langle x^4\rangle \le Z(H)$.

Now consider $m \ge 4$. We shall show that there is an element $y_1 \in H$ such that $H = \langle x, y_1\rangle$ and $\langle x\rangle \cap \langle y_1\rangle = \{1\}$. We proceed by induction on $n$.

If $n = 2$, then we have to consider only the case $\langle x\rangle \cap \langle y\rangle = \langle y^2\rangle$. By Lemma 188.4, $(H/\langle y^2\rangle)'$ is elementary abelian of order $\le 2^3$ and by Lemma 188.3, $H/\langle x^2\rangle \cong E_4$ or $D_8$ so that $\langle x\rangle^H = \langle x, x^y\rangle$ is either equal $\langle x\rangle$ (in which case $\langle x\rangle \trianglelefteq H$) or $\langle x\rangle^H \cong M_{2^{m+1}}$ with $(\langle x\rangle^H)' = \langle y^2\rangle$, $m \ge 4$. We compute:

$$[y, x^2] = [y, x][y, x]^x = [y, x]^2([y, x]^{-1}[y, x]^x) = [y, x]^2[y, x, x].$$

On the other hand,

$$[y, x]^2 \in \langle y^2\rangle \le Z(H) \quad \text{and} \quad [y, x] \in \langle x\rangle^H \quad \text{so that } [y, x, x] \in \langle y^2\rangle.$$

We have proved that $[y, x^2] \in \langle y^2\rangle$, and so $y$ (normalizing $\langle x^2\rangle$) centralizes the subgroup $\langle x^{2^{m-2}}\rangle \cong C_4$. But then $y_1 = x^{2^{m-2}}y$ is of order 2 and $H = \langle x, y_1\rangle$ with $\langle x\rangle \cap \langle y_1\rangle = \{1\}$.

If $n \geq 3$, then we set $\langle x\rangle \cap \langle y\rangle = \langle y^{2^k}\rangle$, $k \geq 1$.

First assume $k \geq 2$. Consider $\bar{H} = H/\langle y^{2^{k+1}}\rangle$. Then $o(\bar{x}) = 2^{m-n+k+1} \geq 2^4$ and $o(\bar{y}) = 2^{k+1}$. By Lemma 188.4, we get $\exp(\bar{H}') \leq 4$, and so $[\bar{x}, \bar{y}]^4 = (\bar{x}^{-1}\bar{x}^{\bar{y}})^4 = 1$. Note that $|\langle \bar{x}, \bar{x}^{\bar{y}}\rangle : \langle \bar{x}\rangle| \leq 2$. If $\langle \bar{x}, \bar{x}^{\bar{y}}\rangle$ is a group of maximal class, then $\exp(\langle \bar{x}, \bar{x}^{\bar{y}}\rangle)' \geq 2^3$, contrary to $\exp(\bar{H}') \leq 4$. Hence either $\langle \bar{x}, \bar{x}^{\bar{y}}\rangle$ is abelian or $\langle \bar{x}, \bar{x}^{\bar{y}}\rangle \cong \mathrm{M}_{m-n+k+2}$. It follows from the above:

$$1 = (\bar{x}^{-1}\bar{x}^{\bar{y}})^4 = \bar{x}^{-4}(\bar{x}^4)^{\bar{y}} \quad \text{and so } (\bar{x}^4)^{\bar{y}} = \bar{x}^4\,,$$

which gives $\bar{x}^4 \in Z(\bar{H})$. Since $\langle \bar{x}^2\rangle \trianglelefteq \bar{H}$, we have

$$\mathrm{cl}(\bar{x}^2, \bar{y}) \leq 2 \quad \text{and} \quad [\bar{x}^2, \bar{y}]^2 = [\bar{x}^4, \bar{y}] = 1\,,$$

which implies

$$(\bar{x}^{2^{m-n}}\bar{y})^{2^k} = \bar{x}^{2^{m-n+k}}\bar{y}^{2^k} = 1 \quad \text{and so } o(\bar{x}^{2^{m-n}}\bar{y}) \leq 2^k$$

and therefore $o(x^{2^{m-n}}y) < 2^n$. Since $H = \langle x, x^{2^{m-n}}y\rangle$, we have by induction that there is $y_1 \in H$ such that $H = \langle x, y_1\rangle$ and $\langle x\rangle \cap \langle y_1\rangle = \{1\}$.

Finally, suppose $k = 1$. Since $\langle x^{2^{m-n}}\rangle \trianglelefteq H$, then $\langle x^{2^{m-n}}, y\rangle$ has a cyclic subgroup of index 2. Because $n \geq 3$, we have that $\langle x^{2^{m-n}}, y\rangle$ is either abelian or $\langle x^{2^{m-n}}, y\rangle \cong \mathrm{M}_{n+1}$, which implies

$$(x^{2^{m-n}}y)^{2^{n-1}} = x^{2^{m-1}}y^{2^{n-1}} = 1\,,$$

and so $o(x^{2^{m-n}}y) < 2^n$ and $H = \langle x, x^{2^{m-n}}y\rangle$. By induction, there is $y_1 \in H$ such that $H = \langle x, y_1\rangle$ and $\langle x\rangle \cap \langle y_1\rangle = \{1\}$.

In all cases, by Lemma 188.4, we get $\mho_2(H) \leq Z(H)$, and so the required property $\mho_2(G) \leq Z(G)$ is completely proved.

If $\exp(G) \leq 4$, then $\mho_1(G) \leq Z(G)$ since $\langle x^2\rangle \trianglelefteq G$ for each $x \in G$, and so $\mathrm{cl}(G) \leq 2$. In general, $\exp(G/\mho_2(G)) \leq 4$, and so $\mathrm{cl}(G/\mho_2(G)) \leq 2$. But $\mho_2(G) \leq Z(G)$ and so $\mathrm{cl}(G) \leq 3$.

Let $x, y \in G$ and $o(x) = 2^m$. Note that $\langle x^2\rangle \trianglelefteq G$ and $\langle x^4\rangle \leq Z(G)$. It follows that:

$$(x^2)^y = x^{2+k2^{m-1}} \text{ with } k = 0, 1\,, \quad \text{and then } x^{-2}(x^2)^y = x^{k2^{m-1}} \in Z(G)\,.$$

Let

$$M_1 = \langle x, x^y\rangle \quad \text{so that } |M_1 : \langle x\rangle| \leq 2 \text{ gives } |M_1'| \leq 2\,.$$

Working in the subgroup $M_1$ of class $\leq 2$, we obtain

$$[x, y]^2 = (x^{-1}x^y)^2 = x^{-2}(x^2)^y c_2\,, \quad \text{where } c_2 \in M_1'\,.$$

Hence we have

$$[x, y]^4 = (x^{-2}(x^2)^y c_2)^2 = (x^{k2^{m-1}}c_2)^2 = 1\,.$$

Let $a, b \in G'$ be such that $a^4 = b^4 = 1$. Set $T = \langle a, b\rangle$. If $\langle a\rangle \cap \langle b\rangle = \{1\}$, then Lemma 188.4 implies that $\mathrm{cl}(T) \leq 3$ and $T'$ is elementary abelian. By Theorem A.1.3 (Hall–Petrescu formula),

$$(ab)^4 = a^4 b^4 c_2^{\binom{4}{2}} c_3^{\binom{4}{3}} = 1\,, \quad \text{where } c_i \in \mathrm{K}_i(T), i = 2, 3\,.$$

By Lemma 188.3, $T/\langle a^2, b^2\rangle \cong \mathrm{E}_4$ *or* $\mathrm{D}_8$, where $\langle a^2, b^2\rangle \leq \mathrm{Z}(T)$ is elementary abelian of order 4, and so $\exp(T) \leq 4$ because $o(ab) \leq 4$.

If $o(a) = o(b) = 4$ and $\langle a\rangle \cap \langle b\rangle = \langle a^2\rangle$, then by Lemma 188.3, $T/\langle a^2\rangle \cong \mathrm{E}_4$ *or* $\mathrm{D}_8$. Suppose that $T/\langle a^2\rangle \cong \mathrm{D}_8$ and $\exp(T) = 2^3$. Then $T \cong \mathrm{Q}_{16}$. On the other hand,

$$\mathrm{cl}(G) \leq 3 \quad \text{and so } T' \leq \mathrm{K}_3(G) \leq \mathrm{Z}(G)\,,$$

which contradicts $T \cong \mathrm{Q}_{16}$. It follows that we get in any case, $\exp(T) \leq 4$. We have proved that $\exp(G') \leq 4$ and our theorem is proved. □

# § 189 2-groups with index of every cyclic subgroup in its normal closure ≤ 4

Recall (see § 144) that a group $G$ is called a BI($l$)-group ($G \in \mathrm{BI}(l)$ for short) if for every $a \in G$,

$$|\langle a\rangle^G : \langle a\rangle| \le l \quad \text{for every } a \in G\,.$$

Clearly, the BI($l$)-property of $G$ is inherited by subgroups and factor-groups of $G$.

The $p$-groups that are BI($p^2$)-groups for $p > 2$ have been classified in § 144. Here we classify 2-groups that are BI($2^2$)-groups using nice results obtained by H. Lv, W. Zhou, and X. Guo, where we have repaired some gaps in the proof of their Lemma 2.7 in [LZG] by proving more general Lemmas 189.7 and 189.8.

Let $G \in \mathrm{BI}(4)$ be a 2-group. Then it was proved here that $\mho_3(G) \le \mathrm{Z}(G)$, $\mho_2(G)$ is abelian (Theorem 189.9), $\mathrm{cl}(G) \le 5$ (Theorem 189.12) and $\exp(G') \le 8$ (Corollary 189.13).

**Lemma 189.1.** *Let $G \in \mathrm{BI}(p^k)$ be a $p$-group. Then for all $a \in G$, $|G : \mathrm{N}_G(\langle a\rangle)| \le p^k$ and $\langle a^{p^k}\rangle \trianglelefteq G$.*

*Proof.* The first part of the lemma is the content of Lemma 144.1. Set $o(a) = p^n$ so that $|\langle a\rangle^G| \le p^{n+k}$. For each $b \in G$, one has $\langle a^b\rangle \le \langle a\rangle^G$ and so

$$|\langle a\rangle\langle a^b\rangle| = p^{2n}/|\langle a\rangle \cap \langle a^b\rangle| \le p^{n+k}\,,$$

which gives

$$|\langle a\rangle \cap \langle a^b\rangle| \ge p^{n-k} \quad \text{and so } \langle a\rangle \cap \langle a^b\rangle \ge \langle a^{p^k}\rangle\,, \quad \langle a^{p^k}\rangle^b = \langle a^{p^k}\rangle\,.$$

The proof is complete. □

**Lemma 189.2.** *Let $G = \langle a, b\rangle \in \mathrm{BI}(4)$ be a 2-group. If $\langle a\rangle \cap \langle b\rangle = \{1\}$, then the following holds:*

(1) $\mho_2(G) \le \mathrm{Z}(G)$,

(2) $|G'| \le 2^3$ *and* $\exp(G') \le 2^2$.

*Proof.* Set $o(a) = 2^m$, $o(b) = 2^n$. Since $G$ is a BI(4)-group, Lemma 189.1 gives $\langle a^4\rangle \trianglelefteq G$ and $\langle b\rangle^{a^4} = \langle b\rangle$. As $\langle a\rangle \cap \langle b\rangle = \{1\}$, we see that $[a^4, b] = 1$, and so $a^4 \in \mathrm{Z}(G)$. Similarly, $b^4 \in \mathrm{Z}(G)$. Firstly, we prove (2) and we need to consider the following two cases:

(i) Suppose that one of $m, n$ is smaller than 3. Without loss of generality, we assume $n \le 2$. Thus, $|\langle b\rangle^G| \le 4$, and so $|G'| \le 2^3$. Indeed, $G/\langle b\rangle^G$ is cyclic, and so if $N$ is a $G$-invariant maximal subgroup of $\langle b\rangle^G$, then $G/N$ is abelian, and so $G'$ is a proper subgroup of $\langle b\rangle^G$. To prove $\exp(G') \le 2^2$, we need only to consider the case $|G'| = 2^3$. Since

$$G' \le \langle a\rangle^G \cap \langle b\rangle^G \quad \text{and} \quad |\langle a\rangle^G : \langle a\rangle| \le 2^2\,, \quad |\langle b\rangle^G : \langle b\rangle| \le 2^2\,,$$

it follows that:

$$G' \cap \langle a\rangle \ne \{1\} \quad \text{and} \quad G' \cap \langle b\rangle \ne \{1\}\,,$$

which together with $\langle a\rangle \cap \langle b\rangle = \{1\}$ implies that $G'$ contains two distinct involutions, and so $\exp(G') \le 2^2$.

(ii) It remains to consider the case $m \ge n \ge 3$. We may assume $|G'| \ge 2^3$. Since $G' \le \langle a\rangle^G \cap \langle b\rangle^G$, it follows that

$$G' \cap \langle a\rangle \ne \{1\} \quad \text{and} \quad G' \cap \langle b\rangle \ne \{1\}\,.$$

Thus,

$$a^{2^{m-1}} \in \langle a\rangle^G \cap \langle b\rangle^G\,, \quad b^{2^{n-1}} \in \langle a\rangle^G \cap \langle b\rangle^G \quad \text{and} \quad \mathrm{E}_4 \cong \langle a^{2^{m-1}}, b^{2^{n-1}}\rangle = N \le \mathrm{Z}(G)\,,$$

where Lemma 189.1 is used. Let $\bar{G} = G/N$ and we use the bar convention. Then we see that

$$\bar{G} = \langle \bar{a}, \bar{b}\rangle\,, \quad \langle \bar{a}\rangle \cap \langle \bar{b}\rangle = \{1\}\,, \quad |\langle \bar{a}\rangle^{\bar{G}} : \langle \bar{a}\rangle| \le 2 \quad \text{and} \quad |\langle \bar{b}\rangle^{\bar{G}} : \langle \bar{b}\rangle| \le 2\,.$$

By Lemma 144.2, $|\bar{G}'| = 2$, and then $|G'| = 2^3$ and $\exp(G') \le 2^2$.

Next we prove (1). By (2), $|G'| \le 2^3$ and $\exp(G') \le 2^2$ so that $|\mho_1(G')| \le 2$. Hence $x^2 \in \mathrm{Z}(G)$ for each $x \in G'$. Let $g_1 = a^i b^j \in G$. By Theorem A.1.3 (Hall–Petrescu formula),

$$(a^i b^j)^4 = a^{4i} b^{4j} c_2^{\binom{4}{2}} c_3^{\binom{4}{3}} c_4^{\binom{4}{4}}\,, \quad \text{where } c_i \in \mathrm{K}_i(G)\,, \quad i = 2, 3, 4\,.$$

Now, by the above,

$$a^{4i} b^{4j} \in \mathrm{Z}(G)\,, \quad c_2^{\binom{4}{2}} = c_2^6 \in \mathrm{Z}(G)\,, \quad c_3^{\binom{4}{3}} = c_3^4 = 1\,, \quad c_4^{\binom{4}{4}} = c_4 \in \mathrm{Z}(G)$$

since $\mathrm{cl}(G) \le 4$. Hence we get $g_1^4 \in \mathrm{Z}(G)$. For any $g = a^{k_1} b^{l_1} \dots a^{k_r} b^{l_r} \in G$, we see that $g^4 \in \mathrm{Z}(G)$, by induction on $r$, as required. □

**Lemma 189.3.** *Let $G \in \mathrm{BI}(4)$ be a 2-group. Then $|\langle a^2\rangle^G : \langle a^2\rangle| \le 2$ for every $a \in G$.*

*Proof.* If $o(a) \le 2$, then there is nothing to prove. So we suppose $o(a) \ge 4$. By Lemma 189.1, $\langle a^4\rangle \trianglelefteq G$. Set $\bar{G} = G/\langle a^4\rangle$ with the bar convention. Then $o(\bar{a}) = 4$ and $\bar{K} = \langle \bar{a}\rangle^{\bar{G}}$ is of order at most $2^4$. If $|\bar{K}| \le 2^2$, then $\bar{K} = \langle \bar{a}\rangle$ and $\langle a\rangle \trianglelefteq G$. Clearly, the lemma is true in this case. Now suppose $|\bar{K}| \ge 2^3$. Then $\bar{K}$ is noncyclic. It follows that $|\bar{K} : \Phi(\bar{K})| \ge 2^2$, which implies $|\mho_1(\bar{K})| \le 4$. Clearly, $\langle \bar{a}^2\rangle^{\bar{G}} \le \mho_1(\bar{K})$. Hence $|\langle \bar{a}^2\rangle^{\bar{G}} : \langle \bar{a}^2\rangle| \le 2$ and then $|\langle a^2\rangle^G : \langle a^2\rangle| \le 2$. □

**Lemma 189.4.** *Let $G = \langle a, b\rangle$ be a noncyclic 2-group such that*

$$\langle a\rangle \cap \langle b\rangle \ne \{1\}\,, \quad o(a) = o(b) = 2^4\,, \quad \textit{and } |\langle a\rangle^G : \langle a\rangle| \le 2\,.$$

*If*

$$|\langle b\rangle^G : \langle b\rangle| \le 2 \textit{ or } |\langle b\rangle^G : \langle b\rangle| \le 4 \quad \textit{and} \quad \langle a^4\rangle \le \langle a\rangle \cap \langle b\rangle\,,$$

*then $o(ab) \le 2^3$.*

*Proof.* Set $c = [a, b]$. It follows from $|\langle a\rangle^G : \langle a\rangle| \le 2$ and $c \in \langle a\rangle^G$ that $c^2 \in \langle a\rangle$ and $(c^2)^a = c^2$. Because $G$ is not cyclic, then $\langle a\rangle \ne \langle b\rangle$, and so $1 \ne |\langle a\rangle \cap \langle b\rangle| \le 2^3$.

(i) Suppose $|\langle a\rangle \cap \langle b\rangle| = 2^3$. Then

$$\langle a^2\rangle = \langle b^2\rangle = \langle a\rangle \cap \langle b\rangle \le Z(G)\,.$$

Since $|\langle a\rangle^G : \langle a\rangle| \le 2$, then $|G/\langle a^2\rangle| \le 2^3$. Set $\bar{G} = G/\langle a^2\rangle$ and use the bar convention. Since $\bar{G}$ is generated by two distinct involutions $\bar{a}$ and $\bar{b}$, we have $\bar{G} \cong E_4$ or $D_8$. In the first case, $\langle a\rangle$ and $\langle b\rangle$ are two distinct abelian maximal subgroups of $G$, and so $|G'| \le 2$. Suppose $\bar{G} \cong D_8$. Let $\bar{S}$ be the cyclic subgroup of order 4 in $\bar{G}$ so that $S$ is an abelian maximal subgroup of $G$. By Lemma 1.1, $|G| = 2|Z(G)|\,|G'|$. Since $|G| = 2^6$ and $2^3 \le |Z(G)| \le 2^4$ (noting that $|Z(\bar{G})| = 2$), it follows that $|G'| \le 4$. Hence $\exp(G') \le 4$ and $\mathrm{cl}(G) \le 3$.

(ii) Assume $|\langle a\rangle \cap \langle b\rangle| = 2^2$. Then

$$\langle a^4\rangle = \langle b^4\rangle = \langle a\rangle \cap \langle b\rangle \le Z(G)$$

and by the hypothesis,

$$2^4 \le |\langle a\rangle^G| \le 2^5 \quad \text{and} \quad |G/\langle a^4\rangle| \le 2^5\,.$$

If $|\langle a\rangle^G| = 2^4$, then $\langle a\rangle \trianglelefteq G$ and since $b$ centralizes $\langle a^4\rangle$, it follows that $b$ centralizes $\langle a\rangle/\langle a^4\rangle$, and so $c = [a, b] \in \langle a\rangle \cap \langle b\rangle \le Z(G)$. It follows that $G' = \langle c\rangle$ is of order $\le 4$, $\mathrm{cl}(G) \le 2$ and $\exp(G') \le 4$.

Now suppose that $|\langle a\rangle^G| = 2^5$. Obviously, $\langle a\rangle^G$ is noncyclic. If $c \in \langle a\rangle$, then $a^b \in \langle a\rangle$ and $\langle a\rangle^G = \langle a\rangle$, contrary to our assumption. Hence $\langle a\rangle^G = \langle a, c\rangle$. Since $a^4 \in \langle a\rangle^G \cap Z(G)$, Theorem 1.2 implies $\langle a\rangle^G \cong C_{2^4} \times C_2$ or $\langle a\rangle^G \cong M_{2^5}$. Set

$$\langle a\rangle^G = \langle a, d\rangle\,, \quad \text{where } d^2 = 1\,, \quad a^d = d \text{ or } a^d = a^{1+8}\,.$$

Then

$$a^b = a^l d\,, \quad \text{where } (2, l) = 1\,, \quad \text{and so } (a^2)^b = a^l(da^l d)\,.$$

If $a^d = a$, then $(a^2)^b = a^{2l}$ and $(a^4)^b = a^{4l}$. If $a^d = a^{1+8}$, then $(a^2)^b = a^l(da^l d) = a^{10l}$ and $(a^4)^b = a^{20l} = a^{4l}$. Since $a^4 \in Z(G)$, we get in any case $(a^4)^b = a^4 = a^{4l}$, and so $l = 1 + 4l_1$ for some integer $l_1$. Hence

$$c = [a, b] = a^{-1}(b^{-1}ab) = a^{-1}a^{1+4l_1}d = a^{4l_1}d \in \Omega_2(\langle a\rangle^G) \cong C_4 \times C_2\,.$$

Since $\Omega_2(\langle a\rangle^G) \trianglelefteq G$, it follows that

$$G/\Omega_2(\langle a\rangle^G) \quad \text{is abelian}\,,\ G' \le \Omega_2(\langle a\rangle^G)\,,$$

and so

$$|G'| \le 2^3\,, \quad \exp(G') \le 4\,, \quad \mathrm{cl}(G) \le 4\,.$$

(iii) Finally, assume $|\langle a\rangle \cap \langle b\rangle| = 2$. Then

$$\langle a^8\rangle = \langle b^8\rangle = \langle a\rangle \cap \langle b\rangle \le Z(G)\,.$$

Set $\bar{G} = G/\langle a^8\rangle$. Then $\langle \bar{a}\rangle \cap \langle \bar{b}\rangle = \{1\}$ and, by the hypothesis,

$$|\langle \bar{a}\rangle^{\bar{G}} : \langle \bar{a}\rangle| \le 2\,, \quad |\langle \bar{b}\rangle^{\bar{G}} : \langle \bar{b}\rangle| \le 2\,.$$

By Lemma 144.2, $|\bar{G}'| \le 2$. It follows $|G'| \le 2^2$, $\exp(G') \le 4$ and $\mathrm{cl}(G) \le 3$.

We have proved that in any case $|G'| \le 2^3$, $\exp(G') \le 4$ and $\mathrm{cl}(G) \le 4$. By the Hall–Petrescu formula (Theorem A.1.3),

$$(ab)^8 = a^8 b^8 c_2^{\binom{8}{2}} c_3^{\binom{8}{3}} c_4^{\binom{8}{4}}\,, \quad \text{where } c_i \in \mathrm{K}_i(G)\,, \quad i = 2, 3, 4\,.$$

Since $a^8 = b^8$ is an involution, $o(c_2) \le 4$, $o(c_3) \le 4$ and $o(c_4) \le 2$ (because $|G'| \le 2^3$), we get $(ab)^8 = 1$ and we are done. □

**Remark 1.** Under the assumptions of Lemma 189.4, there exists $b_1 \in G$ such that $G = \langle a, b\rangle = \langle a, b_1\rangle$ with $o(b_1) < o(b)$.

**Lemma 189.5.** *Let $G = \langle a, b\rangle \in \mathrm{BI}(4)$ be a 2-group with $o(a) = 2^5$, $o(b) \le 2^4$. If $\langle a\rangle \cap \langle b\rangle \ne \{1\}$, then there exists $b_1 \in G$ such that $G = \langle a, b_1\rangle$ with $\langle a\rangle \cap \langle b_1\rangle = \{1\}$ and $o(b_1) < o(b)$.*

*Proof.* First suppose that $o(b) = 2^4$ and set $H = \langle a^2, b\rangle$. By Lemma 189.3, $|\langle a^2\rangle^H : \langle a^2\rangle| \le 2$. If $|\langle a^2\rangle \cap \langle b\rangle| \ge 4$, then Lemma 189.4 implies $o(a^2 b) \le 2^3$. In that case we set $h = a^2 b$ so that $o(h) \le 2^3$ and $G = \langle a, h\rangle$.

Let us consider the case $|\langle a^2\rangle \cap \langle b\rangle| = 2$. Then

$$\langle b^8\rangle = \langle a^2\rangle \cap \langle b\rangle \le Z(G)\,.$$

By Lemma 189.1, $\langle a^4\rangle \trianglelefteq G$ and $a^4$ normalizes $\langle b\rangle$. This gives $[a^4, b] \in \langle a^4\rangle \cap \langle b\rangle = \langle b^8\rangle$. Set

$$c = [a^2, b]\,, \quad z = b^8 \in Z(G)\,.$$

Then

$$[a^4, b] = [a^2 a^2, b] = [a^2, b]^{a^2}[a^2, b] = c^{a^2} c = z^{\epsilon}\,, \quad \text{where } \epsilon = 0, 1\,.$$

Hence $c^{a^2} = c^{-1} z^{\epsilon}$, and so $(c^2)^{a^2} = c^{-2}$. On the other hand, $|\langle a^2\rangle^H : \langle a^2\rangle| \le 2$, and so $\langle a^2\rangle^H$ is isomorphic to one of the following groups:

$$\mathrm{C}_{2^4}\,, \quad \mathrm{C}_{2^4} \times \mathrm{C}_2\,, \quad \mathrm{M}_{2^5}\,.$$

In any case,

$$|\langle a^2\rangle^H : \Omega_3(\langle a^2\rangle^H)| = 2 \quad \text{and so } H/\Omega_3(\langle a^2\rangle^H) \text{ is abelian since } H/\langle a^2\rangle^H \text{ is cyclic}\,.$$

Hence $H' \le \Omega_3(\langle a^2\rangle^H)$, and so $c \in \Omega_3(\langle a^2\rangle^H)$, which implies $c^2 \in \langle a^4\rangle$. By the above,

$$(c^2)^{a^2} = c^2 = c^{-2} \quad \text{and so } c^4 = 1\,.$$

It follows that

$$c \in \Omega_2(\langle a^2\rangle^H) \cong \mathrm{C}_4 \text{ or } \mathrm{C}_4 \times \mathrm{C}_2 \quad \text{and so } H' \le \Omega_2(\langle a^2\rangle^H)\,.$$

We have proved that $|H'| \le 2^3$ and $\exp(H') \le 4$. By Theorem A.1.3 (Hall–Petrescu formula),

$$(a^2b)^8 = a^{16}b^8c_2^{\binom{8}{2}}c_3^{\binom{8}{3}}c_4^{\binom{8}{4}} = a^{16}b^8 = 1\,, \quad \text{where } c_i \in \mathrm{K}_i(H)\,, \quad i = 2, 3, 4\,.$$

Set $h = a^2b$. Then we have again (as in the first paragraph of the proof) $G = \langle a, h\rangle$ with $o(h) \le 2^3$. If $\langle a\rangle \cap \langle h\rangle = \{1\}$, then we set $h = b_1$ and we are done.

Now suppose that $o(b) = 2^3$, where $\langle a\rangle \cap \langle b\rangle \ne \{1\}$. Set $H_1 = \langle a^4, b\rangle$. By Lemma 189.1, $\langle a^4\rangle \trianglelefteq H_1$ and $\langle b\rangle \trianglelefteq H_1$. It follows that $\mathrm{cl}(H_1) \le 2$. If $\langle a^4\rangle \cap \langle b\rangle = \langle b^2\rangle$, then $H_1$ has two distinct cyclic maximal subgroups (of order 8), and so either $H_1 \cong \mathrm{M}_{2^4}$ or $H_1 \cong \mathrm{C}_8 \times \mathrm{C}_2$. In any case, there is an involution $b_1 \in H_1 - \langle a^4\rangle$ so that $G = \langle a, b_1\rangle$ with $\langle a\rangle \cap \langle b_1\rangle = \{1\}$ and $o(b_1) = 2 < o(b)$. If $\langle a^4\rangle \cap \langle b\rangle = \langle b^4\rangle$, then $|H_1'| \le 2$. We have

$$(a^4b)^4 = a^{16}b^4[b, a^4]^{\binom{4}{2}} = 1\,.$$

Since $H_1 = \langle a^4, a^4b\rangle$ and $|H_1 : \langle a^4\rangle| = 4$, it follows that $o(a^4b) = 4$ and $\langle a^4\rangle \cap \langle a^4b\rangle = \{1\}$. Hence in this case we may set $b_1 = a^4b$ and we get $G = \langle a, b_1\rangle$ with $\langle a\rangle \cap \langle b_1\rangle = \{1\}$ and $o(b_1) = 4 < o(b)$.

Finally, assume that $o(b) = 2^2$, where $\langle a\rangle \cap \langle b\rangle = \langle b^2\rangle$. Set again $H_1 = \langle a^4, b\rangle$. By Lemma 189.1, $\langle a^4\rangle \trianglelefteq H_1$ and $\langle b\rangle \trianglelefteq H_1$. But then acting with $\langle a^4\rangle \cong \mathrm{C}_8$ on $\langle b\rangle \cong \mathrm{C}_4$, we see that $\langle a^8\rangle \cong \mathrm{C}_4$ centralizes $\langle b\rangle \cong \mathrm{C}_4$, and so $b_1 = a^8b$ is an involution in $H_1 - \langle a^4\rangle$. Since $|H_1 : \langle a^4\rangle| = 2$, we have $H_1 = \langle a^4, b_1\rangle$, and so $G = \langle a, b_1\rangle$ with $\langle a\rangle \cap \langle b_1\rangle = \{1\}$ and $o(b_1) = 2 < o(b) = 4$. The proof is complete. □

**Lemma 189.6.** *Let $G = \langle a, b\rangle \in \mathrm{BI}(4)$ be a 2-group with $o(a) = 2^5$, $o(b) \le 2^5$. If $\langle a\rangle \cap \langle b\rangle \ne \{1\}$, then there exists $b_1 \in G$ such that $G = \langle a, b_1\rangle$ with $\langle a\rangle \cap \langle b_1\rangle = \{1\}$ and $o(b_1) < o(b)$.*

*Proof.* By Lemma 189.5, we need only to consider the case $o(b) = 2^5$.

Case 1. $G'$ contains a $G$-invariant 4-subgroup $N$.

(i) First assume that $\langle a^{16}\rangle = \langle b^{16}\rangle \le N$. Write $\bar{G} = G/N$. Since

$$N \le G' \le \langle a\rangle^G \cap \langle b\rangle^G\,, \quad \text{then } \langle \bar{a}\rangle^{\bar{G}} = (\langle a\rangle^G N)/N = \langle a\rangle^G/N\,.$$

It follows that

$$|\langle \bar{a}\rangle^{\bar{G}} : \langle \bar{a}\rangle| \le 2 \quad \text{and similarly } |\langle \bar{b}\rangle^{\bar{G}} : \langle \bar{b}\rangle| \le 2\,.$$

Clearly, $o(\bar{a}) = o(\bar{b}) = 2^4$.

If $\langle \bar{a} \rangle \cap \langle \bar{b} \rangle = \{1\}$, then Lemma 144.2 gives $|\bar{G}'| \le 2$, and then

$$|G'| \le 2^3\,, \quad \exp(G') \le 4, \mathrm{cl}(G) \le \ .$$

By the Hall–Petrescu formula,

$$(ab)^{16} = a^{16} b^{16} c_2^{\binom{16}{2}} c_3^{\binom{16}{3}} c_4^{\binom{16}{4}} = 1\,, \quad \text{where } c_i \in \mathrm{K}_i(G)\,, \quad i = 2, 3, 4\,.$$

Thus, $o(ab) \le 16$ and since $G = \langle a, ab \rangle$, we have either $\langle a \rangle \cap \langle ab \rangle = \{1\}$ (and we are done) or $\langle a \rangle \cap \langle ab \rangle \ne \{1\}$, and then, by Lemma 189.5, there exists $b_1 \in G$ such that

$$G = \langle a, b_1 \rangle \quad \text{with } \langle a \rangle \cap \langle b_1 \rangle = \{1\}\,, \quad o(b_1) < o(b)\,.$$

If $\langle \bar{a} \rangle \cap \langle \bar{b} \rangle \ne \{1\}$, then Lemma 189.4 implies that there is $x \in G$ such that $\bar{G} = \langle \bar{a}, \bar{x} \rangle$ and $o(\bar{x}) \le 2^3$. Hence

$$G = \langle a, x, N \rangle = \langle a, x \rangle \quad \text{since } N \le G' \le \Phi(G)\,.$$

Also, $o(x) \le 2^4$ and either $\langle a \rangle \cap \langle x \rangle = \{1\}$ (and we are done) or $\langle a \rangle \cap \langle x \rangle \ne \{1\}$, and then Lemma 189.5 gives that there exists $b_1 \in G$ such that

$$G = \langle a, b_1 \rangle \quad \text{with } \langle a \rangle \cap \langle b_1 \rangle = \{1\}\,, \quad o(b_1) < o(b)\,.$$

(ii) Now suppose $\langle a^{16} \rangle = \langle b^{16} \rangle \ne N$. Then $N_1 = \langle a^{16} \rangle N \cong \mathrm{E}_8$ is normal in $G$ since $\langle a^{16} \rangle \le \mathrm{Z}(G)$. Consider $\bar{G} = G/N_1$ with the bar convention and note that $\bar{a} \trianglelefteq G$, $\bar{b} \trianglelefteq G$, and $o(\bar{a}) = o(\bar{b}) = 2^4$. If $\langle \bar{a} \rangle \cap \langle \bar{b} \rangle \ne \{1\}$, then Lemma 189.4 implies that there is $x \in G$ such that $\bar{G} = \langle \bar{a}, \bar{x} \rangle$ and $o(\bar{x}) \le 2^3$. Hence $G = \langle a, x, N_1 \rangle = \langle a, x \rangle$ since $N_1 \le \Phi(G)$. Also, $o(x) \le 2^4$, and so Lemma 189.5 implies that there exists $b_1 \in G$ such that $G = \langle a, b_1 \rangle$ with $\langle a \rangle \cap \langle b_1 \rangle = \{1\}$ and $o(b_1) < o(b)$. If $\langle \bar{a} \rangle \cap \langle \bar{b} \rangle = \{1\}$, then by Lemma 144.2, $|\bar{G}'| \le 2$ and then $|G'| \le 2^4$. Hence $\exp(G') \le 4$ and $\mathrm{cl}(G) \le 5$ since $N_1 \cong \mathrm{E}_8$. By the Hall–Petrescu formula (Theorem A.1.3),

$$(ab)^{16} = a^{16} b^{16} c_2^{\binom{16}{2}} c_3^{\binom{16}{3}} c_4^{\binom{16}{4}} c_5^{\binom{16}{5}} = 1\,, \quad \text{where } c_i \in \mathrm{K}_i(G)\,, \quad i = 2, 3, 4, 5\,.$$

Thus, $o(ab) \le 2^4$, $G = \langle a, ab \rangle$ and either $\langle a \rangle \cap \langle ab \rangle = \{1\}$ (and we are done) or $\langle a \rangle \cap \langle ab \rangle \ne \{1\}$ and then Lemma 189.5 implies that there is $b_1 \in G$ such that $G = \langle a, b_1 \rangle$ with $\langle a \rangle \cap \langle b_1 \rangle = \{1\}$ and $o(b_1) < o(b)$.

Case 2. $G'$ does not possess a $G$-invariant 4-subgroup.

By Lemma 1.4, $G'$ is cyclic since $G' \le \Phi(G)$. Set $c = [a, b]$ so that $G' = \langle c \rangle$. We have

$$\langle a \rangle^G = \langle a, c \rangle\,, \quad \langle b \rangle^G = \langle b, c \rangle\,, \quad \text{and so } c^4 \in \langle a \rangle \cap \langle b \rangle \le \mathrm{Z}(G) \text{ and } \mathrm{cl}(G) \le 4\,.$$

Suppose that $[G', G] \nleq \mho_2(G')$. Then there is $g \in G$ such that $c^g = c^{-1}c^{4i}$ for some integer $i$. Since $c^4 \in Z(G)$,

$$c^4 = (c^4)^g = c^{-4}c^{16i}, \quad c^8 = c^{16i}, \quad c^{8(2i-1)} = 1 \quad \text{and we get } c^8 = 1 .$$

In what follows we want to show that $c^8 = 1$ and so we may assume $[G', G] \leq \mho_2(G')$.

(1) Assume $|\langle a \rangle \cap \langle b \rangle| = 2^4$ so that

$$\langle a^2 \rangle = \langle b^2 \rangle = \langle a \rangle \cap \langle b \rangle \leq Z(G) .$$

From $1 = [a^2, b] = [a, b]^a[a, b] = c^a c$, we get $c^a = c^{-1}$ and $(c^4)^a = c^{-4} = c^4$, and so $c^8 = 1$.

(2) Suppose $|\langle a \rangle \cap \langle b \rangle| = 2^3$ so that

$$\langle a^4 \rangle = \langle b^4 \rangle = \langle a \rangle \cap \langle b \rangle \leq Z(G) .$$

Then $1 = [a, b^4] = [a, b^2][a, b^2]^{b^2}$, and so $[a, b^2]^{b^2} = [a, b^2]^{-1}$. On the other hand, by Lemma 139.1,

$$\langle [a, b^2] \rangle = \langle [a, b]^2 \rangle = \langle c^2 \rangle .$$

Hence $(c^2)^{b^2} = c^{-2}$, and so $c^4 = (c^4)^{b^2} = c^{-4}$, which gives $c^8 = 1$.

(3) Suppose $|\langle a \rangle \cap \langle b \rangle| = 2^2$ so that

$$\langle a^8 \rangle = \langle b^8 \rangle = \langle a \rangle \cap \langle b \rangle \leq Z(G) .$$

By Lemma 189.1, $\langle b^4 \rangle \trianglelefteq G$ and $\langle b^4 \rangle$ normalizes $\langle a \rangle$ so that

$$[a, b^4] \in \langle a \rangle \cap \langle b^4 \rangle \leq \langle a \rangle \cap \langle b \leq Z(G) .$$

We get

$$1 = [a, b^8] = [a, b^4 b^4] = [a, b^4][a, b^4]^{b^4} = [a, b^4][a, b^4] = [a, b^4]^2 .$$

By our assumption, $[G', G] \leq \mho_2(G')$, and so for each $g \in G$,

$$c^g = cc^{4i} \quad \text{with some integer } i, \quad (c^2)^g = (cc^{4i})^2 = c^2 c^{8i} .$$

Consider the subgroup $H = \langle a, b^2 \rangle$ so that $H' = \langle [a, b^2] \rangle$. By Lemma 139.1,

$$\langle [a, b^2] \rangle = \langle [a, b]^2 \rangle = \langle c^2 \rangle \quad \text{and so } H' = \langle c^2 \rangle .$$

By the above, $[H', H] \leq \mho_2(H')$, and so by Lemma 139.1 applied to $H$,

$$\langle [a, b^4] \rangle = \langle [a, b^2]^2 \rangle = \mho_1(H') = \langle c^4 \rangle .$$

By the above, $[a, b^4]^2 = 1$, and so $c^8 = 1$.

(4) Suppose $|\langle a\rangle \cap \langle b\rangle| = 2$. Since $c^4 \in \langle a\rangle \cap \langle b\rangle$, we get again $c^8 = 1$.

In any case we have $c^8 = 1$, and so $G'$ is cyclic of order $\le 8$. By the Hall–Petrescu formula (Theorem A.1.3),

$$(ab)^{16} = a^{16}b^{16}c_2^{\binom{16}{2}}c_3^{\binom{16}{3}}c_4^{\binom{16}{4}} = 1\,, \quad \text{where } c_i \in \mathrm{K}_i(G)\,, \quad i = 2, 3, 4\,.$$

Thus, $G = \langle a, ab\rangle$ and $o(ab) \le 2^4$. By Lemma 189.5, there exists $b_1 \in G$ such that $G = \langle a, b_1\rangle$ with $\langle a\rangle \cap \langle b_1\rangle = \{1\}$ and $o(b_1) < o(b)$. The Lemma is proved. □

**Lemma 189.7** (Janko). *Let $G = \langle a, b\rangle$ be a noncyclic 2-group such that*

$$\langle a\rangle \cap \langle b\rangle \ne \{1\}\,, \quad o(a) = o(b) = 2^n\,, \quad n \ge 4\,,$$
$$|\langle a\rangle^G : \langle a\rangle| \le 2\,, \quad \textit{and} \quad |\langle b\rangle^G : \langle b\rangle| \le 2\,.$$

*Then $o(ab) \le 2^{n-1}$.*

*Proof.* Set

$$z = a^{2^{n-1}} = b^{2^{n-1}} \in \langle a\rangle \cap \langle b\rangle \le \mathrm{Z}(G)\,.$$

Since $|\langle a\rangle^G : \langle a\rangle| \le 2$, the normal closure $\langle a\rangle^G$ is isomorphic to one of the following groups:

$$\mathrm{C}_{2^n}\,, \quad \mathrm{M}_{2^{n+1}}\,, \quad \mathrm{C}_{2^n} \times \mathrm{C}_2\,,$$

and so

$$(\langle a\rangle^G)' \le \langle z\rangle\,, \quad \mho_1(\langle a\rangle^G) = \langle a^2\rangle \trianglelefteq G\,, \quad |G : \mathrm{N}_G(\langle a\rangle)| \le 2\,, \quad |\langle a\rangle^G : \Omega_{n-1}(\langle a\rangle^G)| = 2$$

(since $\langle a\rangle^G$ has at most two cyclic subgroups of index 2). Therefore $G/\Omega_{n-1}(\langle a\rangle^G)$ is abelian giving that $c = [a, b]$ is contained in $\Omega_{n-1}(\langle a\rangle^G)$, and so $c^{2^{n-1}} = 1$. Similarly, $|\langle b\rangle^G : \langle b\rangle| \le 2$ gives

$$(\langle b\rangle^G)' \le \langle z\rangle\,, \quad \langle b^2\rangle \trianglelefteq G\,, \quad |G : \mathrm{N}_G(\langle b\rangle)| \le 2\,, \quad c \in \Omega_{n-1}(\langle b\rangle^G)\,.$$

We claim that $\mathrm{cl}(G) \le 3$ and $G'$ is either elementary abelian of order $\le 4$ or $G'$ is cyclic of order $\le 2^{n-2}$. Hence we may assume $c \ne 1$ (i.e., $G$ is nonabelian). We have

$$c^a = cz^\epsilon\,, \quad c^b = cz^\eta\,, \quad \text{where } \epsilon, \eta \in \{0, 1\} \quad \text{and so } \langle c, z\rangle \trianglelefteq G \text{ giving } G' \le \langle c, z\rangle\,.$$

If $c^2 = 1$, then $G'$ is elementary abelian of order $\le 4$ and we are done. In what follows we assume that $c^2 \ne 1$. Then

$$c^2 \in \langle a^4\rangle \cap \langle b^4\rangle \le \mathrm{Z}(G) \quad \text{and} \quad \langle c\rangle > \langle z\rangle\,.$$

In this case,

$$\langle c\rangle \trianglelefteq G \quad \text{and so } G' = \langle c\rangle\,, \quad \mathrm{cl}(G) \le 3\,.$$

Suppose, by way of contradiction, that $o(c) = 2^{n-1}$. In that case,

$$o(c^2) = 2^{n-2}\,, \quad c^2 \in \langle a^4\rangle \cap \langle b^4\rangle \le \mathrm{Z}(G), \quad \text{and so } \langle c^2\rangle = \langle a^4\rangle = \langle b^4\rangle\,.$$

In particular, $[a, b^4] = 1$. On the other hand, $\langle b^2\rangle \trianglelefteq G$ and $b^2$ normalizes $\langle a\rangle$ and so, $[a, b^2] \in \langle a\rangle \cap \langle b^2\rangle \le Z(G)$. This gives

$$1 = [a, b^4] = [a, b^2b^2] = [a, b^2][a, b^2]^{b^2} = [a, b^2]^2 .$$

If $[G', G] \not\le \mho_2(G')$, then there is $g \in G$ such that $c^g = c^{-1}c^{4i}$ for some integer $i$. Since $c^2 \in Z(G)$, it follows

$$c^2 = (c^2)^g = (c^g)^2 = c^{-2}c^{8i}, \quad c^{4(2i-1)} = 1 ,$$

and so $c^4 = 1$, contrary to $o(c) = 2^{n-1}$, $n \ge 4$. Thus, $[G', G] \le \mho_2(G')$ and then Lemma 139.1 gives $\mho_1(G') = \langle c^2\rangle = \langle [a, b^2]\rangle$. By the above $[a, b^2]^2 = 1$, and so $c^4 = 1$, contrary to $o(c) = 2^{n-1}$. We have proved that $o(c) \le 2^{n-2}$, and so $\exp(G') \le 2^{n-2}$.

By the Hall–Petrescu formula (Theorem A.1.3),

$$(ab)^{2^{n-1}} = a^{2^{n-1}} b^{2^{n-1}} c_2^{\binom{2^{n-1}}{2}} c_3^{\binom{2^{n-1}}{3}}, \quad \text{where } c_i \in K_i(G), \quad i = 2, 3 .$$

It follows $(ab)^{2^{n-1}} = 1$, completing the proof. □

**Lemma 189.8** (Janko). *Let $G = \langle a, b\rangle \in \mathrm{BI}(4)$ be a noncyclic 2-group with $o(a) = 2^m$, $m \ge 5$, and $o(b) \le o(a)$. If $\langle a\rangle \cap \langle b\rangle \ne \{1\}$, then there exists $b_1 \in G$ such that $G = \langle a, b_1\rangle$ with $\langle a\rangle \cap \langle b_1\rangle = \{1\}$ and $o(b_1) < o(b)$.*

*Proof.* If $m = 5$, then the result follows from Lemma 189.6. Therefore in the sequel we assume $m \ge 6$. We choose an element $b \in G$ of the smallest possible order $o(b) = 2^n$, $n \le m$, such that $G = \langle a, b\rangle$ and $\langle a\rangle \cap \langle b\rangle \ne \{1\}$. This fact will be used many times in the proof.

Set $a' = a^{2^{m-n}}$ so that $o(a') = 2^n$. Since $\langle a\rangle \cap \langle b\rangle \ne \{1\}$ and $G$ is noncyclic, we have $n \ge 2$. Suppose $n = 2$ so that $|\langle a\rangle \cap \langle b\rangle| = 2$. By Lemma 189.1, $\langle a^4\rangle$ (of order $\ge 2^4$) normalizes $\langle b\rangle \cong C_4$, and so $\langle a'\rangle \cong C_4$ centralizes $\langle b\rangle$. It follows that $a'b = i$ is an involution. We get $G = \langle a, i\rangle$, $\langle a\rangle \cap \langle i\rangle = \{1\}$, $o(i) = 2 < o(b)$ and we are done. From now on we assume $n \ge 3$.

First suppose $m - n \ge 2$ and then Lemma 189.1 gives $\langle a'\rangle \trianglelefteq G$ and $a'$ normalizes $\langle b\rangle$. Set $H_0 = \langle a', b\rangle$, where $\mathrm{cl}(H_0) \le 2$ since $H_0' \le \langle a'\rangle \cap \langle b\rangle \le Z(H_0)$. We get

$$(a'b)^{2^{n-1}} = (a')^{2^{n-1}} b^{2^{n-1}} [b, a']^{\binom{2^{n-1}}{2}} = [b, a']^{\binom{2^{n-1}}{2}} ,$$

and note that

$$[b, a'] \in \langle a'\rangle \cap \langle b\rangle \quad \text{and} \quad |\langle a'\rangle \cap \langle b\rangle| \le 2^{n-1} .$$

Assume that $|\langle a'\rangle \cap \langle b\rangle| = 2^{n-1}$. Then $b^2 \in \langle a'\rangle \cap \langle b\rangle \le Z(H_0)$, and so

$$1 = [b^2, a'] = [b, a']^2 \quad \text{and} \quad o(a'b) \le 2^{n-1} .$$

If $|\langle a'\rangle \cap \langle b\rangle| \le 2^{n-2}$, then, by the above, we get again $o(a'b) \le 2^{n-1}$. Because $G = \langle a, a'b\rangle$ and $o(a'b) < 2^n$, we must have $\langle a\rangle \cap \langle a'b\rangle = \{1\}$, and we are done.

In what follows we assume $m-n \leq 1$. Suppose for a moment $m-n = 1$. In this case $a' = a^2$, $o(a') = o(b) = 2^n$, $n \geq 5$, and we consider the subgroup $H = \langle a', b\rangle$, where $\langle a'\rangle \cap \langle b\rangle \neq \{1\}$. If we find in $H$ an element $b_1$ such that $o(b_1) < 2^n$ and $H = \langle a', b_1\rangle$, then $G = \langle a, b_1\rangle$, and so the minimality of $o(b)$ implies $\langle a\rangle \cap \langle b_1\rangle = \{1\}$, and in this case we are done.

We have seen that we need only to consider the case $G = \langle a, b\rangle$ with $m = n \geq 5$, where $o(a) = o(b) = 2^m$ and $\langle a\rangle \cap \langle b\rangle \neq \{1\}$. In view of Lemma 189.6, we may in fact assume $m \geq 6$.

Case 1. $G'$ contains a $G$-invariant 4-subgroup $U$.

(i) First assume $a^{2^{m-1}} = b^{2^{m-1}} \in U$. Let us consider $\bar{G} = G/U = \langle \bar{a}, \bar{b}\rangle$. Because

$$U \leq G' \leq \langle a\rangle^G \cap \langle b\rangle^G, \quad \text{then } |\langle \bar{a}\rangle^{\bar{G}} : \langle \bar{a}\rangle| \leq 2, \quad |\langle \bar{b}\rangle^{\bar{G}} : \langle \bar{b}\rangle| \leq 2.$$

Clearly, $o(\bar{a}) = o(\bar{b}) = 2^{m-1}$.

If $\langle \bar{a}\rangle \cap \langle \bar{b}\rangle = \{1\}$, then Lemma 144.2 gives $|\bar{G}'| \leq 2$, and so

$$|G'| \leq 2^3, \quad \exp(G') \leq 4, \quad \mathrm{cl}(G) \leq 4.$$

By the Hall–Petrescu formula,

$$(ab)^{2^{m-1}} = a^{2^{m-1}} b^{2^{m-1}} c_2^{\binom{2^{m-1}}{2}} c_3^{\binom{2^{m-1}}{3}} c_4^{\binom{2^{m-1}}{4}} = 1, \quad \text{where } c_i \in \mathrm{K}_i(G), \quad i = 2, 3, 4.$$

Hence $o(ab) \leq 2^{m-1}$, and so

$$o(ab) < o(b) \quad \text{which together with } G = \langle a, ab\rangle \text{ gives } \langle a\rangle \cap \langle ab\rangle = \{1\},$$

and we are done.

If $\langle \bar{a}\rangle \cap \langle \bar{b}\rangle \neq \{1\}$, then Lemma 189.7 implies that there is $x \in G$ such that $\bar{G} = \langle \bar{a}, \bar{x}\rangle$ and $o(\bar{x}) \leq 2^{m-2}$. Hence

$$G = \langle a, x, U\rangle = \langle a, x\rangle \quad \text{since } U \leq G' \leq \Phi(G) \text{ and } o(x) \leq 2^{m-1}.$$

Again, the minimality of $o(b) = 2^n = 2^m$ implies $\langle a\rangle \cap \langle x\rangle = \{1\}$ and we are done.

(ii) Suppose $a^{2^{m-1}} = b^{2^{m-1}} \notin U$. Then $\mathrm{E}_8 \cong U_1 = \langle a^{2^{m-1}}\rangle \times U$ since $a^{2^{m-1}}$ is a central involution in $G$. Consider $\bar{G} = G/U_1$ (with bar convention) and note that $\bar{a} \trianglelefteq G$, $\bar{b} \trianglelefteq G$ and $o(\bar{a}) = o(\bar{b}) = 2^{m-1}$.

If $\langle \bar{a}\rangle \cap \langle \bar{b}\rangle \neq \{1\}$, then Lemma 189.7 implies that there is $x \in G$ such that $\bar{G} = \langle \bar{a}, \bar{x}\rangle$ and $o(\bar{x}) \leq 2^{m-2}$. Hence

$$G = \langle a, x, U_1\rangle = \langle a, x\rangle \quad \text{since } U_1 \leq \Phi(G) \text{ and } o(x) \leq 2^{m-1}.$$

The minimality of $o(b) = 2^n = 2^m$ implies $\langle a\rangle \cap \langle x\rangle = \{1\}$, and we are done.

If $\langle \bar{a}\rangle \cap \langle \bar{b}\rangle = \{1\}$, then Lemma 144.2 gives $|\bar{G}'| \leq 2$, and so

$$|G'| \leq 2^4, \quad \exp(G') \leq 4, \quad \mathrm{cl}(G) \leq 5, \quad \exp(\mathrm{K}_3(G)) \leq 2$$

since $K_3(G) \le U_1 \cong E_8$. By the Hall–Petrescu formula,

$$(ab)^{2^{m-1}} = a^{2^{m-1}} b^{2^{m-1}} c_2^{\binom{2^{m-1}}{2}} c_3^{\binom{2^{m-1}}{3}} c_4^{\binom{2^{m-1}}{4}} c_5^{\binom{2^{m-1}}{5}} = 1\,,$$
$$\text{where } c_i \in K_i(G)\,, \quad i = 2, \dots, 5\,.$$

Hence $o(ab) \le 2^{m-1}$, and so

$$o(ab) < o(b) \quad \text{which together with } G = \langle a, ab\rangle \text{ gives } \langle a\rangle \cap \langle ab\rangle = \{1\}\,,$$

and we are done.

Case 2. $G'$ does not possess a $G$-invariant 4-subgroup.

By Lemma 1.4, $G'$ is cyclic since $G' \le \Phi(G)$. Set $c = [a, b]$ so that $G' = \langle c\rangle$. We have

$$\langle a\rangle^G = \langle a, c\rangle\,, \quad \langle b\rangle^G = \langle b, c\rangle\,, \quad \text{and so } c^4 \in \langle a\rangle \cap \langle b\rangle \le Z(G) \text{ and } \mathrm{cl}(G) \le 4\,.$$

Assume, by way of contradiction, that $o(c) \ge 2^{m-1}$. Since $c^4 \in \langle a\rangle \cap \langle b\rangle$, we get

$$|langlea\rangle \cap \langle b\rangle| \ge 2^{m-3} \quad \text{and so } b^8 \in \langle a\rangle \cap \langle b\rangle \le Z(G)\,, \quad [a, b^8] = 1\,.$$

By Lemma 189.1, $\langle b^4\rangle \trianglelefteq G$ and $b^4$ normalizes $\langle a\rangle$, which implies $[a, b^4] \in \langle a\rangle \cap \langle b^4\rangle \le Z(G)$. Hence

$$1 = [a, b^8] = [a, b^4][a, b^4]^{b^4} = [a, b^4]^2\,.$$

Suppose that $[G', G] \nleq \mho_2(G')$. Then there is $g \in G$ such that $c^g = c^{-1}c^{4i}$ for some integer $i$. Since $c^4 \in Z(G)$,

$$c^4 = (c^4)^g = c^{-4}c^{16i}\,, \quad c^8 = c^{16i}\,, \quad c^{8(2i-1)} = 1 \quad \text{and we get } c^8 = 1\,,$$

contrary to $o(c) \ge 2^{m-1}$, $m \ge 6$. Hence $[G', G] \le \mho_2(G')$, and so for each $x \in G$, we have $c^x = cc^{4j}$ ($j$ is an integer) and $(c^2)^x = c^2c^{8j}$. By Lemma 139.1, we get $\langle [a, b^2]\rangle = \langle c^2\rangle$. Setting $H = \langle a, b^2\rangle$, we get $H' = \langle [a, b^2]\rangle = \langle c^2\rangle$. Since, by the above, $[H', H] \le \mho_2(H')$, we may apply Lemma 139.1 on the group $H$ and we get

$$\langle [a, b^4]\rangle = \langle [a, b^2]^2\rangle = \langle c^4\rangle\,.$$

By the above, $[a, b^4]^2 = 1$, and so $c^8 = 1$, contrary to $o(c) \ge 2^{m-1}$, $m \ge 6$.

We have proved that $o(c) \le 2^{m-2}$. Thus,

$$\exp(G') \le 2^{m-2}\,, \quad \exp(K_3(G)) \le 2^{m-3}\,, \quad \exp(K_4(G)) \le 2^{m-4}\,, \quad m \ge 6\,.$$

By the Hall–Petrescu formula,

$$(ab)^{2^{m-1}} = a^{2^{m-1}} b^{2^{m-1}} c_2^{\binom{2^{m-1}}{2}} c_3^{\binom{2^{m-1}}{3}} c_4^{\binom{2^{m-1}}{4}} = 1\,, \quad \text{where } c_i \in K_i(G)\,, \quad i = 2, 3, 4\,.$$

Hence $o(ab) \le 2^{m-1}$, and so

$$o(ab) < o(b) \quad \text{which together with } G = \langle a, ab\rangle \text{ gives } \langle a\rangle \cap \langle ab\rangle = \{1\}\,,$$

and we are done. □

**Theorem 189.9.** *Let $G \in \mathrm{BI}(4)$ be a 2-group. Then $\mho_3(G) \le Z(G)$, and $\mho_2(G)$ is abelian.*

*Proof.* Let $a, b \in G$, where $o(a) = 2^m$, $o(b) = 2^n$. It suffices to prove $[a^8, b] = [a, b^8] = 1$ and $[a^4, b^4] = 1$. Set $H = \langle a, b\rangle$ and without loss of generality we may suppose $m \ge n$. By Lemma 189.2, we need only to consider that $\langle a\rangle \cap \langle b\rangle \ne \{1\}$.

If $m \ge 5$, then by Lemma 189.8, there is $b_1 \in H$ such that

$$H = \langle a, b_1\rangle, \ \langle a\rangle \cap \langle b_1\rangle = \{1\}\,.$$

By Lemma 189.2, $\mho_2(H) \le Z(H)$, and we are done.

Now suppose $4 \ge m \ge n$. It follows from $\langle a\rangle \cap \langle b\rangle \ne \{1\}$ that $[a^8, b] = [a, b^8] = 1$. Set $\langle a^{2^k}\rangle = \langle a\rangle \cap \langle b\rangle$, where $k \ge 1$. If $k \le 2$, then $\langle b^4\rangle \le \langle a^4\rangle$, and so $[a^4, b^4] = 1$. If $k = 3$, then we consider the subgroup $H_1 = \langle a^4, b\rangle$. By Lemma 189.1, $\langle a^4\rangle \trianglelefteq H_1$. Acting with $\langle b\rangle$ on $\langle a^4\rangle \cong C_4$, we see that $b^2$ centralizes $\langle a^4\rangle$, and so $[a^4, b^4] = 1$, and we are done. □

**Proposition 189.10.** *Let $G \in \mathrm{BI}(4)$ be a 2-group. If $|G| = 2^{m+2}$ and $\exp(G) \ge 2^m$, $m \ge 2$, then $|G'| \le 2^3$.*

*Proof.* We may assume that $G$ is nonabelian. Suppose that $\exp(G) = 2^{m+1}$. If $G \cong M_{m+2}$, then $|G'| = 2$. If $G$ is of maximal class, then setting $G' = \langle c\rangle$, we have $c^4 \in Z(G)$, and so $|G| \le 2^5$ and $|G'| \le 2^3$.

Now suppose that $\exp(G) = 2^m$. Then $G$ is not of maximal class, and so if $m \le 4$, a result of O. Taussky implies $|G'| \le 2^3$. Suppose that $m \ge 5$ and let $a \in G$ with $o(a) = 2^m$. Let $b \in G - \langle a\rangle$. By Lemma 189.8, there is $b_1 \in G$ such that

$$\langle a, b\rangle = \langle a, b_1\rangle \quad \text{and} \quad \langle a\rangle \cap \langle b_1\rangle = \{1\}\,.$$

If $G = \langle a, b\rangle$, then Lemma 189.2 implies $|G'| \le 2^3$. If $\langle a, b\rangle < G$, then $o(b_1) = 2$ since $|G| = 2^{m+2}$. Let $b_2 \in G - \langle a, b_1\rangle$. Similarly, $|G'| \le 2^3$ if $G = \langle a, b_2\rangle$. Hence we need only to consider the case where $H_1 = \langle a, b_1\rangle$ and $H_2 = \langle a, b_2\rangle$ are proper subgroups of $G$ and $|H_1| = |H_2| = 2^{m+1}$. Since $H_i \in \mathrm{BI}(4)$ $(i = 1, 2)$, then either $H_i \cong M_{2^{m+1}}$ or $H_i \cong C_{2^m} \times C_2$, and so $|H_i'| \le 2$. By Exercise 1.69, $|G' : (H_1' H_2')| \le 2$, and so $|G'| \le 2^3$. □

**Proposition 189.11.** *Let $G$ be a 2-group with a cyclic subgroup $\langle a\rangle$ of index 2. If $o(a) = 2^n \ge 2^4$, then $G$ is not generated by its elements of order $2^3$.*

*Proof.* We may assume that $G$ is nonabelian. If $G$ is of maximal class, then all elements in $G - \langle a\rangle$ are of order $\le 4$. Hence $G \cong M_{2^{n+1}}$. But then $\Omega_{n-1}(G) \cong C_{2^{n-1}} \times C_2$ contains all elements of order $2^3$. □

**Theorem 189.12.** *Let $G \in \mathrm{BI}(4)$ be a 2-group and $a, b \in G$. Then $|\langle [a, b]\rangle^G| \le 2^4$, $[a, b]^8 = 1$, and hence $\mathrm{cl}(G) \le 5$.*

*Proof.* Let $o(a) = 2^m$, $o(b) = 2^n$ and $c = [a, b]$. Without loss of generality, suppose $m \ge n$. It suffices to prove $|\langle c\rangle^G| \le 2^4$ and $c^8 = 1$.

By Proposition 189.10, $|(\langle b\rangle^G)'| \le 2^3$. If $a \in \langle b\rangle^G$, then $c \in (\langle b\rangle^G)'$ and we have $|\langle c\rangle^G| \le 2^4$ and $c^8 = 1$. So we need to consider only the case $a \notin \langle b\rangle^G$ and similarly, $b \notin \langle a\rangle^G$.

If $\langle a\rangle \cap \langle b\rangle = \{1\}$, we consider $A = \langle a\rangle^G \cap \langle b\rangle^G$. Then $|A| \le 2^4$ for $G \in \mathrm{BI}(4)$. If $|A| \ge 2^3$, then $a^{2^{m-1}}, b^{2^{n-1}} \in A$. In this case, $|\Omega_1(A)| \ge 4$. Hence $\exp(A) \le 2^3$. Clearly, this is also true for $|A| \le 2^2$. Hence, $|\langle c\rangle^G| \le 2^4$ and $c^8 = 1$ since $c \in A$. Therefore, we assume in the sequel that $\langle a\rangle \cap \langle b\rangle \ne \{1\}$.

(1) Suppose that $m \ge 5$. By Lemma 189.8, there exists $b_1 \in H = \langle a, b\rangle$ such that $H = \langle a, b_1\rangle$ and $\langle a\rangle \cap \langle b_1\rangle = \{1\}$. By the argument in the previous paragraph, $|\langle [a, b_1]\rangle^G| \le 2^4$ and $\exp(\langle [a, b_1]\rangle^G) \le 2^3$. Hence $|\langle c\rangle^G| \le 2^4$ and $c^8 = 1$ for $c \in \langle [a, b_1]\rangle^G$.

(2) Suppose that $m \le 3$. If $m \le 2$, then $|\langle a\rangle^G| \le 2^4$, and so $|\langle c\rangle^G| \le 2^3$ since $|\langle b\rangle^G| \le 2^4$, $c \in \langle a\rangle^G \cap \langle b\rangle^G$ and $\langle a\rangle^G \ne \langle b\rangle^G$.

Now suppose $m = 3$. Then $|\langle a\rangle^G| \le 2^5$ and note that $c \in \langle a\rangle^G \cap \langle b\rangle^G$. If $|\langle c\rangle^G| = 2^5$, then $\langle a\rangle^G = \langle b\rangle^G$, a contradiction. Hence $|\langle c\rangle^G| \le 2^4$. If $\langle c\rangle^G$ is a cyclic group of order $2^4$, then $|\langle a\rangle^G : \langle c\rangle^G| = 2$ for $\langle a\rangle^G$ is noncyclic. By Proposition 189.11, $\langle a\rangle^G$ cannot be generated by elements of order $2^3$, which contradicts $o(a) = 2^3$. Hence $\exp(\langle c\rangle^G) \le 2^3$ and then $c^8 = 1$.

(3) Let $m = 4$. If $o(b) \le 2^3$, then $|\langle a\rangle^G \cap \langle b\rangle^G| \le 2^4$ for $b \notin \langle a\rangle^G$, by assumption. So $|\langle c\rangle^G| \le 2^4$ and if $\langle c\rangle^G \cong C_{2^4}$, then $\langle b\rangle^G$ (noncyclic) has a cyclic subgroup of order $2^4$ and index 2. In this case, $o(b) = 2^3$, which gives a contradiction, by Proposition 189.11. It follows that $\exp(\langle c\rangle^G) \le 2^3$, and so $c^8 = 1$. Hence we need to consider only the case that $o(a) = o(b) = 2^4$. Set $H = \langle a, b\rangle$.

(3.1) $\langle a^4\rangle \le \langle a\rangle \cap \langle b\rangle$. If $|\langle a\rangle^H : \langle a\rangle| \le 2$, then, by Lemma 189.4, there exists $b_1 \in H$ such that $H = \langle a, b_1\rangle$ and $o(b_1) \le 2^3$. By the above argument, $c \in \langle [a, b_1]\rangle^G$ with $|\langle [a, b_1]\rangle^G| \le 2^4$ and $c^8 = 1$. So we need only to consider the case $|\langle a\rangle^H : \langle a\rangle| = 4$ and (similarly) $|\langle b\rangle^H : \langle b\rangle| = 4$. In this case we have

$$H = \langle a\rangle^H \langle b\rangle^H = \langle a\rangle^G \langle b\rangle^G \trianglelefteq G\,.$$

Set $\bar{H} = H/\langle a^4\rangle$ with the bar convention and then $|\bar{H}| \le 2^6$. We shall prove $|H'| \le 2^4$.

If $|\bar{H}| = 2^6$, then

$$|\langle \bar{a}\rangle^{\bar{H}} \cap \langle \bar{b}\rangle^{\bar{H}}| = 4\,, \quad |\bar{H}'| \le 4\,.$$

Therefore, $|H'| \le 2^4$. If $|\bar{H}| = 2^5$,then $|H/\langle a^8\rangle| = 2^6$. By Proposition 189.11, $H/\langle a^8\rangle$ is not a 2-group of maximal class. Thus, $|(H/\langle a^8\rangle)'| \le 2^3$ and $|H'| \le 2^4$. The case $|\bar{H}| = 2^4$ would give $\langle a\rangle^H = \langle b\rangle^H$, a contradiction. Thus, $|H'| \le 2^4$.

We claim that $H'$ is not cyclic of order $2^4$. Otherwise, $H' = \langle c\rangle \cong C_{2^4}$. Hence $c^a = c^k$, where $k$ is odd and $1 \le k < 2^4$. Since $\langle a\rangle^H = \langle a\rangle\langle c\rangle$ and $|\langle a\rangle^H| = 2^6$, it follows that $c^4 \in \langle a\rangle$, and so $(c^4)^a = c^{4k} = c^4$. On the other hand, $a^4 \in Z(H)$, and so

$$\begin{aligned}1 &= [a^4, b] = [aa^3, b] = [a, b]^{a^3}[a^3, b] = [a, b]^{a^3}[a, b]^{a^2}[a^2, b]\\ &= [a, b]^{a^3}[a, b]^{a^2}[a, b]^{a}[a, b]\,.\end{aligned}$$

This gives

$$c^{k^3}c^{k^2}c^kc = c^{k^3+k^2+k+1} = 1\,.$$

We get

$$k^3 + k^2 + k + 1 \equiv 1 \pmod{16}$$
$$4k \equiv 4 \pmod{16}\,,$$

and these congruences have no solution for any odd integer $k$ with $1 \le k < 2^4$. Hence $H'$ is not cyclic of order $2^4$, and so $c^8 = 1$ and $|\langle c\rangle^G| \le 2^4$.

(3.2) $\langle a^8\rangle = \langle a\rangle \cap \langle b\rangle$. By Theorem 189.9, $a^8 \in Z(G)$. Set $\bar{G} = G/\langle a^8\rangle$. Since $\bar{H} = \langle \bar{a}, \bar{b}\rangle$ and $\langle\bar{a}\rangle \cap \langle\bar{b}\rangle = \{1\}$, by Lemma 189.2, we have $o(\bar{c}) \le 2^2$. It follows that $c^8 = 1$. Suppose that $|\langle c\rangle^G| = 2^5$ so that

$$o(c) = 2^3\,, \quad |\langle c\rangle^G : \langle c\rangle| = 4\,,$$

and note that

$$c \in \langle a\rangle^G \cap \langle b\rangle^G \quad \text{and} \quad |\langle a\rangle^G| \le 2^6\,, \quad |\langle b\rangle^G| \le 2^6\,.$$

Thus

$$|\langle c\rangle^G \cap \langle a\rangle| \ge 2^3 \quad \text{and} \quad |\langle c\rangle^G \cap \langle b\rangle| \ge 2^3\,.$$

It follows that $a^2, b^2 \in \langle c\rangle^G$ and since

$$|\langle a^2\rangle\langle b^2\rangle| = (|\langle a^2\rangle|\,|\langle b^2\rangle|) : |\langle a^2\rangle \cap \langle b^2\rangle\,| = 2^5\,,$$

we obtain

$$\langle c\rangle^G = \langle a^2\rangle\langle b^2\rangle = \langle a\rangle^G \cap \langle b\rangle^G\,.$$

Then

$$\langle a\rangle^G = \langle a, b^2\rangle \quad \text{and also } \langle b\rangle^G = \langle b, a^2\rangle\,.$$

It follows that

$$H = \langle a, b\rangle = \langle a\rangle^G\langle b\rangle^G \trianglelefteq G\,.$$

By Lemma 189.2, $|\bar{H}'| \le 2^3$ and $\exp(\bar{H}') \le 2^2$. Therefore, $|H'| \le 2^4$ and $H' \trianglelefteq G$, a contradiction (since we have assumed $|\langle c\rangle^G| = 2^5$). Hence we have $|\langle c\rangle^G| \le 2^4$ and $c^8 = 1$, as required. □

**Corollary 189.13.** *Let $G \in \mathrm{BI}(4)$ be a 2-group. Then* $\exp(G') \le 2^3$.

*Proof.* Note that

$$G'' = [\mathrm{K}_2(G), \mathrm{K}_2(G)] \le \mathrm{K}_4(G)\,,$$

and so

$$[G', G''] \le [\mathrm{K}_2(G), \mathrm{K}_4(G)] \le \mathrm{K}_6(G) = \{1\}$$

since, by Theorem 189.12, $\mathrm{cl}(G) \le 5$. Hence $G'$ is of class at most 2. By Theorem 189.12, $G'$ is generated by commutators $[x, y]\,,\quad x, y \in G$, where $o([x, y]) \le 8$.

Let $a, b$ be any elements of order $\leq 8$ in $G'$. If $\langle a\rangle \cap \langle b\rangle = \{1\}$, then Lemma 189.2 implies $\exp(\langle a, b\rangle)' \leq 4$. This gives $(ab)^8 = a^8 b^8 [b, a]^{\binom{8}{2}} = 1$. If $\langle a\rangle \cap \langle b\rangle \neq \{1\}$, then $[b, a^4] = 1$, and so $[b, a]^4 = [b, a^4] = 1$.This gives again $(ab)^8 = a^8 b^8 [b, a]^{\binom{8}{2}} = 1$, and we are done. □

# Appendix 45 Varia II

This is a continuation of Appendix 40.

*1°*. Let us begin with the following obvious:

**Lemma A.45.1.** *Let $N$ be a normal subgroup of a $p$-group $G$ and let $|N/\mho_1(N)| = p^u$ and $|(G/N)/\mho_1(G/N)| = p^v$. Then $|G/\mho_1(G)| \le p^{u+v}$.*

*Proof.* Clearly, $\mho_1(N) \le \mho_1(G)$, so one may assume without loss of generality that $\mho_1(N) = \{1\}$; then $|N| = p^u$. Write $\bar{G} = G/\mho_1(G)$. Then $|\bar{N}| \le p^u$ and $|\bar{G}/\bar{N}| \le p^v$, and the result follows since $|\bar{G}| = |\bar{G}/\bar{N}| \cdot |\bar{N}| \le p^v \cdot p^u = p^{u+v}$. □

**Problem.** Let $G$ be a $p$-group, $p > 2$, and there is a cyclic $L \lhd G$ such that $G/L$ is absolutely regular. Is it true that $G$ is regular?

**Proposition A.45.2.** *If $G$ is a counterexample of minimal order to the problem, then:*
(a) *all proper sections of $G$ are regular so that* $\mathrm{d}(G) = 2$,
(b) $\mho_1(G) = \mathrm{Z}(G)$ *is cyclic of order $p^{e-1}$, where $p^e$* $= \exp(G)$,
(c) *$G/\mathrm{Z}(G)$ is of order $p^p$ and exponent $p$. In particular,* $\mathrm{cl}(G) = p$,
(d) *$\Omega_1(G)$ is of order $p^p$ and exponent $p$ and $L\Omega_1(G) < G$.*

*Proof.* By hypothesis, $G$ is irregular so that $|L| > p$, by Remark 7.2.

Assume that $|G| = p^{p+1}$. Then $G$ is of maximal class (Theorem 7.1 (b)). However, a $p$-group of maximal class has no normal cyclic subgroup of order $p^2$, unless $p = 2$. Thus, $|G| > p^{p+1}$.

Let $H < G$. Then $HL/L$ is absolutely regular as a subgroup of $G/L$. One has $HL/L \cong H/(L \cap H)$, and so, by induction, $H$ is regular since $H \cap L$ is cyclic. Let $R$ be a minimal normal subgroup of $G$. If $R \le L$, then $G/R$ is regular by induction, since $L/R$ is cyclic and $(G/R)/(L/R) \cong G/L$ is absolutely regular. Now let $R \nleq L$. Then $LR/R \cong L$ is cyclic and $(G/R)/(LR/R) \cong G/LR$ is absolutely regular as an epimorphic image of $G/L$. In that case, by induction, $G/R$ is regular. Thus, all sections of $G$ are regular. It follows that $\mathrm{d}(G) = 2$.

By Theorem 7.4 (c), $\mathrm{Z}(G) = \mho_1(G)$ is cyclic of order $p^{e-1}$, where $p^e = \exp(G)$. By Lemma A.45.1 and Theorem 9.8 (a), $|G/\mho_1(G)| = p^p$. Thus, $G/\mathrm{Z}(G)$ is of order $p^p$ and exponent $p$. It follows that $\mathrm{cl}(G) = p$ and $|\mathrm{Z}(G)| > p$ since $|G| > p^{p+1}$, and hence $G$ is not of maximal class.

As $G/L$ is absolutely regular, $G$ has no subgroup of order $p^{p+1}$ and exponent $p$ (indeed, if $H < G$ is such subgroup, then the group $HL/L \cong H/(H \cap L)$ of order $\ge p^p$ and exponent $p$ is not absolutely regular). Since $G$ is not of maximal class, it contains a normal subgroup $M$ of order $p^p$ and exponent $p$ (Theorem 12.1 (a)). Assume that $M < \Omega_1(G)$. Then there is $x \in \Omega_1(G) - M$ of order $p$. Set $H = \langle x, M\rangle$. Then $\exp(H) > p$, by the above, and we conclude that $H$ is irregular (Theorem 7.2 (b)) of order $p^{p+1}$. As $H < G$, we get a contradiction. Thus, $\Omega_1(G) = M$ has order $p^p$. As $\mathrm{d}(G) = 2$, it follows that $ML < G$. Indeed, $|M \cap L| = p$ and $M/(M \cap L) = (M/(M \cap L)) \times L/(M \cap L)$ is not two-generator. □

**Remark 1.** Below Mann's letter dated August 18, 2011 follows: For all primes, there exists a $p$-group $G$ containing a normal cyclic subgroup $N$ such that $G/N$ is absolutely regular but $G$ is not regular. If $p = 2$, any nonabelian metacyclic 2-group is an example. For an odd $p$, let $G$ be a minimal irregular $p$-group of exponent at least $p^3$, class $p$, and $|G/\mho_1(G)| = p^p$. Such groups are constructed in Section 2 of [Man2]. Let $z$ be a generator of $Z(G)$, and let $u$ be an element of $K_{p-1}(G) - K_p(G)$. Then $x = zu$ generates a cyclic subgroup $N$ as required. This solves Problem 1 (see also #2542 (i) in the list of Research Problems and Themes IV).

*2°*. If $G$ is metacyclic, $p > 2$ and $|G'| = p^k$, then $\mathrm{cd}(G) = \{1, p, \dots, p^k\}$ [HetK]. Let us prove the following:

**Proposition A.45.3.** *Let $G$ be a $p$-group with cyclic $G'$. Then there is $\chi \in Irr(G)$ such that $\chi = \mu^G$ for $\mu \in Lin(A)$, where $A$ is abelian of minimal index in $G$. Next,* $\ker(\chi) \cap G' = \{1\}$.

*Proof.* One may assume that $G$ is nonabelian. Since $\bigcap_{\chi \in \mathrm{Irr}_1(G)} \ker(\chi) = \{1\}$, there is $\chi \in \mathrm{Irr}_1(G)$ such that $\Omega_1(G') \nleq \ker(\chi)$; then $G' \cap \ker(\chi) = \{1\}$ since $G'$ is cyclic. There are $A < G$ and $\mu \in \mathrm{Lin}(A)$ such that $\chi = \mu^G$. It follows from

(1) $$\{1\} = G' \cap \ker(\chi) = G' \cap \ker(\mu)_G \geq G' \cap \left(\bigcap_{x \in G} x^{-1}A'x\right)$$

that $\bigcap_{x \in G} x^{-1}A'x = \{1\}$. Since $G'$ is cyclic, one has $A' \trianglelefteq G$, and so $A' = \{1\}$, hence $A$ is abelian of minimal index in $G$ (see Introduction, Theorem 17). □

**Corollary A.45.4.** *Let $G$ be a nonabelian metacyclic 2-group and $K < G'$. Then there is $\chi \in Irr_1(G)$ such that $G' \cap \ker(\chi) = K$. In particular, $2 \in \mathrm{cd}(G)$.*

*Proof.* The first assertion follows from Proposition A.45.3. Let us prove the second assertion. Let $K < G'$ be of index 2. Then $G/K$ is minimal nonabelian (Lemma 65.2 (a)). Let $\chi \in \mathrm{Irr}_1(G/K)$; then $G' \cap \ker(\chi) = K$. Now the result follows from Theorem 17 in Introduction. □

**Exercise 1.** Let $G$ be a nonabelian $p$-group, $p > 2$. Then $G$ is regular $\iff$ any two of its minimal nonabelian subgroups generate the regular subgroup.

**Solution.** Assume that $G$ is irregular. Without loss of of generality one may assume that $G$ is minimal irregular; then $\mathrm{d}(G) = 2$. It suffices to show that $G$ is generated by two minimal nonabelian subgroups. By Theorem 10.28, there is a minimal nonabelian subgroup, say $A$, that is not contained in $\Phi(G)$. Set $F = A\Phi(G)$; then $F \in \Gamma_1$. By Theorem 10.28, there is a minimal nonabelian subgroup, say $B$, not contained in $F$; then $G = BF = \langle B, A, \Phi(G)\rangle$. Set $H = \langle A, B\rangle$. Since $H\Phi(G) = G$, one obtains $H = G$.

**Exercise 2.** If $p > 3$ and all nonabelian subgroups are normal in a $p$-group $G$, then $G$ is regular. (**Hint.** It suffices to show that $G$ has no minimal irregular subgroup. Assume that $H \leq G$ is minimal irregular. Use Exercise 1, Fitting's lemma and Theorem 7.1 (b).)

**Exercise 3.** Suppose that a $p$-group $G$ of order $p^p$ of maximal class and exponent $p > 3$ has an abelian subgroup of index $p$. Then $\alpha_1(G) = p^{p-1}$.

**Hint.** All nonabelian maximal subgroups of $G$ are of maximal class. Working by induction on $|G|$, use Lemma 57.1 and Theorem 9.6 (f).

Mann in a letter dated August 7, 2011 reported that if a $p$-group $G$ possesses a maximal normal abelian subgroup $A$ with cyclic quotient group $G/A$, then $\mathrm{cd}(G) = \{1, p, \dots, p^k\}$, where $p^k = |G/A|$ (see also [Man30, Proposition 11]). This result inspired the following:

**Proposition A.45.5** (= Proposition 148.22). *Suppose that a group $G$ possesses a maximal normal abelian subgroup $A$ such that the quotient group $G/A$ is either cyclic $p$-group or generalized quaternion group. Then there is $\lambda \in Lin(A)$ with $\lambda^G \in Irr(G)$.*

**Hint.** Let $B/A < G/A$ be of order $p$ and $\chi \in \mathrm{Irr}_1(B)$. In that case, $|\mathrm{Irr}_1(\chi_A)| = p$ (Burnside) so, if $\lambda \in \mathrm{Irr}_1(\chi_A)$, then $A = \mathrm{I}_G(\lambda)$.

**Corollary A.45.6.** *If $A$ is a maximal abelian subgroup of a $p$-group $G$ such that $G/A$ is a cyclic $p$-group or a generalized quaternion group, then $A$ is an abelian subgroup of minimal index in $G$.*

**Exercise 4.** Let $A$ be a maximal abelian normal subgroup of a 2-group $G$ such that $G/A$ is a generalized quaternion group. Estimate $\mathrm{b}(G) = \max\{\chi(1) \mid \chi \in \mathrm{Irr}(G)\}$.

**Exercise 5.** Let $N$ be a normal subgroup of a $p$-group $G$ such that $G/N$ is either cyclic or generalized quaternion. Suppose that if $H/N < G/N$ is of prime order, then $\mathrm{b}(H) > \mathrm{b}(N)$ (here $\mathrm{b}(H) = \max\{\chi(1) \mid \chi \in \mathrm{Irr}(H)\}$). Take $\chi \in \mathrm{Irr}(H)$ with $\chi(1) = \mathrm{b}(H)$. If $\chi_N = \phi_1 + \cdots + \phi_p$ is the Clifford's decomposition, then $\phi_1^G \in \mathrm{Irr}(G)$.

**Exercise 6.** Let $G$ be a nonabelian $p$-group. Then $\mathrm{k}(G) \leq (|G| + (p^2 - 1)|G : G'|)/p^2$ with equality $\iff \mathrm{cd}(G) = \{1, p\}$.

**Solution.** One has

$$|G| = |G : G'| + \sum_{\chi \in \mathrm{Irr}_1(G)} \chi(1)^2 \geq |G : G'| + (\mathrm{k}(G) - |G : G'|)p^2 .$$

It follows that $\mathrm{k}(G) \leq (|G| + (p^2 - 1)|G : G'|)/p^2$ with equality $\iff \mathrm{cd}(G) = \{1, p\}$.

**Exercise 7.** Let $x_1, \dots, x_k$ be the system of distinct representatives of the noncentral conjugacy classes of a nonabelian group $G$. Is it true that $\langle x_1, \dots, x_k \rangle = G$?

$3^o$. In this subsection we prove the following:[1]

**Proposition A.45.7.** *Suppose that $G$ is a metacyclic $p$-group with derived subgroup $G'$, $p > 2$, $|G| = p^{2s+e}$, where $e \in \{0, 1\}$, and let $|G : G'| = p^k$. Then $s + 1 \leq k$.*

*Proof.* Set $|G| = p^n(= p^{2s+e})$, $e \in \{0, 1\}$, $|Z(G)| = p^z$ and $|G'| = p^{n-k}$. Let $L$ be a normal cyclic subgroup of $G$ such that $G/L$ is cyclic; then $G' < L$. Let $L \leq A < G$, where $A$ is a maximal abelian subgroup of $G$; then $G/A$ is cyclic as an epimorphic image of $G/L$. By Lemma 1.1, $|A| = |G'|\,|Z(G)|$. By Lemma 148.14,

$$(2)\ \ p^k = |G : G'| = |A| = |G'|\,|Z(G)| = p^{n-k}p^z \Rightarrow k = \frac{n+z}{2} = \frac{2s+e+z}{2} = s + \frac{e+z}{2}..$$

Assume that the theorem is false. Then $k < s + 1$ or, what is the same, $s + \frac{e+z}{2} < s + 1$, and so $e + z < 2$. Then $z = 1$ and $e = 0$ since $z \geq 1$. Thus, $n = 2s$ and, by (2), $2k = n + z = 2s + 1$; a final contradiction. □

$4^o$. The following result was reported by Mann a couple years ago:

**Proposition A.45.8.** *If a $p$-group $G$ is of class $p$, then* $\exp(G/Z(G)) = \exp(G')$.

**Remark 2.** If $G$ is a $p$-group and $x \in G$, then $\mathrm{cl}(\langle x, G'\rangle) < \mathrm{cl}(G)$. Indeed, write $H = \langle x, G'\rangle$. Then $H/K_3(G)$ is abelian, and hence $\mathrm{cl}(H) < \mathrm{cl}(G)$.

**Remark 3.** Let $q$ be a power of $p$. If $G$ is a regular $p$-group, then $x^q = b^q \iff (x^{-1}b)^q = 1$ (Theorem 7.2 (a)).

*Proof of Proposition A.45.8.* If $x, y \in G$, then, by Remark 2,

$$\mathrm{cl}(\langle x, x^y\rangle) = \mathrm{cl}(\langle x, x[x, y]\rangle = \mathrm{cl}(\langle x, [x, y]\rangle \leq \mathrm{cl}(\langle x, G'\rangle) < \mathrm{cl}(G) = p\,.$$

Assume that $x^q \in Z(G)$, where $q$ is a power of $p$. Then $x^q = (x^q)^y = (x^y)^q$; therefore $[x, y]^q = (x^{-1}x^y)^q = 1$ (Remark 3). Since the regular subgroup $G'$ (Remark 3) is generated by elements of order $\leq q$, it follows that $\exp(G') \leq q \leq \exp(G/Z(G))$ (Theorem 7.2 (b)). The other direction is proved in the same way, reading the above backward. □

**Exercise 8.** If $G$ is a $p$-group with regular $\langle x, G'\rangle$ for all $x \in G$, then $\exp(G') = \exp(G/Z(G))$.

**Solution.** Let $x, y \in G$. Then the subgroup $H = \langle x, G'\rangle$ is regular, by hypothesis, and $x, x^y \in H$ since $H \triangleleft G$. Let $q = \exp(G/Z(G))$; then $x^q \in Z(G)$ and so $x^q = (x^q)^y = (x^y)^q \Rightarrow [x, y]^q = (x^{-1}x^y)^q = 1$ (Remark 3). Thus, the regular subgroup $G'$, being generated by commutators of orders dividing $q$, has exponent $\leq q$. As in the proof of Theorem A.45.8, the other direction is proved in the same way, reading the above backward.

As Mann noticed, the result of Exercise 8 is a partial case of [Man4, Theorem 4].

---

1 For a more general result see § 204.

$5^o$. Here another proof of [INS, Theorem 3.3] is offered.

**Theorem A.45.9.** *Suppose that a two-generator $p$-group $G$ has no epimorphic image of order $p^3$ and exponent $p > 2$. Then $G$ is metacyclic.*

*Proof.* One may assume that $G$ is nonabelian. Let $R < G'$ be $G$-invariant of index $p$. Then $\bar{G} = G/R$ is minimal nonabelian (Lemma 65.2 (a)). Assume that $G$ is nonmetacyclic. Then $\bar{G}$ is also nonmetacyclic (Theorem 36.1). By Lemma 65.1,

$$\bar{G} = \langle x, y \mid x^{p^m} = y^{p^n} = z^p = 1,\ z = [x, y],\ [x, z] = [y, z] = 1\rangle .$$

Then $\bar{G}/\langle x^p, y^p\rangle$ is nonabelian of order $p^3$ generated by elements of order $p$, contrary to the hypothesis. Thus, $G$ is metacyclic. □

$6^o$. Recall that $\Omega_n^*(G) = \langle x \in G \mid o(x) = p^n\rangle$. If $G$ is metacyclic, then

$$w(G) = \max\{i \mid |\Omega_i(G)| = p^{2i}\}, \quad R(G) = \Omega_{w(G)}(G) .$$

Then $G/R(G)$ is either cyclic or a 2-group of maximal class (Lemma 1.4).

**Proposition A.45.10.** *Let $G$ be a nonabelian metacyclic $p$-group of exponent $p^e$. If $\Omega_e^*(G) < G$, then $G/R(G)$ is a 2-group of maximal class $\not\cong Q_8$.*

*Proof.* If $p > 2$, then $\Omega_e^*(G) = G$ since $G$ is regular (Theorem 7.2 (b)). Thus, $p = 2$. Let $G$ be not of maximal class. Then $R(G) > \{1\}$ (Lemma 1.4). Assume that $G/R(G)$ is cyclic. In that case, $\Omega_{w(G)}^*(R(G)) = R(G) \Rightarrow G > R(G)$. One has $\Omega_{e-1}(G) < G$ so that $\Omega_e^*(G) = \langle G - \Omega_{e-1}(G)\rangle = G$, a contradiction. Now let $G/R(G)$ be a 2-group of maximal class. If $G/R(G) \cong Q_8$, then $\Omega_{e-1}(G) < G$ so that $\Omega_e^*(G) = G$, a contradiction. Thus, $G/R(G) \not\cong Q_8$. In that case, $|G : \Omega_e^*(G)| = 2$ (apply the above result to a cyclic subgroup $T/R(G)$ of index 2 in $G/R(G)$). □

$7^o$ **Theorem A.45.11.** *Let $G$ be a $p$-group of exponent $p^e > p$. Any two elements of $G$ of distinct orders are permutable $\iff \Omega_{e-1}(G) \le Z(G)$.*

*Proof.* If $\Omega_{e-1}(G) \le Z(G)$, then any two elements of $G$ of different orders are permutable. Therefore, it suffices to prove implication $\Rightarrow$. Assume that $G$ is nonabelian. Write $\Omega = \Omega_{e-1}(G)$ and $H = \Omega_e^*(G)$. Then $[\Omega, H] = \{1\}$, by hypothesis. Also $\Omega \cup H = G$, so that either $\Omega = G$ or $H = G$. If $H = G$, then $\Omega = Z(G)$, and we are done. Now let $\Omega = G$; then $H \le Z(G)$. Let $x \in \Omega^\#$ be of order $< p^e$ and let $y \in H - \Omega$ be of order $p^e$. Then $o(xy) = p^e$, so that $xy \in H \le Z(G)$. As $y \in Z(G)$, it follows that $x \in Z(G)$. Since such elements generate $\Omega$, we get $\Omega \le Z(G)$, and we are done (in that case, $G\Omega \cup H$ is abelian). □

$8^o$. If $G$ is a group, then $\mathrm{n}(G) = |\mathrm{Irr}_1(G)|$, the number of nonlinear irreducible characters of $G$. In this subsection the following theorem from Kazarin's letter, obtained at least 30 years ago, will be proved:

**Theorem A.45.12** (L. Kazarin, personal communication). *Let $G$ be a nonabelian $p$-group with $|G'| = p^k$, $k = 2s + e$, $e \in \{0, 1\}$. If $e = 0$, then $\mathrm{n}(G) \geq (p^2 - 1)s$. If $e = 1$, then $\mathrm{n}(G) \geq (p^2 - 1)s + (p - 1)$.*

*Proof.* Let $|G| = p^m$ and let $N \leq G'$ be $G$-invariant; then $\mathrm{Irr}(G \mid N) \subseteq \mathrm{Irr}_1(G)$.

(i) Suppose that $|N| = p$ and $\mathrm{Lin}(N) = \{1_N, \mu_1, \dots, \mu_{p-1}\}$; then all $\mu_i$ are $G$-invariant. One has $\mathrm{Irr}(\mu_i^G) \subseteq \mathrm{Irr}_1(G)$ and, by reciprocity and the Clifford theorem, $\mathrm{Irr}(\mu_i^G) \cap \mathrm{Irr}(\mu_j^G) = \emptyset$ for $i \neq j$. It follows that $|\mathrm{Irr}(G \mid N)| \geq p - 1$.

(ii) Suppose that $|N| = p^2$. Then

$$p^{m-2}(p^2 - 1) = |G| - |G/N| = \sum_{\chi \in \mathrm{Irr}(G|N)} \chi(1)^2 \equiv |\mathrm{Irr}(G \mid N)| \pmod{p^2 - 1},$$

and so $p^2 - 1 \mid |\mathrm{Irr}(G \mid N)|$. Since $|\mathrm{Irr}(G \mid N)| > 0$, it follows that $|\mathrm{Irr}(G \mid N)| \geq p^2 - 1$.

In view of (i) and (ii), one may assume in what follows that $|G'| = p^k > p^2$. We proceed by induction on $k$. Set $k = 2s + e$, where $e \in \{0, 1\}$. Next, $\mathrm{Irr}_1(G) = \mathrm{Irr}_1(G/N) \cup \mathrm{Irr}(G \mid N)$ ($N$ is as above) is a partition or, what is the same,

$$\mathrm{n}(G) = \mathrm{n}(G/N) + |\mathrm{Irr}(G \mid N)|. \tag{3}$$

(iii) Let $e = 0$ and let $N < G'$ be of order $p^2$. Then $|(G/N)'| = p^{2(s-1)}$ so, by (3), (ii) and induction, one has $\mathrm{n}(G) \geq (p^2 - 1)(s - 1) + (p^2 - 1) = (p^2 - 1)s$.

(iv) Let $e = 1$ and let $N < G'$ be of order $p$. Then $|(G/N)'| = p^{2s}$ hence, by (ii), $\mathrm{n}(G/N) \geq (p^2 - 1)s$. Since $|\mathrm{Irr}(G \mid N)| \geq p - 1$, by (i), one obtains, by (3), $\mathrm{n}(G) \geq (p^2 - 1)s + (p - 1)$. □

**Proposition A.45.13.** *Let $G$ be a group of exponent $p$.*

(a) *If any two subgroups of $G$ of order $p^2$ are permutable, then $G$ is abelian, unless $|G| = p^3$.*

(b) *If $|G| > p^6$ and any two subgroups of $G$ of order $p^3$ are permutable, then $G$ is abelian.*

*Proof.* Assume that $G$ is nonabelian. Let $R \lhd G$ be of order $p$.

(a) All subgroups of $G/R$ having order $p$, are permutable, by hypothesis, so that $G/R$ is elementary abelian. It follows from this that $R$ is the unique minimal normal subgroup of $G$ and so $G$ is extraspecial. Assume that $|G| > p^3$. Let $A < G$ be minimal nonabelian; then $|A| = p^3$. By Proposition 10.17, $\mathrm{C}_G(A) \nleq A$. Let $A' \neq X \leq \mathrm{C}_G(A)$, where $X$ is of order $p$ and let $U, V < A$ of order $p$ be such that $UV \neq VU$. Then $UX, VX < X \times A$ are not permutable of order $p^2$, a contradiction.

(b) By (a), either $|G/R| = p^3$ or $G/R$ is abelian. In the first case, $G$ of order $p^4$ satisfies the condition. Now let $|G| > p^4$. It follows that $R = G'$ is the unique normal subgroup of $G$ of order $p$ so that $G$ is extraspecial.

(b1) Let $|G| = p^7$. Then $G = A_1 * A_2 * A_3$ (central product), where $A_i \cong S(p^3)$, $i = 1, 2, 3$ (Lemma 4.3). Let $U_i$ be noncentral subgroups of order $p$ in $A_i$ ($i = 2, 3$). Set

$H = A_1 \times U_2 \times U_3$. Let $V_i/(U_2 \times U_3)$ $(i = 1, 2)$ be nonpermutable subgroups of order $p$ in $H/(U_2 \times U_3)$. Then $\langle V_1, V_2\rangle = H$ so $V_1V_2 \neq V_2V_1$ (here we use the product formula). Thus, $G$ does not satisfy the hypothesis.

(b2) Now let $|G| > p^7$. Then there is extraspecial $H < G$ of order $p^7$. As $H$ does not satisfy the hypothesis, by (b1), so $G$ does not. □

**Exercise 9.** All noncyclic subgroups of a nonabelian $p$-group $G$, $p > 2$, are normal $\iff$ one of the following holds:
(a) $G$ is a metacyclic minimal nonabelian group,
(b) $G = \Omega_1(G)Z(G)$, where $\Omega_1(G) \cong S(p^3)$ and $Z(G)$ is cyclic, or
(c) $G$ is a 3-group of maximal class and order $3^4$, $|\Omega_1(G)| = 3^2$.

**Hint.** The group $G$ has no elementary abelian subgroup of order $p^3$ (otherwise, $G$ is abelian). Therefore result follows easily from Theorem 13.7.

*9⁰* **Proposition A.45.14.** *If $G$ is a $p$-group and $\chi \in Irr(G)$ is faithful, then $|Z(G)| \mid \chi(1) \cdot |G : G'|$.*

*Proof.* One may assume that $G$ is nonabelian. By assumption, $Z = Z(G)$ is cyclic. By Clifford's theorem, $\chi_Z = \chi(1)\lambda$, where $\lambda \in \mathrm{Lin}(Z)$ is faithful. Let $T$ be a representation of $G$ affording $\chi$ and put $\det(\chi)(g) = \det(T(g))$ for $g \in G$ (here $\det(T(g))$ is the determinant of the matrix $T(g)$). Clearly, $\det(\chi)$ is a linear character of $G$ and its order in the group $\mathrm{Lin}(G)$ divides $|G : G'|$. Let $g \in Z$. Since $T_Z(g) = \lambda(g)I_{\chi(1)}$ is a scalar matrix of size $\chi(1)$, it follows that $\det(\chi_Z) = \lambda^{\chi(1)}$. As we have noted, $\det(\chi)^{|G:G'|} = 1_G$ so that $\lambda^{\chi(1)|G:G'|} = 1_Z$, and therefore $|Z| = o(\lambda) \mid \chi(1)|G : G'|$. □

*10⁰* **Proposition A.45.15.** *Suppose that a nonabelian metacyclic $p$-group $G$ has an abelian subgroup $A$ of index $p$. Let a cyclic $L \lhd G$ be such that $G/L$ is cyclic. If $L \nleq A$, then $G$ is minimal nonabelian.*

*Proof.* One has $G' < L$, $G' \leq A \cap L$ and $G = AL$; then $A/(A \cap L) \cong G/L$ is cyclic. One has $C_G(A \cap L) \geq AL = G$ so that $G' \leq A \cap L \leq Z(G)$. In that case, $G/(A \cap L)$ is a noncyclic abelian group so that $G/Z(G)$ is a noncyclic abelian group with cyclic subgroup $A/Z(G)$ of index $p$ (note that $A/Z(G)$ is an epimorphic image of $A/(A \cap L)$). It follows that $G/Z(G)$ contains two distinct cyclic subgroups $A/Z(G)$ and, say $V/Z(G)$, of index $p$. Then $A$, $V$ are two distinct abelian maximal subgroups of $G$ so that $A \cap V = Z(G)$. Since $d(G) = 2$ and $G/Z(G) \cong E_{p^2}$, it follows that $G$ is minimal nonabelian. □

*11⁰* **Proposition A.45.16.** *If a $p$-group $G$ is such that any its subgroup is a direct factor of its normal closure, then $G/E$ is Dedekindian, where $E$ is a maximal elementary abelian subgroup of $G$.*

*Proof.* Let $E$ be a maximal elementary abelian subgroup of $G$ and $E \leq H < G$. Then $H^G = H \times L$ for some $L < G$. If $L > \{1\}$, there is $x \in L$ of order $p$. In this case, $E \times \langle x\rangle > E$

is elementary abelian, contrary to the choice of $E$. Thus, $L = \{1\}$ so that $H \lhd G$. In particular, $E \lhd G$. By the above, $G/E$ is Dedekindian. □

**Proposition A.45.17.** *A nonabelian metacyclic p-group G satisfies the hypothesis of Proposition A.45.16* $\iff$ *it is minimal nonabelian.*

*Proof.* One may assume that $|G| > p^3$. It is easy to see that $G$ is not a 2-group of maximal class. By Proposition 10.19, $E = \Omega_1(G) \cong \mathrm{E}_{p^2}$. By Proposition A.45.16 and Theorem 1.20, $G/E$ is either abelian or $\cong \mathrm{Q}_8$. Assume that $G/E \cong \mathrm{Q}_8$; then $G$ is not minimal nonabelian and $G$ is a group of Lemma 42.1 (c) so that $\mathrm{Z}(G) \cong \mathrm{C}_4$. One has

$$G = \langle a, b \mid o(a) = o(b) = 8\,, \quad a^4 = b^4\,, \quad a^b = a^{-1} \rangle\,.$$

Here $\mathrm{Z}(G) = \langle b^2 \rangle$. The centralizer $C = \mathrm{C}_G(a) = \langle a, b^2 \rangle$ is abelian of type $(8, 2)$ and $C$ is a unique abelian subgroup of index 2 in $G$. It follows that the subgroup $N = \mathrm{N}_G(\langle b \rangle) \cong \mathrm{M}_{16}$ and $\langle b \rangle^G = N$ so that $G$ does not satisfy the hypothesis. Thus, $G/E$ is abelian so that $G' < E$ is of order $p$. By Lemma 65.2 (a), $G$ is minimal nonabelian. Clearly, any metacyclic minimal nonabelian $p$-group $X$ satisfies the hypothesis of Proposition A.45.16 (indeed, if $C < X$ is nonnormal, it is cyclic and $C^X = C \times X'$). □

*12°* **Proposition A.45.18.** *If every two elements of equal order in a p-group G commute, then G is abelian.*

*Proof.* Assume that $G$ is a counterexample of minimal order. Then $G$ is minimal nonabelian since all its proper subgroups satisfy the hypothesis. Let $\exp(G) = p^e$. As we know, $G$ is generated by elements of some equal order which implies that $G$ is abelian, contrary to the assumption. □

Let $G$ be an arbitrary finite group in which every two elements of equal order commute. Let us prove that $G$ is abelian. Assume that $G$ is a counterexample of minimal order. By induction, $G$ is minimal nonabelian. In view of Proposition A.45.18, $G$ is nonnilpotent. Then $G = P \cdot Q$, where $Q = G' \in \mathrm{Syl}_q(G)$ and $P = \langle x \rangle \in \mathrm{Syl}_p(G)$ is maximal in $G$. If $y \in Q^\#$, then $\langle x, x^y \rangle = G$, a contradiction.

**Exercise 10.** Classify the nonabelian groups $G$ in which any two elements of distinct orders commute. (**Hint.** See Theorem A.45.11.)

**Lemma A.45.19.** *Let G be a p-group of exponent > p. Every element of G of order p is permutable with every element of G of order > p* $\iff$ $\Omega_1(G) < \mathrm{Z}(G)$.

*Proof.* (Compare with the proof of Theorem A.45.11.) Assume that $G$ is nonabelian. Let $\Omega = \Omega_1(G)$ and $H = \mathrm{H}_p(G)$. Then $[\Omega, H] = \{1\}$, by hypothesis. As $G = \Omega \cup H$, it follows that either $\Omega = G$ or $H = G$. If $H = G$, then $\Omega \le \mathrm{Z}(G)$, and we are done. Now let $\Omega = G$; then $H \le \mathrm{Z}(G)$. As $\Omega$ is nonabelian, it contains two noncommuting elements $x, y$ of order $p$. Let $h \in H$ be of order $> p$. Then $[x, hy] = 1$, by hypothesis, and this implies that $[x, y] = 1$, a contradiction. □

**Exercise 11** (Janko). A noncyclic $p$-group $G$ has the property that each of its maximal cyclic subgroups is contained in exactly one maximal subgroup of $G \iff \mathrm{d}(G) = 2$ and $\Phi(G) = \{x^p \mid x \in G\}$.

**Solution.** If $x \in G - \Phi(G)$, then $\langle x \rangle$ is a maximal cyclic subgroup of $G$ and so $\Phi(G)\langle x \rangle$ must be the unique maximal subgroup of $G$ containing $\langle x \rangle$. This gives $|G : (\Phi(G)\langle x \rangle)| = p$ and so $\mathrm{d}(G) = 2$. Let $\langle y \rangle$ be a maximal cyclic subgroup in $\Phi(G)$. Then $\langle y \rangle$ cannot be a maximal cyclic subgroup in $G$, by hypothesis, and so there is $z \in G - \Phi(G)$ such that $z^p = y$. This implies that each element in $\Phi(G)$ is a $p$-th power of an element in $G$ and so $\Phi(G) = \{z^p \mid z \in G\}$. Conversely, if $G$ is a noncyclic two-generator $p$-group and $\Phi(G) = \{z^p \mid z \in G\}$, then each maximal cyclic subgroup $\langle z \rangle$ of $G$ has the property $z \in G - \Phi(G)$ and so $\Phi(G)\langle z \rangle$ is the unique maximal subgroup of $G$ containing $\langle z \rangle$.

*13°* **Exercise 12.** Let $L$ be a maximal cyclic subgroup of a noncyclic $p$-group $G$, $|L| > p$. If for any maximal cyclic subgroup $B \neq L$ of $G$ one has $L \cap B = \{1\}$ then $G \cong \mathrm{D}_{2^n}$.

**Solution.** Let $L < M \leq G$, where $|M : L| = p$. Obviously, $L$ is the unique cyclic subgroup of index $p$ in $M$ and any cyclic subgroup of $M$ is either contained in $L$ or has order $p$. It follows from Theorem 1.2 that $M$ is a 2-group of maximal class. By Exercise 10.10, $G$ is of maximal class. Clearly, $G$ has no cyclic subgroup of order 4 not contained in $L$. It follows that $G$ is dihedral.

A small modification of the above argument shows that if $G$ is a $p$-group and $L < G$ is maximal cyclic of order $> p$, then there is in $G$ a maximal cyclic subgroup $A$ of order $|L|$ such that $|A \cap L| = \frac{1}{p}|L|$, unless $G$ is a 2-group of maximal class.

*14°*. All subgroups of order 2 in the group $\mathcal{H}_{2,2} = \langle a, b \mid a^4 = b^4 = 1,\ b^a = b^3 \rangle$ are characteristic. It is interesting to classify the metacyclic minimal nonabelian $p$-groups (even, more generally, metacyclic $p$-groups) all of whose subgroups of order $p$ are characteristic.

If a metacyclic $p$-group $G = B \cdot A$ of order $p^{m+n}$, where $A, B$ are cyclic of orders $p^m, p^n$ respectively, $n > m$, $A \lhd G$, and $C < G$ is cyclic of order $p^n$, then the subgroup $\Omega_{n-m}(C)$ is characteristic (of order $p^{n-m}$) in $G$. Indeed, if $U < G$ is cyclic of order $p^n$, then, by the product formula, $C \cap U \geq \Omega_{n-m}(C)$. Use this observation to solve the following two exercises.

**Exercise 13.** Let $G = \langle a, b \mid o(a) = p^m,\ o(b) = p^n,\ \langle a \rangle \lhd G,\ \langle a \rangle \cap \langle b \rangle = \{1\},\ m < n \rangle$. Then $\langle b^{p^m} \rangle$ is characteristic in $G$.

It follows that if, in Exercise 13, $p = 2$, then all subgroups of order 2 are characteristic in $G$.

**Exercise 14.** Let $G = \langle a, b \mid o(a) = p^m,\ o(b) = p^n,\ [a, b] = c,\ c^p = 1,\ [a, c] = [b, c] = 1,\ m > n\}$. Is it true that $\langle a^{p^{m-n}} \rangle$ is characteristic in $G$?

**Exercise 15.** Prove that a group $G$ of exponent $p^e$ is abelian $\iff$ $\Omega_e^*(G) \le Z(G)$.

**Exercise 16.** Let $A$ be a maximal abelian subgroup of a nonabelian $p$-group $G$. Prove that there is a minimal nonabelian $B \le G$ such that $|A \cap B| > p$.

**Hint.** Let $A < T \le G$, where $|T : A| = p$. Then any minimal nonabelian subgroup of $T$ satisfies the condition.

**Exercise 17.** Let $G = R(G)$ be a metacyclic $p$-group (see §§ 124, 148), and an $\mathcal{A}_k$-subgroup $H < G$, $k > 1$. Then $H/\Omega_1(G)$ is an $\mathcal{A}_{k-1}(G)$-group (see Remark 148.33).

**Hint.** One has $\Omega_1(G) = \Omega_1(H)$. By Proposition 72.1 (a), $|H'| = p^k$. Then $|(H/\Omega_1(G))'| = p^{k-1}$ so that $H/\Omega_1(G)$ is an $\mathcal{A}_{k-1}$-group (Proposition 72.1 (a) again).

**Exercise 18.** Classify the $p$-groups of order $\ge p^{n+3}$, $n > 2$, all of whose subgroups of index $p^n$ are metacyclic.

**Hint.** Such groups have no elementary abelian subgroup of order $p^3$. Use Theorem 13.7 and § 50.

**Exercise 19.** Study the subgroup structure of the groups $G = \langle a, b \mid a^{p^m} = b^{p^n} = c^p,\ c = [a, b], [a, c] = [b, c] = 1\rangle$.

**Exercise 20.** Study the subgroup structure of the groups $G = \langle a, b \mid a^{p^m} = b^{p^n},\ a^b = a^{1+p^{m-1}},\ m > 1\rangle$.

**Exercise 21.** Given $m > 1$ and $p > 2$, study the subgroup structure of the groups $G = \langle a, b \mid a^{p^m} = b^{p^{m-1}},\ a^b = a^{1+p^{m-1}}\rangle$.

**Exercise 22.** Given $m > 2$, study the subgroup structure of the groups $G = \langle a, b, c \mid a^{2^m} = b^{2^{m-2}} = c^2,\ a^b = a^5,\ a^c = a^{-1},\ [b, c] = 1\rangle$.

A group $H$ is said to be *capable* if there is a group $G$ such that $G/Z(G) \cong H$.

**Exercise 23.** Prove that a capable $p$-group has no normal subgroup $\cong M_{p^n}$.

**Solution.** Let a capable $p$-group $G/Z(G)$ have a normal subgroup $M/Z(G) \cong M_{p^n}$. Using Lemma 1.1, one obtains $|M'| = p$ so that $M' \le Z(G)$ since $M'$, being characteristic in $M \lhd G$, is $G$-invariant. Then $M/Z(G)$ is abelian, a contradiction.

**Exercise 24.** Classify the q-self dual minimal nonabelian $p$-groups.

**Solution.** Let $G$ be a minimal nonabelian q-self dual $p$-group (i.e., any epimorphic image of $G$ is isomorphic to a subgroup of $G$). By hypothesis, all proper epimorphic images of $G$ are abelian. It follows that $G'$ is the unique minimal normal subgroup of $G$ so that $Z(G)$ is cyclic. By Lemma 65.1, $G \in \{S(p^3), D_8, M_{p^n}\}$ ($Q_8$ is not q-self dual).

**Exercise 25.** If $G$ is a nonabelian $p$-group with abelian subgroup of index $p$ and $|G : G'| = p^2$, then $G$ is of maximal class.

**Solution.** One may assume that $|G| > p^3$. By Lemma 1.1, $|Z(G)| = \frac{1}{p}|G : G'| = p$. By induction, $G/Z(G)$ is of maximal class since $Z(G) < G'$. Then $G$ is of maximal class since $|Z(G)| = p$.

**Exercise 26.** Suppose that a group $G$ of exponent $p$ contains an abelian subgroup $A$ of index $p$. Then $\mathrm{cl}(G) \le p - 1$.

**Solution.** One may assume that $G$ is nonabelian; then $Z(G) < A$. We proceed by induction on $|G|$. If $|Z(G)| > p$, there are in $Z(G)$ two distinct subgroups $K$ and $L$ of order $p$. Then $G$ is isomorphic to a subgroup of $(G/K) \times (G/L)$ so, by induction, $\mathrm{cl}(G) \le \max\{\mathrm{cl}(G/K), \mathrm{cl}(G/L)\} \le p-1$. Now let $Z(G)$ be of order $p$. Then $|G : G'| = p|Z(G)| = p^2$ (Lemma 1.1). It follows from Exercise 25 that $G$ is of maximal class. Since $\exp(G) = p$, Theorem 9.5 implies that $|G| \le p^p$ so that $\mathrm{cl}(G) \le p - 1$.

The following exercise (as the previous one) is a partial case of [Hup1, Satz III.10.10] which asserts that the class of any regular $p$-group containing an abelian subgroup of index $p$ does not exceed $p - 1$.

**Exercise 27.** Suppose that a regular $p$-group $G$ contains an elementary abelian subgroup $A$ of index $p$. Then $\mathrm{cl}(G) \le p - 1$. (**Hint.** Repeat word for word the solution of Exercise 26 to prove that $G$ is of maximal class. Use Theorems 9.5 and 9.6 (b,e).)

As a Sylow $p$-subgroup of the symmetric group $S_{p^2}$, $p > 2$, shows, Exercise 27 is not true for irregular $p$-groups. However, the following result holds:

**Exercise 28.** Suppose that a $p$-group $G$ contains an elementary abelian subgroup of index $p$. Then, if $p > 2$, $\mathrm{cl}(G) \le p$.

**Solution.** One may assume that $Z(G)$ is of order $p$ (see the solution of Exercise 26). In that case, $|G : G'| = p^2$ (Lemma 1.1) so that $G$ is of maximal class (Exercise 25). It follows that $\mathrm{cl}(G) \le p$ (Theorem 9.6 (b,e); see also Exercise 9.13).

**Exercise 29.** If a $p$-group $G$ is a $\mathcal{A}_n$-group with regular $G'$, then $\exp(G') \le p^n$.

**Solution.** We proceed by induction on $n$. Let $A, B \in \Gamma_1$ be distinct. Then $A$ is an $\mathcal{A}_u$-group and $B$ is an $\mathcal{A}_v$-group, $u, v \le n - 1$. Since $A'$ and $B'$ are regular, one obtains, by induction, $\exp(A') \le p^{u-1} \le p^{n-1}$ and $\exp(B') \le p^{v-1} \le p^{n-1}$. Write $H = A'B'$. By Theorem 7.2 (b), $\exp(H) \le p^{n-1}$. Then $\exp(G') \le p \cdot \exp(H) \le p^n$ since, by Exercise 1.69, $|G' : H| \le p$.

**Exercise 30.** Suppose that a group $G$ of exponent $p > 2$ contains an abelian subgroup, say $A$, of index $p^2$. Then $\mathrm{cl}(G) \le 2(p - 1)$.

**Solution.** By Exercise 1.6 (a), one may assume that $A \lhd G$. Then $G/A \cong E_{p^2}$. Let $U/A, V/A < G/A$ be distinct of order $p$. By Exercise 26, $\mathrm{cl}(A) \le p - 1$ and $\mathrm{cl}(V) \le p - 1$. Now the result follows, by Fitting's lemma (Introduction, Theorem 21).

**Exercise 31.** Suppose that a group $G$ of exponent $p > 2$ contains an abelian subgroup, say $A$, of index $p^3$. Then $\mathrm{cl}(G) \leq 4(p-1)$.

**Solution.** By Alperin's theorem (see § 39), one may assume that $A \triangleleft G$. Let $U/A$ and $V/A$ be two distinct maximal subgroups of $G/A$. By Exercise 30, one has $\mathrm{cl}(U) \leq 2(p-1)$ and $\mathrm{cl}(V) \leq 2(p-1)$. Then, by Fitting's lemma, $\mathrm{cl}(G) \leq \mathrm{cl}(U) + \mathrm{cl}(V) \leq 4(p-1)$.

**Exercise 32.** Let a group $G$ of exponent $p > 2$ contain a normal abelian subgroup, say $A$, of index $p^n$. Then $\mathrm{cl}(G) \leq 2^{n-1}(p-1)$.

**Hint.** If $H/A$ is maximal in $G/A$, then, by induction, $\mathrm{cl}(H) \leq 2^{n-2}(p-1)$. Apply Fitting's lemma.

**Exercise 33.** If a nonabelian $p$-group $G$ is such that all elements of the set $G - A$ have order $p$ for any maximal abelian subgroup $A < G$, then $\exp(G) = p$.

**Solution.** Let $x \in G - A$ and $x \in B$, where $B \neq A$ is maximal abelian subgroup of $G$. Then all elements of the set $A - B$ have order $p$ and $\langle A - B \rangle = A$. It follows that $\exp(A) = p$, and we conclude that $\exp(G) = p$.

**Exercise 34.** Let $G = \langle a, b \mid a^{p^m} = b^{p^2} = 1,\ b^a = b^{p+1} \rangle$, $m > 2$ be minimal nonabelian and $\mathrm{M}(G)$ its Schur multiplier. Is it true that $|\mathrm{M}(G)| \leq p$?

**Solution.** Since $|G'| = p$ and $\mathrm{d}(G) = 2$, $G$ is metacyclic minimal nonabelian (Lemma 65.2 (a)) and $\Omega_2(G) = \langle a^{p^{m-2}}, b \rangle$ is abelian of type $(p^2, p^2)$. As $G/\Omega_2(G)$ is cyclic and, if $F_1/\Omega_2(G) < G/\Omega_2(G)$ is maximal, then $F$ is abelian of type $(p^{m-1}, p^2)$. If $H_1 \in \Gamma_1 - \{F_1\}$, then $H_1$ as abelian of type $(p^m, p)$ so that $\Omega_2(H_1)$ is abelian of type $(p^2, p)$. Let $\Gamma$ be a representation group of $G$. There is in $\mathrm{Z}(\Gamma) \cap \Gamma'$ a subgroup $M \cong \mathrm{M}(G)$ such that $\Gamma/M \cong G$. Let $\mathcal{F}/M$ and $\mathcal{H}/M$ be distinct maximal subgroups of $\Gamma/M$ with cyclic subgroups of index $p$. Then $|\mathcal{F}'| \leq p$ and $|\mathcal{H}'| \leq p$. Next, $\mathcal{F}', \mathcal{H}'$ are incident of orders $\leq p$ since $\Gamma'$ is cyclic (Theorem 47.2). One has $|M| < |\Gamma'| \leq p^2$ (Exercise 1.69) so that $|M| \leq p$.

**Exercise 35.** Let $G$ be a nonabelian $p$-group of order $> p^3$, $p > 2$. If all subgroups of $G$ of order $p^3$ are isomorphic, then $G = \langle a, b \mid o(a) = o(b) = p^2,\ a^b = a^{1+p} \rangle$ is metacyclic of order $p^4$ and exponent $p^2$.

**Hint.** If there is in $G$ a subgroup $E \cong \mathrm{E}_{p^3}$, then $\exp(G) = p$ and $G$ has no minimal nonabelian subgroups (Lemma 65.1); in that case $G$ is abelian, a contradiction. Application of Proposition 1.3 and Theorem 13.7 allows us to finish the proof.

**Exercise 36.** Let $G$ be a noncyclic metacyclic $p$-group of order $> p^4$. Is it true that $G$ has two nonisomorphic subgroups of order $p^4$?

**Exercise 37.** Suppose that a finite group $G$ has a nontrivial center. If for any $x \in G - \mathrm{Z}(G)$ and $y \in x\mathrm{Z}(G)$ one has $\pi(y) = \pi(x)$, then the number $|\mathrm{Z}(G)|$ is a prime power.

**Solution.** Let $u \in G - Z(G)$ be such that $o(u) = p^\alpha$ for some prime $p$. Assume that $v, w \in Z(G)$, where $o(v)$ is a power of a prime $q$ and $o(w)$ is a power of a prime $r \neq q$. Then for $uv, uw \in uZ(G)$ one has $\pi(uv) \neq \pi(uw)$, a contradiction. Thus, $|Z(G)|$ is a prime power.

**Exercise 38.** Suppose that $G$ is a noncyclic 2-group of order $> 2^3$ and $\exp(G) > 2$. If all subgroups of $G$ of order 8 are isomorphic, then $|G| \leq 2^7$.

**Solution.** It follows easily from Proposition 1.3 that $\exp(G) = 4$, so that $G$ has no elementary abelian subgroup of order 8. Hence, if $G$ is abelian, it is of type $(4, 4)$. If $G$ is nonabelian metacyclic, then $G \cong \mathcal{H}_{2,2}$. If $G$ is nonmetacyclic, there is a metacyclic normal subgroup $N$ such that $G/N$ is isomorphic to a subgroup of $D_8$ (Theorem 50.2). Since $\exp(N) \leq 4$, one has $|N| \leq 2^4$, by the above, and so $|G| = |N|\,|G/N| \leq 2^4 \cdot 2^3 = 2^7$. (The minimal nonmetacyclic group of order $2^5$ satisfies the condition.)

**Exercise 39.** Does there exist a noncyclic metacyclic $p$-group $G$ of order $> p^4$, all of whose subgroups of order $p^4$ are isomorphic?

**Hint.** By Theorem 1.2, $G$ has no cyclic subgroup of index $p$. Let $R \triangleleft G$ be abelian of type $(p, p)$ (Lemma 1.4). If $G/R$ has a subgroup of type $(p, p)$, then $\exp(G) = p^2$; then $|G| < p^5$, a contradiction. Thus, $G/R$ is either cyclic or generalized quaternion group, and so $G$ is a group of Lemma 42.1 (c) and order $2^5$. The last groups, however, do not satisfy the condition.

**Exercise 40.** Classify the metacyclic $p$-groups $G$ of order $> p^5$, all of whose subgroups of order $p^5$ are isomorphic.

**Exercise 41.** Study the metacyclic $p$-groups $G$ all of whose subgroups of the same order have the same class.

**Exercise 42** ( = Remark 10.5). An irregular $p$-group containing a maximal regular subgroup $R$ of order $p^p$, is of maximal class.

**Solution.** Let $R < H \leq G$, where $|H : R| = p$. Then $H$ is irregular so of maximal class. Now the result follows from Exercise 10.10.

**Exercise 43.** Classify the irregular $p$-groups all of whose maximal regular subgroups are either absolutely regular or of exponent $p$.

**Solution.** Let an absolutely regular $F < G$ be maximal regular ($F$ exists since $\exp(G) > p$, by Theorem 7.1 (b)). Let $F < H \leq G$, where $|H : F| = p$; then $H$ is irregular. If any such $H$ is of maximal class, then $G$ is of maximal class (Exercise 10.10). Assume that $H$ is not of maximal class. By Exercise 42, $|F| > p^p$. Then there is $R \triangleleft H$ of order $p^p$ and exponent $p$ (Theorem 12.1 (a)), so that $H = RF$ and $|H| > p^{p+1}$. Let $R < K < H$, where $|K : R| = p$. Then $K$ is irregular, by hypothesis, and we conclude that $K$ is of maximal class (Theorem 7.1 (b) again). It follows from Exercise 10.10 that $H$ is of maximal class, contrary to the assumption. Thus. $G$ is of maximal class. Any irregular $p$-group of maximal class satisfies the hypothesis (Theorem 9.6 (e)).

**Exercise 44.** Let $G$ be a finite group. If any two conjugate prime power subgroups of $G$ not contained in $\Phi(G)$, are permutable, then $G$ is nilpotent.

**Solution.** Assume that $G$ is nonnilpotent. Then it contains a minimal nonnilpotent subgroup $H = PQ$, where $Q = H' \in \mathrm{Syl}_q(H)$ and $P = \langle a \rangle \in \mathrm{Syl}_p(H)$. As $\Phi(G)$ is nilpotent, one has $H \not\leq \Phi(G)$. Consideration of $H \cap \Phi(G)$ shows that $P \not\leq \Phi(G)$. If $P_1 \in \mathrm{Syl}_p(H) - \{P\}$, then $PP_1 \neq P_1P$. Thus, $H$ does not exist, so $G$ is nilpotent.

**Exercise 45.** Let $G$ be a finite group and $N \lhd G$. If any two conjugate prime power subgroups of $G$ that are not contained in $N$ are permutable, then all minimal nonnilpotent subgroups of $G$ are contained in $N$.

**Solution.** Let $G, H, P, P_1$ be as in Exercise 44. Since $P, P_1$ are not permutable, at least one of them, say $P$, is contained in $N$. Since $N \lhd G$ and $P \sim P_1$ in $H$ so in $G$, one has $P_1 \leq N$. It follows from $\langle P, P_1 \rangle = H$ that $H \leq N$. We suggest to the reader to prove that $G/N$ is nilpotent.

**Exercise 46.** Deduce Exercise 44 from Exercise 45.

**Exercise 47.** If a noncyclic $p$-group $G$ of order $> p^3$ and exponent $> p$ has no abelian subgroup of type $(p^2, p)$, then $G$ is a group of maximal class.

**Solution.** Let $L < G$ be cyclic of order $p^2$. If $\mathrm{C}_G(L) = L$, then $G$ is of maximal class (Proposition 1.8). It follows from Theorems 9.5 and 9.6 that $p = 2$ (consider $G_1$). Now let $\mathrm{C}_G(L) > L$. Then $\Omega_1(\mathrm{C}_G(L)) = \Omega_1(L)$, so that $\mathrm{C}_G(L)$ is cyclic. Let $\mathrm{C}_G(L) < B \leq G$ be such that $|B : \mathrm{C}_G(L)| = p$. It is easily seen (see Theorem 1.2) that $B$ is of maximal class. Now the result follows from Exercise 10.10. Use Theorems 9.5 and 9.6 to finish the classification. Is it true that $G \cong \Sigma_9$?

**Exercise 48.** Let $G$ be a $p$-group with cyclic subgroup $T$ of index $p$. Is it true that if $A, B < G$ are nonincident, then $A \cap B$ has index $p$ either in $A$ or in $B$?

**Exercise 49.** If $n$ is such that $\mathrm{Z}_n(G)$ is regular, where $G$ is an irregular $p$-group, then either $\mathrm{e}_n(G) = 0$ or $\mathrm{e}_n(G) \equiv 1 \pmod p$ (use Theorem 7.2).

A noncyclic $p$-group, in which the intersection of any two distinct subgroups of order $p^2$ has order $p$, we call a $(*)$-group.

**Exercise 50.** The $(*)$-group $G$ of order $> p^3$ has no proper subgroup $\cong \mathrm{E}_{p^3}$.

**Hint.** Let $\mathrm{E}_{p^3} = E < G$ and $E < H \leq G$. Then $H = \langle x, E \rangle$. Let $x \in R < H$, where $|R| = p^2$. Then $E = S \times (R \cap E)$ and $S \cap R = \{1\}$, a contradiction.

**Exercise 51.** Is it true that if a $p$-group $G$ is a noncyclic abelian $(*)$-group of order $> p^3$, then it has a cyclic subgroup of index $p$?

**Exercise 52.** If $G$ is a noncyclic $(*)$-group, $p > 2$, then one and only one of the following holds:
(a) $|G| \leq p^3$,
(b) $G = \Omega_1(G)C$, where $\Omega_1(G) \cong S(p^3)$ and $C$ is cyclic of order $> p$,
(c) $G$ has a cyclic subgroup of index $p$,
(d) $p = 3$, $G$ is a 3-group of class 3 and order $3^4$, $G \not\cong \Sigma_9 \in \mathrm{Syl}_3(\mathrm{S}_9)$.

**Hint.** By Exercise 50 and Theorem 13.7, one of the following holds: (i) $G = \Omega_1(G)C$, where $\Omega_1(G) \cong S(p^3)$ and $C$ is cyclic of order $> p$; (ii) $G$ is a 3-group of maximal class; (iii) $G$ is metacyclic. Any group from (i) satisfies the condition. Suppose that $G$ is metacyclic. Let $H = \Omega_2(G)$. Then $p^3 \leq |H| \leq p^4$. If $|H| = p^3$, then $G$ is as in (c). Now let $|H| = p^4$. Then there are in $H$ two distinct cyclic subgroups $A, B$ of order $p^2$ such that $A \cap B = \{1\}$, a contradiction. Let $G$ be a 3-group of maximal class. Then the fundamental subgroup $G_1$ of $G$ is of order $3^3$, so that $|G| = 3^4$. It is easy to see that $G \not\cong \Sigma_9 \in \mathrm{Syl}_3(\mathrm{S}_9)$.

**Exercise 53.** Let a $p$-group $G$ of order $> p^{p+1}$ be neither absolutely regular nor of maximal class. If $H \in \Gamma_1$ is absolutely regular, then $G = H\Omega_1(G)$, where $\Omega_1(G)$ is of order $p^p$ and exponent $p$ (Theorem 12.1 (b)). Then all members of the set $\Gamma_1$ not containing $\Omega_1(G)$ are absolutely regular.

**Solution.** Let $F \in \Gamma_1$ be such that $\Omega_1(G) \not\leq F$ and assume that $F$ is not absolutely regular ($\Omega_1(G) \not\leq \Phi(G)$ since $\Omega_1(G) \not\leq H \in \Gamma_1$). Then $F \cap \Omega_1(G) = \Omega_1(F)$ is of order $p^{p-1}$ and exponent $p$. It follows from Theorem 12.1 (a) that $F$ is of maximal class. Then, by Theorem 12.12 (b), $G/\mathrm{K}_p(G)$ is of order $p^{p+1}$ and exponent $p$. In that case, the set $\Gamma_1$ has no absolutely regular member, a contradiction.

**Exercise 54.** Let $H$ be a proper subgroup of maximal class of a $p$-group $G$ and suppose that all subgroups of $G$ containing $H$ are two-generator. Then $G$ is of maximal class.

**Solution.** Let $H < M$ with $|M : H| = p$. Then $M$ is of maximal class, by Theorem 12.12 (a). Now the result follows from Exercise 10.10.

**Exercise 55.** If $G$ is a nonmetacyclic minimal nonabelian $p$-group of exponent $p^e > p$, then $\Omega_e^*(G) = G$. (**Hint.** One has $\exp(G/G') = p^e$.)

*15°*. Below, the list of Schur multipliers $\mathrm{M}(G)$ of the nonabelian groups $G$ of order 16 is presented. We assume that $G$ has no cyclic subgroup of prime index (for such groups Schur multipliers are known). The presented information is taken from [Kar1, Table 1]; see also [Tah].
- $G = \mathrm{C}_2 \times \mathrm{D}_8$. $\mathrm{M}(G) \cong \mathrm{E}_8$.
- $G = \mathrm{C}_2 \times \mathrm{Q}_8$. $\mathrm{M}(G) \cong \mathrm{E}_4$.
- $G \cong \mathcal{H}_{2,2} = \langle a, b \mid o(a) = o(b) = 4,\ a^b = a^3 \rangle$. $\mathrm{M}(G) = \mathrm{C}_2$. (See Remark 148.53.)
- $G$ is nonmetacyclic minimal nonabelian of order 16. $\mathrm{M}(G) \cong \mathrm{E}_4$.
- $G = \mathrm{D}_8 * \mathrm{C}_4$ (central product of order 16). $\mathrm{M}(G) \cong \mathrm{E}_4$.

Note that if $G = S(p^3)$, then $\mathrm{M}(G) \cong \mathrm{E}_{p^2}$ (Schur).

The Schur multipliers of some interesting $p$-groups are computed (see [Kar2]), among them metacyclic $p$-groups (see also Theorem 47.2) and $\Sigma_{p^n}$.

*16⁰* **Exercise 56.** Suppose that $G$ is a noncyclic $p$-group of exponent $> p^2$. If, for distinct cyclic subgroups $A$ and $B$ of $G$ of the same fixed order $> p^2$ one has $|A \cap B| \le p$, then $G$ is a 2-group of maximal class.

**Solution.** One may assume that $A$ is a maximal cyclic subgroup of $G$ of order $> p^2$. Let $A < C \le G$, where $|C : A| = p$. Then $A$ is the unique cyclic subgroup of $C$ of index $p$. It follows from Theorem 1.2 that $C$ is a 2-group of maximal class. By Exercise 10.10, $G$ is also of maximal class.[2]

**Exercise 57.** Let $R$ be a normal subgroup of order $p^p$ and exponent $p$ of an irregular $p$-group $G$ and suppose that, whenever $R_1$ is a normal subgroup of $G$ containing $R$ as a subgroup of index $p$, then $R_1$ is of maximal class. Is it true that $R_1 = G$?

**Exercise 58.** Let $G$ be a 2-group. If $A < B \le G \Rightarrow \mathrm{d}(A) \le \mathrm{d}(B)$, then all two-generator subgroups of $G$ are metacyclic.

**Solution.** Let $B \le G$ be two-generator. If $A$ is maximal in $B$, then $\mathrm{d}(A) \le 2$, by hypothesis, so that $B$ is metacyclic (Corollary 36.6).

**Exercise 59.** Classify the $p$-groups all of whose subgroups are permutable with all maximal cyclic subgroups.

**Solution** (Mann). Let $H, C < G$ be any subgroup and a cyclic subgroup, respectively. Embed $C$ is a maximal cyclic subgroup $D$. By assumption, $HD = DH$. Let $|D : C| = p^i$. Then $HD$ contains a subgroup $E$ of index $p^i$ which contains $H$, and then $E = HC$ (the case $p^i \ge |HD : H|$, in which $C \le H$ is ignored). Thus $H$ permutes with all cyclic subgroups, therefore with all subgroups, and $G$ is modular.

**Exercise 60.** Let $A$ be a nonnormal subgroup of a $p$-group $G$. Set $B = \mathrm{N}_G(A)$, $C = \mathrm{N}_G(B)$ and $p^k = |C : B|$. Then $B$ contains exactly $p^k$ subgroups conjugate with $A$ in $C$ and so normalizing $A$. In particular (Matsuyama; see [Mat, 2.1.5], where another proof is given), some conjugate subgroup $A^x$ different from $A$ normalizes $A$.

**Exercise 61.** Let $G$ be a $p$-group. If $A < B \le G$, where $|B : A| = p \Rightarrow |B : B'| > |A : A'|$, then $G$ is abelian.

**Solution.** Assume that $B \le G$ is minimal nonabelian. If $A < B$ is maximal, then $|B : B'| = |A : A'| (= |A|)$, a contradiction. Thus, $G$ has no minimal nonabelian subgroup so it is abelian.

2 One may assume that $A$ is fixed.

**Exercise 62.** Let $G$ be a $p$-group, $p > 3$. If $A < B \leq G$ with $|B : A| = p \Rightarrow \mathrm{d}(A) \leq \mathrm{d}(B)$, then $G$ is regular.

**Solution.** Suppose that $G$ is irregular. Let $B \leq G$ be minimal irregular. Then $\mathrm{d}(B) = 2$ so all maximal subgroups of $B$ are two-generator, by hypothesis. By Theorem 9.8 (a), $|B/\mho_1(B)| \geq p^p \geq p^5$ since $p \geq 5$. By Theorem 5.8 (b), there is in $B/\mho_1(B)$ a maximal subgroup $A/\mho_1(B)$ that is not generated by two elements. Then $\mathrm{d}(A) \geq 3 > 2 = \mathrm{d}(B)$, contrary to the hypothesis. Thus, $B$ does not exist so that $G$ is regular.

**Exercise 63.** If a $p$-group $G$, $p > 2$, contains a normal cyclic subgroup $L$ such that $|G/L| = p^{p-1}$, then $G$ is regular.

**Solution.** Assume that $G$ is irregular. Then, by Theorem 7.1 (b), $|L| > p$. By Exercise 9.1 (b) and Lemma 1.4, $G$ is not of maximal class. Therefore, by Theorem 12.1 (a), there is $R \triangleleft G$ of order $p^p$ and exponent $p$. Since $|R \cap L| = p$, one has $G = RL$, by the product formula. Next, $G/(R \cap L) = R/(R \cap L) \times L/(R \cap L)$, $|R/(R \cap L)| = p^{p-1}$. It follows that

$$\mathrm{cl}(G) \leq 1 + \mathrm{cl}(G/(R \cap L)) = 1 + \mathrm{cl}(R/(R \cap L)) \leq 1 + (p-2) = p - 1\,,$$

and so $G$ is regular (Theorem 7.1 (b)), contrary to the assumption.

**Exercise 64.** Suppose that a $p$-group $G$, $p > 3$, contains a normal abelian subgroup $A$ of rank 2 such that $|G/A| = p^{p-2}$. Is it true that $G$ is regular? What will be, provided $A$ is nonabelian metacyclic? (See Lemma A.45.1.)

**Exercise 65.** Study the irregular 3-groups containing a metacyclic minimal nonabelian subgroup of index 3.

**Exercise 66** (Compare with [Att]). Let $G$ be a nonabelian $p$-group with $|\mathrm{Z}_2(G)| = p^2$. Then all maximal subgroups of $G$, except one, have centers of order $p$ (such $p$-groups are classified in § 145.)

**Solution.** Obviously, $|G| > p^3$. All $G$-invariant subgroups of orders $\leq p^2$ are contained in $\mathrm{Z}_2(G)$, and so $\mathrm{Z}_2(G)$ is the unique $G$-invariant subgroup of order $p^2$. Set $F = \mathrm{C}_G(\mathrm{Z}_2(G))$; then $F \in \Gamma_1$. Take $H \in \Gamma_1 - \{F\}$ and assume that $|\mathrm{Z}(H)| > p$; then $\mathrm{Z}(H)$ is $G$-invariant of order $\geq p^2$. It follows that $\mathrm{Z}_2(G) \leq \mathrm{Z}(H)$, and so $\mathrm{C}_G(\mathrm{Z}_2(G)) \geq FH = G$, a contradiction. Thus, $|\mathrm{Z}(H)| = p$.

**Exercise 67.** Classify the nonabelian $p$-groups $G$ having exactly one faithful irreducible character.

**Solution.** The center $\mathrm{Z}(G)$ is cyclic. Let $L \triangleleft G$ be of order $p$. Then $\mathrm{Irr}(G \mid L) = \{\chi\}$, by hypothesis. Set $|G| = p^m$ and $\chi(1) = p^k$. Then

$$p^m = |G| = \chi(1)^2 + |G/L| = p^{2k} + p^{m-1} \Rightarrow p^{m-1}(p-1) = p^{2k}\,.$$

It follows that $p = 2$ and $m - 1 = 2k$. If $\lambda \in \mathrm{Irr}(L)$ is nonprincipal, then $\lambda^G = \chi(1)\chi$, by reciprocity and Clifford's theorem. Conversely, if $p = 2$ and a nonabelian 2-group

$G$ has a unique minimal normal subgroup, say $L$, and $\mathrm{Irr}(G)$ has a member of degree $\sqrt{\frac{1}{2}|G|}$, then $G$ satisfies the condition.

**Exercise 68.** Classify the nonabelian $p$-groups with exactly two faithful irreducible characters.

The following argument has been used in our book for partial groups.

**Exercise 69 (P. Hall. Reported by Mann).** Suppose that a group $\bar{G}$ (not necessarily finite) is generated by a set $\mathcal{M}$ of cyclic subgroups such that $\bigcap_{\bar{A}\in\mathcal{M}} \bar{A} = \bar{L} > \{\bar{1}\}$. Then there is no a group $W$ such that $W/\mathrm{Z}(W) \cong \bar{G}$.

**Solution.** Let $A/\mathrm{Z}(W) = \bar{A} \in \mathcal{M}$; then $A$ is abelian. By hypothesis, $T = \bigcap_{\bar{A}\in\mathcal{M}} A = L > \mathrm{Z}(W)$. On the other hand, $\mathrm{C}_W(T) \geq \langle A \mid \bar{A} \in \mathcal{M}\rangle = W$, i.e., $T \leq \mathrm{Z}(W)$, a contradiction.

**Exercise 70 (Reported by Mann).** If $G = \Sigma_{p^2} = U \operatorname{wr} V$, where $|U| = |V| = p > 2$, then there is no $p$-group $W$ such that $W/\mathrm{Z}(W) \cong G$, i.e., $G$ is not capable.

**Solution.** Since $\mho_1(G) > \{1\}$ is contained in all cyclic subgroups of $G$ of order $p^2$ and these subgroups generate $G$ (indeed, there are in $G$ two distinct regular subgroups of order $p^p$ and exponent $p^2$, and each of these subgroups is generated by its cyclic subgroup of order $p^2$), the result follows from Exercise 69. For $p = 2$, the result does not hold (for example, $G = \mathrm{D}_8 = \mathrm{C}_2 \operatorname{wr} \mathrm{C}_2$, $W = \mathrm{Q}_{16}$).

**Exercise 71.** Let $p > 2$, $G \cong \Sigma_{p^2}$ and let $\Gamma$ be a representation group of $G$. Then $\mathrm{M}(G) < \mathrm{Z}(\Gamma)$, where $\mathrm{M}(G)$ is the Schur multiplier of $G$. (**Hint.** Use Exercise 70.)

**Exercise 72 ( = Exercise 9.13 (Blackburn)).** Let $p > 2$ and $G$ a $p$-group such that $G/\mathrm{K}_{p+1}(G) \cong \Sigma_{p^2}$. Then $\mathrm{K}_{p+1}(G) = \{1\}$.

**Solution.** Assume that $\mathrm{K}_{p+1}(G) > \{1\}$; then $\mathrm{cl}(G) > p$. Consider the quotient group $\bar{G} = G/\mathrm{K}_{p+2}(G)$; then $\mathrm{cl}(\bar{G}) > p$. In that case, $\mathrm{Z}(\bar{G}) = \overline{\mathrm{K}_{p+1}(G)}$ (otherwise, $\mathrm{cl}(\bar{G}) \leq p$). Then $\bar{G}/\mathrm{Z}(\bar{G}) \cong \Sigma_{p^2}$, contrary to Exercise 70.[3]

**Exercise 73.** Let $H$ be a $p$-group of maximal class and order $p^{p+1}$ in which the number $\mathrm{e}_p(H)$ of subgroups of order $p^p$ and exponent $p$ is $< p$. Then there is no a $p$-group $G$ of maximal class and order $> p^{p+1}$ such that $G/\mathrm{K}_{p+1}(G) \cong H$.

**Solution.** Assume that such $G$ exists; then $|G| \geq p^{p+2}$ and $\mathrm{K}_{p+1}(G) < \Phi(G)$ so that $\mathrm{d}(G) = 2$. Let $\Gamma_1 = \{G_1, G_2, \ldots, G_{p+1}\}$, where $G_1$ is the fundamental subgroup of $G$. Then, for $i > 1$, $G_i/\mathrm{K}_p(G_i)$ is of order $p^p$ and exponent $p$ so that $\mathrm{e}_p(G/\mathrm{K}_{p+1}(G)) = p$, a contradiction since $\mathrm{K}_p(G_i) = \mathrm{K}_{p+1}(G)$.

**Exercise 74.** Let $H < G$ be a $p$-group of maximal class and order $p^{p+1}$. Then, if $G$ is of maximal class, then $\mathrm{e}_p(H) \equiv 0 \pmod{p}$.

---

3 This result also follows from Theorem 12.9.

**Solution.** Assume that $e_p(H) \not\equiv 0 \pmod p$. Let $H < M \le G$, where $|M : H| = p$. In that case some subgroup of $H$ of order $p^p$ and exponent $p$ is normal in $M$. As $M$ is of maximal class, this is a contradiction (Theorem 9.6 (c)).

**Exercise 75.** Let $G$ be an irregular $p$-group. If the set $\Gamma_1$ possesses exactly one absolutely regular member, then $G$ is of maximal class. (**Hint.** Use Theorems 12.1 (b) and 13.6, and Exercise 53.)

**Exercise 76** (see Proposition A.81.1). Suppose that a $p$-group $G$, $p > 2$, has a cyclic subgroup $Z$ of index $p^3$. Prove that then either $G$ is metacyclic or possesses a normal abelian subgroup of index $p^2$. If $G$ is metacyclic, it possesses a normal abelian subgroup of index $p^3$. (**Hint.** Use Theorem 13.7.)

**Exercise 77** (see Proposition A.81.2). Suppose that a 2-group $G$ has a cyclic subgroup $Z$ of index 8. Is it true that $G$ has a normal abelian subgroup of index 8?

**Exercise 78.** Suppose that a $p$-group $G$ satisfies the following condition: whenever $A, B < G$ are cyclic and $A \cap B = \{1\}$, then $[A, B] = \{1\}$. Then one and only one of the following holds:

(a) $G$ is abelian,

(b) $G = Q \times E$, where $Q \cong Q_{2^n}$ and $\exp(E) \le 2$.

**Solution.** Assume that $G$ is nonabelian. Then there is minimal nonabelian $A \le G$. Suppose that $A \not\cong Q_8$. If $A$ is metacyclic, then $A = B \cdot C$ (semidirect product with kernel $C$), where $B, C < G$ are cyclic and $B \cap C = \{1\}$ (by Lemma 65.1). However $[B, C] \ne \{1\}$, a contradiction. If $A$ is nonmetacyclic, then $A = \langle B, C \rangle$, where $B, C$ are cyclic and $B \cap C = \{1\}$. However $[B, C] \ne \{1\}$, a contradiction. Thus, all minimal nonabelian subgroups of $G$ are $\cong Q_8$. In that case, by Corollary A.17.3, $G = Q \times E$, where $Q \cong Q_{2^n}$ and $E$ is elementary abelian. If $B < G$ is cyclic of order $> 2$, then $\Omega_1(Q) < B$. Thus, any two cyclic subgroups of order $> 2$ have intersection $> \{1\}$. Since $\Omega_1(G) \le Z(G)$, our $G$ satisfy the hypothesis.

**Exercise 79** (Compare with Exercise 10.10). Let $H < G$ be a subgroup of order $p^p$ and exponent $p$ in a $p$-group $G$. Suppose that $|G : H| > p$. If, whenever $H < M \in \Gamma_1$, then $M$ is of maximal class. Prove that $G$ is also of maximal class.

**Solution.** Let $H \le N < M$, where $|M : N| = p$. As all subgroups of $G$ of order $p|N|$ containing $N$ are of maximal class (see Theorem 9.6), then $G$ is of maximal class, by Exercise 10.10.

**Exercise 80.** Study the nonabelian $p$-groups all of whose nonlinear irreducible characters are induced from cyclic subgroups.

**Exercise 81.** Is it true that there does not exist a nonabelian 2-group $G$ such that, whenever $A < B \le G$, where $A$ is maximal abelian in $G$ and $|B : A| = 2$, then all elements of the set $B - A$ are involutions?

**Exercise 82.** Prove that there does not exist an irregular $p$-group $G$, $p > 2$, such that, whenever $A < B \leq G$, where $A$ is maximal regular in $G$ and $|B : A| = p$, then all elements of the set $B - A$ have order $p$.

**Solution.** Assume that this is false. Let $E < G$ be of exponent $p$ of maximal possible order and let $E \leq R < G$, where $R$ is maximal regular. Then $E = \Omega_1(R)$ is characteristic in $R$. Let $R < B \leq G$, where $|B : R| = p$; then $E \triangleleft B$. Take $x \in B - R$; then $o(x) = p$, by hypothesis. Put $H = \langle x, E\rangle$; then $|H : E| = p$ and $H \leq B$. One has $H \cap R = E$ so that $H - E = H - R$. Since $H - E = H - R \subseteq B - R$, all elements of the set $H - E$ have order $p$. Since $H = E \cup (H - E)$ and $\exp(E) = p$, one obtains $\exp(H) = p$, a contradiction since $H > E$.

*17°* **Exercise 83.** A metacyclic 2-group $G$ has no two distinct subgroups $\cong E_4$, unless it is of maximal class $\not\cong Q_{2^n}$. (**Hint.** Use Lemma 1.4 and Proposition 10.19.)

**Exercise 84.** Let $H < G$ be a metacyclic subgroup of order $2^{2e}$ and exponent $2^e > 2$ of a 2-group $G$. Study the structure of $G$ provided $N_G(H)$ is metacyclic.

**Exercise 85.** If a two-generator $p$-group $G$ of class 2 has the cyclic derived subgroup $G'$, then $|G| \geq |G'|^3$, and this estimate is the best possible.

**Hint.** Clearly, $G'$ is isomorphic to a subgroup of the Schur multiplier $M(G/G')$ of the (abelian) group $G/G'$. Let $G/G'$ be abelian of type $(p^m, p^n)$, $m \geq n$; then $M(G/G') \cong C_{p^n}$ (Schur; see [BZ, Chapter 6]) so that $|G'|^2 \leq p^{2n} \leq |G/G'| \Rightarrow |G'|^3 \leq |G|$. If $G$ is nonabelian of order $p^3$, then $|G'|^3 = |G|$.

**Exercise 86.** Let a group $G$ be metabelian, $a \in G'$ and $x, y \in G$. Then $[a, x, y] = [a, y, x]$.

**Solution** (Mann). The elements of $G$ induce on $G'$, by conjugation, a commutative group of automorphisms. This group spans a commutative subring of the endomorphism ring of the abelian group $G'$. In particular, if $x, y \in G$, then the endomorphisms $-1 + x$ and $-1 + y$ commute. In that case,

$$[a, x] = a^{-1}a^x = a^{-1+x} \Rightarrow [a, x, y] = a^{(-1+x)(-1+y)} = a^{(-1+y)(-1+x)} = [a, y, x]\,.$$

**Exercise 87.** If an abelian $A < G$ is soft, then $C_G(a) = A$ for some $a \in A$.

**Solution.** Assume that the conclusion is false. As $N_G(A)$ is a unique subgroup of $G$ containing $A$ as a subgroup of index $p$, it follows that $N_G(A) \leq C_G(a)$ for all $a \in A$. Therefore, $A < Z(N_G(A)) = C_G(A)$, a contradiction.

**Exercise 88.** Study the nonmetacyclic $p$-groups $G$ such that the normalizer of any their nonnormal noncyclic metacyclic subgroup is metacyclic.

**Exercise 89.** Present a $p$-group of minimal possible order all of whose minimal nonabelian subgroups are nonnormal.

**Hint.** In any $p$-group of maximal class and order $\geq p^5$ with abelian subgroup of index $p$ all minimal nonabelian subgroup are nonnormal.

**Exercise 90.** Let $R \cong \mathrm{E}_{p^2}$ be a subgroup of a $p$-group $G$. If $\mathrm{N}_G(R)$ is metacyclic so is $G$.

**Solution.** If $R$ is a characteristic subgroup in $\mathrm{N}_G(R)$, then $G = \mathrm{N}_G(R)$ is metacyclic. If $R$ is not characteristic in $\mathrm{N}_G(R)$, then $p = 2$ and $\mathrm{N}_G(R) \cong \mathrm{D}_8$ (Proposition 10.19). Then $G$ is a 2-group of maximal class (Remark 10.5), so metacyclic (Proposition 1.6 and Theorem 1.2).

**Exercise 91.** If $G$ is a metacyclic $p$-group, $p > 2$, such that $|G/\mathrm{Z}(G)| = p^4$, then $G$ is an $\mathcal{A}_2$-group.

**Solution.** By Lemma 65.1, $G$ is not minimal nonabelian so that $|G'| > p$ (Lemma 65.2 (a)). Therefore, by Corollary 148.2, the set $\Gamma_1$ has no abelian member so that $G/\mathrm{Z}(G)$ has no cyclic subgroup of index $p$. Thus, $\exp(G/\mathrm{Z}(G)) = p^2$. Let $T/\mathrm{Z}(G) < G/\mathrm{Z}(G)$ be of index $p$; then $T/\mathrm{Z}(G)$ is abelian of type $(p^2, p)$. In that case, $T$ is either abelian or minimal nonabelian (check!). As $\mathrm{Z}(G) < \Phi(G)$, it follows that all members of the set $\Gamma_1$ are either abelian or minimal nonabelian hence $G$ is an $\mathcal{A}_2$-group.

**Exercise 92.** If a $p$-group $G, p > 2$, has no subgroup $\cong S(p^3)$, then $\Omega_1(G)$ is elementary abelian. (See Lemma 30.4 (a).) In particular, the same conclusion holds if all minimal nonabelian subgroups of $G$, $p > 2$, are metacyclic.

**Solution.** Let $E \triangleleft G$ be elementary abelian of maximal order. Assume that there is an element $x \in G - E$ of order $p$ and set $H = \langle x, E\rangle$. By Theorem 10.1, $H$ is nonabelian. By Lemma 57.1, there is $a \in E$ such that $A = \langle a, x\rangle$ is minimal nonabelian. As $\Omega_1(A) = A$, it follows that $A \cong S(p^3)$ (Lemma 65.1), a contradiction.

**Exercise 93.** Let $\mathrm{e}_n(G)$ ($\mathrm{e}'_n(G)$) be the number of subgroups (normal subgroups) of order $p^n$ and exponent $p$ in a $p$-group $G$. Prove that if $G$ is an irregular $p$-group, then $\mathrm{e}_n(G) \equiv 1 \pmod p$ and $\mathrm{e}'_n(G) \equiv 1 \pmod p$ for all $n < p$. (This is not a new result; see §§ 12, 13, 18.)

**Solution** (Mann). As $\mathrm{e}_n(G) \equiv \mathrm{e}'_n(G) \pmod p$, it suffices to show that $\mathrm{e}'_n(G) \equiv 1 \pmod p$. Write $K = \mathrm{Z}_n(G)$; then $K$ is regular (Theorem 7.1 (b)). In that case, $\mathrm{e}_n(K) = \mathrm{e}_n(\Omega_1(K)) \equiv 1 \pmod p$ (Kulakoff). All $G$-invariant subgroups of order $p^n$ and exponent $p$ are contained in $\Omega_1(K)$. However $\mathrm{e}'_n(G) \equiv \mathrm{e}_n(\Omega_1(K)) \equiv 1 \pmod p$.

**Exercise 94.** Suppose that a two-generator 2-group $G$ contains an abelian subgroup $A$ of index 2 and $G$ is neither abelian nor minimal nonabelian. If $B$ is another maximal abelian subgroup of $G$, then $G/\mathrm{Z}(G)$ is of maximal class and $|B| = 2|\mathrm{Z}(G)|$, unless $|G : B| = 2$ (then $G/\mathrm{Z}(G) \cong \mathrm{E}_4$).

**Exercise 95.** Study the nonabelian $p$-groups $G$ such that all subgroups of $G$ nonincident with $\Phi(G)$ are abelian.

**Hint.** Let $\Phi(G) < H < G$, where $|H : \Phi(G)| = p$. If $H$ is nonabelian, there is a minimal nonabelian $A < H$ such that $A \nleq \Phi(G)$.

**Exercise 96.** Let $G$ be a metacyclic $p$-group of order $p^5$ with $|G'| = p^2$. Prove that $G$ has $p$ isomorphic $\mathcal{A}_1$-subgroups.

**Solution.** By Corollary 65.3, $G$ is an $\mathcal{A}_2$-group; then $\alpha_1(G) \geq p$ and $G$ is not a 2-group of maximal class. If $X$ is a metacyclic $\mathcal{A}_1$-group of order $p^4$, then either $X \cong \mathrm{M}_{p^4}$ or $\exp(X) = p^2$. Write $R = \Omega_1(G)$; then $R \cong \mathrm{E}_{p^2}$ (Lemma 1.4) and $G/R$ is nonabelian so that $R < \Phi(G)$. Set $A = \mathrm{C}_G(R)$. If $A = G$, then all $\alpha_1(G) \geq p$ minimal nonabelian subgroups have exponent $p^2$, so are isomorphic. Now let $A \in \Gamma_1$. Assume that $A$ is abelian; then $A$ is a unique abelian subgroup of index $p$ in $G$ since $|G'| > p$. In that case, all $\alpha_1(G) = p$ minimal nonabelian, say $H$, subgroups are $\cong \mathrm{M}_{p^4}$ (consider $H \cap A$). Now let $A$ be nonabelian; then $\exp(A) = p^2$ and $\alpha_1(G) = p + 1$. All $\alpha_1(G) - 1 = p$ minimal nonabelian subgroups $\neq A$ are $\cong \mathrm{M}_{p^4}$.

**Exercise 97.** Suppose that $G$ is a $p$-group of exponent $p^e$, $p > 2$, such that $\Omega_k^*(G)$ is abelian for all $k \leq e$. Prove that $G$ is abelian. (**Hint.** Prove that $G$ has no minimal nonabelian subgroup.)

**Exercise 98.** If $G$ is a metacyclic $p$-group such that $\mathrm{cd}(G) = \{1, p\}$, then $G$ has an abelian subgroup of index $p$. If $p > 2$, then $G$ is minimal nonabelian; see § 148. (**Hint.** This follows from Theorem 22.5.)

**Exercise 99.** If for every element $x$ of a $p$-group $G$, $p > 2$, one has $\mathrm{cl}(\langle x\rangle^G) \leq \frac{1}{2}(p-1)$, then $G$ is regular.

**Solution.** Take $x, y \in G$; then, by hypothesis, $\mathrm{cl}(\langle x, y\rangle) \leq \mathrm{cl}(\langle x\rangle^G) + \mathrm{cl}(\langle y\rangle^G) \leq p - 1$ so that $\langle x, y\rangle$ is regular (Theorem 7.2 (b)); then $G$ is regular.

**Exercise 100.** Classify the nonabelian 2-groups $G$ in which any two noncommuting elements of the same order generate a $p$-group of maximal class. (**Hint.** For $p = 2$, see § 90.)

**Exercise 101.** Suppose that $G$ is a modular $p$-group. Then all minimal nonabelian subgroups of $G$ are metacyclic. (**Hint.** Subgroups of modular $p$-groups are modular. Use Lemma 65.1.)

**Exercise 102.** Let a $p$-group $G$ of exponent $p^e > p > 2$ be such that $\Omega_e^*(G)$ is abelian. Prove that $\Omega_e^*(G)$ is the unique maximal normal abelian subgroup of $G$.

**Solution.** Clearly, $\Omega_e^*(G)$ is a maximal abelian subgroup of $G$. Assume that $A \neq \Omega_e^*(G)$ is a maximal abelian normal subgroup of $G$. Then $\mathrm{cl}(A\Omega_e^*(G)) = 2$ (Fitting's lemma) so $A\Omega_e^*(G)$ is regular since $p > 2$. In that case, $\Omega_e^*(A\Omega_e^*(G)) = A\Omega_e^*(G) > \Omega_e^*(G)$, a contradiction.

**Exercise 103.** Find all possible values of the minimal degree $\delta(G)$ of faithful representations by permutations of a $p$-group $G$ of maximal class and order $p^n$ (for example, $\delta(Q_{2^n}) = 2^n$, $\delta(D_{2^n}) = \delta(SD_{2^n}) = 2^{n-1}$).

**Exercise 104.** Classify the nonmetacyclic 2-groups $G$ all of whose maximal metacyclic subgroups are of maximal class.

**Hint.** Assume that $G$ has a normal subgroup $R \cong E_4$. Let $R \le M < G$, where $M$ is maximal metacyclic in $G$. As $M$ is of maximal class, it follows that $|M| \le 2^3$.

A $p$-group $X$ is said to be a $(**)$-group if it contains a normal absolutely regular subgroup $A$ such that $X/A$ is absolutely regular. Any $p$-group of maximal class is a $(**)$-group.

**Exercise 105.** If all maximal $(**)$-subgroups of an irregular $p$-group $G$ are of maximal class, $p > 2$, then $G$ is also of maximal class.

**Solution.** Let $T \le G$, where $T$ is a $(**)$-subgroup of maximal possible order; then $T$ is of maximal class, by hypothesis. Let $H < T$ be absolutely regular of index $p$ (Theorems 9.5 and 9.6) and let $H < T_1 \le G$, where $|T_1 : H| = p$. Since $|T_1| = |T|$ and $T_1$ is a $(**)$-group, it follows that $T_1$ is of maximal class. Then $G$ is of maximal class, by Exercise 10.10.

**Exercise 106.** All noncyclic subgroups of a non-Dedekindian metacyclic $p$-group $G$ are normal $\iff$ one of the following holds:
(a) $G$ is minimal nonabelian,
(b) $\Omega_1(G) \cong E_{p^2}$ and $G/\Omega_1(G) \cong Q_8$ (in that case, $G$ is a group from Lemma 42.1 (c)),
(c) $G \cong Q_{16}$.

**Solution.** If $G$ has no normal abelian subgroup of type $(p, p)$, then $G$ is a 2-group of maximal class (Lemma 1.4 and Theorem 1.2); moreover, $G \in \{Q_8, Q_{16}\}$. Now let $R \triangleleft G$ be abelian of type $(p, p)$. Then $G/R$ is either abelian or $\cong Q_8$. In the first case, $|G'| = p$ so that $G$ is minimal nonabelian (Lemma 65.2 (a)); such $G$ satisfies the hypothesis. Now let $G/R \cong Q_8$. In that case $R = \Omega_1(G)$. Such $G$ satisfies the hypothesis as well (see Lemma 42.1).

**Exercise 107.** Show that a metabelian group $G$ of exponent $p > 2$ satisfies $\mathrm{cl}(G) \le (p-1) \cdot \mathrm{d}(G)$.

**Solution.** The subgroup $G' = \Phi(G)$ is abelian. If $\Phi(G) < H < G$ with $|H : \Phi(G)| = p$, then $\mathrm{cl}(H) \le p-1$ (Exercise 26). Let $G/\Phi(G) = (H_1/\Phi(G)) \times \cdots \times (H_d/\Phi(G))$, where $|H_i : \Phi(G)| = p$ for all $i \le \mathrm{d}(G)$. Then result follows from Fitting's lemma since $\mathrm{cl}(H_i) \le p-1$ for $i = 1, \dots, \mathrm{d}(G)$.

**Exercise 108.** Let $G$ be a $p$-group of maximal class and order $p^n$ with abelian subgroup $A$ of index $p$. The number of distinct centralizers of noncentral elements of $G$ is equal to $p^{n-2} + 1$.

**Solution.** If $x \in A - Z(G)$, then $C_G(x) = A$ is abelian. If $x \in G - A$, then $|C_G(x)| = p^2$. In that case $G - Z(G) = \bigcup_{x \in G - Z(G)}(C_G(x) - Z(G))$. The number of centralizers $\neq A$, is equal to $\frac{|G|-|A|}{p^2-p} = \frac{p^n - p^{n-1}}{p(p-1)} = p^{n-2}$.

It is easy to prove, arguing in a similar manner, that if a 3-group $G$ of maximal class has a nonabelian fundamental subgroup and $|G| = 3^n$ (in that case $n > 4$), then the number of distinct centralizers of noncentral elements is equal to the cardinality of any maximal set of pairwise noncommuting elements, and that number is $3^{n-2} + 4$.

**Exercise 109.** Study the $\mathcal{A}_n$-groups of prime power order, $n > 1$, that are not covered by the nonabelian members of the set $\Gamma_1$.

**Hint.** Let $d(G) > 2$. If $D \in \Gamma_2$ is nonabelian, then $G$ is covered by all members of the set $\Gamma_1$ containing $D$, contrary to the hypothesis. Thus, all members of the set $\Gamma_2$ are abelian so that $\Phi(G) \leq Z(G)$. It follows easily that then $d(G) = 3$. Any two generator $p$-group $G$ satisfies the hypothesis $\iff$ there is an abelian $A \in \Gamma_1$.

**Exercise 110.** If a modular 2-group $G$ (see § 73) has a section $\cong Q_8$, then it has a subgroup $\cong Q_8$.

**Solution.** We proceed by induction on $|G|$. Let $L \triangleleft K \leq G$, where $K/L \cong Q_8$. Bearing in mind our aim, one may assume, without loss of generality, that $K = G$; then $L \triangleleft G$. We also may assume that $L > \{1\}$. Let $|L| = 2$. By Theorem 1.2, $G$ has no cyclic subgroup of index 2 hence it is not of maximal class so $|G : G'| = 8$ (Proposition 1.6). Next, $G$ has exactly 3 involutions. If $G/G' \cong E_8$, then $G$ contains a proper minimal nonabelian subgroup, say $Q$, and, by the above, $Q \cong Q_8$ (indeed, $D_8$ has $5 > 3$ involutions). Now let $d(G) = 2$. Then $G$ is minimal nonabelian (Lemma 65.2 (a)). If $G$ is metacyclic and its subgroup $R \neq G'$, $R \neq L$ is of order 2, then $G/R \cong D_8$ is nonmodular, a contradiction. If $G$ is nonmetacyclic and $R < Z(G)$ is of order 2 with $R \neq G', L$, then $G/R \cong D_8$, a contradiction again. Now let $|L| > 2$. Let $S < L$ be $G$-invariant of index 2. Then, by the above, $G/S$ contains a subgroup $T/S \cong Q_8$. By induction, $T$ contains a subgroup $Q_0 \cong Q_8$.

**Exercise 111** ([Schm5], Theorem 2.3.8). If a modular 2-group $G$ has a section $\cong Q_8$, then $G$ is Dedekindian.

**Solution.** We use induction on $|G|$. By Exercise 110, $G$ contains a subgroup $Q \cong Q_8$. Assume that $G$ is non-Dedekindian. Then, by Iwasawa's Theorem 73.15, there is a $G$-invariant abelian subgroup $A < G$ such that $G = \langle A, t\rangle$ and $a^t = a^{1+2^s}$ for all $a \in A$ and some integer $s \geq 2$. Since $G$ is nonabelian, $\exp(A) > 2^s \geq 4$ (moreover, $\Omega_s(A) \leq Z(G)$). Assume that $Q < H < G$. Then, by induction, $H$ is Dedekindian so $\exp(H) = 4 < 8 \leq \exp(A)$, which is a contradiction. Thus, $G = H = QA$. Since $\Omega_2(A) \leq Z(G)$, we get $|Q \cap A| = 2$. Then $G/A \cong Q/(Q \cap A) \cong E_4$ is noncyclic, a final contradiction.

**Exercise 112** ([Schm5], Lemma 2.3.9). Let $L$ be a cyclic quasinormal subgroup of a $p$-group $G$ and $|L| = \exp(G) = p^e$. Then $L_G > \{1\}$.

**Solution.** Assume that $L_G = \{1\}$. Then there is a cyclic $U < G$ that does not normalize $\Omega_1(L)$. In particular, $L \cap U = \{1\}$. Write $H = LU$; then $|H| \le p^{2e}$. In that case, $\Omega_1(L)$ is not normal in $H$ so that $L_H = \{1\}$. It follows that there is $x \in H$ such that $L \cap L^x = \{1\}$. We have $|LL^x| = p^{2e} \ge |H|$ so that $H = LL^x$. Thus, $H$ is a product of two permutable conjugate (in $H$) subgroups $L$ and $L^x$, which is impossible in view of Lemma A.28.8. Thus, $L_G > \{1\}$.

**Exercise 113.** Suppose that a generalized quaternion group $Q$ is quasinormal in a 2-group $G$ and $\exp(Q) = \exp(G)$. Is is true that $Q_G > \{1\}$?

**Exercise 114.** Is the following assertion true?: If a 2-group $Q$ of maximal class is quasinormal in a 2-group $G$ such that $\exp(Q) = \exp(G)$, then $Q_G > \{1\}$.

**Exercise 115.** Let $G = Q \times E$, where $Q \cong \mathrm{Q}_8$ and $\exp(E) \mid 2$. Prove that all subgroups of $G$ are normal.

**Solution.** One may assume that $E > \{1\}$. Let $\{1\} < H < G$. If $\exp(H) = 2$, then $H \le \Omega_1(G) = \mathrm{Z}(G)$, and so $H \lhd G$. If $\exp(H) = 4$, then $G' = \mho_1(G) < H$, and $H \lhd G$.

**Exercise 116.** Let $G = Q_1 \times Q_2$, where $Q_1 \cong Q_2 \cong \mathrm{Q}_8$. Then $G$ has a nonnormal subgroup of order 4.

**Solution.** Let $D$ be the diagonal subgroup of $G$; then $D$ is not normal in $G$. The subgroup $D \cong \mathrm{Q}_8$ is generated by cyclic subgroups of order 4. Therefore, one of these subgroups is not normal in $G$.

**Exercise 117.** Classify the nonabelian $p$-groups without special sections.

**Solution** (Janko). By hypothesis, $G$ has no nonabelian sections of order $p^3$ so it is modular. Every special $p$-group has a nonabelian section of order $p^3$. Hence, as $G$ has no nonabelian section of order $p^3$, it has no special section. If a modular 2-group has a section $\cong \mathrm{Q}_8$, it is Dedekindian (see Exercise 111). Thus, $G$ satisfies the hypothesis $\iff$ is is non-Dedekindian modular $p$-group.

**Exercise 118.** Characterize the $p$-groups of exponent $> p$ in which any two abelian subgroups of order $p^2$ are permutable.

**Exercise 119.** Present two modular 2-groups $A, B$ such that $A \times B$ is nonmodular.

**Solution.** Let $A \cong \mathrm{Q}_8 \cong B$, $G = A \times B$ and $D$ be the diagonal subgroup of $G$; then $D \cong \mathrm{Q}_8$. Assume that $G$ is modular. Let $X < D$ and $Y < A$ be of order 4. By assumption, $XY = YX$ is modular. By remark, following Lemma 73.2, $XY$ is metacyclic. By Proposition 73.3, $XY$ is abelian. Thus, any subgroup of order 4 in $A$ centralizes any subgroup of order 4 in $D$ so that $A$ centralizes $D$, a contradiction.

Let $G$ be as in Exercise 119 and let $L < \mathrm{Z}(G)$ be of order 2, $A' \neq L \neq B'$. Set $\bar{G} = G/L$; then $\bar{G} = \bar{A} * \bar{B}$ is extraspecial of order $2^5$. Let $\bar{C} < \bar{B}$ be cyclic of order 4. Then $\bar{H} = \bar{A} * \bar{C}$ is nonmodular since it contains two noncommuting involutions.

**Exercise 120.** Do there exist two modular $p$-groups $A$ and $B$, $p > 2$, such that $G = A \times B$ is nonmodular?

**Solution.** Let $G = A \times B$, where $A \cong \mathrm{M}_{p^3} \cong B$. Assume that $G$ is modular. Let $L < Z(G)$ be such that $A' \neq L \neq B'$ and let $\bar{G} = G/L$; then $\bar{G} = \bar{A} * \bar{B}$ is modular and extraspecial of order $p^5$. We get $|\Omega_1(\bar{G})| = |\bar{G}/\mho_1(\bar{G})| = p^4$. Since $\bar{G}$ has no abelian subgroups of order $p^4$ (use Lemma 1.4 to justify this assertion), the subgroup $\Omega_1(\bar{G})$ of exponent $p$ is nonabelian so it is nonmodular, a contradiction.

**Exercise 121.** Let $\phi : G \to G_0$ be a lattice isomorphism of a $p$-group $G$ onto a $p$-group $G_0$. If $Q < G$ is quasinormal, then $Q^\phi$ is also quasinormal in $G_0$.

**Exercise 122.** Suppose that $G$ is a nonabelian $p$-group such that $D = \langle H' \mid H \in \Gamma_1 \rangle < G'$. Then $G/D$ is minimal nonabelian; in particular, $\mathrm{d}(G) = 2$.

**Solution.** One has $D < G' \leq \Phi(G)$. If $H \in \Gamma_1$, then $H/D$ is abelian. Since $G/D$ is nonabelian, by hypothesis, it is minimal nonabelian so that $\mathrm{d}(G) = 2$. In particular, if $\mathrm{d}(G) > 2$, then $D = G'$.

**Exercise 123.** Let $G$ be a 3-group of maximal class and order $\geq 3^5$ and $\phi \in \mathrm{Aut}(G)$ be an automorphism of order 3. Study the structure of the holomorph $\mathcal{H} = \langle \phi, G \rangle$.

**Exercise 124.** Let $G = \Omega_1(G)$ be an irregular $p$-group. Then there is $H \in \Gamma_1$ such that $\Omega_1(H) = H$.

**Solution.** Let $\Omega_1(M) = M < G$ be such that $M$ is as large as possible. If $M \in \Gamma_1$, we are done. Otherwise, let $M < H \in \Gamma_1$. Let $K = \Omega_1(H)$; then $K \triangleleft G$, $M \leq K$ and $|G : K| > p$. Take $x \in G - K$ of order $p$. Write $S = \langle x, K \rangle$; then $\Omega_1(S) = S$ and $|G| > |S| = p|K| \geq p|M| > |M|$, contrary to the choice of $M$. Thus, $M \in \Gamma_1$.

**Exercise 125.** Let $G$ be a $p$-group, $p > 2$. Suppose that $A \in \Gamma_1$ is absolutely regular and $R \triangleleft G$ is of exponent $p$ and order $p^p$. Then one of the following holds:
(a) $\Omega_1(G) = R$,
(b) $G$ is of maximal class and order $p^{p+1}$.

**Solution.** (Compare with the proof of Theorem 12.1 (b).) Assume that $\Omega_1(G) > R$. Let $x \in G - R$ be of order $p$ and write $H = \langle x, R \rangle$; then $|H| = p^{p+1}$. Assume that $H$ is not of maximal class. Then $H$ is regular (Theorem 7.1 (b)) so that $\exp(H) = p$ (Theorem 7.2 (b)). In that case $A \cap H$ is not absolutely regular, a contradiction. Thus, $H$ is of maximal class and so irregular (Theorem 9.5). Assume that $H < G$. By Theorem 9.6 (c), any subgroup of $G$ containing $H$ as a subgroup of index $p$ is not of maximal class. Let $H < T \leq G$, where $|T : H| = p$. Then, by Theorem 12.12 (b), $T/\mho_1(T)$ is of order $p^{p+1}$ and exponent $p$ (Theorem 12.12). In that case, $A \cap T$, the maximal subgroup of $T$, is not absolutely regular, a contradiction. Thus, $G = H$ is of maximal class and order $p^{p+1}$.

**Exercise 126.** Classify the nonabelian $p$-groups, $p > 2$, all of whose nonnormal subgroups are 2-uniserial. (See § 131.)

**Exercise 127.** (a) Let $N$ be a proper nonabelian normal subgroup of a $p$-group $G$, $C_G(N) < N$. If $a \in G - N$, then $x \to x^a$ is an outer automorphism of $N$.
(b) Show that any nonabelian group $L$ of order $p^3$ admits an outer $p$-automorphism $\phi$ such that $|C_L(\phi)| = p^2$.

**Solution.** (a) Assume the this is false. Then there is $b \in N$ such that $x^a = x^b$ for all $x \in N$. It follows that $x^{ab^{-1}} = x$ for all $x \in N$, i.e., $ab^{-1} \in C_G(N)$. By hypothesis, $ab^{-1} \in N \Rightarrow a \in Nb = N$, contrary to the choice of $a$.
(b) We consider only case $p < 5$. (i) Let $p = 2$ and $G = SD_{2^4}$. Let $a \in G$ be of order 8 and let $L \in \Gamma_1$ be nonabelian. Then, by (a), $a$ induces on $L$ an outer automorphism of order 4 with $|L \cap \langle a \rangle| = 4$ fixed points. (ii) Let $p = 3$, $G = \Sigma_9 \in \mathrm{Syl}_3(S_9)$, let $L \in \Gamma_1$ be nonabelian, $A \in \Gamma_1$ elementary abelian and $a \in A - \Phi(G)$. Then $a$ induces on $L$ an outer automorphism of order 3 with $|L \cap A| = 3^2$ fixed points. Since $L$ can be chosen to be isomorphic with $S(3^3)$ and $M_{3^3}$, we are done.

**Exercise 128** (Zassenhaus [Zas2]). A finite group $G$ is abelian $\iff$ $N_G(A) = C_G(A)$ for all primary abelian $A \leq G$.

**Solution.** Assume that $G$ is nonabelian. Let $M \leq G$ be minimal nonabelian. If $M$ is nonnilpotent and $A = M'$, then $M \nleq C_G(A)$ but $M \leq N_G(A)$ so that $N_G(A) \neq C_G(A)$, a contradiction. If $M$ is primary and $A$ is maximal in $M$, then we obtain a contradiction as above. Thus, $M$ does not exist so $G$ is abelian.

**Exercise 129.** Suppose that $G$ is a nonabelian $p$-group with $G' \leq \Omega_1(Z(G))$. Then any two noncommuting elements of $G$ generate a minimal nonabelian subgroup.

**Solution.** Let $x, y \in G$ be not commute. Set $H = \langle x, y \rangle$. By hypothesis, $H' \leq G' \leq \Omega_1(Z(G))$ so that $H' \leq H \cap Z(G) \leq Z(H)$ is of exponent $p$. Since $d(H) = 2$, it follows from Lemma 65.2 (a) that $H$ is minimal nonabelian.

**Exercise 130.** Classify the nonabelian $p$-groups $G$ of exponent $p^e > p$ in which any two distinct cyclic subgroups of order $p^e$ generate $G$.

**Solution.** If $G$ has a cyclic subgroup of index $p$, then $G \in \{Q_8, M_{p^{e+1}}\}$ (Theorem 1.2). Next assume that $G$ has no cyclic subgroup of index $p$. Let $C < M < G$, where $C$ is cyclic of order $p^e$ and $|M : C| = p$. By hypothesis, $M$ has exactly one cyclic subgroup of index $p$ so by Theorem 1.2, $p = 2$ and $M$ is of maximal class. By Exercise 10.10, $G$ is of maximal class so it has a cyclic subgroup of index 2, contrary to the assumption.

**Exercise 131.** Suppose that a $p$-group $G$ of class $> 2$ has no normal subgroup of class 2. Prove that $G$ has only one maximal normal abelian subgroup. (**Hint.** Use Fitting's lemma.)

**Exercise 132.** Suppose that an irregular $p$-group $G$, $p > 2$, possesses a subgroup of order $p^p$ and exponent $p$. If all such subgroups are of maximal class, then one of the following holds:

(a) $G = H\Omega_1(G)$, where $H \in \Gamma_1$ is absolutely regular and $|\Omega_1(G)| = p^p$,
(b) $G$ is of maximal class.

**Solution.** Assume that $G$ is not of maximal class. Then it contains a normal subgroup $R$ of order $p^p$ and exponent $p$ (Theorem 12.1 (a)). By hypothesis, $R$ is of maximal class. Let $L < R$ be a $G$-invariant abelian subgroup of type $(p, p)$ and $C_G(L) = H$; then $R \nleq H$ and $H \in \Gamma_1$. Assume that $H$ has a $G$-invariant subgroup $R_1$ of order $p^p$ and exponent $p$; then $L \nleq R_1$ since $R_1$ is of maximal class. In that case, $LR_1$ is of exponent $p$ and order $> p^p$ (Theorem 7.1 (b)). If $L < R_2 < LR_1$, where $R_2$ is of order $p^p$, then $R_2$ is not of maximal class, contrary to the hypothesis. Thus, $H$ has no $G$-invariant subgroup of order $p^p$ and exponent $p$ so it is absolutely regular, by Theorems 13.5 and 12.1 (b). By Theorem 12.1 (b), $R = \Omega_1(G)$. Since $R \nleq H$, we get $G = HR$.

**Exercise 133.** Suppose that a $p$-group $G$ of order $> p^p$ contains a subgroup of order $p^p$ and exponent $p$, and all such subgroups have centers of order $p$. Then one of the following holds:
(a) $G = \Omega_1(G)H$, where $|\Omega_1(G)| = p^p$ and $H$ is absolutely regular maximal subgroup of $G$,
(b) $G$ is of maximal class.

**Solution.** By hypothesis, $p > 2$. Let $e_p(G)$ be the number of subgroups of order $p^p$ and exponent $p$ in $G$. Let $R \cong E_{p^2}$ be $G$-invariant (Lemma 1.4). Set $H = C_G(R)$. Then $H$ has no subgroup of order $p^p$ and exponent $p$ so that $H$ is absolutely regular (Theorem 12.1 (a)). It follows from $H < G$ that $H \in \Gamma_1$. By Theorems 7.2 (b) and 12.1 (b), $\Omega_1(G)$ is of order $p^p$ and exponent $p$ and so $G = H\Omega_1(G)$.

**Exercise 134** ([Pas], Lemma 2.1). Suppose that any nonnormal subgroup of a non-Dedekindian $p$-group $G$ has order $\leq p^k$. Then $|G'| \leq p^k$.

**Solution.** Assume that $|G'| > p^k$. By Proposition 1.23, there is in $G'$ a $G$-invariant subgroup $L$ of index $p$ such that $G/L$ is non-Dedekindian. Then there is in $G/L$ a nonnormal subgroup $H/L$. In that case, $|H| \geq p|L| = |G'| > p^k$, contrary to the hypothesis.

**Exercise 135.** Given $k$, the intersection of all subgroups of a $p$-group $G$ that have order $p^k$ is contained in $\Phi(G)$.

**Solution.** Let $A < H \leq G$, where $|A| = p^k$ and $|H| = p^{k+1}$. Then the intersection of all subgroups of $H$ of order $p^k$ coincides with $\Phi(H) \leq \Phi(G)$.

**Exercise 136.** Suppose that a $p$-group $G$ contains a normal subgroup $R$ of order $p$ such that $G/R$ is a metacyclic minimal nonabelian group. Prove that $G$ is either minimal nonabelian or an $\mathcal{A}_2$-group.

**Solution.** If $R \nleq \Phi(G)$, then $G = R \times A$, where $A \in \Gamma_1$, so $G$ is an $\mathcal{A}_2$-group. Now suppose that $R \leq \Phi(G)$; then $d(G) = d(G/R) = 2$. Let $H \in \Gamma_1$. One has $R < H$ so $H/R$ is abelian. Since $G/R$ is metacyclic, we get $d(H/R) = 2$. If $d(H) = 2$, then $H$ is either abelian or minimal nonabelian (Lemma 65.2 (a)). If $d(H) = 3$, then $H = R \times F$, where $F$ is maximal in $H$. Since $F \cong H/R$ is abelian, then $H$ is abelian. Thus, all members of the set $\Gamma_1$ are either abelian or minimal nonabelian so that $G$ is either an $\mathcal{A}_1$- or an $\mathcal{A}_2$-group.

**Exercise 137.** Suppose that a $p$-group $G$ contains a normal subgroup $R$ of order $p$ such that $G/R$ is a metacyclic $\mathcal{A}_n$-group. Is it true that $G$ is either $\mathcal{A}_n$- or an $\mathcal{A}_{n+1}$-group?

**Exercise 138.** Let a $p$-group $G$ be neither abelian nor minimal nonabelian. Write $p^k = \min\{|A| \mid A < G \text{ is minimal nonabelian}\}$. Prove that the intersection of all nonabelian subgroups of $G$ of order $p^k$ (all these subgroups are minimal nonabelian) is contained in $\Phi(G)$. (Compare with Exercise 135.)

**Solution.** Let $A < H \le G$, where $A$ is minimal nonabelian of order $p^k$ and $|H| = p^{k+1}$. It is easily checked that $H$ is an $\mathcal{A}_2$-subgroup. Clearly, $d(H) \le 3$. The number of abelian maximal subgroups in $H$ is either 0, 1, or $p+1$. If $d(H) = 2$, then $\alpha_1(H) \in \{p, p+1\}$, so the intersection of minimal nonabelian subgroups of $H$ coincides with $\Phi(H) \le \Phi(G)$. If $d(H) = 3$, then $\alpha_1(H) \in \{p^2+p+1, p^2+p, p^2\}$. Since any subgroup of $H$ of index $p^2$ is contained in at most $p+1 (< p^2)$ maximal subgroups of $H$, it follows that the intersection of all minimal nonabelian subgroups of $H$ has index $p^3$ in $H$ so coincides with $\Phi(H) \le \Phi(G)$.

**Exercise 139.** Suppose that a $p$-group $G$ contains a nonabelian subgroup of order $p^k$. Is it true that the intersection of all nonabelian subgroups of $G$ of order $p^k$ is contained in $\Phi(G)$?

**Exercise 140.** Suppose that a nonabelian $p$-group contains exactly one maximal subgroup, say $H$, of exponent $> p$. Prove that
(a) $G$ is irregular of exponent $p^2$,
(b) $d(G) = 2$, $G' = \Phi(G)$,
(c) if $p = 2$, then $G = D \times E$, there $D \cong D_8$ and $\exp(E) \mid 2$.

**Solution.** Since $|\Gamma_1| \ge p+1 \ge 3$, the set $\Gamma_1$ contains two distinct members, say $A$ and $B$, of exponent $p$. We have $G = AB$ so that $\Omega_1(G) = G$ is of exponent $> p$; then $G$ is irregular (Theorem 7.2 (b)) of exponent $p^2$ and $G' = \Phi(G)$. Let $y \in H$ be of order $p^2$ and $x \in A - H$. Since $U = \langle x, y\rangle$ is of exponent $p^2$ and $U \nleq H$, we get $U = G$ so that $d(G) = 2$. Now let $p = 2$. Then $Z(G) = A \cap B$ is elementary abelian of index 4 and so $G = D \times E$, where $D \cong D_8$ and $\exp(E) \le 2$ (check!).

$18^o$ **Exercise 141.** Let $G$ be a $p$-group and let $H < G$ be metacyclic of order $p^{2n}$ and exponent $p^n$, $p > 2$. If $N_G(H)$ is metacyclic, then $H \trianglelefteq G$. (**Hint.** $H = \Omega_n(N_G(H))$ is characteristic in regular $N_G(H)$ (Theorem 7.2 (c).)

**Exercise 142.** Suppose that a nonabelian q-self dual $p$-group has the cyclic derived subgroup. Prove that $Z(G)$ is cyclic.

**Solution.** Assume that $Z(G)$ is noncyclic. Then there is in $Z(G)$ a subgroup $R$ of order $p$ such that $R \nleq G'$. In that case, $|(G/R)'| = |G'|$. Since $G$ is self dual, there is $H \in \Gamma_1$ with $|H'| = |G'|$ ($H \cong G/R$), contrary to Proposition 72.1 (b). Indeed, assume that $|G'| = p^k$; then $G$ and $H$ are $\mathcal{A}_k$-group which is impossible.

**Exercise 143.** If a $p$-group $G$ has two distinct maximal subgroups, say $A$ and $B$, with cyclic $A'$ and $B'$, then it is metabelian.

**Solution.** Since $A'$ and $B'$ are $G$-invariant cyclic, they lie in $Z(G')$, and so $A'B' \leq Z(G')$. By Exercise 1.69, $|G' : A'B'| \leq p$, so $G'$ is abelian.

**Exercise 144.** Find the norm of a minimal nonabelian $p$-group. (For example, the norm of $G = M_{p^n}$ coincides with its center (indeed, if $A < G$ is nonnormal and $C \in \Gamma_1$ is cyclic not containing $A$, then $C$ does not normalize $A$).

**Exercise 145.** Let $G$ be a group with $|G'|$ a prime and $H < G$ with $H \nleq Z(G)$. Then $H \lhd G \iff G' < H$. (**Hint.** If $[G, H] > \{1\}$, then $G' = [G, H] \leq H$.)

**Exercise 146.** Classify the 2-groups with three distinct nonabelian Dedekindian maximal subgroups (see Appendix 245).

**Exercise 147.** Classify the modular $p$-groups that are special.

**Exercise 148.** Classify the modular 2-groups that are metacyclic.

**Exercise 149.** Let $G$ be a nonabelian $p$-group. Suppose that $N_G(A)$ is abelian for all nonnormal abelian $A < G$. Describe the structure of $G$.

**Exercise 150** (Q. H. Zhang). Given a nonabelian $p$-group $G$, the intersection of all nonabelian subgroups of $G$ equals the intersection of all its $\mathcal{A}_1$-subgroups.

**Solution.** Let $M$ be the intersection of all nonabelian subgroups of $G$ and $N$ the intersection of all its $\mathcal{A}_1$-subgroups. Clearly, $M \leq N$. Conversely, let $H \leq G$ be nonabelian and let $K \leq H$ be minimal nonabelian. Then $N \leq K \leq H$, i.e., $N$ is contained in all nonabelian subgroups of $G$ so that $N \leq M$. Thus, $M = N$.

**Exercise 151.** Let $\Theta$ be a group theoretic property inherited by subgroups. Check the following assertion: Given a $p$-group $G$ that in not a $\Theta$-group, the intersection of minimal non-$\Theta$-subgroups of $G$ equals the intersection of all its subgroups containing minimal non-$\Theta$-subgroups.

**Exercise 152.** Find the number of irreducible characters of degree $p$ of a $p$-group of maximal class.

**Solution.** Let $G$ be a $p$-group of maximal class and let $N$ be the least normal subgroup of $G$ such that $G/N$ has an abelian subgroup of index $p$; then $N < \Phi(G)$. Take $\chi \in \mathrm{Irr}_1(G)$ of degree $p$. Then there are $H \in \Gamma_1$ and $\mu \in \mathrm{Lin}(H)$ such that $\chi = \mu^G$ (Blichfeldt). In that case, $\ker(\chi) = \ker(\mu^G) = \ker(\mu)_G \geq (H')_G = H'$ since $H' \lhd G$.

Since $G/H'$ has an abelian subgroup $H/H'$ of index $p$, we get $N \leq H'$ (here we use Exercise 9.1 (b)). Thus, the kernel of any nonlinear irreducible character of $G$ of degree $p$ contains $N$. Conversely, all nonlinear irreducible characters of $G/N$ have degree $p$, by Ito's theorem on degrees (see Introduction, Theorem 17). Thus, the desired number equals the number $\mathrm{n}(G/N) = |\mathrm{Irr}_1(G/N)|$ of nonlinear irreducible characters of the quotient group $G/N$. Setting $|G/N| = p^k$, we obtain $\mathrm{n}(G/N) = \frac{p^k - p^2}{p^2} = p^{k-2} - 1$, where $p^k - p^2 = |G/N| - |G/G'|$ is the sum of squares of degrees of nonlinear irreducible characters of $G/N$; that number equals $\mathrm{n}(G/N) \cdot p^2$.

**Exercise 153.** A nonabelian $p$-group $G$ has an irreducible character of degree $p \iff$ there is $N \lhd G$ such that $G/N$ is nonabelian with abelian subgroup of index $p$.

**Solution.** If $N$ exists, there is $\chi \in \mathrm{Irr}_1(G/N)$. By Theorem 17 in Introduction, $\chi(1) = p$. If $\psi \in \mathrm{Irr}_1(G)$ is of degree $p$, then $G/\ker(\psi)$ has an abelian subgroup of index $p$ [BZ, Theorem 18.1].

**Exercise 154.** Suppose that $G$ is a $p$-group with $|G'| = p^k > p$ and assume that $G$ has no nonnormal subgroups of order $p^{k+1}$. Then all minimal normal subgroups of $G$ are contained in $G'$.

**Solution.** Assume that a $G$-invariant $L < G$ of order $p$ is not contained in $G'$; then $LG' = L \times G'$ and $|(G/L)'| = |G'| = p^k$. By Theorem 1.23, $G/L$ contains a nonnormal subgroup $H/L$ of order $\geq p^k$. We get a contradiction since $|H| > p^k$.

**Exercise 155.** Suppose that $A, B \in \Gamma_1$ are distinct $\mathcal{A}_r$- and $\mathcal{A}_s$-subgroups of a metacyclic $p$-group $G$, $r < s$. Then $G$ is an $\mathcal{A}_{s+1}$-group.

**Solution.** By Theorem 72.1, $|A'| = p^r$, $|B'| = p^s$. As $G'$ is cyclic, we get $A' < B'$. By Exercise 1.69, $|G'| \leq p^{s+1}$. Let $A' \leq L < B'$, where $|B' : L| = p$; then $|L| \geq p^{s-1}$. Then the nonabelian $B/L < G/L$ so $G/L$ is not an $\mathcal{A}_1$-group. It follows that $|(G/L)'| > p$, and we conclude that $|G'| > p^s$. Thus, $|G'| = p^{s+1}$ so, by Theorem 72.1, $G$ is an $\mathcal{A}_{s+1}$-group.

**Exercise 156.** If a nonabelian $p$-group $G$, $p > 2$, with the cyclic derived subgroup $G'$ contains an abelian subgroup $A$ of index $p$, then $|G'| = p$.

**Solution.** By Lemma 1.1, $A/\mathrm{Z}(G) \cong G'$ is cyclic (see the proof of Lemma 1.1). Therefore, $G/\mathrm{Z}(G)$ is either abelian of type $(p^n, p)$ or $\cong \mathrm{M}_{p^{n+1}}$ (Theorem 1.2). Then $G/\mathrm{Z}(G)$ has two distinct cyclic subgroups $A/\mathrm{Z}(G)$ and $B/\mathrm{Z}(G)$ of index $p$. The subgroups $A, B$ are abelian so that $A \cap B = \mathrm{Z}(G)$, and hence $|G'| = \frac{1}{p}|G : \mathrm{Z}(G)| = p$. It follows that if $\mathrm{d}(G) = 2$ (in particular, if $G$ is metacyclic), then $G$ is an $\mathcal{A}_1$-group.

**Exercise 157.** Let $G$ be a special 2-group and, for some $a \in G$ of order 4, there is $b \in G$ such that $b^2 = a^2$ and $[a, b] \neq 1$. Then $\langle a, b\rangle \cong \mathrm{Q}_8$.

**Exercise 158.** Describe a Sylow 2-subgroup of the automorphism group of a nonabelian Dedekindian 2-group.

**Exercise 159.** Let $G$ be a minimal nonabelian $p$-group. If any two distinct maximal cyclic subgroups are permutable and generate $G$, then $G \cong Q_8$.

**Solution.** Let $A < G$ be nonnormal maximal cyclic and $x \in G - N_G(A)$. By hypothesis, $G = AA^x$, which is impossible in view of Lemma A.28.8. Thus, $A$ does not exist so that $G \cong Q_8$.

**Exercise 160.** Suppose that all maximal abelian subgroups of a $p$-group $G$ have order $p^\nu$ and all its minimal nonabelian subgroups have order $p^{\nu+1}$. Then

(a) if $A < B \leq G$, where $A$ is maximal abelian and $|B : A| = p$, then $B$ is minimal nonabelian,

(b) if $B \leq G$ be minimal nonabelian, then all maximal subgroups of $B$ are maximal abelian in $G$.

**Solution.** (a) Let $B \leq G$ be as in the statement. Let $B_1 \leq B$ be minimal nonabelian. Then $|B_1| = p^{\nu+1} = |B|$ so $B = B_1$ is minimal nonabelian.

(b) All maximal subgroups of $B$ are abelian of order $p^\nu$. If $A \leq A_1 < G$, where $A_1$ is maximal abelian, then $|A_1| = p^\nu = |A|$ so $A = A_1$ is maximal abelian.

**Exercise 161.** Suppose that $H$ is an absolutely regular subgroup of a $p$-group $G$ with $|\Omega_1(H)| = p^{p-1}$ and $|H| \geq p^{p+1}$. If all subgroups of $G$ containing $H$ as a subgroup of index $p$, are absolutely regular, then $G$ is either absolutely regular or of maximal class. (**Hint.** Use Theorem 12.1 (a).)

**Exercise 162.** Let $G \not\cong Q_8$ be a $p$-group of maximal class. Prove that the norm $\mathcal{N}(G)$ of $G$ has order $\leq p^2$. (See §§ 143, 171.) (**Hint.** By Corollary 140.8, $\mathcal{N}(G) \leq Z_2(G)$.)

**Exercise 163.** If $H$ is a quasinormal subgroup of index 4 in a 2-group $G$, then $H \trianglelefteq G$.

**Solution.** Suppose that this is false. Then $H/H_G$ is quasinormal but nonnormal of index 4 in $G/H_G \cong D_8$, a contradiction since a nonnormal subgroup of order 2 is not quasinormal in $D_8$.

**Exercise 164.** Let $H$ be a quasinormal subgroup of index $p^2$ in a $p$-group $G$, $p > 2$. Then $H \trianglelefteq G$.

**Exercise 165.** Let $H$ be a subgroup of maximal class and order $> p^{p+1}$ of a $p$-group $G$. If for all $x \in G$ of order $p$, the subgroup $D = \langle x, H\rangle$ is of maximal class, then $G$ is of maximal class.

**Solution.** Assume that $G$ is not of maximal class. Then there is in $G$ a subgroup $T$ containing $H$ as a subgroup of index $p$ and such that $T$ is not of maximal class (Exercise 10.10). By Lemma 12.3, $c_1(H) \equiv 1 + p + \cdots + p^{p-2} \pmod{p^p}$. By Theorem 13.2 (a), $c_1(T) \equiv 1 + p + \cdots + p^{p-1} \pmod{p^p}$. It follows that there is $x \in T - H$ of order $p$. Then $T = \langle x, H\rangle$ is of maximal class, contrary to the choice of $T$.

**Exercise 166.** If a $p$-group $G$ has only one irreducible character, say $\chi$, of degree $p$, then $p = 2$ and $G/\ker(\chi)$ is nonabelian of order 8.

**Solution.** One may assume that $\ker(\chi) = \{1\}$. Then, by [BZ, Theorem 18.1], there is in $G$ an abelian subgroup $A$ of index $p$. By Ito's theorem on degrees, $\mathrm{cd}(G) = \{1, p\}$. Let $|G| = p^{n+k}$, $|G'| = p^k$. Then $p^{n+k} = p^n + p^2$. It follows that $p^{n-2}(p^k - 1) = 1$ so that $n = 2$, $p^k = 2$, and hence $|G| = 2^n \cdot 2^k = 8$.

**Exercise 167.** Study the $p$-groups having exactly two distinct irreducible characters of degree $p$ (for example: $\mathcal{H}_{2,2}$ for $p = 2$).

**Exercise 168.** Classify the $p$-groups having $t \leq p$ irreducible characters of degree $p$.

Let $G$ be a nonabelian $p$-group. A subgroup $H < G$ is said to be a *centralizer* provided there is $x \in G - \mathrm{Z}(G)$ such that $\mathrm{C}_G(x) = H$. The centralizers cover $G$, and there are at least $p + 1$ centralizers in $G$, by Lemma 116.3. There are in $G$ at least $p + 1$ maximal abelian subgroups. If $x \in G - \mathrm{Z}(G)$, then any maximal abelian subgroup of $\mathrm{C}_G(x)$ is a maximal abelian subgroup in $G$.

**Exercise 169.** If a nonabelian $p$-group $G$ has exactly $p + 1$ centralizers, then all these centralizers have index $p$ in $G$.

**Exercise 170.** A nonabelian $p$-group $G$ has exactly $p + 1$ maximal abelian subgroups $\iff G = M\mathrm{Z}(G)$, where $M$ is minimal nonabelian.

**Exercise 171.** Classify the $p$-groups $G$ all of whose subgroups have multipliers of order $\leq p$. (**Hint.** Describe the abelian subgroups of $G$.)

**Exercise 172.** Suppose that $G$ is a nonabelian 2-group of order 16. Then:
(a) if $c_2(G) = 1$, then $G \cong D_{16}$,
(b) if $c_2(G) = 2$, then $G \in \{M_{16}, D_8 \times C_2\}$,
(c) if $c_2(G) = 3$, then $G \in\cong SD_{16}$,
(d) if $c_2(G) = 4$, then either $G \cong \langle a, b \mid a^4 = b^2 = 1,\ c = [a, b],\quad c^2 = 1,\ [a, c] = [b, c] = 1\rangle$ or $G = Q_8 * C_4$,
(e) if $c_2(G) = 5$, then $G \cong Q_{16}$,
(f) if $c_2(G) = 6$, then $G \in \{Q_8 \times C_2, \mathcal{H}_{2,2}\}$. (**Hint.** Use Theorem 1.17 (b).

**Exercise 173.** If $G$ is a $p$-group of order $> p^p$ such that $|\mathrm{H}_p(G)| = p^p$. Then $G$ is of maximal class and order $p^{p+1}$.

**Solution.** Let $\mathrm{H}_p(G) < F$, where $|F : \mathrm{H}_p(G)| = p$. Then $\mathrm{H}_p(F) = \mathrm{H}_p(G) < F$ so that $F$ is irregular. Thus, $\mathrm{H}_p(G)$ is a maximal regular subgroup of $G$. It follows that $G$ is of maximal class (Remark 10.5). Assume that $|G| > p^{p+1}$. Then $\exp(G_1) > p$, $G_1 \geq \mathrm{H}_p(G)$ and $|G_1| > p^p$, a contradiction. Thus, $|G| = p^{p+1}$.

**Exercise 174.** Let $G = \Omega_1(G)$ be irregular, $p > 2$. Then $G$ contains a subgroup of order $\geq p^p$ and exponent $p$.

**Solution.** Let $E \lhd G$ be of exponent $p$ of maximal order and let $x \in G - E$ be of order $p$. Set $H = \langle x, E\rangle$. By the choice of $E$, one has $\exp(H) > p$ so $H$ is irregular. Then $|H| \geq p^{p+1}$ so $|E| = \frac{1}{p}|H| \geq p^p$.

**Exercise 175.** If all subgroups of a $p$-group $G$ are two-generator, then one of the following holds:
(a) $G$ is metacyclic,
(b) $G$ is a 3-group of maximal class $\not\cong \Sigma_9$,
(c) $G \cong S(p^3)$,
(d) $G$ is a group of maximal class and order $p^4$, $p > 2$, $G$ has no subgroup $\cong \mathrm{E}_{p^3}$.

**Hint.** If $p = 2$, then $G$ is metacyclic (Corollary 36.6). If $p > 2$, use Theorem 13.7.

**Exercise 176.** Suppose that $p$-group $G$, $p > 2$, contains exactly one proper nonmetacyclic minimal nonabelian subgroup. Then one of the following holds:
(a) $\Omega_1(G) \cong \mathrm{E}_{p^3}$ and $\Omega_1(G) \in \Gamma_1$,
(b) $G$ contains a nonabelian subgroup $A \cong S(p^3)$ such that $\mathrm{C}_G(A)$ is cyclic.

**Solution.** Let $E < G$ be a normal elementary abelian subgroup of maximal order. Assume that $\Omega_1(G) > E$. Let $x \in \Omega_1(G) - E$ be of order $p$; then $H = \langle x, E\rangle$ is nonabelian (Theorem 10.1). By Lemma 57.1, there is in $E$ an element $a$ such that $A = \langle a, x\rangle$ is minimal nonabelian. It follows from $\Omega_1(A) = A$ that $A \cong S(p^3)$ (Lemma 65.1). Assume that $\mathrm{C}_G(A)$ is noncyclic. Then there is in $\mathrm{C}_G(A) - A$ an element $y$ of order $p$. In that case, $A \times \langle y\rangle$ contains $p^2$ minimal nonabelian subgroups $\cong A$ so is nonmetacyclic, a contradiction. Thus, $E = \Omega_1(G)$. Assume that $|E| < p^3$. Then $G$ is a 3-group of maximal class. By Theorem 9.6, $|G| = 3^4$ and $G$ is minimal nonmetacyclic, a contradiction. Thus, $E \cong \mathrm{E}_{p^3}$ and $|G| = p^4$.

**Exercise 177.** The following conditions for a noncyclic abelian $p$-group $G$ are equivalent:
(a) any maximal cyclic subgroup of $G$ is a direct factor,
(b) $\exp(G) \leq p^2$.

**Solution.** One has (b) $\Rightarrow$ (a) (Introduction, Lemma 4 (b)). Let us prove the reverse implication. Assume that $C < G$ is cyclic of order $p^3$ and $A < G$ is of order $p$, $A \not\leq C$. Write $H = C \times A$. Let $U < H$ be cyclic of order $p^2$ such that $U \not\leq C$. Then $H/U$ is cyclic of order $p^2$ so that $U$ is not a direct factor of $H$ (otherwise, $\exp(H) = p^2$). It follows that $U$ is not a direct factor of $G$ (here we apply the modular law). Thus, $C$ does not exist so that $\exp(G) \leq p^2$.

**Exercise 178.** Suppose that a $p$-group $G$ has a subgroup of order $p^k$ and exponent $p$. Is it true that if every subgroup of $G$ of order $p^k$ and exponent $p$ has a normal complement, then $\exp(G) = p$?

**Hint.** Let us consider case $k = 2$. Let $A \le G$ be minimal nonabelian. Assume that $A$ has a normal abelian subgroup $R$ of type $(p, p)$. Then $R$ has no normal complement in $A$ so in $G$. Thus, $A$ does not exist. It follows that $G$ is abelian and then $\exp(G) = p$.

**Exercise 179.** If $A$ is a proper minimal nonabelian subgroup of a metacyclic $p$-group $G$, $|G'| > p^2$, then $A \cap G' > A'$.

**Solution.** By Lemma 65.2 (a), $|G'| > p$. Let $G' < L < G$, where $L$ is cyclic such that $G/L$ is cyclic. Assume that $A \cap G' = A'$; then $A \cap L = A'$. On one side, $AL/L$, as a subgroup of $G/L$, is cyclic. On the other side, $AL/L \cong A/A'$ is noncyclic, a contradiction.

**Exercise 180.** Study the s-self dual (q-self dual) $p$-groups $G$ with cyclic derived subgroup.

**Exercise 181.** Let $G$ be a the nonabelian two-generator $p$-group. If $G' < C < G$, where $C$ is cyclic, then $G$ is metacyclic.

**Solution.** Let $L < G'$ be of index $p$. Then $G/L$ is minimal nonabelian (Lemma 65.2 (a)) and metacyclic since $G'/L$ is not maximal cyclic in $G/L$ (Lemma 65.1). Then, by Theorem 36.1, $G$ is metacyclic.

**Exercise 182.** Study the $p$-groups $G$ with cyclic $G'$ such that, whenever $A < G$ is abelian, there is $N \triangleleft G$ satisfying $G/N \cong A$.

**Exercise 183.** Prove that the $p$-groups $G$ such that, whenever $A < B < G$, then $A^G < B^G$ are Dedekindian.

**Solution.** Assume that $A < G$ is nonnormal. Then $A < A^G$. Taking $B = A^G$, one has $A < B < G$ and $A^G = B^G$, a contradiction. Thus, $G$ is Dedekindian.

**Exercise 184** ([Zha3]). Suppose that a nonabelian 2-group $G$ has no abelian subgroup of index 2 and all nonabelian two-generator subgroups of $G$ are maximal in $G$. Then $G$ is isomorphic to a Sylow 2-subgroup of the simple Suzuki group Sz(8).

**Solution.** By hypothesis, all nonabelian two-generator subgroups of $G$ are $\mathcal{A}_1$-groups, and these subgroups are maximal in $G$. All nonabelian maximal subgroups are two-generator since they contain $\mathcal{A}_1$-subgroups. It follows that $G$ is an $\mathcal{A}_2$-group. Now the result follows from § 71.

**Exercise 185.** Let $G$ be a two-generator $p$-group and $H \in \Gamma_1$. Is it true that if $H/\mathrm{K}_n(G)$ is of maximal class for some $n > 1$, then $G/\mathrm{K}_n(G)$ is of maximal class?

**Exercise 186.** Study the $p$-groups $G$ of exponent $p^e > p$, whose proper subgroup $\Omega_e^*(G)$ is of maximal class.

**Hint.** By Exercise 9.28(ii), $e = 2$. By Theorem 7.2 (b), $G$ is irregular. By Lemma 12.3 (b) and Theorem 13.2 (b), $G$ is of maximal class. Then $|G| \le p^{2(p-1)+1}$ (Theorem 9.6). In that case, $p > 2$.

**Exercise 187.** Classify the irregular $p$-groups in which the intersection of any two distinct maximal regular subgroups of $G$ is absolutely regular.

**Exercise 188.** Classify the $p$-groups of order $> p^{p+2}$ all of whose subgroups of order $p^{p+2}$, except one, are of maximal class. (**Hint.** Use Theorems 12.1 (a) and 13.6.)

**Exercise 189** (Janko). If a group $G$ of exponent $p^e$ contains only one maximal subgroup, say $H$, of exponent $p^e$, then $\mathrm{d}(G) = 2$.

**Solution.** Let $h \in H$ be of order $p^e$ and $g \in G - H$. Write $F = \langle h, g\rangle$ and assume that $F < G$. Let $F \le H_1 \in \Gamma_1$. As $H_1 \ne H$ and $\exp(H_1) = p^e$, we get a contradiction, Thus, $F = G$ so that $\mathrm{d}(G) = 2$.

**Exercise 190.** Let $G$ be a group of exponent $p^e$. Suppose that $M_1, \dots, M_k$ are all members of the set $\Gamma_1$ having exponent $p^e$. Is it true that if $k \le p$, then $\mathrm{d}(G) = 2$?

**Exercise 191.** Classify the nonabelian metacyclic $p$-groups (i) all of whose minimal nonabelian subgroups have no cyclic subgroups of index $p$, (ii) of exponent $p^e$ covered by cyclic subgroups of order $p^e$.

# Appendix 46 On Zsigmondy primes

This section is written by M. Roitman from the University of Haifa. We follow the exposition in [Roi1] of some known results related to Zsigmondy primes; generally, the proofs here are more detailed than in [Roi1]. Recall that if $a$ and $n$ are natural numbers greater than 1, then a prime $p$ is called a *Zsigmondy prime* for $\langle a, n\rangle$ if $p$ does not divide $a$, and $a$ is of order $n$ modulo $p$ (see [Fei1], and Theorem A.46.3 below). If $p$ is a Zsigmondy prime for $\langle a, n\rangle$, then $n \mid p-1$ by Fermat's theorem; thus $p \geq n+1$. A Zsigmondy prime $p$ for $\langle a, n\rangle$ is called a *large Zsigmondy prime* for $\langle a, n\rangle$ if $p > n+1$ or $p^2 \mid a^n - 1$. Clearly, as remarked in [Fei1], a prime $p$ is a large Zsigmondy prime for $\langle a, n\rangle$ if and only if $p$ is a Zsigmondy prime for $\langle a, n\rangle$ such that $|a^n - 1|_p > n+1$. Zsigmondy primes are used in finite group theory (see, e.g., [Art]). For applications of large Zsigmondy primes to finite groups see [FZ].

We now recall some basic properties of cyclotomic polynomials, which we will use below (see, e.g., [Rib], Preliminaries, Section 1). For $n \geq 1$ the cyclotomic polynomial $\Phi_n(X)$ is defined as $\Phi_n(X) = \prod_{i=1}^{\varphi(n)}(X - \epsilon_i)$, where $\epsilon_1, \dots, \epsilon_{\varphi(n)}$ are the primitive $n$-th roots of unity and $\varphi$ is Euler's totient function. If $n > 2$, then $\Phi_n(a) > 0$ for all real $a$, since $\Phi_n(X)$ is a monic polynomial over rational integers with no real roots. Moreover, if $n > 1$ and $a > 1$ are natural numbers, then $\Phi_n(a) = \prod_{i=1}^{\varphi(n)} |a - \epsilon_i| > (a-1)^{\varphi(n)} \geq 1$.

Since $X^n - 1 = \prod_{d:\, d|n} \Phi_d(X)$, every Zsigmondy prime for $\langle a, n\rangle$ divides $\Phi_n(a)$.

Except for these basic facts, this section is self-contained.

**Lemma A.46.1.** *Let $a > 1$ and $n = q^i r$ be integers, where $q$ is a prime, $i \geq 1$, and $r$ is a natural number not divisible by $q$. Let $b = a^{q^{i-1}}$. Then $\Phi_n(a) > [b^{q-2}(b-1)]^{\varphi(r)}$.*

*Proof.* We have

$$\Phi_n(a) = \frac{\Phi_r(a^{q^i})}{\Phi_r(a^{q^{i-1}})} = \frac{\Phi_r(b^q)}{\Phi_r(b)} = \frac{\prod_{i=1}^{\varphi(r)}(b^q - \epsilon_i)}{\prod_{i=1}^{\varphi(r)}(b - \epsilon_i)} > \left(\frac{b^q - 1}{b+1}\right)^{\varphi(r)},$$

where the $\epsilon_i$ s are roots of unity. Since $b^q - 1 \geq b^{q-2}(b^2 - 1)$, the lemma follows. □

**Remark 1.** In the setting of the previous lemma, $\Phi_n(a) \geq \left(\frac{b^q+1}{b+1}\right)^{\varphi(r)}$. Indeed, let $1 \leq j \leq \varphi(r)$, and set $\epsilon_j = \alpha + i\beta$, where $\alpha$ and $\beta$ are real numbers such that $\alpha^2 + \beta^2 = 1$. We have

$$\left|\frac{b^q - \epsilon_j}{b - \epsilon_j}\right|^2 = \frac{(b^q - \alpha)^2 + \beta^2}{(b-\alpha)^2 + \beta^2} = \frac{b^{2q} + 1 - 2b^q\alpha}{b^2 + 1 - 2b\alpha} = b^{q-1}\left(1 + \frac{b^{q+1} + \frac{1}{b^{q-1}} - b^2 - 1}{b^2 + 1 - 2b\alpha}\right)$$

(in the previous calculation we just got rid of $\alpha$ in the numerator). We see that $\left|\frac{b^q-\epsilon_j}{b-\epsilon_j}\right|^2$ is an increasing function of $\alpha$ in the interval $[-1, 1]$ such that $\alpha(-1) = \left(\frac{b^q+1}{b+1}\right)^2$.

**Proposition A.46.2.** *Let $a > 1$ and $n > 1$ be natural numbers. Let $q$ be a prime factor of $\Phi_n(a)$. Then $q$ is not a Zsigmondy prime for $\langle a, n\rangle \iff q \mid n$. In this case, $q$ is* *the largest prime factor of $n$, and $n = q^i r$, where $r$ is a natural number dividing $q - 1$; moreover, if $q^2$ divides $\Phi_n(a)$, then $n = 2$. Conversely, if $n = 2$ and there are no Zsigmondy primes for $\langle a, 2\rangle$, then $q = 2$, and $q^2 = 4$ divides $\Phi_2(a) = a + 1$.*

*Thus, if there are no Zsigmondy primes for $\langle a, n\rangle$, then $\Phi_n(a)$ is a power of $q$; also if $n > 2$ or $q > 2$, then $\Phi_n(a) = q$.*

*Proof.* If $q \mid n$, then $n$ does not divide $q - 1$, so $q$ is not a Zsigmondy prime for $\langle a, n\rangle$ (for this direction we do not need the assumption that $q$ is a prime factor of $\Phi_n(a)$).

Conversely, assume that $q$ is not a Zsigmondy prime for $\langle a, n\rangle$. Then there exists a prime factor $p$ of $n$ such that $a^{n/p} \equiv 1 \pmod q$. Since $\Phi_n(X) \mid \frac{X^n-1}{X^{n/p}-1}$, we see, putting $c = a^{n/p}$, that $q \mid \frac{c^p-1}{c-1}$. We have $\frac{c^p-1}{c-1} = \sum_{i=0}^{p-1} c^i \equiv p \pmod q$. Thus $q \mid p$, which implies that $q = p$. Hence $q \mid n$, and by the previous argument, $a^{n/p} \not\equiv 1 \pmod q$ for each prime factor $p \neq q$ of $n$. It follows that the order of $a \pmod q$ is of the form $r := \frac{n}{q^i}$ for some $i \geq 1$, so $\frac{n}{q^i} \mid q - 1$. Thus $q$ is the largest prime factor of $n$.

Let $c = a^{n/q}$.

If $q > 2$ let $d = c - 1$, so $q \mid d$. We have

$$\frac{a^n - 1}{a^{n/q} - 1} = \frac{c^q - 1}{c - 1} = \frac{(1 + d)^q - 1}{d} = q + \sum_{i=2}^{q-1} \binom{q}{i} d^{i-1} \equiv q \pmod{q^2}$$

since $q > 2$, so the binomial coefficients $\binom{q}{i}$ are divisible by $q$ for $2 \leq q - 1$. Thus $q^2$ does not divide $\Phi_n(a)$.

If $q = 2$ and $n > 2$, then $n$ is a power of 2 and $n \geq 4$. We have $\frac{c^2-1}{c-1} = c+1$. Since $a$ is odd and $\frac{n}{2}$ is even, we obtain $c = a^{n/2} \equiv 1 \pmod 4$. Thus $c + 1 \equiv 2 \pmod 4$. Hence 4 does not divide $\Phi_n(a)$ as claimed.

If $n = 2$, then $q = 2$ since $q$ is the largest prime factor of $n$. Also $2 \mid \Phi_2(a) = a + 1$. Since $a + 1 \geq 3$ and $a + 1 = \Phi_2(a)$ is a power of 2, we see that $4 \mid \Phi_2(a)$. □

**Theorem A.46.3** (Zsigmondy's Theorem). *Let $a > 1$ and $n > 1$ be natural numbers. There exists a prime $q$ such that $a$ is of order $n$ modulo $q$ with exactly the following exceptions:* (1) $n = 2$, $a = 2^s - 1$, where $s \geq 2$, and (2) $n = 6$, $a = 2$.

*Proof.* Assume condition (1). Let $q$ be a Zsigmondy prime for $\langle a, n\rangle$. Since the order of $a$ modulo $q$ is 2, we see that $a \equiv -1 \pmod q$. Hence $2^s \equiv 0 \pmod q$, so $q = 2$. Thus $a \equiv 1 \pmod 2$, a contradiction.

Assume condition (2). If $q$ is a Zsigmondy prime for $\langle a, n\rangle$, then $q \geq n + 1 = 7$ and $q \mid 2^6 - 1 = 63$, so $q = 7$. However, $2^3 \equiv 1 \pmod 7$, a contradiction.

Assume now that there are no Zsigmondy primes for $\langle a, n\rangle$.

If $n = 2$, then by Proposition A15.2, $\Phi_n(a) = a + 1 = 2^s$ for some integer $s \geq 2$, and case (1) holds.

Assume that $n > 2$. By Proposition A15.2, $\Phi_n(a) = q$, where $q$ is the largest prime factor of $n$.

If $q = 2$, then $n = 2^s$ for some positive integer $s$, and $\Phi_n(a) = a^{2^{s-1}} + 1 > 2$, a contradiction.

Thus $q \geq 3$. Let $n = q^i r$, where $r$ is an integer not divisible by $q$. Set $b = a^{q^{i-1}}$. By Lemma A15.1, $b^{q-2} < \Phi_n(a) = q$. This implies that $q = 3$ and $b = 2$, so $a = 2$ and $i = 1$. Since 7 is a Zsigmondy prime for $\langle 2, 3\rangle$, we see that $n > 3$. Since $i = 1$ we have $n = qr$; also $r \mid q - 1$ by Proposition A15.2. Thus $n$ divides $q(q-1) = 6$. It follows that $n = 6$. □

For the present proof we have used a proof of Zsigmondy's Theorem based on [Art] due to Y. Berkovich and G. Freiman.

We now turn to the proof of Feit's results on Zsigmondy primes.

**Proposition A.46.4.** *Let $a, n$ be natural numbers greater than 1. Let $q$ be the largest prime factor of $n$. Assume that there exist Zsigmondy primes for $\langle a, n\rangle$, but no large Zsigmondy primes. Then $n + 1$ is the unique Zsigmondy prime for $\langle a, n\rangle$, and $\Phi_n(a) = q^i(n + 1)$, where $i \geq 0$. Moreover, if $n > 2$ or $q > 2$, then $i = 0, 1$.*

*Proof.* Let $p$ be the Zsigmondy prime for $\langle a, n\rangle$, thus $p \geq n + 1$. Since $p$ is not a large Zsigmondy prime, we see that $p = n + 1$, thus $n + 1$ is the unique Zsigmondy prime for $\langle a, n\rangle$. Since each Zsigmondy prime for $\langle a, n\rangle$ divides $\Phi_n(a)$, and $n + 1$ is not a large Zsigmondy prime, we see that $n + 1 \mid \Phi_n(a)$, but $(n + 1)^2$ does not divide $\Phi_n(a)$. By Proposition A15.2, each prime different from $q$ that divides $\Phi_n(a)$ is a Zsigmondy prime for $\langle a, n\rangle$, and so it is equal to $n + 1$. It follows that $\Phi_n(a) = q^i(n + 1)$, as stated. By Proposition A15.2, if $n > 2$ or $q > 2$, then $i = 0, 1$. □

**Lemma A.46.5.** *For $n \geq 1$, $n \neq 6$ we have $2^{\varphi(n)} \geq n$; thus, for all $n \geq 1$, we have $2^{\varphi(n)} \geq \frac{2}{3}n$. Moreover, for odd $n > 3$ we have $2^{\varphi(n)} > 3n$.*

*Proof.* Let $n > 1$. First let $n$ be odd. Then $2^{\varphi(n)} \equiv 1 \pmod{n}$, so $2^{\varphi(n)} = kn + 1$ for some integer $k \geq 1$. It follows that $2^{\varphi(n)} > n$.

If $k = 1$, let $s = \varphi(n)$. Thus $n = 2^s - 1$, so $s > 1$, and $(2^s - 2, n) = 1$. The $s$ numbers $2^i$ for $0 \leq i \leq s - 1$ are coprime with $n$ and are between 1 and $n$. Since $s = \varphi(n)$, it follows that $2^s - 2 = 2^i$ for some $i \in \{0, \dots, s - 1\}$. Hence $2^{s-1} - 1 = 2^{i-1}$, so $i - 1 = 0$, $s = 2$ and $n = 2^s - 1 = 3$. It follows that for odd $n > 3$ we have $k > 1$, and since $k$ is odd we infer that $2^{\varphi(n)} > 3n$.

Let $n \neq 6$ be even. Set $n = 2^i m$, where $m$ is odd and $i \geq 1$. If $m = 1$, then $2^{\varphi(n)} = 2^{2^{i-1}} \geq 2^i = n$. If $i = 1$, then $m > 3$ since $n \neq 6$. Thus, by the previous paragraph, $2^{\varphi(n)} = 2^{\varphi(m)} > 2m = n$. Finally, if $i > 1$ and $m > 1$, then $m \geq 3$ since $m$ is odd, so $\varphi(m) \geq 2$. As $2^{\varphi(m)} > m$, we obtain

$$2^{\varphi(n)} = 2^{2^{i-1}\varphi(m)} \geq 2^{i\varphi(m)} \geq 2^{i+\varphi(m)} = 2^i 2^{\varphi(m)} > 2^i m = n$$

(since $xy \geq x + y$ for real $x \geq 2, y \geq 2$). □

**Remark 2.** For an integer $n > 1$, we have equality $2^{\varphi(n)} = n$ if and only if $n = 2$ or 4. Indeed, assume that $2^{\varphi(n)} = n$. Let $s = \varphi(n)$. Thus $n = 2^s$, so $\phi(n) = 2^{s-1}$. Hence $2^s = n = 2^{2^{s-1}}$, so $s = 2^{s-1}$. It follows that $s = 1, 2$, that is, $n = 2, 4$. The converse is trivial.

**Theorem A.46.6.** *[Fei1, Theorem A] If $a$ and $n$ are integers greater than 1, then there exists a large Zsigmondy prime for $\langle a, n\rangle$, with exactly the following exceptions: (a) $n = 2$ and $a = 2^s 3^t - 1$ for integers $s \geq 0$ and $t = 0, 1$, with $s \geq 2$ if $t = 0$, (b) $a = 2$ and $n = 4, 6, 10, 12$ or $18$, (c) $a = 3$ and $n = 4$ or $6$, and (d) $a = 5$ and $n = 6$.*

*Proof.* It is easy to show in each case that there are no large Zsigmondy primes for $\langle a, n\rangle$. For example, assume (a). Let $p$ be a Zsigmondy prime for $\langle a, n\rangle$. Since the order of $a$ modulo $p$ is 2, we see that $p$ divides $a + 1 = 2^s 3^t$. As $p \geq n + 1$, we obtain $p \geq 3$, so that $p = 3$ and $t \geq 1$; by (a), $t = 1$. Hence $p$ is not a large Zsigmondy prime, so there are no large Zsigmondy primes in case (a).

We also check the case $a = 2, n = 18$ in (b). Let $p$ be a Zsigmondy prime for $\langle 2, 18\rangle$. Hence $p \mid 2^9 + 1$. We have $2^9 + 1 = (2^3 + 1)(2^6 - 2^3 + 1)$. Since $p \geq n + 1 = 19$, we see that $p \mid 2^6 - 2^3 + 1$. We have $2^6 - 2^3 + 1 = 2^3(2^3 - 1) + 1 = 57$. Hence $p = 19$ is not a large Zsigmondy prime: $p = n + 1$, and $p^2$ does not divide $a^n - 1$. Thus there are no large Zsigmondy primes for $\langle 2, 18\rangle$.

For the converse, by Zsigmondy's Theorem A15.3, we may assume that there exist Zsigmondy primes for $\langle a, n\rangle$, but no large Zsigmondy primes. By Proposition A15.4, $n+1$ is the unique Zsigmondy prime for $\langle a, n\rangle$, and $\Phi_n(a)$ equals either $n+1$ or $q^i(n+1)$, where $q$ is the largest prime factor of $n$ and $i \geq 0$. Since $n > 1$, it follows that $n$ is even.

First let $n = 2$. Then $\Phi_2(a) = a + 1 = q^i(n + 1) = 2^i \cdot 3$, so case (a) holds.

Let $n > 2$.

(i) Let $n$ be a power of 2: $n = 2^i, i \geq 2$.

By Proposition A15.4, $\Phi_n(a)$ equals either $n + 1 = 2^i + 1$ or $2(n + 1) = 2(2^i + 1)$.

If $\Phi_n(a) = n + 1$, then $a^{2^{i-1}} = 2^i$ since $\Phi_{2^i}(a) = a^{2^{i-1}} + 1$. As $2^i \leq a^i \leq a^{2^{i-1}} = 2^i$, we obtain $a = 2$, $2^{i-1} = i$, and $i = 2$, that is, $n = 4$ (case (b)).

If $\Phi_n(a) = 2(n + 1)$, then $a^{2^{i-1}} + 1 = 2(2^i + 1)$; thus $a^{2^{i-1}} = 2^{i+1} + 1$, and $a > 1$ is odd, so $a \geq 3$. If $i = 2$, then $a = 3$ and $n = 4$ (case (c)). If $i > 2$, then $2^{i-1} \geq i + 1$, so $a^{2^{i-1}} \geq 3^{i+1} > 2^{i+1} + 1$, a contradiction.

(ii) Assume that $n$ is not a power of 2. Let $q$ be the largest prime factor of $n$, thus $q$ is odd. Let $n = q^i r$, where $r$ is an integer not divisible by $q$. Since $n$ is even, $r$ is even as well. Set $b = a^{q^{i-1}}$.

(ii-1) First assume that $\Phi_n(a) = q(n + 1)$.

By Lemma A15.1,

$$\left[b^{q-2}(b-1)\right]^{\varphi(r)} < \Phi_n(a) = q(rq^i + 1) \leq q[rq(b-1) + 1] \leq (b-1)q(rq+1)\,.$$

Hence

$$b^{(q-2)\varphi(r)} < q(rq+1)\,. \tag{1}$$

By Proposition A15.2, $r \mid q - 1$; hence $q(rq + 1) \leq q((q - 1)q + 1) < q^3$, so

$$b^{(q-2)\varphi(r)} < q^3\,. \tag{2}$$

If $q = 3$, then $r = 2$ since $r$ is even and $r$ divides $q - 1 = 2$. Also $a^{3^{i-1}} = b < q(rq+1) = 21$ by (1), thus $i = 1, 2$. Since $n = 3^i \cdot 2$ and $q(n+1) = \Phi_n(a) = b^2 - b + 1$, we see that if $i = 1$, then $n = 6$ and $b = a = 5$ (case (d)); if $i = 2$, then $n = 18$ and $b = 8$, $a = 2$ (case (b)).

Assume now that $q > 3$.

If $r > 2$, then $\varphi(r) \geq 2$. By (2),

$$b^{2(q-2)} < q^3 . \tag{3}$$

Since the function $\frac{2^{2x}}{x^3}$ is increasing for $x \geq 3$, and the inequality $2^{2(m-2)} > m^3$ holds for $m = 6$, we obtain that this inequality holds for every integer $m \geq 6$. As $q > 3$ is odd, by (3) we obtain that $q = 5$. Since the inequality (3) does not hold for $b = 3$ and $q = 5$, it follows that $b = 2$, $i = 1$, $n = 5r$. Since $r > 2$ is even and $r \mid q - 1$, we obtain $r = 4$. Thus $n + 1 = 21$ is not a prime, a contradiction.

Let $r = 2$.

We have $\Phi_n(a) = \Phi_{2q^i}(a) = \frac{b^q+1}{b+1}$. Since $q \mid \Phi_n(a)$, we obtain that $b + 1 \equiv b^q + 1 \equiv 0 \pmod{q}$. Thus $b \geq q - 1$. By (1), $(q-1)^{q-2} < q(2q+1)$. But for $q \geq 5$ we have $(q-1)^{q-2} \geq (q-1)^3 > q(2q+1)$, since $(q-1)^3 - q(2q+1) = q^3 - 5q^2 + 2q - 1 > 0$, a contradiction.

(ii-2) Assume that $\Phi_n(a) = n + 1$.

Similarly to the inequality (1), we have

$$\left(b^{q-2}\right)^{\varphi(r)} < rq + 1 . \tag{4}$$

Dividing by $2^{\varphi(r)}$ and using the inequality $2^{\varphi(r)} \geq \frac{2}{3}r$ (Lemma A15.5), we obtain

$$\left(\frac{b^{q-2}}{2}\right)^{\varphi(r)} < \frac{3}{2}q + 1 . \tag{5}$$

Hence $\frac{b^{q-2}}{2} < \frac{3}{2}q + 1$, so

$$b^{q-2} < 3q + 2 . \tag{6}$$

Since the function $\frac{2^x}{x}$ is increasing for $x \geq 2$, and the inequality $2^{m-2} \geq 3m + 2$ holds for $m = 7$, we obtain that this inequality holds for every integer $m \geq 7$. Hence by (6), we have $q = 3, 5$.

Let $q = 3$. Let $n = 2^j 3^i$, where $j \geq 1$ (since $r$ is even), and $i \geq 1$ (since $3 \mid n$). Since $r = 2^j \neq 6$, we have by Lemma A15.5 that $\varphi(r) \geq r$, so the estimate in (6) can be improved to obtain that $b^{q-2} < 2q + 2$, that is, $b < 8$. Since $b = a^{q^{i-1}}$, we see that $i = 1$. We have $\Phi_n(a) = \frac{\Phi_{2^j}(a^3)}{\Phi_{2^j}(a)} = \frac{c^3+1}{c+1}$, where $c = a^{2^{j-1}}$. Thus $c^2 - c + 1 = n + 1$, that is, $c(c-1) = 2^j \cdot 3$. Hence $c = 3$ or $c - 1 = 3$. If $c = 3$, then $j = 1$, $a = 3$, $n = 6$ (case (c)). If $c - 1 = 3$, then $4 = c = 2^j$, $j = 2$, $a = 2$, $n = 12$ (case (b)).

Let $q = 5$. By (5) we obtain $2^{2\varphi(r)} = \left(\frac{2^{q-2}}{2}\right)^{\varphi(r)} < \frac{3}{2} \cdot 5 + 1 = 8.5$. Thus $4^{\varphi(r)} < 9$, which implies $\varphi(r) = 1$, $r = 2$ since $r$ is even. By (6), we have $b^3 < 17$, so $b = a = 2$, $i = 1$, and $n = q^i r = 10$ (case (b)). □

In what follows, we will prove [Fei1, Theorem B], but we first need some auxiliary results.

**Lemma A.46.7.** *Let $a > 1$ and $n > 2$ be natural numbers. Then $\Phi_n(a) > a^{\varphi(n)/2}$.*

*Proof.* If $n$ is a power of 2, $n = 2^m$, then $\Phi_n(a) = a^{2^{m-1}} + 1 > a^{2^{m-1}} = a^{\varphi(n)}$. Otherwise, let $q$ be an odd prime factor of $n$. Set $n = q^i r$, where $r$ is an integer not divisible by $q$. Set $b = a^{q^{i-1}}$. By Lemma A15.1, we have

$$\Phi_n(a) > \left(b^{q-2}\right)^{\varphi(r)} \geq \left(b^{\frac{q-1}{2}}\right)^{\varphi(r)} = a^{\varphi(n)/2}$$

since $\varphi(n) = q^{i-1}(q-1)\varphi(r)$. □

**Lemma A.46.8** ([Fei1], Lemma 2.5). *If $n \geq 1$, then $\varphi(n) \geq \frac{1}{2}\sqrt{n}$.*

*Proof.* If $p$ is an odd prime and $m \geq 1$, then$\varphi(p^m) = (p-1)p^{m-1} \geq \sqrt{p^m}$ since $(p-1)^2 p^{m-2} \geq (p-1)^2 p^{-1} \geq 1$, so $((p-1)p^{m-1})^2 \geq p^m$. Also $\varphi(2^m) = 2^{m-1} \geq \frac{1}{2}\sqrt{2^m}$. Since $\varphi(mn) = \varphi(m)\varphi(n)$ for $m, n$ coprime, the lemma follows. □

**Corollary A.46.9.** *For any natural numbers $a > 1, n > 2$, we have $\Phi_n(a) > a^{\frac{\sqrt{n}}{4}}$.*

**Theorem A.46.10** ([Fei1], Theorem B). *Let $N \geq 1$ be an integer. Then for all but finitely many pairs of integers $\langle a, n\rangle$ with $a > 1$ and $n > 2$, there exists a Zsigmondy prime $p$ with $|a^n - 1|_p > nN + 1$.*

*Proof.* Let $a > 1$ and $n > 2$ be natural numbers such that there are no Zsigmondy primes $p$ for $\langle a, n\rangle$ for which $|a^n - 1|_p > nN + 1$. Since for each Zsigmondy prime $p$ it holds $p \equiv 1 \pmod{n}$, there are at most $N$ Zsigmondy primes such that $p \leq nN + 1$. Let $q$ be the largest prime factor of $n$. By Proposition A15.2, $q^2$ does not divide $\Phi_n(a)$, and each prime factor of $p$ of $\Phi_n(a)$ such that $p \neq q$ is a Zsigmondy prime. Since $\Phi_n(a) \mid a^n - 1$ and $|a^n - 1|_p \leq nN + 1$ for each Zsigmondy prime $p$, it follows that $\Phi_n(a) \leq q(nN + 1)^N$. By Corollary A15.9, this implies $a^{\frac{\sqrt{n}}{4}} < n(nN + 1)^N$. □

Wedderburn used Zsigmondy's theorem in order to prove that any finite division ring is commutative. We reproduce one of Wedderburn's proofs, slightly revised. Let $D$ be a finite division ring of dimension $n$ over its prime subfield $\mathrm{GF}(p)$. First assume that there is a Zsigmondy prime $q$ for $\langle p, n\rangle$. By Cauchy's Lemma there exists an element $g \in D - \{0\}$ of order $q$. Let $F$ be the subring of $D$ generated by $g$. Let $m = [F : \mathrm{GF}(p)]$, so $m \leq n$. Since $g^{p^m - 1} = 1$, we have $q \mid p^m - 1$. Thus $m = n$, so $D = F$ is commutative.

If there are no Zsigmondy primes for $\langle p, n\rangle$ then, by Zsigmondy's Theorem, either $n = 2$, or $n = 6$ and $p = 2$. If $n = 2$, then $D$ is commutative, since it is the subring generated by an arbitrary element in $D - \mathrm{GF}(F)$. If $n = 6$ and $p = 2$, then the order of $2 \pmod{9}$ is 6. By Sylow's Theorem, $D - \{0\}$ contains a subgroup $G$ of order 9 (which is abelian). Let $F$ be the subring of $D$ generated by $G$, and let $m = [F : \mathrm{GF}(p)]$, thus $9 \mid 2^m - 1$. It follows that $m = 6$, and $D = F$ is commutative.

# Appendix 47 The holomorph of a cyclic 2-group

Let $C = \langle g \rangle$ be a cyclic 2-group of order $2^n$ and $A = \mathrm{Aut}(C)$ the automorphism group of $C$. Denote by $\mathcal{H} = A \cdot C$ the natural semidirect product of $A$ and $C$ with kernel $C$. The group $\mathcal{H}$ is said to be the *holomorph* of $C$. If $n = 1$, then $A = \{1\}$ so $\mathcal{H} \cong C$. If $n = 2$, then $A \cong \mathrm{C}_2$ and $\mathcal{H} \cong \mathrm{D}_8$, the dihedral group of order 8. In what follows we assume that $n > 2$. In that case, $A = \langle \alpha \rangle \times \langle \beta \rangle$, where $o(\alpha) = 2$ and $o(\beta) = 2^{n-2}$ so that $A$ is abelian of type $(2, 2^{n-2})$. Clearly, $\mathrm{C}_{\mathcal{H}}(C) = C$.

One can choose $\alpha$ and $\beta$ so that

(1) $$g^{\alpha} = g^{-1}, \quad g^{\beta} = g^5.$$

Obviously, $\Omega_1(A) \cong \mathrm{E}_4$. Let $\Omega_1(A) = \{1, \alpha, \sigma, \tau\}$, where $\tau = \beta^{2^{n-3}}$ and $\sigma = \alpha\tau$. Set

(2) $$U = \langle C, \alpha \rangle, \; S = \langle C, \sigma \rangle, \; T = \langle C, \tau \rangle.$$

The subgroups $U, S, T$ are $\mathcal{H}$-invariant nonabelian of order $2^{n+1}$ with cyclic subgroup $C$ of index 2. If $C < K < \mathcal{H}$ with $|K : C| = 2$, then $K \in \{U, S, T\}$ since $\Omega_1(\Omega_1(A)C/C) \cong \mathrm{E}_4$.

One has $U \cong \mathrm{D}_{2^{n+1}}$. Therefore, since $R = \Omega_1(A) \cdot C$ is not of maximal class (indeed, $\mathrm{d}(R) = 3$ since $\Phi(R) = \mho_1(C)$). By Exercise 10.10, $T$ is not of maximal class. It is easy to show that $S \cong \mathrm{SD}_{2^{n+1}}$. Therefore, $U, S, T$ are pairwise nonisomorphic. As $\mathcal{H}/C$ is abelian, it follows that $U, S, T \triangleleft \mathcal{H}$. Since the nonabelian subgroup $T$ is not of maximal class, it follows from Theorem 1.2 that $T \cong \mathrm{M}_{2^{n+1}}$. Thus,

(3) $$g^{\alpha} = g^{-1}, \quad g^{\sigma} = g^{-1+2^{n-1}}, \quad g^{\tau} = g^{1+2^{n-1}}.$$

As all involutions of $\mathcal{H}$ lie in the subgroup $R = \Omega_1(A) \cdot C$ and $\mathrm{c}_1(D) = 2^n + 1$, $\mathrm{c}_1(S) = 2^{n-1} + 1$, $\mathrm{c}_1(T) = 3$, $\mathrm{c}_1(C) = 1$, one has, by Hall's enumeration principle,

(4) $$\mathrm{c}_1(\mathcal{H}) = \mathrm{c}_1(R) = \mathrm{c}_1(U) + \mathrm{c}_1(S) + \mathrm{c}_1(T) - 2\mathrm{c}_1(C) = 3 \cdot 2^{n-1} + 3.$$

It follows that $\Omega_1(\mathcal{H}) = R$ so that $R$ is characteristic in $\mathcal{H}$.

One has $\mathcal{H}/\mho_1(C) \cong \mathrm{C}_2 \times A$, which is abelian of type $(2, 2, 2^{n-2})$. Since $\mho_1(C) = \Phi(R) < \Phi(\mathcal{H})$, it follows that $\mathrm{d}(\mathcal{H}) = 3$.

As $\mathrm{c}_2(U) = 1$, $\mathrm{c}_2(S) = 2^{n-2} + 1$, $\mathrm{c}_2(T) = 2$, $\mathrm{c}_2(C) = 1$, and $U \cup S \cup T = R$, one obtains

$$\mathrm{c}_2(R) = \mathrm{c}_2(U) + \mathrm{c}_2(S) + \mathrm{c}_2(T) - 2\mathrm{c}_2(C) = 1 + (2^{n-2} + 1) + 2 - 2 = 2^{n-2} + 2\,.$$

Let $I, J < A(\cong \mathrm{Aut}(C))$ be distinct cyclic subgroups of order 4 (by assumption, $n > 3$). Next, $\Omega_2(A) = IJ$, $\mathrm{c}_2(CI) = \mathrm{c}_2(CJ) = 6$, all subgroups of $\mathcal{H}$ of order 4 are contained in $\Omega_2(A) \cdot C$. One has $CI \cap CJ \cap R = T$ and $\Omega_2(A)C/T \cong \mathrm{E}_4$. Therefore,

$$\mathrm{c}_2(\mathcal{H}) = \mathrm{c}_2(\Omega_2(A) \cdot C) = \mathrm{c}_2(R) + \mathrm{c}_2(IC) + \mathrm{c}_2(JC) - 2\mathrm{c}_2(T) = 2^{n-2} + 2 + 2 \cdot 6 - 2 \cdot 2 = 2^{n-2} + 10.$$

Let $n \geq k > 2$. Then $c_k(U) = 1 = c_k(S)$, $c_k(T) = 2$ so that

$$c_k(R) = c_k(U) + c_k(S) + c_k(T) - 2c_k(C) = 1 + 1 + 2 - 2 = 2\,.$$

Arguing as above and using the equality $c_k(\mathcal{H}) = c_k(\Omega_k(A) \cdot C)$, it is possible to compute $c_k(\mathcal{H})$ for $k > 2$.

Write

$$X = \langle C, \alpha, \beta^2 \rangle = \langle U, \beta^2 \rangle\,, \quad Y = \langle C, \beta \rangle,\ Z = \langle C, \alpha\beta \rangle\,.$$

Then $X, Y, Z \in \Gamma_1(\mathcal{H})$, $Y, Z$ are metacyclic, $d(X) = 3$ since $d(X/\mho_1(C)) = 3$.

Since $U \lhd \mathcal{H}$ is of maximal class (see (2)), we get, since $\mathcal{H}/U' = \mathcal{H}/\mho_1(C)$ is abelian of type $(2^n, 2, 2)$,

(5) $$\mho_1(U) = U' = \mathcal{H}'.$$

The subgroup $T$ has two cyclic subgroups $C = \langle g \rangle$ and $C_1 = \langle g\tau \rangle$ of index 2 (note, that $c_n(G) = 2$, by the above). Since $C, T \lhd \mathcal{H}$, it follows that $C_1 \lhd \mathcal{H}$. The quotient group $\mathcal{H}/C_1$ is noncyclic since the group $\mathcal{H}$ is nonmetacyclic. Next,

(6) $$\alpha, \sigma, \tau \notin C_1.$$

Write

(7) $$X_1 = \langle \alpha, C_1 \rangle\,, \quad Y_1 = \langle \sigma, C_1 \rangle.$$

Considering the action of $\alpha$ on $C \cap C_1$ and taking into account that $C_1$ has index 2 in $X_1$ and $Y_1$, we conclude that $X_1$ and $Y_1$ are of maximal class.

Since $C \cap C_1 = \mho_1(C) = \mathcal{H}'$ (see (5)), the quotient group $\mathcal{H}/C_1$ is abelian. By (7), $C_1 \cap A = \{1\}$ so that $\mathcal{H} = A \cdot C_1$. Since all involutions in $\Omega_1(A)$ do centralize $C_1$, we get $C_{\mathcal{H}}(C_1) = C_1$.

Write

(8) $$U_1 = \langle \alpha, \beta^2, C_1 \rangle = \langle X_1, \beta^2 \rangle\,, \quad S_1 = \langle \beta, C_1 \rangle.$$

Since $C \nleq U_1$, $C \nleq S_1$, the subgroups $X, Y, Z, U_1, S_1$ are pairwise distinct maximal subgroups of $\mathcal{H}$. Thus, all maximal subgroups of $\mathcal{H}$ of exponent $2^n$ are listed. Next,

(9) $$W = A \cdot \mathcal{H}' \in \Gamma_1 - \{X, Y, Z, U_1, S_1\}\,, \quad d(W) = 3.$$

Let $D_0 < T$ be a noncyclic maximal subgroup ($D_0$ is abelian of type $(2^{n-1}, 2)$); then $D_0 \lhd \mathcal{H}$ since $D_0$ is characteristic in $T \lhd G$. Write

(10) $$W_1 = A \cdot D_0 \in \Gamma_1 - \{X, Y, Z, W, U_1, S_1\}.$$

Thus, all $7 = 2^3 - 1$ maximal subgroups of $\mathcal{H}$ are listed. Since $W_1' \leq \mathcal{H}'$ and $\mathcal{H}'$ is cyclic, we see that $W' < D_0$ so $d(W_1) > 2$. Considering $W_1 \cap Y$ and taking into account that $Y$ is metacyclic, we conclude that $d(W_1) = 3$. Thus, the numbers of generators of maximal subgroups of $\mathcal{H}$ are 2, 2, 2, 2, 3, 3, 3.

Assume that $E_8 \cong E \triangleleft \mathcal{H}$; then $E < \Omega_1(\mathcal{H}) = R = \Omega_1(A) \cdot C = EC$, $|E \cap C| = 2$, and we conclude that $R/(E \cap C)$ is abelian, i.e., $|R'| \le 2 < 2^{n-1} = |U'|$, which is a contradiction since $U < R$. Thus, $\mathcal{H}$ has no normal subgroup $\cong E_8$. However, $\mathcal{H}$ has a subgroup $\Omega_1(A) \times \Omega_1(C) \cong E_8$ which is non-$\mathcal{H}$-invariant.

Next, we note that $\exp(\mathcal{H}) = 2^n$. This is a consequence of the following:

**Proposition A.47.1.** *Let a $p$-group $X = A \cdot B$ be a semidirect product of two cyclic groups $A$ and $B$ with kernel $B$. Then* $\exp(X) = \max\{|A|, |B|\}$.

*Proof.* Set $|A| = p^f$, $|B| = p^e$. If $p > 2$, our claim follows from regularity of $X$ (Theorems 7.1 (c) and 7.2 (b)). Next we assume that $p = 2$. We have $\Phi(X) = \Phi(A) \cdot \Phi(B)$ and $|\Phi(A)| = 2^{f-1}$, $|\Phi(B)| = 2^{e-1}$. Using induction on $e + f$, one obtains $\exp(\Phi(X)) = \max\{2^{f-1}, 2^{e-1}\}$. Since $\exp(X) \le 2 \cdot \exp(\Phi(X))$, we get $\exp(X) \le \max\{2^e, 2^f\}$. Since the reverse inequality obviously holds, we are done. □

As $\mathcal{H} = U \cup S \cup T$ and $\exp(U) = \exp(S) = \exp(T)$, it follows easily from Proposition A.47.1 (see definitions of $U, S, T$) that $\exp(\mathcal{H}) = 2^n$.

**Exercise 1.** Let $S$ be the Sylow $p$-subgroup of the holomorph of the cyclic group of order $p^n$, $p > 2$.

(a) Prove that $\exp(S) = p^n$.

(b) Find $c_k(S)$ for all $k \le n$.

(c) Describe all maximal subgroups of $S$. (**Hint.** The group $S$ is regular, by Theorem 7.1 (c). Use Theorem 7.2.)

**Exercise 2.** Let $\mathcal{H}$ be the Sylow $p$-subgroup of the holomorph of the group $C_{p^n}$, $n > 2$. Describe the minimal nonabelian subgroups of $\mathcal{H}$. (**Hint.** Use Lemma 57.1.)

# Appendix 48 Some results of R. van der Waall and close to them

In this appendix we present some results of R. van der Waall.

Recall that a $p$-group is *modular* $\iff$ any two its subgroups are permutable or, what is the same, all subgroups are quasinormal.

Let $\mathcal{M}_2$ be the set of all nonabelian quaternion-free 2-groups (see § 56). If $p > 2$, we denote by $\mathcal{M}_p$ the set of all nonabelian modular $p$-groups (see § 73).

**Theorem A.48.1** ([Waa3], Theorem 2). *An $\mathcal{M}_p$-group, $p > 2$, contains a characteristic subgroup of index $p$.*

In particular, a nonabelian modular $p$-group, $p > 2$, contains a noncentral proper characteristic subgroup [Waa6, Theorem 1.2.7].

It follows from Ward's Theorem 56.1 that a non-Dedekindian modular 2-group contains a characteristic subgroup of index 2. Thus, any non-Dedekindian modular $p$-group ($p$ is arbitrary) contains a characteristic maximal subgroup.

**Proposition A.48.2** ([Waa5]). *Every nonabelian section of a $p$-group $G$, $p > 2$, contains a characteristic subgroup of index $p$ $\iff$ $G \in \mathcal{M}_p$.*

If $G$ is a nonabelian modular $p$-group, $p > 2$, and $\bar{L}$ is its nonabelian section, then $\bar{L}$ is modular, and so possesses a characteristic subgroup of index $p$ (Theorem A.48.1). If $G$ is a nonmodular $p$-group, $p > 2$, it has a section $\cong S(p^3)$ that has no characteristic subgroup of index $p$. If $G$ is a nonmodular 2-group, it has a section $\cong D_8$ that has a characteristic subgroup of index 2.

**Theorem A.48.3** ([Waa1], Theorem 1). *Let $p > 2$ and $G$ be a group of order $p^n$ with cyclic derived subgroup $G'$. Then $p^{[n/2]+1} \le |G/G'|$.*

A particular case of this theorem we prove now. For the general case, see § 204.

**Corollary A.48.4.** *Suppose that $G$ is a metacyclic $p$-group with the derived subgroup $G'$ of index $p^k$, $p > 2$, $|G| = p^{2s+e}$, where $e \in \{0, 1\}$. Then $s + 1 \le k$.*

*Proof.* Set $|G| = p^n = p^{2s+e}$, $|Z(G)| = p^z$, $|G'| = p^{n-k}$. Let $L$ be a normal cyclic subgroup of $G$ such that $G/L$ is cyclic; then $G' < L$. Let $L \le A < G$, where $A$ is a maximal abelian subgroup of $G$; then $G/A$ is cyclic as an epimorphic image of $G/L$. By Lemma 1.1, $|A| = |G'|\,|Z(G)|$. By Lemma 148.14,

$$(2)\quad p^k = |G : G'| = |A| = |G'|Z(G)| = p^{n-k}p^z \Rightarrow k = \frac{n+z}{2} = \frac{2s+e+z}{2} = s + \frac{e+z}{2}.$$

Assume that the theorem is false. Then $k < s + 1$ or, what is the same, $s + \frac{e+z}{2} < s + 1$, and so $e + z < 2$. Then $z = 1$ and $e = 0$ since $z \ge 1$. Thus, $n = 2s$ and, by (2), $2k = n + z = 2s + 1$, a contradiction. □

**Theorem A.48.5** ([Waa1]). *Let $G$ be a noncyclic $p$-group, $p > 2$, such that $Z(\Omega_1(G))$ and $Z(\Phi(G))$ are cyclic. Then $G = Z(G)E$, where $Z(G)$ is cyclic and $E$ is extraspecial.*

*Proof.* Clearly, $G$ is nonabelian. By Lemma 1.4, $\Phi(G)$ is cyclic so $G$ is regular (Theorem 7.1 (c)). Let $\Phi(G) \le L < G$, where $L$ is maximal cyclic. We have $\Phi(G) = G'\mho_1(G)$. Let $L < T \le G$, where $|T : L| = p$. Then, by Theorem 1.2, $T = L\Omega_1(T)$. Such subgroups as $T$ generate $G$ since $G/L$ is elementary abelian. It follows that $G = L\Omega_1(G)$ and $\Omega_1(G)$ is of exponent $p$, by Theorem 7.2 (b). Therefore, $|L \cap \Omega_1(G)| = p$ so $L \cap \Omega_1(G) = \Omega_1(L)$ and $\Omega_1(G)/\Omega_1(L) \cong G/L$ is elementary abelian. By hypothesis, $Z(\Omega_1(G))$ is of order $p$ so that $\Omega_1(G)$ is extraspecial. Since $L$, $\Omega_1(G)$ are normal in $G$, the subgroup $G/\Omega_1(L)$ is abelian so $\Omega_1(L) = G'$. Now the result follows from Lemma 4.3 (indeed, $\Omega_1(G)$ is a central product of nonabelian groups of order $p^3$). □

In is interesting to describe those $p$-groups that occur as $G$-invariant subgroups of $\Phi(G)$, where $G$ is a $p$-group. It is easily seen that all abelian $p$-groups have the above property (which we call the $\Phi$-property). There is only one nonabelian group of order $p^4$ that has $\Phi$-property.

By Lemma 1.4, there is no $p$-group $G$ such that $N \cong \Phi(G)$ for a nonabelian $p$-group $N$ with cyclic $Z(N)$.

Allenby [Alle] showed that if $N$ is a $G$-invariant subgroup of $\Phi(G)$, then there exists a group $X$ such that $N \cong \Phi(X)$. This solves a problem of B.H. Neumann. All groups of order 32 that have $\Phi$-property are listed in [WaaN].

Suppose that the $p$-group $P$ satisfies the following condition:

(∗) Whenever $G$ is a group containing $P$ as a Sylow $p$-subgroup, where $p$ is a minimal prime divisor of $|G|$, then $G$ is $p$-nilpotent. All (∗)-groups of order dividing 32 are determined in [Waa7]. Below we present an infinite family of (∗)-groups (this example is due to A.J. Wong; see [Hup1, Kap. IV. Theorem 3.5]).

**Exercise 1** (= Remark 148.53). The group $P \cong M_{2^n}$ is a (∗)-group. Moreover, if $p$ is a minimal prime divisor of $|G|$ and $P \cong M_{p^n}$ is a Sylow $p$-subgroup of a group $G$, then $G$ is $p$-nilpotent.

**Solution.** Assume that $G$ is a counterexample of minimal order. Then all those containing $P$ proper subgroups of $G$ are $p$-nilpotent. By Theorem A.52.2, there is in $G$ a minimal nonnilpotent subgroup $H = Q_1 \cdot H'$, where $E_{p^2} \cong H' \in \mathrm{Syl}_p(H)$ and a cyclic $Q_1 \in \mathrm{Syl}_q(H)$, a prime $q > p$. Since $q \mid p^2 - 1$ (see [BJ1, Lemma A.22.1]), we get $p = 2$ and $q = 3$. As in the proof of Theorem A.52.3, one has $H' \lhd G$. Write $\bar{G} = G/H'$. Since Sylow 2-subgroup $\bar{P}$ of $\bar{G}$ is cyclic, $\bar{G}$ has a normal 2-complement $\bar{T}$, by Theorem A.52.1. Thus, $G = PT$ is 2-solvable. Write $T = T_1H'$, where $T_1$ is a $2'$-Hall subgroup of $T$. By Frattini and Schur–Zassenhaus, $G = N_G(T_1)T_1H' = N_G(T_1)H'$ so that $P = H'(P \cap N_G(T_1))$, by the modular law. As $\Omega_1(P) = H'$, we get $|G : N_G(T_1)| = p$. As $T_1$ is characteristic in $N_G(T_1) \lhd G$, one obtains $T_1 \lhd G$, and we are done since $T_1$ is a $2'$-Hall subgroup of $G$.

**Exercise 2.** Let $p$ be the minimal prime divisor of $|G|$. If $P \in \mathrm{Syl}_p(G)$ is abelian of type $(p^{a_1}, \ldots, p^{a_n})$, where $a_1 < \cdots < a_n$, then $G$ is $p$-nilpotent.

**Exercise 3.** If $G$ is a 2-group and $\Phi(G)$ is nonabelian of order $2^4$, then $\Phi(G) = D \times C$, where $|C| = 2$ and $|D| = 8$.

**Solution.** By Lemma 1.4, $\mathrm{Z}(\Phi(G)) \cong \mathrm{E}_4$ so that $\exp(\Phi(G)) = 4$. Assume that $\mathrm{d}(\Phi(G)) = 2$. Then $\Phi(G)$ is metacyclic (Theorem 44.12) so that $\Phi(G) \cong \mathcal{H}_{2,2}$. If $L < \Phi(G)$ is of order 2 and $L \neq \Phi(G)'$, then $L$ is characteristic in $\Phi(G)$ so normal in $G$ and $\Phi(G)/L$ is nonabelian of order 8, which is impossible (Lemma 1.4). Thus, $\mathrm{d}(\Phi(G)) = 3$. Let $M < \Phi(G)$ be minimal nonabelian. Then $\Phi(G) = M\mathrm{Z}(\Phi(G))$. If $R \leq \mathrm{Z}(\Phi(G))$ and $R \neq \mathrm{Z}(M)$, then $\Phi(G) = M \times R$, and the result follows.

# Appendix 49 Kegel's theorem on nilpotence of $\mathrm{H}_p$-groups

A finite group $G$ is called an $\mathrm{H}_p$-group if $G$ possesses an automorphism $\alpha$ of prime order $p$ such that for all $g \in G$ the equality

$$(*) \qquad g^{1+\alpha+\dots+\alpha^{p-1}} = gg^{\alpha}\dots g^{\alpha^{p-1}} = 1$$

holds. (Note that if for some $g \in G$, $g^{\alpha} = g$, then $(*)$ gives $g^p = 1$.)

**Exercise 1.** Suppose that $G$ is an $\mathrm{H}_p$-group with respect to automorphism $\alpha$ of order $p$. Let $W = \langle \alpha, G \rangle$ be the extension of $G$ by $\langle \alpha \rangle$. Prove that all elements of the set $W - G$ have the same order $p$. It follows that if $u \in W - G$, then $u$ induces the automorphism $\alpha$ of $G$ satisfying $(*)$.

**Exercise 2.** Let $G$ and $\alpha$ be as in $(*)$, $N$ is $\alpha$-invariant subgroup of $G$. Then $G/N$ is an $\mathrm{H}_p$-group.

**Exercise 3** (Burnside). $\mathrm{H}_2$-groups are abelian.

**Exercise 4.** Let $G = (F, H)$ be a Frobenius group with kernel $H$. Then $H$ is an $\mathrm{H}_p$-group for any prime divisor $p$ of $|F|$.

**Exercise 5.** Let $G$ and $\alpha$ be as in $(*)$, $q \in \pi(G) - \{p\}$. Then there is an $\alpha$-invariant $Q \in \mathrm{Syl}_q(G)$.

In [HTh] Hughes and Thompson have proved the following result:

**Theorem A.49.1.** *If $G$ is an $\mathrm{H}_p$-group but not a $p$-group, then $G$ is generated by all elements of order distinct from $p$. In addition, $G$ possesses a characteristic nilpotent $p'$-subgroup $K$ such that $G/K$ is a $p$-group and so $G$ is solvable.*

**Exercise 6.** Let $G$ and $\alpha$ be as in $(*)$, $q \in \pi(G) - \{p\}$. Using Exercise 5 and considering $N_G(Q_0)$, where $Q_0$ runs over all nonidentity characteristic subgroups of $Q$ (see [Hup1, Kap IV, Hauptsatz 6.2]), prove that if $G$ is a minimal counterexample to Theorem A.49.1, to $|\pi(G)| = 2$.

In [Keg2] O.H. Kegel proves the following important result:

**Theorem A.49.2.** *If a finite group $G$ admits an automorphism $\alpha$ of prime order $p$ such that for all $g \in G$ equality $(*)$ holds, then $G$ is nilpotent.*

*Proof.* Suppose that the theorem is false. Let $G$ be a counterexample of minimal order. Since $G$ is not nilpotent, it follows that $G$ is not a $p$-group and therefore we may use Theorem A.49.1. Let $H$ be a semidirect product of $G$ with $\langle s \rangle \cong C_p$, where $s$ induces on $G$ (by conjugation) the automorphism $\alpha$. By Theorem A.49.1, $G$ has a nilpotent normal

$p'$-subgroup $K \neq \{1\}$ so that $G/K \neq \{1\}$ is a $p$-group and $G$ is generated by elements of order distinct from $p$. Let $P_1$ be a Sylow $p$-subgroup of $H$ such that $P_1 \geq \langle s \rangle$.

(a) $G$ has an $\langle s \rangle$-invariant Sylow $p$-subgroup $P$ and an $\langle s \rangle$-invariant subgroup of order $p$ contained in $Z(P)$.

Indeed, since $G \lhd H$, $P = P_1 \cap G \in \mathrm{Syl}_p(G)$ and $P$ is $\langle s \rangle$-invariant. In this case, $P \neq \{1\}$ because, otherwise, $G$ would be nilpotent. Take a subgroup of order $p$ in $Z(P_1) \cap P$ that is obviously $\langle s \rangle$-invariant.

(b) Each proper $\langle s \rangle$-invariant subgroup of $G$, say $M$, and each factor-group $G/N$ with $\{1\} < N \trianglelefteq G$ being $\langle s \rangle$-invariant, is nilpotent.

Let $L \in \{M, G/N\}$. If $s$ induces on $L$ the identity automorphism, then it follows from $(*)$ that $o(x) \mid p$ for all $x \in L$, and therefore $L$ is a $p$-group, so nilpotent. If $s$ induces of $L$ an automorphism of order $p$ (for $L = G/N$, see Exercise 2), then $L$ is nilpotent, by induction.

(c) Let us prove that $Z(G) = \{1\}$.

Clearly, $Z(G)$, being characteristic in $G$, is $s$-invariant. Therefore, $G/Z(G)$ is nilpotent, by (b). But then $G$ would be also nilpotent, contrary to the assumption.

(d) The nilpotent characteristic subgroup $K \neq \{1\}$ of $G$ is abelian.

Assume, by way of contradiction, that $(\{1\} <)Z(K) < K$. In that case, $PZ(K)$, being a proper $\langle s \rangle$-invariant subgroup of $G$, is nilpotent, by (b), and so $P$ centralizes $Z(K)$. But then $\{1\} < Z(K) \leq Z(G)$, contrary to (c).

(e) $P$ is cyclic of order $p$.

By (a), there is an $\langle s \rangle$-invariant subgroup $\langle t \rangle$ of order $p$ in $Z(P)$. If $\langle t \rangle < P$, then $\langle t \rangle \cdot K$ is a proper $\langle s \rangle$-invariant subgroup of $G$ and so, by (b), $K\langle t \rangle$ is nilpotent. But then $t$ centralizes $K$ and so $t \in Z(G)$, contrary to (c).

(f) There is an element $w \in G - K$ such that $o(w) \neq p$ and $p \mid o(w)$.

By Theorem A.49.1, $G$ is generated by elements of order $\neq p$ and so the existence of $w \in G - K$ follows.

Finally, $k = w^p \neq 1$ and $k \in K$ (here $w$ is as in (f)). Hence $w = kg = gk$ with $o(g) = p$. By (d) and (e), $C_G(k) \geq \langle K, g \rangle = G$ and so $k \in Z(G)$, contrary to (c). Hence minimal counterexamples do not exist, and so the theorem is proved. □

Theorem A.49.1 is a generalization of Thompson's theorem on nilpotence of kernels of Frobenius groups.

**Exercise 7** ([Keg2]). Let $\Sigma$ be a nontrivial partition of a group $G$ such that, whenever $A \in \Sigma$, then $A^x \in \Sigma$ for all $x \in G$. Then one of the following hold:

(a) $G = (F, N)$ is a Frobenius group with the kernel $N$ and a complement $F$ such that $\Sigma = \{N, F^x \mid x \in F\}$, or

(b) any component $A \in \Sigma$ is an $H_p$-group for some prime $p$ and $N_G(A) > A$.

**Solution**. Let $A \in \Sigma$. If $N_G(A) = A$, then, by Frobenius' Theorem, $G = (A, N)$ is a Frobenius group with a complement $A$. By [BZ, Lemma 3 and Theorem 10.6], $A$ has no nontrivial partition, and this completes the proof of (a). Now assume that $G$ is not a Frobenius group. Then $\mathrm{N}_G(A) > A$ for all $A \in \Sigma$. Let $A < B \leq \mathrm{N}_G(A)$ be such that $|B : A| = p$, a prime. Then there is a subset $\Sigma_1$ in $\Sigma$ such that

$$B = A \cup \left( \bigcup_{C \in \Sigma_1} (B \cap C) \right)$$

is a nontrivial partition. Thus, all elements in the set $B - A$ have order $p$ so that $A$ is an $\mathrm{H}_p$-group, by Exercise 1.

# Appendix 50 Sufficient conditions for 2-nilpotence

$1^o$ *Introduction.* Recall that an element $x$ of a finite group $G$ is said to be *real* if $x$ and $x^{-1}$ are conjugate in $G$ (write $x \sim x^{-1}$). If $H < G$ and $x \in H$ is real in $H$, then $x$ is real in $G$, but the converse assertion is not true (for example, a real element $x$ of order $> 2$ is not real in $C_G(x)$). In what follows the expression '$x$ is real' denotes '$x$ is real in $G$'. If $x \in G$ is real, then all elements of the cyclic subgroup $\langle x \rangle$ are also real. Note that involutions are real in any group.

Obviously,

$$(*) \qquad x \in G \text{ of order 4 is real} \iff x \notin Z(N_G(\langle x \rangle)).$$

**Remark 1.** We use $(*)$ to classify the minimal nonabelian 2-groups $G$ containing a real element $x$ of order 4. Write $X = \langle x \rangle$. Then $x^y = x^{-1}$ for some $y \in G$ so that the subgroup $H = \langle x, y \rangle$ is nonabelian, and we conclude that $G = H$, $X = \langle x \rangle \triangleleft G$ and $G/X$ is cyclic so $G$ is metacyclic. Assume that $X < Y < G$, where $Y$ is cyclic. Then $X \le \Phi(Y) \le \Phi(G) = Z(G)$, and so $x$ is not real. Thus, $X$ is a normal maximal cyclic subgroup of $G$ such that $G/X$ is cyclic, and the converse is also holds as the above argument shows. It follows that either $G = M_{2^n}$, $n > 3$, or $G \cong Q_8$, or $G = \langle a, b \mid a^4 = b^{2^n} = 1, \quad a^b = a^{-1} \rangle$ (see Lemma 65.1). A minimal nonabelian 2-group $W$ has no real element of order $> 4$. Indeed, assume that $x \in W$ is real of order $2^n > 2^2$. Then an element $z \in \langle x \rangle$ of order 4 lies in $\Phi(\langle x \rangle) \le \Phi(W) = Z(W)$, so nonreal, a contradiction.

This section is inspired by the following theorem:

**Theorem A.50.1** ([IN2], Theorem B). *Suppose that a group $A$ of odd order acts on a 2-group $S$. If $A$ centralizes all real elements of $S$ of order $\le 4$, then $A$ acts trivially on $S$.*

Note that the proof of Theorem A.50.1 in [IN2] is character-theoretic. It is asked in [IN2] if it is possible to produce character-free proof of that theorem. Our proof of more general Theorem A.50.6 is character-free. Another elementary proof has appeared in [Mar] (I saw it on the day when this note was finished). Thus, three entirely distinct proofs of Theorem A.50.1 are produced!

A group $G$ is said to be *$p$-nilpotent* if it has a normal $p$-complement. All sections of $p$-nilpotent groups are also $p$-nilpotent.

Theorem A.50.1 is a generalization, in the case $p = 2$, of the following:

**Theorem A.50.2** (compare with [Hup1], Satz IV.5.12). *Let $p$ be a prime divisor of the order of a group $G$ and $S \in Syl_p(G)$.*

(a) *If $p > 2$ and every prime power cyclic $p'$-subgroup of $G$ centralizes $\Omega_1(S)$, then $G$ is $p$-nilpotent.*
(b) *If $p = 2$ and every prime power cyclic $2'$-subgroup of $G$ centralizes $\Omega_2(S)$, then $G$ is 2-nilpotent.*

In what follows $p$ and $q$ are distinct primes.

Theorem A.50.2 is an easy consequence of the following three known theorems. Theorem A.50.3, a fundamental result of finite group theory, was proved by Frobenius.

**Theorem A.50.3** ( = [Isa5], Theorem 9.18). *If $G$ is not $p$-nilpotent, then there is a nonnilpotent $H \leq G$ such that $H = QP$, where $H' = P \in \mathrm{Syl}_p(H)$ and $Q \in \mathrm{Syl}_q(G)$ is cyclic.*

**Theorem A.50.4** (O. Schmidt; see Appendix 22.). *Suppose that $G$ is a minimal nonnilpotent group. Then $G = PQ$, where $\langle y\rangle = P \in \mathrm{Syl}_p(G)$, $y^q \in \mathrm{Z}(G)$, $Q = G' \in \mathrm{Syl}_q(G)$.*

(a) *If $Q$ is abelian, then it is a minimal normal subgroup of $G$ and $|Q| = q^b$, where $b = \min\{n > 0 \mid q^n \equiv 1 \pmod p\}$.*

(b) *If $Q$ is nonabelian, it is special (i.e., $\mathrm{Z}(Q) = Q' = \Phi(Q)$ is elementary abelian), $Q \cap \mathrm{Z}(G) = \mathrm{Z}(Q)$. In particular, $Q/\Phi(Q)$ is a minimal normal subgroup of $G/\Phi(Q)$. If $q > 2$, then $\exp(Q) = q$. Next, $\mathrm{N}_Q(P) = \mathrm{Z}(Q)$.*

**Remark 2.** It follows from Theorem A.50.3 that if $G$ is not $p$-nilpotent, it contains a minimal nonnilpotent subgroup $F = QP$ such that $F' = P \in \mathrm{Syl}_p(F)$.

**Lemma A.50.5** (= Lemma 65.1). *Let $G$ be a minimal nonabelian $p$-group. Then*

$$\mathrm{d}(G) = 2\,, \quad \mathrm{Z}(G) = \Phi(G)\,, \quad |G'| = p\,, \quad |\Omega_1(G)| \leq p^3\,.$$

*If $|\Omega_1(G) \neq p^3$, then $G$ is metacyclic.*

*Proof of Theorem A.50.2.* Assume that $G$ is a counterexample of minimal order. Then, by Remark 2, there is in $G$ a non-$p$-nilpotent minimal nonnilpotent subgroup

$$F = Q_1P_1\,, \text{ where } P_1 = F' \in \mathrm{Syl}_p(F)\,, \quad Q_1 = \langle y\rangle \in \mathrm{Syl}_q(F)\,, \quad y^q \in \mathrm{Z}(F)\,.$$

If $p > 2$, then $\exp(P_1) = p$; if $p = 2$, then $\exp(P_1) \leq 4$ with equality provided $P$ is nonabelian (Theorem A.50.4). Therefore, by hypothesis, $Q_1$ centralizes $P_1$ so $F$ must be nilpotent, a contradiction. □

Let $G$ be a group and $p > 2$ divides $|G|$. If, whenever $x \in G$ is a $p$-element, then $\mathrm{C}_G(x) \geq \mathrm{O}^p(G) = \langle a \in G \mid p \notin \pi(o(a))\rangle$, then $G$ is $p$-nilpotent. This follows from Theorem A.50.2.

$2^o$ *Main results.* The main result of this subsection is the following:

**Theorem A.50.6.** *Let $S$ be a Sylow 2-subgroup of a finite group $G > S$. If every prime power cyclic $2'$-subgroup of $G$ centralizes all real elements of $S$ of orders $\leq 4$, then $G$ is 2-nilpotent.*

This is a consequence of the following key result:

**Lemma A.50.7.** *Let $G = Q \cdot S$ be a minimal nonnilpotent group, where $G' = S \in \mathrm{Syl}_2(G)$ is nonabelian and $Q = \langle y\rangle \in \mathrm{Syl}_q(G)$ is cyclic, a prime $q > 2$. Then $S$ contains a real element of order 4.*

*Proof.* Take $a \in S$ of order 4; then $a \in S - \Phi(S)$ since $S$ is special. Set $H = \langle a^{y^i} \mid i = 0, 1, \ldots, q-1\rangle$, where $y$ is a generator of $Q \in \mathrm{Syl}_q(G)$. Then $H$ is a minimal $Q$-invariant subgroup of $S$ containing $a$. Since $S/\Phi(S)$ is abelian, we get $H\Phi(S)/\Phi(S) \triangleleft G/\Phi(S)$. By Theorem A.50.4 (b), $S/\Phi(S)$ is a minimal normal subgroup of $G/\Phi(S)$, and this implies that $H\Phi(S) = S$, and we conclude that $H = S$. Since $S$ is nonabelian, it follows that the elements $a$ and $b = a^{y^i}$ for some $i$ are not permutable. Write $M = \langle a, b\rangle$; then $M$ is nonabelian, $\mathrm{d}(M) = 2$, so that $|M/\Phi(M)| = 4$. Next, $\Phi(M)$, being a subgroup of $\Phi(S) = \mathrm{Z}(S)$, is elementary abelian. Since $a^2 \in \Phi(S) = \mathrm{Z}(S)$, we get $a^2 = b^2$. Let us consider the quotient group $\bar{M} = M/\langle a^2\rangle$.

If involutions $\bar{a}$ and $\bar{b}$ are permutable, then $|M| = 2^3$ so that $M \cong \mathrm{Q}_8$ (indeed, the nonabelian subgroup $M$ of order 8 has two distinct cyclic subgroups $\langle a\rangle$ and $\langle b\rangle$ of order 4). In this case all elements of order 4 are real in $M$ so in $G$. Now assume that involutions $\bar{a}$ and $\bar{b}$ are not permutable. Then $\bar{M} \cong \mathrm{D}_8$ since $\mathrm{cl}(\bar{M}) = 2$, and we conclude that $|M| = 2^4$.

Suppose that $M$ is metacyclic. Since $\exp(M) = \exp(S) = 4$, then $M = \langle u, v \mid u^4 = v^4 = 1,\quad u^v = u^{-1}\rangle \cong \mathcal{H}_{2,2}$; then the element $u$ is real in $M$ so in $G$.

Now suppose that $M$ is nonmetacyclic. Since $M$ is of class 2, we get $|M/M'| = 8$, by Taussky's Proposition 1.6. In that case, $M/M'$ is abelian of type $(4, 2)$ since $\mathrm{d}(M) = 2$, and we conclude that $M$ is minimal nonabelian (Lemma 65.2 (a)). Then $\Omega_1(M)$ of order $2^3$ (Lemma A.50.5) has index 2 in $M$ hence $\Omega_1(M) \nleq \mathrm{Z}(S)$, and so there is an involution $c \in M$ such that $c \notin \mathrm{Z}(S) = \Phi(S)$. Write $F = \langle c^{y^i} \mid i = 0, 1, \ldots, q-1\}$. Then the subgroup $F$ is $Q$-invariant so, as with $H$ in the first paragraph of the proof, one has $F\Phi(S) = S$ and $F = S$. Since $S$ is nonabelian, the involutions $c$ and $c' = c^{y^i}$ for some $i$ are not permutable. It follows that $D = \langle c, c'\rangle$ is dihedral. Since $\exp(S) = 4$, we conclude that $D \cong \mathrm{D}_8$. In that case the element $cc'$ is real of order 4 in $D$ so in $G$. The proof is complete. □

**Exercise 1.** Let $G = Q \cdot P$ be minimal nonnilpotent $\{q, 2\}$-group, $P = G' \in \mathrm{Syl}_2(G)$ is nonabelian. Then $P$ contains a metacyclic minimal nonabelian subgroup of order $\le 16$. If $P$ has no minimal nonabelian subgroup of order 16, it has a subgroup $\cong \mathrm{Q}_8$.

**Solution.** By Lemma A.50.7, $P$ contains a real element $a$ of order 4. Therefore, there is $b \in P$ such that $a^b = a^{-1}$. Since $o(b) \le \exp(P) = 4$, the subgroup $H = \langle a, b\rangle$ is metacyclic of order $\le 16$. If $|H| = 16$, it is not of maximal class, and therefore $|H'| = 2$ (Taussky). It follows that $H$ is minimal nonabelian. The last assertion follows from what has just been proved and Ward's Theorem 56.1 since $P$ has no characteristic subgroup of index 2.

Now we are ready to prove Theorem A.50.6.

*Proof of Theorem A.50.6.* Assume that $G$ is not 2-nilpotent. Then it contains a minimal nonnilpotent subgroup $M = Q \cdot P$, where $M' = P \in \mathrm{Syl}_2(M)$ and $Q \in \mathrm{Syl}_q(M)$ is cyclic (Remark 2). One may assume that $P \le S$, where $S$ is as in the statement of the theorem. Since $Q$ centralizes $\mathrm{Z}(P)$, the subgroup $P$ is nonabelian (Theorem A.50.4), and $P/\mathrm{Z}(P)$

is a minimal normal subgroup of $M/\mathrm{Z}(P)$, and it follows that $\mathrm{C}_P(Q) = \mathrm{Z}(P)$. By Lemma A.50.7, there is in $P$ a real element $a$ of order 4. By hypothesis, $Q$ centralizes $a$. Since $a \in P - \mathrm{Z}(P)$, we get a contradiction. □

We claim that Theorem A.50.6 implies Theorems A.50.1 and A.50.2. To justify this assertion, let us deduce, for example, Theorem A.50.1 from Theorem A.50.6. We use the same notation as in the statement Theorem A.50.1. Let $G = A \cdot S$ be the natural semidirect product with kernel $S \in \mathrm{Syl}_2(G)$. Assume that the action of $A$ on $S$ is nontrivial; then $G$ is not 2-nilpotent. By Remark 2, there is in $G$ a minimal nonnilpotent subgroup $M = Q \cdot P$, there $M' = P \in \mathrm{Syl}_2(M)$ and $Q \in \mathrm{Syl}_q(M)$ for some prime $q > 2$. By the theorem of Schur–Zassenhaus, there is $x \in G$ such that $Q^x \le A$. Set $H = M^x$. Then $H$, by Theorem A.50.6, is 2-nilpotent, so nilpotent, a contradiction.

It is fairly surprising that the authors of [IN2] and [Mar], managed to prove Theorem A.50.1 without using Remark 2. I am sure that it is possible to prove almost all $p$-nilpotence criteria with help of either Remark 2 or Tate's theorem [BZ, Exercise 7.23]). In contrast to Theorem A.50.1, we do not assume in Theorem A.50.6 that $G$ has a $2'$-Hall subgroup and its Sylow 2-subgroup is normal.

**Corollary A.50.8** (compare with [IN2], Theorem C). *Let $S \in \mathrm{Syl}_2(G)$ and suppose that every cyclic $2'$-subgroup centralizes all involutions of $S$. If $S$ has no real elements of order 4, then $G$ is 2-nilpotent.*

*Proof.* Assume that $G$ is not 2-nilpotent. Then $G$ contains a minimal nonnilpotent subgroup $M = Q \cdot P$, where $M' = P \in \mathrm{Syl}_2(M)$ and $Q \in \mathrm{Syl}_q(M)$ is cyclic (Remark 2). One may assume that $P \le S$, where $S \in \mathrm{Syl}_2(G)$. Since $Q$ centralizes $\mathrm{Z}(P)$, by hypothesis, $P$ is nonabelian, and so $\mathrm{N}_P(Q) = \mathrm{Z}(P)$ (Theorem A.50.4 (b)). Then, by Lemma A.50.7, there is in $P$ a real element $a$ of order 4, contrary to the hypothesis. □

**Supplement 1 to Theorem A.50.6.** *Let $S \in \mathrm{Syl}_2(G)$ contain a real element of order 4. If every prime power cyclic $2'$-subgroup of $G$ centralizes all real elements of $S$ of order 4, then $G$ is 2-nilpotent.*[1]

*Proof.* Assume that $G$ is not 2-nilpotent. Then there is in $G$ a minimal nonnilpotent subgroup $H = Q{\cdot}P$, where $H' = P \in \mathrm{Syl}_2(H)$ and $Q \in \mathrm{Syl}_q(G)$ is cyclic. One may assume that $P \le S$. It follows from the proof of Theorem A.50.6 that $P$ has no real elements of order 4 so $P$ is abelian (Lemma A.50.7). By Theorem A.50.4 (a), $\exp(P) = 2$. Let $x \in S$ be real of order 4 and let $y \in S$ be such that $x^y = x^{-1}$. By hypothesis, $x$ centralizes all Sylow $q$-subgroups of $H$. Since $H$ is generated by its Sylow $q$-subgroups, it follows that $x$ centralizes $H$. Write $P_1 = \langle x, P\rangle$. If $a \in P_1$ is of order 4, then $a = xc$ or $x^3c$, where $c \in P$. Then $a^y = a^{-1}$. Thus, all elements of $P_1$ of order 4 are real and so are centralized by $Q$. Note that the abelian subgroup $P_1$ of type $(4, 2, \dots, 2)$ is generated

1 In contrast to Theorem 50.6, we do not assume here that all $2'$-elements of $G$ centralize all involutions of $S$.

by elements of order 4. It follows that $Q$ centralizes $P_1$. In that case, $H = QP$ is abelian, a contradiction. □

**Supplement 2 to Theorem A.50.6.** *Let $S \in Syl_2(G)$. Suppose that all $2'$-subgroups of the group $G$ are permutable with all cyclic subgroups of $S$ generated by real elements of order $\leq 4$. Then $G$ is 2-nilpotent.*

*Proof.* Assume that $G$ is not 2-nilpotent. Then there is in $G$ a minimal nonnilpotent $\{2, q\}$-subgroup $H = Q \cdot P$, where $P = H' \in \mathrm{Syl}_2(H)$ and $Q \in \mathrm{Syl}_q(G)$, $q > 2$ (Remark 2).

(i) Assume that $P$ is abelian; then $\exp(P) = 2$ (Lemma A.50.5 (a)). In that case, $\mathrm{N}_H(Q) = Q$ and $|P| > 2$. Let $L < P$ be of order 2. Then $LQ = QL$ is of order $2|Q|$ so $\mathrm{N}_H(Q) \geq LQ > Q$, a contradiction.

(ii) Thus, $P$ is nonabelian. By Lemma A.50.7, there is in $P$ a cyclic subgroup $X$ generated by a real element of order 4. In that case, $\mathrm{N}_P(Q) = Z(P)$. As $XQ = QX$ and $X \in \mathrm{Syl}_2(XQ)$ is cyclic, we get $X < \mathrm{N}_P(Q)$ (Burnside) so that $\mathrm{N}_P(Q) \not\leq Z(P)$, contrary to what has just been proven. Thus, $H$ does not exist, so $G$ is 2-nilpotent, by Remark 2. □

**Remark 3.** Let $p > 2$ be the least prime divisor of $|G|$ and $S \in \mathrm{Syl}_p(G)$. Suppose that, whenever $X \leq S$ is of order $p$ and $Y < G$ is a $p'$-subgroup, then $XY = YX$. To prove that $G$ is $p$-nilpotent, it suffices to prove that $G$ has no minimal nonnilpotent subgroup $H$ with $H' \in \mathrm{Syl}_p(H)$ (see Supplement 2 to Theorem A.50.6).

**Remark 4.** Let $p > 2$, $G > S$, $S$, $X$ and $Y$ be as in Remark 3. If $G$ is not $p$-nilpotent, there is in $S$ an abelian subgroup $R$ of type $(p, p)$ and a $p'$-subgroup $T$ such that $RT \neq TR$. Indeed, otherwise, there is minimal nonnilpotent subgroup $H = Q \cdot P$, where $P = H' \in \mathrm{Syl}_p(H)$. One may assume that $P \leq S$. If $RQ = QR$ for any abelian subgroup $R$ of type $(p, p)$ in $P$, then $Q \lhd QR$ (Sylow). Since $P$ is generated by such subgroups as $R$, we get $Q \lhd QP$ so $QP$ is nilpotent, a contradiction.

**Remark 5.** Next we improve slightly [G-S, Corollary 6]. Let $n \geq 1$ be fixed and $S \in \mathrm{Syl}_p(G)$. If $p > 2$, we suppose that $\Omega_1(S) \leq Z_n(G)$, the $n$-th member of the upper central series of $G$. If $p = 2$, we suppose that all real elements of orders $\leq 4$ are contained in $Z_n(G)$. Assume that $G$ is not $p$-nilpotent. Then there is in $G$ a minimal nonnilpotent subgroup $M = Q \cdot P$, where $P = M' \in \mathrm{Syl}_p(M)$ and $Q \in \mathrm{Syl}_q(M)$ is cyclic. One may assume that $P \leq S$. If $\exp(P) = p$, then $P = \Omega_1(P)$ so, by hypothesis, $P \leq \Omega_1(S) \leq Z_n(G)$; then $Q$ centralizes $P$, a contradiction. Thus, $p = 2$ and $\exp(P) = 4$. In that case $P$ has a real element, say $x$, of order 4 (Lemma A.50.7). If $x \in Z_n(G)$, then $x$ centralizes $Q$, which is not the case since $C_P(Q) = Z(P)$ does not contains $x$. Thus, $M$ does not exist so $G$ is $p$-nilpotent, by Remark 2.

**Exercise 2.** If $G$ is a metacyclic 2-group that has no nonabelian sections of order 8, then $\exp(\Omega_n(G)) \leq 2^n$.

**Exercise 3.** If all nontrivial characteristic subgroups of a nonabelian $p$-group $G$ have the same order, then there is in $G$ only one nontrivial characteristic subgroup.

**Solution.** By hypothesis, all nontrivial characteristic subgroups of $G$ have index $|G : \Phi(G)| > p$. Assume that $U \neq \Phi(G)$ is a nontrivial characteristic subgroup of $G$. Then $|\Phi(G)| = |U\Phi(G)| < |G| \Rightarrow U\Phi(G) = \Phi(G) \Rightarrow U = \Phi(G)$. (It is easy to prove that $G$ is special.)

**Exercise 4.** Classify the minimal nonabelian $p$-groups $G$ such that $\mathrm{Z}(G) < \mathcal{N}(G)$, where $\mathcal{N}(G)$ is the norm of $G$.

**Exercise 5.** Find the norms of the primary $\mathcal{A}_2$-groups.

**Exercise 6.** Find the norms of the metacyclic $p$-groups.

**Exercise 7.** Let $G$ be a group of even order, $P \in \mathrm{Syl}_2(G)$. Let $\mathcal{E}(P)$ be the set of all elementary abelian and special subgroups of $P$. Suppose that for any $S \in \mathcal{E}(P)$ such that $\mathrm{N}_G(S)$ is not a 2-group (if $S$ does not exist, then $G$ is 2-nilpotent, by Remark 2), one of the following holds:

(a) if $S$ is abelian, then any $2'$-element of prime power order from $\mathrm{N}_G(S)$ centralizes some involution from $S$,

(b) if $S$ is nonabelian, then any $2'$-element of prime power order from $\mathrm{N}_G(S)$ centralizes some element of $S$ of order 4.

Then $G$ is 2-nilpotent.

**Hint.** Assume that $G$ is non-2-nilpotent. Then there is in $G$ a minimal nonnilpotent subgroup $H$ with $S = H' \in \mathrm{Syl}_2(H)$. Mimic the proof of Theorem A.50.6. (In view of Lemma A.50.7, this result generalizes Theorem A.50.6.)

**Old problem.** Study the 2-groups all of whose elements are real. (This will be $\Longleftrightarrow$ all irreducible characters of $G$ are real-valued; see [Dor, Lemma 23.2, Theorem 23.3 (a)].) Moreover, study the 2-groups all of whose elements of order 4 are real.

# Appendix 51 Varia III

This is a continuation of Appendices 40 and 45. In some exercises we offer new proofs of several results which have proved in the previous volumes.

*1°* **Exercise 1.** If derived subgroups of two distinct maximal subgroups of a $p$-group $G$ are cyclic, then $G'$ is abelian.

**Solution.** Let distinct $H, F \in \Gamma_1$ have cyclic derived subgroups. Then $H', F'$, being $G$-invariant cyclic, centralizes $G'$ so $H'F' \leq \mathrm{Z}(G')$. By Exercise 1.69, $|G' : H'F'| \leq p$ hence $G'$ is abelian.

**Exercise 2** (= Remark 10.5). Let $H$ be a proper subgroup of a $p$-group $G$. If $\mathrm{N}_G(H)$ is of maximal class, so is $G$.

**Solution.** If $|H| = p^3$, the result follows from Proposition 10.17. Now let $|H| > p^3$. Clearly, $\mathrm{Z}(G) < H$. By induction, $G/\mathrm{Z}(G)$ is of maximal class, and the result follows since $|\mathrm{Z}(G)| = p$.

*2°*. In this subsection nonnilpotent groups are considered. A group $G$ is said to be *Lagrangian* (in short: an L-group) if for any $d \mid |G|$ it contains a subgroup of index $d$. By Hall–Chunikhin's theorem, L-groups are solvable. All supersolvable groups are L-groups. The group $\mathrm{S}_4$ and its representation groups are L-groups (these groups have no Sylow tower so are nonsupersolvable).

**Proposition A.51.1.** *If all proper subgroups of an L-group G are L-groups, then G has an ordered Sylow tower of the same type as supersolvable groups (such groups are called dispersive).*

*Proof.* Let $G$ be a counterexample of minimal order. By induction, all proper subgroups of $G$ are dispersive but $G$ is nondispersive. Therefore, if $p$ is the minimal prime divisor of $|G|$, then $G$ is not $p$-nilpotent. Then, by Frobenius' normal $p$-complement theorem, there is in $G$ a minimal nonnilpotent subgroup $H = Q \cdot P$, where $P = H' \in \mathrm{Syl}_p(H)$. Since $H$ has no subgroup of index $p$, it is not an L-group, a contradiction. □

In the case of nonnilpotent groups, minimal nonnilpotent subgroups play the same role as minimal nonabelian subgroups in the case of $p$-groups.

*3°* **Proposition A.51.2.** *If a p-group G of order $p^4$ has the derived subgroup of order $p^2$, then G is of maximal class. In particular, if X is of order $p^5$ and $|X : X'| = p^2$, then all epimorphic images of X of order $p^4$ are of maximal class. (**Hint.** As $\mathrm{d}(X) = 2$, we get $\mathrm{Z}(X) \leq \Phi(X) = X'$.)*

**Proposition A.51.3.** *Suppose that a nonabelian two-generator p-group G of order* $> p^3$ *has the cyclic Frattini subgroup. Then G has a cyclic subgroup of index p.*

*Proof.* If $H \in \Gamma_1$, then $|H : \Phi(G)| = p$, so that $H$ is metacyclic. Thus, all maximal subgroups of $G$ are metacyclic. If $G$ is nonmetacyclic, then it is minimal nonmetacyclic, contrary to Theorems 66.1 and 69.4. Thus, $G$ is metacyclic. Then $\Phi(G) = \mho_1(G)$ so that $\mho_1(G) = \langle a^p \rangle$ for some $a \in G$, and hence $|G : \langle a \rangle| = p$. □

**Exercise 3.** Suppose that $G$ is a $p$-group of order $p^{m+2}$ with $|G : G'| = p^2$, $m \geq 1$. If, for all $i = 1, \ldots, m-1$, there is in $G'$ only one $G$-invariant subgroup of index $p^i$, then $G$ is of maximal class. If, in addition, $m > 2$, then $p = 2$.

**Solution.** We proceed by induction on $|G|$. By hypothesis, $\mathrm{Z}(G)$ is cyclic and $\mathrm{d}(G) = 2$ so that $\mathrm{Z}(G) \leq \Phi(G) = G'$. One may assume that $m > 1$. Let $L \leq \mathrm{Z}(G) \cap G'$ be of order $p$. By induction, $G/L$ is of maximal class. It remains to show that $\mathrm{Z}(G) = L$. Assume that this is false. Then $\mathrm{Z}(G) \cong \mathrm{C}_{p^2}$. Assume that $R \lhd G$ is abelian of type $(p, p)$. Then $R\mathrm{Z}(G)/\Omega_1(\mathrm{Z}(G)$ is central subgroup $\cong \mathrm{E}_{p^2}$ so $G/\Omega_1(\mathrm{Z}(G)$ is not of maximal class, a contradiction. Then, by Lemma 1.4, $G' = \Phi(G)$ is cyclic, and, by Proposition A.51.3, $G$ has a cyclic subgroup $C$ of index $p$. By Theorem 1.2, $G$ is a 2-group of maximal class.

**Exercise 4.** Find the norm of primary $\mathcal{A}_1$- and $\mathcal{A}_2$-groups (see § 175).

**Exercise 5.** Let $G$ be a nonabelian $p$-group, $p > 2$, with exactly one nontrivial characteristic subgroup. If $\mathrm{d}(G) = 3$, then $|G| \leq p^6$.

**Solution.** It follows from $\mathrm{cl}(G) = 2$ that $G$ is regular and one has $\Omega_1(G) = \mho_1(G) = \Phi(G) = \mathrm{Z}(G)$ hence $G$ is special and $|G'| = p^k$, where $k \leq \frac{1}{2} \cdot 3(3-1) = 3$, and the result follows.

**Exercise 6.** Classify the minimal nonabelian $p$-groups $G$ that have no subgroups $A$ of order $p^3$ and $B$ of order $p^2$ such that $A \cap B = \{1\}$.

**Exercise 7.** A noncyclic metacyclic $p$-group $G$ is such that the set $G - \Phi(G)$ contains an element of order $p$ $\iff$ it has a cyclic subgroup of index $p$ and $\not\cong \mathrm{Q}_{2^n}$.

**Solution.** If $G$ has no normal abelian subgroups of type $(p, p)$, then result follows from Lemma 1.4 and Theorem 1.2. Now let $G$ have a normal subgroup $R \cong \mathrm{E}_{p^2}$. If $|G| = p^3$, then $G \in \{\mathrm{D}_8, \mathrm{M}_{p^3}\}$. Now let $|G| > p^3$; then $R = \Omega_1(G) \cong \mathrm{E}_{p^2}$. In that case, $\Phi(G)$ contains only one subgroup of order $p$, so is cyclic (see Proposition 10.17). As $\mho_1(G) = \Phi(G)$, then $G$ has a cyclic subgroup of index $p$, and the result follows from Theorem 1.2.

**Exercise 8** (Janko). Let $G$ be a non-Dedekindian $p$-group such that the normal closures of all its nonnormal cyclic subgroups are minimal nonabelian. If $p > 2$ and $A$ is a maximal normal abelian subgroup of $G$, then $G/A$ is cyclic.

**Solution.** Let $A$ be a maximal normal abelian subgroup of $G$ and let $p > 2$. Since the normal closure in $G$ of each cyclic subgroup of $A$ is abelian, it follows that each cyclic subgroup of $A$ is $G$-invariant. But then each subgroup of $A$ is $G$-invariant. Let $B/A$ be a normal subgroup of order $p$ in $G/A$. By Theorem 125.1 applied on the subgroup $B$, $A$ is of exponent $p^e$, $e \geq 2$, and there is $b \in B - A$ such that $a^b = a^{1+p^{e-1}}$ for all $a \in A$. Suppose that $G/A$ is noncyclic. Then there is a subgroup $C/A$ of order $p$ in $G/A$, which is distinct from $B/A$. Applying Theorem 125.1 on the subgroup $C$, we see that there is $c \in C - A$ such that $a^c = a^{1+p^{e-1}}$ for all $a \in A$. Note that $BC/A \cong \mathrm{E}_{p^2}$ (recall that $B \lhd G$). In that case, $bc^{-1} \in G - A$ centralizes $A$, a contradiction.

**Exercise 9.** Let $G$ be a regular $p$-group of exponent $p^w$ such that all the indices of the series $G > \mho_1(G) > \cdots > \mho_{w-1}(G) > \{1\}$ are equal to $p^d$; then $|G| = p^{dw}$. Find the number of cyclic subgroups $C$ in $G$ of order $p^w$ with the same $\Omega_1(C)$.

**Exercise 10.** Is it true that there does not exist a $p$-group $G = A \times B$ such that $A < G$, $B < G$ and $A$ is characteristic in $G$?

**Solution** (Mann). This is true. Let $Z \leq \mathrm{Z}(B)$ be of order $p$ and let $\phi \in \mathrm{Hom}(A, Z)$ be nontrivial. Then there is $\alpha \in \mathrm{Aut}(G)$ such that $\alpha(a) = a\phi(a) \notin A$ for all $a \in A^\#$ and $\phi(b) = b$ for all $b \in B$.

**Exercise 11.** If a metacyclic $p$-group is not generated by elements of the same order, then $G/R(G) \cong \mathrm{SD}_{2^n}$ (see § 124).

Recall that $\Omega_n^*(G) = \langle x \in G \mid o(x) = p^n \rangle$.

**Exercise 12.** A $p$-group $G$ of maximal class is not generated by elements of the same order $\iff \Omega_1(G) < G$ and $\Omega_2^*(G) < G$. (**Hint.** See the solution of Exercise 13.)

**Exercise 13** (Mann). If a $p$-group $G$ of maximal class is not generated by elements of the same order, then $p = 2$ and $G \cong \mathrm{SD}_{2^n}$.

**Solution.** By Theorem 13.19 (a), one has $G = G_1 \cup \Omega_1(G) \cup \Omega_2^*(G)$. If $p = 2$, then $G \cong \mathrm{SD}_{2^n}$ (Exercise 11). If $p > 2$, then either $G = \Omega_1(G)$ or $G = \Omega_2^*(G)$ (Lemma 116.3 (a)).

**Exercise 14.** Let $G$ be a $p$-group and $H \in \Gamma_1$. Suppose that, whenever $x \in G - H$, then $o(x) < p^p$. Then $\Omega_i^*(G) = G$ for some $i \in \{1, \ldots p-1\}$.

**Solution.** One has $H \cup \Omega_1(G) \cup \Omega_2^*(G) \cdots \cup \Omega_{p-1}^*(G) = G$, and therefore one of the summands coincides with $G$ (see Lemma 116.3 (a)). But $H < G$.

**Exercise 15.** Let $\mathcal{M}$ be the set of those $p$-groups of exponent $p^{p+1}$ that are not generated by elements of the same order. Let $G \in \mathcal{M}$. Setting $H_i = \Omega_i^*(G)$, $i = 1, \ldots, p+1$, one obtains $G = \bigcup_{i=1}^{p+1} H_i$ and the subgroup $D = \bigcap_{i=1}^{p+1} H_i$ has index $p^2$ in $G$, $H_i \in \Gamma_1$ for all $i$; moreover, $H_i \cap H_j = D$ (see Lemma 116.3 (b)). Therefore, all elements of the set $H_i - D$ have the same order $p^i$, $i = 1, 2, \ldots, p+1$. If $G$ is a minimal element of the set $\mathcal{M}$, then $\mathrm{d}(G) = 2$.

**Solution.** Let $G \in \mathcal{M}$. Set $H_i = \Omega_i^*(G)$, $i = 1, \ldots, p+1$. Then $G = H_1 \cup \cdots \cup H_{p+1}$ so that, by Lemma 116.3 (b), $H_i \in \Gamma_1$ for $i = 1, \ldots, p+1$ and, if $D = \bigcup_{i=1}^{p+1} H_i$, then $|G : D| = p^2$.

Assume that there is $x \in H_j - D$ of order $p^i$ for some $i \neq j$. Then $x \in H_i \cap H_j = D$, a contradiction. It follows that all elements of the set $H_j - D$ have the same order $p^j$. In particular, if $p = 2$, then consideration of $H_1$ shows that the subgroup $D$ is abelian (Burnside). Now assume that $G$ is a minimal element of the set $\mathcal{M}$ and $\mathrm{d}(G) > 2$. Let $U \in \Gamma_1 - \{G_1, \ldots, G_{p+1}\}$. Then $U$ is not generated by elements of the same order (if, for example, $U = \Omega_k^*(U)$, then $G = \Omega_k^*(G)$, which is impossible) so that $U \in \mathcal{M}$. But $|U| < |G|$, contrary to the choice of $G$. Thus, $\mathrm{d}(G) = 2$.

**Exercise 16.** Study the nonabelian metacyclic 2-groups $G$ without real elements of order 4.

**Exercise 17.** Let $G$ be a noncyclic metacyclic $p$-group, $p > 2$. Then $G = AB$, where $A, B < G$ are cyclic and $A \cap B = \{1\}$.

**Solution.** The quotient group $\bar{G} = G/R(G)$ is cyclic (see § 124). If $G = R(G)$, the assertion is obvious. Therefore, one may assume that $\bar{G} > \{\bar{1}\}$. If a cyclic $C < G$ is such that $CR(G) = G$, then $|C| = \exp(G)$. Set $A = C \cap R(G)$. Then $R(G) = AB$, where $B$ is cyclic and $A \cap B = \{1\}$. It follows that $G = CB$ with $C \cap B = \{1\}$.

**Exercise 18.** Let $p > 2$ be the least prime divisor of the order of a group $G$, let $P \in \mathrm{Syl}_p(G)$ be metacyclic. Then $G$ is $p$-nilpotent.

**Hint.** Assuming that $G$ is a minimal counterexample, we see that $G = Q \cdot P$ is minimal nonnilpotent, where $P = G' \in \mathrm{Syl}_p(G)$ is $\cong \mathrm{E}_{p^2}$ and $Q \in \mathrm{Syl}_q(G)$ is cyclic, $q > p$ is a prime, $p^2 \equiv 1 \pmod{q}$, which is impossible.

**Exercise 19.** Every irregular $p$-group, $p > 3$, contains a maximal subgroup that is not two-generator. (**Hint.** Use Theorem 9.8 (a).)

**Exercise 20.** If $G$ is a nonabelian group, then $D = \bigcap_{\chi \in \mathrm{Irr}_1(G)} \ker(\chi) = \{1\}$.

**Solution.** Assume that $D > \{1\}$. If $\chi \in \mathrm{Irr}_1(G)$ and $\lambda \in \mathrm{Lin}(G)$, then $\lambda\chi \in \mathrm{Irr}_1(G)$. If $x \in D$, then

$$\lambda(x)\chi(1) = \lambda(x)\chi(x) = (\lambda\chi)(x) = (\lambda\chi)(1) = \lambda(1)\chi(1)$$

so that $\lambda(x) = 1$. Thus, $x \in \bigcap_{\chi \in \mathrm{Irr}(G)} \ker(\chi) = \ker(\rho_G) = \{1\}$ (here $\rho_G$ is the regular character of $G$), and hence $D = \{1\}$.

**Exercise 21.** Study the capable (see § 21) extraspecial $p$-groups.

**Exercise 22.** Any modular $p$-group is powerful (see § 26).

**Solution.** Let $G$ be a modular $p$-group. If $p > 2$, then $G/\mho_1(G)$ is modular of exponent $p$ so abelian; in that case, $G$ is powerful. Now let $p = 2$. One may assume that $\exp(G) > 2$. Write $\bar{G} = G/\mho_2(G)$. Then $\bar{G}$ is modular of exponent 4. As we know (see Proposition 73.5), $\bar{G}$ is abelian, and hence $G$ is powerful.

**Exercise 23.** Classify the minimal nonabelian 2-groups $G$ containing a real element of order 4.

**Solution.** Let $x \in G$ be real of order 4. There is $y \in G$ such that $x^y = x^{-1}$. Then the nonabelian subgroup $\langle x, y\rangle = G$ so $G$ is metacyclic. Let $G = \langle u, v \mid o(u) = 2^a,\ o(v) = 2^b,\ u^v = u^{1+2^{a-1}}\rangle$. If $a, b > 2$, then $G$ has no real element of order 4 since $\Omega_2(G) \leq Z(G)$. If $a = 2$, then $u$ is real of order 4. Now let $a > 2$. Then $b \leq 2$. If $b = 1$, then any generator of the noncentral cyclic subgroup of order 4 is real. The case $b = 2$ we leave to readers.

**Exercise 24.** Suppose that $G$ is an extraspecial 2-group. If $\Omega_1(G) < G$, then $G \cong Q_8$.

**Solution.** Suppose that $|G| > 8$. Then either

$$G \cong G^1 = D_8 * \cdots * D_8 \text{ or } G \cong G^2 = Q_8 * D_8 * \cdots * D_8\,.$$

Clearly, $\Omega_1(G^1) = G^1$. To prove that $\Omega_1(G^2) = G^2$, it suffices to prove that $\Omega_1(D_8 * Q_8) = D_8 * Q_8$. Let $U, V < Q_8$ be a distinct cyclic of order 4. Then $\Omega_1(D_8 * U) = D_8 * U$ and $\Omega_1(D_8 * V) = D_8 * V$. Now $(D_8 * U)(D_8 * V) = D_8 * Q_8$.

$4^o$. If $G$ is an extraspecial 2-group, then $|G : \Omega_2^*(G)| \leq 2$ with equality only if $G \cong D_8$ (check!).

**Exercise 25.** Classify the nonabelian $p$-groups $G$ having exactly $p+1$ pairwise distinct principal series. Is it true that $G$ is of maximal class? (**Answer.** Yes. Use Lemma 9.1.)

**Exercise 26.** If $G$ is a nonabelian Dedekindian group, then $G = Q \times E \times A$, where $Q \cong Q_8$, $\exp(E) \leq 2$ and $A$ is abelian of odd order.

**Solution.** As $G$ is nilpotent, it suffices to assume that $G$ is a $p$-group. Any Dedekindian minimal nonabelian $p$-group is isomorphic to $Q_8$. Therefore, by Corollary A.17.3, $G = R \times E$, where $R \cong Q_{2^n}$ end $\exp(E) \leq 2$. If $n > 3$, then $Q_{2^n}$ contains a nonnormal subgroup of order 4. Therefore, $n = 3$.

**Exercise 27.** If $G$ is a two-generator nonabelian group of exponent 3, then $|G| = 3^3$.

**Hint.** Assume that $|G| > 3^3$. Let $K \triangleleft G$ be of index $3^4$. To obtain a contradiction, it suffices to assume that $K = \{1\}$; then $|G| = 3^4$. Since $G$ is not minimal nonabelian (Lemma 65.1), one has $|Z(G)| = 3$. Since $G/Z(G)$ is nonabelian, then $\mathrm{cl}(G) = 3$, i.e., $G$ is of maximal class. However, this contradicts Theorem 9.5.

**Exercise 28.** If $G$ is a minimal nonabelian $p$-group all of whose nonnormal subgroups are cyclic, then either $|G| = p^3$ or $G$ is metacyclic.

**Solution.** One may assume that $|G| > p^3$ and $G$ is nonmetacyclic (otherwise, $G$ satisfies the condition); then $\Omega_1(G) \cong E_{p^3}$ (Lemma 65.1). Let

$$G = \langle a, b \mid o(a) = p^m,\ o(b) = p^n,\ [a, b] = d,\ d^p = [a, d] = [b, d] = 1\rangle\,.$$

If one of the numbers $m, n$ is $> 1$, the one of subgroups $\langle a, b^{p^{n-1}}\rangle$, $\langle a^{p^{m-1}}, b\rangle$ is nonnormal and noncyclic. Thus, $m = n = 1$; then $|G| = p^3$.

**Exercise 29** ([Hup1] Satz IV.2.7). Let $p$ be the minimal prime divisor of the order of a group $G$, $P \in \mathrm{Syl}_p(G)$ is abelian of type $(p^{e_1}, \dots, p^{e_k})$, where $e_1 < \cdots < e_k$. Then $G$ is $p$-nilpotent.

**Hint.** Let $G$ be a counterexample of minimal order. Then, whenever $P \le T < G$, then $T$ is $p$-nilpotent, by induction. Set $N = \mathrm{N}_G(P)$. If $N < G$, then $P \le \mathrm{Z}(\mathrm{N}_G(P))$ so $G$ is $p$-nilpotent (Burnside's Theorem A.52.1). Now let $N = G$; then $P \lhd G$. Since $\mathrm{Aut}(P)$ has no elements of prime order $q > p$ (Theorem 6.9), it follows that $P \le \mathrm{Z}(G)$ so $G$ is $p$-nilpotent (Theorem A.52.1 again).

**Exercise 30.** Classify the noncyclic $p$-groups $G$ of order $> p^2$ satisfying the following condition: Whenever $N > \{1\}$ is a proper normal subgroup of $G$, then it has only one $G$-invariant subgroup of index $p$.

**Solution.** Let us prove that $G$ is of maximal class. We proceed by induction on $|G|$. Clearly, $\mathrm{Z}(G)$ is cyclic. One may assume that $|G| > p^3$. Assume that $|\mathrm{Z}(G)| > p$. Let $L \le \mathrm{Z}(G)$ be of order $p^2$ and let $R \lhd G$ be abelian of type $(p, p)$ (Lemma 1.4). By induction, $G/\Omega_1(L)$ is of maximal class. However, $G/\Omega_1(L)$ has the central subgroup $RL/\Omega_1(L)$ of type $(p, p)$, a contradiction. Thus, $|\mathrm{Z}(G)| = p$ and $G/\mathrm{Z}(G)$ is of maximal class, so is $G$.

**Exercise 31.** Let $G \cong \mathcal{H}_{2,2}$ and let $\phi : G \to G_0$ be a lattice isomorphism. Prove that $G_0 \cong G$.

**Hint.** Clearly, $\mathrm{d}(G_0) = \mathrm{d}(G) = 2$. Let us prove that $G_0$ is nonabelian. Assume that this is false. Then all maximal cyclic subgroups of $G_0$ have order 4. However, $G$ has a maximal cyclic subgroup of order 2, and this is a contradiction. Prove that if $G_0$ is an $\mathcal{A}_1$-group, it is metacyclic, and then $G_0 \cong G$. If $G_0$ is not an $\mathcal{A}_1$-group, it has a proper $\mathcal{A}_1$-subgroup and then, by Theorem 12.12 (a), $\mathrm{d}(G_0) = 3 > 2 = \mathrm{d}(G)$, a final contradiction.

**Exercise 32.** Let $G$ be a nonabelian metacyclic $p$-group of order $p^4$ and exponent $p^2$, $p > 2$. Is it true that $G$ is lattice isomorphic to an abelian group of type $(p^2, p^2)$?

**Exercise 33.** Let $G$ be a nonmetacyclic minimal nonabelian group of order 16 and let $\phi : G \to G_0$ be a lattice isomorphism. Prove that $G_0 \cong G$.

**Solution.** One has $\Gamma_1 = \{E, F, H\}$, where $E \cong \mathrm{E}_8$ and $F, H$ are abelian of type $(4, 2)$. Then $E^\phi \cong E$, $F^\phi \cong H^\phi \cong H$. As $\mathrm{d}(G) = 2$ and $\mathrm{d}(E) = 3$, then $G_0$ is minimal nonabelian so that $G_0 \cong G$.

**Exercise 34.** Classify the $p$-groups lattice isomorphic to the nonmetacyclic minimal nonabelian group of order $p^4$. $p > 2$.

**Exercise 35.** Classify the $\mathrm{D}_8$-free minimal nonabelian 2-groups.

**Solution.** A metacyclic group $G = \langle a, b \mid o(a) = 2^m, \quad o(b) = 2^n, \quad a^b = a^{1+2^{m-1}}\rangle$ is $D_8$-free $\iff$ $m > 2$. If a nonmetacyclic group $G = \langle a, b \mid o(a) = 2^m, \quad o(b) = 2^n, \quad c = [a, b], \quad c^2 = 1, \quad [a, c] = [b, c] = 1\rangle$, then $G/\langle a^2, b^2\rangle \cong D_8$. Thus, $G$ is metacyclic.

**Exercise 36.** Classify the $S(p^3)$-free minimal nonabelian $p$-groups, $p > 2$.

*5°.* If $G > \{1\}$ is a $p$-group, then the inequality $\exp(G/\Omega_1(G)) < \exp(G)$ is true. If $G > \{1\}$ is a regular $p$-group of exponent $> p$, then $p \cdot \exp(\mho_1(G)) = \exp((G)$.

**Proposition A.51.4.** *If* $G > \{1\}$ *is a metacyclic p-group, then* $\exp(\Phi(G)) < \exp(G)$.

*Proof.* (See also Lemma 148.40.) If $R(G) = \{1\}$, then $G$ is either cyclic or a 2-group of maximal class, and the result is obvious. Let $R(G) > \{1\}$. Then $G/R(G)$ is either cyclic or a 2-group of maximal class. Let $\exp(G) = p^e$. In the first case, $\Omega_{e-1}(G) \in \Gamma_1$ has exponent $p^{e-1}$, and the result follows since $\Phi(G) < \Omega_{e-1}(G)$. If $G/R(G)$ is a 2-group of maximal class and $\Phi/R(G) = \Phi(G/R(G))$, then $\Phi = \Phi(G)$ and $\exp(\Phi) < \exp(G)$. □

**Exercise 37.** Classify the $p$-groups $G$ whose Frattini subgroup is maximal cyclic in $G$.

**Exercise 38.** Modular $p$-groups $G$, $p > 2$, are regular.

**Solution.** If $x, y \in G$, then $\langle x, y\rangle$ is metacyclic (see remark following Proposition 73.2), so regular. It follows that $G$ is regular.

**Exercise 39.** Study the 2-groups all of whose nonnormal subgroups are either cyclic or of maximal class.

**Hint.** Assume that $G$ has a normal subgroup $R$ of type $(2, 2)$. Let $A/R < G/R$ be nonnormal. Then $A \cong D_8$ so $|A/R| = 2$. Thus, $G/R$ is either Dedekindian or all of its nonnormal subgroups have order 2. See Theorem 1.25.

**Exercise 40.** If a non-Dedekindian 2-group $G$ is an extension of a group $L$ of order 2 by a nonabelian Dedekindian 2-group, then $G' \le Z(G)$.

**Solution.** One may assume that $|G'| > 2$; then $L < G'$ hence $|G'| = 4$ so that $G$ is non-Dedekindian (Theorem 1.20). Then, by Passman's Theorem 1.23, there is a $G$-invariant $R < G'$ of order 2 such that $G/R$ is non-Dedekindian. It follows that $R \ne L$, and hence $G' = L \times R \le Z(G)$.

**Exercise 41.** Let $G$ be an abelian group of type $(p^{c_1}, \dots, p^{c_n})$, $c_1 > c_i$ for $i > 1$. Then there is no $p$-group $W$ such that $W/Z(W) \cong G$.

**Exercise 42.** Let $G$ be a nonabelian Dedekindian 2-group. Show that there is no a 2-group $W$ such that $W/Z(W) \cong G$.

**Exercise 43.** Let $G$ be a nonabelian 2-group such that $|\Phi(G)| = 2$ and $\Omega_2^*(G) = G$. Show that there is no a 2-group $W$ such that $W/Z(W) \cong G$.

**Exercise 44.** If $G$ is a nonabelian $p$-group such that all subgroups of minimal nonabelian subgroups of $G$ are $G$-invariant, then $G$ is Dedekindian (use Corollary A.17.3).

**Exercise 45.** Classify the $p$-groups $G$ that are $\mathcal{A}_4$-groups and such that $|G'| = p$.

**Exercise 46.** Present a classification of the minimal nonmetacyclic $p$-groups with $|G'| = p$ independent of §§ 66 and 69.

**Exercise 47** (Reported by B. M. Schein). Let $\delta(G)$ be the minimal degree of a representation of a group $G$ by permutations. Then $\delta(G) = |G|$ if and only if one of the following holds:

(a) $G$ is cyclic,

(b) $G \cong \mathrm{E}_4$,

(c) $G$ is a generalized quaternion group.

**Exercise 48.** (a) Suppose that a $p$-group $G$ has only one normal subgroup, say $N$, of order $p^2$ and $N \nleq \mathrm{Z}(G)$. If $H \in \Gamma_1 - \{\mathrm{C}_G(N)\}$, then $\mathrm{Z}(H)$ is of order $p$.

(b) Suppose that a $p$-group $G$ has only one noncyclic normal subgroup, say $N$, of order $p^2$ and $N \nleq \mathrm{Z}(G)$. If $H \in \Gamma_1 - \{\mathrm{C}_G(N)\}$, then $\mathrm{Z}(H)$ is cyclic. (**Hint.** Use Theorem 1.17 (b))

**Solution.** (a) Set $F = \mathrm{C}_G(N)$; then $F \in \Gamma_1$. Let $H \in \Gamma_1 - \{F\}$ and assume that $\mathrm{Z}(H)$ is of order $> p$. Then there is in $\mathrm{Z}(H)$ a $G$-invariant subgroup $R$ of order $p^2$. By hypothesis, $R = N$; then $\mathrm{C}_G(N) \geq FH = G$, a contradiction.

**Exercise 49.** Classify the nonabelian metacyclic $p$-groups $G$ all of whose minimal nonabelian subgroups are q-self dual.

**Exercise 50.** Classify the modular minimal nonabelian $p$-groups. Prove that the group $G = \langle a, b \mid o(a) = o(b) = p^m > p^2, \quad a^b = a^{1+p^{m-1}} \rangle$ is modular.

**Exercise 51** (Janko). If a group $G$ of exponent $p^e$ contains only one maximal subgroup, say $H$, of exponent $p^e$, then $\mathrm{d}(G) = 2$.

**Solution.** Let $h \in H$ be of order $p^e$ and $g \in G - H$. Write $F = \langle h, g \rangle$ and assume that $F < G$. Let $F \leq H_1 \in \Gamma_1$. As $H_1 \neq H$ and $\exp(H_1) = p^e$, we get a contradiction, Thus, $F = G$ so that $\mathrm{d}(G) = 2$.

**Exercise 52.** Is it true that the nonabelian metacyclic group of order 16 and exponent 4 is capable? The same question for the nonabelian metacyclic group of order $p^4$ and exponent $p^2$, $p > 2$.

**Exercise 53.** Is the following assertion true?: A $p$-group $G$ is capable $\iff$ $\mathrm{Z}(\Gamma) \cong \mathrm{M}(G)$, where $\Gamma$ is a representation group of $G$ and $\mathrm{M}(G)$ is the Schur multiplier of $G$.

**Exercise 54.** Classify the metacyclic $p$-groups admitting a nontrivial partition.

**Solution.** Let $G$ be a metacyclic $p$-group admitting a nontrivial partition. It is easy to check that if $G$ is abelian, then $G \cong \mathrm{E}_{p^2}$. Now let $G$ be nonabelian. By Theorem 164.2, $\mathrm{H}_p(G) < G$. It follows that $\Omega_1(G) = G$ hence $G$ is a 2-group of maximal class (Proposition 10.19) so that $G$ is dihedral (Theorem 1.2).

**Second solution of Exercise 54.** There is $C \lhd G$ such that $G/C$ is cyclic; then $C$ is maximal cyclic in $G$. Assume that $G$ is nonabelian; then $|C| > p$. Let $U/C \leq G/C$ be of order $p$. Then $U$ admits a cyclic partition so that $U$ is dihedral. By Exercise 10.10, $G$ is of maximal class, so dihedral.

**Exercise 55.** Describe a nonmetacyclic representation group of the abelian group $A$ of type $(p^2, p^2)$. (**Hint.** By [BZ, Theorem 6.6.4], $\mathrm{M}(A) \cong \mathrm{C}_{p^2}$.)

**Exercise 56.** Describe a nonmetacyclic representation group of the abelian group of type $(p^m, p^n)$.

**Exercise 57.** Let $G$ be a minimal nonabelian $p$-group of order $> p^3$ and exponent $p^e (> p)$. Then $\mathrm{H}_p(G) = G$ (here $\mathrm{H}_p(G) = \langle x \in G \mid o(x) > p \rangle$).

**Solution.** By Lemma 65.1, $\Omega_1(G) < G$. Therefore $\mathrm{H}_p(G) = \langle G - \Omega_1(G) \rangle = G$. Moreover, if $\exp(G) = p^e > p$ for our $G$, then $\Omega_e^*(G) = G$ since $\Omega_{e-1}(G) < G$.

**Exercise 58** (D. Taunt). Classify the nonnilpotent solvable groups having only one nontrivial characteristic subgroup.

**Hint.** The socle $S$ of $G$ is characteristic so $S = \mathrm{F}(G)$, the Fitting subgroup of $G$. In that case $S$ is an elementary abelian $p$-group. It is easy to see that all minimal normal $p$-subgroups of $G$ have the same order.

**Exercise 59.** Let $\Gamma$ be a representation group of a finite group $G$. Then there is $M \leq \Gamma' \cap \mathrm{Z}(\Gamma)$ such that $\Gamma/M \cong G$. Prove that $M \leq \Phi(\Gamma)$. (**Hint.** If $X$ is a group, then $X' \cap \mathrm{Z}(X) \leq \Phi(X)$.)

**Exercise 60.** (a) Construct the character table of the group $\mathcal{H}_{2,2} = \langle a, b \mid a^4 = b^4 = 1, \ a^b = a^3 \rangle$.
(b) Find all $G$ satisfying $\mathrm{X}(G) = \mathrm{X}(\mathcal{H}_{2,2})$, where $\mathrm{X}(G)$ is the character table of $G$.
(c) Let $p > 2$ and $H$ be a nonabelian metacyclic group of order $p^4$ and exponent $p^2$. Find all groups $G$ such that $\mathrm{X}(G) = \mathrm{X}(H)$.

**Exercise 61.** Let $G$ be a $p$-group of maximal class. Describe the groups $H$ such that $\mathrm{X}(H) = \mathrm{X}(G)$.

**Exercise 62.** Classify the $p$-groups $G$ such that $\mathrm{X}(G) = \mathrm{X}(\mathrm{M}_{2^n})$. Moreover, study the $p$-groups $G$ having the same character table as a minimal nonabelian $p$-group.

**Exercise 63.** Let $p$ be the least prime divisor of the order of a group $G$, $\mathrm{M}_{p^n} \in \mathrm{Syl}_p(G)$. Then $G$ is $p$-nilpotent. (**Hint.** See Appendix 52.)

**Exercise 64.** Is it true that if $X(G) = X(M_{2^4})$, then $G \cong M_{2^4}$?

**Exercise 65.** Let $G$ be a group of maximal class and order $2^n$, $n > 3$. Is it true that if $X(G) = X(H)$, then $H \cong G$?

**Exercise 66.** Is it true that an abelian $p$-group of rank 2 is capable $\iff$ it is homocyclic?

**Exercise 67.** Deduce Burnside's normal $p$-complement theorem (see Theorem A.52.1) from Frobenius' normal $p$-complement theorem (see Theorem A.52.2).

**Solution** (Mann). Let $P \in \mathrm{Syl}_p(G)$ be such that $P \le Z(N_G(P))$. We proceed by induction on $|G|$. Then all proper subgroups of $G$ containing $P$ are $p$-nilpotent. If $N_G(S) < G$ for all nonidentity $S \le P$, then $G$ is $p$-nilpotent, by Theorem A.52.2. Thus, there is in $G$ a nonidentity $G$-invariant $p$-subgroup $S$; then $S \le P$, by hypothesis $S < P$. Write $C = C_G(S)$; then $C$ is normal in $G$. If $C < G$, then it has a normal $p$-complement $C_1$, and $C_1 \lhd G$. Set $\bar{G} = G/C_1$. The group $\bar{G}$ satisfied the hypothesis;[1] it has a normal $p$-complement $\bar{H}$. In that case, $H$ is a normal $p$-complement of $G$. Now let $C = G$. Then again, by induction, $\bar{G}$ has a normal $p$-complement $H$. By Frobenius' Theorem A.52.2, $H = S \times F$. Since $F$ is characteristic in $H \lhd G$, one has $F \lhd G$, so $F$ is a normal $p$-complement of $G$.

**Exercise 68** ([Isa19], Lemma 10.3). Let $G$ be a $p$-group and let $A \in \Gamma_1$ be elementary abelian. If $A = \langle x\rangle^G$ for some $x \in A$, then $G$ is of maximal class.

**Solution.** It suffices to show that $|G : G'| = p^2$ (then $G$ must be of maximal class). Assume that $|G : G'| > p^2$. Then $|G : G'\langle x\rangle^G| > p = |G : A|$, a contradiction.

**Exercise 69.** Let $G$ be a $p$-group and let $A \in \Gamma_1$ be abelian. Is it true that if $A = \langle x\rangle^G$ for some $x \in A$ such that $x^p \in G'$, then $G$ is of maximal class?

$6^o$. The following result due to T. Yoshida (see [Isa19, Theorem 10.1]) is important:

**Theorem A.51.5.** *Let $P \in Syl_p(G)$ have no epimorphic image isomorphic to $\Sigma_{p^2} = C_p \operatorname{wr} C_p$ and let $N = N_G(P)$. Then Sylow $p$-subgroups of the quotient groups $G/G'$ and $N/N'$ are isomorphic.*[2]

**Exercise 70.** Let $P = Q \times A$, where $Q \cong Q_8$ and $A$ is abelian of exponent $> 2$. Then $P$ has a section $\cong D_8$.

**Solution.** Let $C \le A$ be of order 4 and $H = Q \times C$. Let $L < H$ be of order 2 such that $Z(Q) \ne L \ne \Omega_1(C)$. Set $\bar{H} = H/L$. Then $\bar{H} = \bar{Q} * \bar{C}$ has a subgroup $\cong D_8$ (see Appendix 16).

1 See paragraph following Theorem 52.3; see also the proof of Theorem 52.3.
2 In particular, $N_G(P)$ is $p$-nilpotent $\iff$ $G$ is.

**Exercise 71.** Is the following assertion true?: Let $P \in \mathrm{Syl}_p(G)$ be modular. If $\mathrm{N}_G(P)$ is $p$-nilpotent, so is $G$. (**Hint.** Use Theorem A.51.5 and the nonmodularity of $\mathrm{D}_8$ and $\mathrm{S}(p^3)$.)

**Exercise 72.** Let $P = M \times A \in \mathrm{Syl}_p(G)$, where $M$ is modular and $A$ is elementary abelian. Suppose that $\mathrm{N}_G(P)$ is $p$-nilpotent. Is it true that $G$ is $p$-nilpotent?

**Exercise 73.** If $G$ is a $p$-group of order $> p^p$ such that $|\mathrm{H}_p(G)| = p^p$, then $G$ is of maximal class and order $p^{p+1}$.

**Solution.** Let $\mathrm{H}_p(G) < F$, where $|F : \mathrm{H}_p(G)| = p$. Then $\mathrm{H}_p(F) = \mathrm{H}_p(G) < F$ so that $F$ is irregular (Theorem 7.2 (b)). By Theorem 7.1 (b), $F$ is of maximal class. It follows that $G$ is of maximal class (Exercise 10.10). By Theorem 9.6 (c), $|G| = p^{p+1}$.

**Exercise 74.** If any epimorphic image of a $p$-group $G$ has at most $1 + p$ subgroups of order $p$, then $G$ is metacyclic.

**Solution.** One may assume that $G' > \{1\}$. By hypothesis, $\mathrm{d}(G) = 2$. Let $L < G'$ be $G$-invariant of index $p$. Then $G/L$ is minimal nonabelian (Lemma 65.2 (a)). By Lemma 65.1, $G/L$ is metacyclic, and so $G$ is metacyclic (Theorem 36.1).

**Exercise 75.** Let $G$ be a $p$-group of maximal class and order $> p^{p+1}$, $p > 3$. Prove that $\mathrm{d}(G_1) > 2$, where $G_1$ is the fundamental subgroup of $G$. (**Hint.** Use Theorems 5.8 (b) and 9.8 (a).)

**Exercise 76.** Find $\alpha_1(M \times \mathrm{C}_{p^n})$, where $M$ is a minimal nonabelian $p$-group.

**Exercise 77** (See § 208). Suppose that $G$ is a non-Dedekindian $p$-group. If $\mathrm{Z}(G)$ is contained in any nonnormal maximal cyclic subgroup of $G$, then $G$ is a generalized quaternion group of order $> 8$.

**Solution.** By hypothesis, $\mathrm{Z}(G)$ is cyclic. Assume that $G \not\cong \mathrm{Q}_{2^n}$. Then $G$ contains two distinct subgroups $U, V$ of order $p$. Let $U \le S < G$ and $V \le T < G$, where $S, T$ are maximal cyclic subgroups. If $S \triangleleft G$, then $\mathrm{Z}(G) \cap S > \{1\}$ so that $U \le \mathrm{Z}(G)$. If $S$ is not normal in $G$, then $\mathrm{Z}(G) < S$, by hypothesis, so that $U \le \mathrm{Z}(G)$. Similarly, $V \le \mathrm{Z}(G)$. In that case $\mathrm{Z}(G)$ is noncyclic since $U \times V \le \mathrm{Z}(G)$. Thus, $G$ contains only one subgroup of order $p$ so that $G \cong \mathrm{Q}_{2^n}$, $n > 3$ (Proposition 1.3).

**Exercise 78.** Classify the nonabelian $p$-groups $G$ such that, whenever $A, B < G$ are distinct minimal nonabelian, then $A \le \mathrm{N}_G(B)$ and $B \le \mathrm{N}_G(A)$.

**Solution.** By Theorem 10.28, all minimal nonabelian subgroups of $G$ are normal. By the same theorem, all nonabelian subgroups of $G$ are normal. Such $G$ are classified in [FA].

**Exercise 79.** Find the norm of a $p$-group of maximal class, $p > 2$.

**Exercise 80.** Is it true that all quasinormal subgroups of a $p$-group $G$ of maximal class are normal, unless $G \cong \mathrm{M}_{p^3}$?

**Exercise 81.** Let $E$ be an elementary abelian subgroup of a $p$-group $G$, $|E| \le p^4$, $p > 2$. Prove that there is in $E^G$ a $G$-invariant subgroup $\cong E$. (**Hint.** Apply Lemma 1.4 and Theorems 10.4 and 10.5 to $E^G$. According to [Kon1], a similar result is true for $|E| = p^5$.)

**Exercise 82.** If a noncyclic $p$-group $G$ has only one proper regular subgroup of order $p^{p+1}$, then $G$ is of maximal class and order $p^{p+2}$.

**Solution.** Below we use Proposition 1.3 and Theorems 5.3–5.5. By Proposition 1.3, $s_{p+1}(G) > 1$ so that $G$ has an irregular subgroup of order $p^{p+1}$. Let $p > 2$. Then $s_{p+1}(G) \equiv 1 + p \pmod{p^2}$. In this case, the number of irregular subgroups of order $p^{p+1}$ in $G$ is $\equiv p \pmod{p^2}$. But irregular groups of order $p^{p+1}$ are of maximal class (Theorem 7.1 (b)). It follows from Theorem 13.6 that $G$ is of maximal class. Let $G_1$ be the fundamental subgroup of $G$. Assume that $|G_1| > p^{p+1}$. Then $s_{p+1}(G_1) > 1$, a contradiction since all subgroups of $G_1$ are regular. Thus, $|G| = p^{p+2}$. Now let $p = 2$. Assume that $G$ is not of maximal class. Then, by Theorem 5.5, $s_{2+1}(G) \equiv 1 + 2 \pmod 4$ so that the number of irregular (= of maximal class) subgroups of order $2^{2+1}$ in $G$ is $\equiv 2 \pmod 4$. By Theorem 5.4, $G$ is a 2-group of maximal class and order $2^4$.

**Exercise 83.** If a minimal nonabelian $p$-group $G$ is a monolith, then either $|G| = p^3$ or $G \cong M_{p^n}$, $n > 3$.

**Exercise 84.** [Kor, Lemma 2.1] If a 2-group $G$ is such that $\mathrm{cl}(G) = 2$ and $\exp(G) = 4$, then $\Phi(G) \le \Omega_1(Z(G))$.

**Solution.** It follows from $[a^2, b^2] = [a, b]^4 = 1$ for all $a, b \in G$ that squares of any two elements of $G$ commute. Therefore, $\Phi(G) = \mho_1(G)$ is elementary abelian. Assume that $\Phi(G) \nleq Z(G)$. Then there are $t, x \in G$ such that $[t^2, x] \ne 1$. As $\mathrm{cl}(G) = 2$, we have $(tx)^2 = t^2x^2[x, t]$ so that

$$[x, t] = x^2t^2(tx)^2 \Rightarrow [x, t^2] = [x, t]^2 = (x^2t^2(tx)^2)^2 = x^4t^4(tx)^4 = 1\,,$$

contrary to the choice of $t$ and $x$.

**Exercise 85.** Is there a true analog of Exercise 84 for $p$-groups of class 2 and exponent $p^2$ provided $p > 2$?

**Exercise 86.** Study the $p$-groups $G$ of exponent $p^e > p$, satisfying $|\Omega_k(G)/\Omega_{k-1}(G)| = |G/\mho_1(G)|$ for all $k \in \{1, \dots, e\}$.

**Exercise 87.** Is it true that there is no nonabelian group $G$ of exponent $p$ such that $E^G$ is nonabelian for any non-$G$-invariant $E \cong E_{p^2}$?

**Exercise 88.** Let $\Sigma$ be a nontrivial partition of a $p$-group $G$. It is easy to see that that $|\Sigma| = c_1(G)$. Classify the $p$-groups $G$ for which $|\Sigma| \le c_1(G) - 1$ for an appropriate $\Sigma$.

**Exercise 89.** A $p$-group $G$ is metacyclic $\iff$ $G/G''$ is metacyclic (see § 36).

**Exercise 90.** (i) Study the $p$-groups $G$ such that $G/G''$ is of maximal class.

(ii) If $G$ is a minimal nonabelian 2-group of order $2^{2m}$ and exponent $2^m$, $m \geq 3$, then $G$ is modular.

**Solution.** (ii) Assume that $G$ is nonmodular. Then there is $L \lhd M \leq G$ such that $M/L \cong \mathrm{D}_8$. As $G$ is minimal nonabelian and $M$ is nonabelian, we get $M = G$. Then $G/\mho_2(G) \cong \mathcal{H}_{2,2}$ is nonmodular, a contradiction since $G' \leq \mho_2(G)$.

**Exercise 91.** The group $G = \Sigma_{p^n}$, $n > 1$, has the nontrivial Schur multiplier. (**Hint.** See Theorem 21.12.)

**Exercise 92.** If $A, B \lhd G$, then $\mathrm{dl}(AB) \leq \mathrm{dl}(A) + \mathrm{dl}(B)$.

**Exercise 93.** Prove that a group $G$ is nilpotent $\iff$ any two its conjugate elements of prime power order generate a nilpotent subgroup.

**Hint.** If a group is nonnilpotent, it contains a minimal nonnilpotent subgroup, say $H$. Show that $H$ is generated by two of its conjugate elements of prime power order.

**Exercise 94.** If $H$ is a maximal absolutely regular subgroup of an irregular $p$-group $G$, then $|H/\mho_1(H)| = p^{p-1}$. (**Hint.** Use Theorem 9.8 (a).)

**Exercise 95.** Classify the $p$-groups of exponent $p$ all of whose subgroups, having derived subgroup of order $p$, are minimal nonabelian.

**Solution.** Let $M < G$ be minimal nonabelian. Then $|M| = p^3$ and $\mathrm{C}_G(M) < G$ so $G$ is of maximal class (Proposition 119.17). Let us prove that $G$ possesses an abelian subgroup of index $p$. One may assume that $|G| > p^4$. Let $R \lhd G$ be abelian of order $p^2$ and set $A = \mathrm{C}_G(R)$; then $|G : A| = p$. Assume that $A$ is nonabelian. In that case $A$ contains a minimal nonabelian subgroup $K$; then $|K| = p^3$ and $R \nleq K$. In that case, $|(KR)'| = |K'| = p$ and $K$ is not minimal nonabelian, a contradiction. Thus, $A$ is abelian.

**Exercise 96.** Suppose that all minimal nonabelian subgroups of a $p$-group $G$ have order $p^3$. If all of those subgroups of $G$ that have a derived subgroup of order $p$ are minimal nonabelian, then $G$ is of maximal class with an abelian subgroup of index $p$. The converse is also true.

**Exercise 97.** Let $R \cong \mathrm{E}_{p^2}$ be a central subgroup of a $p$-group $G$. Suppose that for any $L < R$ of order $p$, the quotient group $G/L$ is modular. It is true that $G$ is modular?

**Exercise 98.** Classify the $p$-groups $G$ containing a maximal subgroup $H$ such that all maximal cyclic subgroups of $G$ are not contained in $H$.

**Solution** (Janko). We prove that $G$ is cyclic. Let $x \in H$. Then either there is $h \in H$ such that either $h^p = x$ or $\langle x \rangle$ is a maximal cyclic subgroup of $H$. In the first case $x \in \mho_1(G)$. In the second case, there is $g \in G$ such that $g^p = x$, so again $x \in \mho_1(G)$. Thus, $H \leq \mho_1(G)$. As, in our case, $G/\mho_1(G)$ is cyclic, it follows that $G$ is cyclic.

**Exercise 99.** The number of absolutely regular subgroups of order $p^p$ in a $p$-group $G$ is $\equiv 0 \pmod p$, unless $G$ is either absolutely regular or a 2-group of maximal class.

**Hint.** One may assume that $G$ is irregular. Use Theorem 13.5 and Sylow's theorem.

**Exercise 100.** Let $A$ be a nilpotent maximal subgroup of a group $G$ and suppose that all Sylow subgroups of $A$ are regular. Is it true that $G$ has a nonidentity normal Sylow subgroup? (**Hint.** Use Wielandt's Theorem A.52.4 (a).)

**Exercise 101.** Find all nontrivial characteristic subgroups of the group $G = \mathcal{H}_{2,2}$.

**Exercise 102.** Find all nontrivial characteristic subgroups of the nonabelian metacyclic group of order $p^4$ and exponent $p^2$, $p > 2$.

**Exercise 103.** Classify the minimal nonabelian 2-groups that have a section $\cong \mathrm{Q}_8$.

**Exercise 104.** Classify the minimal nonabelian $p$-groups without characteristic subgroup of index $p$.

**Exercise 105.** Let $p > 2$ and suppose that any maximal abelian subgroup of rank 2 coincides with its centralizer in a nonabelian $p$-group $G$ having noncyclic center $\mathrm{Z}(G)$. Prove that $\Omega_1(G) \cong \mathrm{E}_{p^2}$.

**Solution.** Let $R \cong \mathrm{E}_{p^2}$ be a subgroup of $\mathrm{Z}(G)$ and assume that $x \in G - R$ is of order $p$. Take $a \in R^\#$ and let $\langle a, x\rangle \le A < G$, where $A$ is a maximal abelian subgroup of $G$ of rank 2. Then $R \le \mathrm{C}_G(A)$ and $R \not\le A$, contrary to the hypothesis. Thus, $x$ does not exist so that $\Omega_1(G) = R = \mathrm{E}_{p^2}$.

**Exercise 106** (Mann). Suppose that all minimal nonabelian subgroups of a nonabelian group $G$ of exponent $p$ are normal. Then either $|\Phi(G)| = p$ or $G$ is of maximal class and order $p^4$.

**Solution.** Assume that $|G| > p^4$. All nonabelian subgroups of $G$ are normal (Theorem 10.28), so that if $A < G$ is minimal nonabelian, then $G/A$ is elementary abelian. It follows that $\Phi(G) < A$ (see Lemma 1.4), so that $|G : \Phi(G)| > p^2$ since $|G| > p^4$. Assume that $\mathrm{C}_G(A) \not\le A$, and let $C < \mathrm{C}_G(A)$ be of order $p$ such that $C \not\le A$. Then the intersection of all minimal nonabelian subgroups in $A \times C$ has order $p$ so that $|\Phi(G)| = p$ (such $G$ satisfies the condition). Now let $\mathrm{C}_G(A) < A$. Then $G$ is of maximal class (Proposition 10.17) and order $p^4$ (here we use Proposition 1.3 and Exercise 9.1 (b)), contrary to the assumption.

**Exercise 107.** Let $G$ be a noncyclic $p$-group of exponent $p^e > p$ and let $A < G$ be a cyclic subgroup of order $p^e$. Suppose that for any cyclic $B < G$ of order $> p$ such that $A, B$ are nonincident, one has $A \cap B = \{1\}$. Then $p = 2$ and $G$ is dihedral.

**Solution.** If $A < G$ is unique cyclic of order $p^e$, then $G$ is dihedral, by Theorems 1.17 (b) and 1.2 and Exercise 10.10. Let $A < H \le G$ be such that $|H : A| = p$. Then, by Theorem 1.2, $H$ is dihedral. By Exercise 10.10, $G$ is of maximal class. By Theorem 1.2, $G$ is dihedral.

**Exercise 108.** Let $G$ be a noncyclic $p$-group of exponent $p^e > p^2$ and let $A < G$ be a cyclic subgroup of order $p^e$. If for any cyclic $B < G$ of order $p^e$, $B \neq A$, one has $|A \cap B| \leq p^{e-2}$, then $p = 2$ and $G$ is of maximal class.

**Exercise 109.** If $G$ is a $p$-group such that $A < B < G$ implies $A^G < B^G$, then $G$ is Dedekindian.

**Solution.** Assume that $G$ is non-Dedekindian. Then there is in $G$ a maximal nonnormal subgroup $A$. Taking $B = A^G$, we get $A < B < G$ and $A^G = B = B^G$, contrary to the hypothesis.

**Exercise 110.** Let a $p$-group $G$ be an $\mathcal{A}_n$-group, $n > 1$. Then there are minimal nonabelian $A, B < G$ such that $A \cap B \not\leq \mathrm{Z}(A)$ and $A \cap B \not\leq \mathrm{Z}(B)$.

**Solution.** Let $A < G$ be minimal nonabelian of the least order and let $A < H \leq G$, where $|H : A| = p$; then $H$ is an $\mathcal{A}_2$-subgroup. Take a nonabelian $B < H$, $B \neq A$ (see § 76). Then $A \cap B$ is maximal in $A$ and in $B$ so such $A, B$ satisfy the condition.

**Exercise 111.** Let $G$ be a noncyclic $p$-group. Suppose that a cyclic $L < G$ of order $> p$ is such that $\mathrm{C}_G(L) > L$ is cyclic. Then $G$ is a 2-group of maximal class.

**Solution.** Write $F = \mathrm{C}_G(L)$ and let $F < H \leq G$ be such that $|H : F| = p$. Then $L \leq \Phi(F) \leq \Phi(H)$ and $L \not\leq \mathrm{Z}(H)$, by hypothesis. It follows from Theorem 1.2 that $p = 2$ and $H$ is of maximal class. By Exercise 10.10, $G$ is of maximal class.

**Exercise 112.** Let $L > \{1\}$ be a cyclic subgroup of a $p$-group $P$. Suppose that $P$ is neither cyclic nor a 2-group of maximal class. Given $k$, the number of cyclic subgroups in $P$ of order $p^k|L| < |P|$ containing $L$ is $\equiv 0 \pmod{p}$.

**Solution.** One may assume that $|G| > p^3$. We use induction on $|P|$. Given $L < T \leq P$, let $\alpha(T)$ denote the number of cyclic subgroups of $T$ of order $p^k|L|$ containing $L$. Assume that $\alpha(P) \not\equiv 0 \pmod{p}$. Take $L < M < P$, where $M$ is cyclic of order $p^k|L|$. It follows that $L \leq \Phi(M) \leq \Phi(P)$. Let $C = \mathrm{C}_P(L)$; then $\alpha(C) = \alpha(P)$. Therefore, without loss of generality, one may assume that $C = P$; then $L \leq \mathrm{Z}(P)$. One may assume that $P$ has no cyclic subgroup of index $p$ (Theorem 1.2). By Hall's enumeration principle (see Theorem 5.2), $\alpha(P) = \sum_{T \in \Gamma_1,\, L<T} \alpha(T) \pmod{p}$. Then, if $L < T \in \Gamma_1$ and $T$ is not a 2-group of maximal class, then $\alpha(T) \equiv 0 \pmod{p}$, by induction. If $T$ is a 2-group of maximal class, then $\alpha(T) \equiv 1 \pmod{2}$. As the number of members of the set $\Gamma_1$ that are 2-groups of maximal class and contain $L$ is a multiple of 2 (prove!), it follows from the obtained results and the formula for $\alpha(P)$ that $\alpha(P) \equiv 0 \pmod{p}$.

**Exercise 113.** Let $A$ and $B$ be distinct minimal nonabelian subgroups of index $p$ in a $p$-group $G$. If $\mathrm{d}(G) = 3$, then $\Phi(G) \leq \mathrm{Z}(G)$.

**Solution.** Let $C = A \cap B$; then $|A : C| = p = |B : C|$. In that case, $Z(A) = \Phi(A) < C$ and $Z(B) < C$. Note that $Z(A) = \Phi(A) = \Phi(G)$ since $|G : \Phi(A)| = p^3 = |G : \Phi(G)|$. Similarly, $Z(B) = \Phi(G)$. Then $C_G(\Phi(G)) \geq AB = G$, as required.

**Exercise 114.** Let $A$ be an abelian maximal subgroup of a finite group $G$. Is it true that $G$ has a nonidentity normal Sylow subgroup? (**Hint.** (By Herstein, $G$ is solvable so $|G : A|$ is a prime power (Galois).)

**Exercise 115.** Let a nonabelian 2-group $G = C \cdot E$ (semidirect product with kernel $E \cong E_8$, $C \cong C_4$). Show that $G$ contains a nonmetacyclic minimal nonabelian subgroup of order 16. ([**Hint.**] (Use Lemma 57.1.))

**Solution.** If $C_G(E) > E$, any minimal nonabelian subgroup of $G$ has order 16. Now let $C_G(E) = E$. Let $x$ be a generator of $C$. By Lemma 57.1, there is $a \in E$ such that $H = \langle a, x\rangle$ is minimal nonabelian. Clearly, $|H| = 16$ and $|\Omega_1(H)| = 8$ so $H$ is nonmetacyclic.

**Exercise 116.** If all noncyclic subgroups of a minimal nonabelian $p$-group $G$ of order $> p^3$ are normal, then it is metacyclic.

**Solution.** Assume that $G$ is nonmetacyclic; then $\Omega_1(G) \cong E_{p^3}$. As all subgroups of $\Omega_1(G)$ of order $p^2$ are $G$-invariant, it follows that $\Omega_1(G) \leq Z(G)$. Let

$$G = \langle a, b \mid o(a) = p^m,\ o(b) = p^n,\ n > 1,\ [a, b] = c,\ o(c) = p,\ [a, c] = [b.c] = 1\rangle .$$

Then a noncyclic subgroup $A = \langle a, b^p\rangle$ is not $G$-invariant, a contradiction.

**Exercise 117.** Classify the nonmodular $p$-groups all of whose proper sections are modular. (**Answer.** $D_8$ and $S(p^3)$). See Appendices 24 and 59.)

**Exercise 118.** Suppose that $G$ is a nonabelian metacyclic $p$-group. Study the structure of $G$ provided for every $L \triangleleft G$ of order $p$ there is $A < G$ such that $G/L \cong A$.

**Exercise 119.** (a) Classify the minimal nonabsolutely regular $p$-groups.
(b) Classify the $p$-groups all of whose maximal subgroups are either absolutely regular or of maximal class.

**Solution.** (a) If there is in $G$ a subgroup $R$ of order $p^p$ and exponent $p$, then $G = R$. If $R$ does not exist, then $G$ is of maximal class (Theorem 12.1 (a)). By Theorems 9.5 and 9.6, $|G| = p^{p+1}$ and $|\Omega_1(G)| = p^{p-1}$. (**Hint.** (b) Apply Theorems 12.1, 9.6 and 9.5.)

**Exercise 120.** Suppose that a $p$-group $G$ is such that, whenever $A, B < G$ with $A \cap B = \{1\}$, then $AB = A \times B$. Then one and only one of the following holds:
(a) $G$ is abelian,
(b) $G = Q \times E$, where $Q \cong Q_{2^n}$ and $\exp(E) \leq 2$.

**Solution.** All abelian $p$-groups satisfy the hypothesis. Suppose that $G$ is nonabelian. Let $A \leq G$ be minimal nonabelian. It follows from Lemma 65.1 that $A \cong Q_8$. Now the result follows from Corollary A.17.3.

**Exercise 121.** Classify the nonabelian $p$-groups $G = \Omega_1(G)$ such that $\langle A, x\rangle$ is minimal nonabelian for any maximal abelian $A < G$ and $x \in G - A$ of order $p$.

**Solution** (Janko). Assume that there is in $G$ a subgroup $E \cong E_{p^3}$ and let $E \le A$, where $A < G$ is maximal abelian. If $x \in G - E$ is of order $p$, then $M = \langle x, A\rangle$ is not minimal nonabelian since $|\Omega_1(M)| > p^3$ (Lemma 65.1). Thus, $E$ does not exist. Assume that $R \triangleleft G$ is abelian of type $(p, p)$ and $R \le A$ where $A$ is a maximal abelian normal subgroup in $G$. Let $x \in G - A$ be of order $p$. Then the minimal nonabelian subgroup $H = \langle x, A\rangle \in \{S(p^3), D_8\}$ since $|\Omega_1(H)| > p^2$ and $G$ has no subgroup $\cong E_{p^3}$. We conclude that $A \cong E_{p^2}$. Therefore. it follows from $C_G(A) = A$ and $A \triangleleft G$ that $|G| = p^3$ so that $G \in \{S(p^3), D_8\}$. If $G$ has no normal abelian subgroup of type $(p, p)$, it is a 2-group of maximal class (Lemma 1.4), and it is easy to check that such $G$ does not exist (Theorem 1.2).

**Exercise 122.** Classify the groups of exponent $p$ all of whose subgroups of order $> p^2$ are quasinormal.

**Exercise 123.** Classify the minimal nonabelian $p$-groups $G$ satisfying the following condition: For any $L \triangleleft G$ of order $p$ there is $A < G$ such that $G/L \cong A$.

**Exercise 124.** Suppose that $L$ is a quasinormal cyclic subgroup of a $p$-group $G$, $p > 2$, $C_G(L) = L$ and $|L| > p^2$. Is it true that $G$ is metacyclic?

**Solution.** Write $N = N_G(L)$. Then $N/L$ is cyclic so $\Omega_1(N) = E_{p^2}$. Since all subgroups of $G$ of order $p$ normalize $L$, it follows that $\Omega_1(G) = \Omega_1(N) \cong E_{p^2}$, and so $G$ is either metacyclic or a 3-group of maximal class (Theorem 13.7). Assume that $G$ is a 3-group of maximal class. As $|L| > 3^2$, one has $L < G_1$, where $G_1$ is the fundamental subgroup of $G$ (Theorem 13.19). Since $Z(G_1) < L$, this is a contradiction since $Z(G_1)$ is noncyclic. Thus, $G$ is metacyclic.

**Exercise 125.** Let $G = A \times B$, where $A$ and $B$ are irregular. Is it true that all maximal subgroups of $G$ are irregular?

**Solution.** Assume that $H \in \Gamma_1$ is regular. Then $A, B \nleq H$. Write $K = H \cap B$. It follows from $G = BH$ that $A \cong G/B \cong H/K$ is irregular, $\Rightarrow$ $H$ is irregular, contrary to the assumption.

**Exercise 126.** Let $G = A * B$ be the central product of two irregular $p$-groups $A$ and $B$. Is it true that all maximal subgroups of $G$ are irregular?

**Exercise 127.** Is it true that all maximal subgroups of $G = \Sigma_{p^n} \in \mathrm{Syl}_p(S_{p^n})$, $n > 2$, are irregular?

**Solution.** The group $G = \Sigma_{p^{n-1}} \operatorname{wr} C_p$ is the standard wreath product. Then $\Phi(G)$ contains a subgroup $\cong \Sigma_{n-1}(G)$ (check!) which is irregular, Thus, all maximal subgroups of $G$ are irregular.

**Exercise 128.** Let $G = R \operatorname{wr} C_p$, where $R$ is an irregular $p$-group. Is it true that all members of the set $\Gamma_1$ are irregular?

**Exercise 129.** Check that all minimal nonmetacyclic groups of order $2^5$ are $\mathcal{A}_2$-groups.

**Exercise 130.** Study the nonabelian $p$-groups all of whose proper nonabelian subgroups have the same (i) Frattini subgroup, and (ii) derived subgroup.

**Exercise 131.** Let $G = D_8 * Q_8$ be the extraspecial group of order $2^5$. Does there exist a characteristic subgroup of index 2 in $G$?

**Exercise 132.** Study the $p$-groups $G$ such that the normal closure $H^G$ is of maximal class for any nonnormal $H < G$. Describe the $\mathcal{A}_1$-subgroups of $G$.

**Hint.** We show that in the case under consideration $p = 2$. Assume, by way of contradiction, that $p > 2$. By hypothesis, there is $H \lhd G$ of maximal class. Let $L \cong E_{p^2}$ be a $G$-invariant subgroup of $H$ (Theorem 1.17 (b) and Proposition 1.3). If $U \neq Z(H)$ is a subgroup of $L$ of order $p$, then $U^G = L$ is not of maximal class, a contradiction. Thus, $p = 2$. Now assume that $A$ is a minimal nonabelian subgroup of $G$ and assume that there is in $A$ a nonnormal cyclic subgroup $L$ of order $> 2$. Then $L^A = L \times A'$ is abelian of type $(|L|, 2)$. As a 2-group of maximal class has no subgroup $\cong L^A$, we get a contradiction. Thus, all nonnormal cyclic subgroups of $A$ have order 2. It follows that $A \in \{D_8, M_{2^n}\}$.

**Exercise 133.** Study the $p$-groups $G$ in which any nonnormal subgroup is contained in a two-generator maximal subgroup of $G$. (In fact, this is a fairly difficult problem.)

**Exercise 134.** Prove that the metacyclic Frattini subgroup of a 2-group $G$ is powerful.

**Solution.** It suffices to show that $\Phi(G)/\mho_2(\Phi(G))$ is abelian (see § 26). As $\mho_2(\Phi(G))$ is $G$-invariant, one may assume without loss of generality that $\mho_2(\Phi(G)) = \{1\}$. Then $|\Phi(G)| \leq 2^4$. If $|\Phi(G)| = 2^3$, then $\Phi(G)$ is abelian (Lemma 1.4). Now let $|\Phi(G)| = 2^4$. Assume that $\Phi(G)$ is nonabelian. Then all subgroups of $\Phi(G) \cong \mathcal{H}_{2,2}$ of order 2 are characteristic, so normal in $G$. Let $L \neq \Phi(G)'$ be a subgroup of order 2 in $\Phi(G)$. Then $\Phi(G)/L$ is a normal nonabelian subgroup of order 8 in $G/L$, contrary to Lemma 1.4.

**Exercise 135.** Let $\phi \in \operatorname{Aut}(G)$ be of order $n$ and $\operatorname{GCD}(n, |G|) = 1$. If $a^{1+\phi+\cdots+\phi^{n-1}} = 1$ for all $a \in G$, then $C_G(\phi) = \{1\}$.

**Exercise 136.** Classify the nonabelian $p$-groups $G$, containing two distinct elementary abelian subgroups of index $p$. (**Answer.** $G = S \times E$, where $S \in \{S(p^3), D_8\}$ and $E$ is elementary abelian; see § 245.)

**Exercise 137.** Find the number of faithful irreducible characters in a 2-group $G = A * B$, where $A$, $B$ are groups of maximal class and orders $2^m$, $2^n$, respectively, $A \cap B = Z(A)$.

**Exercise 138.** Suppose that a 2-group $G$ of order $> 2^4$, which is not of maximal class, possesses a subgroup $H$ of maximal class and index 2. Then it contains only one normal abelian subgroup of type $(2, 2)$.

**Solution.** Assume that $R, R_1 < G$ are distinct $G$-invariant abelian subgroups of type $(2, 2)$. As $H$ has no normal subgroup of type $(2, 2)$, it follows that $D = RR_1 \cong \mathrm{D}_8$ (indeed, consider $D \cap H$). Let $C < H$ be cyclic of index 2; then $C \lhd G$. One has $C \cap R = C \cap R_1 = \Omega_1(C)$ and $H \cap D = \Omega_2(C)$. The subgroups $T = CR$ and $T_1 = CR_1$ are either abelian of type $(|C|, 2)$ or $\cong \mathrm{M}_{2|C|}$. It follows that $\Omega_2(C)$ centralizes $R$ and $R_1$ so also $RR_1 = D$. As $\Omega_2(C) < D$ is central (in $D$) of index 2, it follows that $D$ is abelian, a final contradiction.

**Exercise 139.** If all noncyclic abelian subgroups of a non-Dedekindian $p$-group $G$ are normal, then $G$ has no subgroup $\cong \mathrm{E}_{p^3}$.

**Exercise 140.** The following conditions for a nonabelian $p$-group $G$ are equivalent:
(a) $G$ has exactly $p + 1$ maximal abelian subgroups,
(b) $|G : \mathrm{Z}(G)| = p^2$,
(c) $G = S\mathrm{Z}(G)$, where $S$ is minimal nonabelian.

**Exercise 141.** Any group of order $> p^p$ and exponent $p$ is covered by abelian subgroups of order $p^3$. (**Hint.** Use Proposition 1.8 and Theorem 9.5.)

**Exercise 142.** Study the 2-groups $G$ all of whose elements of order 4 are real but $G$ contains a nonreal element (for example: $G = \mathrm{Q}_{2^4}$).

**Exercise 143** (see Appendix 97). Suppose that a 2-group $G$ of order $2^{n+1}$ is neither cyclic nor of maximal class and $n > 2$. Then the number $t$ of central subgroups $D$ of order 2 such that $G/D$ is of maximal class and order $2^n$, is even.

**Solution.** Let $D$ be as in the statement. If $T/D < G/D$ is cyclic of index 2, then $T$ is abelian. If $D = \mathrm{Z}(G)$, then $D < G'$ so $G$ is of maximal class (Taussky). Therefore, one may assume that $\mathrm{Z}(G) > D$; then $|\mathrm{Z}(G)| = 4$. By Lemma 1.1, $|G : G'| = 2|\mathrm{Z}(G)| = 8$. Assume that $\mathrm{Z}(G)$ is cyclic. Then there is in $G$ a normal subgroup $R \cong \mathrm{E}_4$. In that case, $\mathrm{Z}(G/D) \cong \mathrm{E}_4$ so $G/D$ is not of maximal class, a contradiction. Thus, $\mathrm{Z}(G) \cong \mathrm{E}_4$. By Proposition 1.6 (Taussky), $D \not\leq G'$. If $D_1 = G' \cap \mathrm{Z}(G)$, then $G/D_1$ is not of maximal class (Proposition 1.6). If $D_1, D_2, D$ are all subgroups of order 2 in $\mathrm{Z}(G)$, then $D_2 \not\leq G'$ since $\mathrm{Z}(G) \cap G' = D_1 \neq D_2$. If $G/D_2$ is not of maximal class, then $|G/D_2 : (G/D_2)'| = 8$ and so $|G : G'| > 8$ since $D_2 \not\leq G'$, a contradiction. Thus, $G/D_2$ is of maximal class; then $t = 2$.

**Exercise 144.** Let $G$ be a two-generator $p$-group such that $G' \cong \mathrm{C}_{p^n}$, $p > 2$, $n > 1$. Then $|H'| = p^{n-1}$ for all $H \in \Gamma_1$.

**Hint.** Let $L < G'$ be of index $p^2$. Then $(G/L)' \cong C_{p^2}$. If $H \in \Gamma_1$, then $|(H/L)'| \geq p$ (Corollary 148.2 (a)) so $|H'| \geq p|L| = p^{n-1}$. If $K < G'$ is of index $p$, then $G/K$ is minimal nonabelian, by Lemma 65.2 (a), so $|H'| \leq |K| = p^{n-1}$.

**Exercise 145.** If $p$-groups $G$, $G_0$ have isomorphic lattices of normal subgroups and $G_0$ is of maximal class, then $G$ is also of maximal class. (**Hint.** Use Exercise 9.1 (b).)

**Exercise 146.** Study the $p$-groups whose lattices of normal subgroups are isomorphic with lattices of normal subgroups of metacyclic $p$-groups.

**Exercise 147.** Classify the minimal nonabelian 2-groups $G$ all of whose elements are real.

**Solution.** Obviously, $\exp(Z(G)) = 2$ and so $\exp(G) = 4$ since $Z(G) = \mho_1(G)$. Let $X = \langle x \rangle$, where $x \in G$ is real of order 4. Then $X \lhd G$ since $x$ together with inverting $x$ an element $y \in G$ generate $G$. It follows that $G$ is metacyclic and $X \not\leq Z(G) = \Phi(G)$. Since $\exp(G) = 4$, it follows that $|G| \leq 2^4$. If $|G| = 2^4$, then $G$ contains a nonreal element (namely, a generator of nonnormal cyclic subgroup of order 4). Thus, $G$ is nonabelian of order 8.

*7°* **Proposition A.51.6.** *Let $G$ be a nonabelian metacyclic $p$-group of order $p^{2n}$ and exponent $p^n$, let $A \lhd G$ be cyclic such that $G/A$ is cyclic. If $A \leq H \in \Gamma_1$, then $H$ is characteristic in $G$.*

*Proof.* By Theorem 1.17 (b), $c_n(G) \equiv 0 \pmod p$. Therefore, there is in $G$ a normal cyclic subgroup $B \neq A$ of order $p^n$. As $G$ is nonabelian, $A \cap B > \{1\}$. Therefore, $AB \leq H$. Thus, all $G$-invariant cyclic subgroups of $G$ of order $p^n$ generate a proper characteristic subgroup $F < G$. Let $\phi \in \mathrm{Aut}(G)$ and assume that $H^\phi \neq H$. One has $F^\phi = F$. Therefore, $H/F$ and $H^\phi/F$ are distinct subgroups of equal order in a cyclic subgroup $G/F$, a contradiction. □

**Exercise 148.** Study the $p$-groups $G$ such that $G/K_3(G)$ is metacyclic. (**Hint.** Use Theorem 36.1.)

*8°* **Exercise 149.** Let $G$ be a metacyclic group of order $p^{2e}$ and exponent $p^e$. If $e > 1$ and $i < e$, then $|\Omega_i(G)| = p^{2i}$ and $|G/\mho_i(G)| = p^{2i}$.

**Exercise 150** (A. Saeidi, personal communication; see #1764). For a nonabelian $p$-group $G$ the following conditions are equivalent:

(a) $G$ has $< p$ nonlinear irreducible characters of each degree,

(b) $G$ is extraspecial.

**Solution.** Let $\mathrm{cd}(G) = \{1, p^{a_1}, \dots, p^{a_k}\}$, where $a_1 < \cdots < a_k = r$. Then

$$|G| = |G:G'| + \sum_{\chi\in\mathrm{Irr}_1(G)} \chi(1)^2 \le |G:G'| + (p-1)\sum_{i=1}^{r} p^{2i}$$
$$= |G:G'| + (p-1)p^2\frac{p^{2r}-1}{p^2-1} = |G:G'| + p^2\frac{p^{2r}-1}{p+1}.$$

It follows from $p^{2r} \le \frac{1}{p}|G|$ that

$$|G| \le |G:G'| + p^2\frac{\frac{1}{p}|G|-1}{p+1} = |G:G'| + \frac{p}{p+1}|G| - \frac{p^2}{p+1}.$$

Assume that $|G'| \ge p^2$. Then

$$|G| \le \frac{|G|}{p^2} + \frac{p}{p+1}|G| - \frac{p^2}{p+1} \Rightarrow |G|\left(1 - \frac{1}{p^2} - \frac{p}{p+1}\right) < 0 \Rightarrow p^2 - p - 1 < 0,$$

which is impossible. Thus, $|G'| = p$ since $G$ is nonabelian. Let $|G| = p^m$, $|\mathrm{Irr}_1(G)| = s$, $\mathrm{cd}(G) = \{1, p^t\}$. Then

$$p^m = p^{m-1} + sp^{2t} \Rightarrow p^{m-1}(p-1) \le (p-1)p^{2t} \Rightarrow t \ge \frac{1}{2}(m-1).$$

Next, $p^{2t} \le |G:\mathrm{Z}(G)| \le p^{m-1} \Rightarrow t \le \frac{1}{2}(m-1)$, and we conclude that

$$t = \frac{1}{2}(m-1), \quad |\mathrm{Z}(G)| = p = |G'| \Rightarrow \mathrm{Z}(G) = G'.$$

If $x, y \in G$, then, as $\mathrm{cl}(G) = 2$, one obtains $1 = [x,y]^p = [x, y^p]$ so that $\mho_1(G) \le \mathrm{Z}(G)$. It follows that $\exp(G/\mathrm{Z}(G)) = p$ so that $\mathrm{Z}(G) = \Phi(G) = G'$ is of order $p$, and this implies that $G$ is extraspecial, i.e., (a) $\Rightarrow$ (b). The reverse implication is known.

**Exercise 151.** Prove that there does not exist a $p$-group $G$ containing a proper subgroup $H$ of maximal class such that $\mathrm{N}_G(H)/H'$ is two-generator abelian.

**Solution.** Let $H < T \le G$, where $|T:H| = p$; then $T/H'$ is two-generator abelian. In that case, $\mathrm{d}(T) = 2$ since $H' = \Phi(H) \le \Phi(T) \le \Phi(\mathrm{N}_G(H))$. It follows that $T$ is of maximal class (Theorem 12.12 (a)). By Exercise 10.10, $G$ is of maximal class. In that case, $T/H'$ is nonabelian (of order $p^3$), a contradiction.

**Exercise 152.** (a) Let $G$ be a regular $p$-group, let $H \triangleleft G$ be of exponent $p$ and $G = \langle x\rangle \cdot H$ (semidirect product with kernel $H$). Then $x^p$ centralizes $H$.
(b) (Mann–Poznick–Fradkin [MP-F, Proposition 2]) Let $G$ be a group of exponent $p$ and order $\le p^{p-1}$. Then all $p$-automorphisms of $G$ have order $p$.

**Solution.** (a) Let $y \in H^\#$. Then $[x,y]^p = 1 \Rightarrow [x^p, y] = 1$ (Theorem 7.2 (e)) so that $x^p \in \mathrm{C}_G(y)$. Since $y \in H^\#$ is arbitrary, we are done.

(b) Let $\alpha \neq \mathrm{id}_G$ be a $p$-automorphism of $G$, and consider the semidirect product $H = \langle\alpha\rangle \cdot G$ with kernel $G$. Then $H/G$ is cyclic and $|G| \leq p^{p-1}$ implies $G \leq Z_{p-1}(H)$ so that $\mathrm{cl}(H) \leq p-1$. It follows that $H$ is regular (Theorem 7.1 (b)). By (a), $\alpha^p$ centralizes $G$ which implies that $o(\alpha) = p$.

**Exercise 153.** Classify the nonabelian $p$-groups containing an abelian normal subgroup $A$ such that $G/A$ is cyclic and all maximal abelian subgroups of $G$ are normal.

**Exercise 154.** Classify the $p$-groups of order $> p^3$ all of whose abelian subgroups of order $p^3$ are isomorphic.

**Exercise 155.** Study the $p$-groups of order $> p^3$ all of whose noncyclic abelian subgroups of order $p^3$ are isomorphic.

**Exercise 156.** Find the number of faithful irreducible characters in a group $G$ of maximal class and order $2^n$.

**Solution.** One has $|\mathrm{Irr}_1(G)| = \frac{|G|-4}{4} = 2^{n-2} - 1$. As all nonfaithful irreducible characters of $G$ contain $Z(G)$ in their kernels, there are exactly $\frac{2^{n-1}-4}{4} = 2^{n-3} - 1$ nonfaithful nonlinear irreducible characters. It follows that the desired number equals $(2^{n-2} - 1) - (2^{n-3} - 1) = 2^{n-3}$.

**Exercise 157** (Theorem 12.12 (a)). Let a two-generator $p$-group $G$ contain a subgroup $M$ of maximal class and index $p$. Then $G$ is of maximal class.

**Solution.** Let $|G| = p^4$; then $M$ is nonabelian of order $p^3$. As $G$ is not of maximal class, one has $G = MC_G(M)$ (Proposition 10.17). In that case, $d(G) = 3$, a contradiction. Now let $|G| > p^4$. Write $R = Z(M)$; then $R \triangleleft G$ is of order $p$. By induction, $G/R$ is of maximal class. We are done if $R = Z(G)$. Assume that $R < Z(G)$. Then $Z(G) \not\leq M$ so that $G = MZ(G)$. In that case, $G/R = (M/R) \times (Z(G)/R)$ hence $d(G/R) = 3 > 2 = d(G)$, a contradiction.

**Exercise 158.** Study the $p$-groups $G$ with a proper absolutely regular $H_p$-subgroup. (**Hint.** Use Theorem 12.1 (b) to prove that $G$ is of maximal class.)

**Exercise 159.** Let $G$ be a nonabelian regular $p$-group of exponent $p^e > p$. Then there is in $G$ an element $x$ of order $p^e$ such that $x \in G - A$, where $A < G$ is a maximal abelian.

**Solution.** The intersection of all maximal abelian subgroups of $G$ coincides with $Z(G)$. Therefore, if our $x \in G - Z(G)$, there is in $G$ a maximal abelian subgroup $A$ such that $x \notin A$.

**Exercise 160.** Let $G$ be a two-generator $p$-group, $p > 2$, such that $G' \cong E_{p^3}$. Then there is $H \in \Gamma_1$ with $|H'| \leq p$.

**Hint.** Let $L < G'$ be $G$-invariant of order $p$. Then $(G/L)' \cong E_{p^2}$. By Proposition 145.5, there is in $G/L$ an abelian maximal subgroup $H/L$. In that case, $H' \leq L$ so that $|H'| \leq p$.

A $p$-group $G$ is said to be *monotone* if $A < B \le G \Rightarrow \mathrm{d}(A) \le \mathrm{d}(B)$. The $p$-groups all of whose proper subgroups are metacyclic are monotone.

**Exercise 161.** Classify the monotone $p$-groups of maximal class.

**Hint.** The 2-groups of maximal class are monotone. A 3-group $G$ of maximal class is monotone $\iff G \not\cong \Sigma_9$. Let $G$ be a $p$-group of maximal class and order $> p^3$, $p > 3$. Then all maximal subgroups of $G$ are two-generator. It follows that $|G/\mho_1(G)| = p^3$ (Theorems 9.8 (a) and 5.8 (b)) so that $G$ is of order $p^4$ with $\Omega_1(G) \cong S(p^3)$ (see Theorems 9.5 and 9.6).

**Exercise 162.** Let $H$ be a subgroup of maximal class of a 2-group $G$. If $\mathrm{N}_G(H)$ is metacyclic, then $G$ is of maximal class. (**Hint.** Use Proposition 10.19 and Remark 10.5.)

**Exercise 163.** Suppose that the intersection of any two distinct conjugate subgroups of a $p$-group $G$, $p > 2$, is either absolutely regular or of maximal class. If $A, B < G$ are distinct conjugates such that $D = A \cap B$ is of maximal class, then $|D| \le p^{p+1}$.

**Solution.** Assume that $|D| > p^{p+1}$; then $G$ is not of maximal class (Theorem 9.6). By Theorem 12.1 (a), there is $R \triangleleft G$ of order $p^p$ and exponent $p$; then $R \not\le D$ (Theorem 9.6 (c)) so that $|R \cap D| \le p^{p-1}$. By hypothesis, $G/R$ is abelian (Theorem 1.20), so is $D/(R \cap D) \cong RD/R$. It follows that $|D/(R \cap D)| = p^2$ (recall that $D$ is of maximal class), so that $|D| = p^2 |R \cap D| \le p^{p-1} \cdot p^2 = p^{p+1}$, contrary to the assumption.

**Exercise 164.** Classify the $p$-groups $G$ of exponent $> p$ in which the intersection of any two distinct nonnormal cyclic subgroups of the same order is equal to $\{1\}$. Is it true that all cyclic subgroups of order $> p$ are normal in $G$?

**Solution for** $p > 2$. We assume that there is in $G$ a nonnormal cyclic subgroup $A$ of order $> p$. Let $A < H \le G$, where $|H : A| = p$. Then, by Theorem 1.2, $H$ is either abelian of type $(p^n, p)$, $n > 1$, or $\cong \mathrm{M}_{p^{m+1}}$, $m > 1$. In that case, $H$ has exactly $p$ cyclic subgroups of index $p$, say $A = A_1, \dots, A_p$. Then, by hypothesis, $A_2, \dots, A_p$ are $G$-invariant so, in view of $p > 2$, $H$ is also normal in $G$. It follows that $A \triangleleft G$, a contradiction. We suggest to the reader to consider the more difficult case $p = 2$.

9° **Exercise 165.** A minimal nonabelian group $G$ of exponent $p^e$ is generated by two elements of order $p^e$.

**Solution.** One may assume that $e > 1$ and $|G| > p^3$. Then $G' < \Omega_{e-1}(G) < G$ so that $\langle G - \Omega_{e-1}(G) \rangle = G$. Thus, $G$ is generated by elements of order $p^e$. Since $\mathrm{d}(G) = 2$, the result follows.

**Exercise 166.** Let $G$ be a nonabelian $p$-group, $p > 2$. If $\mathrm{C}_G(M) < M$ for any nonabelian $M \le G$, then either $G$ is of maximal class or $\Omega_1(G)$ is elementary abelian.

**Solution.** Let $E \triangleleft G$ be elementary abelian of maximal order. Assume that $E < \Omega_1(G)$. Then there is $x \in G - E$ of order $p$. Set $H = \langle E, x \rangle$. By Theorem 10.1, $H$ is nonabelian.

Therefore, by Lemma 57.1, there is $a \in E$ such that $A = \langle a, x \rangle$ is minimal nonabelian. It follows from $\Omega_1(A) = A$ that $A \cong S(p^3)$. By hypothesis, $C_G(A) < A$ so that $G$ is of maximal class (Proposition 10.17). We suggest to prove that $G$ has an abelian subgroup of index $p$.

**Exercise 167.** Let $H < G$ be a subgroup of maximal class of a $p$-group $G$. Suppose that $N_G(H/\Phi(H))$ is metacyclic. Is it true that $G$ is of maximal class? (**Answer.** Yes, by Exercise 151.)

**Exercise 168.** Classify the irregular $p$-groups of order $> p^{p+1}$ in which the intersection of any two distinct maximal subgroups is absolutely regular.

**Solution.** Any irregular $p$-group of maximal class satisfies the hypothesis (see Theorems 9.5 and 9.6). Now let $G$ be not of maximal class. Then there is $R \lhd G$ of order $p^p$ and exponent $p$ (Theorem 12.19 (a). By hypothesis, there is in $G/R$ only one maximal subgroup so it is cyclic. Let $H/R = \Omega_1(G/R)$. One of the following holds: (i) $R = \Omega_1(G)$, or (ii) $\exp(H) = p$. In case (i), $G$ satisfies the hypothesis (in that case, $G$ is an $L_p$-group; see §§ 17 and 18). Next we consider case (ii). Let $U/H$ be maximal in $G/H$ and let $V \in \Gamma_1$ be such that $H \nleq V$. Then $U \neq V$ and $U \cap V$ is not absolutely regular. Thus, case (ii) is impossible so that $G$ is either of maximal class or an $L_p$-group.

*10°* **Exercise 169** ([Hup1], Satz 16.5 (b)). The group $E = E_{p^p}$ has no an automorphism of order $p^2$.

**Solution.** Assume that $\alpha \in \mathrm{Aut}(E)$ is of order $p^2$ and set $G = \langle \alpha \rangle \cdot E$. Then $\Omega_1(G) = \langle \alpha^p \rangle \cdot E = K$. Assume that $K$ is irregular. Then $K$ is of maximal class (Theorem 7.1 (b)). By Exercise 10.10, $G$ is of maximal class, a contradiction since $|G/G'| > p^2$. Thus, $K$ is regular. Then $\alpha^p$ centralizes $E$ (Exercise 152 (a)), a contradiction.

**Remark 1.** Let $G$ be a $p$-group, $R = Z_2(G) \cong E_{p^2}$, $H = C_G(R) < G$; then $H \in \Gamma_1$. Suppose that $\Omega_1(H)$ is of order $p^k$ and exponent $p$. If $G$ possesses an elementary abelian subgroup $E$ of order $p^k$, then $\Omega_1(H) \cong E$. Indeed, assume that $E \nleq H$; then $H \cap E \cong E_{p^{k-1}}$ has index $p$ in $\Omega_1(H)$. If $R \leq E$, then $C_G(R) \geq HE = G$, a contradiction. Thus, $R \nleq E$. In that case, $\Omega_1(H) = (H \cap E)R \cong E_{p^k}$ since $[H \cap E, R] = \{1\}$.

**Exercise 170.** If $H$ is a quasinormal subgroup of a $p$-group $G = \Omega_1(G)$, then $H \lhd G$. In particular, all quasinormal subgroups of $\Sigma_{p^n}$ are normal.

**Exercise 171.** Study the non-Dedekindian $p$-groups covered by nonnormal subgroups.

**Exercise 172.** Given $n$ and $p > 2$, prove that there is a $p$-group $G$ of maximal class containing a cyclic subgroup $C$ of order $p^n$ such that $C_G = \{1\}$, where $C_G = \bigcap_{x \in G} C^x$.

**Solution.** This is trivial for $n = 1$ (see Proposition 1.3). Now we let $n > 1$. Let $|G| = p^{n(p-1)+1}$ and let $G_1$ be the fundamental subgroup of $G$; then $\exp(G_1) = p^n$, $|G_1| =$

$p^{n(p-1)}$ and $|\mho_{n-1}(G_1)| = p^{p-1} > p$. Therefore, there is in $\mho_1(G_1)$ a subgroup $L = \langle x \rangle$ of order $p$ such that $L \nleq Z(G)$. By Theorem 7.2 (c), there is $y \in G_1$ such that $y^{p^{n-1}} = x$. The cyclic subgroup $C = \langle y \rangle$ possesses the required property.

**Exercise 173.** Let $A$ be an abelian subgroup of order $p^p$ and rank $p - 1$ of a $p$-group of maximal class $G$, $p > 2$. Is it true that $\Omega_1(A) \leq G_1$?

**Exercise 174.** Let $H$ be a subgroup of index $p^n$ in a $p$-group $G$ and $H_G = \{1\}$. Then $\exp(G) \leq p^n$. Estimate $d(G)$. (**Hint.** The group $G$ is isomorphic to a subgroup of $\Sigma_{p^n} \in \mathrm{Syl}_p(S_{p^n})$.)

**Exercise 175.** Let $H \in \Gamma_1$ be a subgroup of maximal class of a $p$-group $G$. If $G$ is not of maximal class, then all maximal subgroups of $H$ are $G$-invariant. (**Hint.** Use Remark 10.5.)

**Exercise 176.** Study the regular $p$-groups of exponent $p^e$ covered by cyclic subgroups of order $p^e$.

**Answer.** The group $G$ satisfies this condition $\iff$ $|G| = |\Omega_1(G)|^e$. If $G$ is a $p$-group of maximal class, $p > 2$, $|G_1| = p^{e(p-1)}$, $e > 1$, then its fundamental subgroup $G_1$ satisfies the condition.

**Exercise 177.** Let $G_0$ be a metacyclic minimal nonabelian $p$-group

$$G_0 = \langle a, b \mid a^{p^m} = b^{p^n} = 1, \quad m > n, \quad m > 2, \quad a^b = a^{1+p^{m-1}} \rangle \cong M_p(m, n).$$

Classify the $p$-groups $G$ satisfying $\mathcal{L}_N(G) \cong \mathcal{L}_N(G_0)$.

It is interesting to study the finite groups $G$ satisfying $\mathcal{L}_N(G) \cong \mathcal{L}_N(G_0)$ for a given $p$-group $G_0$. If $G, G_0$ are groups with identical character tables, then $\mathcal{L}_N(G) \cong \mathcal{L}_N(G_0)$ (see § 178). This explains the importance of the lattice $\mathcal{L}_N(G)$. A lattice isomorphism of groups $G, G_0$ does not imply isomorphism of the lattices $\mathcal{L}_N(G)$ and $\mathcal{L}_N(G_0)$ (for example: $\mathcal{L}_N(Q_8) \cong \mathcal{L}_N(D_8)$, but $\mathcal{L}(Q_8) \ncong \mathcal{L}(D_8)$).

**Exercise 178.** Is it true that a $p$-group $G$ is modular if all its maximal cyclic subgroups are quasinormal? (**Hint.** Use the modular law.)

**Exercise 179.** Suppose that $G \ncong \Sigma_9$ is a 3-group of maximal class. Prove that if $A < G$ is such that $A \cap Z(G) = \{1\}$, then $A$ is cyclic. (**Hint.** Otherwise, $G$ contains a subgroup $\cong E_{3^3}$, contrary to Exercise 9.13.)

**Exercise 180.** Let $G$ be a $p$-group of order $> p^3$. Any two distinct maximal subgroups of $G$ have cyclic intersection $\iff$ there is in $G$ a cyclic subgroup of index $p$. (**Hint.** Use Lemma 1.4.)

**Exercise 181.** Let $G$ be a nonabelian group of order $> p^4$ and exponent $p$. If all nonabelian subgroups of $G$ of order $p^4$ are of maximal class, then $G$ is of maximal class with abelian subgroup of index $p$. (**Hint.** Use Proposition 10.17 and Exercise 10.10.)

**Exercise 182.** Present a minimal nonabelian $p$-group which is a product of two nonnormal subgroups.

**Exercise 183.** Classify the nonabelian $p$-groups $G$ in which the intersection of any two distinct maximal abelian subgroups is elementary abelian.

**Solution** (Janko). This is $\iff$ $\mathrm{Z}(H)$ is elementary abelian for any nonabelian $H \le G$. Indeed, let $H \le G$ be nonabelian and let $A, B < H$ be distinct maximal abelian and let $A \le U < G$, $B \le V < G$, where $U$, $V$ are maximal abelian in $G$; then $U \ne V$ since $\langle A, B\rangle$ is nonabelian. In that case, $A \cap B \le U \cap V$ is elementary abelian. Conversely, let $U \cap V$ be elementary abelian. As $\mathrm{Z}(H) \le A \cap B$, it follows that $\mathrm{Z}(H)$ is elementary abelian. Now let $\mathrm{Z}(H)$ be not elementary abelian for some nonabelian $H \le G$. If $A, B < H$ are as above, then $\mathrm{Z}(H) \le A \cap B$ is not elementary abelian. If $U$, $V$ are as above, then $U \cap V \ge A \cap B$ is not elementary abelian, a contradiction.

**Exercise 184.** Suppose that all normal subgroups of a $p$-group $G$ of index $p^2$ have exponent $p$ and $\mathrm{d}(G) > 2$. Prove that $\exp(G) = p$.

**Solution** (Mann). Assume that there is $x \in G$ of order $> p$. Then $H = \langle x\rangle^G \in \Gamma_1$. Let $y \in G - H$ and $K = \langle x, y\rangle$. Assume that $K \le L \in \Gamma_1$ (clearly, $K < G$ since $\mathrm{d}(G) > 2 = \mathrm{d}(K)$). Then $y \in L$ and $H = \langle x\rangle^G = L$, a contradiction since $y \notin H$.

**Exercise 185.** Let $G \cong \mathcal{H}_{2,2}$ be a nonabelian metacyclic group of order 16 and exponent 4. Is it true that $|\mathrm{Aut}(G)| = 2^5$?

**Hint.** One has $G = \langle a, b \mid a^4 = b^4 = 1,\quad a^b = a^3\rangle$. If $X = \langle x\rangle < G$ and $Y = \langle y\rangle < G$ and $X$ is normal of order 4 and $Y$ is nonnormal of order 4, then the pair $\{x, y\}$ satisfies the same relations as $a$, $b$. There are $2 \cdot 2$ possibilities for $x$ and $2 \cdot 4 = 8$ possibilities for $y$.

**Exercise 186.** Let $p > 2$ and $G$ be a nonabelian $p$-group of exponent $> p$. If all minimal nonabelian subgroups of $G$ have exponent $p$, then $\mathrm{H}_p(G)$ is a unique maximal abelian normal subgroup of $G$.

**Solution** (See Mann's commentary to Problem 115 yielding a more exact result). Let $A$ be a maximal normal abelian subgroup of $G$ and $x \in G - A$. Then, by Lemma 57.1, there is $a \in A$ such that $M = \langle a, x\rangle$ is minimal nonabelian. By hypothesis, $\exp(M) = p$ so that all elements of the set $G - A$ have order $p$. As $A$ is generated by all elements of $G$ of order $\exp(G) > p$, it follows that $A$ is uniquely determined and $A = \mathrm{H}_p(G)$.

**Exercise 187.** Study the nonabelian $p$-groups $G$ such that $M \cap \Phi(G) \le \Phi(M)$ for each minimal nonabelian $M \le G$.

**Hint.** As $\Phi(G)$ has no minimal nonabelian subgroup, it is abelian. Assume that $\Phi(G) < H$, where $H$ is nonabelian of order $p|\Phi(G)|$. Let $M \le H$ be minimal nonabelian. Then $M \cap \Phi(G) > \Phi(M)$, a contradiction. Thus, $H$ is abelian. In that case, $\Phi(G) \le \mathrm{Z}(G)$. Conversely, if $\Phi(G) \le \mathrm{Z}(G)$, then $G$ satisfies the condition.

**Exercise 188.** Classify the $p$-groups all of whose maximal subgroups, except one, are of maximal class. (**Hint.** Use Theorem 13.6.)

**Exercise 189.** Given $n > 1$, describe all minimal nonabelian subgroups of the Sylow $p$-subgroup of the holomorph of $C_{p^n}$.

**Exercise 190.** Study the primary $\mathcal{A}_n$-groups, $n > 3$, all of whose $\mathcal{A}_3$-subgroups are metacyclic (see § 202.)

**Exercise 191.** Let $G$ be a $p$-group of maximal class and order $> p^{p+1}$, $p > 3$. Then $d(G_1) > 2$.

**Solution.** By Theorem 1.2, $d(G_1) > 1$. Assume that $d(G_1) = 2$. Then, by Theorem 9.6 (e), all members of the set $\Gamma_1$ are two-generator. Write $\bar{G} = G/\mho_1(G)$; then $|\bar{G}| = p^p$ and all maximal subgroups of $\bar{G}$ are two-generator, contrary to Theorem 5.8 (b).

The same argument shows that if $G$ is a $p$-group of maximal class and order $p^{p+1}$, $p > 3$, at least one member of the set $\Gamma_1$ is not a two-generator.

**Exercise 192.** Any metacyclic $p$-group $G$ is generated by elements of equal order, unless $p = 2$ and $G/R(G) \cong SD_{2^n}$ (see §§ 124 and 148).

**Exercise 193.** Suppose that, whenever $A$ is a minimal nonabelian subgroup of a $p$-group $G$, there is $B < G$ of maximal class containing $A$ as a subgroup of index $p$. Describe the structure of $G$. (**Hint for $p = 2$.** Every minimal nonabelian subgroup of $G$ has order 8. Apply Theorem 90.1.)

**Exercise 194.** Suppose that any maximal subgroup of a $p$-group $G$ is generated by elements of equal order but $G$ is not. Prove that $G$ is neither regular nor minimal irregular.

**Solution.** Let $\exp(G) = p^e$. Since $\Omega_e^*(G) < G$, it follows that $G$ is irregular (Theorem 7.2 (b)). Assume that $G$ is minimal irregular and let $H \in \Gamma_1$ be of exponent $p^e$. Then $H = \Omega_e^*(H) = \Omega_e^*(G)$. If $F \in \Gamma_1 - \{H\}$, then $\exp(F) = p^{e-1}$ so, being regular, $F = \Omega_{e-1}^*(F) = \Omega_{e-1}(G)$. It follows that $G = F \cup H$ so that $G \in \{F, H\}$, by Lemma 116 (a), a contradiction.

**Exercise 195.** There is no noncyclic $p$-group such that exponents of its maximal subgroups are pairwise distinct. (**Hint.** If $G$ is a $p$-group and $H \in \Gamma_1$, then $\exp(H) \in \{\frac{1}{p}\exp(G), \exp(G)\}$.)

**Exercise 196.** Prove that a 2-group $G$ all of whose metacyclic sections are abelian, must be abelian. (**Hint.** Consider a minimal nonabelian subgroup of $G$.)

**Exercise 197.** Prove that there is only one noncyclic 2-group $G$ all of whose maximal subgroups, except one, have exponent 2, namely, $G \cong D_8$. If $G$ has three pairwise distinct maximal subgroups $A, B, C$, which are of exponent 2, then $\exp(G) = 2$.

**Solution.** Obviously, $G$ is nonabelian. If $G$ is minimal nonabelian, then $G \cong D_8$ (see Lemma 65.1). Otherwise, the set $\Gamma_1$ has two distinct nonabelian members, say $A$, $B$ (Exercise 1.6). Then $\exp(A) > 2$, $\exp(B) > 2$, a contradiction. Now let $A, B, C$ be of exponent 2 and assume that $G$ is nonabelian. Then $A \cap B = B \cap C = C \cap A = Z(G)$. Therefore, $A \cup B \cup C = G$ so that $\exp(G) = 2$.

**Exercise 198.** Let $G_0 \cong M_{p^{n+1}}$, $n > 2$. Classify the $p$-groups $G$ satisfying $\mathcal{L}_N(G) \cong \mathcal{L}_N(G_0)$.

Note that if $p > 2$, then $\mathcal{L}_N(M_{p^3}) \cong \mathcal{L}_N(S(p^3))$.

**Exercise 199.** If $G$ is a homocyclic group of exponent $p^e$, then all maximal cyclic subgroups of $G$ have order $p^e$.

**Solution.** One may assume that $d(G) > 1$. We use induction on $|G|$. Let $C < G$ be maximal cyclic and assume that $|C| < p^e$. Then $\exp(G/\Omega_1(C)) = p^e$. Therefore, there is a cyclic $L < G$ of order $p^e$ such that $\Omega_1(C) \nleq L$. By Exercise 4 in Introduction, $G = L \times H$, where $H$ is homocyclic. Write $\bar{G} = G/L$. Then, by induction, $\bar{C}$ is a direct factor of the homocyclic group $\bar{H} \cong H$, a contradiction since $|C| < p^e = \exp(H)$.

**Exercise 200.** Let $H$ be a subgroup of a $p$-group $G$ such that $N_G(H)$ is of maximal class. Then $G$ is of maximal class.

**Solution.** This coincides with Remark 10.5. Here we present an alternate proof. If $H$ is characteristic in $N_G(H)$, then $N_G(H) = G$, so $G$ is of maximal class. Now let $H$ be not characteristic in $N_G(H)$. Then $|N_G(H) : H| = p$, so the subgroup $N_G(H)$ of maximal class is a unique subgroup of $G$ containing $H$ as a subgroup of index $p$. Now the result follows from Exercise 10.10.

**Exercise 201.** Classify the $p$-groups of maximal class covered by cyclic subgroups of equal order.

**Solution.** Let a $p$-group $G$ of maximal class and exponent $p^e$ be covered by cyclic subgroups of order $p^e$. One may assume that $e > 1$. Then, by Theorem 13.19, $e = 2$. It is easily seen that $|G| > p^{p+1}$, unless $G \cong Q_8$. All elements of the set $G - G_1$ have order $p^2$. It follows that $\Omega_1(G) = \Omega_1(G_1)$ and hence (the fundamental subgroup) $G_1$ is covered by cyclic subgroups of order $p^2$. Then $|G_1| = p^{2(p-1)}$. We have $|G| = p^{2(p-1)+1} = p^{2p-1}$ and $|\Omega_1(G)| = p^{p-1}$, and such a $G$ satisfies the hypothesis.

**Exercise 202.** An irregular $p$-group $G = \Omega_1(G)$ is generated by subgroups of order $p^p$ and exponent $p$.

**Solution.** Let $x \in G$ be of order $p$ and let $R < G$ be of exponent $p$ of maximal order containing $x$. Assume that $|R| < p^p$. Set $N = N_G(R)$. Then $R = \Omega_1(N)$ (Theorems 7.1 (b) and 7.2 (b)) so $R$ is characteristic in $N$, and we conclude that $N = G$ so $\Omega_1(G) = R < G$, a contradiction since $R$ is regular (Theorem 7.1 (b)).

Two groups $G$, $G_0$ such that $\mathcal{L}_N(G) \cong \mathcal{L}_N(G_0)$ are called $N$-lattice isomorphic. If $\phi$ : $\mathcal{L}_N(G) \to \mathcal{L}_N(G_0)$ is a lattice isomorphism and $R \lhd G$, then instead of $R^\phi$ we write $R_0$. In that case, $\mathcal{L}_N(G/R) \cong \mathcal{L}_N(G_0/R_0)$. Let $G \cong \mathrm{M}_{p^n}$ and $G_0$ be abelian of type $(p^n, p)$. Then $\mathcal{L}(G) \cong \mathcal{L}(G_0)$ but $\mathcal{L}_N(G) \ncong \mathcal{L}_N(G_0)$.

**Exercise 203.** If groups $G$, $G_0$ are $N$-lattice isomorphic and $G_0 \cong \mathrm{E}_{p^2}$, then $G \cong G_0$.

**Solution.** We do not assume that $G$ is a $p$-group. Let $L_0, R_0 < G_0$ be distinct subgroups of order $p$ in $G_0$ and $L$, $R$ corresponding members of $\mathcal{L}_N(G)$. Then $G = L \times R$ and $L$, $R$ are minimal normal subgroups of $G$. Since $|\mathcal{L}_N(G)| = |\mathcal{L}_N(G_0)| = p + 1 > 2$, it follows that $G \cong G_0$ (if exactly one of subgroups $L$, $R$ is nonabelian, then $|\mathcal{L}_N(G)| = 2 < p+1$.)

**Exercise 204.** If $G$, $G_0$ are $N$-lattice isomorphic and $G_0 \cong \mathrm{E}_{p^n}$, $n > 1$, then $G \cong G_0$.

**Solution.** In view of Exercise 203, one may assume that $n > 2$. Let $G_0 = R_{01} \times R_{02} \times \cdots \times R_{0n}$, where all direct factors $R_{0i}$ are of order $p$. Then $G = R_1 \times R_2 \times \cdots \times R_n$, where $R_i$ corresponds to $R_{0i}$. Obviously, all $R_i$ are simple. Let $H_0 \cong \mathrm{E}_{p^{n-2}}$ and let $H$ be the corresponding member of $\mathcal{L}_N(G)$. By Exercise 203, $G/H \cong \mathrm{E}_{p^2}$. It follows that all $R_i$ have order $p$ so that $G \cong \mathrm{E}_{p^n} \cong G_0$.

**Exercise 205.** Let $G$ be a nonabelian $p$-group and let $p^e$ be a maximal exponent of a minimal nonabelian subgroup of $G$. If all subgroups of $G$ of order $\leq p^e$ are $G$-invariant, then $G$ is a Dedekindian 2-group, $G = Q \times E$, where $Q$ is a generalized quaternion group and $\exp(E) \leq 2$.

**Solution.** Any minimal nonabelian subgroup of $G$ is Dedekindian so $\cong \mathrm{Q}_8$. By Corollary A.17.3, $G = Q \times E$, where $Q \cong \mathrm{Q}_{2^n}$ and $\exp(E) \leq 2$. In that case, $e = 2$. It follows that $Q \cong \mathrm{Q}_8$ so that $G$ is Dedekindian (Theorem 1.20).

**Exercise 206.** Let $G$ and $G_0$ be $p$-groups such that $\mathcal{L}_N(G) \cong \mathcal{L}_N(G_0)$. Is it true that $\mathrm{d}(G) = \mathrm{d}(G_0)$ and $\mathrm{d}(\mathrm{Z}(G)) = \mathrm{d}(\mathrm{Z}(G_0))$? (See Exercise 204.)

**Exercise 207.** Let $G$ be a nonabelian metacyclic group of order $2^4$ and exponent $2^2$ and let $G_0$ be a nonmetacyclic minimal nonabelian group of order $2^4$ and exponent $2^2$. Prove that $G$ and $G_0$ are N-lattice isomorphic.[3]

**Exercise 208.** Classify the $p$-groups that are N-lattice isomorphic with minimal nonabelian $p$-groups.

**Exercise 209.** Classify the $p$-groups that are N-lattice isomorphic with extraspecial $p$-groups.

**Exercise 210.** Let $a_1, \dots, a_p$ be pairwise noncommuting elements of a $p$-group $G$, $n > 1$. Then there exists $a_{p+1} \in G - \{a_1, \dots, a_p\}$ such that elements $a_1, \dots, a_{p+1}$ are pairwise noncommuting.

---

3 According to Isaacs' report, these groups have the same character tables.

**Solution.** By § 116, $C_G(a_1) \cup \cdots \cup C_G(a_p) = M \subset G$. Then any $a_{p+1} \in G - M$ is not permutable with all $a_1, \ldots, a_p$.

**Exercise 211.** If a special $p$-group $G$ possesses an abelian subgroup $A$ of index $p$, then $|G| = p|Z(G)|^2$.

**Solution.** By Lemma 1.1, $A/Z(G) \cong G'$. As $G' = Z(G)$, the result follows.

**Exercise 212.** Prove that the minimal nonmetacyclic group of order $2^5$ has an abelian subgroup of index 2.

**Exercise 213.** Let $G_0$ be a minimal nonmetacyclic group of order $2^5$ and let $G$ be a 2-group such that $\mathcal{L}_N(G) \cong \mathcal{L}_N(G_0)$. Is it true that $G \cong G_0$?

**Exercise 214.** If a $p$-group $G$ of exponent $p^e > p > 2$ has no normal subgroup $\cong E_{p^3}$, then $|G| \le p^{2e+1}$ with equality if $G$ is a 3-group of maximal class. (**Hint.** Use Theorem 13.7.)

**Exercise 215.** Does there exist an irregular $p$-group $G$ satisfying $\mathcal{L}_N(G) \cong \mathcal{L}_N(G_0)$ for some regular $p$-group $G_0$ of order $> p^{p+1}$? (See Appendix 42.)

Let $A_0$ be an abelian $p$-group and let a $p$-group $A$ be such that $\mathcal{L}_N(A) \cong \mathcal{L}_N(A_0)$. Let us prove that $A$ is abelian. One may assume that $A_0$ is noncyclic. Let $L_0, M_0 < A_0$ be of order $p$, and $L, M$ be corresponding (normal) subgroups of $A$. Then, by induction, $A/L, A/M$ are abelian so that $A$ is abelian since $L \cap M = \{1\}$. Is it true that $A \cong A_0$?

**Exercise 216.** Let $G, G_0$ be groups of exponent $p$ satisfying $\mathcal{L}_N(G) \cong \mathcal{L}_N(G_0)$. Is it true that $\mathrm{cl}(G) = \mathrm{cl}(G_0)$?

**Exercise 217** ([FO3]). (a) Let $G$ be a $p$-group of maximal class and order $p^n$ with abelian subgroup $A$ of index $p$. Let $\mathcal{M}$ be a maximal set of pairwise noncommuting elements in $G$. Then $|\mathcal{M}| = p^{n-2} + 1$.

(b) Let $G$ be a 3-group of maximal class and order $3^n > 3^3$ and let $\mathcal{M}$ be as in (a). Then $|\mathcal{M}| = 3^{n-2} + 1$ if $G_1$ is abelian and $3^{n-2} + 4$ if $G_1$ is nonabelian.

**Solution.** (a) Let $x \in G - A$. Then $C_G(x)$ is of order $p^2$ and $C_G(x) - Z(G)$ has exactly $p^2 - p$ elements. Therefore, $|\mathcal{M}| - 1 = \frac{|G-A|}{p^2-p} = p^{n-2}$ so that $|\mathcal{M}| = p^{n-2} + 1$.

(b) Note that $G_1$ is either abelian or minimal nonabelian. If $x \in G - G_1$, then $|C_G(x)| = 3^2$. Therefore, as in (a), $|\mathcal{M} \cap (G - G_1)| = 3^{n-2}$. If $G_1$ is abelian, then $|\mathcal{M}| = 3^{n-2} + 1$. If $G_1$ is nonabelian, it is minimal nonabelian so that $|\mathcal{M} \cap G_1| = 3 + 1 = 4$, and the result follows in that case.

**Exercise 218.** (i) If $G, G_0$ are extraspecial $p$-groups of equal order, then $\mathcal{L}_N(G) \cong \mathcal{L}_N(G_0)$.

(ii) Let $G_0$ be an extraspecial $p$-group. Classify the $p$-groups satisfying $\mathcal{L}_N(G) \cong \mathcal{L}_N(G_0)$.

**Exercise 219.** Let $G_0$ be a special 2-group such that $|\Omega_1(G_0)| = 4$. Study the $p$-groups satisfying $\mathcal{L}_N(G) \cong \mathcal{L}_N(G_0)$.

**Exercise 220.** Let $G_0$ be a holomorph of a cyclic $p$-group. Classify the $p$-groups $G$ satisfying $\mathcal{L}_N(G) \cong \mathcal{L}_N(G_0)$.

**Exercise 221.** Let $G_0$ be a representation group of the group $\mathrm{E}_{p^n}$. Classify the $p$-groups $G$ satisfying $\mathcal{L}_N(G) \cong \mathcal{L}_N(G_0)$.

**Exercise 222.** Let $G_0$ be the central product of two $p$-groups of maximal class with amalgamated centers. Classify the $p$-groups $G$ satisfying $\mathcal{L}_N(G) \cong \mathcal{L}_N(G_0)$.

**Exercise 223.** A 2-group $G$ is metacyclic $\iff$ each subgroup of $G$ of order $\leq 2^5$ is metacyclic. (**Hint.** By Theorem 66.1, $G$ has no minimal nonmetacyclic subgroup.)

**Exercise 224.** If all abelian subgroups of maximal order of a nonabelian $p$-group of order $> p^3$ are cyclic, then $G$ is a 2-group of maximal class.

**Solution.** Assume that $G$ is not a 2-group of maximal class. Then there is $G$ a normal subgroup $R \cong \mathrm{E}_{p^2}$. Let $A < G$ be abelian of maximal order; then $A$ is cyclic, by hypothesis, and $|A| > p^2$ (Proposition 1.8). Write $H = RA$. Then $\mathrm{C}_H(R)$ is noncyclic abelian of order $|A|$, a contradiction. Thus, $R$ does not exist so $G$ is a 2-group of maximal class.

**Exercise 225.** If each minimal nonabelian subgroup $A$ of a nonabelian 2-group $G$ satisfies $\exp(\Omega_1(A)) = 2$, then $\Omega_1(G)$ is elementary abelian. (**Hint.** Consider $\langle x, y\rangle$, where $x, y \in G$ are distinct involutions.)

**Exercise 226.** Let $G$ be a regular $p$-group of exponent $p^e > p$ such that $\mho_i(G) = \Omega_{e-i}(G)$ for $i \leq e$. If $x \in G^\#$, then there is a cyclic $C \leq G$ of order $p^e$ such that $x \in C$.

**Solution.** Let $o(x) = p^k < p^e$. Then $x \in \mho_{e-k}(G)$. Therefore, there is $a \in G$ such that $a^{p^{e-k}} = x$ (see Theorem 7.2 (c)). In that case, $x \in \langle a\rangle$, where $o(a) = p^e$.

**Exercise 227.** If a two-generator $p$-group $G$ has the abelian Frattini subgroup, then the number of abelian subgroups of index $p^2$ in $G$ is $\equiv 1 \pmod p$.

**Solution.** Let $a_k(X)$ be the number of abelian subgroups of index $p^k$ in a $p$-group $X$. If $H \in \Gamma_1$, then $a_1(H) \equiv 1 \pmod p$ (Exercise 1.6). By Hall's enumeration principle, $a_2(G) \equiv \sum_{H\in\Gamma_1} a_1(H) \equiv \sum_{H\in\Gamma_1} 1 \equiv |\Gamma_1| \equiv 1 \pmod p$.

**Exercise 228.** (a) Let $G$ be an irregular $p$-group of maximal class such that $\mathrm{H}_p(G) < G$. Then $|G : \mathrm{H}_p(G)| = p$.
(b) If $G$ is a metacyclic $p$-group of order $> p^2$ such that $\mathrm{H}_p(G) < G$, then $G \cong \mathrm{D}_{2^n}$.

**Exercise 229.** If all absolutely regular subgroups of a $p$-group $G$, $p > 3$, are abelian, then $\Omega_1(G)$ is abelian. (**Hint.** Let $E$ is maximal normal elementary abelian subgroup of $G$. Use Theorem 10.1 and Lemmas 57.1 and 65.1.)

**Exercise 230.** Suppose that a $p$-group $G, p > 2$, contains exactly one proper nonmetacyclic minimal nonabelian subgroup. Then one of the following holds:
(a) $\Omega_1(G)$ is elementary abelian,
(b) $G$ contains a nonabelian subgroup $A$ of order $p^3$ and exponent $p$ such that $C_G(A)$ is cyclic.

**Solution.** Let $E < G$ be a normal elementary abelian subgroup of maximal order. Assume that $\Omega_1(G) > E$. Let $x \in \Omega_1(G) - E$ be of order $p$; then the subgroup $\langle x, A\rangle$ is nonabelian (Theorem 10.1). By Lemma 57.1, there is in $E$ an element $a$ such that $A = \langle a, x\rangle$ is minimal nonabelian. It follows from $o(x) = o(a) = p$ that $\Omega_1(A) = A$ hence $A \cong S(p^3)$. Assume that $C_G(A)$ is noncyclic. Then there is in $C_G(A) - A$ an element $y$ of order $p$. In that case, $A \times \langle y\rangle$ contains $p^2$ minimal nonabelian subgroups $\cong A$, so nonmetacyclic, a contradiction.

**Exercise 231.** Suppose that a nonabelian $p$-group $G$ contains an abelian subgroup of index $p$. Then
(a) if $x \in G - A$, then $C_A(x) = Z(G)$,
(b) any two distinct maximal abelian subgroups of $G$ have intersection $Z(G)$.

**Exercise 232.** Let a $p$-group $G = M \times E$, where $M$ is minimal nonabelian and $E > \{1\}$ is elementary abelian. Prove that all minimal, nonabelian subgroups of $G$ are isomorphic.

**Solution.** Let $H < G$ be minimal nonabelian and assume that there is $L \leq H \cap E$ of order $p$. Then $G = L \times T$, where $T \in \Gamma_1$. By the modular law, $H = L \times (H \cap T)$ is abelian, a contradiction. Thus, $H \cap E = \{1\}$, and so $G = H \times E$. Next, $H \cong G/E \cong M$.

**Exercise 233.** Let $p > 2$. If all minimal nonabelian subgroups of a nonabelian $p$-group $G$ have order $> p^3$, then $\Omega_1(G)$ is abelian.

**Solution.** Let $E < G$ be normal elementary abelian of maximal possible order and assume that there is $x \in G - E$ of order $p$. Set $A = \langle x, E\rangle$. By Theorem 10.1, $A$ is nonabelian. Therefore, by Lemma 57.1, there is $a \in E^\#$ such that

$M = \langle a, x\rangle$ is minimal nonabelian. It follows from $\Omega_1(M) = M$ that $|M| = p^3$ (Lemma 65.1), contrary to the hypothesis. Thus, $\Omega_1(G) = E$.

**Exercise 234.** Suppose that a $p$-group $G$ admits a nontrivial partition $\Sigma$. Describe orders of members of the set $\Sigma - \{H_p(G)\}$.

**Exercise 235.** Given $n$, find the nontrivial partition of $E_{p^n}$ with minimal number of components.

**Exercise 236.** Let $p > 2$. Suppose that $G$ is a $p$-group such that, whenever $A \leq G$ is minimal nonabelian, then $C_G(A) < A$. Prove that then $G$ is either of maximal class or $\Omega_1(G)$ is abelian.

**Exercise 237.** Let $G$, $G_0$ be two $p$-groups of exponent $p^e$. Suppose that $c_i(G) = c_i(G_0)$ for all $i \leq e$. Is it true that $G \cong G_0$ provided (a) $G$ and $G_0$ are minimal nonabelian, (b) $G$ is minimal nonabelian and $|G| = |G_0|$?

**Exercise 238.** Prove that a nonabelian two-generator $p$-group $G$ in which $G'$ is a maximal cyclic subgroup of $G$, is nonmetacyclic.

**Solution.** Let $L < G'$ have index $p$. Then $G/L$ is minimal nonabelian (Lemma 65.2 (a)) and $G'/L$ is a maximal cyclic subgroup of order $p$ in $G/L$. It follows from Lemma 65.1 that $G/L$ is nonmetacyclic. Then $G$ is also metacyclic.

**Exercise 239.** Does there exist a $p$-group admitting a partition all of whose components (i) are pairwise non-isomorphic, (ii) have pairwise distinct orders?

The following result is known.

**Exercise 240.** Suppose that a nonabelian $p$-group $G$ possesses an abelian subgroup of index $p$. If $|G : G'| = p^2$, then $G$ is of maximal class.

**Solution.** To prove this, we proceed by induction on $|G|$. One may assume that $|G| > p^3$. By Lemma 1.1, $A/Z(G) \cong G'$ so that $|Z(G)| = p$. By induction, $G/Z(G)$ is of maximal class. In that case $G$ is of maximal class since $Z(G)$ is of order $p$.

The following result was reported by Qinhai Zhang.

**Exercise 241.** The following conditions are equivalent for an abelian $p$-group $G$ of order $p^n$:
(a) all subgroups of index $p$ are characteristic in $G$,
(b) all subgroups of index $p$ of $G$ are pairwise nonisomorphic,
(c) all quotient groups of order $p^{n-1}$ are pairwise nonisomorphic,
(d) $G$ is cyclic.

**Solution.** It is obvious that $(d) \Rightarrow (b) \Rightarrow (a)$ and $(d) \Rightarrow (c)$. It remains to prove the implications $(a) \Rightarrow (d)$ and $(c) \Rightarrow (d)$. Let $a_1 \in G$ be of maximal order. If $G$ is not cyclic, one can write $G = \langle a_1 \rangle \times \langle a_2 \rangle \times \cdots \times \langle a_s \rangle$, where $s \geq 2$.

Let $M = \langle a_1 \rangle \times \langle a_2^p \rangle \times \langle a_3 \rangle \times \cdots \times \langle a_s \rangle$ be a subgroup of index $p$ in $G$. Take $\sigma \in \mathrm{Aut}(G)$ such that $a_1 \mapsto a_1 a_2$ and $a_i \mapsto a_i$ for $2 \leq i \leq s$. Since $M^\sigma \neq M$, $M$ is not characteristic in $G$. Thus, $(a) \Rightarrow (d)$.

Let $o(a_i) = p^{m_i}$ and $N = \langle a_2^{p^{m_2-1}} \rangle$; then $|G/N| = p^{n-1}$. Take $\delta \in \mathrm{Aut}(G)$ such that $a_2 \mapsto a_2 a_1^{p^{m_1-m_2}}$ and $a_i \mapsto a_i$ for $i \neq 2$. Then $N^\delta = K := \langle a_2^{p^{m_2-1}} a_1^{p^{m_1-1}} \rangle$ is of order $|N| = p$. As $G/N$ and $G/K$ are abelian of type $(p^{m_1-1}, p^{m_2}, \dots, p^{m_s})$, we get $G/N \cong G/K$. Thus, $(c) \Rightarrow (d)$.

**Exercise 242.** Is it true that a nonabelian $p$-group $G$ is metacyclic $\iff$ all of its metabelian subgroups are metacyclic?

**Hint.** Let $E$ be an elementary abelian subgroup of $G$ of maximal order and let $E \leq A < G$, where $A$ is maximal abelian. Let $A < B \leq G$, where $|B : A| = p$. Then $B$ is metacyclic, by hypothesis. It follows that $|E| \leq p^2$. Applying Theorem 13.7, we see that, provided $p > 2$, $G$ is metacyclic. For $p = 2$, use § 50.

**Exercise 243.** Classify the nonabelian $p$-groups $G$ such that, whenever $H \leq G$ is nonabelian, then $|H/H'| = p^2$.

**Solution.** Let $A$ be a maximal normal abelian subgroup of $G$ and $A < B \leq G$ be such that $|B/A| = p$. It follows from $|B/B'| = p^2$ that $B$ is of maximal class. One may assume that $|G| > p^3$. Let $L \triangleleft G$ be of index $p^4$. Then $G/L$ has an abelian subgroup $C/L$ of index $p$; then $|C/C'| \geq p^3$. It follows that $C$ is abelian. As $|G/G'| = p^2$, the group $G$ is of maximal class. Any $p$-group of maximal class with an abelian subgroup of index $p$ satisfies the hypothesis.

**Exercise 244.** Classify the nonabelian $p$-groups $G = \Omega_1(G)$ all of whose subgroups of index $\leq p^2$ are two-generator.

**Exercise 245.** Classify the nonabelian $p$-groups $G$ such that, whenever $C < B \leq G$, where $C$ is maximal cyclic and $|B : C| = p$, then $B \cong \mathrm{M}_{p|C|}$.

**Exercise 246.** Study the nonabelian $p$-groups $G$ such that for any $x \in \mathrm{Z}_2(G) - \mathrm{Z}(G)$ and any $y \in G - \mathrm{C}_G(x)$, the subgroup $\langle x, y \rangle$ is minimal nonabelian.

**Exercise 247.** Study the nonabelian $p$-groups $G$ such that, whenever $A < G$ is maximal abelian and $x \in A - \mathrm{Z}(G)$, then $\mathrm{C}_G(x) = A$.

# Appendix 52 Normal complements for nilpotent Hall subgroups

Suppose that a finite group $G$ has a normal Hall subgroup $N$. Let $\pi = \pi(|G : N|)$ be the set of prime divisors of $|G : N|$ and $q$ a prime divisor of $|N|$. There is in $G$ a $\pi$-Hall subgroup $H$ so that $G = HN$ and $H \cap N = \{1\}$ (Schur–Zassenhaus). Next, all maximal $(\pi \cup q)$ subgroups of $G$ are conjugate [BZ, Lemma 14.C.1 (b)], so are the $(\pi \cup q)$-Hall subgroups of $G$. In particular, if $H$ is maximal in $G$, then $N$ is a $q$-subgroup. We see that existence of a normal Hall subgroup yields a lot of useful information on the subgroup structure of $G$. Therefore, it is not surprising that many papers are devoted to establishing criteria guaranteeing the existence of such subgroups. A fairly complete exposition of this theme is contained in [Hup1, Kap. IV]. This section is also devoted to this theme and contains some additional information.

Let $p$ be a prime. A group $G$ is said to be *$p$-nilpotent* if it has a normal $p'$-Hall subgroup, say $T$ (the subgroup $T$ is said to be a *normal $p$-complement* of $G$). If $\pi$ is a set of primes, then a group $G$ is said to be *$\pi$-nilpotent* if it is $p$-nilpotent for all $p \in \pi$. In the last case, $G$ has a nilpotent $\pi$-Hall subgroup (Schur–Zassenhaus) and a normal $\pi'$-Hall subgroup $N$ so that $G = H \cdot N$ (semidirect product with kernel $N$). All sections of $\pi$-nilpotent groups are $\pi$-nilpotent.

In this appendix we present a number of conditions guaranteeing $\pi$-nilpotence. The stated below two classical results are the best-known $p$-nilpotence criteria.

**Theorem A.52.1** (Burnside; see [Hup1], Hauptsatz IV.2.6, and [Isa1], Theorem 9.13). *If $P \in \mathrm{Syl}_p(G)$ lies in the center of its normalizer $\mathrm{N}_G(P)$, then $G$ is $p$-nilpotent.*

The original Burnside's proof of Theorem A.52.1 uses the transfer homomorphism of $G$ in $P$.

**Exercise 1.** Let $p$ be the least prime divisor of $|G|$.
(a) If $P \in \mathrm{Syl}_p(G)$ is cyclic, then $G$ is $p$-nilpotent.
(b) If $P \in \mathrm{Syl}_p(G)$ is abelian of type $(p^n, p)$, $n > 1$, then $G$ is $p$-nilpotent.

**Hint.** In both cases, $q > p$ does not divide $|\mathrm{Aut}(P)|$. Apply theorem A.52.1.

The second criterion, Frobenius' Theorem A.52.2, we state in a form (equivalent to original one) that is more suitable for applications.

We recall some known results (see Lemma I in Appendix 56; see also Appendix 22). A group $G$ is said to be *minimal nonnilpotent* if it is nonnilpotent but all its proper subgroups are nilpotent. Then $G = P \cdot Q$ (semidirect product), where $G' = Q \in \mathrm{Syl}_q(G)$, $P \in \mathrm{Syl}_p(G)$, $p$ and $q$ are distinct primes. If $Q$ is abelian, then $\exp(Q) = q$. In the proof of Theorem A.52.6 the following additional information is used. If $Q$ is nonabelian, then $\exp(Q) \le q^2$ with equality only for $q = 2$. O. Schmidt [Sch4] was the first who investigated minimal nonnilpotent groups. Lemma I in Appendix 56 contains more full information on the structure of $G$.

Below we use the following fundamental result (see [BZ, Lemma 14.C.1]):

**Theorem A.52.2** (Frobenius; see [Hup1], Satz IV.5.8, and [Isa1], Theorem 9.18). *A group $G$ is not $p$-nilpotent $\iff$ it contains a minimal nonnilpotent subgroup $H$ such that $H' \in \mathrm{Syl}_p(H)$.*

In particular, $G$ is $p$-nilpotent $\iff$ $\mathrm{N}_G(A)$ is $p$-nilpotent for all $p$-subgroups $A > \{1\}$.

Theorem A.52.3 shows that Theorem A.52.1 is a consequence of Theorem A.52.2 and some results described above. Note that below we do not try to produce a proof that is easier than Burnside's original proof. Some generalizations of Theorem A.52.1 are also proved in this section.

Below we also use the following theorem (see [BZ, Lemma 14.C.1]). If $G$ is a $\pi$-solvable group and $q$ a prime, then all maximal $(\pi \cup q)$-subgroups are solvable and conjugate in $G$ (and so are Hall subgroups); a partial case of this result was mentioned above.

**Theorem A.52.3.** *Theorem A.52.2 $\Rightarrow$ Theorem A.52.1.*

*Proof.* Assume that $G$ is a counterexample of minimal order. Let $P \in \mathrm{Syl}_p(G)$ be such that $P \le \mathrm{Z}(\mathrm{N}_G(P))$; then $P$ is the nonnormal abelian subgroup of $G$ (otherwise, $G$ is $p$-nilpotent, by Theorem A.52.2). By induction, if $P \le H < G$, then $H$ is $p$-nilpotent since $\mathrm{N}_H(G) = \mathrm{N}_G(P) \cap H$ is $p$-nilpotent. Therefore, if $P_0 \le P$, then either $\mathrm{N}_G(P_0)$ is $p$-nilpotent or $P_0 \lhd G$. If $P_G = \{1\}$, then $G$ is $p$-nilpotent, by Theorem A.52.2. Therefore, one may assume that $P_0 = P_G > \{1\}$. As $\mathrm{N}_{G/P_0}(P/P_0) = \mathrm{N}_G(P)/P_0$ is $p$-nilpotent, then $G/P_0$ has a normal $p$-complement $T/P_0$, by induction; then $T \cap P = P_0$ and $G = PT$ is $p$-solvable. Let $F$ be a $p'$-Hall subgroup of $T$ (Schur–Zassenhaus); then $F$ is a $p'$-Hall subgroup of $G$. One has $T = FP_0$, $G = FP$; in that case, $G$ contains a $\{p, q\}$-Hall subgroup $K_q > P$ for all $q \in \pi(F)$. Let $Q = K_q \cap F \in \mathrm{Syl}_q(K_q) \subseteq \mathrm{Syl}_q(G)$. If $Q \lhd K_q$ for any choice of $K_q$, then $P$ normalizes $F$, contrary to the assumption. Therefore one may assume that $K_q$ is not $p$-nilpotent for some $q \in \pi(F)$. Then, by induction, $K_q = G$, and $G$ is solvable, by Burnside's two-prime theorem. The $p$-length of $G$ is $\mathrm{l}_p(G) = 1$ (Hall–Higman). As $P$ is not normal in $G$, it follows that $\mathrm{O}_q(G) = Q_G > \{1\}$, and then $PQ_G \lhd G$. By assumption, $Q_G < Q$. By Frattini's argument, $G = \mathrm{N}_G(P)PQ_G = \mathrm{N}_G(P)Q_G \Rightarrow P < \mathrm{N}_G(P)$. Let $Q_1 \in \mathrm{Syl}_q(\mathrm{N}_G(P))$. As $P < \mathrm{N}_G(P) < G$, one has $Q_1 \lhd \mathrm{N}_G(P)$, by induction, and therefore $\mathrm{N}_G(P) = P \cdot Q_1$ (semidirect product with kernel $Q_1$). It follows from $G = \mathrm{N}_G(P)Q_G = PQ_1Q_G$ that $Q_1Q_G \in \mathrm{Syl}_q(G)$. However, $P \in \mathrm{Syl}_p(G)$ normalizes $Q_1$ and $Q_G$ so $Q_1Q_G \in \mathrm{Syl}_p(G)$, and we conclude that $G = P(Q_1Q_G)$ is $p$-nilpotent. Thus, $G$ is not a counterexample. □

In his letter dated February 15, 2015, Mann wrote that it is possible to produce a shorter proof of Theorem A.52.3 using the following fact: If $L \lhd G$, $P \in \mathrm{Syl}_p(G)$, $N = \mathrm{N}_G(P)$, then $\mathrm{N}_{G/L}(PL/L) = NL/L$. Let us prove the last equality. Write $\mathrm{N}_{G/L}(PL/L) = H/L$. It follows from $N \le H$ that $\mathrm{N}_H(P) = N$. By Frattini's lemma, $H = \mathrm{N}_H(P)PL = \mathrm{N}_H(P)P = NL$ which implies $\mathrm{N}_{G/L}(PL/L) = H/L = NL/L$. Using this, let us prove Theorem A.52.3. As in the proof of Theorem A.52.3, one may assume that $G = PQ$, where $Q \in \mathrm{Syl}_p(G)$. Let $N = \mathrm{N}_G(P)$. Note that $\mathrm{l}_p(G) = 1$ ($P$ is abelian). One may assume that $Q_1 = \mathrm{O}_q(G) > \{1\}$ (otherwise, $P \lhd G$). By the above, $\mathrm{N}_{G/Q_1}(PQ_1/Q_1) = NQ_1/Q_1 < G/Q_1$. Then $NQ_1/Q_1 \cong N/(N \cap Q_1)$ is $p$-nilpotent so, by induction, $Q/Q_1 \lhd G/Q_1 \Rightarrow Q \lhd G$.

The following assertion follows easily from Theorem A.52.2:

**Proposition A.52.4** (Wong [Won2] for $p = 2$; see also [Hup1], Satz IV.3.5). *If $p$ is a minimal prime divisor of $|G|$, $P \in \mathrm{Syl}_p(G)$ and*

$$P \cong \mathrm{M}_{p^n} = \langle a, b \mid a^{p^{n-1}} = b^p = 1,\ a^b = a^{1+p^{n-2}}, \quad n > 2 \text{ and } n > 3 \text{ for } p = 2\rangle\,,$$

*then $G$ is $p$-nilpotent.*

*Proof.* Assume that $G$ is a counterexample of minimal order. Then there is in $G$ a minimal nonnilpotent subgroup $H = Q \cdot P_0$, where $P_0 = H' \in \mathrm{Syl}_p(H)$ and $Q \in \mathrm{Syl}_q(H)$, $q > p$ is a prime. It follows from $P_0 \cong \mathrm{E}_{p^2}$ that $p^2 \equiv 1 \pmod q$ so that $p = 2$, $q = 3$. Set $N = \mathrm{N}_G(P_0)$. Then $P < N \Rightarrow H < N$ so that $N$ is not $p$-nilpotent. By induction, $N = G$ so that $P_0 \lhd G$. By Burnside, $G/P_0$ has a normal $p$-complement $T/P_0$ so that $G = TP$ and $T \cap P = P_0$. Let $V$ be a $p'$-Hall subgroup of $T$. By Schur–Zassenhaus and Frattini's argument, $G = N_G(V)T = \mathrm{N}_G(V)VP_0 = \mathrm{N}_G(V)P_0$. Since $V$ is not normal in $G$, it follows that $|G : \mathrm{N}_G(V)| = p^2$. Then, by the modular law, $P = P_0(P \cap \mathrm{N}_G(V))$ so that $\Omega_1(P) = P\Omega_1(P \cap \mathrm{N}_G(V))$ has order $p^3 > p^2 = \Omega_1(P)$, a contradiction. □

**Remark 1.** Let an abelian $p$-group $P$ of type $(p^{e_1}, \dots, p^{e_n})$, $e_1 < \cdots < e_n)$ be a Sylow $p$-subgroup of $G$. If $p$ is the least prime divisor of $|G|$, then $G$ is $p$-nilpotent. This follows from Theorems 6.9 and A.52.1.

Next we use the following important results due to Wielandt and Thompson, respectively; see [Hup1], Satz IV.8.1, Hauptsatz IV.6.2):

**Theorem A.52.5.** *Let $G$ be a group and $p \in \pi(G)$.*
(a) *(Wielandt). If $P \in \mathrm{Syl}_p(G)$ is regular and $P$ is a direct factor of* $\mathrm{N}_G(P)$, *then $G$ is $p$-nilpotent.*
(b) *(Thompson) Let $p$ be an odd prime divisor of $|G|$ and $P \in \mathrm{Syl}_p(G)$. If $G$ is not $p$-nilpotent, then there exists a characteristic subgroup $P_0 > \{1\}$ of $P$ such that* $\mathrm{N}_G(P_0)$ *is not $p$-nilpotent.*

Theorems A.52.5 (a,b) are deep generalizations of Theorems A.52.1 and A.52.2, respectively (the second one only for $p > 2$).

**Definition 1.** A $p$-group $P$ is said to be a $(*)$-group if, whenever $P \in \mathrm{Syl}_p(G)$ and $\mathrm{N}_G(P)$ is $p$-nilpotent, so is $G$.

**Definition 2.** A $p$-group $P$ is said to be *good* if it satisfies the following conditions:
(1) $\Omega_1(P) \le \mathrm{Z}(P)$ and, if $p = 2$, then $\Omega_2(P)$ is abelian, and

(2) the quotient group $P/R$ is a $(*)$-group for every noncyclic $R \leq \Omega_1(Z(P))$.

So defined property, generally speaking, is not inherited by epimorphic images (see paragraph preceding Exercise 2). The group $P = M_{p^{n+1}}$, $n > 2$, is not good since $\Omega_1(P) \nleq Z(P)$. The group $Q_8$ is not good since $\Omega_2(Q_8)$ is nonabelian.

Let $G$ be a metacyclic minimal nonabelian $p$-group. If $G$ has no cyclic subgroup of index $p$ and $|G| > 2^4$ for $p = 2$, it is good.

Let $G$ be a nonabelian good $p$-group. It follows from commutativity of $\Omega_2(G)$ that $Z(G)$ is noncyclic (Proposition 1.3). Any metacyclic $p$-group of exponent $2^e$ and order $2^{2e}$ without a nonabelian section of order $2^3$ is good (see § 148); then $e > 2$.

Let a minimal nonabelian $p$-group, $p > 2$,

$$P = \langle a, b \mid o(a) = o(b) = p^2, \quad [a, b] = c, \quad o(c) = p, \quad [a, c] = [b, c] = 1\rangle$$

of order $p^5$. Set $R = \langle a^p, b^p\rangle (\cong E_{p^2})$. Then $P/R \cong S(p^3)$ is a $(*)$-group (Theorem A.52.5 (a)) hence $P$ is good.

Let a minimal nonabelian group

$$P = \langle a, b \mid a^{p^m} = b^{p^n} = 1,\ m > 1,\ n > 2,\ a^b = a^{1+p^{m-1}}\rangle$$

be metacyclic of order $p^{m+n}$. Then $P$ is a good $p$-group since $\Omega_1(P) < Z(P)$, $\Omega_2(P)$ is abelian of type $(p^2, p^2)$ and, if (noncyclic) $R \leq \Omega_1(G)$, then $G' < R$ so that $G/R$ is abelian hence is a $(*)$-group (Theorem A.52.1). However $G/\mho_1(\langle b\rangle) \cong M_{p^{m+1}}$ is not good. (If $m = n = 2$, then $P$ is not good since $\Omega_2(G) = G$ is nonabelian.)

**Exercise 2.** Classify the good minimal nonabelian $p$-groups.

In what follows we use the following Wielandt's $D_\pi^n$-theorem [Wiel5] (the notation $D_\pi^n$ is due to P. Hall): If a group contains a nilpotent $\pi$-Hall subgroup, then all its maximal $\pi$-subgroups are conjugate.

Wielandt (see [Wiel2] or [Hup1], Satz IV.7.3) has proved that if $H$ is a Hall but not a Sylow subgroup of a group $G$ and $N_G(P) = H$ for all $P \in \mathrm{Syl}(H)$, then $H$ has a normal complement in $G$, i.e., $G$ is $\pi$-nilpotent.

**Theorem A.52.6.** *Let $p$ be a minimal prime divisor of the order of a finite group $G$ and let $P \in \mathrm{Syl}_p(G)$ be good. If $N_G(P)$ is $p$-nilpotent, so is $G$.*

*Proof.* One may assume that $P$ is not normal in $G$ (otherwise, there is nothing to prove further). One has to show that $G$ is $p$-nilpotent.

Assume that $G$ is a counterexample of minimal order. Then, by Theorem A.52.2, there is in $G$ a minimal nonnilpotent subgroup $S = Q_1 \cdot S'$ (semidirect product), where $S' \in \mathrm{Syl}_p(S)$, $Q_1 \in \mathrm{Syl}_q(S)$ is cyclic and a prime $q > p$. One may assume that $S' \leq P \in \mathrm{Syl}_p(G)$. By Lemma A.22.1, $S'$ is noncyclic. By induction:

(i) If $P \leq T < G$, then $T$ is $p$-nilpotent since $N_T(P) = T \cap N_G(P)$ is $p$-nilpotent.

By Lemma A.22.1, $\exp(S') = p$ if $p > 2$, and in that case $S'$ is abelian. If $p = 2$, then $S \leq \Omega_2(P)$, and so $S'$ is abelian again (Definition 2); then, by Lemma A.22.1, $\exp(S') = 2$, and so, by Definition 2, $S' \leq Z(P)$.

(ii) We claim that $S' \lhd G$. Indeed, assume that $N = N_G(S') < G$. As $P < N$, the subgroup $N$ is $p$-nilpotent, and this is a contradiction since the non-$p$-nilpotent subgroup $S \leq N$. Since $S'(\leq Z(P))$ is noncyclic, $P/S'$ is a $(*)$-group (Definition 2).

Write $\mathcal{N} = N_G(P)$. As we have noted, $\mathcal{N} < G$. By hypothesis, $\mathcal{N}$ is $p$-nilpotent. However $N_{G/S'}(P/S') = \mathcal{N}/S' < G/S'$ is $p$-nilpotent so is $G/S'$ since $P/S'$ is a $(*)$-group. Let $T/S'$ be the (normal) $p'$-Hall subgroup of $G/S'$; then $G = PT$ with $P \cap T = S'$ and $T \lhd G$. The group $G$ is $p$-solvable so it contains a $\{p \cup q\}$-Hall subgroup $K$, and one may assume that $K = PQ$, where $Q \in \mathrm{Syl}_q(G)$ (see [BZ, Lemma 14.C.1]). Since all maximal $\{p \cup q\}$-subgroups of $G$ are conjugate [BZ, Lemma 14.C.1 (b)], one may assume, in addition, that $S \leq K$ so that $K$ is not $p$-nilpotent. Therefore, as $|G : K|$ is a $p'$-number, we get $(PQ =)K = G$, by (i) (in particular, $S < G$); then $G$ is solvable, by Burnside's two-prime theorem. One has to prove that $Q \lhd G$.

Write $C = C_G(S')$; then $C \lhd G$ ($C \neq G$ since $S \not\leq C$). By Definition 2, $P \leq C$. Since $P$ is not normal in $G$, it follows that $P$ is not normal in $C$ (otherwise, $P$ is characteristic in $C$) so that $C > P$ is nonnilpotent. By (i), $C$ has a characteristic $p'$-Hall subgroup $C_1$ (in fact, $C_1$ is a $q$-subgroup since $G$ is a $\{p, q\}$-group); then $\{1\} < C_1 \lhd G$, and hence $C_1 < Q$ since $Q$ is not normal in $G$, by assumption. Then:

(iii) $PC_1 < G$. Write $\bar{G} = G/C_1$. In that case, $\bar{P} \cong P$ is good.

Assume that $PC_1 \lhd G$. Then $\bar{P} = \overline{PC_1} \lhd \bar{G}$. In that case, $\bar{G}$ has a normal $p'$-Hall subgroup $\bar{Q}$ (see the paragraph following the proof of Theorem A.52.3). Thus, $Q \lhd G$, and we are done in this case.

Next assume that:

(iv) $PC_1$ is not normal in $G$.

(v) By the paragraph following the proof of Theorem A.52.3, $N_{\bar{G}}(\bar{P}) = \mathcal{N}C_1/C_1$ is $p$-nilpotent. Then, by induction, $\bar{G}$ is $p$-nilpotent, i.e., $\bar{Q} \lhd \bar{G} \Rightarrow Q \lhd G$ since $C_1 < Q$, i.e., $G$ is not a counterexample. □

**Remark 2.** Let $P$ be a Sylow subgroup of a group $G$. We claim that $P$ is a direct factor of $PC_G(P)$. Without loss of generality, one may assume that $PC_G(P) = G$. Then $C_G(P) \leq N_G(P) \Rightarrow P \lhd G$ so that $C_G(P) \lhd G$. As $|G : C_G(P)|$ is a power of $p$, any $p'$-element of $G$ centralizes $P$, and now $p$-nilpotence of $G$ follows from Theorem A.52.2.

**Theorem A.52.7.** *Let $H$ be a $\pi$-Hall subgroup of a group $G$ and all Sylow subgroups of $H$ are either good or $(*)$-groups. Let $\pi_0 = \{p \in \pi \mid P \in \mathrm{Syl}_p(G)$ is good$\}$. Suppose, in addition that, whenever $p \in \pi_0$ and $q \in \pi' \cap \pi(G)$, then $p < q$. If $N_G(H)$ is $\pi$-nilpotent so is $G$.*

*Proof.* As $N_G(H)$ is $\pi$-nilpotent, then $H$ is nilpotent. It will suffice to show that $G$ is $p$-nilpotent for every prime $p \mid |H|$. Assume that $G$ is a counterexample of minimal order; then $H$ is not normal in $G$. By induction, if $H \leq T < G$, then $T$ is $\pi$-nilpotent. By assumption, there is a prime $p \mid |H|$ such that $\mathrm{N}_G(P)$ is not $p$-nilpotent for $P \in \mathrm{Syl}_p(H)$. Therefore, since $H < \mathrm{N}_G(P)$ ($<$ since $\mathrm{N}_G(P)$ is not $\pi$-nilpotent and $H$ is nilpotent), we must have $\mathrm{N}_G(P) = G$. Indeed, if $\mathrm{N}_G(P) < G$, then $\mathrm{N}_G(P)$ is $\pi$-nilpotent so $p$-nilpotent; then $G$ is $p$-nilpotent, by Theorems A.52.5 (a) and A.52.6, contrary to the choice of $P$. Thus, $P \lhd G$. Write $T = P\mathrm{C}_G(P)$; then $T$ is $G$-invariant and $H < T$ ($<$ since $H$ is not $G$-invariant).

Assume that $T < G$. Then $T$ is $\pi$-nilpotent, by induction. Let $D$ be the $\pi'$-Hall subgroup of $T$ so that $T = H \cdot D$ (semidirect product with kernel $D$); then $D \lhd G$ and $D > \{1\}$ since $H < T \lhd G$. Write $\bar{G} = G/D$. By Schur–Zassenhaus' theorem and Frattini's argument, one has $G = \mathrm{N}_G(H)T = \mathrm{N}_G(H)HD = \mathrm{N}_G(H)D$. As $\mathrm{N}_G(H)$ is $\pi$-nilpotent, the group $\bar{G} = \mathrm{N}_G(H)D/D \cong \mathrm{N}_G(H)/(\mathrm{N}_G(H) \cap D)$ is $\pi$-nilpotent, so has the normal $\pi'$-Hall subgroup $\bar{Q}$. Then $Q$, the inverse image of $\bar{Q}$ in $G$, is a normal complement to $H$ in $G$ so that $G = H \cdot Q$ (semidirect product with kernel $Q$), being $\pi$-nilpotent, is not a counterexample.[1]

Thus, $T = G$; then $P\mathrm{C}_G(P) = T = G$. In that case, $P$ is a direct factor of $G$ (Remark 23) so $G$, being $p$-nilpotent, is not a counterexample. The proof is complete. □

**Corollary A.52.8** (Carter [Car1]). *Let $H < G$ be a nilpotent $\pi$-Hall subgroup of a group $G$ all of whose Sylow subgroups are regular. If $\mathrm{N}_G(H)$ is $\pi$-nilpotent, so is $G$.*

Indeed, by Theorem A.52.5 (a), regular $p$-groups are $(*)$-groups so the result follows from Theorem A.52.7.

**Corollary A.52.9.** *Let $H$ be a proper nilpotent $\pi$-Hall subgroup of a group $G$ and suppose that all Sylow subgroups of $H$ are either good or $(*)$-groups. Suppose that $\pi_0 = \{p \in \pi \mid P \in \mathrm{Syl}_p(G)$ is good$\}$ and, if $p \in \pi_0$ and $q \in \pi' \cap \pi(G)$, then $p < q$. If $H \leq T < G$, implying $\pi$-nilpotence of $T$, then one of the following holds:*
(a) *$G$ is $\pi$-nilpotent,*
(b) *$G = Q \cdot H$ (semidirect product with kernel $H$), where $Q \in \mathrm{Syl}_q(G)$ is cyclic, $q$ a prime and $\mho_1(Q) \leq \mathrm{Z}(G)$.*

*Proof.* Let $H$ be not normal in $G$. Then $\mathrm{N}_G(H)$ is $\pi$-nilpotent, by hypothesis. In that case, by Theorem A.52.7, $G$ is $\pi$-nilpotent, i.e., we have case (a). As $\pi$-nilpotence is inherited by subgroups, all containing $H$ subgroups of $G$ are $\pi$-nilpotent.

Now assume that $G$ is not $\pi$-nilpotent. Then $H \lhd G$, by Theorem A.52.7. By the Schur–Zassenhaus theorem, there is $Q < G$ such that $G = Q \cdot H$ (semidirect product with kernel $H$) and $Q$ is not normal in $G$. Let $M < Q$ be maximal. Then, by hypothesis, $H, M \lhd MH$ so that $MH = M \times H$. Thus, all maximal subgroups of $Q$ centralize $H$ so that $Q$ is not generated by its maximal subgroups, and we conclude that $Q$ is a cyclic $q$-group for some prime $q \notin \pi$. By the above, $\mho_1(Q)$, the unique maximal subgroup

1 Here one can use the same argument as in the paragraph following the proof of Theorem A.52.3.

of $Q$, centralizes $H$, and so $\mho_1(Q) \leq Z(G)$. A group with such structure satisfies the hypothesis. □

**Exercise 3.** Suppose that $H$ is a proper abelian $\pi$-Hall subgroup of a nonnilpotent group $G$ such that, whenever $H \leq T < G$, then $T$ is $\pi$-nilpotent. Then one of the following holds:
(a) $G$ is $\pi$-nilpotent,
(b) $G = Q \cdot H$ (semidirect product with kernel $H$), where $Q \in \mathrm{Syl}_q(G)$ is cyclic, $q$ a prime and $\mho_1(Q) \leq Z(G)$.

*Proof.* This follows from Corollary A.53.9. □

**Proposition A.52.10.** *Let $H$ be a nilpotent $\pi$-Hall subgroup of a group $G$, $2 \notin \pi$. Suppose that, whenever $H \leq T < G$, then $T$ is $\pi$-nilpotent. Then one of the following holds:*
(a) *$|G : H|$ is a prime power and $G$ is solvable,*
(b) *$G$ is $\pi$-nilpotent.*

*Proof.* (i) Suppose that $G$ is $\pi$-solvable and $|G : H|$ is not a prime power. Then there is in $G$ a $\pi'$-Hall subgroup $F$. For every prime $q \mid |F|$, there is $Q \in \mathrm{Syl}_q(G)$ such that $K_q = HQ = QH < G$. By hypothesis, $Q \triangleleft K_q$. Without loss of generality, one may assume that $Q \leq F$. Since $|K_q \cap F| = |Q|$, one has $K_q \cap F = Q$. Thus, for every choice of a prime $q \mid |F|$, $H$ normalizes some $Q \in \mathrm{Syl}_q(F)$. It follows that $F \triangleleft G$ so $G$ is $\pi$-nilpotent, and we have case (b). If $G$ is $\pi$-solvable and not $\pi$-nilpotent, then $|G : H|$ is a prime power, by the above.

(ii) If $|G : H|$ is a prime power, then $G$ is solvable, by Wielandt's Theorem A.28.14.

(iii) Now assume that $G$ is nonsolvable: then $|G : H|$ is not a prime power. We have to prove that $G$ is $\pi$-nilpotent. By (i), it suffices to show that $G$ is $\pi$-solvable. Assume that $G$ is a counterexample of minimal order. Then $G$ is not $p$-nilpotent for some $p \in \pi$. By Theorem A.52.5 (b), there is in $P \in \mathrm{Syl}_p(H)$ a nonidentity characteristic subgroup $P_0$ such that $\mathrm{N}_G(P_0)$ is not $p$-nilpotent. Then $H < \mathrm{N}_G(P_0)$, and therefore $\mathrm{N}_G(P_0) = G$, i.e., $P_0 \triangleleft G$. Then the pair $H/P_0 < G/P_0$ satisfies the hypothesis. By induction, $G/P_0$ is $\pi$-solvable, so is $G$. It is easy to see that $G$ is as in (a). □

**Remark 3.** Suppose that a nilpotent subgroup $H$ is maximal in a group $G$ and all Sylow subgroups of $H$ are either $(*)$-groups or good. Let $\pi_0 = \{p \in \pi \mid P \in \mathrm{Syl}_p(G) \text{ is good}\}$. Suppose, in addition that, whenever $p \in \pi_0$ and $q \in \pi_0' \cap \pi(G)$, then $p < q$. Then $G$ is solvable. To prove, we use induction on $|G|$. If $H$ is a Hall subgroup of $G$, then either $H \triangleleft G$ or $\mathrm{N}_G(H) = H$. In the first case $G/H$ is of prime order, so $G$ is solvable. In the second case, $G = H \cdot Q$ is a semidirect product with kernel $Q \triangleleft G$, by Theorem A.52.7. Then $G$ is $\pi$-solvable, where $\pi = \pi(H)$. In that case, by [BZ, Lemma 14.C.1], for any $q \in \pi(Q)$, one has $H < T \leq G$, where $T$ is a $(\pi \cup q)$-Hall subgroup of $G$. It follows that $T = G$. Since $G$ is solvable [Wiel2], we are done in this case. Now suppose that $H$ is not a Hall subgroup of $G$. In that case, there is a prime $p \mid |H|$ such that $P \in \mathrm{Syl}_p(H)$

is not a Sylow subgroup of $G$. Then $N_G(P) = G$, and $G/P$ is solvable, by induction. In that case, $G$ is solvable.

**Exercise 4** (Thompson [Tho2]). If a group $G$ has a maximal subgroup that is nilpotent of odd order, then $G$ is solvable. (**Hint.** Use Theorem A.52.10.)

**Exercise 5.** Let $H = P \times F$ be a nilpotent maximal subgroup of a group $G$, where $P \in \mathrm{Syl}_2(H)$ is either good or a $(*)$-group. Then $G$ is solvable.

**Exercise 6.** Suppose that $P \in \mathrm{Syl}_2(G)$ is nonabelian Dedekindian. If $N_G(P)$ is 2-nilpotent, so is $G$.

**Problem.** Let $P \in \mathrm{Syl}_2(G)$ have no nonabelian sections of order 8. Is it true that if $G$ is not 2-nilpotent, then there exists a characteristic subgroup $P_0 > \{1\}$ of $P$ such that $N_G(P_0)$ is not 2-nilpotent.

# Appendix 53 *p*-groups with large abelian subgroups and some related results

This appendix, presenting a survey of some papers of the $p$-group theorists from Shanxi Normal University, was written using material presented by Qinhai Zhang.

Chinese mathematicians L.K. Hua and H.F. Tuan many years ago wrote a number of interesting papers devoted, in particular, to *Anzahl* theorems on $p$-groups. Their investigations were continued by modern Chinese $p$-group theorists, mainly from Shanxi Normal University.

The following identities are used in computations in metabelian groups:

**Theorem A.53.0.** *Let $G$ be a metabelian group and $a, b \in G$. For any positive integers $i$ and $j$, let*

$$[ia, jb] = [a, b, \underbrace{a, \dots, a}_{i-1}, \underbrace{b, \dots, b}_{j-1}]$$

*(in particular, $[1a, 1b] = [a, b]$). Then:*

(a) *For any positive integers $m$ and $n$, one has*

$$[a^m, b^n] = \prod_{i=1}^{m}\prod_{j=1}^{n}[ia, jb]^{\binom{mi}{\binom{n}{j}}}\,.$$

(b) *Let $n$ be a positive integer. Then:*

$$(ab^{-1})^n = a^n \prod_{i+j\le n}[ia, jb]^{\binom{n}{i+j}} b^{-n}\,.$$

The following result is basic:

**Theorem A.53.1** ([? ]). *Suppose that $G$ is a nonabelian two-generator $p$-group having an abelian subgroup $A$ of index $p$, $|G/Z(G)| > p^2$. Then $G/Z(G)$ is of maximal class.*

*Proof.* One may assume that $G$ is not an $\mathcal{A}_1$-group; then $|G/Z(G)| > p^2$. Let $b \in G - A$, and $a \in A - \Phi(G)$. Then $G = \langle a, b\rangle$. Let $\mathrm{cl}(G) = c$. Since $G$ is not minimal nonabelian, we get $c > 2$ (Theorem 102.1). For $2 \le i \le c$, one has $\mathrm{K}_i(G) = \langle [a, b, (i-2)b], [b, a, (i-2)b], \ \mathrm{K}_{i+1}(G)\rangle$ [Suz1, vol. 2, page 42, Theorem 3.8] (here $[a, b, (i-2)b] = [a, (i-1)b]$, $[b, a, (i-2)b] = [b, a, \underbrace{b, \dots, b}_{i-2}]$). Note that $G' < A$ and $A$ is abelian. Since $G'$ is abelian, we get

$$1 = [xx^{-1}, b] = [x, b][x, b, x^{-1}][x^{-1}, b] = [x, b][x^{-1}, b] \quad \text{for all } x \in G'\,.$$

Hence $[x, b]^{-1} = [x^{-1}, b]$ for all $x \in G'$. Noting that $[b, a] = [a, b]^{-1}$, we get $[[b, a], b] = [[a, b]^{-1}, b] = [a, b, b]^{-1}$. By induction on $c$ (the equality in the previous sentence is the induction basis), one has $[b, a, (i-2)b]^{-1} = [a, b, (i-2)b] = [a, (i-1)b]$. Hence, for $2 \le i \le c$, one obtains

$$\mathrm{K}_i(G) = \langle [a, (i-1)b], \ [a, (i-1)b]^{-1}, \ \mathrm{K}_{i+1}(G)\rangle = \mathrm{K}_i(G) = \langle [a, (i-1)b], \ \mathrm{K}_{i+1}(G)\rangle.$$

In particular, $K_c(G) = \langle [a, (c-1)b] \rangle$ is cyclic. It follows from $[a, (c-1)b] \in K_c(G) \le Z(G)$ and $b^p \in A$ that $[a, (c-1)b]^p = [[a, (c-2)b], b]^p = [a, (c-2)b, b^p] = 1$, and hence $|K_c(G)| = p$. Since $cl(G/K_{i+1}(G)) = i$, we get, by the above, $|K_i(G/K_{i+1}(G))| = p$ for $2 \le i \le c$. Hence $|K_i(G)| = p^{c-i+1}$ for $2 \le i \le c$. In particular, since $K_2(G) = G'$, we get $|G'| = p^{c-2+1} = p^{c-1}$. By Lemma 1.1, $|G| = p|G'||Z(G)|$. Hence $|G/Z(G)| = p|G'| = p^c$. Since $cl(G/Z(G)) = c - 1$, the quotient group $G/Z(G)$ of order $p^c$ is of maximal class, completing the proof. □

**Exercise 1** ([? ]). Suppose that $G$ is a nonabelian two-generator $p$-group having an abelian subgroup $A$ of index $p$. Let $M \in \Gamma_1$ be nonabelian. Then $Z(M) = Z(G)$.

**Solution.** Since $Z(G) \le \Phi(G) < M$, then $Z(G) \le Z(M)$. By Lemma 1.1, $|G| = p|G'||Z(G)|$ and $|M| = p|M'||Z(M)|$. Hence

$$|G'/M'| = p|Z(M)/Z(G)| \tag{1}$$

so that $G' > M'$ and we conclude that $G/M'$ is nonabelian.

Let $\bar{G} = G/M'$. Then $\bar{G}$ is not abelian since $G'/M' \ne 1$. Since the two-generator $p$-group $G/M'$ has two distinct abelian maximal subgroups $A/M'$ and $M/M'$, it follows that it is minimal nonabelian hence $|G'/M'| = p$ (Lemma 65.1). It follows from the last equality and (1) that $Z(M) = Z(G)$.

**Theorem A.53.2.** *Suppose that $G$ is a $p$-group, $p > 2$, and $N$ is a $G$-invariant subgroup of $G'$ of order $p$ such that $G/N$ is minimal nonabelian; then* $d(G) = 2$ *so that* $|\Gamma_1| = p + 1$.

(a) *If $G/N \cong M_p(m, n)$, then $G$ has $p$ pairwise distinct minimal nonabelian subgroups of index $p$.*
(b) *If $G/N \cong M_p(m, 1, 1)$, then $G$ has $p - 1$ pairwise distinct minimal nonabelian subgroups of index $p$.*
(c) *If $G/N \cong M_p(m, n, 1)$, where $m \ge n \ge 2$, then $G$ has no minimal nonabelian subgroup of index $p$.*

*In particular, in parts (a) and (b), $G$ is an $\mathcal{A}_2$-group since all its nonabelian maximal subgroups are minimal nonabelian.*

*Proof.* By Lemma 65.1,

$$d(G/N) = 2\,, \quad |(G/N)'| = |G'/N| = p \Rightarrow |G'| = |G/N|\,|N| = p^2$$

so that $G$ is not minimal nonabelian. Let

$$\mathcal{M} = \{H \mid \text{ where } H \in \Gamma_1 \text{ is nonabelian}\}\,, \quad \mathcal{N} = \{H \mid \text{ where } H \in \Gamma_1 \text{ is abelian}\}\,.$$

Since $d(G) = 2$, one has $|\mathcal{M}| + |\mathcal{N}| = |\Gamma_1| = p + 1$. Since $G$ is not minimal nonabelian, $|\mathcal{N}| < p + 1$. It follows from Exercise 1.6 (a) that $|\mathcal{N}| \le 1$ (indeed, if $|\mathcal{N}| > 1$, then $G$ is minimal nonabelian, which is not a case). Hence $|\mathcal{M}| \ge p \ge 3$.

For all $H \in \mathcal{M}$, the subgroup $H/N$ is maximal in $G/N$. Since $G/N$ is minimal nonabelian, we conclude that $H/N$ is abelian and hence $H' \le N$. Since $H' \ne \{1\}$, we get $N = H' \le \Phi(H)$, which implies $\mathrm{d}(H) = \mathrm{d}(H/N)$.

(a) If $G/N \cong \mathrm{M}_p(m, n)$, then $\mathrm{d}(H/N) \le 2$ for all maximal subgroups $H$ of $G$. Hence $\mathrm{d}(H) = 2$ for all $H \in \mathcal{M}$. Therefore, since $\mathrm{d}(H) = 2$, $H$ is minimal nonabelian for all $H \in \mathcal{M}$ (Lemma 65.2 (a)). Since $|\mathcal{M}| \ge p \ge 3$, $G$ has $p > 2$ pairwise distinct minimal nonabelian subgroups of index $p$.

(b) If $G/N \cong \mathrm{M}_p(m, 1, 1)$, write $M/N = \Omega_{m-1}(G/N)$; then $M \in \Gamma_1$ and $\mathrm{d}(M) \ge \mathrm{d}(M/N) = 3$. Since $\mathrm{d}(H/N) \le 2$ for all $H/N \nleq M/N$, one obtains $\mathrm{d}(H) = 2$ for all $H \in \mathcal{M} - \{M\}$ (indeed, $N = H' \le \Phi(H)$). Hence $H$ is minimal nonabelian for all $H \in \mathcal{M} - \{M\}$ (Lemma 65.2 (a)). Since $|\mathcal{M}| \ge p \ge 3$, $G$ has $p - 1 \ge 2$ pairwise distinct minimal nonabelian subgroups of index $p$.

(c) If $G/N \cong \mathrm{M}_p(m, n, 1)$, where $m \ge n \ge 2$, then, since $\mathrm{d}(\Phi(G/N)) = 3$, one has $\mathrm{d}(H/N) = 3$ for all maximal subgroups $H/N$ of $G/N$. Hence $\mathrm{d}(H) = 3$ for all $H \in \mathcal{M}$. In that case, $H$ is not minimal nonabelian for all $H \in \mathcal{M}$, i.e., $G$ has no minimal nonabelian subgroup of index $p$. □

**Exercise 2.** Suppose that $G$ is a 2-group and $N$ is a $G$-invariant subgroup of $G'$ of order 2 such that $G/N$ is minimal nonabelian.
(a) If $G/N \cong \mathrm{M}_2(m, n)$, then $G$ has 2 distinct minimal nonabelian subgroups of index 2.
(b) If $G/N \cong \mathrm{M}_2(m, 1, 1)$, then $G$ has a minimal nonabelian subgroup of index 2.
(c) If $G/N \cong \mathrm{M}_2(m, n, 1)$, where $m \ge n \ge 2$, then $G$ has no minimal nonabelian subgroup of index 2.

**Exercise 3.** If a $p$-group $G$ is an $\mathcal{A}_n$-group with regular $G'$, then $\exp(G') \le p^n$.

**Solution.** We proceed by induction on $n$. This is true for $n = 1$ (Lemma 65.1). Let $A, B \in \Gamma_1$ be distinct. Then $A, B$ are $\mathcal{A}_2$- and $\mathcal{A}_s$-subgroups, where $r, s \le n - 1$ so that $|A'|, |B'| \le p^{n-1}$, by induction, and we conclude that $\exp(A'B') \le p^{n-1}$ (Theorem 7.2 (b)). Next, $A', B' \trianglelefteq G$. By Exercise 1.69, $|G' : A'B'| \le p$ hence $\exp(G') \le p \exp(A'B') \le p^n$. (By Proposition 72.1, if a $p$-group $G$ is a metacyclic $\mathcal{A}_n$-group, then $|G'| = p^n$.)

**Proposition A.53.3.** *Suppose that $G$ is a $p$-group, $N \le \mathrm{Z}(G)$ and $G/N$ is an $\mathcal{A}_1$-group. Then $G = H * N$, where $\mathrm{d}(H) = 2$ and there exists a subgroup $K \le \mathrm{Z}(H) \cap H'$ with $\mathrm{d}(K) \le 3$ such that $H/K$ is an $\mathcal{A}_1$-group.*

*Proof.* Let $G/N = \langle aN, bN\rangle$ and $H = \langle a, b\rangle$. Then $\mathrm{d}(H) = 2$ and $G = H * N$. Hence $H/(H \cap N) \cong G/N$ is an $\mathcal{A}_1$-group. Set $K = \Phi(H')\mathrm{K}_3(H)$. It follows that $K \le H \cap N \le \mathrm{Z}(H)$. and $\exp(H'/K) = p$. Therefore, since $\mathrm{d}(H) = 2$, and $K < H'$, $H/K$ is an $\mathcal{A}_1$-group, by Lemma 65.2 (a). It follows from $\mathrm{K}_3(H) \le N \le \mathrm{Z}(G)$ that $\mathrm{K}_4(H) = 1$. Therefore, the subgroup $H' = \langle [a,b], [a,b,a], [a,b,b] \rangle$ [Suz1, vol. 2, page 42, Theorem 3.8] is abelian. It follows that $\mathrm{d}(K) \le \mathrm{d}(H') \le 3$. □

*Proof.* Let $G/N = \langle aN, bN \rangle$ and $H = \langle a, b \rangle$. Then $\mathrm{d}(H) = 2$ and $G = HN$. Hence $H/(H \cap N) \cong G/N$ is an $\mathcal{A}_1$-group. It follows that $\Phi(H')\mathrm{K}_3(H) \le H \cap N \le \mathrm{Z}(H)$. Let $K = \Phi(H')\mathrm{K}_3(H)$. Since $\mathrm{d}(H) = 2$, $H/K$ is an $\mathcal{A}_1$-group. Since $\mathrm{K}_3(H) \le \mathrm{Z}(G)$, $\mathrm{K}_4(H) = 1$. Hence $H' = \langle [a,b], [a,b,a], [a,b,b] \rangle$ is abelian. It follows that $\mathrm{d}(K) \le \mathrm{d}(H') \le 3$. □

**Corollary A.53.4.** *Let $G$ be a 2-group. If $N \le \mathrm{Z}(G) \cap G'$ and $G/N \cong \mathrm{Q}_8$, then $N = \{1\}$.*

*Proof.* It follows from $N \le G' \le \Phi(G)$ that $\mathrm{d}(G) = \mathrm{d}(G/N) = 2$. If $N < H \in \Gamma_1$, then $H$ is abelian. It follows that $G$ is minimal nonabelian so that $|N| \le |G'| = p$. As $G/N$ is nonabelian, we get $N < G'$; hence $N = \{1\}$. □

**Remark 1.** Let $G$ be a 2-group. If $N \le \mathrm{Z}(G) \cap G'$ and $G/N \in \{\mathrm{Q}_{2^n}, \mathrm{SD}_{2^n}\}$, then $N = \{1\}$. Indeed, by Proposition 1.6, $G$ is a 2-group of maximal class since $|G : G'| = 4$. Assume that $N > \{1\}$. Let $R < N$ be $G$-invariant of index 2. Then, by Theorem 1.2, $G/R$ is not of maximal class, a contradiction.

**Exercise 4.** Let $G$ be a $p$-group. If $N \le \mathrm{Z}(G) \cap G'$ and $G/N \cong \mathrm{M}_{p^n}$, then $N = \{1\}$. (**Hint.** Prove that $G$ is minimal nonabelian.)

**Exercise 5 ([WZ]).** Suppose that the centers of all nonabelian subgroups of a $p$-group $G$ coincide. If $x, y \in G - \mathrm{Z}(G)$ and $[x,y] = 1$, then $\mathrm{C}_G(x) = \mathrm{C}_G(y)$.

**Solution.** Assume that $\mathrm{C}_G(x) \ne \mathrm{C}_G(y)$. One may assume, without loss of generality, that there is $z \in \mathrm{C}_G(x) - \mathrm{C}_G(y)$. As the subgroup $H = \langle x, y, z \rangle$ is nonabelian (by the choice, $yz \ne zy$), it follows that $x \in \mathrm{Z}(H) = \mathrm{Z}(G)$, a contradiction. Therefore, $\mathrm{C}_G(x) = \mathrm{C}_G(y)$.

**Exercise 6.** Suppose that $G$ is a metacyclic $p$-group and $A$ a subgroup of $G$ such that $A$ is not contained in $\Phi(G)$. Then the following statements are equivalent:
(a) $G' \le A$,
(b) $A \lhd G$,
(c) $A\Phi(G') \lhd G$.

*Proof.* Implications $(a) \Longrightarrow (b)$ and $(b) \Longrightarrow (c)$ are obvious.

$(b) \Longrightarrow (a)$: Since $A$ is not contained in $\Phi(G)$ and $\mathrm{d}(G) = 2$, there are $a \in A - \Phi(G)$ and $b \in G - A\Phi(G)$ such that $G = \langle a, b \rangle$. Since $A \lhd G$, it follows that $G/A = \langle Ab \rangle$ is cyclic so that $G' \le A$.

$(c) \Longrightarrow (a)$: Since $A\Phi(G') \lhd G$, then $G' \le A\Phi(G')$ by the equivalence of (1) and (2). Then $G' = \Phi(G')(G' \cap A) = G' \cap A$, by the modular law. Therefore $G' \le A$. □

**Exercise 7.** Assume that $G$ is a $p$-group and $M \in \Gamma_1$. Then $G' = [G - M, M] = \langle [x,y] \mid x \in G - M,\ y \in M] \rangle$.

**Solution.** Let $N = [G - M, M]$. Clearly, $N \lhd G$ and $N \le G'$. Let $\bar{G} = G/N$. It suffices to prove that $\bar{G}$ is abelian. Since $[\overline{G - M}, \bar{M}] = \bar{1}$, one obtains $|C_{\bar{G}}(\bar{M})| > 1 + |\overline{G - M}| > \frac{1}{2}|\bar{G}|$. Hence $C_{\bar{G}}(\bar{M}) = \bar{G}$ so that $\bar{M} \le Z(\bar{G})$. Since $\bar{G}/\bar{M} \cong G/M$ is cyclic (of order $p$), it follows that $\bar{G}$ is abelian, as required.

Next we list some results presenting some main achievements of $p$-group theorists from Shanhai Normal University, China.

(1) P.J. Li [Li1. Li2] has classified the $p$-groups of class $> 2$ all of whose proper subgroups have class $\le 2$. This is a deep generalization of $\mathcal{A}_n$-groups, $n \le 2$.
(2) The $p$-groups all of whose nonnormal subgroups have orders $\le p^3$ were classified in [ZGQX, ZS, ZLS]. Recall that D. Passman, in Theorem 1.25, has classified the non-Dedekindian $p$-groups all of whose subgroups of order $> p$ are normal (this result has a lot of applications in the book).
(3) The $p$-groups all of whose nonnormal subgroups are abelian (so called *metahamiltonian* $p$-groups) were classified in [AZ] and [FA]. This is also a deep generalization of $\mathcal{A}_n$-groups, $n \le 2$. (See also § 228 where some results on metahamiltonian $p$-groups are proved.)
(4) The $p$-groups containing a minimal nonabelian subgroup of index $p$, are classified in the series [QYXA, ALQZ, QXA, AHZ, QZGA], thereby solving Problem 239. The proof is surprisingly complicated. Undoubtedly, this is a basic result.
(5) The $p$-groups containing a cyclic subgroup of index $p^3$ are classified in [ZL2] (see also [Nei] and [Tit]).
(6) The $p$-groups such that $G/Z(G)$ is a minimal nonabelian $p$-group are classified in [LQC, QZ, QH2, QZ2].
(7) The $p$-groups all of whose subgroups of index $p^3$ are abelian (= $\mathcal{A}_3$-groups) are classified in [ZZLS]. (Of course, primary $\mathcal{A}_1$- and $\mathcal{A}_2$-groups, satisfy the above condition.)
(8) Finite $p$-groups of class 3 all of whose proper subgroups have class at most 2 are classified in [SQG].

The $p$-group theorists from Shanhai Normal University have solved a number of research problems posed in this book, in particular, problems 46, 87, 116(i,ii), 239, 279, 511, 535, 538, 768, 692, 704, 706, 1476, 725, 824, 869, 1278, 1305, 1347, 1381, 1392, 1459, 2279, and 2347. The following problems were solved but their solutions have not published: 49, 56, 436 (b), 451, 736, 903, 946, 1016, 1484, and 1576. A few problems were solved in part.

For more detailed information, see papers listed in Bibliography.

# Appendix 54 On Passman's Theorem 1.25 for $p > 2$

Passman has classified the $p$-groups all of whose subgroups of order $> p$ are normal (Theorem 1.25). In this appendix one assumes that only abelian subgroups of order $> p$ are normal. It appears that this does not affect the conclusion. The last result follows easily from Kazarin's Theorem 63.4. But for $p > 2$, if the group is nonmetacyclic, Blackburn's Theorem 13.7 is used.

**Remark 1.** Suppose that $G$ is a nonabelian $p$-group all of whose subgroups of order $> p$ are normal. Suppose that $E \cong \mathrm{E}_{p^3}$ is a subgroup of $G$. If $L < E$ is of order $p$ and $U$, $V$ are distinct subgroups of $E$ of order $p^2$ containing $L$, then $U, V \triangleleft G$, and so $L = U \cap V \triangleleft G$. It follows that $L \leq Z(G)$ and so $E \leq Z(G)$. Let $Z < G$ be cyclic; then the noncyclic group $ZE/Z \cong E/(E \cap Z)$ contains two distinct subgroups $X/Z$ and $Y/Z$ of order $p$. The subgroups $X$, $Y$ are distinct noncyclic abelian subgroups of order $> p$, so normal in $G$. Then $Z = X \cap Y \triangleleft G$; hence all cyclic subgroups of $G$ are normal and therefore $G$ is Dedekindian, contrary to the hypothesis. Thus, $G$ has no subgroup $\cong \mathrm{E}_{p^3}$. We will consider only case $p > 2$. Let $L \triangleleft G$ be of order $p$. By hypothesis, all subgroups of $G/L$ are normal so $G/L$ is abelian (Theorem 1.20). It follows that $L = G'$, so $G'$ is the unique minimal normal subgroup of $G$. By Theorem 13.7, $G$ is either metacyclic or $G = \Omega_1(G)C$, where $\Omega_1(C) \cong S(p^3)$. If $G$ is metacyclic, it is minimal nonabelian (Lemma 65.2 (a)), and (the monolith) $G \cong \mathrm{M}_{p^n}$. In the second case, (the monolith) $G = \Omega_1(G) * C$ (Lemma 4.3).

**Exercise 1.** Classify the nonabelian $p$-groups all of whose abelian subgroups of order $> p$ are normal. (**Answer.** Such groups are classified in Theorem 1.25.)

Now we are ready to prove the following:

**Theorem A.54.1** (see § 180). *Suppose that all noncyclic abelian subgroups of a nonabelian p-group G, p > 2, are normal. Then one of the following holds:*

(a) $|G| = p^3$,

(b) *$G$ is a 3-group of maximal class and order $3^4$ such that $|\Omega_1(G)| = 3^2$,*

(c) *$G = E * C$, there $E = \Omega_1(G) \cong S(p^3)$ and $C$ is cyclic of order $> p$,*

(d) *$G$ is metacyclic minimal nonabelian,*

(e) *$G$ is metacyclic and $\Omega_1(G) \nleq Z(G)$. Then $H = \mathrm{C}_G(\Omega_1(G))$ is either abelian or minimal nonabelian. All subgroups of $H/\Omega_1(G)$ are normal in $G/\Omega_1(G)$. If $A < G$ is abelian such that $A \nleq H$, then $A$ is cyclic. If $B < G$ is minimal nonabelian, then either $B \leq H$ or $B \cong \mathrm{M}_{p^n}$.*

*Proof.* One may assume that $|G| > p^3$. As in Remark 1, $G$ has no normal subgroup $\cong \mathrm{E}_{p^3}$. In view of Theorem 13.7, one has to consider the following cases:

(i) If $G = EC$, where $E = \Omega_1(G) \cong S(p^3)$ and $C$ is cyclic of order $> p$ and index $p^2$ in $G$, then, as in the proof of Remark 1, one obtains $|G'| = p$. By Lemma 4.3, $G = EZ(G)$, where $Z(G)$ is cyclic. If $A < G$ is noncyclic abelian, then either $G' < A$ or $G' \times \Omega_1(A) \cong E_{p^3}$. In the first case, $A \lhd G$, the second case is impossible, by the above.

(ii) Now let $G$ be a 3-group of maximal class. By the above, $G$ has no subgroup $\cong E_{3^3}$. If $|G| = 3^4$, then $G$ satisfies the hypothesis $\iff$ $|\Omega_1(G)| = 3^2$, by Exercise 9.1 (b). Assume that $|G| > 3^4$. Let $G_1 < G$ be the fundamental subgroup of $G$ (see Theorems 9.5 and 9.6). Then $G_1$ is metacyclic and $|\Omega_2(G_1)|$ is of order $3^4$ and exponent $3^2$. Since $3+1$ maximal subgroups of $\Omega_2(G_1)$ are noncyclic abelian, only one of them is $G$-invariant, by Exercise 9.1 (b), a contradiction. Thus, $G$ is as in (b).

(iii) Now let $G$ be metacyclic. If $G$ contains a nonabelian subgroup of order $p^3$, then $G \cong M_{p^3}$ is of order $p^3$ (Proposition 10.19), contrary to the assumption. Thus, $G$ has no such subgroup.

(iii1) Suppose that $\Omega_1(G) \le Z(G)$. Then $G/\Omega_1(G)$ is abelian (Theorem 1.20). In that case, $|G'| = p$ since $G'$ is cyclic so that $G$ is minimal nonabelian (Lemma 65.2 (a)), and all such $G$ satisfy the hypothesis (indeed, all noncyclic abelian subgroups of $G$ contain $\Omega_1(G) > G'$).

(iii2) Let $\Omega_1(G) \nleq Z(G)$. Then, by (iii1), $H = C_G(\Omega_1(G)) \in \Gamma_1$ is either abelian or minimal nonabelian. Let $A < G$ be noncyclic abelian. Then $\Omega_1(G) \le A$. In that case, $H = C_G(\Omega_1(G)) \ge A$. By hypothesis, $A \lhd G$. Thus, all abelian subgroups of $G$ not contained in $H$ are cyclic. If $U/\Omega_1(G) \le H/\Omega_1(G)$ is cyclic, then $U \lhd G$ since $U$ is noncyclic abelian. In that case, all subgroups of $H/\Omega_1(G)$ are normal in $G/\Omega_1(G)$. Now let $\Omega_1(G) \nleq Z(G)$. Then any minimal nonabelian subgroup of $G$ is $\cong M_{p^n}$. Assume that this is false. By Theorem 10.28, there is in $G$ a minimal nonabelian subgroup $A$ such that $A \nleq H$. By Proposition 10.19, $|A| > p^3$. Then $\Omega_1(A) = \Omega_1(G) = \Omega_1(H)$. By Lemma 65.1, $\Omega_1(G) = \Omega_1(A) \le Z(A)$. In that case, $\Omega_1(G) \le Z(G)$, contrary to the hypothesis. □

**Problem 1.** Classify the $p$-groups all of whose subgroups (abelian subgroups) of order $> p^2$ are normal.

**Problem 2.** Study the nonmetacyclic $p$-groups $G$ (i) all of whose maximal metacyclic subgroups are normal, (ii) such that $N_G(M)$ is metacyclic for every nonnormal metacyclic $M < G$.

**Problem 3.** Classify the $p$-groups all of whose minimal nonabelian subgroups have a cyclic subgroup of index $p$.

**Problem 4.** Study the $p$-groups $G$ such that $N_G(M)$ is an $\mathcal{A}_2$-subgroup for any minimal nonabelian $M < G$.

**Problem 5.** Study the $p$-groups in which any nonnormal abelian subgroup is contained in the center of its normalizer.

**Problem 6.** Classify the groups of Theorem A.54.2 (e) up to isomorphism.

**Problem 7.** Prove an analog of Theorem A.54.2 for $p = 2$.

# Appendix 55 On $p$-groups with the cyclic derived subgroup of index $p^2$

The $p$-groups $G$ with the cyclic derived subgroup $G'$ are not classified yet. In the following theorem the case $|G : G'| = p^2$ is considered.

**Exercise 1.** (a) If a $p$-group $G$ of order $p^4$ has the derived subgroup of order $p^2$, then $G$ is of maximal class.

(b) If a $p$-group $G$ is such that $|G : G'| = p^2$ and $|G| > p^4$, then $\mathrm{cl}(G) \geq 3$.

**Solution.** Clearly, $\mathrm{d}(G) = 2$.

(a) By Lemma 65.1, $G$ is not minimal nonabelian. Let $H < G$ be minimal nonabelian; then $|G : H| = p$ and $H$ is of maximal class. Now the result follows from Theorem 12.12 (a).

(b) If $L < G'$ is $G$-invariant of index $p^2$, then, by (a), $\mathrm{cl}(G) \geq \mathrm{cl}(G/L) \geq 3$.

**Theorem A.55.1.** *Suppose that the derived subgroup of a nonabelian $p$-group $G$ is cyclic and has index $p^2$. Then:*

(a) $|G| = p^3$,

(b) $p = 2$ *and $G$ is of maximal class.*

*In particular, $G$ is of maximal class.*

*Proof.* In view of Proposition 1.6 (Taussky), one may assume that $p > 2$ (otherwise, we have case (b)). Assume that $|G| = p^m > p^3$. As $G/G' \cong \mathrm{E}_{p^2}$, one has $G' = \Phi(G)$ and, clearly, $G'$ is a unique normal subgroup of index $p^2$ in $G$. Let $A, B < G$ be $G$-invariant subgroups of the same index $> p^2$ in $G$. Since $A, B < \Phi(G) = G'$ and $G'$ is cyclic, one has $A = B$. It follows that $G$ has exactly one normal subgroup of index $p^i$ for $i = 2, \dots, m-1$. By Lemma 9.1, $G$ is of maximal class. Since $G$ has no normal noncyclic subgroup of order $p^2$, it follows that $p = 2$ (Lemma 1.4). □

*Second proof.* Each maximal subgroup of $G$ has a cyclic subgroup $G' = \Phi(G)$ of index $p$. Therefore $G$ is either metacyclic or minimal nonmetacyclic. If $G$ is metacyclic, then it follows from $|G/G'| = p^2$ that either $G \cong \mathrm{M}_{p^3}$ or $G$ is a 2-group of maximal class. If $G$ is minimal nonmetacyclic, then $G \cong \mathrm{S}_{p^3}$ (see §§ 66 and 69). □

*Third proof (Janko).* One may assume that $|G| > p^3$. In view of Taussky, one may assume that $p > 2$. As each member of the set $\Gamma_1$ has a cyclic subgroup $G'$ of index $p$, its derived subgroup is of order $\leq p$ (Theorem 1.2); therefore, $|G'| \leq p^2$ (Exercise 1.69). Thus, $|G| = p^4$ (here we use assumption $|G| > p^3$). By Lemma 1.1, $|\mathrm{Z}(G)| = p$ so that $G$ is of maximal class since $G/\mathrm{Z}(G)$ is nonabelian. By Lemma 1.4, $G$ has a normal abelian subgroup $R$ of type $(p, p)$, and this implies that $R = G'$ since $G'$ is the unique normal subgroup of $G$ of index $p^2$, and this is a contradiction since $G'$ is cyclic. Thus, in the case under consideration, $|G| = p^3$. It follows after the third sentence that $G$ is of maximal class, by Exercise 1 (a).) □

**Exercise 2.** Suppose that a nonabelian two-generator $p$-group $G$ of order $> p^3$ has cyclic Frattini subgroup. Then $G$ has a cyclic subgroup of index $p$.

**Solution.** If $H \in \Gamma_1$, then $H$ has a cyclic subgroup $\Phi(G)$ of index $p$, so metacyclic. Assume that $G$ is not metacyclic. Then $G$ is minimal nonmetacyclic. This contradicts Theorems 66.1 and 69.1. Thus, $G$ is metacyclic. Then $\Phi(G) = \mho_1(G)$. Therefore, $\mho_1(G) = \langle a^p \rangle$ for some $a \in G$, and hence $|G : \langle a \rangle| = p$ since $|G : \mho_1(G)| = p^2$. (The condition $|G| > p^3$ is essential.)

**Exercise 3.** Use § 87 to classify the $p$-groups $G$ such that $G/G'$ is abelian of type $(p^n, p)$.

**Exercise 4.** Let $N < \Phi(G)$ be a cyclic normal subgroup of a $p$-group $G$. Use § 87 to classify the $p$-groups $G$ such that $G/N$ is either abelian of type $(p^n, p)$ or $\cong \mathrm{M}_{p^n}$.

**Problem 1.** Classify the $p$-groups with the cyclic derived subgroup of index $p^3$.

**Problem 2.** Classify the two-generator $p$-groups whose Frattini subgroup is abelian of type $(p^n, p)$.

**Problem 3.** Study the $p$-groups $G$ such that the derived subgroup $G'$ is abelian of index $p^2$.

# Appendix 56 On finite groups all of whose $p$-subgroups of small orders are normal

This appendix was inspired by Isaacs' letter (April 2007), where the following result is proved: if a finite group $G$ is not 2-nilpotent (then $4 \mid |G|$) and all its subgroups of order 4 are $G$-invariant, then a Sylow 2-subgroup of $G$ is elementary abelian and normal. An analog of this result for some prime $p$ that is, in particular, the minimal prime divisor of $|G|$, is proved in Theorem A.56.1.

In Theorem A.56.4, the non-$p$-nilpotent groups $G$ of orders divisible by $p^3$, where $p$ is a minimal prime divisor of $|G|$, in which all noncyclic subgroups of order $p^3$ are quasinormal, are classified.

In Theorem A.56.7, the non-$p$-nilpotent groups $G$ of orders divisible by $p^3$, where $p$ is a minimal prime divisor of $|G|$, all of whose noncyclic abelian subgroups of order $p^3$ are quasinormal, are classified.

As a rule, $p$ is the minimal prime divisor of $|G|$. A group $G$ is said to be *$p$-nilpotent* if it has a normal $p'$-Hall subgroup. By $\pi(G)$ we denote the set of prime divisors of $|G|$.

Recall that a group $G$ is said to be *minimal nonnilpotent* (*minimal nonabelian*) if it is nonnilpotent (nonabelian) but all its proper subgroups are nilpotent (abelian). All considered below minimal nonabelian groups have prime power order.

A subgroup $Q$ of a group $G$ is said to be *quasinormal* if $QF = FQ$ for all $F \leq G$. If $S \in \mathrm{Syl}_p(G)$ is quasinormal in $G$ and $x \in G$ is arbitrary, then $SS^x = S$ so that $S^x = S$ and $S \lhd G$. Moreover, quasinormal subgroups are subnormal [Ore], however, we do not use this fact in what follows. The groups all of whose subgroups are quasinormal (these groups are termed *modular*), are classified by Iwasawa (see also §73). In view of Iwasawa's classification, it is natural to study the groups in which only part of subgroups are quasinormal. The main difficulty with quasinormal subgroups is due to the fact that the intersection of two quasinormal subgroups, generally speaking, is not quasinormal.

In Lemmas I–VI some results used below are stated.

**Lemma I** (= Lemma A.22.1). *Let $H$ be a minimal nonnilpotent group. Then $|H| = p^\alpha q^\beta$, where $p, q \in \pi(H)$ are distinct and $\alpha, \beta$ are positive integers. Let $P_0 \in \mathrm{Syl}_p(H)$, $Q \in \mathrm{Syl}_q(H)$.*

(a) *One of the subgroups $P_0$, $Q$ (say $P_0$) is normal and coincides with $H'$.*

(b) *If $P_0$ is abelian, then $P_0 \cap Z(H) = \{1\}$, $P_0$ is a minimal normal subgroup of $H$ so that $P_0$ is elementary abelian, and $\alpha$ is the order of $p \pmod q$.*

(c) *The subgroup $Q$ is cyclic and $|Q : (Q \cap Z(H))| = q$.*

(d) *If $P_0$ is nonabelian, then $P_0$ is special and $|P_0/P_0'| = p^b$, where $b$ is the order of $p$ $\pmod q$ (so that, if $p < q$, then $b > 1$), $P_0 \cap Z(H) = Z(P_0) = P_0' = Z(P_0)$, $P_0/P_0'$ is a minimal normal subgroup of $H/P_0'$. If $p > 2$, then $\exp(P_0) = p$.*

In what follows, the notation introduced in Lemma I for any minimal nonnilpotent group $H$ is retained.

Subgroup $Q$ of Lemma I plays a key role below. Indeed, if $H$ is as in the lemma, then $Q$ controls the set of all relevant normal or quasinormal $p$-subgroups of the group $H$. For example, if $|P_0/P_0'| > p$ and $x \in P_0 - P_0'$, then $Q$ does not normalize the subgroup $\langle x \rangle$.

**Lemma II** (= Proposition 10.17). *If $B$ is a nonabelian subgroup of order $p^3$ in a $p$-group $G$ and $C_G(B) < B$, then $G$ is of maximal class.*

**Lemma III** ( = Exercise 1.8a = Lemma 65.1). *Let $G$ be a minimal nonabelian $p$-group. Then $G = \langle a, b \rangle$ and one of the following holds:*

(a) $a^{p^m} = b^{p^n} = c^p = 1,\ [a, b] = c,\ [a, c] = [b, c] = 1,\quad |G| = p^{m+n+1},\ G = \langle b \rangle \cdot (\langle a \rangle \times \langle c \rangle) = \langle a \rangle \cdot (\langle b \rangle \times \langle c \rangle)$ *(semidirect products with kernels in brackets), $G$ is nonmetacyclic, $Z(G) = \langle a^p, b^p, c \rangle$.*

(b) $a^{p^m} = b^{p^n} = 1,\ a^b = a^{1+p^{m-1}},\quad |G| = p^{m+n}$ *and $G = \langle b \rangle \cdot \langle a \rangle$ (semidirect product with kernel $\langle a \rangle$), $G$ is metacyclic, $Z(G) = \langle a^p, b^p \rangle$.*

(c) $a^4 = 1,\ a^2 = b^2,\ a^b = a^{-1},\quad G \cong Q_8$.

The following lemma is a strong form of the Frobenius normal $p$-complement theorem that is used below.

**Lemma IV.** *If a group $G$ is non-$p$-nilpotent, then it contains a minimal nonnilpotent subgroup $H$ such that $H' \in Syl_p(H)$.*

Indeed, by [Isa5, Theorem 9.18], there is in $G$ a nonnilpotent subgroup $T = Q \cdot P$, where $P$ is a normal Sylow $p$-subgroup of $T$ and $Q \in \mathrm{Syl}_q(T)$ is cyclic, $q \in \pi(G) - \{p\}$. If $H \leq T$ is minimal nonnilpotent, then $H$ is as in the conclusion of Lemma IV.

Next we use the following lemma freely.

**Lemma V** (= Theorem 34.8). (a) *If a 2-group $G$ of order $> 4$ has a cyclic subgroup of index 2, then $Aut(G)$ is a 2-group, unless $G \cong Q_8$.*

(b) *(Theorem 1.16) Let $G$ be a $p$-group and $|G : \Phi(G)| = p^d$. Then*

$$|Aut(G)| \mid (p^d - 1)(p^d - p) \dots (p^d - p^{d-1})|\Phi(G)|^d = |\Phi(G)|^d |Aut(E_{p^d})|\,.$$

*In particular, if $|G| = p^3$ and a prime $> p$ divides $|Aut(G)|$, then $G$ is either $Q_8$ or $E_{p^3}$.*

The following assertion follows immediately from the basic theorem of abelian $p$-group theory:

**Lemma VI.** *If $G$ is a noncyclic abelian $p$-group of exponent $> p$, then any element of $G$ of order $p$ is contained in an appropriate abelian subgroup of $G$ of type $(p^2, p)$.*

Our first result is the following:

**Theorem A.56.1.** *Let $p$ be a prime divisor of the order of a non-$p$-nilpotent group $G$. Then, by Lemma IV, there is in $G$ a minimal nonnilpotent subgroup $H = Q \cdot P_0$, where*

*$H, P_0 = H'$ and $Q$ are such as in Lemma I. Suppose, in addition, that $p \not\equiv 1 \pmod q$; then $P_0$ is noncyclic, by Lemma I(b). If all subgroups of $G$ of order $p^2$ are $H$-invariant and permutable with all their $G$-conjugates, then $P \in \mathrm{Syl}_p(G)$ is abelian of type $(p, p)$ and $G$-invariant.*

*Proof.* One may assume that $P_0 \le P \in \mathrm{Syl}_p(G)$.

If $|P| = p^2$, then $P = P_0$, and $P < H$. By the above, $P \cong \mathrm{E}_{p^2}$. Now, $PP^x = P^xP$ for all $x \in G$ implies that $P^x = P$ so that $P \triangleleft G$, and we are done.

Next we assume that $|P| > p^2$.

(i) Suppose that $P_0$ is abelian; then $\exp(P_0) = p$ and $P_0$ is a minimal normal subgroup of $H$, by Lemma I(b). Assume that $|P_0| > p^2$. Then every subgroup of $P_0$ of order $p$, being the intersection of two appropriate $H$-invariant (by hypothesis) subgroups of order $p^2$, is $H$-invariant, contrary to Lemma I(b). Thus, $|P_0| = p^2$. If $P_0 < T \le P$, where $|T : P_0| = p$ ($T$ exists since $|P| > p^2 = |P_0|$), then there is in $T$ a maximal subgroup $P_1 \neq P_0$. As $P_1$ is $Q$-invariant, then $P_0 \cap P_1 < P_0$ is also $Q$-invariant so $H$-invariant of order $p$, contrary to Lemma I(b). Thus, if $P_0$ is abelian, then $T$ does not exist so that $P = P_0 \cong \mathrm{E}_{p^2}$, $P \triangleleft G$, as desired.

In what follows we assume that $P_0$ is nonabelian; then $P_0$ is special, $P_0/P_0'$ is a minimal normal subgroup of $H/P_0'$ (Lemma I(d)). In that case, $\exp(P_0) \le p^2$.

(ii) Suppose that $\exp(P_0) = p^2$; then $p = 2$ (Lemma I). Let $L < P_0$ be cyclic of order 4; then $L \triangleleft H$, by hypothesis, and so $LP_0'/P_0'$ is a normal subgroup of order $2 < |P_0/P_0'|$ in $H/P_0'$, contrary to Lemma I(d).

(iii) Now let $\exp(P_0) = p$. As $P_0$ is nonabelian, one has $p > 2$. Let $L = \langle a \rangle \times \langle b \rangle$, where $a \in (P_0')^\#$ and $b \in P_0 - P_0'$. Then $L$, being of order $p^2$, is $H$-invariant. One has $|L \cap P_0'| = o(a) = p$. In this case, $LP_0'/P_0'$ is normal of order $p < |P_0/P_0'|$ in $H/P_0'$, contrary to Lemma I(d).

Thus, $P = P_0 \cong \mathrm{E}_{p^2}$ is normal in $G$, completing the proof. □

**Corollary A.56.2.** *Let $p$ be a prime divisor of the order of a non-$p$-nilpotent group $G$ such that $\mathrm{GCD}(p - 1, |G|) = 1$. If all subgroups of $G$ of order $p^2$ are quasinormal, then $P \in \mathrm{Syl}_p(G)$ is isomorphic to $\mathrm{E}_{p^2}$ and normal in $G$.*

**Corollary A.56.3.** *Let $G$ be a non-$p$-nilpotent group, where $p$ is the minimal prime divisor of $|G|$, and suppose that every subgroup of order $p^2$ is quasinormal in $G$. Then $p = 2$ and $G$ has a normal Sylow 2-subgroup $P \cong \mathrm{E}_4$.*

*Proof.* By Lemma IV, $P \in \mathrm{Syl}_p(G)$ is noncyclic. By Corollary A.56.2, $P \cong \mathrm{E}_{p^2}$ is normal in $G$. Since $G$ is not $p$-nilpotent, there is in $G$ a minimal nonnilpotent subgroup $H$ as in Lemma I. By Lemma I(b), $p^2 \equiv 1 \pmod q$ so that $p = 2$ since $p < q$. □

**Remark 1.** Let $p$ be the minimal prime divisor of $|G|$. If $p > 2$ and all subgroups of $G$ of order $p$ are quasinormal, then $G$ is $p$-nilpotent. If $p = 2$ and all cyclic subgroups of $G$ of orders 2 and 4 are quasinormal, then $G$ is 2-nilpotent. Indeed, assume that $H$ is a non-$p$-nilpotent minimal nonnilpotent subgroup of $G$. If $\exp(H') = p$, then not all subgroups of $H$ of order $p$ are quasinormal in $H$, so in $G$. If $\exp(H') = 4$, then not all cyclic subgroups of $H$ of orders 2 and 4 are quasinormal in $H$, so in $G$. Here we have used Lemma I. Therefore, $H$ does not exist, and so $G$ is $p$-nilpotent, by Lemma IV.

**Theorem A.56.4.** *Let $G$ be a non-$p$-nilpotent group of order divisible by $p^3$, where $p$ is the minimal prime divisor of $|G|$, and suppose that every noncyclic subgroup of $G$ of order $p^3$ is quasinormal. Then $P \in \mathrm{Syl}_p(G)$ is $G$-invariant and $P \in \{\mathrm{E}_{p^3}, \mathrm{Q}_8\}$.*

*Proof.* Being noncyclic (Lemma IV), $P \in \mathrm{Syl}_p(G)$ has a noncyclic subgroup of order $p^3$. This is true if $G$ has an abelian subgroup of type $(p, p)$. If $G$ has no such subgroup, it is a 2-group of maximal class (Lemma 1.4); then any of its nonabelian subgroups of order 8 are such as asserted. By Lemma IV, $G$ has a non-$p$-nilpotent minimal nonnilpotent subgroup $H$. Let $H$, $Q$ and $P_0 \leq P$ be as in Lemma I; then $P_0$ of order $> p$ is either of exponent $p$ or special (Lemma I).

(i) Suppose that $|P| = p^3$. Then $P \in \mathrm{Syl}_p(G)$, being quasinormal, is normal in $G$. In that case, $F = QP \geq H$ is nonnilpotent so that $P \in \{\mathrm{E}_{p^3}, S(p^3), \mathrm{Q}_8\}$ (Lemma V). Assume that $P \cong S(p^3)$. In this case, $|\mathrm{Aut}(P)|$ divides $(p^2-1)(p-1)p^3$ (Lemma V(b)). Therefore, $q \mid p+1$. Now $q > p \Rightarrow q = p+1 \Rightarrow p = 2$, a contradiction. Thus, $P \in \{\mathrm{E}_{p^3}, \mathrm{Q}_8\}$.

In what follows we assume that $|P| > p^3$.

(ii) If $R < P$ is noncyclic of order $p^3$, then $RQ < G$ and $R \in \mathrm{Syl}_p(RQ)$ is quasinormal in $RQ$ hence $R \lhd RQ$. Thus, $Q$ normalizes all noncyclic subgroups of $P$ of order $p^3$. It follows easily that $Q$ centralizes all abelian subgroups of $P$ of type $(p^2, p)$ (Lemma V).

(iii) Assume that $P$ is abelian (of order $> p^3$, by assumption). In this case, $\exp(P_0) = p$ (Lemma I(b)). If $\exp(P) > p$, then $Q$ also centralizes all abelian subgroups of type $(p^2, p)$ from $P$ (see the last sentence of the previous paragraph) so $Q$ also centralizes $\Omega_1(P)$ (see the last sentence in (ii) and Lemma VI). As $P_0 \leq \Omega_1(P)$, we get a contradiction.

Now let $\exp(P) = p$ and let $X < P_0$ be of order $p$. Then $X$, being the intersection of all containing $X$ ($Q$-invariant) subgroups of $P$ of order $p^3$, is itself $Q$-invariant, contrary to Lemma I(b) applied to $H$.

Thus, $P$ is nonabelian (of order $> p^3$). The above argument shows that $P$ has no subgroup $\cong \mathrm{E}_{p^4}$.

(iv) We claim that $|P_0| > p^2$. Assume, however, that $|P_0| = p^2$. Then $p^2 \equiv 1 \pmod{q}$ (Lemma I(b)) so that $p = 2$, $q = 3$ since $p < q$. Assume that $L < P$ is $H$-invariant of order 4 such that $L \nleq P_0$. Then $L \cap P_0 = \{1\}$ since $L \cap P_0 < P_0$ is $H$-invariant; in that case, $|LP_0| = 16$. If $L < U < LP_0$, then $U$ of order 8 is $H$-invariant, by hypothesis, so that $U \cap P_0 < P_0$ is $H$-invariant and, by the product formula, has order 2, contrary to Lemma I(b). Thus, $P_0$ is the unique $H$-invariant subgroup of $P$ of order 4. Let $P_0 < S < T \leq P$, where $|S : P_0| = |T : S| = 2$ (recall that $|P| > 8$, by assumption). By Lemma VI, $T$ is nonabelian since $T$ is noncyclic and $\exp(T) > 2$ (see the previous paragraph). Let $V < T$ be noncyclic maximal. If $P_0 \nleq V$, then $P_0 \cap V$ is $H$-invariant of order 2, contrary to Lemma I(b). Thus, all noncyclic maximal subgroups of $T$ contain $P_0$. Assume that $T$ contains only one noncyclic maximal subgroup. Then $T$ is either abelian of type (8, 2) or $T \cong \mathrm{M}_{16}$ (Theorem 1.2). Let $A < T$ be noncyclic maximal; then $A$ is abelian of type (4, 2). In that case, $Q$ centralizes $A$ (Lemma V(b)), a contradiction since $P_0 < A$. Thus, $T$ contains two distinct noncyclic maximal subgroups. Assume that $T$ has a cyclic subgroup of index 2. Then $T$ is of maximal class (Theorem 1.2). Two ($Q$-invariant) noncyclic maximal subgroups of $T$ generate $T$ so that $T$ is $Q$-invariant and $Q$ centralizes $T > P_0$ (Lemma V(a)), a contradiction. Thus, all maximal subgroups of $T$ are noncyclic, and this implies $\Phi(T) = P_0 \cong \mathrm{E}_4$ so that $\mathrm{d}(T) = 2$. Application of Lemma II shows that $T$ is minimal nonabelian (see also Theorem 12.12 (a)). Then any maximal subgroup of $T$ is $Q$-invariant and one of them, say $A$, is abelian of type (4, 2) (Lemma III) which is centralized by $Q$ (Lemma VI). It follows that $Q$ stabilizes the chain $\{1\} < A < T$ so $Q$ centralizes $T > P_0$, a contradiction.

Thus, if $|P_0| = p^2$, then $p = 2$, $T$ does not exist; hence $|P| = 8$. By (i), $P \cong \mathrm{E}_8$ (indeed, $\mathrm{E}_4 \cong P_0 < P$) completing this case.

In what follows we assume that $|P_0| > p^2$.

Assume that $P_0$ is abelian; then, as above, $P_0 \cong \mathrm{E}_{p^3}$. Let $P_0 < T \leq P$, where $|T : P_0| = p$. If $V < T$ is maximal and $V \neq P_0$, then $V \cap P_0 < P_0$ is $H$-invariant of order $p^2$, contrary to Lemma I(b). Thus, $P_0$ is nonabelian so it is special (Lemma I(d)).

(v) Let $|P_0| = p^3$ and $P_0 < T \leq P$, where $|T : P_0| = p$ so that $|T| = p^4$. Let $S \neq P_0$ be noncyclic maximal in $T$ ($S$ exists since $T \ncong \mathrm{M}_{p^4}$). Then $S \cap P_0 > P_0'$ is $Q$-invariant of order $p^2 < |P_0|$, contrary to Lemma I(d). Thus, if $|P_0| = p^3$, then $T$ does not exist; hence $P = P_0 \triangleleft G$ and $P \in \{\mathrm{E}_{p^3}, \mathrm{Q}_8\}$.

Next we assume that $|P_0| > p^3$ (by the paragraph preceding (v), $P_0$ is nonabelian).

(vi) Assume that $p > 2$. Then $\exp(P_0) = p$ (Lemma I). Let $M < P_0$ be minimal nonabelian. By Lemma III, $M \cong S(p^3)$. As $\mathrm{C}_H(Q) = Q\mathrm{Z}(P_0)$ is abelian, it follows that $QM$ is nonnilpotent. Then $p^2 \equiv 1 \pmod{q}$; hence $p = 2$ since $q > p$, contrary to the assumption.

Thus, $p = 2$.

(vii) Let $|P_0'| = 2$; then $P_0$ is extraspecial so that it is a central product of nonabelian subgroups of order 8, all of whose, except possibly one, are $\cong \mathrm{D}_8$. Since $|P_0| \geq 2^5$, there is in $P_0$ a subgroup $D \cong \mathrm{D}_8$. Then $Q$ centralizes $D$ (Lemma V(a)) so that $Q$ centralizes a cyclic subgroup, say $L$, of order 4 from $D$. In that case, $LP_0'/P_0'$ is a (proper) normal subgroup of $H/P_0'$ of order 2, contrary to Lemma I(d).

Thus, $|P_0'| > 2$.

Now we are ready to complete the proof. If $L < P_0$ is cyclic of order 4 and $X < P_0'$ is of order 2 such that $X \nleq L$ ($X$ exists since $P_0$ is elementary abelian of order > 2), then $Q$ centralizes the abelian subgroup $XL = X \times L$ of type $(4, 2)$. It follows that $Q$ centralizes $P_0'L$ and so $P_0'L/P_0'$ is a normal subgroup of $H/P_0'$ of order 2, contrary to Lemma I(d). □

**Corollary A.56.5.** *Let $G$ be a non-$p$-nilpotent group of order divisible by $p^3$, where $p > 2$ is a minimal prime divisor of $|G|$, and suppose that every noncyclic subgroup of $G$ of order $p^3$ is normal. Then the subgroup $P \cong \mathrm{E}_{p^3}$ is a $G$-invariant Sylow $p$-subgroup of $G$.*

**Corollary A.56.6.** *Let $G$ be a non-2-nilpotent group of order divisible by 8 and suppose that all its noncyclic subgroups of order 8 are $G$-invariant. Then $P \in \{\mathrm{E}_8, \mathrm{Q}_8\}$ is a normal Sylow 2-subgroup of $G$.*

**Problem 1.** Let $p$ be a minimal prime divisor of the order of a non-$p$-nilpotent group $G$ and suppose that $p^4$ divides $|G|$. Study the $p$-structure of $G$ provided all subgroups of order $p^4$ are normal.

**Problem 2.** Let $p$ be the minimal prime divisor of the order of a group $G$. Study the structure of $G$ if all proper subgroups of $G$ are either $p$-closed or $p$-nilpotent. (A group is said to be *$p$-closed* if its Sylow $p$-subgroup is normal.)

**Problem 3.** Study the $p$-groups $G$ such that, whenever $C < G$ is cyclic and $\phi \in \mathrm{Aut}(G)$, then $CC^{\phi} = C^{\phi}C$.

# Appendix 57 $p$-groups with a 2-uniserial subgroup of order $p$ and an abelian subgroup of type $(p, p)$

Let $L$ be a proper subgroup of a $p$-group $G$ and let $\mathcal{C}(L, G)$ be the number of maximal chains connecting $L$ with $G$ (a chain $L = L_0 < L_1 < \cdots < L_n = G$ is said to be *maximal* if all its indices are equal to $p$). Then

$$(*) \qquad \mathcal{C}(L, G) = \sum_{H \in \Gamma_1,\, L \leq H} \mathcal{C}(L, H)$$

(see Exercises 5.6 and 5.7). It follows from $(*)$, by induction on $|G : L|$, that $\mathcal{C}(L, G) \equiv |\Gamma_1| \equiv 1 \pmod p$. Therefore, it is natural to study the structure of a $p$-group $G$ containing a proper subgroup $L$ such that $\mathcal{C}(L, G) = 1$. To obtain interesting results, we have to assume that $|L|$ is small. This question was considered in [BlaH] and Theorem 134.5. Let us describe the approach in [BlaH].

The following definitions are due to Blackburn and Hethelyi [BlaH]. Let $L$ be a proper subgroup of a $p$-group $G$, $|G : L| \geq p^n$. Then $L$ is said to be *$n$-uniserial* in $G$, if for any $i \leq n$ there is in $G$ only one subgroup of $G$ of order $p^i|L|$ containing $L$. If $L$ is $n$-uniserial for $p^n = |G : L|$, it is called *uniserial* in $G$; in this case, there is only one maximal chain connecting $L$ with $G$, i.e., $\mathcal{C}(L, G) = 1$. If $L$ is nonnormal of index $p^2$ in $G$, then it is uniserial. The uniseriality of a 'small' $L < G$ is a restrictive condition as shown by [BlaH] and the results of this appendix.

In § 131 the $p$-groups containing a 2-uniserial subgroup of order $p$ are classified (Theorem 131.1; for $p > 2$ this result was obtained in [BlaH, Theorem 2.3]). Below we offer another proof of Theorem 131.1. In conclusion we treat the $p$-groups with uniserial elementary abelian subgroups of order $p^2$ and reduce their classification to classification of the $p$-groups with a cyclic subgroup of index $p^2$ or $p^3$ (such $p$-groups are classified).

**Exercise 1.** If the identity subgroup is 2-uniserial in a $p$-group $G$, then $G$ is cyclic.

**Hint.** The group $G$ has exactly one subgroup of every order $p$ and $p^2$. Now the result follows from Proposition 1.3.

**Exercise 2.** If $L$ is a 2-uniserial subgroup of a $p$-group $G$, then the quotient group $\mathrm{N}_G(L)/L$ is cyclic. (**Hint.** Use Exercise 1.)

Theorem A.57.1 in the case when $L$ of order $p$ is uniserial has also appeared in an old paper of the first author.

**Theorem A.57.1** (Blackburn–Hethelyi [BlaH] for $p > 2$ and Janko for $p = 2$). *Suppose that $L$ is a 2-uniserial subgroup of order $p$ in a noncyclic $p$-group $G$ of order $> p^2$. Then one of the following holds:*

(a) *$G$ is abelian with cyclic $G/L$,*
(b) *$G \cong M_{p^n}$. In that case, $|L^G| = p^2$,*
(c) *$G$ is of maximal class. In that case, $|G : L^G| = p$,*
(d) *$p = 2$ and $|\Omega_2(G)| = 2^3$, $G$ is a group from Lemma 42.1 (c). In that case, $|L^G| = p^2$.*

*Proof.* Let $|G| = p^3$. If $G$ is abelian, then $G/L$ is cyclic. If $G$ is nonabelian, then $G \ncong Q_8$ and $L \neq G'$. Next we assume that $|G| > p^3$.

If $L \lhd G$, then $G$ is as in (a), by Exercise 1. Next we assume that $L$ is not normal in $G$. Then $N_G(L) < G$ is as in (a). Write $C = C_G(L)(= N_G(L))$. Then $C/L$ is cyclic (Exercise 1). If $C$ is cyclic, then $G = C$ is cyclic since $L$ is characteristic in $C = N_G(L)$. Next we assume that the subgroup $L$ is nonnormal in $G$; then $G$ is non-Dedekindian.

If $|C| = p^2$, then $G$ is of maximal class (Proposition 1.8), i.e., $G$ is as in (c). By Theorem 9.6 (f), some $G$ satisfy the hypothesis for an appropriate $L$ (if $L$ is not contained in the fundamental subgroup $G_1$ of $G$, then $L$ is 2-uniserial, even uniserial). In that case, $|G : L^G| = p$. Next we also assume that $G$ is not a 2-group of maximal class.

If $G$ is of maximal class and order $p^4$, then its fundamental subgroup $G_1$ is abelian of type $(p^2, p)$ and, if $L < G_1$, then $L^G = \Phi(G)$ so $L$ is not 2-uniserial. Clearly, if $G$ is of maximal class of order $> p^3$, then $L \nleq G_1$. Next we also assume that $G$ is not of maximal class.

Next assume that $|C| > p^2$. Then $C$ is (noncyclic) abelian with a cyclic subgroup of index $p$ and $E = \Omega_1(C) \cong E_{p^2}$ is a proper subgroup of $C$ (see Exercise 2). By assumption, $C < G$. Write $N = N_G(E)$. By hypothesis, $N/E$ has only one subgroup of order $p$ so that $N/E$ is either cyclic or a generalized quaternion group (Proposition 1.3). As $E$ is characteristic in $N$ (recall that $C$ of order $> p^2$ has a cyclic subgroup of index $p$), it follows that $N = G$. If $N/E = G/E$ is cyclic, then $G$ is as in (b) since $\Omega_1(N) = E$, in view of the existence of $C$. Next assume that $G/E = N/E$ is a generalized quaternion group. Then $|\Omega_2(G)| = 2^3$, i.e., $G$ is as in (d) (Lemma 42.1). □

**Exercise 3.** Let $L < G$ be as in Theorem A.57.1 and $G$ in order $> p^3$ is nonabelian. If $L < H \in \Gamma_1$, then $L$ is 2-uniserial in $H$. Using this and induction, present alternate proof of Theorem A.57.1.

It is easy to show that groups (a), (b) and (d) of Theorem A.57.1 satisfy its hypothesis. But not all $p$-groups of maximal class satisfy its hypothesis (for example, $Q_{2^n}$). However, if $|C_G(L)| = p^2$, then $G$ satisfies the hypothesis. If $L$ is a noncentral subgroup of order 2 in a group of Theorem A.57.1 (d), then it is 2-uniserial but not uniserial since $G/\Omega_2(G)$ has $> 1$ subgroups of order 2. Therefore, if a $p$-group $G$, $p > 2$, contains an uniserial subgroup of order $p$, then $G$ is one of groups (a,b,c) of that theorem. This coincides with [BlaH, Theorem 2.3].

Below we consider a $p$-group $G$ containing a 2-uniserial subgroup $A \cong E_{p^2}$ (see Problem 389). Write $N = N_G(A)$ and assume that $|G| > p^4$.

(i) Suppose that $|N| = p^3$. As $N < G$, it follows that $A$ is not characteristic in $N$, and we conclude that $N$ is not abelian of type $(p^2, p)$. If $N$ is nonabelian, then

$C_G(A) = A$, and so $G$ is of maximal class (Proposition 1.8). It remains to consider the case when $N \cong E_{p^3}$. In that case, $C_G(N) \leq C_G(A) = N_G(A) = N$. Such $G$, for $p = 2$, are described in § 51.

(ii) Now suppose that $|N| > p^3$, then the identity subgroup is 2-uniserial in $N/A$. It follows from Exercise 1 that $N/A$ is cyclic. If $\Omega_1(N) = A$, then $A$ is characteristic in $N$ so that $N = G$; $G$ is as in (a) or (b) of Theorem A.57.1 (here we use Proposition 1.3). Now let $A < \Omega_1(N)$. Then $C_G(A) \geq \Omega_1(N)$ since $N/A$ is cyclic of order $> p$, and we conclude that $E = \Omega_1(N) \cong E_{p^3}$ and $N = EC$, where $C$ is cyclic of index $p^2$ in $N$. Write $N_1 = N_G(E)$. By hypothesis, $N_1/E$ has only one subgroup of order $p$ so that $N_1/E$ is either cyclic or $\cong Q_{2^n}$. In both cases, $E = \Omega_1(N_1)$ is characteristic in $N_1$, and this implies that $N_1 = G$. If $G/E$ is cyclic, then $G$ has a cyclic subgroup of index $p^2$. Such $G$ are classified in § 74 (for $p = 2$, the defining relations are given; in the general case, see [Nin]). Now let $G/E \cong Q_{2^n}$. In this case, $G$ has a cyclic subgroup of index $2^3$. Indeed, if $T/E < G/E$ is cyclic of index 2 and $T = EZ$, where $Z$ is cyclic, then $|T : Z| = 2^2$ so that $|G : Z| = |G : T|\,|T : Z| = 2^3$ (see [Tit] and [ZL2]).

**Problem 1.** Classify the $p$-groups containing a 2-uniserial cyclic subgroup of order $p^2$.

**Problem 2.** Given $n > 2$, study the $\mathcal{A}_n$-groups all of whose minimal nonabelian subgroups are uniserial (2-uniserial).

**Problem 3.** Classify the noncyclic $p$-groups all of whose maximal cyclic subgroups are uniserial (2-uniserial).

**Problem 4.** Study the nonmetacyclic $p$-groups all of whose maximal metacyclic subgroups are uniserial (2-uniserial).

**Problem 5.** Study the non-Dedekindian $p$-groups all of whose maximal nonnormal subgroups are uniserial (2-uniserial).

Some additional information on 2-uniserial and uniserial $p$-groups for $p$ odd is contained in [BlaH]. There are also posed three related problems.

**Problem 6.** Classify the $p$-groups containing a 2-uniserial cyclic subgroup of order $p^2$.

**Problem 7.** Given $n > 2$, study the $\mathcal{A}_n$-groups all of whose minimal nonabelian subgroups are uniserial (2-uniserial).

# Research problems and themes IV

I am convinced that the special problems in all their complexity constitute the stock and core of mathematics; and to master their difficulties requires on the whole the harder labor.
– H.Weyl, The Classical Groups, Princeton Univ. Press, 1973.

The heart of mathematics is its problems.
– P. Halmos

Theories change, but the groups remain.
– M. Hall and J.K. Senior, The Groups of Order $2^n$, $n \leq 6$.

A well-chosen problem can isolate an essential difficulty in a particular area, serving as a particular benchmark against which progress in this area can be measured.
– P. Erdos

A scientific person will never understand why he should believe opinions only because they are written in a certain book. Furthermore, he will never believe that the results of his own attempts are final.
– A. Einstein

Science will stagnate if it is made to serve practical goals.
– A. Einstein

The gradation of all science is to cover the greatest number of empirical facts by logical deduction from the smallest number of hypotheses or axioms.
– A. Einstein

You are never sure whether or not a problem is good unless you actually solve it.
– M. Gromov

This is the fourth list of research problems and themes written by the first author. As in the previous lists, I have used numerous constructive remarks and comments by Zvonimir Janko and Avinoam Mann. Qinhai Zhang sent me a great number of remarks. This allowed me to improve this list considerably. As in the previous lists, short information about the solved problems is presented after their statements. Only a few problems from this list were solved since it was started in the very beginning of 2011.

In many cases we suggest finding a solution to a problem for $p$-groups of class two or groups of exponent $p$. These partial cases may appear fairly difficult. In many cases it will be useful at first to solve some problems for some partial cases of $p$-groups ($\mathcal{A}_n$-groups with small $n$, or metacyclic or regular $p$-groups).

Most items in this list, as in the previous ones, are devoted to subgroup structures of $p$-groups. Many problems are posed in order to investigate the impact of minimal nonabelian subgroups on the structure of a group. Any item beginning with 'Study', we consider as a theme for research. Many items from this and the previous lists make sense for nonnilpotent groups.

These lists show how little we know (or how much we do not know) about $p$-groups.

2292. Classify the pairs $H, G$ of nonisomorphic (having distinct orders) $p$-groups and satisfying $|\mathrm{Aut}(H)|_p = |\mathrm{Aut}(G)|_p$. Compare $|H|$ and $|G|$. The case when $H < G$, presents a special interest. (One has $\mathrm{Aut}(\mathrm{D}_8) \cong \mathrm{D}_8$, $|\mathrm{Aut}(\mathrm{E}_8)|_2 = 8$. Compare with #211, which was solved by B. Sambale.)
2293. Given $n$, does there exist a $p$-group of exponent $> p$ that contains $\geq n$ maximal subgroups of exponent $p$ having pairwise distinct orders? The same problem for maximal elementary abelian subgroups.
2294. Classify the $p$-groups $G$ such that $|H : H_G| \leq p$ for all (i) $H < G$, (ii) abelian $H < G$, (iii) cyclic $H < G$, (iv) metacyclic $H < G$, (v) regular $H < G$, (vi) nonabelian $H < G$ with cyclic $H'$, (vii) minimal nonabelian $H < G$, (viii) nonabelian $H < G$. (Example: 2-groups of maximal class.)
2295. Describe all representation groups of a $p$-group $G = M \times C$, where $M$ is nonabelian metacyclic and $C$ is cyclic (see Theorem 47.2).
2296. Classify the $p$-groups $G$ containing two distinct $F, H \in \Gamma_1$ such that, whenever $x \in F - (F \cap H)$ and $y \in H - (F \cap H)$, then $\langle x, y\rangle$ is (i) either abelian or minimal nonabelian, (ii) of maximal class, (iii) metacyclic, (iv) of class $\leq 2$, (v) regular, (vi) $G$-invariant (six problems).
2297. Study the irregular $p$-groups $G = AB$, where $A < G$ is absolutely regular and $B < G$ is cyclic. (Example: $G$ is of maximal class.)
2298. Classify the metacyclic 2-groups which cannot be presented in the form $G = AB$, where $A, B < G$ and $A \cap B = \{1\}$. (All metacyclic $p$-groups, $p > 2$, do not satisfy the above condition.)
2299. Given $e > 2$, study the irregular $p$-groups $G$ of exponent $\geq p^e$ such that $\mathrm{c}_k(G) = p^{p-1}$ for some $k \in \{2, \dots, e\}$ (see Theorem 13.2 (b)).
2300. A group $G$ is said to be *capable* if $G \cong H/\mathrm{Z}(H)$ for some group $H$ (R. Baer; see also § 21). Study the capable $p$-groups (i) having an abelian subgroup of index $p$, (ii) of maximal class, (iii) that are special.
2301. Study the nonabelian $p$-groups all of whose subgroups of class 2 are minimal nonabelian.
2302. Classify the 2-groups $G$ such that $|\mathrm{N}_G(C) : C| \leq 4$ for all nonnormal cyclic $C < G$ (for $p > 2$, see [ZG1]).
2303. Study the $p$-groups $G$ with $H \in \Gamma_1$ such that $\langle h, x\rangle$ is $G$-invariant for all $h \in H^\#$ and $x \in G - H$. (This is solved by the second author; see § 246.)
2304. Let $H$ be a nonabelian absolutely regular $p$-group of exponent $p^e$ and order $p^{(p-1)e}$. Does there exist a $p$-group $G$ of maximal class satisfying $G_1 \cong H$ (here $G_1$ is the fundamental subgroup of $G$)?
2305. Study the nonabelian $p$-groups $G$ such that $|\mathrm{Z}(G/\ker(\chi))| = p$ for all $\chi \in \mathrm{Irr}^\#(G)$.

2306. Find $\alpha_1(G)$, the number of minimal nonabelian subgroups of $G$, where $G = A \times B$ and $A, B$ are minimal nonabelian $p$-groups.

2307. Find the maximal $n$ such that there is a $p$-group $G$ of exponent $p$ and order $p^n$ such that $p^2$ does not divide $\exp(\mathrm{Aut}(G))$.

2308. Study the nonmetacyclic $p$-groups all of whose maximal cyclic (metacyclic) subgroups are nonnormal.

2309. Classify the special $p$-groups $G$ satisfying $|\mathrm{cd}(G)| \le 3$.

2310. Study the $p$-groups covered by normalizers (normal closures) of nonnormal cyclic subgroups.

2311. Study the $p$-groups of exponent $> p$ all of whose cyclic subgroups of order $> p$ are quasinormal (see § 63).

2312. Let $G$ be a $p$-group. Set $\mathcal{N}_1(G) = \mathcal{N}(G)$, $\mathcal{N}_2(G)/\mathcal{N}_1(G) = \mathcal{N}(G/\mathcal{N}_1(G))$, and so on (here $\mathcal{N}(G)$ is the norm of $G$). Study the $p$-groups $X = \mathcal{N}_2(X)$.

2313. Study the two-generator $p$-groups with a derived subgroup of order $p^2$, describe their maximal abelian and minimal nonabelian subgroups.

2314. Describe the Schur multipliers and the representation groups of minimal nonabelian $p$-groups.

2315. Study the $p$-groups all of whose nonnormal abelian subgroups are cyclic (according to a letter dated 1/9/2012, the second author has solved this problem.)

2316. Study the 2-groups all of whose maximal subgroups are products of two cyclic subgroups.

2317. Does there exist an irregular $p$-group $G$, $p > 2$, such that, whenever $A, B < G$ are distinct minimal irregular, then $A \cap B = \{1\}$?

2318. Study the irregular $p$-groups $G$ such that, whenever $A < B \le G$, where $A$ is maximal regular and $|B : A| = p$, then $B$ is minimal irregular.

2319. Given $c$ and $p > 2$, does there exist a $p$-group $G$ of class $c$ such that $\mathrm{Aut}(P)$ is a $p$-group?

2320. Study the $p$-groups $G$, $p > 2$, such that $G/G''$ is (i) of maximal class, (ii) special.

2321. Study the $p$-groups that are representation groups of some of their proper epimorphic images.

2322. Let $G$ be a group of order $p^m > p^{p+1}$ and let $\mu_{m-1}(G)$ be the number of $D \lhd G$ such that $G/D$ is of maximal class and order $p^{m-1}$. Is it true that $p \mid \mu_{m-1}(G)$, unless $G$ is of maximal class?

2323. Classify the $p$-groups $G$ such that for any (i) $H \in \Gamma_1$ there is $L \lhd G$ satisfying $G/L \cong H$, (ii) $L \le \mathrm{Z}(G)$ of order $p$ there is $H \in \Gamma_1$ satisfying $G/L \cong H$. (For a partial cases, see [Yin1] and § 192.)

2324. Study the $p$-groups $G$ such that, whenever $A < B \le G$ are nonabelian, then $\mathrm{Z}(A) \le \mathrm{Z}(B)$.

2325. Study the $p$-groups $G$ containing a subgroup $H$ of maximal class and order $p^4$ such that $\mathrm{C}_G(H) < H$ and $|\mathrm{N}_G(H) : H| = p$.

2326. Classify the $p$-groups $G$ such that $|\mathrm{Z}(H)| \le p^2$ for all $H \in \Gamma_1$.

2327. Study the irregular $p$-groups all of whose nonnormal subgroups are (i) absolutely regular, (ii) metacyclic (see § 16).

2328. Study the special 2-groups without real elements of order 4 (all of whose elements of order 4 are real).

2329. Estimate the minimal index of a normal abelian subgroup of a $p$-group of maximal class containing an abelian subgroup of index $p^4$.

2330. Study the $p$-groups $G$ such that, whenever $A < G$ with $A_G = \{1\}$, then $A$ is cyclic.

2331. Study the $p$-groups $G$ such that, whenever $A \neq B$ are minimal nonabelian subgroups of $G$ and $x \in A - B$, $y \in B - A$, then $\langle x, y\rangle$ is minimal nonabelian. (This is solved by the second author; see § 149 and [Jan38].)

2332. Study the irregular $p$-groups $G$ of order $p^{np}$ such that $|\Omega_i(G)| = |G/\mho_{n-i}(G)| = p^{ip}$ for $i = 1, \dots, n$.

2333. Study the set $\mathcal{E}$ of elements $a$ of an irregular $p$-group $G$ such that, whenever $y \in G$, then $\langle a, y\rangle$ is regular.

2334. Study the $p$-groups $G$ such that $|G/\ker(\chi)| = p^{\mu}\chi(1)^2$ for all $\chi \in \mathrm{Irr}^{\#}(G)$, $\mu \leq 2$ (see #1.)

2335. Describe the structure of two-generator $p$-groups $G$ such that $G/\mathrm{Z}(G)$ is metacyclic of order $p^{2e}$ and exponent $p^e$.

2336. Suppose that $G$ is a $p$-group such that $\mathrm{Aut}(G)$ is of exponent $p$. Is it possible to estimate $\exp(G)$? (This is solved by Mann; see § 150.)

2337. Study the special 2-groups $G$ satisfying $|G : \Omega_1(G)| = 2$.

2338. Study the $p$-groups $G$ containing a 2-uniserial cyclic subgroup $L$ of order $\leq p^3$ (see § 131 and Appendix 61).

2339. Does there exist a $p$-group $G$ admitting a partition all of whose components are of maximal class?

2340. Study the nonabelian $p$-groups $G$ such that, whenever $A < B \leq G$, where $A$ is maximal abelian and $|B : A| = p$, then $B$ is minimal nonabelian (see § 148).

2341. Let $A$ be an abelian subgroup of a nonabelian $p$-group $G$. Is it possible that $\mathrm{Aut}(A)$ and $\mathrm{Aut}(G)$ are isomorphic? Consider in detail the case $|G : A| = p$.

2342. Study the nonabelian $p$-groups $G$ such that, whenever $A < G$ is maximal abelian, then (i) $\mathrm{Z}(\mathrm{N}_G(A))$ is cyclic, (ii) $\mathrm{N}_G(A)$ is of class 2.

2343. (Mann) Let $H$ be a metacyclic 2-group. Classify the 2-groups $G$ satisfying $\mathrm{s}_1(G) = \mathrm{s}_1(H)$ and $\mathrm{s}_2(G) = \mathrm{s}_2(H)$ (see § 124).

2344. Study the $p$-groups all of whose nonabelian maximal subgroups are two-generator.

2345. Classify the nonmetacyclic 2-groups $G$ containing a metacyclic subgroup of index 2.

2346. Study the two-generator $p$-groups with special Frattini subgroup (see § 60).

2347. Study the two-generator 2-groups $G$ with cyclic $G'$. In particular, study the subgroup structure of the representation groups of the abelian group of type $(2^n, 2^m)$ and find the number of these representation groups.

2348. Given $e > 1$, study the $p$-groups of exponent $p^e$ such that $|N_G(L)| = p^{e+1}$ for any cyclic $L < G$ of order $p^e$.
2349. Given $n > p + 1$, study the $p$-groups $G$ admitting an irredundant covering $G = A_1 \cup \cdots \cup A_n$ such that $|A_i \cap A_j| \leq p$ for $i \neq j$.
2350. (*By an idea of N. Mazza*) Study the $p$-groups $G$ admitting an automorphism of order $p$ fixing all elements of $\Phi(G)$.
2351. Classify the $p$-groups $G$ such that $A \cap B$ is Dedekindian for any nonincident $A, B < G$.
2352. Given a special $p$-group $G$ of order $p^n$ and $|G'| = p^2$, estimate minimal and maximal orders of maximal abelian subgroups of $G$.
2353. Given a $p$-group $G$, does there exist a $p$-group $W$ such that $W/\Omega_1(W) \cong G$?
2354. Study the special $p$-groups $G$ such that all their normal subgroups are incident with $Z(G)$.
2355. Classify the special $p$-groups with an abelian subgroup of index $p$.
2356. Study the $p$-groups of exponent $p^e > p^2$, all of whose cyclic subgroups of order $p^e$ are maximal abelian.
2357. Study the irregular $p$-groups with exactly one maximal subgroup of exponent $> p$.
2358. Does there exist a special $p$-group $G$ (group of exponent $p$) such that $\mathrm{Aut}(G)$ is abelian?
2359. Classify the special $p$-groups $G$ all of whose maximal abelian subgroups have orders $\leq p^2|Z(G)|$ (see § 20.)
2360. Study the $p$-groups without characteristic maximal abelian subgroups.
2361. Study the nonabelian $p$-groups $G$ in which the center of any nonabelian member of the set $\Gamma_1$ is either cyclic or abelian of type $(p, p)$.
2362. Study the $p$-groups $G$ in which the center of the centralizer of any noncyclic abelian subgroup $A$ coincides with $A$.
2363. Classify the special $p$-groups $G$ such that $|\mathrm{Aut}(G)|_p \leq |G|$ (compare with #2256).
2364. Given $n > 1$, present a $p$-group $G$ satisfying $k(A) = k(B)$ for all $A, B \in \Gamma_1$ (here $k(A)$ is the class number of $A$).
2365. Study the $p$-groups $G$ such that $C_G(M) \cong M_{p^n}$ for any cyclic $M < G$ of order $p^2$ and some $n > 3$.
2366. Given a $p$-group $G$, decide if there is a $p$-group $H$ such that $H/Z(H)$ contains a subgroup $\cong G$.
2367. Study the $p$-groups $G$ such that, if $A \in \{\Phi(G), G', Z(G), \mho_1(G), \Omega_1(G)\}$, then there is $N \lhd G$ such that $G/N \cong A$ (five problems).
2368. Study the $p$-groups $G$ such that $H_G$ is cyclic (abelian, Dedekindian) for all nonnormal $H < G$.
2369. Classify the primary $\mathcal{A}_n$-groups, $n > 1$, all of whose $\mathcal{A}_1$- and $\mathcal{A}_2$-subgroups have cyclic centers (for a partial case, see § 238).

2370. Study the nonabelian $p$-groups $G$ such that $A \cap B$ is cyclic for any distinct maximal abelian subgroups $A, B < G$ (this is solved by the second author for $p = 2$; see Theorem 91.1).

2371. Study the metacyclic $p$-groups with given subgroup (element) breadth (see #1703).

2372. Study the primary $\mathcal{A}_n$-groups, $n > 1$, such that, whenever $A, B < G$ are distinct $\mathcal{A}_1$-subgroups, then $A \cap B$ is maximal either in $A$ or in $B$.

2373. Study the $p$-groups $G$ containing a proper nonabelian metacyclic subgroup $H$ of order $p^4$ and exponent $p^2$ such that (i) $C_G(H) < H$, (ii) $C_G(x) < H$ for all $x \in H - Z(H)$.

2374. Study the irregular $p$-groups in which the intersection of any two distinct maximal regular subgroups is (i) absolutely regular, (ii) has exponent $p$.

2375. Write $p^{\delta_n} = \max\{|G'| \mid p\text{-group } G \text{ is an } \mathcal{A}_n\text{-group}\}$. Is it true that the set of $n$ such that $\delta_{n+1} = 1 + 2\delta_n$ is finite?

2376. Let $G$ be a capable $p$-group (see #2300) and let $W$ be a $p$-group of the least order such that $W/Z(W) \cong G$. The subgroup $W$ is said to be a *covering* of $G$. Next we ask the reader to study coverings of some capable $p$-groups only. (i) Given a metacyclic $p$-group, study the structure of its coverings. (ii) Study the coverings of an abelian $p$-group. (iii) Describe coverings of minimal nonmetacyclic $p$-groups.

2377. Let $G = AB$ be a nonmetacyclic 2-group, where $A$ and $B$ are cyclic. Describe (i) the subgroup, normal and power structures of $G$, (ii) the representation groups of $G$, (iii) $\text{Aut}(G)$, (iv) the 2-groups that are lattice isomorphic with $G$.

2378. Study the irregular $p$-groups $G$, $p > 2$, such that, whenever $M < H \leq G$ with $|H : M| = p$, then $H$ is irregular for any maximal absolutely regular $M < G$.

2379. Let a $p$-group $G = AB$ and let $p^a$, $p^b$ be the maximal orders of subgroups of exponent $p$ in $A$, $B$, respectively. Estimate the maximal possible order of a subgroup of exponent $p$ in $G$.

2380. Study the irregular $p$-groups all of whose maximal regular subgroups are normal.

2381. Study the 2-groups all of whose elements of order $> 2$ are nonreal.

2382. Study the irregular $p$-groups all of whose regular subgroups are metabelian (see #3370).

2383. Study the $p$-groups all of whose nonfaithful nonlinear irreducible characters have degree $p$.

2384. Study the $p$-groups $G$ containing a maximal subgroup $H$ such that any abelian subgroup of $H$ is not a maximal abelian subgroup of $G$.

2385. Study the $p$-groups $G$ of exponent $> p$ such that $H_p(G) < G$ and $G = U \cdot H_p(G)$ (semidirect product with kernel $H_p(G)$ and complement $U$).

2386. Describe all characteristic subgroups of $\mathcal{A}_n$-groups, $n \leq 3$ (see [ZZLS]).

2387. Classify the $p$-groups of exponent $p$ all of whose nonabelian subgroups of order $p^3$ are nonnormal.

2388. Let $G = Z_1 \dots Z_n$, where all $Z_i$ are $G$-invariant cyclic. Then $\mathrm{cl}(G) \le n$, by Fitting's lemma (see Introduction, Theorem 1.21). Find all $n$ for which the equality is possible.

2389. Classify the $p$-groups $G$ of order $p^n$ and exponent $p$ with minimal possible $\alpha_1(G)$ (= the number of minimal nonabelian subgroups in $G$).

2390. Study the $p$-groups $G$ such that all non-$G$-invariant subgroups of $H$ have equal order for all $H \in \Gamma_1$.

2391. Study the $p$-groups all of whose maximal abelian subgroups are homocyclic.

2392. Study the $p$-groups $G$ such that, for all nonabelian $H \in \Gamma_1$, whenever $C < H$ is non-$G$-invariant, then $C$ is cyclic.

2393. Classify the $p$-groups possessing a soft cyclic subgroup.

2394. Study the $p$-groups (special $p$-groups, regular $p$-groups) of order $p^n$ with minimal possible class number.

2395. Classify the $p$-groups (special $p$-groups) of exponent $> p > 2$, all of whose $\mathcal{A}_1$-subgroups have order $p^3$ (see § 90 for $p = 2$ and Mann's comment to #115).

2396. Does there exist a nonabelian $p$-group of exponent $> p$ all of whose maximal abelian subgroups and $\mathcal{A}_1$-subgroups have equal order?

2397. Study the irregular $p$-groups all of whose nonnormal regular subgroups have regular normalizers (this is a partial case of #2380).

2398. Study the $p$-groups all of whose nontrivial characteristic subgroups have exponent $p$.

2399. Study the $p$-groups $G$ in which every maximal cyclic subgroup, say $C$, is contained in at most $p + 1$ subgroups of $G$ of order $p|C|$.

2400. Classify the 2-groups all of whose nonnormal subgroups are either cyclic, or abelian, or of type $(2, 2)$, or 2-groups of maximal class (see §§ 174 and 175 where this problem is solved by the second author).

2401. Classify the $p$-groups (groups of exponent $p$) in which any two nonabelian subgroups of equal order are isomorphic.

2402. Study the $p$-groups $G$ all of whose minimal nonabelian subgroups containing $\Phi(G)$ generate $G$ (see #2502).

2403. Classify the $p$-groups $G$ all of whose subgroups have (i) element breadth $\le 1$, (ii) subgroup breadth $\le 1$ (see #1703).

2404. Study the nonabelian $p$-groups all of whose nonabelian subgroups with cyclic derived subgroups are metacyclic.

2405. Study the nonabelian $p$-groups $G$ such that $\langle A, B\rangle$ is an $\mathcal{A}_2$-subgroup for any distinct minimal nonabelian $A, B < G$.

2406. Study the $p$-groups $G$ with a nonabelian $H \in \Gamma_1$ such that $H \cap A \le \mathrm{Z}(H)$ for all maximal abelian $A < G$ not contained in $H$.

2407. Given $n > 1$, study the primary $\mathcal{A}_n$-groups, $n > 2$, all of whose (i) maximal subgroups, (ii) maximal subgroups except one, are $\mathcal{A}_{n-1}$-groups (see §§ 76 and 100–102).

2408. Describe the automorphism groups of $\mathcal{A}_2$-groups (see [Kur] and [Mal6]).

2409. Study the $p$-groups $G$ such that $G/C_G(K)$ is abelian for any $G$-class $K$.

2410. (i) Given a homocyclic $p$-group $G$, describe the group of all automorphisms of $G$ leaving fixed all elements of $G$ of order $p$. (ii) The same problem for an arbitrary abelian $p$-group $G$.

2411. Estimate the number of groups $G$ of order $p^n$ such that $G$ (i) is two-generator, (ii) is of class 2, (iii) is metabelian, (iv) has an abelian (cyclic) $\Phi(G)$ ($\mho_1(G)$, $G'$, $Z(G)$), (v) is absolutely regular, (vi) is regular, (vii) is of maximal class, (viii) is of exponent $p$, (ix) satisfies $|G'| = p^m$, (x) satisfies $\exp(\Omega_i(G)) = p^i$ for all $p^i \leq \exp(G)$, (xii) satisfies $\exp(G') = p$, (xii) $|G/K_3(G)| = p^3$ for $p > 2$, (xiii) satisfies $\Omega_1(G) = G$.

2412. Classify the $p$-groups $G$ with $|\mathrm{cd}(G)| = |\mathrm{Irr}_1(G)|$ (for example: $M_{p^4}$, see [BCH]).

2413. Study the special $p$-groups of exponent $> p$ (i) all of whose minimal nonabelian subgroups are isomorphic, (ii) without normal cyclic subgroup of order $p^2$.

2414. Study the regular $p$-groups $G$ all of whose maximal subgroups are characteristic.

2415. Study the irregular $p$-groups $G$ such that $G/\ker(\chi)$ is an $\mathcal{A}_1$-group for any $\chi \in \mathrm{Irr}_1(G)$.

2416. Given $p > 7$, does there exist an irregular $p$-group $G$ of maximal class having a subgroup $E \cong E_{p^{p-3}}$ but not having a normal subgroup $\cong E$? (See Theorems 10.4, 10.5 and § 219.)

2417. (Partial cases of a well-known problem) Study the $p$-groups (two-generator $p$-groups, $p$-groups of maximal class) $G, p > 2$, such that $\mathrm{Aut}(G)$ is a $p$-group (see [Hor1, Hor2]).

2418. Classify the nonabelian $p$-groups $G$ satisfying $|H| = p|H'|\,|Z(H)|$ for all nonabelian $H \leq G$.

2419. Study the irregular $p$-groups $G$ such that for any minimal nonabelian $R < G$ there is a cyclic $C < G$ such that $\langle R, C\rangle = G$.

2420. Classify the $p$-groups in which any two nonnormal cyclic subgroups of equal order are conjugate (see § 234 and Appendix 94).

2421. (P.G. Kontorovich) Study the $p$-groups covered by proper characteristic subgroups of index $> p$.

2422. (Kazarin) A group $G$ is an SR-group if all its elements are real and $\chi\tau$ is multiplicity-free for any $\chi, \tau \in \mathrm{Irr}(G)$. Let a 2-group $G$ be an SR-group. (i) Is it true that $\mathrm{dl}(G)$ is bounded? (ii) What is the structure of $G$ of derived length $\leq 3$?

2423. Classify the $p$-groups such that $A^G \cap B^G = A \cap B$ for any nonnormal nonincident $A, B < G$.

2424. Given $d$, study the $d$-generator $p$-groups $G$, in which any minimal system of generators consists of pairwise noncommuting elements.

2425. Describe $\mathrm{Aut}(M \times C)$, where $M$ is a minimal nonabelian $p$-group and $C$ is a cyclic $p$-group.

2426. Study the $p$-groups in which any $\mathcal{A}_1$-subgroup has order $\leq p^4$.

2427. (Isaacs) Is it true that the character table of a $p$-group determines the minimal index of its abelian subgroup?
2428. Find the numbers of orders of maximal abelian subgroups and minimal nonabelian subgroups in $\Sigma_{p^n} \in \mathrm{Syl}_p(\mathrm{S}_{p^n})$ and $\mathrm{UT}(n.p) \in \mathrm{Syl}_p(\mathrm{GL}(n,p))$.
2429. Present an irregular $p$-group $G$ of order $p^{p+1}$ such that $|\mathrm{Aut}(G)|$ is as small as possible.
2430. Study the irregular $p$-groups $G$ satisfying $|\mho_1(G)| = p$.
2431. Study the $p$-groups $G$ such that (i) $\mathrm{X}_1(G) = \mathrm{X}_1(\Sigma_{p^n})$, (ii) $\mathrm{X}_1(G) = \mathrm{X}_1(\mathrm{UT}(n,p))$, where $\mathrm{X}_1(G)$ is the first row of the character table of $G$ (see [T-V]), (iii) $\mathrm{X}_1(G) = \mathrm{X}_1(S)$, where $S$ is a representation group of $\mathrm{E}_{p^n}$.
2432. Classify the 2-groups $G$ containing a nonabelian metacyclic subgroup $H$ of order 16 and exponent 4 (in the notation of this book, it is $\mathcal{H}_{2,2}$) that normalizes all subgroups of order $\leq 8$ nonincident with it.
2433. Study the $p$-groups $G$ satisfying $|\mathrm{cd}(\Gamma)| = |\mathrm{cd}(G)|$, where $\Gamma$ is a representation group of $G$.
2434. Classify the minimal nonabelian and maximal subgroups of the group $\Sigma_{p^3} \in \mathrm{Syl}_p(\mathrm{S}_{p^3})$.
2435. Suppose that $H$ is an irregular $p$-group of order $p^{p+1}$, $p > 2$, containing exactly $p$ subgroups of order $p^p$ and exponent $p$. Does there exist a $p$-group $G$ of maximal class and order $> p^{p+1}$ such that the quotient group $G/\mathrm{K}_{p+1}(G) \cong H$?
2436. Study the $p$-groups $G$ such that $|G : A^G| = p$ for every nonnormal (i) abelian, (ii) minimal nonabelian, (iii) cyclic $A < G$.
2437. Study the $p$-groups $G$ such that $G/\mathrm{K}_3(G)$ is the nonmetacyclic minimal nonabelian group of order $p^4$.
2438. Given $n$, do there exist $p$-groups $H < G$ such that $|G : H| = p^n$ and $\mathrm{Aut}(G) \cong \mathrm{Aut}(H)$? (See #211 and § 126.)
2439. Study the $p$-groups $G$ such that, whenever $K < H \in \Gamma_1$, then $K^G \cap H = K^H$.
2440. Study the group of those automorphisms of a nonabelian $p$-group $G$ that stabilize all (i) $\mathcal{A}_1$-subgroups of $G$, (ii) all subgroups of $G$ of order $\leq p^2$.
2441. Study the $p$-groups $G$ such that, whenever $A < G$ is nonnormal cyclic, then the normal closure $A^G$ is of maximal class. Is it true that $p = 2$? (This is solved by the second author; see Theorem 223.1.)
2442. Study the subgroup structure of the metacyclic 2-groups $G$ of exponent $2^e$ and order $2^{2e}$ (see § 148).
2443. Given a 2-group $G$, does there exist a 2-group $W$ containing $G$ as a normal subgroup and such that all elements of $W$ of order 4 are real in $W$? If the set of such $W$ is nonempty, estimate the minimal possible value of $|W : G|$.
2444. Given $e > 1$, study the $p$-groups $G$ of exponent $p^e$ satisfying $\Phi(G) = \Omega_e^*(G)$.
2445. Study the nonabelian $p$-groups $G$ such that $|G : \mathrm{N}_G(A)| \leq p$ for all cyclic $A < H$, where $H$ runs over all minimal nonabelian subgroups of $G$ (see § 125).
2446. Given $n$, study the groups of exponent $2^n$, all of whose elements are real.

2447. Classify the $p$-groups $G$ such that (i) $G = \langle Z(H) \mid H \in \Gamma_1 \rangle$, (ii) $G = \langle \mathcal{N}(H) \mid H \in \Gamma_1 \rangle$ (here $\mathcal{N}(H)$ is the norm of $H$). (iii) Is it true that the derived length of any group from (i) or (ii) is bounded?

2448. Study the irregular $p$-groups, all of whose nonabelian regular subgroups are two-generator.

2449. Present an irregular $p$-group all of whose maximal regular subgroups are nonnormal.

2450. Study the $p$-groups $G$ in which, for each $n$, $K_n(G)$, the $n$-th member of the lower central series of $G$ is a unique $G$-invariant subgroup of its order.

2451. Study the $p$-groups $G$ such that, whenever $A, B < G$ are nonabelian, then (i) $|A'| = |B'| \Rightarrow |A| = |B|$, (ii) $|A| = |B| \Rightarrow |A'| = |B'|$.

2452. Study the non-Dedekindian $p$-groups $G$ such that, whenever $A < G$ is nonnormal, then $A < B < G$, where $B$ is minimal nonabelian.

2453. Classify the $p$-groups all of whose maximal subgroups, except one, are modular (see § 73).

2454. Study the $p$-groups $G$ such that a Sylow $p$-subgroup of $\mathrm{Aut}(G)$ is abelian (see #1906).

2455. Study the metacyclic $p$-groups $H$ such that, for some two-generator $p$-group $G$, one has (i) $H \cong \Phi(G)$, (ii) $H \cong G'$, (iii) $H \cong \mho_1(G)$.

2456. Describe the representation groups of $\Gamma$, where $\Gamma$ is a representation group of $E_{p^n}$.

2457. Study the $p$-groups all of whose maximal cyclic subgroups are (i) normally complemented, (ii) complemented (for example: $D_{2^n}$).

2458. Study the groups $G$ of order $p^m$ such that for any $n < m$ there is in $G$ a characteristic subgroup of order $p^n$ (see § 148.) Is it true that there is a maximal chain of subgroups of $G$ all of whose members are characteristic?

2459. Suppose that all maximal subgroups of a nonabelian $p$-group $G$ are characteristic. Is it true that then: (i) $d(G)$ is bounded? (ii) $p = 2$?

2460. Study the nonabelian $p$-groups $G$ such that, whenever $A < B \le G$, where $A$ is maximal abelian subgroup of $G$ and $|B : A| = p$, then $|B'| = p$ ($d(B) = 2$).

2461. Study the $p$-groups $G$ satisfying $|G' : \Phi(G)'| = p$ (see § 111).

2462. Classify the special $p$-groups (i) that have a special representation group, (ii) whose Schur multiplier is trivial.

2463. Suppose that a nonabelian $p$-group $G$ has a maximal elementary abelian subgroup of order $p^e$. Is it true that the orders of elementary abelian subgroups of $G$ are bounded? (For $e = 2$, see [GM] and § 134.)

2464. Study the $p$-groups $G$ and $G_0$ such that $\mathcal{L}(G) \cong \mathcal{L}_N(G_0)$ (here $\mathcal{L}(G)$, $\mathcal{L}_N(G_0)$ are the lattices of subgroups of $G$ and normal subgroups of $G_0$, respectively).

2465. Study the 2-groups in which the number of subgroups $\cong E_8$ is even.

2466. Study the $p$-groups all of whose nonnormal abelian subgroups have orders $\le p^3$ (see [ZS] and [ZLS]).

2467. Study the $p$-groups with cyclic center all of whose maximal subgroups have noncyclic centers.

2468. Study the nonabelian two-generator $p$-groups $G$ with metacyclic $G/Z(G)$.

2469. Study the $p$-groups $G$ such that, whenever $H < G$ is nonnormal cyclic, then $H^G$ is either abelian or minimal nonabelian.

2470. Study the $p$-groups whose $\mho$- and $\Omega$-series coincide.

2471. Study the $p$-groups all of whose minimal nonabelian subgroups, except one, have order $p^3$ and exponent (i) $p$, (ii) $p^2$ (see Mann's commentary to #115).

2472. Classify the $p$-groups $G$ with $|G'| = p^k$, $k = 2s + e$, $e \in \{0, 1\}$, satisfying $\mathrm{n}(G) = \frac{1}{2}(p^2 - 1)k + (p - 1)e$, where $\mathrm{n}(G) = |\mathrm{Irr}_1(G)|$ (see Kazarin's Theorem A.45.12).

2473. Let $G$ be an arbitrary finite group with $\Phi(G) > \{1\}$. Present a proof independent of the Sylow theorem that $Z(\Phi(G)) > \{1\}$.

2474. Study the $p$-groups $G$ such that all subgroups of $G$ nonincident with $\Phi(G)$ (i) are abelian, (ii) are either abelian or minimal nonabelian.

2475. Classify the $p$-groups $G$ such that $\mathrm{Aut}(G) \cong \mathrm{Aut}(A)$, where the $p$-group $A$ is abelian (minimal nonabelian); see [Kur] and [Mal6].

2476. Study the $p$-groups $G$ such that $\mathrm{H}_p(G) = \mho_1(G)$ (the strong inclusion $\mathrm{H}_p(G) < \mho_1(G)$ impossible).

2477. (Old problem) Study the $p$-groups which are products of $n$ pairwise permutable cyclic subgroups. State a necessary and sufficient condition when such groups are powerful (see § 26). Estimate the derived length of such $p$-groups.[1]

2478. Does there exist an irregular $p$-group in which any two distinct conjugate minimal irregular subgroups have intersection $\{1\}$?

2479. Given $k > 1$, estimate the class of the $p$-groups $G$ all of whose metabelian subgroups have class $\leq 2$ (see § 163).

2480. Find the minimal value of $\mathrm{k}(G)$, where $G$ runs over all two-generator groups of order $p^n$.

2481. Study the $p$-groups $G$ such that, whenever $H \in \Gamma_1$ is nonabelian, then $H = M \times A$, where $M \leq H$ is minimal nonabelian and $A$ is abelian. Consider in detail the case when $M \cong Q_8$.

2482. Study the capable 2-groups having a normal dihedral subgroup.

2483. Study the nonabelian $p$-groups $G$ such that, whenever $A < G$ is maximal abelian, there is $\mu \in \mathrm{Lin}(A)$ satisfying $\mu^G \in \mathrm{Irr}(G)$.

2484. Study the $p$-groups $G$ such that, whenever $H < G$ is maximal nonnormal, then $|G : \mathrm{N}_G(H)| = p$.

2485. Classify the $p$-groups $G$ such that $\mathrm{Aut}(G)$ acts transitively on the set of maximal cyclic subgroups.

---

1 Commentary of Mann. Let $p > 2$ and let a $p$-group $G = C_1 \dots C_n$, where $C_i = \langle x_i \rangle$ and any two factors permutable. Then $C_i C_j$ is metacyclic for all $i, j$ (Theorem 9.11 or Corollary 36.8) so that $[x_i, x_j] \in \mho_1(G)$. Since, in the case under consideration, $G' = \langle [x_i, x_j], \quad i, j \in \{1, \dots, n\} \rangle$, we get $G' \leq \mho_1(G)$, i.e., $G$ is powerful. This is a result from the unpublished PhD Thesis of B.W. King.

2486. Study the $p$-groups $G$ all of whose maximal elementary abelian subgroups are maximal abelian.
2487. Study the two-generator groups $G$ of exponent $p$ with abelian $\Phi(G)$.
2488. Study the $p$-groups $G$ with noncyclic centers all of whose central subgroups are characteristic. Is it true that $d(Z(G))$ is bounded? Classify the special $p$-groups satisfying the above condition.
2489. Classify the $p$-groups with a normal homocyclic subgroup of index $\leq p^2$.
2490. Study the $p$-groups $G$ with the special $\mathrm{Inn}(G)$.
2491. Classify the $p$-groups $G$ containing a nontrivial subgroup $H$ such that all subgroups of $G$ that are nonincident with $H$ are $G$-invariant.
2492. Study the $p$-groups in which the normal closures of all nonnormal subgroups (cyclic subgroups, abelian subgroups) have equal order (compare with § 62).
2493. Study the $p$-groups in which the normal closure of each metacyclic subgroup has the cyclic derived subgroup.
2494. Study the special 2-groups satisfying (i) $G = \Omega_1(G)$, (ii) $G = \Omega_2^*(G)$.
2495. Study the nonabelian $p$-groups $G$ such that $A \cap G' = A'$ ($A \cap \Phi(G) = \Phi(A)$) for any nonabelian $A < G$ nonincident with $G'$ (with $\Phi(G)$).
2496. Study the $p$-groups $G$ such that $\langle x, y\rangle$ is of maximal class for all noncommuting $x, y \in G - G'$ ($x, y \in G - \Phi(G)$). See § 90.
2497. Study the nonabelian $p$-groups $G$ containing a minimal nonabelian subgroup $A$ such that all maximal subgroups of $A$ are self centralizing in $G$.
2498. Study the 2-groups with metacyclic Frattini (derived) subgroup.
2499. Study the $p$-groups $G$, $p > 2$, with $\mathrm{H}_p(G) < G$ and $|\mathrm{H}_p(G)'| = p$ (see [GMS1] and Appendix 91).
2500. Given a $p$-group $H$, does there exist an overgroup $G > H$ such that all subgroups of $G$ that are $\cong H$, are non-$G$-invariant and conjugate?
2501. Study the $p$-groups $G$ in which all $\mathcal{A}_2$-subgroups have cyclic centers.
2502. Classify the $p$-groups $G$ all of whose subgroups containing $\Phi(G)$ as a subgroup of index $p$ are minimal nonabelian (this was solved by the second author; see § 165).
2503. Given $n > 1$, does there exist a nonabelian group of exponent $p$ admitting a nontrivial partition by subgroups of order $p^n$?
2504. Given $n$, present a group (i) of order $p^n$ having maximal possible derived length, (ii) of minimal order having derived length $n$.
2505. Given a nonabelian $p$-group $G$, is it true that the Schur multiplier of $G$ is characteristic in some representation group of $G$?
2506. Study the $p$-groups $G$, $p > 2$, whose proper $\mathrm{H}_p$-subgroup is abelian (regular).
2507. Let $G$ be a $p$-group and $\Gamma$ a $p$-group containing a normal subgroup $A$ such that $A \leq \Gamma'$ and $\Gamma/A \cong G$. Find a necessary and sufficient condition under which $|A|$ is bounded. Compare $|A|$ and $|\mathrm{M}(G)|$.
2508. (Old problem) Study the $p$-groups $G$ admitting a fixed-point-free automorphism. Consider in detail the case when $\exp(G) = p$.

2509. Classify the $p$-groups in which the normalizers of all nonnormal cyclic subgroups are metacyclic.

2510. Study the 3-groups $G$ whose $H_3$-subgroup is minimal nonabelian.

2511. Does there exist a noncyclic $p$-group all of whose subgroups of index $p$ and order $p$ are characteristic?

2512. Let a group $A$ be abelian of type $(p, p^2, \dots, p^n)$. Find the minimal $n = n(p)$ such that a Sylow $p$-subgroup of $\mathrm{Aut}(A)$ is irregular.

2513. Study the $p$-groups $G$ with irregular $\Phi(G)$ ($\mho_1(G)$, $\Omega_1(G)$) such that $\Phi(H)$ ($\mho_1(H)$, $\Omega_1(H)$) is abelian for any $H \in \Gamma_1$.

2514. Study the irregular $p$-groups $G$ such that a Sylow $p$-subgroup of $\mathrm{Aut}(G)$ is regular.

2515. Study a $p$-group $G$ such that some of its representation groups contain a subgroup $\cong G$.

2516. Study the irregular $p$-groups $G = \Omega_1(G)$ such that, whenever $R$ is a maximal absolutely regular subgroup of $G$ and $x \in G - R$ is of order $p$, then the subgroup $\langle x, R \rangle$ is irregular.

2517. Classify the nonmodular $p$-groups all of whose nonquasinormal subgroups are cyclic (see § 16 and ##1572,768).

2518. For all $n$, find $\mathrm{s}_n(G)$ and $\mathrm{c}_n(G)$, where $G = \Sigma_{p^2} * \Sigma_{p^2}$ is the central product of order $p^{2p+1}$ (here $\Sigma_{p^2} \in \mathrm{Syl}_p(\mathrm{S}_{p^2})$).

2519. Classify the abelian $p$-groups $G$ such that $\exp(\Gamma) = \exp(G)$, where $\Gamma$ is a representation group of $G$.

2520. Describe the automorphism groups of minimal nonmetacyclic 2-groups.

2521. Classify the $p$-groups $G$ such that $\mathcal{L}_N(G) = \mathcal{L}(M)$, where a $p$-group $M$ is (i) minimal nonabelian, (ii) metacyclic, (iii) of maximal class, (iv) $\Sigma_{p^n}$, (v) abelian.

2522. Let $G$ be a $p$-group with $\mathrm{d}(G) = d > 2$, $G = \langle a_1, \dots, a_d \rangle$. Describe the structure of $G$, if for different $r, s \in \{1, \dots, d\}$, the subgroup $\langle a_r, a_s \rangle$ is minimal nonabelian.

2523. Study the $p$-groups in which any two subgroups of equal order (of different orders) are permutable.

2524. Does there exist a $p$-group $G$ such that $\exp(\mathrm{M}(G)) > \exp(G)$? Here $\mathrm{M}(G)$ is the Schur multiplier of $G$.

2525. Study the $p$-groups all of whose nonabelian maximal subgroups are special (see [Cos]).

2526. Study the $p$-groups $G$ satisfying $|\mathrm{Aut}(G) : \mathrm{Inn}(G)|_p = p$.

2527. Classify the $p$-groups $G$ such that for some distinct $A, B \in \Gamma_1$ and any $a \in A - B$, $b \in B - A$, the subgroup $\langle a, b \rangle$ is $G$-invariant.

2528. Study the $p$-groups, $p > 2$, with the proper special $H_p$-subgroup.

2529. Classify the nonmodular $p$-groups in which the quasinormal (normal) closure of any nonquasinormal cyclic subgroup is abelian.

2530. (Old problem) Let a $p$-group $G = AB$, where $A, B < G$. Study the structure of $G$ provided $\exp(G) = \exp(A)\exp(B)$ (for some results on this topics, see [Man31]).[2]

2531. Classify the $p$-groups admitting an automorphism of order $p$ with exactly $p^2$ fixed points (see §§ 51 and 77).

2532. Study the nonabelian $p$-groups $G$ satisfying (i) $\mathrm{cl}(G) = \mathrm{cl}(\Phi(G))$, (ii) $\mathrm{cl}(G) = \mathrm{cl}(\mho_n(G))$ for a given $n$.

2533. Study the $p$-groups in which any two-generator subgroup has cyclic derived subgroup.

2534. Does there exist a nonabelian $p$-group $G$ satisfying (i) $\mathrm{Aut}(G/Z(G)) \cong \mathrm{Aut}(G)$, (ii) $\mathrm{Aut}(G/G') \cong \mathrm{Aut}(G)$, (iii) $\mathrm{Aut}(G/\Phi(G)) \cong \mathrm{Aut}(G)$, (iv) $\mathrm{Aut}(G) \cong \mathrm{Aut}(G')$, (v) $\mathrm{Aut}(G) \cong \mathrm{Aut}(Z(G))$, (vi) $\mathrm{Aut}(G) \cong \mathrm{Aut}(\Phi(G))$?

2535. Study the two-generator $p$-groups $G$ with $|G : Z(G)| = p^3$.

2536. Study the subgroup, normal and power structures of the $p$-group $G = E \operatorname{wr} C$ of order $|E|^{|C|}|C|$ (the regular wreath product), where $C$ is cyclic and $E$ is elementary abelian. Describe $\mathrm{Aut}(G)$ and $M(G)$.

2537. Does there exist a $p$-group (irregular $p$-group) all of whose maximal abelian (regular) subgroups are pairwise nonisomorphic?

2538. Study the $p$-groups in which any two nonincident subgroups with cyclic derived subgroups have a cyclic intersection.

2539. Study the irregular $p$-groups all of whose absolutely regular subgroups are abelian.

2540. Study the nonabelian $p$-groups $G$ all of whose nonconjugate maximal abelian subgroups $A, B$ satisfy $\langle A, B\rangle = G$ (see § 212).

2541. Study the $p$-groups $G$ such that, whenever $A, B < G$ are nonincident, then $\langle A, B\rangle \le N_G(A \cap B)$.

2542. Study the irregular $p$-groups $G$ containing a normal (i) cyclic subgroup $N$ such that $G/N$ is absolutely regular,[3] (ii) absolutely regular subgroup $N$ such that $G/N$ is cyclic.

2543. Given $n$ and a $p$-group $E = \Omega_n(E)$, does there exist a $p$-group $G$ such that $G > \Omega_n(G) \cong E$?

2544. Describe $\mathrm{Aut}(M \times E)$, where $M$ is a 2-group of maximal class and $\exp(E) = 2$.

2545. Study the $p$-groups $G$ such that $\Phi(G)$ is irregular of order (i) $p^{p+2}$, (ii) $p^{p+3}$.

2546. (A partial case of #860) Classify the nonabelian groups of exponent $p$ covered by nonabelian subgroups of order $p^3$.

---

**2** As Mann informed me, if $A, B$ are abelian, then B. Howlett [How] has proved that $\exp(G) \le \exp(A)\exp(B)$, and this result is best possible. Indeed, the wreath product $A \operatorname{wr} B = B \cdot C$, where $C$ is its base, has exponent $\exp(A)\exp(B)$.).

**3** As Mann has noted, the set of such $G$ is nonempty; see Remark 1 in Appendix 45.

2547. Classify the $p$-groups $G$ of exponent $> p$ in which any two cyclic subgroups of the same order $> p$ are automorphic. (Two subgroups $A, B$ of $G$ are said to be *automorphic* if $B = A^{\phi}$ for some $\phi \in \mathrm{Aut}(G)$.)

2548. Suppose that $H \cong G'$ for some $p$-group $G$. Is it true that $H \cong \Phi(W)$ ($H \cong \mho_1(W)$) for some $p$-group $W$?

2549. (M.L. Lewis [Lew]) Let $G$ be a nonabelian $p$-group. Suppose that for any $x \in G - \mathrm{Z}(G)$, all elements of the coset $x\mathrm{Z}(G)$ are conjugate. Is it true that $|\mathrm{Z}(G)| \leq |G|^{2/3}$?

2550. Classify the nonabelian groups of exponent $p$ all of whose nonabelian subgroups of order $p^4$ are nonnormal.

2551. Classify the $p$-groups admitting an irredundant covering $G = A_1 \cup A_2 \cup \cdots \cup A_n$ such that exactly two of its components, say $A_1$ and $A_2$, have intersection $A_1 \cap A_2 > \{1\}$.

2552. Study the $p$-groups $G$ of exponent $> p^2$ such that, whenever $C$ is a nonmaximal cyclic subgroup of $G$, then $C$ is contained in exactly $p$ cyclic subgroups of order $p|C|$ (see Lemma A.69.6).

2553. Study the $p$-groups $G$ such that $|\mathrm{N}_G(A)/\mathrm{C}_G(A)| = |\mathrm{Aut}(A)|_p$ for all maximal cyclic $A < G$.

2554. Study the nonabelian 2-groups containing exactly one normal subgroup of index 8. (A Sylow 2-subgroup of the simple Suzuki group Sz(8) satisfies this condition.)

2555. Study the nonabelian $p$-groups all of whose cyclic subgroups are TI-subgroups. (A more general result is proved by the second author; see § 187. See also § 63.)

2556. Study the $p$-groups $G$ in which any epimorphic image of $G$ is isomorphic to an appropriate subgroup (normal subgroup) of $G$.

2557. Study the subgroup, normal and power structures of the $p$-group $G = H \operatorname{wr} C$ of order $|H|^{|C|}|C|$, where $H$ is homocyclic and $C$ is cyclic.

2558. Study the $p$-groups with trivial Schur multiplier all of whose maximal subgroups have nontrivial Schur multipliers (see #1827).

2559. Study the minimal irregular $p$-groups all of whose maximal subgroups, except one, are of exponent $p$.

2560. Classify the 2-groups $G$ containing a normal abelian subgroup $R \cong \mathrm{E}_4$ such that $G/R$ is of maximal class (for a partial case, see § 67).

2561. (Old problem) Let a $p$-group $G = AB$ for $A, B < G$. Estimate $\mathrm{dl}(G)$ in terms $\mathrm{dl}(A)$ and $\mathrm{dl}(B)$.

2562. Classify the 2-groups admitting an automorphism of order 4 with exactly two fixed points (see § 77).

2563. Study the regular $p$-groups of order $p^n$ having maximal possible class.

2564. Study the $p$-groups $G = A * B$ (central product) such that $A, B$ are characteristic in $G$ (for example: $G = \mathrm{Q}_8 * \mathrm{C}_4$ of order 16).

2565. Given a $p$-group $E$ of derived length $n$ (of exponent $p^n$, of class $n$), does there exist a $p$-group $G$ of derived length $> n$ (of exponent $> p^n$, of class $> n$) such that $G/G^{(n)} \cong E$ ($G/\mho_n(G) \cong E$, $G/\mathrm{K}_{n+1}(G) \cong E$)?

2566. Classify the $p$-groups $G$ in which all nonnormal $H < G$ satisfying $|H|^2 < |G|$ have the same order (see § 210).

2567. Study the $p$-groups $G$ such that $\mho_n(H) = \mho^n(H)$ for all $H \leq G$ and $n$.

2568. Classify the $p$-groups $G$ such that for any maximal cyclic $C < G$, all maximal subgroups of $G$ containing $C$, are of class $\leq 2$.

2569. Study the nonabelian $p$-groups $G$ containing a maximal subgroup $H$ such that all minimal nonabelian subgroups of $G$ not contained in $H$, have order $p^3$.

2570. Study the $p$-groups $G$ such that, whenever nonabelian subgroups $A < B \leq G$, then $\mathrm{d}(A) \leq \mathrm{d}(B)$ (see § 233).

2571. Is it true that the Burnside two-generator group of exponent 4 and order $2^{12}$ is capable?

2572. Given a group $E$ of exponent $p$, does there exist a $p$-group $G$ of exponent $> p$ such that $G/\mathrm{H}_p(G) \cong E$?

2573. Study the $p$-groups all of whose nonlinear irreducible characters of equal degree have an equal number of zeros.

2574. Does there exist $n$ and $p$ such that, if $\Omega_n(G)$ is regular, so is a $p$-group $G$?

2575. Study the $p$-groups $G$ such that, whenever $A < F < G$, there is $H < G$, $H \neq A$, such that $F \cap H = A$.

2576. Let $G$ be a $p$-group of order $p^m$ and let $\mathrm{s}_t(G) = \max\{\mathrm{s}_i(G) \mid i \in \{1, 2, \ldots, m-1\}$. Find $t$ for $\Sigma_{p^n}$ and $\mathrm{UT}(n, p)$.

2577. Classify the special $p$-groups all of whose maximal subgroups are special (this was solved by J. Cossey [Cos]. Moreover, the $p$-groups all of whose maximal subgroups are special, are classified [Cos]).

2578. Suppose that a $p$-group $G = F * H$ (central product) is such that $F \cap H = \mathrm{Z}(F) = \mathrm{Z}(H)$. Study the group $A = \langle \alpha \in \mathrm{Aut}(G) \mid F^\alpha = F,\ H^\alpha = H \rangle$. Consider in detail the case when $A = \mathrm{Aut}(G)$.

2579. Classify the $p$-groups all of whose metabelian subgroups are metacyclic (for $p > 2$, see Exercise A.45.192).

2580. (Old problem) Let $\phi : G \to G_0$ be a lattice isomorphism of $p$-groups. Study the embedding of $N^\phi$ in $G_0$ for $N \lhd G$.

2581. Classify the $p$-groups $G$ such that, whenever $A, B < G$ are distinct conjugates, then either $A \leq \mathrm{N}_G(B)$ or $B \leq \mathrm{N}_G(A)$.

2582. Study the $p$-groups $G$ all of whose irreducible characters assume $\leq p$ distinct nonzero values.

2583. Study the $p$-groups in which every subgroup of order $p^2$ is contained in a nonabelian subgroup of order $p^3$.

2584. Classify the 2-groups $G$ such that, whenever $A, B < G$ are distinct conjugates, then $A \cap B$ is either cyclic or a generalized quaternion group. (This problem was solved by the second author; see § 184. Compare with #1572, also solved by the second author.)

2585. Does there exist a $p$-group of class $p$ with irregular (i) $\Phi(G)$, (ii) $\mho_1(G)$, (iii) $\mho_2(G)$, (iv) $G'$?

2586. Estimate the minimal value of $|\mathrm{Aut}(G)|_p$, where $G$ runs over all nonabelian groups of order $p^n$ and exponent $p$.

2587. Study the irregular $p$-groups $G$ such that there exists a regular $p$-group $H$ satisfying $\mathrm{c}_k(G) = \mathrm{c}_k(H)$ ($\mathrm{s}_k(G) = \mathrm{s}_k(H)$) for all $k$ (for example: $G = \mathrm{H}_{2,2}$, $H = \mathrm{C}_4 \times \mathrm{C}_4$).

2588. Let a $p$-group $W = E \cdot G$ (semidirect product with kernel $G$), where $E \cong \mathrm{E}_{p^2}$. Study the structure of $G$ provided $\mathrm{C}_G(E)$ is of order $p$.

2589. Study the $p$-groups $G$ containing a subgroup of index $p$ and exponent $p$ such that the Schur multiplier of $G$ is trivial.

2590. Given $n > 1$, classify the $p$-groups $G$ such that $\Omega_1(G) \cong \Sigma_{p^n} \in \mathrm{Syl}_p(\mathrm{S}_{p^n})$. (If $n = 2$, then $G \cong \Sigma_{p^2}$, by Theorem 13.2 (a) and Exercise 9.3.)

2591. Classify the nonabelian $p$-groups $G$ such that for any $H \in \Gamma_1$ there are $\leq p$ minimal nonabelian subgroups not contained in $H$ (see § 76).

2592. Study the $p$-groups $G$, $p > 2$, such that $\Phi(H) = \mho_1(H)$ for all $H \leq G$.

2593. Classify the $p$-groups in which each subgroup is normal in its normal closure.

2594. Classify the $p$-groups with cyclic derived subgroup containing a maximal abelian subgroup of order $p^4$ (see § 148).

2595. Estimate $|\mathrm{Aut}(G)|_p$, where $G$ is a metabelian group of maximal class and order $p^n$.

2596. Let $\mathcal{M}_p$ be the set of all irregular $p$-groups of exponent $p^2$ containing a subgroup of exponent $p$ and index $p$. Given $n$, does there exist $G \in \mathcal{M}_p$ such that $|\mho_1(G)| = p^n$?

2597. Classify the $p$-groups all of whose maximal subgroups, except one, are special (see [Cos]).

2598. Study the $p$-groups $G$ such that, whenever $A, B < G$ are distinct conjugate subgroups, then $\langle A, B\rangle = A^G$.

2599. Study the $p$-groups $G > \mathrm{H}_p(G)$, $p > 2$, such that, whenever $x \in G - \mathrm{H}_p(G)$, there is $h \in \mathrm{H}_p(G)$ such that $\langle x, h\rangle \cong S(p^3)$.

2600. Study the $p$-groups $G$ such that, whenever $A < G$ is maximal abelian, then $A/\mathrm{Z}(G)$ is maximal abelian in $G/\mathrm{Z}(G)$.

2601. Let $G = AB$ be of exponent $p$ and $A, B$ be abelian with $A \cap B = \{1\}$. Find the maximal possible value of $\mathrm{cl}(G)$ in terms of $|A|$, $|B|$.

2602. Study the irregular $p$-groups $G$ satisfying $|\Omega_1(G)| = |G/\mho_1(G)| = p^p$.

2603. Classify the $p$-groups $G$ all of whose $\mathcal{A}_1$-subgroups contain $\Phi(G)$. (See #2502. According to a letter of the second author dated 28/2/2012, he solved this problem; see § 167.)

2604. Classify the two-generator $p$-groups $G$ with elementary abelian subgroup $\Phi(G)$.

2605. Study the $p$-groups $G$ such that all their subgroups containing $\Phi(G)$ as a subgroup of index $p$ are special.

2606. Is it true that for any nonabelian $p$-group $G$ there is a minimal nonabelian subgroup $A \leq G$ such that $A \cap \mathrm{Z}(G) > \{1\}$ (according to the second author, the answer is 'yes'; see Theorem 200.1).

2607. Study the $p$-groups $G$ such that $\mathrm{cl}(G) = \mathrm{cl}(A) + \mathrm{cl}(B)$ for any distinct $A, B \in \Gamma_1$.

2608. Study the $p$-groups $G$ possessing a normal subgroup $A$ and such that all subgroups of $G$ containing $A$ as a subgroup of index $p$, except one, are of maximal class (see Exercise 10.10).

2609. Classify the $p$-groups all of whose maximal absolutely regular subgroups are (i) of class $\leq 2$, (ii) maximal regular.

2610. Study the $p$-groups $G$ such that, whenever $A < G$ is nonnormal, then $|A : (A \cap G')| = p$.

2611. Study the group of those automorphisms of a $p$-group $G$ that induce the identity on the quotient group $G/G'$.

2612. Study the $p$-groups $G = AB$ such that $A$ and $B$ are characteristic in $G$ and $A$ is a minimal supplement of $B$ in $G$ and $B$ is a minimal supplement of $A$ in $G$.

2613. Given $n > 1$, study the $p$-groups $G$ such that $G/\mho_n(G) \cong \Sigma_{p^n} \in \mathrm{Syl}_p(\mathrm{S}_{p^n})$ (see #2565).

2614. Study the $p$-groups $G$ such that, whenever $H < G$ is nonnormal, there is a cyclic $C < H$ such that $C^G = H^G$.

2615. (Burnside) Let $G$ be a $p$-group and $A \leq \mathrm{Aut}(G)$ be a subgroup all those automorphisms of $G$ which leave fixed all $G$-classes. Study the structure of $A$.

2616. Estimate the maximal number of $\mathcal{A}_1$-subgroups in $p$-groups $G$ with $\mathrm{d}(G) = d$ and $|G| = p^n$. Is it true that if $n$ is sufficiently large, then $\exp(G) = p$? (See § 76.)

2617. Study the nonabelian $p$-groups $G$ such that for any maximal abelian $A < G$ there is an abelian $B < G$ satisfying $G = AB$.

2618. (Compare with Lemma 57.1) Study the nonabelian $p$-groups $G$ such that, whenever $A < G$ is maximal abelian, then, whenever $x \in G - A$, there is $a \in A$ such that $\langle a, x \rangle$ is minimal nonabelian.

2619. Classify the $p$-groups $G$ such that, whenever $A, B < G$ are nonnormal of equal order, then $A \leq \mathrm{N}_G(B)$ and $B \leq \mathrm{N}_G(A)$.

2620. Classify the 2-groups $G$ such that $G/L$ is of maximal class for some cyclic $L \lhd G$.

2621. A 2-group is said to be a $\mathcal{D}_n$-group if all of its subgroups of index $2^n$ are Dedekindian but it contains a non-Dedekindian subgroup of index $2^{n-1}$ and $G$ is not an $\mathcal{A}_n$-group. (i) Classify the $\mathcal{D}_n$-groups for $n \leq 3$ (see Appendix 90 for $n = 1$, § 245 for $n = 2$, Theorem A.17.2 and Corollary A.17.3.). (ii) Classify the 2-groups all of whose subgroups of index 2 (of index 4), except one, are Dedekindian (see #3849).

2622. Let $A$ be a maximal normal abelian subgroup of a nonabelian $p$-group $G$. Find a sufficient condition for $\mathrm{b}(G) = \max\{\chi(1) \mid \chi \in \mathrm{Irr}(G)\} = |G : A|$.

2623. Study the $p$-groups (special $p$-groups) all of whose proper nonabelian epimorphic images are special.

2624. Classify the nonabelian $p$-groups all of whose minimal nonabelian subgroups are nonmetacyclic of order $p^4$.

2625. Find the minimal order of a $p$-group $G$ whose Frattini subgroup (derived subgroup, $\mho_1(G)$) is irregular.

2626. Study the $p$-groups whose automorphism group is metacyclic.

2627. Study the $p$-groups such that, whenever $H < G$ is nonnormal and $K$ is the characteristic closure of $H$ in $G$, then $|G : K| \le p$ (compare with § 62).

2628. (R. van der Waall) Let $R$ be a $p$-group such that there is no $p$-group $G$ with $R = \Phi(G)$. Does this imply that there is no $p$-group $H$ such that $R$ is a proper $H$-invariant subgroup of $\Phi(H)$? (See [Alle].)[4]

2629. Study the nonabelian $p$-groups $G$ such that, whenever $A, B < G$, where $A$ is minimal nonabelian and nonincident with $B$, then $B$ is permutable with all subgroups of $A$.

2630. Study the irregular $p$-groups $G$ with noncyclic center, $p > 2$, all of whose proper epimorphic images of order $\frac{1}{p}|G|$ are regular.

2631. Classify the $p$-groups $G$ containing a nonabelian subgroup $R$ of order $p^3$ such that $C_G(R)$ is cyclic (see [ZG2]).

2632. Given $p$-groups $A$ and $B$, $\exp(A) = p$, does there exist a $p$-group $G$ satisfying $\Omega_1(G) = A$ and $G/A \cong B$?

2633. Classify the special $p$-groups $G$ of exponent $p^2$ such that all $p$-groups that are lattice isomorphic to $G$ are also special.

2634. Let $A$ be a maximal normal abelian subgroup of a $p$-group $G$. Compare $\alpha_1(G/A)$ and $\alpha_1(G)$ (see Appendix 87).

2635. Classify the non-Dedekindian $p$-groups all of whose nonnormal maximal cyclic subgroups (nonabelian subgroups with cyclic derived subgroups) have normalizers with a cyclic derived subgroup.

2636. Given $n > 1$, construct a nonabelian $p$-group $G$ such that $\Omega_n(G)$ is abelian of type $(p, p^2, \dots, p^n)$.

2637. Classify the metacyclic $p$-groups $G$ such that all $p$-groups that are lattice isomorphic with $G$ are $\cong G$ (see §§ 124 and 148).

2638. Study the $p$-groups covered by nonnormal (i) subgroups, (ii) cyclic subgroups, (iii) abelian subgroups, (iv) subgroups of class 2, (v) metabelian subgroups, (vi) minimal nonmetacyclic subgroups, (vii) minimal nonabelian subgroups (see #860).

2639. Study the $p$-groups $G$ such that, whenever $\chi, \tau \in \mathrm{Irr}_1(G)$ and $\chi(1) \le \tau(1)$, then $|\mathrm{T}_\chi| \ge |\mathrm{T}_\tau|$ (here $\mathrm{T}_\chi = \{a \in G \mid \chi(a) = 0\}$).

2640. Is it true that the set of natural numbers $n > 2$ such that there is a primary $\mathcal{A}_n$-group $G$ with $\alpha_{n-1}(G) = 1$, is finite? (See [ZZLS] for $n = 3$.)

2641. (I. Schur) Given an abelian $p$-group $G$, find the number of its representation groups. Consider in detail the case when $G$ is homocyclic.

2642. Classify the normal subgroups in (i) $\Sigma_{p^n} \in \mathrm{Syl}_p(\mathrm{S}_{p^n})$, (ii) $\mathrm{UT}(n, p) \in \mathrm{GL}(n, p)$.

---

4 R. van der Waall [Waa2] has found the groups $H$ of order $p^4$ for which there is a finite group $G$ such that $H \lhd G$ and $H \le \Phi(G)$. Next, in [WaaN] all 2-groups of order $2^5$, which are Frattini subgroups of appropriate finite groups, are classified.

2643. Does there exist a $p$-group all of whose maximal subgroups are isomorphic and characteristic?

2644. Study the $p$-groups all of whose subgroups of class 2 have abelian subgroups of index $p$.

2645. Study the nonmodular $p$-groups $G$ such that $Z(G)$ is noncyclic and $G/L$ is modular for each $L \leq Z(G)$ of order $p$.

2646. Classify the $p$-groups $G$ such that, whenever $H < G$ satisfies $|H|^2 < |G|$, then $H$ is quasinormal in $G$.

2647. Find the number of normal subgroups of index $p^k$, $k \in \{2, 3, 4\}$, in $\Sigma_{p^n}$ and $\mathrm{UT}(n, p) \in \mathrm{Syl}_p(\mathrm{GL}(n.p))$.

2648. Let a $p$-group $G \cong A \times A$. Describe (i) the members of the set $\Gamma_1$, (ii) the structure of $\mathrm{Aut}(G)$ in the terms of $A$.

2649. Suppose that any maximal subgroup of a $p$-group $G$ is generated by elements of equal order but $G$ is not. Is it true that (i) $\mathrm{d}(G)$ is bounded? (ii) Study the structure of $G$ provided, in addition, $\exp(G) = p^{p+1}$, in detail.

2650. Classify the $p$-groups, all of whose maximal subgroups have cyclic Frattini subgroups (derived subgroups). See §§ 66 and 69.

2651. Classify the 2-groups with a nonabelian Dedekindian subgroup of index 2 (see Appendix 90 and § 245).

2652. (Inspired by Isaacs' letter) Let $G, G_0$ be $p$-groups such that $\mathcal{L}_N(G) = \mathcal{L}_N(G_0)$. Suppose that $G_0$ contains an abelian subgroup of index $p$. Is it possible to find the minimal index of an abelian subgroup in $G$?

2653. Study the $p$-groups all of whose maximal subgroups, except one, are two-generator (see § 70 and 113).

2654. Classify the $p$-groups $G$ such that, whenever $F, H < G$ have equal order, then (i) $\mathrm{k}(F) = \mathrm{k}(H)$, (ii) $\mathrm{cl}(F) = \mathrm{cl}(H)$, (iii) $\mathrm{dl}(F) = \mathrm{dl}(H)$ (three problems).

2655. Study the $p$-groups $G$ such that $\mathrm{cl}(\Gamma) = \mathrm{cl}(G)$ for all representation groups $\Gamma$ of $G$.

2656. Study the groups of exponent $p$ having only one representation group.

2657. Study the $p$-groups all of whose representation groups have cyclic derived subgroups. (All metacyclic $p$-groups satisfy this condition; see Theorem 47.2.)

2658. Classify the $p$-groups $G$ such that $G/\mathcal{N}(G)$ is (i) abelian, (ii) of maximal class, (iii) extraspecial, (iv) minimal nonabelian (v) metacyclic, (vi) absolutely regular, (vii) elementary abelian, (viii) abelian of type $(p, p)$ (here $\mathcal{N}(G)$ is the norm of $G$).

2659. Study the irregular $p$-groups with a maximal regular subgroup of order $p^{p+1}$.

2660. Study the $p$-groups $G$ such that, whenever $A, B \lhd G$ are of the same order, then $G/A \cong G/B$.

2661. Study the $p$-groups $G$ such that $[x, y] \in \mho_1(\langle x, y\rangle)$ for all $x, y \in G$.

2662. Classify the irregular 3-groups all of whose regular subgroups have cyclic derived subgroups.

2663. Given a $p$-group $G_0$, study the $p$-groups $G$ such that there is a bijection $\phi : G \to G_0$ satisfying $C_G(x) \cong C_{G_0}(x^\phi)$ for all $x \in G - Z(G)$.

2664. Given $n$, does there exist a group $G$ of order $p^n$ such that $\mathrm{Aut}(G)$ has an abelian subgroup $H$ of order $\geq p^n$?

2665. Given $n$, does there exist a $p$-group $G$ containing a subgroup $H$ of index $p^n$ such that $k(H) \geq k(G)$?

2666. Study the $p$-groups $G$ such that $G/\mho_1(\mho_1(G))$ is special (see #740).

2667. Classify the nonmodular $p$-groups $G$ in which any two nonpermutable subgroups generate $G$.

2668. Study the $p$-groups $G$ in which any minimal nonabelian subgroup $A$ contains a maximal subgroup $B$ such that $N_G(B) = A$.

2669. Given $n$, estimate the minimal index of a $G$-invariant subgroup of exponent $p$ in a $p$-group $G$ provided that $G$ possesses a subgroup of exponent $p$ and index $p^n$. Consider in detail the cases $n = 2, 3$.

2670. Let $X_1(G) = X_1(H)$, where $\exp(H) = p$ and $G$ is a $p$-group. Is it true that $\exp(G)$ is unbounded? (Here $X_1(G)$ is the first column of the character table of $G$.)

2671. Let $G$ be a $p$-group of maximal class. Study the $p$-groups $H$ of order $|G|$ such that $k(H) \leq k(G)$. Is it true that $H$ is of maximal class?

2672. Study the $p$-groups $G$ such that for any nonnormal $A, B < G$ one has $|A^G| = |B^G|$.

2673. Let $X_1(G) = X_1(G_0)$, where $G$ is a $p$-group and $G_0$ is an absolutely regular $p$-group, $p > 2$. Is it true that $G$ may be irregular?

2674. Study the $p$-groups $G$ that are connected with any maximal cyclic subgroup by only one maximal chain.

2675. Given $n > 3$, classify the $p$-groups $G$ with $|G'| = p^n$ without nonnormal subgroups of order $> p^n$ (see [Pas] and Theorem 1.23, #2279, [ZLS], and [ZS]).

2676. Given $n > 2$, construct a $p$-group of exponent $p^n$ containing maximal abelian subgroups of exponents $p, p^2, \dots, p^n$.

2677. (Janko) Classify the $p$-groups of exponent $\geq p^3$ all of whose cyclic subgroups of order $\geq p^3$ are normal (this is solved in § 229).

2678. Classify the irregular $p$-groups all of whose maximal regular subgroups are two-generator.

2679. Is it true that for any $p$-group $H$ there is a two-generator $p$-group $G$ such that $H$ is isomorphic to a subgroup of $\Phi(G)$? (Without restriction on $d(G)$ this is true.)

2680. Study the $p$-groups $G$ such that, whenever a noncyclic $A < G$, then there is in $A$ a maximal subgroup which is $G$-invariant.

2681. Find a necessary and sufficient condition for a direct product of two regular $p$-groups to be regular.

2682. Given $n$, does there exist a $p$-group having exactly $n$ faithful irreducible characters?

2683. Classify the $p$-groups all of whose minimal nonabelian subgroups are pairwise non-isomorphic (for example: $SD_{16}$. See § 206).

2684. Study the irregular $p$-groups that are lattice isomorphic to regular $p$-groups (see Appendix 42).

2685. Let $G = A_1 \cup \cdots \cup A_n$ and $G = B_1 \cup \cdots \cup B_n$ be two distinct nontrivial partitions of a $p$-group $G$ with minimal possible number of components. Let $|A_1| \geq \cdots \geq |A_n|$ and $|B_1| \geq \cdots \geq |B_n|$. Is it true that (i) $|A_i| = |B_i|$ for all $i$, (ii) if $\exp(G) > p$, then $|A_1| = |\mathrm{H}_p(G)| = |B_1|$?

2686. Study the nonabelian $p$-groups $G$ such that, whenever $A < G$ is minimal nonabelian and $x \in G - A$, then $\mathrm{C}_A(x) = \mathrm{Z}(A)$.

2687. Study the $p$-groups $G$ of order $p^{p+1}$ such that $p^2$ does not divide $\exp(\mathrm{Aut}(G))$.

2688. Let the Schur multiplier of a $p$-group $G$ be of order $p$. Is it possible to estimate $\mathrm{d}(G)$?

2689. Classify the $p$-groups $G$ containing a proper special subgroup $H$ such that all subgroups of $G$ containing $H$ as a subgroup of index $p$ are special.

2690. Study the nonabelian two-generator $p$-groups $G$ such that $G'$ is a maximal cyclic subgroup of $G$.

2691. Study the $p$-groups $G$ with a cyclic derived subgroup of order $p^k$ such that $\mathrm{cd}(G) = \{1, p, \ldots, p^k\}$ (see [HetK], [Man30] and § 148).

2692. Let $G$ be a 2-group. Suppose that $H \in \Gamma_1$ has no $G$-invariant elementary abelian subgroup of order 8. Describe the structure of $H$.

2693. (Isaacs; see [INS, Theorem A]) Classify the 2-groups with exactly one nonlinear rational-valued irreducible character.

2694. Find the number of rational-valued nonlinear irreducible characters for a nonabelian metacyclic 2-group $G$ with given defining relations.

2695. Study the $p$-groups $G$ such that, whenever $H < G$ is noncyclic, then all proper characteristic subgroups of $H$ are normal in $G$.

2696. (i) The intersection of normalizers of all nonquasinormal subgroups of a group $G$ is said to be the *quasinorm* of $G$. Study the structure of the $p$-groups with nonabelian quasinorm (see §§ 143 and 171). (ii) Compare norms and quasinorms of a $p$-group.

2697. Classify the characteristic subgroups of an abelian $p$-group of given type.

2698. Let $G$ be a nonabelian metacyclic $p$-group. Study the nonmetacyclic $p$-groups $H$ having the same character table as $G$.[5]

2699. Study the nonabelian $p$-groups $G$ all of whose minimal nonabelian subgroups not contained in $\Phi(G)$ have order $p^3$.

2700. Classify the $p$-groups all of whose proper subgroups have cyclic derived subgroups (this was solved by the second author; see §§ 137 and 139).

2701. Given $k > 1$, does there exist a special $p$-group all of whose nonabelian subgroups of index $p^k$ are special? (See #2577 and [Cos].)

5 If $G = \langle a, b \mid a^4 = b^4 = 1, a^b = a^{-1} \rangle$ and $H = \langle a, b \mid a^4 = b^2 = 1, a^b = ac, c^2 = 1, [a, c] = [b, c] = 1 \rangle$, then $G$ and $H$, according to Isaacs' information, have identical character tables.

2702. Study the $p$-groups $G$ all of whose normal subgroups of index $> p$ are pairwise nonisomorphic.

2703. Let $G$ and $G_0$ be two $p$-groups, $\Gamma_1 = \{F_1, \dots, F_n\}$ and $\Gamma_1^0 = \{H_1, \dots, H_n\}$ (the set of maximal subgroups of $G_0$). Suppose that $F_i \cong H_i$ for all $i \le n$. Compare the structures of $\Phi(G)$ and $\Phi(G_0)$.

2704. Study the $p$-groups $G$ in which $\Phi(G)$ is the unique normal subgroup of its order. The same problem with $\mho_1(G)$ (with $G'$) instead of $\Phi(G)$.

2705. Estimate the order of the Schur multiplier of a $p$-group of maximal class and order $p^n$, $p > 2$.

2706. Find the number of special groups $G$ of order $p^n$ with the center of order $p^2$.

2707. Study the representation groups of the $p$-groups $G$ with $|\Phi(G)| = p$.

2708. Study the groups $G$ of exponent $p$ whose representation groups are irregular (are of exponent $p$).

2709. Study the abelian $p$-groups $G$ such then $\Gamma/\mathrm{Z}(\Gamma) \cong G$, where $\Gamma$ runs through all representation groups of $G$.

2710. Study the nonabelian $p$-groups, $p > 2$, in which any $p + 1$ pairwise noncommuting elements generate either a regular subgroup or a subgroup of maximal class.

2711. Let $p$-groups $G$ and $W$ be lattice isomorphic and let $G'$ be cyclic. Describe the structure of $W'$.

2712. Study the nonabelian $p$-groups $G$ such that any $M \in \Gamma_1$ contains the normalizers of all its non-$G$-invariant subgroups.

2713. Given $n$, classify the $p$-groups $G$ all of whose proper subgroups of order $p^n$ are pairwise nonisomorphic.

2714. Study the nonabelian nonmetacyclic 2-groups all of whose nonabelian metacyclic subgroups are of maximal class.

2715. Study the 2-groups containing only one subgroup $\cong \mathrm{SD}_{2^n}$.

2716. Study the $p$-groups all of whose maximal subgroups, except one, (i) have cyclic derived subgroup (this is solved by the second author; see § 146), (ii) are metabelian.

2717. Given $n > 2$, does there exist a $p$-group admitting a partition that contains $n$ components of pairwise distinct orders?

2718. Study the $p$-groups $G$ satisfying $\mho_1(\mho_1(G)) > \mho_2(G)$ and estimate $|\mho_1(\mho_1(G)) : \mho_2(G)|$.

2719. Classify the nonabelian $p$-groups all of whose nonnormal nonabelian subgroups have (i) cyclic center, (ii) cyclic derived subgroup.

2720. Does there exist a non-absolutely regular $p$-group, $p > 2$, all of whose maximal absolutely regular subgroups are pairwise nonisomorphic?

2721. Study the subgroup and normal structure of the $p$-groups $G$, $p > 2$, satisfying $|\Omega_1(G)| = p^p$ (for $p = 2$, see § 82).

2722. Study the $p$-groups $G$ containing exactly $p$ subgroups $\cong \Sigma_{p^2} (\in \mathrm{Syl}_p(\mathrm{S}_{p^2})$ (see Theorem 30.10).

2723. Study the $p$-groups $G$ in which the kernels of their nonlinear irreducible characters form a chain.

2724. Study the $p$-groups $G$ containing a proper subgroup $H$ such that (i) $G = \langle H^\alpha \mid \alpha \in \mathrm{Aut}(G)\rangle$, (ii) $G = \bigcup_{\alpha\in\mathrm{Aut}(G)} H^\alpha$.

2725. Study the nonabelian $p$-groups $G$ containing a proper subgroup $H$ such that, whenever $\chi \in \mathrm{Irr}_1(G)$, then there exists $\lambda \in \mathrm{Irr}(H)$ with $\chi = \lambda^G$.

2726. Study the $p$-groups $G$ containing a self-centralizing subgroup $\cong \mathrm{C}_{p^3}$.

2727. Study the $p$-groups $G$ satisfying $\mathrm{H}_p(G) = \mho_1(G)$.

2728. Given $k \geq 4$, study the $p$-groups $G$ all of whose normal subgroups of order $p^k$ are two-generator (for $k = 3$, see § 10 and Theorems 13.7, 50.1, and 50.2).

2729. (Isaacs [Isa8]) Does there exist a $p$-group $G$ admitting a nontrivial partition all of whose components are nonabelian?

2730. Study the $p$-groups admitting a nontrivial irredundant covering by $\leq 2p$ subgroups (see § 116).

2731. Does there exist a capable $p$-group (see #2300) with a trivial Schur multiplier?[6]

2732. Let $a(X)$ be the number of maximal abelian subgroups of a nonabelian $p$-group $X$. Classify the $p$-groups $G$ that are not covered by $a(G) - 1$ maximal abelian subgroups.

2733. Study the $p$-groups in which the normalizer of any maximal cyclic subgroup is abelian.

2734. (Old problem inspired by Fitting's lemma and treated by I.D. Macdonald [Macd1]) Given $n$, study the $p$-groups of class $2n$ all of whose maximal subgroups have class $n$.

2735. Does there exist an irregular $p$-group of exponent $p^e$ covered by cyclic subgroups of order $p^e$? (See #580.)

2736. (Old problem) Study the $p$-groups $G$ such that $|G|$ does not divide $|\mathrm{Aut}(G)|$.

2737. Study the $p$-groups containing exactly one two-generator maximal subgroup.

2738. Study the absolutely regular $p$-groups, $p > 3$, all of whose representation groups are absolutely regular.[7]

2739. (i) Study the nonabelian $p$-groups of exponent $p$ in which any two subgroups (nonabelian subgroups) of order $p^3$ are permutable. (ii) Study the $p$-groups $G$ containing a nontrivial subgroup $N$ such that any two subgroups of $G$ nonincident with $N$ are permutable (see § 161).

2740. Classify the $p$-groups $G$ such that for any two nonincident $A, B < G$, one has $A \cap B = A_G \cap B_G$.

---

6 Mann's solution. There is no such $p$-groups. Suppose that $G$ is capable. and write $G = H/\mathrm{Z}(H)$. Then $H' \cap \mathrm{Z}(H) > \{1\}$, but $H' \cap \mathrm{Z}(H)$ is a subgroup of the Schur multiplies $\mathrm{M}(G)$ of $G$ (see Karpilovsky's book on Schur multipliers).

7 If $G$ is a metacyclic $p$-group, $p > 3$, then all its representation groups are absolutely regular; if, in addition, $G$ is nonabelian, this is also true for $p = 3$ (see Theorem 47.2).

2741. Let

$$\{1\} = A_0 < A_1 < \cdots < A_{n-1} < A_n = G, \quad \{1\} = B_0 < B_1 < \cdots < B_{n-1} < B_m = G$$

be two series of $G$-invariant subgroups such that $A_{i+1}/A_i$ and $B_{j+1}/B_j$ are maximal $G$-invariant abelian (elementary abelian, of exponent $p$) subgroups of $G/A_i$ and $G/B_j$, respectively, $i = 0, 1, \dots, n-1$ and $j = 0, 1, \dots, m-1$. Estimate $|n - m|$ (three problems).

2742. Describe all automorphisms of order $p$ of a homocyclic $p$-group $G$ and describe the group $\Omega_1(P)$, where $P \in \mathrm{Syl}_p(\mathrm{Aut}(G))$.

2743. Study the pairs of $p$-groups $G$, $G_0$ such that $\exp(G) = p < \exp(G_0)$ and the character tables $\mathrm{X}(G) = \mathrm{X}(G_0)$.

2744. Classify the $p$-groups in which the normal closure of any cyclic subgroup (i) is either abelian or an $\mathcal{A}_1$-subgroup (see §§ 223 and 224), (ii) has the cyclic center (this is solved by the second author; see § 236).

2745. Let $\delta_2(G)$ be the number of two-generator members of the set $\Gamma_1$ of maximal subgroups of a $p$-group $G$. Find all possible values of $\delta_2(G)$, where $G$ runs over all groups of exponent $p$ (if $|G| > p^3$, then $p \mid \delta_2(G)$, by Theorem 5.8 (b)).

2746. Does there exist a special $p$-group $G$ of exponent $p^2$, $p > 2$, $|G'| > p$, such that $G/\mho_1(G)$ is special?

2747. Study the $p$-groups $G$ with noncyclic $\mathrm{Z}(G)$ all of whose epimorphic images of order $\frac{1}{p}|G|$ are special.

2748. Study the $p$-groups having the same character tables as $\mathcal{A}_n$-groups, $n \le 2$.

2749. Classify the nonabelian $p$-groups $G$ such that $\mathrm{C}_G(x)$ is abelian for every $x \in G$ of order $> p$ (as noted by the second author, $G$ satisfies this condition $\iff$ the center of any nonabelian subgroup of $G$ has exponent $p$).

2750. Classify the $p$-groups $G$ such that $H_G$ is either cyclic or abelian of type $(p, p)$ for any nonnormal $H < G$.

2751. (A partial case of an old problem) Study the groups $G$ of exponent $p$ admitting a fixed-point-free automorphism.

2752. Present a nonabelian absolutely regular $p$-group of exponent $> p > 3$ all of whose representation groups are not absolutely regular (see Theorem 47.2).

2753. Classify the $p$-groups $G$ such that $\mathrm{C}_G(x) < M$ for all $x \in M - \mathrm{Z}(M)$, where $M$ runs over all minimal nonabelian subgroups of $G$ (this is solved by the second author; see Theorem 92.2).

2754. Study the $p$-groups $G$ such that, for any nonnormal cyclic $C < G$, (i) $C_G = \{1\}$, (ii) $|C_G| \le p$, (iii) $|C : C_G| = p$ (see § 63).

2755. Study the $p$-groups all of whose subgroups of index $p^2$ have cyclic derived subgroups.

2756. Classify the $p$-groups all of whose noncyclic subgroups are quasinormal.

2757. Classify the $p$-groups in which the intersection of any two distinct conjugate cyclic subgroups is $= \{1\}$ (this was solved by the second author; see § 187).

2758. Given $n > 1$, does there exist a $p$-group $G$ satisfying $\alpha_2(G) = n$? (See [ZLS] and Appendix 53.)
2759. Study the $p$-groups all of whose nonnormal subgroups of index $\geq p^3$ are abelian (see ##768, 1278 and [ZZLS]).
2760. Study the irregular $p$-groups $G = \Omega_1(G)$ all of whose maximal subgroups of exponent $p$ are normal.
2761. Study the special 3-groups all of whose representation groups are irregular (for $p > 3$ this is impossible, by Theorem 7.1 (b)).
2762. Classify the $p$-groups $G$ of class $> 2$ without normal subgroups of class 2.[8]
2763. Study the nonabelian $p$-groups $G$ such that, whenever $x \in G - Z(G)$, then $C_G(x)$ is metacyclic.
2764. (Isaacs–Passman) Given $n > 1$, study the $p$-groups $G$ satisfying $\mathrm{cd}(G) = \{1, p^n\}$ (for $n = 1$, see Theorem 22.5).
2765. (Inspired by Isaacs' Problem 2427) Let $G$, $G_0$ be $p$-groups such that $\mathcal{L}_N(G) \cong \mathcal{L}_N(G_0)$ and suppose that the nonabelian $G_0$ has an abelian subgroup of index $p$. Is it possible to estimate the minimal index of an abelian subgroup in $G$?
2766. Study the $p$-groups such that $\exp(G) = \exp(\Gamma)$, where $\Gamma$ runs over all representation groups of $G$ (consider in detail the case when $\exp(G) = p$).
2767. Study the $p$-groups in which every abelian subgroup is contained in a subgroup of class 2 (see § 151).
2768. Given $k \in \{2, \dots, d(G) - 1\}$, does there exist a $p$-group $G$, all of whose subgroups of index $p^k$, containing $\Phi(G)$, are special? (See [Cos] where this problem is solved for $k = 1$.)
2769. Classify the $p$-groups $G$ such that $\mathrm{cd}(H) \subseteq \{1, p\}$ for all $H \in \Gamma_1$.
2770. Suppose that a $p$-group $G$ possesses a metacyclic subgroup of index $p^n$. Estimate the minimal index in $G$ of a normal metacyclic subgroup. Consider in detail $n = 2, 3$ (see § 39).
2771. Given $n > p^2 + p + 1$, does there exist a $p$-group $G$ satisfying $\alpha_1(G) = n$? (See § 76.)
2772. A subgroup $H$ of a $p$-group $G$ is said to be *$p$-characteristic* if $H^\alpha = H$ for all $p$-automorphisms $\alpha$ of $G$. Study the $p$-groups all of whose normal subgroups are $p$-characteristic.
2773. Study the $p$-groups $G$ with $p$-closed ($p$-nilpotent) $\mathrm{Aut}(G)$.
2774. Classify the $p$-groups in which any subgroup is either metacyclic or of class $\leq 2$.
2775. Study the $p$-groups $G$ such that $\Phi(H) = H'$ for all nonabelian $H \leq G$.
2776. Study the special $p$-groups $G$ such that $Z(G) = Z(H)$ for all $H \in \Gamma_1$.
2777. Study the groups of exponent $p$, all of whose lattice isomorphic groups are isomorphic.

---

8 Any $p$-group of maximal class of order $\geq p^5$ with abelian subgroup of index $p$ satisfies this condition.

2778. Study the $p$-groups $G$ containing a normal irregular subgroup $M$ of maximal class such that $G/M$ is cyclic of order $p^n$ (for $n = 1$, see Theorem 12.12).

2779. Classify the $p$-groups $G$ such that $|H : \mathrm{Z}(H)| \le p^3$ for all $H \in \Gamma_1$ (see also § 151).

2780. Study the $p$-groups of class 2 having (i) a representation group of class 2, (ii) the trivial Schur multiplier.

2781. Study the normal, power and subgroup structures of the irregular $p$-groups $G$ with cyclic $\mho_1(G)$ and satisfying $|G/\mho_1(G)| = p^p$ (see ##136, 594).

2782. Let $G$ be a $p$-group all of whose maximal subgroups are characteristic. Is it true that (i) all maximal subgroups of its representation groups are characteristic, (ii) $p = 2$?

2783. (Passman, Ito) Study the $p$-groups having a faithful irreducible character of degree $p^2$.

2784. Study the $p$-groups containing a normal $\mathcal{A}_1$-subgroup of index $p^2$ (see [Mal6] and [Kur]).

2785. Given $p > 2$, does there exist a group of order $p^p$ and exponent $p > 2$ all of whose representation groups have exponent $p$? (See [Joh]. If such group, say $H$, exists, then there is no irregular group $G$ of order $p^{p+1}$ such that $G/\mho_1(G) \cong H$.)

2786. Study the $p$-groups $G$ containing a normal subgroup $N > \{1\}$ such that some representation group of $G/N$ is isomorphic to $G$. Is it true that $N$ is abelian?

2787. Classify the $p$-groups $G$ such that, whenever $H < G$ is minimal nonabelian and $x \in H$, then $|G : \mathrm{C}_G(x)| \le p$ (this is solved by the second author; see § 217).

2788. Study the irregular $p$-groups $G$ all of whose minimal irregular subgroups are isomorphic to $\Sigma_{p^2}$.

2789. Find all $n$ such that, whenever a $p$-group $G$ contains an abelian subgroup of index $p^n$, then there is in $G$ a normal subgroup of index $p^n$ of class $\le 2$.

2790. Study the $p$-groups $G$ such that each $H \in \Gamma_{\mathrm{d}(G)-1}$ is complemented.

2791. Study the $p$-groups such that $\mathrm{C}_G(x)$ is abelian for all (i) $x \in G - G'$, (ii) $x \in G - \Phi(G)$.

2792. Study the $p$-groups $G$ with $M \in \Gamma_1$ such that $|\mathrm{C}_M(x)| \le p^2$ for each $x \in G - M$.

2793. Study the $p$-groups containing $p + 1$ conjugate classes of maximal abelian subgroups (a group is covered by maximal abelian subgroups).

2794. Given $p > 2$, study the metabelian $p$-groups $G$ such that $G/\mathrm{K}_p(G)$ is of maximal class.

2795. Study the $p$-groups (groups of exponent $p$) covered by nonnormal subgroups.

2796. Study the $p$-groups all of whose $\mathcal{A}_1$-subgroups have nonabelian centralizers (this is solved by the second author; see § 233).

2797. Given $n$, classify the $p$-groups in which the intersection of any two nonincident subgroups of indices $p^n$ and $p^{n+1}$ is (i) abelian, (ii) absolutely regular, (iii) metacyclic.

2798. Classify the irregular $p$-groups containing only one maximal regular subgroup which is absolutely regular (see Appendix 75).

2799. (Janko) Study the $p$-groups all of whose cyclic subgroups of order $p^2$ are normal (see §§ 229 and 230).

2800. Study the $p$-groups of maximal class, $p > 2$, all of whose representation groups are of maximal class. Is it true that if a $p$-group $G$ of maximal class has a representation group of maximal class, then all its representation groups are of maximal class?

# Bibliography


[AHZ] L. J. An, R. F. Hu and Q. H. Zhang, Finite $p$-groups with a minimal nonabelian subgroup of index $p$ (IV), *J. Algebra Appl.* **14**(2015), no. 3, 1550020 (54 pages).

[Alle] R. B. J. T. Allenby, Normal subgroups contained in Frattini subgroups are Frattini subgroups, *Proc. Amer. Math. Soc.* **78** (1980), 315–318.

[Alp1] J. L. Alperin, On a special class of regular $p$-groups, *Trans. Amer. Math. Soc.* **106** (1963), 77–99.

[Alp2] J. L. Alperin, Large abelian subgroups of $p$-groups, *Trans. Amer. Math. Soc.* **117** (1965), 10–20.

[Alp3] J. L. Alperin, Centralizers of abelian normal subgroups of $p$-groups, *J. Algebra* **1** (1964), 110–113.

[AK1] B. Amberg and L. Kazarin, On the rank of a finite product of two $p$-groups, Groups–Korea 94, Kim/Johnson, W. de Gruyter, Berlin (1995), 1–8.

[ALQZ] L. J. An, L. L. Li, H. P. Qu and Q. H. Zhang, Finite $p$-groups with a minimal nonabelian subgroup of index $p$ (II), Science China, *Math.* **56** (2013), 1–17.

[AQXY] L. J. An, H. P. Qu, M. Y. Xu and C. S. Yang, Quasi-NC groups, *Comm. Algebra* **36**(2008), no. 11, 4011–4019.

[Art] E. Artin, The orders of of the linear groups, *Comm. Pure and Appl. Math.* **8** (1955), 355–365.

[AS] M. Aschbacher and S. D. Smith, The Classification of Quasithin Groups 1, Mathematical Surveys and Monographs, Volume 111, AMS, 2004.

[Att] M. S. Attar, A necessary condition of nonabelian $p$-groups with second centre of order $p^2$, *Indian J. Pure Appl. Math.* **42**(3) (2011), 183–186.

[AZ] L. J. An and Q. H. Zhang, Finite metahamiltonian $p$-groups, *J. Algebra*, online published in http://dx.doi.org/10.1016/j.jalgebra.2014.12.004

[BW] A. H. Baartman and J. J. Woeppel, The automorphism group of a $p$-group of maximal class with abelian maximal subgroup, *Fund. Math.* **93**, 1 (1976). 41–46.

[Bae2] R. Baer, Partitionen endlicher Gruppen, *Math. Z.* **75** (1961), 333–372.

[Bae3] R. Baer, Gruppen mit Hamiltonschen Kern, *Comp. Math.* **2** (1935), 241–246.

[Baer10] R. Baer, Groups with abelian norm quotient group, *Amer. J. Math.* **61**, 3 (1939), 700–708.

[BauBl] G. Baumslag and N. Blackburn, Groups with cyclic upper central factors, *Proc. London. Math. Soc.* **(3) 10** (1960), 531–544.

[Bei2] B. Beisigel, Die Automorphismengruppen homozyklisher $p$-Gruppen, *Arch. Math.* **29**, 4 (1977), 363–366.

[Ber3] Y. Berkovich, A generalization of theorems of Ph. Hall and Blackburn and an application to non-regular $p$-groups, *Math. USSR Izv.* **35** (1971), 815–844.

[Ber25] Y. Berkovich, Finite Groups. Selected Topics, Parts I, II, in preparation.

[Ber26] Y. Berkovich, Maximal abelian and minimal nonabelian subgroups of some finite two-generator $p$-groups especially metacyclic, in preparation.

[Ber30] Y. Berkovich, Finite $p$-groups with few minimal nonabelian subgroups. With an appendix by Z. Janko, *J. Algebra* **297**, 1 (2006), 62–100.

[Ber43] Y. Berkovich, Groups of Prime Power Order, vol. 1, Walter de Gruyter, Berlin, 2008.

[Ber53] Y. Berkovich, Some remarks on subgroups of maximal class of a finite $p$-group, *J. Algebra and Applications* **14**, 1550043 (2015) (pages 20).

[Ber54] Y. Berkovich, Finite groups all of whose small subgroups are normal, J. Algebra and Applications 13, 6 (2014), 1450006, 10pp.

[Ber55] Y. Berkovich, Sufficient conditions for 2-nilpotence of a finite groups, *J. Algebra and Applications* **12** 4 (2012), pp7.

[Ber56] Y. Berkovich, On the theorem of Mann about $p$-groups all of whose nonnormal subgroups are elementary abelian, Isr. J. Math., to appear.

[Ber57] Y. Berkovich, Alternate proofs of the theorem of G. A. Miller of 2-groups all of whose proper subgroups are Dedekindian, manuscript.

[BCH] Y. Berkovich, D. Chillag, and M. Herzog, Finite groups in which the degrees of the nonlinear irreducible characters are distinct, Proc. Amer. Math. Soc. 115 (1992), 955–959.

[BI] Y. Berkovich and I. M. Isaacs, $p$-supersolvability and actions of $p$-groups stabilizing certain subgroups, *J. Algebra* **414** (2014), 82–94.

[BJ1] Y. Berkovich and Z. Janko, Groups of Prime Power Order, vol. 2, Walter de Gruyter, Berlin, 2008.

[BJ2] Y. Berkovich, and Z. Janko, Structure of finite $p$-groups with given subgroups, *Contemporary Mathematics* **402** (2006), 13–93.

[BJ3] Y. Berkovich, and Z. Janko, On subgroups of finite $p$-groups, *Israel J. Math.* **171** (2009), 29–49.

[BJ4] Y. Berkovich, and Z. Janko, Groups of Prime Power Order, vol. 3, Walter de Gruyter, Berlin, 2011.

[BerZhang] Y. Berkovich and Q. H. Zhang, $\mathcal{A}_1$- and $\mathcal{A}_2$-subgroups of finite $p$-groups, J. Algebra Appl. 13, 2(2013), 1350095, 6pp.

[BZ] Y. Berkovich and E. Zhmud, Characters of Finite Groups, Parts 1, 2, Translations of Mathematical Monographs 172, 181, American Mathematical Society, Providence, RI, 1998, 1999.

[BT] E. Best and O. Taussky, A class of groups, *Proc. Roy. Irish. Acad. Sect. A.* **47** (1942), 55–62.

[Bla3] N. Blackburn, On a special class of $p$-groups, *Acta Math.* **100** (1958), 45–92; (1958), 422–427.

[Bla5] N. Blackburn, Generalizations of certain elementary theorems on $p$-groups, *Proc. London Math. Soc.* **11** (1961), 1–22.

[Bla7] N. Blackburn, Finite groups in which the nonnormal subgroups have nontrivial intersection, *J. Algebra* **3** (1966), 30–37.

[Bla13] N. Blackburn, Groups of prime-power order having an abelian centralizer of type $(r, 1)$, *Monatshefth. Math.* **99** (1985), 1–18.

[BlaDM] N. Blackburn, M. Deaconescu and A. Mann, Equilibrated groups, *Proc. Cambridge Philos. Soc.* **120**, 2 (1996), 579–588.

[BlaH] N. Blackburn and L. Hethelyi, Some further properties of soft subgroups, *Arch. Math.* **65 (5)** (1997), 365–371.

[Blac2] S. R. Blackburn, Groups of prime power order with derived subgroup of prime order, *J. Algebra* **219** (1999), 625–657.

[Carl] J. Carlson, Maximal elementary abelian subgroups of rank 2, *J. Group Theory* **10** (2007), 5–13.

[Car1] R. W. Carter, Normal complements of nilpotent self-normalizing subgroups, **78** (1962), 149–150.

[Che] Y. Cheng, On finite $p$-groups with cyclic commutator subgroups, *Arch. Math.* **39** (1982), 295–298.

[ChM] D. Chillag and A. Mann, Nearly odd order and nearly real finite groups, Comm. Algebra (1998), 2041–2064.

[Cos] J. Cossey, Special maximal subgroups of $p$-groups, *Bull. Aust. Math. Soc.* **89**, 3 (2014), 415–419.

[CT] G. Corsi Tani, Automorphisms fixing every normal subgroup of a $p$-group, *Bull. Un. Mat. Ital. A* **64** (1985), 245–252.

[Cre1] E. Crestani, On 3-generated 2-groups whose maximal subgroups are 2-generated, *Comm. Alg.* **39** (2001). 881–887.

[Cre] E. Crestani, Monotone 2-groups, PhD thesis, Univ. of Padova.

[CrM] E. Crestani and F. Menegazzo, Monotone 2-groups, *J. Group Theory* **15** (2012), 359–383.

[Dar] R. Dark and M. L. Newell, On condition for commutators to form a subgroup, *J. London Math. Soc.* **17** (1978), 251–252.

[Ded] R. Dedekind, Über Gruppen, deren sämtliche Teiler Normalteiler sind, *Math. Ann.* **48** (1897), 548–561.

[Dor] L. Dornhoff, Group Representation Theory, Part A, Marcel Dekker, New York, 1971.

[ERNS] S. Evans-Riley, M. F. Newman and C. Schneider, On the soluble length of groups with prime-power order, *Bull. Aust. Math. Soc.* **59** (1999) 343–346.

[FA] X. G. Fang and L. J. An, The classification of finite metahamiltonian $p$-groups, submitted.

[Fei1] W. Feit, On large Zsigmondy primes, *Proc. Amer. Math. Soc.* **102** (1988), 29–36.

[FO3] R. Fouladi and R. Orfi, Maximal subsets of pairwise noncommuting elements of some $p$-groups of maximal class, *Bull. Aust. Math. Soc.* **84** (2011), 447–451.

[FT] W. Feit and J. G. Thompson, Solvability of groups of odd order, *Pacific J. Math.* **13** (1963), 775–1029.

[FZ] W. Feit and G. Seitz, On finite rational groups and related topics, *Illinois J. Math.* **33** (1988), 103–131.

[F-A1] G. A. Fernandez-Alcober, An introduction to finite $p$-groups: regular $p$-groups and groups of maximal class, *Math. Contemp.* **20** (2001), 155–226.

[GMS1] N. Gavioli, A. Mann, and C. Scoppola, Two applications of the Hughes subgroup of finite group, Ischia Group Theory 2006, World Scientific, Singapore, 2007, 138–146.

[GMS2] N. Gavioli, V. Monti, and C. Scoppola, Pro-$p$-groups with waists, *J. Algebra* **351** (2012), 130–137.

[Gil] J. D. Gillam, A note on finite metabelian $p$-groups, *Proc. Amer. Math. Soc.* **25** (1970), 189–190.

[Gla1] G. Glauberman, Large abelian subgroups of finite $p$-groups, *J. Algebra* **196** (1997), 301–338.

[Gla2] G. Glauberman, Large abelian subgroups of groups of prime exponent, *J. Algebra* **237** (2001), 735–768.

[Gla5] G. Glauberman, Large subgroups of small class in finite $p$-groups, *J. Algebra* **272** (2004), 128–153.

[Gla6] G. Glauberman, Centrally large subgroups of finite $p$-groups, *J. Algebra* **300** (2006), 480–508.

[Gla7] G. Glauberman, Existence of normal subgroups in finite $p$-groups, *J. Algebra* **319** (2008), 800–805.

[Gla8] G. Glauberman, A $p$-group with no normal large abelian subgroup, *Contemp. Math.* **524**, AMS (2010), 61–65.

[GM] G. Glauberman and N. Mazza, $p$-groups with maximal elementary abelian subgroup of rank 2, *J. Algebra* **323** (2010), 1729–1737.

[Gol] Y. A. Golfand, On groups all of whose subgroups are nilpotent, *Dokl. Akad. Nauk SSSR* **125** (1948), 1313–1315.

[Golo1] M. I. Golovanov, Certain classes of finite $p$-groups, Proc. Sem. Algebraic Systems, No. 2, Krasnoyarsk (1970), 5–24.

[Golo2] M. I. Golovanov, Finite groups with condition $|I_m(G)| = p^{m-1}$, Some questions on the theory groups and rings, Krasnoyarsk (1973), 58–82.

[G-S] J. Gonsalez-Sanchez, A $p$-nilpotency criterion, *Arch. Math.* **94** (2010), 201–205.

[Gor1] D. Gorenstein, Finite Groups, Harper and Row, N. Y., 1968.

[Gor3] D. Gorenstein (Editor), Reviews on Finite Groups, Amer. Math. Soc., Providence, RI, 1974.

[GLS] D. Gorenstein, R. Lyons and R. Solomon, The Classification of the Finite Simple Groups, Part 1, Chapter G: General Group Theory, AMS, Providence, RI, 1995.

[GLS5] D. Gorenstein, R. Lyons and R. Solomon, The Classification of the Finite Simple Groups, Number 5, Providence, RI, 2002.

[Grov1] J. R. J. Groves, On minimal irregular $p$-groups, *J. Aust. Math. Soc.* **16** (1973), 78–89.

[Grov2] J. R. J. Groves, On direct products of regular $p$-groups, *Proc. Amer. Math. Soc.* **37** (1973), 377–379.

[Grov3] J. R. J. Groves, Some criteria for the regularity of a direct product of regular $p$-groups, *J. Aust. Math. Soc. Ser A* **24** (1977), 35–49.

[Grun2] O. Grün, Über das direkte Produkt regulärer $p$-Gruppen, *Arch. Math.* **5** (1954), 241–243.

[Grun4] O. Grün, Einige Sätze über Automorphismen abelschen $p$-Gruppen, *Abh. Math. Sem. Univ. Hamburg* **24** (1960), 54–58.

[HalM1] M. Hall, The Theory of Groups, Macmillan, New York, 1959.

[HS] M. Hall and J. K. Senior, On Groups of Order $2^n$, ($n \leq 6$), Macmillan, New York, 1964.

[Hal1] P. Hall, A contribution to the theory of groups of prime power order, *Proc. London Math. Soc.* **36** (1933), 29–95.

[Hal2] P. Hall, On a theorem of Frobenius, *Proc. London Math. Soc.* **40** (1936), 468–501.

[Hal3] P. Hall, The classification of prime power groups, *J. reine angew. Math.* **182** (1940), 130–141.

[Hal4] P. Hall, Nilpotent groups, Can. Math. Congr., Alberta, 1957.

[Hal5] P. Hall, On groups of automorphisms, *J. Math.* **182** (1940), 194–204.

[Hal7] P. Hall. Verbal and marginal subgroups, *J. reine angew. Math.* **182** (1940), 156–167.

[Hal8] P. Hall, The Eulerian functions of a group, *Quart. J. Math.* **7** (1936), 134–151.

[Har] K. Harada, On some 2-groups of normal 2-rank 2, *J. Algebra* **20**, 1 (1972), 90–93.

[Haw] T. O. Hawkes, On the automorphism group of a 2-group, *Proc. London Math. Soc.* **26** (1973), 207–225.

[Hei3] H. Heineken, Nilpotente Gruppen, deren sämtliche Normalteiler charakteristisch sind, *Arch. Math.* **33** (1979/80), 497–503.

[HL1] H. Heineken and H. Liebeck, On $p$-groups with odd order automorphism groups, *Arch. Math.* **24** (1973), 465–471.

[Hel1] G. T. Helleloid, A survey of automorphism groups of finite $p$-groups, arXiv.mathGR/0610294 v2 25 Oct 2006, 1–20.

[HK] M. Herzog and G. Kaplan, Large cyclic subgroups contain non-trivial normal subgroups, *J. Group Theory* **4** (2001), 247–253.

[Het2] L. Hethelyi, On subgroups of $p$-groups having soft subgroups, J. London Math. Soc. (2) 41 (1990), 425–427.

[Het3] L. Hethelyi, Some additional properties of subgroups of $p$-groups having soft subgroups, *Studia Sci. Math. Hungar.* **29** (1994), 317–320.

[Het4] L. Hethelyi, Some remarks on 2-groups having soft subgroups, *Studia Sci. Math. Hungar.* **27** (1992), 295–299.

[Het5] L. Hethelyi, Soft subgroups of $p$-groups, *Ann. Univ. Sci. Budapest Eótvós Sect. Math.* **27** (1984), 81–85.

[HetK] L. Hethelyi and B. Külshammer, Characters, conjugacy classes and centrally large subgroups of $p$-groups of small rank, *J. Algebra* **340** (2011), 199–210.

[HKS1] L. Hethelyi and B. Külshammer and B. Sambale, Conjugacy classes and characters of finite $p$-groups, *Comm. Alg.* **39** (2011), 657–685.

[HP] W. M. Hill and D. B. Parker, The nilpotence class of the Frattini subgroup, *Isr. J. Math.* **15** (1973), 211–215.

[HW] W. M. Hill and C. R. B. Wright, Normal subgroups contained in the Frattini subgroup, *Proc. Amer. Math. Soc.* **35**)1972), 413–415.

[HogK] G. T. Hogan and W. P. Kappe, On the $H_p$-problem for finite $p$-groups, *Proc. Amer. Math. Soc.* **20** (1969), 450–454.

[Hop2] C. Hopkins, Nonabelian groups whose groups of automorphisms are abelian, *Ann. Math.* **29** (1927/28), 508–520.

[Hor1] M. V. Horosevskii, The automorphism groups of finite $p$-groups. (Russian) *Algebra i Logika* **10** (1971), 81–86.

[Hor2] M. V. Horosevskii, The automorphism group of wreath products of finite groups, *Sibirsk. Mat. Z.* **14** (1973), 651–659.

[Hor3] M. V. Horosevskii, Automorphisms of finite groups. (Russian) *Mat. Sb. (N. S.)* **93** (1974), 576–587.

[How] R. B. Howlett, On the exponent of certain factorizable groups, *J. London Math. Soc.* **231** (1985), 265–271.

[HI] R. B. Howlett and I. M. Isaacs, On groups of central type, *Math. Z.* **179** (1982), 555–569.

[HTh] D. R. Hughes and J. G. Thompson, The $H_p$-problem and the structure of $H_p$-groups, *Pacif. J. Math.* **9** (1959), 1097–1101.

[Hum] K. Hummel, The order of the automorphism group of a central product, *Proc. Amer. Math. Soc.* **47** (1975), 37–40.

[Hup1] B. Huppert, Endliche Gruppen, Bd. 1, Springer, Berlin, 1967.

[HupB] B. Huppert and N. Blackburn, Finite Groups II, Springer, Berlin, 1982.

[Isa1] I. M. Isaacs, Character Theory of Finite Groups, Acad. Press, N. Y., 1976.

[Isa5] I. M. Isaacs, Algebra: a graduate course, Brooks/Cole, 1994.

[Isa6] I. M. Isaacs, Commutators and commutator subgroup, *Amer. Math. Monthly* **84** (1977), 720–722.

[Isa8] I. M. Isaacs, Equally partitioned groups, *Pacific J. Math.* **49** (1973), 109–116.

[Isa14] I. M. Isaacs, Large orbits in nilpotent action, *Proc. Amer. Math. Soc.* **127** (1999), 45–50.

[Isa19] I. M. Isaacs, Finite Group Theory, AMS, New York, 2008.

[Isa21] I. M. Isaacs, Semipermutable $\pi$-subgroups, *Arch. Math (Basel)* **102**, 1 (2014), 1–6.

[IN2] I. M. Isaacs and G. Navarro, Normal $p$-complements and fixed elements, *Arch. Math.* **95** (2010), 207–211.

[INS] I. M. Isaacs, G. Navarro and J. Sangroniz, $p$-groups having few almost-rational irreducible characters, *Israel J. Math.* (2012), to appear.

[Ish] K. Ishikawa, On finite $p$-groups which have only two conjugacy lengths, *Isr. J. Math.* **129** (2002), 119–123.

[Ito1] N. Ito, On the degrees of irreducible representations of a finite group, *Nagoya Math. J.* **3** (1951), 5–6.

[Ito10] N. Ito, Über eine Frattini-Gruppe duale Bildung, *Nagoya Math. J.* **9** (1955), 123–127.

[IM] N. Ito and A. Mann, Counting classes and characters of groups of prime exponent, *Israel J. Math.* **156** (2006), 205–220.

[Iwa] K. Iwasawa, Über die endlichen Gruppen und die Verbände ihrer Untergruppen, *J. Univ. Tokyo* **4** (1941), 171–199.

[Jan8] Z. Janko, Finite 2-groups with exactly four cyclic subgroups of order $2^n$, *J. reine angew. Math.* **566** (2004), 135–181.

[Jan27] Z. Janko, Finite 2-groups all of whose maximal cyclic subgroups of composite order are self-centralizing, *J. Group Theory* **10** (2007), 1–4.

[Jan36] Z. Janko, Finite nonabelian 2-groups all of whose minimal nonabelian subgroups are isomorphic to $M_{2^{n+1}} = \langle a, t \mid a^{2^n} = t^2 = 1,\ a^t = a^{1+2^{n-1}} \rangle$, $n \geq 3$ fixed, manuscript.

[Jan37] Z. Janko, Finite nonabelian 2-groups all of whose minimal nonabelian subgroups are metacyclic and have exponent 4, J. Algebra, to appear.

[Jan38] Z. Janko, Finite $p$-groups with many minimal nonabelian subgroups, *J. Algebra* **367** (2012), 363–370.

[Jan39] Z. Janko, Finite $p$-groups all of whose nonabelian maximal subgroups have the largest possible centers, manuscript.

[Jan42] Z. Janko, Finite nonabelian $p$-groups having exactly one maximal subgroup with a noncyclic center, *Arch. Math* **96** (2011), 101–103.

[Jan47] Z. Janko, Finite nonabelian $p$-groups all of whose nonabelian maximal subgroups are either metacyclic or minimal nonabelian, *Arch. Math.* **96** (2011), 105–107.

[Jan48] Z. Janko, Finite $p$-groups all of whose maximal subgroups, except one, have its derived subgroup of order $\leq p$, *Glas. Mat.* **48** (67) (2012), 325–332.

[Jan49] Z. Janko, Finite $p$-groups all of whose maximal subgroups, except one, have cyclic derived subgroup, *J. Algebra and its Applications* **14**, 1 (2015), 1450080, 8pp.

[Jan50] Z. Janko, Finite $p$-groups with exactly two sizes of conjugate classes, manuscript.

[Jan51] Z. Janko, Finite $p$-groups with $A \cap B$ being maximal in $A$ or $B$ for any two nonincident subgroups $A$ and $B$, *Glas. Math.* **49** (69), 1 (2014), 113–118.

[Jan52] Z. Janko, Finite nonabelian $p$-groups all of whose subgroups are q-self dual, *J. Algebra and Appl.* **6**, 1 (2014), 1450008, 7pp.

[Jan53] Z. Janko, Finite $p$-groups in which the normal closure of each nonnormal cyclic subgroup is nonabelian, *Glas. Mat.*, to appear.

[Jan54] Z. Janko, Finite $p$-groups of exponent $p^e$ all of whose cyclic subgroups of order $p^e$ are normal, *J. Algebra* **416** (2014), 274–286.

[Jan55] Z. Janko, Classification of finite $p$-groups with cyclic intersection of any two distinct conjugate subgroups, *Glas. Math.*, to appear.

[Joh] D. L. Johnson, A property of finite $p$-groups with trivial multiplicator, *Amer. J. Math.* **98** (1976), 105–108.

[Kal1] L. Kaloujnine, La structure des $p$-groupes de Sylow des groupes symetriques finis, *Ann. Sci Ecole Norm. Super.* **65** (1968), 239–276.

[Kar1] G. Karpilovsky, The Schur Multiplier, Clarendon Press, Oxford, 1986.

[Kar2] G. Karpilovsky, Group Representations, vol. 2, North-Holland, Amsterdam, 1993.

[Kar3] G. Karpilovsky, Projective Representations of Finite Groups, Dekker, 1985.

[Kaz1] L. S. Kazarin, Groups with certain conditions for normalizers of subgroups, *Uchen. zapiski Perm Univ.* **218** (1969), 268–279 (Russian).

[Kaz2] L. S. Kazarin, On some classes of finite groups, *Soviet Math. Dokl.* **12, 2** (1971), 549–553.

[Kaz3] L. S. Kazarin, Groups with restrictions on normalizers of subgroups, *Izv. vuzov (mathematics)* **2** (1973), 41–50 (Russian).

[Kaz4] L. S. Kazarin, On product of two nilpotent groups, *Questions of group theory and homological algebra* (1981), 62–66; II, ibid (1982), 47–49 (Russian).

[KazSi] L. S. Kazarin and V. Sidel'nikov, On the automorphism group of the Suzuki $p$-algebra, *Math. Notes* **80** (2006), 500–508.

[Keg1] O. H. Kegel, Produkte nilpotenter Gruppen, *Arch. Math.* **12** (1961), 90–93.

[Keg2] O. H. Kegel, Die Nilpotenz der $H_p$-Gruppen, *Math. Z.* **75** (1960), 373–376.

[Keg3] O. H. Kegel, Sylow-Gruppen und Subnormalteiler endlichen Gruppen, *Math. Z.* **78** (1062), 205–221.

[Khu1] E. I. Khukhro, Nilpotent Groups and their Automorphisms, W. de Gruyter, Berlin, 1993.
[Khu2] E. I. Khukhro, $p$-automorphisms of finite $p$-groups, CUP, Cambridge, 1998.
[Kim2] I. Kiming, Structure and derived length of finite $p$-groups possessing an automorphism of $p$-power order having exactly $p$ fixed points, *Math. Scand.* **62** (1988), 153–172.
[Kin1] B. W. King, Normal subgroups of groups of prime-power order, in "Proc. 2nd Int. Con. Theory Groups," Lecture Notes in Math., **372**, Springer, Berlin (1973), 401–408.
[Kin3] B. W. King, Presentation of metacyclic groups, *Bull. Aust. Math. Soc.* **8** (1973), 101–131.
[Kirt1] J. Kirtland, Finite groups with subgroups not contained in Frattini subgroup are permutable, *Arch. Math.* **97** (2011), 399–406.
[Klu] F. L. Kluempen, The power structure of 2-generator 2-groups of class two, *Algebra Colloq.* **9**, 3 (2002), 287–302.
[Kno] H. G. Knoche, Über den Frobeniusschen Klassenbegriff in nilpotenten Gruppen, I, II, *Math. Z.* **55** (1951), 71–83; ibid **59** (1953), 8–16.
[Kon1] M. Konvisser, Embedding of abelian subgroups in $p$-groups, *Trans. Amer. Math. Soc.* **153** (1971), 469–481.
[KonJ] M. Konvisser and D. Jonah, Counting abelian subgroups of $p$-groups. A projective approach, *J. Algebra* **34** (1975), 309–330.
[Kor] I. Korchagina (Capdeboscq), Ranks of $p$-groups, *Proc. Amer. Math. Soc.* **140**, 11 (2012), 3769–3775.
[Kur] J. Kurdics, Automorphisms of a minimal nonabelian $p$-group with odd $p$ I, *Glas. Math. Set III* **46** (66), 2 (2011), 367–383.
[KS] H. Kurzweil und B. Stellmacher, Theorie der endlichen Gruppen. Eine Einführung, Springer, 1998.
[Laf1] T. J. Laffey, The minimum number of generators of a finite $p$-group, *Bull. London Math. Soc.* **5** (1973), 288–290.
[Laf2] T. J. Laffey, Bounding the order of a finite $p$-group, *Proc. Roy. Ir. Acad.* 80a **2** (1980), 131–134.
[Laf3] T. J. Laffey, A lemma on finite $p$-groups and some consequences, *Proc. Camb. Philos. Soc.* **75** (1974), 133–137.
[Laf4] T. J. Laffey, Centralizers of elementary abelian subgroups in finite $p$-groups, *J. Algebra* **51** (1978), 88–96.
[Lang] S. Lang, Algebra, GTM 211, Springer, 2002.
[Leo] A. Leone, Finite minimal non-KC-groups, *Matematiche* **38** (1987), 191–200.
[Leon1] Y. K. Leong, Finite 2-groups of class two with cyclic centre, *J. Aust. Math. Soc. Ser. A* **27** (1979), 125–140.
[Leon2] Y. K. Leong, Odd order nilpotent groups of class two with cyclic centre, J. Aust. Math. Soc. **17** (1974), 142–153.
[Lev1] F. W. Levi, Groups in which the commutator operations satisfy certain algebraic conditions, *J. Indian Math. Soc.* **6** (1942), 87–97.
[Lew] M. L. Lewis, On a $p$-group Camina pairs, *J. Group Theory* **15** (2012), 469–483.
[LP] P. J. Li, A classification of finite $p$-groups whose proper subgroups are of class $\leq 2$ (I), (II), *J. Algebra Appl.* **12**, 3 (2013), 1250171–1–22, 1250171–1–29.
[LW] H. G. Liu and Y. L. Wang, The automorphism group of a generalized extraspecial $p$-group, *Sci. China Math.* **53** (2010), 315–334.
[LM] P. Longobardi and M. Maj, On $p$-groups of breadth two, *Algebra Colloq.* **6** (1999), 121–124.
[LQC] L. L. Li, H. P. Qu and G. Y. Chen, Central extension of inner abelian $p$-groups (I), *Acta Math. Sinica* (*Chin. Ser.*) **53**, 4 (2010), 675–684.

[LQZ] P. J. Li, Haipeng Qu and Jiwen Zeng, Finite $p$-groups whose proper subgroups are of class $\leq n$ (manuscript).

[LubM] A. Lubotzky and A. Mann, Powerful $p$-groups 1, *J. Algebra* **105** (1987), 484–505.

[LZG] Heng Lv, Wei Zhou and Xiuyun Guo, Finite 2-group with index of every cyclic subgroup in its normal closure no greater than 4, *J. Algebra* **342** (2011), 256–264.

[LZY] Heng Lv, Wei Zhou and Dapeng Xu, Some finite $p$-groups with bounded index of every cyclic subgroup in its normal closure, *J. Algebra* **338** (2011), 169–179.

[LZX] Heng Lv, Wei Zhou and Haijing Xu, Finite $p$-groups with small subgroups generated by two conjugate elements, manuscript.

[Macd1] I. D. Macdonald, Generalizations of a classical theorem on nilpotent groups, *Illinois J. Math.* **8** (1964), 556–570.

[Macd3] I. D. Macdonald, Commutators and their products, *Amer. Math. Monthly* **93** (1986), 440–444.

[Macd5] I. D. Macdonald, The breadth of finite $p$-groups, *Proc. Roy. Soc. Edinburgh, Sect. A* **78** (1977/78), 1–39.

[Macd7] I. D. Macdonald, Computer results on Burnside groups, *Bull. Aust. Math. Soc.* **9** (1973), 433–438.

[Macd9] I. D. Macdonald, Some examples in the theory of groups, Mathematical Essays dedicated to A. J. Macintyre, Ohio Univ. Press, Athens, Ohio (1970), 263–269.

[Macd10] I. D. Macdonald, On cyclic commutator subgroups, *J. London Math. Soc.* **38** (1963), 419–422.

[Macd14] I. D. Macdonald, A computer application to finite $p$-groups, *J. Aust. Math. Soc., Ser. A* **17** (1974), 102–112.

[Mal2] I. Malinowska, $p$-automorphisms of finite $p$-groups: problems and questions, *Advances in Group Theory* (2002), 111–127.

[Mal6] I. Malinowska, On the structure of the automorphism group of a minimal nonabelian $p$-group (metacyclic case), *Glas. Mat.* **47** (67) (2012), 153–164.

[Mal7] I. Malinowska, Finite groups all of whose small subgroups are pronormal, manuscript.

[Man2] A. Mann, Regular $p$-groups, *Israel J. Math* **10** (1971), 471–477.

[Man3] A. Mann, Regular $p$-groups, II, *Israel J. Math.* **14** (1973), 294–303.

[Man4] A. Mann, Regular $p$-groups, III, *J. Algebra* **70** (1981), 89–101.

[Man5] A. Mann, The power structure of $p$-groups I, *J. Algebra* **42** (1976), 121–135; II, *J. Algebra* **318** (2007), 953–956.

[Man6] A. Mann, Regular $p$-groups and groups of maximal class, *J. Algebra* **42** (1976), 136–141.

[Man7] A. Mann, Conjugacy classes in finite groups, *Israel J. Math.* **31** (1978), 78–84.

[Man11] A. Mann, The number of generators of finite p-groups, *J. Group Theory* **8** (2005), 317–338.

[Man15] A. Mann, Generators of $p$-groups, Proc. Groups-St. Andrews 1985, Cambridge (1986), 273–281.

[Man16] A. Mann, Some applications of powerful $p$-groups, Proc. Groups-St. Andrews 1989, Cambridge (1991), 370–385.

[Man18] A. Mann, Finite groups with maximal normalizers, *Illinois J. Math.* **12** (1968), 67–75.

[Man20] A. Mann, Some questions about $p$-groups, *J. Aust. Math. Soc. (Series A)* **67** (1999), 356–379.

[Man21] A. Mann, The derived length of $p$-groups, *J. Algebra* **224** (2000), 263–267.

[Man22] A. Mann, Finite $p$-Groups, in preparation.

[Man24] A. Mann, On the power structure of some $p$-groups, *Circ. Mat. Palermo II* **23** (1990), 227–235.

[Man25] A. Mann, Groups with few class sizes and the centralizer equality subgroup, *Isr. J. Math.* **142** (2004), 367–380.

[Man29] A. Mann, On characters-classes duality and orders of centralizers, *Cont. Math.* **402** (2006), 215–217.

[Man30] A. Mann, Normally monomial $p$-groups, *J. Algebra* **300** (2006), 2–9.

[Man31] A. Mann, On the exponent of the product of two groups, *Rend. Sem. Mat. Univ. Padova* **115** (2006), 205–207.

[Man33] A. Mann, The number of subgroups of metacyclic groups, *Contemporary Mathematics* **524** (2010), 93–95.

[Man34] A. Mann, On skew $p$-groups, *Publ. Math. Debrecen* **69** (2006), 353–360.

[Man36] A. Mann, Corrigendum The number of generators of finite p-groups, *J. Group Theory* **14** (2011), 329–331.

[Man37] A. Mann, More on normally monomial $p$-groups, manuscript.

[MM] A. Mann and C. Martinez, The exponent of finite groups, *Arch. Math.* **67** (1996), 8–10.

[MP-F] A. Mann and F. Poznick-Fradkin, Subgroups of powerful groups, *Israel I. Math.* **138** (2003), 19–28.

[Mar] Z. Marcinjak, Fixed elements in 2-groups revisited, *Arch. Math.* **97** (2011), 207–208.

[Mat] H. Matsuyama, Solvability of groups of order $2^a p^b$, *Osaka J. Math.* **10** (1973), 375–378.

[Maz2] V. D. Mazurov, 2-groups with an automorphism of odd order fixing all involutions, *Algebra and Logika* **8, 6** (1969), 874–885 (Russian).

[McK1] S. McKay, Finite $p$-groups, Queen Mary Math. Notes 18, London, 2000.

[McK2] S. McKay, On the structure of a special class of $p$-groups, *Quart. J. math. (2)* **38** (1987), 489–502.

[Men] F. Menegazzo, Automorphisms of $p$-groups with cyclic commutator subgroup, *Rend. Sem. Mat. Padova* **90** (1993), 81–101.

[Mie1] R. J. Miech, metabelian $p$-groups of maximal class, *Trans. Amer. Math. Soc.* **152** (1970), 331–373.

[Mie2] R. J. Miech, Some $p$-groups of maximal class, *Trans. Amer. Math. Soc.* **189** (1974), 1–47.

[Mie3] R. J. Miech, On $p$-groups with a cyclic commutator subgroup, *J. Aust. Math. Soc.* **20** (1975), 178–198.

[Mie4] R. J. Miech, The metabelian $p$-groups of maximal class, *Trans. Amer. Math. Soc.* **236** (1978), 93–119.

[Mie5] R. J. Miech, The metabelian $p$-groups of maximal class, II, *Trans. Amer. Math. Soc.* **272** (1982), 465–484.

[Mil8] G. A. Miller, On the holomorph of the cyclic group of order $p^m$, *Trans. Amer. Math. Soc.* **9** (1908), 232–236.

[Mil9] G. A. Miller, The groups in which every subgroup is either abelian or Hamiltonian, *Trans. Amer. Math. Soc.* **8** (1907), 25–29.

[MilM] G. Miller and H. Moreno, Nonabelian groups in which every subgroup is abelian, *Trans. Amer. Math. Soc.* **4** (1903), 398–404.

[Mill] W. H. Mills, The automorphisms of the holomorph of a finite abelian group, *Trans. Amer. Math. Soc.* **85** (1956), 1–34.

[MLC] E. Morgado Morales and M. Lazo Cortis, On the Sylow $p$-groups of the automorphism group of a finite homocyclic $p$-group, *Rev. Cienc. Mat.* **6** (1985), 35–44 (Spanish).

[Mori2] M. Morigi, Power automorphisms of finite $p$-groups, *Comm. Algebra* **70** (1999), 4853–4877.

[Mori4] M. Morigi, On $p$-groups with abelian automorphism group, *Rend. Sem. Mat. Univ. Padova* **92** (1994), 47–58.

[Mul] O. Müller, On $p$-automorphisms of finite $p$-groups, *Arch. Math.* **32** (1979), 533–538.

[NRSB] L. A. Nazarova, A. V. Roiter, V. V. Sergeichuk and V. M. Bondarenko, Application of modules over a dyad to the classification of finite $p$-groups that have an abelian subgroup of index $p$ and to the classification of pairs of mutually annihilating operators (Russian), Investigations on the theory of representations, *Zap. Nauchn. Sem. LOMI* **28** (1972), 69–92.

[Nei] L. I. Neikirk, Groups of order $p^m$ which contain cyclic subgroups of order $p^{m-3}$, *Trans. Amer. Math. Soc.* **6** (1905), 316–325.

[Nek2] K. G. Nekrasov, On some 2-groups with a small noncyclic Frattini subgroup, Algebraic and logical constructions, *Tver*, 1994, 53–65 (Russian).

[Nin] Y. Ninomiya, Finite $p$-groups with cyclic subgroups of index $p^2$, *Math. J. Okayama Univ.* **36** (1994), 1–21.

[Ore] O. Ore, Contributions to the theory of groups of finite order, *Duke Math. J.* **5** (1938), 431–460.

[Pas] D. S. Passman, Nonnormal subgroups of $p$-groups, *J. Algebra* **15, 3** (1970), 352–370.

[Pas1] D. S. Passman, Character theory and group rings, *Contemp. Math.* **524** (2010), 139–148.

[Paz] G. Pazdersky, Prime power groups which are cyclic extensions of elementary abelian groups, *Math. Nachr.* **97** (1980), 57–68.

[QW] G. H. Qian, Y. M. Wang, A note on character kernels in finite groups of prime power order, *Arch. Math.* **9**, 3 (2008), 193–199.

[Qu1] H. P. Qu, Finite $p$-groups all of whose maximal normal abelian subgroups are soft, *Sci. China Math.* **53**, 11 (2010), 3037–3040.

[QZ2] H. P. Qu and Q. H. Zhang, A central extension of minimal nonabelian $p$-groups (II), *Acta. Math. Sinica* **53:5** (2010), 933–944.

[QH2] H. P. Qu and R. F. Hu, A central extension of minimal nonabelian $p$-groups (III), *Acta. Math. Sinica* **53:6** (2010), 1051–1064.

[QZ] H. P. Qu and L. F. Zheng, A central extension of minimal nonabelian $p$-groups (IV), *Acta. Math. Sinica* **54:5** (2011), 739–752 (in Chinese).

[QZGA] H. P. Qu, L. P. Zhao, J. Gao and L. J. An, Finite $p$-groups with a minimal nonabelian subgroup of index $p$ (V), *J. Algebra Appl.* **13**(2014), no. 7, 1450032(35 pages).

[QXA] H. P. Qu, M. Y. Xu and L. J. An, Finite $p$-groups with a minimal nonabelian subgroup of index $p$ (III), *Sci. China Ser. A* **58**(2015), no. 4, 763–780.

[QYXA] H. P. Qu, S. S. Yang, M. Y. Xu and L. J. An, Finite $p$-groups with a minimal nonabelian subgroup of index $p$ (I), *J. Algebra* **358** (2012), 178–188.

[Rea1] E. W. Read, On the Schur multiplier of a wreath product, *Illinois J. Math.* **20** (1976), 456–466.

[Rea2] E. W. Read, On the centre of a representation group, *J. London Math. Soc. (2)* **16** (1977), 43–50.

[Red1] L. Redei, Das schiefe Produkt in der Gruppentheorie, *Comment. Math. Helvet.* **20** (1947), 225–267.

[Red2] L. Redei, Die endlichen einstufig nichtnilpotenten Gruppen, *Publ. Math. Debrecen* **4** (1956), 303–324.

[Rib] P. Ribenboim, Catalan's conjecture, Academic Press, New York, 1994.

[Rob] D. J. S. Robinson, A course of the theory of groups, NY, Springer, 1996.

[Rock] D. M. Rocke, $p$-groups with abelian centralizers, *Proc. London Math. Soc.* **(3) 30** (1975), 55–75.

[Roi1] M. Roitman, On Zsigmondy primes, *Proc. Amer. th. Soc.* **125** (1997), 1913–1919.

[Ron] C. Ronse, On centralizers of involutions in 2-groups, *Math. Ser. Cambridge Philos. Soc.* **86** (1979), 199–204.

[Rus2] D. J. Rusin, The 2-groups of rank 2, *J. Algebra* **149** (1992), 1–31.

[Sh] S. Shahriari, On normal subgroups of capable groups, *Arch. Math.* **48** (1987), 193–198.
[Sam1] B. Sambale, Fusion systems of bicyclic 2-groups, arXiv: 1206.2962v1 [mat.GR] 123 June 2012.
[Sam2] B. Sambale, Exponent and $p$-rank of finite $p$-groups and applications, *Arch. Math.* **103** (2014), 11–20.
[ST] J. Sangroniz and J. Tent, 2-groups with few rational conjugate classes, *J. Algebra* **338** (2011), 114–121.
[Schm3] P. Schmid, On the automorphism group of extraspecial 2-groups, *J. Algebra* **234** (2000), 492–506.
[Schm5] R. Schmidt, Subgroup Lattices of Groups, Expositions in Mathematics 14, W. de Gruyter, Berlin, 1994.
[Sch4] O. Schmidt, Groups with two classes of nonnormal subgroups (Russian), Proc. Seminar on group theory (1938), 7–26.
[Scot] W. R. Scott, Group theory, Prentice Hall, 1964.
[Sha1] A. Shalev, The structure of finite $p$-groups: effective proof of coclass conjectures, *Invent. Math.* **115** (1994), 315–345.
[Sha2] A. Shalev, Finite $p$-groups, in *"Finite and locally finite groups"*, Kluwer Acad. Publ., Dordrecht, 1995, 401–450.
[SQG] Q. W. Song, H. P. Qu and X. Y. Guo, Finite $p$-groups of class 3 all of whose proper subgroups have class at most 2, Algebra Coll. **17** (2010), 191–201.
[Son] Q. W. Song, Finite two-generator $p$-groups with cyclic derived group, manuscript.
[SS] E. G. Straus and G. Szekeres, On a problem of D. R. Hughes, *Proc. Amer. Math. Soc.* **9** (1958), 157–158.
[Suz1] M. Suzuki, Group Theory I, II, Springer, Berlin, 1982, 1986.
[Suz2] M. Suzuki, Structure of a Group and the Structure of its Lattice of Subgroups, *Erg. Math.* 10, Berlin: Springer 1956.
[Sze] G. Szekeres, On finite metabelian $p$-groups with two generators, *Acta Sci. Math. Szeged* **21** (1960), 270–291.
[Tah] K. Tahara, On the second cohomology groups of semidirect products, *Math. Z.* **129** (1972), 365–379.
[Tes] L. Teschke, Über die Normalteiler der $p$-Sylowgruppe der symmetrische Gruppe von Grade $p^m$, *Math. Nachr.* **87** (1979), 197–212.
[Tho2] J. G. Thompson, Finite groups with fixed-point-free automorphisms of prime order, *Proc. Nat. Acad. Sci. USA* **45** (1959), 578–581.
[Tho3] J. G. Thompson, Nonsolvable finite groups all of whose local subgroups are solvable, 1, *Bull. Amer. Math. Soc.* **74** (1968), 383–437.
[Tho4] J. G. Thompson, Fixed points of $p$-groups acting on $p$-groups, *Math. Z.* **86** (1964), 12–13.
[Tho5] J. G. Thompson, Centralizers of elements in $p$-groups, *Math. Z.* **96** (1967), 292–293.
[Tit] G. N. Titov, Groups containing a cyclic subgroup of index $p^3$, *Mat. Zametki* **28** (1980), 17–24.
[Tua1] H. F. Tuan, A theorem about $p$-groups with abelian subgroup of index $p$, *Acad. Sinica Science Record* **3** (1950), 17–23.
[Tua2] H. F. Tuan, An Anzahl theorem of Kulakoff's type for $p$-groups, *Sci. Rep. Nat. Tsing-Hua Univ. A* **5** (1948), 182–189.
[T-V] H. P. Tong-Viet, Symmetric groups are determined by their character degrees, *J. Algebra* **334** (2011), 275–284.
[V-LF1] A. Vera-Lopez and G. A. Fernandez-Alcober, Centralizers of small order in $p$-groups of maximal class, *Comm. Algebra* **20 (40)** (1992), 1051–1059.

[V-LF2] A. Vera-Lopez and G. A. Fernandez-Alcober, The conjugacy vector of a $p$-group of maximal class, *Israel J. Math.* **86** (1994), 233–255.

[V-LLO] A. Vera-Lopez and L. Ortiz de Elguea, The conjugacy-vectors of all relative holomorphs of an elementary abelian group of order 16, *Portugal Math.* **47**, 3 (1990), 243–257.

[V-LS] A. Vera-Lopez and J. Sangroniz, Classification of all relative holomorphs of an abelian group of type $C_4 \times C_4$, *Rev. Acad. Cienc. Zaragoza (2)* **43** (1988), 7–12 (Spanish).

[Ver1] L. Verardi, On groups whose non-central elements have the same finite number of conjugates, *Boll. U. M. I. (7)* **2-A** (1988), 391–400.

[Ver2] L. Verardi, A class of finite groups of exponent $p$ in which every normal subgroup is characteristic, *Boll. Un. Mat. Ital. B (6)* **4** (1988), 307–317.

[Waa1] R. W. van der Waall, On finite $p$-groups whose commutator subgroups are cyclic, *Indag. Math.* **35** (1973), 342–345.

[Waa2] R. W. van der Waall, Certain normal subgroups of the Frattini subgroup of a finite group, *Indag. Math.* **36** (1974), 382–386.

[Waa3] R. W. van der Waall, On modular $p$-groups, *J. Algebra* **25** (1973), 125–127.

[Waa4] R. W. van der Waall, On modular $p$-groups II, *Indag. Math.* **37** (1975), 79–82.

[Waa5] R. W. van der Waall, A note on modular $p$-groups, *Indag. Math.* **35** (1973), 47–48.

[Waa6] R. W. van der Waall, On monomial groups, *J. reine angew. Math.* **264** (1973), 103–134.

[Waa7] R. W. van der Waall, On $p$-nilpotent forcing groups, *Indag. Math.* **2** (1991), 367–384.

[WaaN] R. W. van der Waall and C. H. W. M. de Nijs, On the embedding of a finite group as a Frattini subgroup, *Bull. Belg. Math. Soc. Simon Stevin* **2** (1995), 519–527.

[Wam] J. W. Wamsley, The deficiency of metacyclic groups, *Proc. Amer. Math. Soc.* **24** (1970), 724–730.

[WZ] Lifang Wand and Qinhai Zhang, Finite $p$-groups whose nonabelian subgroups have the same center, *J. Group Theory* **17** (2014), 689–703.

[Web2] U. H. M. Webb, The number of stem covers of an elementary abelian $p$-group, *Math. Z.* **182**, 3 (1983), 327–337.

[Web3] U. H. M. Webb, The occurrence of groups as automorphisms of nilpotent $p$-groups, *Arch. Math.* **37**, 6 (1981), 481–498.

[Web4] U. H. M. Webb, On the rank of a $p$-group of class 2, *Canad. Math. Bull.* **26** (1983), 101–105.

[Web5] U. H. M. Webb, The Schur multiplier of a nilpotent group, *Trans. Amer. Math. Soc.* **291** (1985), 755–763.

[Web6] U. H. M. Webb, The using graphs to investigate the automorphism groups of nilpotent groups, Graphs and applications, Wiley, New York (1985), 223–247.

[Weir2] A. Weir, The Sylow subgroups of the symmetric groups, *Proc. Amer. Math. Soc.* **6** (1955), 534–541.

[Wie2] J. Wiegold, The Schur multiplier: an elementary approach, Groups — St. Andrews, 1981, *London Math. Soc. Lect. Notes* **71**, 137–154, Cambridge, 1982.

[Wie3] J. Wiegold, The multiplicator of direct product, Quart. J. Math. Oxford (2) (1971), 103–105.

[Wiel2] H. Wielandt, Über Produkte von nilpotenten Gruppen, *Illinois J. Math.* **2** (1958), 611–618.

[Wiel5] H. Wielandt, Zum Satz von Sylow. II. *Math. Z.* **71** (1959), 401–402.

[Wil3] B. Wilkens, 2-groups of breadth 3, *J. Algebra* **318** (2007), 202–224.

[Wil5] B. Wilkens, Finite $p$-groups not characteristic in any $p$-group in which they are properly contained, *Israel J. Math.* **166** (2008), 97–112.

[Wil6] B. Wilkens, $p$-groups without noncharacteristic normal subgroups, *Israel J. Math.* **172**, 357–369.

[Wol] B. Wolf, A note on $p'$-automorphisms of $p$-groups $P$ of maximal class centralizing the center of $P$, *J. Algebra* **190**, 1 (1997), 163–171.

[Won2] W. J. Wong, On finite groups whose 2-Sylow subgroups have cyclic subgroups of index 2, *L. Australian Math. Soc.* **4** (1964), 90–112.

[Wri] D. Wright, Degree of minimal permutation representation of covering groups of abelian groups, *Amer. J. Math.* **96** (1974), 578–592.

[Xu1] M. Y. Xu, The power structure of finite $p$-groups, *Bull. Aust. Math. Soc.* **36**, 1 (1987), 1–10.

[Xu3] M. Y. Xu, A theorem on metabelian $p$-groups and some consequences, *Chin. Ann. Math.*, Ser. B5 (1984), 1–6.

[Xu4] M. Y. Xu, Regular $p$-groups and their generalizations, manuscript.

[XZ1] M. Y. Xu and Q. H. Zhang, A classification of metacyclic 2-groups, *Algebra Colloq.* **13** (2006), 25–34.

[XZA] M. Y. Xu, Q. H. Zhang and L. J. An, Finite $p$-groups all of whose proper nonabelian subgroups are generated by two elements, *J. Algebra* **319, 9** (2008), 3603–3620.

[Yin1] J. H. Ying, Finite groups with all subgroups isomorphic to quotient groups, *Arch. Math.* **24** (1973), 561–570.

[Yin2] J. H. Ying, Finite groups with restricted quotient structures, *Bull. Inst. Math. Acad. Sinica* **8** (1980), 1–14.

[Zas2] H. Zassenhaus, A group-theoretic proof of a theorem of Maclagan Wedderburn, *Proc. Glasgow Math. Assoc.* **1** (1852), 53–63.

[ZL2] Q. H. Zhang and P. J. Li, Finite $p$-groups with a cyclic subgroup of index $p^3$, *J. Math. Res. Appl.* **32**, 5(2012), 505–529.

[Zha2] Q. H. Zhang, A characterization of metacyclic $p$-groups by counting subgroups, ICA-Proceedings, September 30, 2011, 713–720.

[Zha3] Q. H. Zhang, A characterization of smallest Suzuki 2-group, *Acta. Math. Sinica, English Series* **24**:12 (2008), 2011–2014.

[ZG1] X. H. Zhang and X. Y. Guo, Finite $p$-groups whose non-normal cyclic subgroups have small index in their normalizers, *J. Group Theory* **15** (2012), 641–659.

[ZG2] L. B. Zhao and X. Y. Guo, Finite $p$-groups with exactly one $\mathcal{A}_1$-subgroup of given structure of order $p^3$, *Acta Math. Sin. Engl. Ser.* **29**, No. 11 (2013), 2099–2110.

[ZGQX] Q. H. Zhang, X. Q. Guo, H. P. Qu and M. Y. Xu, Finite groups which have many normal subgroups, *J. Korean Math. Soc.* **46**, 6 (2009), 1165–1278.

[ZZLS] Q. H. Zhang. L. B. Zhao, M. M. Li and Y. Q. Shen, Finite $p$-groups all of whose subgroups of index $p^3$ are abelian, *Commun. Math. Stat.* **3**, 1 (2015), 69–162.

[ZLS] Q. H. Zhang, X. X. Li and M. J. Su, Finite $p$-groups whose nonnormal subgroups have orders $\leq p^3$, manuscript.

[ZS] Q. H. Zhang and M. J. Su, Finite 2-groups whose nonnormal subgroups have orders $\leq 2^3$, *Front. Math. China* **7**, 5 (2012), 971–1003.

[ZW] Q. H. Zhang and J. J. Wei, The intersection of subgroups of finite groups, *Arch. Math.* **96** (2011), 9–17.

[Zhm5] E. M. Zhmud, The isomorphisms of the lattice of normal subgroups of a finite nilpotent group, *Vestnik Kharkov Univ.* **26** (1967), 3–17 (Russian).

# Author index

# Subject index

**N**

**W**



**Y**



**Z**

## De Gruyter Expositions in Mathematics

**Volume 60**
Benjamin Fine, Anthony Gaglione, Alexei Myasnikov, Gerhard Rosenberger, Dennis Spellman
*The Elementary Theory of Groups,, 2014*
ISBN 978-3-11-034199-7, e-ISBN 978-3-11-034203-1, Set-ISBN 978-3-11-034204-8

**Volume 59**
Friedrich Haslinger,
*The d-bar Neumann Problem and Schrödinger Operators, 2014*
ISBN 978-3-11-031530-1, e-ISBN 978-3-11-031535-6, Set-ISBN 978-3-11-031536-3

**Volume 58**
Oleg K. Sheinman,
*Current Algebras on Riemann Surfaces, 2012*
ISBN 978-3-11-026452-4, e-ISBN 978-3-11-026452-4, Set-ISBN 978-3-11-916387-3

**Volume 57**
Helmut Strade
*Simple Lie Algebras, Completion of the Classification, 2012*
ISBN 978-3-11-026298-8, e-ISBN 978-3-11-026301-5, Set-ISBN 978-3-11-916682-9

**Volume 56**
Yakov Berkovich, Zvonimir Janko
*Groups of Prime Power Order 3, 2011*
ISBN 978-3-11-020717-0, e-ISBN 978-3-11-025448-8, Set-ISBN 978-3-11-218909-2

**Volume 55**
Rainer Picard, Des McGhee
*Partial Differential Equations, 2011*
ISBN 978-3-11-025026-8, e-ISBN 978-3-11-025027-5, Set-ISBN 978-3-11-218895-8

**Volume 54**
Edgar E. Enochs, Overtoun M. G. Jenda
*Relative Homological Algebra, 2011*
ISBN 978-3-11-021522-9, e-ISBN 978-3-11-021523-6, Set-ISBN 978-3-11-173442-2

Bei Fragen zur Produktsicherheit wenden Sie sich bitte an:
If you have any questions regarding product safety,
please contact:

Walter de Gruyter GmbH
Genthiner Straße 13
10785 Berlin
productsafety@degruyterbrill.com